T0145386

# Springer Series on Bio- and Neurosystems

Volume 7

**Series editor**

Nikola K. Kasabov, Auckland University of Technology, Auckland, New Zealand

The *Springer Series on Bio- and Neurosystems* publishes fundamental principles and state-of-the-art research at the intersection of biology, neuroscience, information processing and the engineering sciences. The series covers general informatics methods and techniques, together with their use to answer biological or medical questions. Of interest are both basics and new developments on traditional methods such as machine learning, artificial neural networks, statistical methods, nonlinear dynamics, information processing methods, and image and signal processing. New findings in biology and neuroscience obtained through informatics and engineering methods, topics in systems biology, medicine, neuroscience and ecology, as well as engineering applications such as robotic rehabilitation, health information technologies, and many more, are also examined. The main target group includes informaticians and engineers interested in biology, neuroscience and medicine, as well as biologists and neuroscientists using computational and engineering tools. Volumes published in the series include monographs, edited volumes, and selected conference proceedings. Books purposely devoted to supporting education at the graduate and post-graduate levels in bio- and neuroinformatics, computational biology and neuroscience, systems biology, systems neuroscience and other related areas are of particular interest.

More information about this series at http://www.springer.com/series/15821

Nikola K. Kasabov

# Time-Space, Spiking Neural Networks and Brain-Inspired Artificial Intelligence

 Springer

Nikola K. Kasabov
Knowledge Engineering and Discovery
  Research Institute (KEDRI)
Auckland University of Technology
Auckland, New Zealand

ISSN 2520-8535          ISSN 2520-8543  (electronic)
Springer Series on Bio- and Neurosystems
ISBN 978-3-662-58607-5          ISBN 978-3-662-57715-8  (eBook)
https://doi.org/10.1007/978-3-662-57715-8

Printed on acid-free paper

This Springer imprint is published by the registered company Springer-Verlag GmbH, DE part of
Springer Nature
The registered company address is: Heidelberger Platz 3, 14197 Berlin, Germany

*Time lives inside us and we live inside Time.*
Vasil Levski-Apostola (1837–1873)
Bulgarian Educator and Revolutionary

*To my mother Kapka Nikolova
Mankova-Kasabova (1920–2012) and my
father Kiril Ivanov Kasabov (1914–1996),
who gave me the light of life, and for those
who came earlier in time; to my family,
Diana, Kapka and Assia, who give me the
light of love; and to those who will come later
in time; I hope they will enjoy the light of life
and the light of love as much as I do.*

# Foreword

Professor Furber is ICL Professor of Computer Engineering in the School of Computer Science at the University of Manchester, UK. After completing his education at the University of Cambridge (BA, MA, MMath, Ph.D.), he spent the 1980s at Acorn Computers, where he was a principal designer of the BBC Micro and the ARM 32-bit RISC microprocessor. As of 2018, over 120 billion variants of the ARM processor have been manufactured, powering much of the world's mobile computing and embedded systems. He pioneered the development of SpiNNaker, a neuromorphic computer architecture that enables the implementation of massively parallel spiking neural network systems with a wide range of applications.

The last decade has seen an explosion in the deployment of artificial neural networks for machine learning applications ranging from consumer speech recognition systems through to vision systems for autonomous vehicles. These artificial neural systems differ from biological neural systems in many important aspects, but most notably in their use of neurons with continuously varying outputs where biology predominantly uses spiking neurons—neurons that emit a pure electro-chemical unit impulse in response to recognising an input pattern. The continuous output of the artificial neuron can be thought of as representing the mean firing rate of its biological equivalent, but in using rates rather than spikes, the artificial network loses the ability to access the detailed spatio-temporal information that can be conveyed in a time sequence of spikes. Biological systems can clearly access this information, but how they use it effectively remains a mystery to science.

Nik Kasabov has done as much as anyone to begin to unlock the secrets of the biological spatio-temporal patterns of spikes, and in this book, he reveals what he has learnt about those secrets and how he has applied that knowledge in exciting new ways. This is deep knowledge, and if we can harness such knowledge in brain-inspired AI systems, then the explosion in AI witnessed over the last decade will look like a damp squib in comparison with what is to follow. This book is not just a record of past work, but also a guidebook for an exciting future!

Steve Furber
CBE, FRS, FREng
Computer Science Department
University of Manchester, UK

# Preface

Everything exists and evolves within time–space and time–space is within everything, from a molecule to the universe. Understanding the complex relationship between time and space has been one of the biggest scientific challenges of all times, including the understanding and modelling the time–space information processes in the human brain and understanding life. This is the strive for *deep knowledge* that has always been the main goal of the human race.

Now that an enormous amount of time–space data is available, science needs new methods to deal with the complexity of such data across domain areas. Risk mitigation strategies from health to civil defence often depend on simple models. But recent advances in machine learning offer the intriguing possibility that disastrous events, as diverse as strokes, earthquakes, financial market crises, or degenerative brain diseases, could be predicted early if the patterns hidden deeply in the intricate and complex interactions between spatial and temporal components could be understood. Although such interactions are manifested at different spatial or temporal scales in different applications or domain areas, the same information-processing principles may be applied.

A radically new approach to modelling such data and to obtaining *deep knowledge* is needed that could enable the creation of faster and significantly better machine learning and pattern recognition systems, offering the realistic prospect of much more *accurate* and *earlier* event prediction, and a better understanding of causal time–space relationships.

The term time–space coined in this book has two meanings:

– The problem space, where temporal processes evolve in time;
– The functional space of time, as it goes by.

This book looks at evolving processes in time–space. It talks about how *deep learning* of time–space data is achieved in the human brain and how this results in *deep knowledge*, which is taken as inspiration to develop methods and systems for deep learning and deep knowledge representation in spiking neural networks (SNN). And furthermore, how this could be used to develop a new type of artificial intelligence (AI) systems, here called brain-inspired AI (BI-AI). In turn, these BI-AI systems can help us understand better the human brain and the universe and for us to gain new deep knowledge.

BI-AI systems adopt structures and methods from the human brain to intelligently learn time–space data. BI-AI systems have six main distinctive features:

(1) They have their structures and functionality inspired by the human brain; they consist of spatially located neurons that create connections between them through deep learning in time–space by exchanging information—spikes. They are built of spiking neural networks (SNNs), as explained in Chaps. 4–6 in the book.
(2) Being brain-inspired, BI-AI systems can achieve not only deep learning, but *deep knowledge* representation in time–space.
(3) They can manifest cognitive behaviour.
(4) They can be used for knowledge transfer between humans and machines as a foundation for the creation of symbiosis between humans and machines, ultimately leading to the integration of human intelligence and artificial intelligence (HI+AI) as discussed in the last chapter of the book.
(5) BI-AI systems are universal data learning machines, being superior to traditional machine learning techniques when dealing with time–space data.
(6) BI-AI systems can help us understand, protect and cure the human brain.

At the more technical level, the book presents background knowledge, new generic methods for SNN, evolving SNN (eSNN) and brain-inspired SNN (BI-SNN) and new specific methods for the creation of BI-AI systems for modelling and analysis of time–space data across applications.

I strongly believe that progress in information sciences is mostly an evolutionary process, that is, building up on what has already been created. In order to understand the principles of deep learning and deep knowledge, SNN and BI-AI, to properly apply them to solve problems, one needs to know some basic science principles established in the past, such as epistemology by Aristotle, perceptron by Rosenblatt, multilayer perceptron by Rumelhart, Amari, Werbos and others, self-organising maps by Kohonen, fuzzy logic by Zadeh, quantum principles by Einstein and Rutherford, von Neumann computing and Atanassoff ABC machine and of course the human brain. All these principles are briefly covered in the book, giving a proper foundation for a better understanding of SNN and BI-AI and how they can be used to understand the time–space puzzles of nature and life and to gain new, deep knowledge.

I have been lucky to meet and talk with some of the pioneers in the fields, such as Shun-ichi Amari, Teuvo Kohonen, Walter Freeman, John Taylor, Lotfi Zadeh, Takeshi Yamakawa, Steve Grossberg, John Andreae, Janus Kacprzyk, Steve Furber, to mention only few of them, who gave me inspiration to go deep in this research. My humble view is that we should not forget our pioneers and teachers who gave us the light of knowledge.

Some of the new methods presented in the book are developed by the author and have already appeared partially in various publications in collaboration with my students and colleagues in the period 2005–2018. I would like to acknowledge the contribution of my colleagues and postdoctoral fellows Lubica Benuskova, Michail Defoin-Platel, Enmei Tu, Zeng-Guang Hou and his students Nelson and James, Jie Yang and his students Lei Zhou and Chengie Gu, Giacomo Indiveri, Qun Song, Paul Pang, Israel Espinosa, Weiqi Yan, Denise Taylor, Grace Wang, Valery Feigin, Rita Krishnamurthi, Carlo Morabito, Nadia Mammone, Veselka Boeva, Marley Vellasco, Andreas Koenig, Mario Fedrizzi, Plamen Angelov, Dimitar Filev, Petia Georgieva, Georgi Bijev, Petia Koprinkova, Chrisina Jayne, Seiichi Ozawa, Cesare Alippi, and many others.

I was privileged to have a large number of Ph.D. students in this period who also contributed to publications used in this book. I acknowledge the contribution of my Ph.D. students Maryam Doborjeh, Neelava Sengupta, Zohre Doborjeh, Anne Abbott, Kaushalya Kumarasinghe, Akshay Gollohalli, Clarence Tan, Vinita Kumar, Wei Cui, Vivienne Breen, Fahad Alvi, Reggio Hartono, Elisa Capecci, Nathan Scott, Norhanifah Murli, Muhaini Othman, Paul Davidson, Kshitij Dhoble, Nuttapod Nuntalid, Linda Liang, Haza Nuzly, Maggie Ma, Gary Chen, Harya Widiputra, Raphael Hu, Stefan Schliebs, Anju Verma, Peter Hwang, Snejana Soltic, Vishal Jain, Simei Wysosky, Liang Goh, Raphael Hu, Gary Chen and others. Special acknowledgement to Helena Bahrami who helped me with the references and the formatting of each of the 22 chapters.

During my long-time work on various topics included in this book and during the writing of the book, I have received a tremendous support and help from my wife Diana and my daughters Kapka and Assia. I thank them and love them!

I did some work on SNN while on a visiting professorship, funded by EU Funding named after the great scientist **Maria Salomea Skłodowska-Curie (b.1867–d.1934)**. My fellowship was hosted by the Institute for Neuroinformatics (INI) at ETH and University of Zurich, working in collaboration with Giacomo Indiveri. I am grateful for this wonderful opportunity named after a remarkable scientist.

I did all the work on the book while maintaining my research, teaching and administrative duties at Auckland University of Technology (AUT). I acknowledge the generous funding and support I have received from this vibrant University since my appointment in 2002, and still continuing. As the Founding Director of the Knowledge Engineering and Discovery Research Institute (KEDRI) at AUT for 16 years now, that allowed me to take a leadership in research, I have been helped tremendously by the KEDRI Administrative Manager Joyce D'Mello. I acknowledge the support and the excellent work by the team of the Springer Series of Bio- and Neurosystems—the series editorial manager Leontina, and also Arun Kumar, Sabine and the whole team involved in this series.

If I have to summarise the philosophy of this book in one sentence as a moto, I would say:

*Inspired by the oneness in nature in time–space, we aim to achieve oneness in data modelling using brain-inspired computation.*

August 2018

Nikola K. Kasabov
Fellow IEEE, Fellow RSNZ,
Fellow IITP NZ, DVF RAE UK

Director
Knowledge Engineering and Discovery
Research Institute (KEDRI)
Auckland University of Technology
Auckland, New Zealand

# About the Book Content by Topics and Chapters and The Pathway of Knowledge

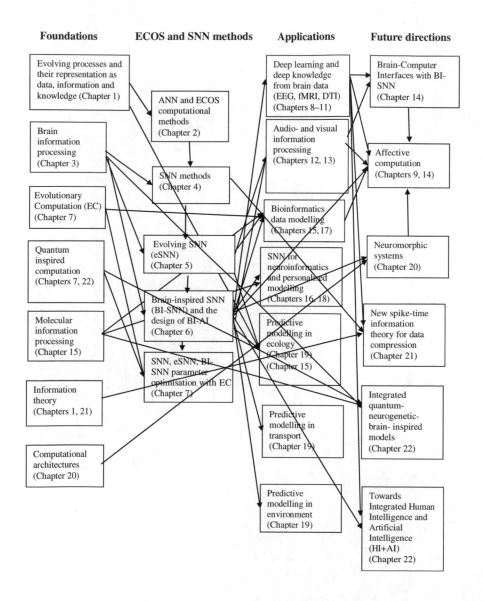

# Contents

## Part VIII   Future Development in BI-SNN and BI-AI

## 20   From von Neumann Machines to Neuromorphic Platforms ...... 661

# About the Author

**Nikola K. Kasabov** is Professor of neural networks and knowledge engineering and Director of the Knowledge Engineering and Discovery Research Institute (KEDRI) at the Auckland University of Technology (AUT), New Zealand. Born in Bulgaria, he has worked previously at the TU Sofia, University of Essex and University of Otago. He is fellow of IEEE, Fellow of the Royal Society (Academy) of New Zealand (RSNZ), Distinguished Fellow of the Royal Academy of Engineering UK and Visiting Professor at several universities, including: Shanghai Jia-Tong University; ETH and University of Zurich; RGU Scotland UK; University of Trento; University of Kaiserslautern;

Universities of Twente and Maastricht. Prof Kasabov originated methods and systems for intelligent information processing, including: evolving connectionist systems, hybrid neuro-fuzzy systems, evolving- and brain–inspired spiking neural network architectures, quantum-inspired methods, methods for personalised modelling in bio and neuroinformatics, published in more than 600 works. He is Past President of the International Neural Network Society (INNS) and the current President of the Asia-Pacific Neural Network Society (APNNS). Prof Kasabov has received the INNS Ada Lovelace and Gabor Awards, APNNS Outstanding Achievements Award, RSNZ Medal, AUT Medal, Honourable Fellowship of the Bulgarian and the Greek Computer Societies, Pavlikeni Honourable Citizenship and other awards. He has been the editor of the Springer Handbook of Bio-/Neuro-informatics published by Springer in 2014 and of the related book series *Springer Series on Bio- and Neurosystems*.

# Part I
# Time-Space and AI. Artificial Neural Networks

# Chapter 1
# Evolving Processes in *Time-Space*.
# Deep Learning and Deep Knowledge
# Representation in Time-Space.
# Brain-Inspired AI

This chapter presents the challenges to information sciences when dealing with complex evolving processes in time-space. The emphasis here is on processes/ systems that evolve/develop/unfold/change in time-space and what characterises them. To model such processes, to extract *deep knowledge* that drives them and to trace how this knowledge changes over time, are among the main objectives of the brain-like approach that we take in this book by using SNN. And before going to SNN in the next chapters, we introduce how evolving processes can be represented as data, information and knowledge, and more specifically, what is *deep knowledge* that we will target to achieve through *deep learning* in SNN.

This chapter consists of the following sections:

1.1. Evolving processes in time-space.
1.2. Characteristics of evolving processes: Frequency, energy, probability, entropy, and information.
1.3. Light and sound.
1.4. Evolving processes in Time-Space and Direction.
1.5. From data and information to knowledge.
1.6. Deep learning and deep knowledge representation in time-space. How deep?
1.7. Statistical, computational modelling of evolving processes.
1.8. Brain-inspired AI (BI-AI).
1.9. Chapter summary and further readings for deeper knowledge.

## 1.1 Evolving Processes in Time-Space

*Time* is defined in the Oxford Dictionary as "The indefinite continued progress of existence, events, etc., in past, present and future regarded as a whole …". *Time* has been studied for many years by the most prolific scientists and cosmologists [1, 2].

© Springer-Verlag GmbH Germany, part of Springer Nature 2019
N. K. Kasabov, *Time-Space, Spiking Neural Networks and Brain-Inspired Artificial Intelligence*, Springer Series on Bio- and Neurosystems 7,
https://doi.org/10.1007/978-3-662-57715-8_1

*Space* is defined in the Oxford Dictionary as "A continuous, unlimited area of expanse which may or may not contain objects ...".

Science aims at understanding Nature and the humanity. Processes in Nature are evolving in both space and time (Fig. 1.1). To understand them humans create models, initially only mental models, as at the time of Aristotle (4c BC) and now mathematical and computational models to extract information and knowledge, and more specifically *deep knowledge* as defined here.

### 1.1.1  What Are Evolving Processes?

We call *evolving processes* or *evolving systems* those that change, develop, unfold in time. Most evolving processes evolve both in time-space. Evolving spatio-temporal processes are characterised by sometimes complex interaction between space and time components in a continuous manner. This interaction may change over time. Such processes may also interact with other processes in the environment. It may not be possible to determine in advance the course of inter-action, unless we discover the important features, the spatio-temporal patterns and rules that drive such processes and their evolution in time.

Evolving spatio-temporal processes are difficult to model because some of their evolving rules (laws) may not be known a priori, they may dynamically change due

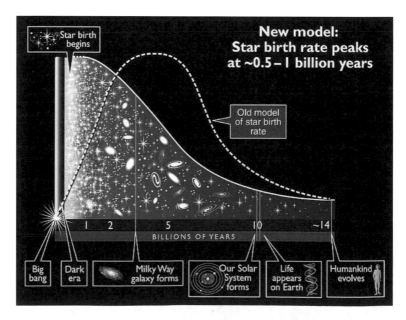

**Fig. 1.1** All processes in Nature are evolving in time-space, from the emergence of the universe to life and the human brain (after [43])

to unexpected perturbations, and therefore they may not be strictly predictable in a longer term. Thus, modelling of such processes is a challenging task with a lot of practical applications in life sciences and engineering.

When processes are evolving, their models need to be evolving too, i.e. to trace the dynamics of the processes and to adapt to changes in these processes over time. For example, a speech recognition system has to be able to adapt to various new accents, and to learn new languages incrementally. A system that models cognitive tasks of the human brain, needs to be adaptive, as all cognitive processes are evolving by nature. (We never stop learning!) In bioinformatics, gene expression modelling systems have to be able to adapt to new information that would define how for example a gene could become inhibited by another gene, the latter being triggered by a third gene, etc. There is an enormous number of tasks from life sciences where the processes evolve and change over time.

It would not be an overstatement to say that everything in nature evolves in time-space. But what are the rules, the laws that drive these processes, and how these rules change over time, how do they evolve? If we knew these rules, we could create computational models that can evolve in a similar manner as the real evolving processes, and use these models to make predictions and to better understand the processes. But if we do not know these rules, we can still try to uncover them from the data collected from these processes using machine learning. This was not possible during the time of Aristotle (4 century BC), but it is possible now as it is demonstrated in the book.

The term "evolving" is used here in a broader sense than the term "evolutionary". The latter is related to a population of individual systems traced over generations [3–5], while the former, as it is used in this book, is mainly concerned with the development of the structure and functionality of an individual system in space and/or time during its lifetime [6]. An evolutionary (population/generation) optimisation of the parameters of this system can be applied as well.

## 1.1.2 Evolving Processes in Living Organisms

The most obvious example of an evolving process is life, defined in the Concise Oxford English Dictionary (1983) as "a state of functional activity and continual change peculiar to organized matter, and especially to the portion of it constituting an animal or plant before death, animate existence, being alive". Continual change in space and time, along with certain stability, is what characterizes life. Modelling living systems requires that the continuous changes are represented in the model, i.e. the model adapts in a life-long mode and at the same time preserves features and principles that are characteristic to the process. The "stability–plasticity" dilemma is a well-known principle of life that is also widely used in connectionist computational models [7].

Perhaps, the most complex information system evolved so far is the human brain. Many interrelated evolving processes are observed at different "levels" of brain functionality (Fig. 1.2).

At the quantum level, particles are in a complex evolving state in space and time, being at several locations at the same time, which is defined by probabilities. General evolving rules are defined by several principles, such as entanglement, superposition, etc. [8, 9] (see also Chaps. 7 and 22).

At a molecular level, RNA and protein molecules, for example, evolve and interact in a continuous way based on the DNA information and on the environment. The central dogma of molecular biology constitutes a general evolving rule, but there are specific rules for different species and individuals. Different spatio-temporal folding and unfolding of proteins in a 3D space define different functions cells in the same organism—Fig. 1.3 [9, 10] (for details see Chap. 15).

At the cellular level (e.g. a neuronal cell) all the metabolic processes, the cell growing, cell division etc., are evolving processes in time-space. At the level of cell ensembles, or at a biological neural network level, an ensemble of cells (neuros) operates in a concert, defining the function of the ensemble or the network through learning, for instance—perception of sound, perception of an image, or learning languages. An example of a general evolving rule is the Hebbian learning rule [11] where neurons create connections between them in space when they are activated together in time [9].

In the human brain, complex dynamic interactions between groups of neurons can be observed when certain cognitive functions are performed, e.g. speech and language learning, visual pattern recognition, reasoning, and decision making [9]. When a person is performing a task brain activities are observed in different parts of the brain over time—Fig. 1.4 (see Chap. 3 for details).

At the level of population of individuals, species evolve through evolution A biological system evolves its structure and functionality through both lifelong learning of an individual and the evolution of populations of many such individuals

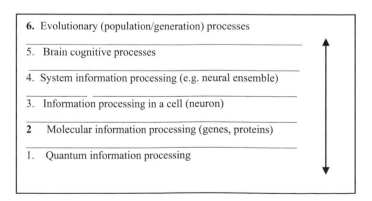

**Fig. 1.2** Many interrelated evolving processes are observed at different "levels" of brain functionality (after [30, 42])

**Fig. 1.3** Evolving processes at a molecular level: Different spatio-temporal folding and unfolding of proteins in a 3D space define different functions of cells in the organism (after [9, 17, 30])

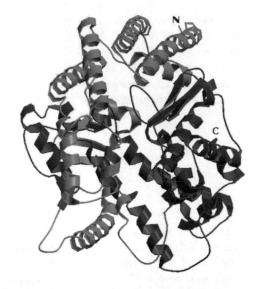

**Fig. 1.4** When a person is performing a task brain activities are observed in different spatially located parts of the brain at different times (after [43])

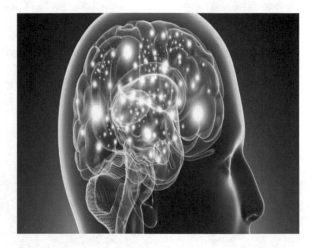

[4, 5]. In other words, an individual is a result of the evolution of many generations of populations, as well as a result of its own developmental lifelong learning processes. The Mendelian and Darwinian rules of evolution have inspired the creation of computational modelling techniques called evolutionary computation (EC) [5, 12] (see Chap. 7 for details).

Interaction in time-space is what makes a living organism a complex one, and that is also a challenge for computational modelling. For example, there are complex interactions between genes in a genome, and between proteins and DNA. There are complex interactions between the genes and the functioning of each neuron, a neural network, and the whole brain. Abnormalities in some of these interactions are known to have caused brain diseases and many of them are unknown at present. An example of interactions between genes and neuronal functions is the observed dependence between long-term potentiation (learning) in the synapses and the expression of the immediate early genes and their corresponding proteins such as Zif/268 [13]. Genetic reasons for several brain diseases have been already discovered, where some genes are expressed at a later time in live through interactions with other genes in the genome (see Chaps. 15 and 16).

### 1.1.3  Spatio-temporal and Spectro-temporal Evolving Processes

The physical interaction between parts of the earth is measured as spatio-temporal seismic data (Fig. 1.5) but what are these deep patterns of interaction in time-space that would trigger an earthquake? (see Chap. 19 for details).

A sound signal represents a spectro-temporal evolving process in time, e.g. music as shown in Fig. 1.6. as a wave form in time (see Chaps. 12 and 13).

Several sources of signals located at different locations, represent a spatio/spectro-temporal process.

The processes of buying/selling shares on the stock market are spatio-temporal, sometimes presented as only spectro- temporal, i.e. the change of the stock prices in time.

To properly model and understand evolving processes, it is important to first understand their characteristics as discussed in the next section.

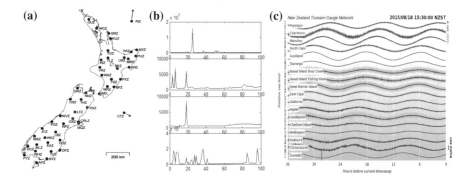

**Fig. 1.5** Geophysical processes are both spatio-temporal and spectro-temporal: **a** Spatially located seismic sites in New Zealand. **b** Temporal seismic activities at four selected seismic sites (spatially located) around Christchurch area manifest different frequency (spectral) characteristics. **c** Sea level at different harbours of New Zealand over time demonstrate both spatial and spectral characteristics (from: http://www.geonet.co.nz)

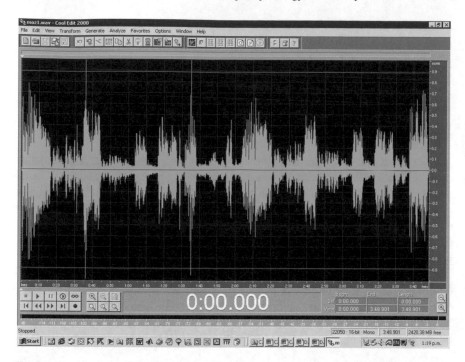

**Fig. 1.6** A wave form of a segment from Mozart's music, represented as intensity of the sound over time

## 1.2   Characteristics of Evolving Processes: Frequency, Energy, Probability, Entropy and Information

Evolving processes are characterised by common characteristics, the most important ones being frequency, entropy, energy and information as explained below.

*Frequency:* Frequency, is defined as the number of a signal/event changes over a period of time (seconds, minutes, centuries, etc.). Some processes have stable frequencies (they are periodic), but other—change their frequencies over time. Different processes are characterised by different frequencies, defined by their physical parameters. Usually, a process is characterised by a spectrum of frequencies. For example, different frequency spectrums are observed as brain activities (e.g. alpha, beta, gamma and delta waves), speech signals, image and video data, seismic processes, music, quantum processes, etc.

Frequency reflects on the changes in the signal (the data) in time. Evolving processes can manifest different behaviour, such as:

- Random: there is no rule that governs changes of the process in time and the process is not predictable.

– Chaotic: the process is predictable but only in a short time ahead, as the changes of the process at a time moment depends on the process changes at previous time moments via a non-linear function.
– Quasy-periodic: the process is manifesting similarity of its changes over time, but slightly modified each time.
– Periodic: the process repeats same patterns of changes over time and is fully predictable (there are fixed rules that govern the process and the rules do not change over time).

Many complex processes in engineering, social sciences, physics, mathematics, economics and other sciences are evolving by nature and can be analysed using the above classification. Some dynamic time series in nature manifest chaotic behaviour, i.e. there are some vague patterns of repetition over time, and the time series are approximately predictable in the near future, but not in the long run [14–17]. Chaotic processes are usually described by mathematical equations that use some parameters to evaluate the next state of the process from its previous states. Simple formulae may describe a very complicated behaviour over time: e.g. a formula that describes fish population growth $F(t+1)$ is based on the current fish population $F(t)$ and a parameter g [14]:

$$F(t+1) = 4gF(t)(1 - F(t)) \tag{1.1}$$

When $g > 0.89$, the function becomes chaotic.

A chaotic process is defined by evolving/changing rules, so that the process lies on the continuum of "orderness" somewhere between random processes (not predictable at all) and quasi-periodic processes (predictable in a longer time-frame, but only to a certain degree). Modelling a chaotic process in reality, especially if the process changes its rules over time, is a task for an adaptive system that captures the changes in the process in time, e.g. the value for the parameter g from the formula above.

All problems from engineering, economics and social sciences that are characterised by evolving processes require continuously adapting models to model them. A speech or sound recognition system (Chaps. 12 and 13), an image recognition system (Chaps. 12 and 13), a multimodal information processing system, a stock prediction system, an intelligent robot, a system that predicts the emergence of insects based on climate (Chap. 19), etc. should always adjust its structure and functionality for a better performance over time. This book offers one approach to achieving this using spiking neural networks (SNN).

Everything is evolving, living organisms for sure, but what are the evolving rules, the laws that govern these processes? Are there any common evolving rules for every material item and for every living organism, along with their specific evolving rules? And what are the specific rules? Do these rules change over time, i.e. do they evolve as well? These are questions that we will address in this book to certain degree, as the process of addressing these issues is also evolving, with our improved understanding of both the processes and the methods that we can use to deal with them.

An evolving process, characterised by its evolving, governing rules, manifests itself in a certain way and produces data that in many cases, can be measured. Through analysis of this data, one can extract patterns of relationship, rules that describe the processes at a certain time, but do they describe the evolving process in the future?

Here, in addition to frequency, we introduce other main characteristics of evolving processes, used in the chapters of the book.

*Energy*

Energy is a major characteristic of any object and organism. It is a quantitative entity that they need to do some work, to move, to heat, to stay alive. There are many aspects of energy under this general definition, including quantum, physical, chemical, thermal, biological etc.

The Albert Einstein's most celebrated energy formula defines energy E that is needed to move and accelerate an object from a stationary position as depending on the mass of the object $m$ and the speed of light $c$:

$$E = m . c^2 \qquad (1.2)$$

The speed of light is used as a constant. It is appr. 300,000 km/s. Energy is associated with mass and the speed of light in vacuum.

Some characteristics of light are important to note as they are used in some of the methods in this book and discussed in the next section.

*Probability, entropy and information*

Having data measuring an evolving process, the question is how do we measure the information contained in the data? There are several ways to define and to measure information depending on the processes. One way is to use a measure of changes in a process called *entropy*, calculated with the use of a measure of uncertainties in these changes called *probability*, as explained below.

The formal theory of probability relies on the following three axioms, where p(E) is the probability of an event E to happen and p(¬E) is the probability of an event not to happen. E1, E2, …, Ek is a set of mutually exclusive events that form an universe U of all possible events, also called problem space:

Axiom 1. $0 \le p(E) \le 1$
Axiom 2. $\sum p(Ei) = 1$, $E1 \cup E2 \cup \cdots \cup Ek = U$, U-problem space
Corollary: $p(E) + p(\neg E) = 1$
Axiom 3. $p(E1 \vee E2) = p(E1) + p(E2)$, where E1 and E2 are mutually exclusive events.

Probabilities are defined as:

- Theoretical—some rules are used to evaluate a probability of an event.
- Experimental—probabilities are learned from data and experiments, e.g. throw dice 1000 times and measure how many times the event "getting the number 6" has happened.
- Subjective—probabilities are based on common sense human knowledge, such as defining that the probability of getting number 6" after throwing dice is 1/6th, without really throwing it at all.

A random variable $x$ is characterized at any moment of time by its *uncertainty* in terms of what value this variable will take in the next moment—its *entropy*. A measure of uncertainty $h(x_i)$ can be associated with each random value $x_i$ of a random variable $x$, and the total uncertainty $H(x)$, called *entropy*, measures our lack of knowledge, the seeming disorder in the space of the variable x:

$$H(X) = \sum_{i=1,\ldots,n} p_i \cdot h(x_i), \tag{1.3}$$

where: $p_i$ is the probability of the variable x taking the value of $x_i$; $h(x_i) = \log(1/p_i)$.

The following axioms for the entropy $H(x)$ apply:

- monotonicity: if $n > n'$ are number of events (values) that a variable x can take, then $Hn(x) > Hn'(x)$, so the more values x can take, the greater the entropy.
- additivity: if x and y are independent random variables, then the joint entropy H(x, y), meaning H(x AND y), is equal to the sum of H(x) and H(y).

The following log function satisfies the above two axioms:

$$h(x_i) = \log(1/p_i) \tag{1.4}$$

If the log has a basis of 2, the uncertainty is measured in [bits], and if it is the natural logarithm ln, then the uncertainty is measured in [nats].

$$H(X) = \sum_{i=1,\ldots,n} (p_i \cdot h(x_i)) = -c \cdot \sum_{i=1,\ldots,n} (p_i \cdot \log p_i), \tag{1.5}$$

where c is a constant.

Based on the *Claude Shannon's* measure of uncertainty—*entropy,* we can calculate an overall probability for a successful prediction for all states of a random variable x, or the predictability of the variable as a whole:

$$P(x) = 2^{-H(x)} \tag{1.6}$$

The max entropy is calculated when all n values of a random variable x are equiprobable, i.e. they have the same probability 1/n—a uniform probability distribution:

$$H(X) = - \sum_{i=1,\ldots,n} p_i \cdot \log p_i \leq \log n \qquad (1.7)$$

*Joint entropy* between two random variables x and y (for example, an input and an output variable in a system) is defined by the formulas:

$$H(x, y) = - \sum_{i=1,\ldots,n} p\left(x_i \, AND \, y_j\right) \cdot \log p\left(x_i \, AND \, y_j\right) \qquad (1.8)$$

$$H(x, y) \leq H(x) + H(y) \qquad (1.9)$$

*Conditional entropy*, i.e. measuring the uncertainty of a variable y (output variable) after observing the value of a variable x (input variable), is defined as follows:

$$H(y|x) = - \sum_{i=1,\ldots,n} p\left(x_i, y_j\right) \cdot \log p\left(y_j|x_i\right) \qquad (1.10)$$

$$0 \leq H(y|x) \leq H(y) \qquad (1.11)$$

Entropy can be used as a measure of the *information* associated with a random variable *x*, its uncertainty, and its predictability.

The *mutual* entropy between two random variables, also simply called *information*, can be measured as follows:

$$I(y; x) = H(y) - H(y|x) \qquad (1.12)$$

The process of on-line information entropy evaluation is important as in a time series of events, after each event has happened, the entropy changes and its value needs to be re-evaluated.

*Bayesian* conditional probability is calculated using the following formula, which represents the conditional probability between two events C and A in terms of event A to happen if the event C has happened (Tamas Bayes, 18 century):

$$p(A|C) = p(C|A) \cdot p(A)/p(C) \qquad (1.13)$$

It follows from the equations:

$$p(A \wedge C) = p(C \wedge A) = p(A|C)p(C) = p(C|A)\,p(A) \qquad (1.14)$$

*Measuring information as correlation between variables*

Correlation coefficients represent the relationship between variables. For every variable $x_i$ ($i = 1, 2, ..., d_1$) its correlation coefficients $Corr(x_i, y_j)$ with all other variables, including output variables $y_j$ ($j = 1, 2, ..., d_2$), are calculated. The following is the formula to calculate the Pearson correlation between two variables x and y based on n values for each of them:

$$Corr = SUM_i((x_i - Mx)(y_i - My))/[(n-1)Stdx . Stdy], \qquad (1.15)$$

where: Mx and My are the mean values of the two variables x and y, and Stdx and Stdy are their respective standard deviations.

*Measuring the level (the value) of information carried in a variable (its importance)*

The *t-test* and the *SNR* methods evaluate how important a variable is to discriminate samples belonging to different classes. For a case of two class problem, a SNR ranking coefficient for a variable x is calculated as an absolute difference between the mean value M1x of the variable for class 1 and the mean M2x of this variable for class 2, divided to the sum of the respective standard deviations:

$$SNR\_x = abs(M1x - M2x)/(Std1x + Std2x) \qquad (1.16)$$

A similar formula is used for the t-test:

$$Ttest\_x = abs(M1x - M2x)/(Std1x^2/N1 + Std2x^2/N2) \qquad (1.17)$$

where: N1 and N2 are the numbers of samples in class 1 and class 2 respectively.

*Transformation of data from one information space to another*

A set of variables measured to carry information for an evolving process form the *problem space, or the information space, the variables representing the dimensions of the space.* These variables can be used to create another set of variables in a new information space, that retains the main information from the original problem space but potentially reduces the dimensionality of the space into a smaller set of variables. Two of the most common transformations are Principal Component Analysis (PCA) and Linear Discriminant Analysis (LDA).

*Principal Component Analysis (PCA)*

PCA aims at finding a representation of a problem space X defined by its variables $X = \{x1, x2, ..., xn\}$ into another orthogonal space having a smaller number of dimensions defined by another set of variables $Z = \{z1, z2, ..., zm\}$, such that every data vector **x** from the original space is projected into a vector **z** of the new space, so that the distance between different vectors in the original space X is maximally preserved after their projection into the new space Z.

*Linear Discriminant Analysis (LDA)*

*LDA* is a transformation of classification data from the original space into a new space of LDA coefficients that has an objective function to preserve the distance between the samples using also the class label to make them more distinguishable between the classes.

## 1.3 Light and Sound

Light and Sound are of a special importance as they, first, affect how we perceive the world, and second, the way we perceive them can be used as inspiration for SNN architectures and for brain-inspired AI that deal with visual and audio information (Chaps. 12 and 13).

*Light* is important electromagnetic radiation that is characterised by certain frequencies and energy. Figure 1.7 shows a spectrum of electromagnetic radiations with *light* being part of it.

Visible light is having wavelengths in the range of 400–700 nanometres (nm), or $4.00 \times 10^{-7}$ to $7.00 \times 10^{-7}$ m, between the infrared (with longer wavelengths) and the ultraviolet (with shorter wavelengths). This wavelength means a frequency range of roughly 430–750 terahertz (THz). The speed of light is used as an universal constant. It is 299,792,458 m/s.

The primary properties of visible light are: intensity, propagation direction, frequency or wavelength spectrum, polarization, energy.

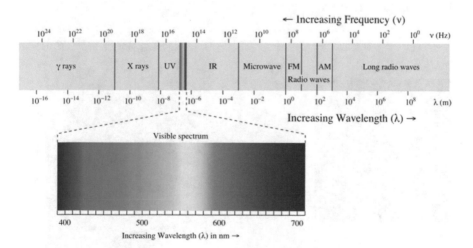

**Fig. 1.7** The frequencies and the wavelengths of electromagnetic radiation, visible *light* being part of a small spectrum [43]

Light has the properties of both:

– electromagnetic waves characterised by frequencies;
– quantum particles, called 'photons'—that is the energy transferred from the light.

When white light illuminates an object or a face, the reflected light at different pixels may have different brightness as the reflected light has different frequencies (see Fig. 1.8). Different brightness means different frequencies of the wave that reaches our retinas. The brightest spots activate earliest the corresponding cells and they send the first signals (spikes) to the brain. This principle is used in some SNN models described in Chaps. 4, 5 and 6 as Rank Order Coding.

The human brain perceives visual information as a trajectory of activation of brain areas in time-space (Chap. 3). The creation of computational models for visual information processing is a subject of computer vision. The way visual information is perceived in the human brain is discussed in Chap. 3 and used in Chap. 13.

In Chaps. 12 and 13 of the book SNN models are developed for both visual and audio information processing and for their integration.

*Sound* is an oscillation under pressure, that is spread as waves in a medium. Sound waves are characterised by:

– Frequency,
– Amplitude
– Speed
– Direction.

Sound that is perceptible by humans has frequencies from about 20 Hz to 20,000 Hz (see Fig. 1.9).

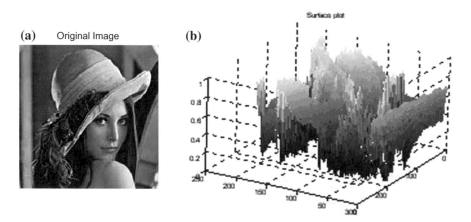

**Fig. 1.8** Original image (**a**) is represented as different intensities of brightness that represent the different frequencies of reflecting light at these pixels—(**b**). And this is how it activates the retina, first brighter pixels are perceived as shown on the z axis of the figure (**b**)

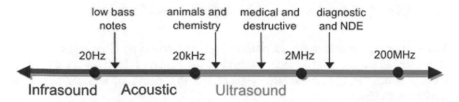

**Fig. 1.9** Some sound frequencies with their approximate ranges for different uses (from [43])

In air, corresponding wavelengths of sound waves range from 17 m to 17 mm. Sometimes speed and direction are combined as a velocity vector; wave number and direction are combined as a wave vector and power of the signal at different frequencies over time is represented as spectrum (Fig. 1.10). A power spectrum represents frequencies of the signal and their power in time. Figure 1.10 shows a spectrogram of the Mozart's music from Fig. 1.6, representing the power of frequencies (on the y axis) over time (on the x axis) as spectro-temporal data.

The way sound is perceived in the human brain is discussed in Chap. 3 and used in Chap. 13.

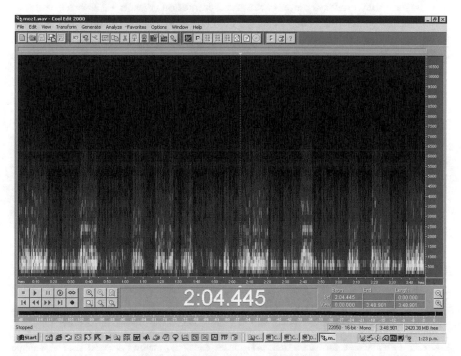

**Fig. 1.10** A spectrogram of the Mozart's music from Fig. 1.6, representing the power of frequencies (on the y axis) over time (on the x axis) as spectro-temporal data

## 1.4    Evolving Processes in Time-Space and Direction

Many evolving processes (in addition to light and sound as discussed above) are characterised by direction (or orientation) in which the signals or the waves spread. Examples are the spread of brain signals and the spread of seismic signals as illustrated below.

Deep learning trajectories of time-space directed connections are created during learning and recall in the brain as discussed in Chap. 3. Chapter 11 introduces a method for modelling time-space and direction on the case study of fMRI (functional Magnetic Resonance Image) and DTI (Difussion Tensor Image) data (see also Fig. 1.11).

Figure 1.11 shows orientational information from a DTI image. Left image shows an axial slice of a single subject's DTI data, registered to structural and MNI standard space. The right image shows a close-up of the right posterior corpus callosum. Directions corresponding to each colour are as follows: Red—left to right or right to left; green—anterior to posterior or posterior to anterior and blue— superior to inferior or inferior to superior (see Chap. 11).

Before an earthquake happens, tectonic pressure, measured at one seismic centre, causes a pressure at another, etc. a chain of such reactions eventually manifested as an earthquake at a final place. Detecting time–space and direction of changes of seismic data may enable a better earthquake prediction. Figure 1.12 shows the map of New Zealand seismic centers and the created map of the direction of changes in these seismic data as edges of a graph developed in a SNN in Chap. 19 of the book. Spike-time learning in SNN allows for directions of changes in the data to be

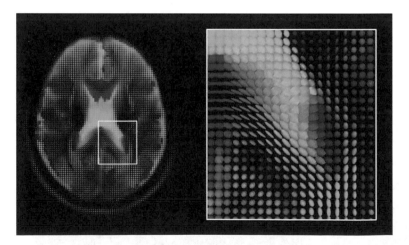

**Fig. 1.11** Orientation information from DTI image. Left image shows an axial slice of a single subject's DTI data, registered to structural and MNI standard brain template. The Right image shows a close-up of the right posterior corpus callosum. Directions corresponding to each colour are as follows: Red—left to right or right to left; green—anterior to posterior or posterior to anterior and blue—superior to inferior or inferior to superior

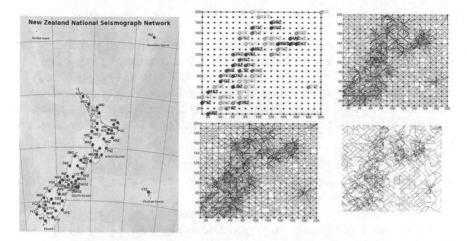

**Fig. 1.12**  Before an earthquake happens, tectonic pressure measured at one seismic centre, causes a pressure at another, also measured there etc. as a chain reaction that eventually manifests as an earthquake at a final place. Detecting the direction of changes of the seismic data may enable a better earthquake prediction. Left figure shows the map of New Zealand seismic centers. Right figures show created maps of direction of seismic changes in the corresponding centres as edges of graphs representing deep knowledge as a result of deep learning in Brain-inspired SNN (Chap. 19)

learned as directed connections between spiking neurons, showing which event happens first (a neuron Ni spikes) and which one follows (a neuron Nj spikes after).

Spike-time learning rules, such as STDP (Spike-Time Dependent Plasticity), to learn time-space and direction of events and changes in the data are discussed in Chaps. 4, 5 and 6 and applied across several applications in other chapters of the book.

Chapter 19 discusses also the detection of radio signals in space-time and direction from objects in the Universe called Pulsars. It also discusses recognition of fast moving objects in time-space and direction.

## 1.5   From Data and Information to Knowledge

Generally speaking, *data* are raw entities: numbers, symbols etc., e.g., 36.

*Information* is labelled, understood, interpreted data, e.g., the temperature of the human body is 36 °C.

*Knowledge* is the understanding of a human, the way we do things, interpretable information in different situations, general information; e.g.:

IF the human temperature is between 36 and 37 °C,
THEN it is most likely that human body is in a healthy state.

Some basic ways to represent data, information and knowledge of evolving processes are presented in this section, while next section discusses ways to represent deep knowledge, both acquired by humans and incorporated in a computer system.

The ultimate goal of information processing is the creation of knowledge. The process of knowledge acquisition from Nature is a continuous process that will never end. This knowledge is then used to understand Nature, to preserve it. From data and information, to knowledge discovery and back. This is what science is concerned with (Fig. 1.13). As shown in Fig. 1.13, modelling evolving processes requires a sequence of procedures that involve dealing with data, information and knowledge, e.g.:

- Searching for *data*: Observe phenomena; collect data; store data;
- Analyse data and extract *information* (e.g. pre-process data, filter, select features, visualise, label data);
- Create a model (learning, reasoning, validation);
- Extract *knowledge* (create/extract rules; reasoning with the knowledge—deductive, inductive);
- Adapt the model (accommodate new data and knowledge).

Extracting knowledge through observation of evolving processes has a long history. At the beginning, there was a school of learning that assumed that understanding of nature and its knowledge representation and articulation would not change with time. Aristotle was perhaps the most pronounced philosopher and encyclopaedist of this school.

**Aristoteles (384-322 BC)** was a pupil of Plato and teacher of Alexander the Great. He is credited with the earliest introduction of formal logic. Aristoteles introduced the theory of *deductive reasoning*.

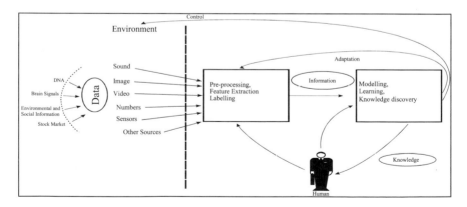

**Fig. 1.13** From data and information to knowledge representation through computational modelling (after [42])

*Example:*

*All humans are mortal (i.e. IF human THEN mortal)*
*New fact: Socrates is a human*
*Deducted inference: Socrates is mortal.*

Aristoteles introduced *epistemology,* which is based on the study of particular phenomena which leads to the articulation of knowledge (rules, formulas) across sciences: botany, zoology, physics, astronomy, chemistry, meteorology, psychology, etc. [18, 19]. According to Aristotle this knowledge was not supposed to change. In places, Aristotle went too far in deriving 'general laws of the universe from simple observations and over-stretched the reasons and conclusions. Because he was perhaps the philosopher most respected by European thinkers during and after the Renaissance, these thinkers, along with institutions, often took Aristotle's erroneous positions, such as defining inferior roles of women in society, which held back science and social progress for a long time.

Over many years after Aristotle, the logic he introduced was further developed into logic systems and rule based systems as a foundation of knowledge-based systems and AI. But this happened due to pioneers in programming analytical devices.

Perhaps the first one was the brilliant British mathematician **Ada Lovelace (1815–1852)** who is considered not only the first programmer, but the first person who demonstrated that an analytical device cannot only be used to crunch numbers, but to deal with symbols as well.

Based on symbolic representation several knowledge representation and reasoning theories and models were developed [6], such as:

- Relations and implications, e.g.: A $\rightarrow$ (implies) B.
- Propositional (true/false) logic, e.g.: IF (A and B) or C THEN D.
- Boolean logic (George Boole).
- Predicate logic: PROLOG.
- Probabilistic logic: e.g. Bayes formula: $p(A/C)) = p(C/A) \cdot p(A)/p(C)$, where p (A/C denotes the conditional probability for an event A to happen if event C has already happened.
- Rule based systems, expert systems, e.g. MYCIN [6].

All above knowledge representations could not deal well with uncertainty of events. Human cognitive behaviour and reasoning is not always based on exact numbers and fixed rules. In 1965 **Lotfi Zadeh (1920–2018)** introduced fuzzy logic [20, 21] that represents information uncertainties and tolerance in a linguistically expressed rules. He introduced fuzzy rules, containing fuzzy propositions and fuzzy inference.

Fuzzy propositions can have truth values between true (1) and false (0), e.g. the proposition "washing time is short" is true to a degree of 0.8 if the time is 4.9 min, where *Short* is represented as a fuzzy set with its membership function—see Fig. 1.14.

**Fig. 1.14** Fuzzy sets representing fuzzy terms of short, medium and long washing time, used to articulate and implement fuzzy rules, such as: IF Washing load is Small THEN Time of washing is Short

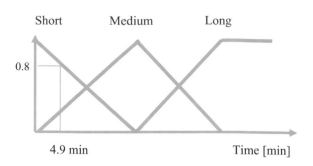

Fuzzy rules can be used to represent human knowledge and reasoning, e.g.

*IF washing load is small*
*THEN washing time is short.*

Fuzzy inference systems calculate exact outputs based on input data and a set of fuzzy rules. However, fuzzy rules need to be articulated in the first instance, they need to change, adapt, evolve through learning, to reflect the way human knowledge evolves. And that is what artificial neural networks (ANN) can do as discussed in Chap. 2. In principle, logic systems and rules, while useful, could be too rigid in some cases to represent the uncertainty in the natural phenomena and some cognitive behaviour. They are often difficult to articulate, and in principle not adaptive to change. 24 centuries after Aristotle, ANN can automate the process of knowledge discovery as they can learn from data and represent the essence of that as rules.

We call the rules discussed above "flat rules", as they represent only single events represented as "flat" vectors of features and there is no time or space of series of events defined in their relationship in time-space.

## 1.6  Deep Learning and Deep Knowledge Representation in Time-Space. How Deep?

### 1.6.1  Defining Deep Knowledge in Time-Space

In contrast to the "flat rules" as discussed in the previous section, *deep knowledge,* as introduced here, represents an informative spatio-temporal pattern of events that happen in space and time in their interaction. This pattern constitutes knowledge that can be interpreted for a better understanding of evolving processes in time-space and for predicting future events.

Continuous learning of time-space data, to capture dynamically changing and informative patterns, 'hidden' deep in time and space, and to predict future events, has been a fundamental science challenge. We call this here *deep learning in time-space.* Inspired by the deep learning capabilities of the human brain, we introduce here the concept of *deep knowledge in time-space*. This is also related to concept formation from multimodal data.

The concept of deep knowledge has been previously studied in different aspects [22–24]. In [23] deep knowledge is defined as '…knowledge that is concerned with underlying meanings and principles; integration of facts and feelings with previously acquired knowledge; fundamental knowledge with general applicability, such as the laws of physics, which can be used in conjunction with other deep knowledge to link evidence and conclusions …'.

Here we define deep knowledge in time-space in a brain-inspired, computational way and that is how it is used in the rest of the book.

Let is consider a set of events E = {E1, E2, …, En}. Each event Ei is defined as:

$$Ei = (Fi, Si, Ti, Pi), \qquad (1.18)$$

where: Fi is a function that defines an operation; Si is the space for function operation; Ti is the time of the function operation; Pi is probability of the function operation to take place.

An *event* could be a simple change in the value of a variable (e.g. increase above a threshold), or a complex cognitive process (e.g. reading a sentence), or an earthquake, etc.

*Time* can be in the *past*, in the *present* or in the *future*.

*Deep knowledge* can be represented in several ways. One way to represent deep knowledge is through deep rules as explained below.

Events Ei and Ej of an evolving process are represented by corresponding functions Fi and Fj, by spatial locations Si and Sj, by times of the events Ti and Tj, by probabilities of the events to happen Pi and Pj, and also by the strength of the relationship (association) between the events Wi, j:

$$\mathbf{W} = \{Wi,j\}, \quad i = 1, \dots, n; \quad j = 1, \dots, n. \qquad (1.19)$$

All parameters of an event can be represented as *crisp* or as *fuzzy* values with corresponding membership functions (see Fig. 1.14), e.g.:

– Location is around Si;
– Time is about Ti;
– Probability is about Pi (see fuzzy probabilities in [6, 21]);
– Strength is around Wi, j; or strength is High.

A hypothetical example of deep knowledge represented as a *deep fuzzy rule* is given below:

*IF* (*event E*1 : *function F*1, *location around S*1, *time about T*1, *probability about P*1)
  *AND* (*strength W*1, 2,)
  (*event E*2 : *function F*2, *location around S*2, *time about T*2, *probability about P*2)
  *AND* (*strength W*2, 3,)
  (*event E*3 : *function F*3, *location around S*3, *time about T*3, *probability about P*3)
  *AND* . . .

  . . .

  (*event En* : *function Fn*, *location around Sn*, *time about Tn*, *probability about Pn*)
  *THEN* (*An informative pattern Q from the measured evolving process is recognised*,
  *that may be used to predict a future event*)

$$(1.20a)$$

The fuzzy rule above allows for the event/task/process Q to be recognised even if only partial match of new data is entered and the rule is applied. This is a brain-inspired principle. For example, we end up with crisp movements as a result of the activation of slightly different clusters of neurons at slightly different times in their sequence, as a reaction to crisp of fuzzy stimuli.

As a partial case, no fuzzy terms will be used, but crisp ones, e.g. the following *deep crisp rule*:

  *IF* (*event E*1 : *function F*1, *location S*1, *time T*1)
    *AND* (*strength W*1, 2,)
    (*event E*2 : *function F*2, *location S*2, *time T*2)
    *AND* (*strength W*2, 3,)
    (*event E*3 : *function F*3, *location S*3, *time T*3)          $(1.20b)$
    *AND*. . .

    . . .

    (*event En* : *function Fn*, *location Sn*, *time Tn*)
    *THEN* (*Task/event Q is recognised*)

Crisp rules would be a case when activities of single neurons are measured in the brain at exact milliseconds time.

Deep knowledge is characterised by the following features:

(1)  It represents informative patterns of multimodal data, deep in time (theoretically unconstrained) and in space (when dealing with spatio-temporal data);
(2)  The knowledge is adaptable in an incremental, theoretically 'life-long' way;
(3)  The knowledge is not restricted by fixed structures;
(4)  The knowledge is obtained in supervised-, unsupervised or semi-supervised modes;

(5) The knowledge is interpretable for a better understating of the data and the processes that generated it;

(6) The knowledge can be used for early and accurate future event prediction.

Deep knowledge is what the human brain learns and manifests all the time, exemplified by:

- Listening or/and playing musical pieces;
- Playing a game;
- Visual perception;
- Predicting the movement of a predator;
- All sorts of cognition;
- Decision making;
- Consciousness;
- …and everything else the brain does.

## 1.6.2    How Deep?

Deep knowledge acquired in the human brain is manifested from hundreds of events in time and space activity of the brain, to hundreds of thousands, depending on the chosen scale to represent this knowledge. In terms of how deep in time the knowledge is represented, it can be represented at every month, or at every 100 ms or even at every single millisecond. In terms of how deep in space the knowledge can be represented, it can be at every large brain area or at every small neuronal cluster, or event at every single neuron.

We can define here the terms *time resolution* and *space resolution* of deep knowledge representation. Other terms that relate to time and space resolution are *temporal and spatial depth* of knowledge. Indeed, how deep knowledge representation can be or should be?

In terms of brain data, deep knowledge that represents an informative pattern of brain activity, can be discovered at a time resolution of a millisecond for various lengths of the brain signals. For example for a time length of 5 min, the events are totalling to 300,000 (300,000 ms), or for 1 s, totalling to 1000 ms, or for a time resolution of 500 ms, totalling to 600 events for a 5 min duration. And this is the temporal dept of the deep knowledge. The spatial dept of this knowledge will be defined by the spatial resolution of the measured brain activities and the size of the areas each measurement (e.g. an EEG channel or fMRI voxel) covers.

Listening to music (e.g. the 2 min piece by Mozart in Fig. 1.10) and measuring EEG brain activities every millisecond, will result in a deep spatio-temporal pattern of brain activities of 120,000 events. What about learning to play and then perform a musical piece without looking at notes for 4 h? That will involve tens of millions of brain activity events at a millisecond resolution. And this is deep knowledge representation that musicians learn.

In terms of seismic environmental data (see Figs. 1.5 and 1.12), deep knowledge that represents a pattern of seismic activity that is detected before an earthquake, can be discovered at a time resolution of a second totalling to 3153,600 time-point events for one year pattern before the earthquake, or to 1800 events 5 h before the earthquake, or to 60 events 1 min before the earthquake. And this is the dept in time of deep seismic knowledge related to risk of earthquake. The spatial dept of this knowledge will be defined by the spatial resolution of the measured seismic activities and the size of the areas each measurement (seismic variable) covers.

Groups of events that happen in a similar time at a similar place can be integrated together in larger knowledge "granules" (chunks of information), thus the term *deep knowledge granularity*.

Optimal spatio-temporal resolution and spatio-temporal depth of the patterns and granularity of knowledge representation for a given task are difficult to define. They vary across tasks and problems and are often restricted by the measured data as illustrated in some of the chapters of this book.

Illustration of deep rules as a result of deep learning in the human brain are given in Chap. 3 (extracted from data measuring brain activities). In Chaps. 6, 8, 10, 13, 18, 19 and in other chapters of the book deep rules are extracted from a deep trained brain-inspired SNN using time-space data.

### 1.6.3   Examples of Deep Knowledge Representation in This Book

Some elements of deep knowledge are manifested in computational models and systems, some of them presented in the book, such as:

- Hidden Markov Models (next section);
- Deep brain EEG and fMRI patterns representing brain perception or cognitive activities (Chaps. 8–11);
- Gene-regulatory networks in Bioinformatics and Neurogenetics (Chap. 17);
- Deep personalised patterns related to individual stroke prediction (Chap. 18)
- Deep climate patterns related to ecological events (Chap. 19);
- Deep geological patterns related to earthquake events (Chap. 19).

## 1.7   Statistical, Computational Modelling of Evolving Processes

*Computational modelling* of evolving processes aims at the development of mathematical and computational models that capture the essence of the dynamics of the processes and facilitate acquiring of knowledge.

## 1.7.1   Statistical Methods for Computational Modelling

Here some of the most popular methods are presented, also used in other chapters of the book for a comparative analysis between their performance and the performance of new methods based on SNN.

*Hidden Markov Models (HMM)* are techniques for modelling the temporal structure of a time series signal, or of a sequence of events [25]. It is a probabilistic pattern matching approach which models a sequence of patterns as the output of a random process. A HMM consists of an underlying Markov chain.

$$P(q(t+1)|q(t), q(t-1), q(t-2), \ldots, q(t-n)) \approx P(q(t+1)|q(t)), \qquad (1.20)$$

where $q(t)$ is state q sampled at a time t.

*Multiple linear regression methods (MLR)*

The purpose of multiple linear regression is to establish a quantitative relationship between a group of p predictor variables (X) and a response, y. This relationship is useful for:

- Understanding which predictors have the greatest effect.
- Knowing the direction of the effect (i.e., increasing x increases/decreases y).
- Using the model to predict future values of the response when only the predictors are currently known.

A linear model takes its common form of:

$$y = X A + b \qquad (1.21)$$

where: p is the number of the predictor variables; y is an n-by-1 vector of observations; X is an n-by-p matrix of regressors; A is a p-by-1 vector of parameters; b is an n-by-1 vector of random disturbances. The solution to the problem is a vector, $A'$ which estimates the unknown vector of parameters.

*Support vector machines*

This is a statistical learning technique introduced by Vapnik [26, 27] which first transforms the data from the original space to a higher dimensional space where data belonging to different classes (outputs) can be discriminated by a hyperplane defined by a set of bordering new data points called support vectors. This is illustrated in Fig. 1.15.

*Evaluating the error and accuracy of the computational models*

One way is to use the least squares solution, so that the model approximates the data with the least root mean square error (RMSE) as follows:

**Fig. 1.15** SVM hyperplane

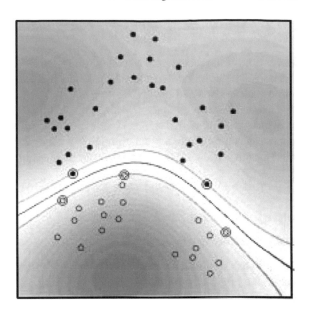

$$RMSE = SQRT(SUM\,i = 1, 2, \ldots, n((y_i - y_i')2)/n) \qquad (1.22)$$

where: $y_i$ is the desired value from the data set corresponding to an input vector $x_i$; $y_i'$ is the value obtained through the model for the same input vector $x_i$ and $n$ is the number of the samples (vectors) in the data set.

Another error measure is also used to evaluate the performance of a regression model—a non-dimensional error index (NDEI)—the RMSE divided to the standard deviation of the data set:

$$NDEI = RMSE/Std \qquad (1.23)$$

A popular method to measure the accuracy of a computational model is the area under the curve (AUC, or also called ROC)—Fig. 1.16, with a value of 1.0 being the best and 0.5 being the worst.

### 1.7.2  Global, Local and Transductive ("Personalised") Modelling [28]

Most of learning models and systems in artificial intelligence developed and implemented so far, are based on inductive inference methods, where a model (a function) is derived from data representing the problem space and this model is further applied on new data. The model is usually created without taking into

**Fig. 1.16** ROC curve is used to measure the accuracy of a computational model, with 1.0 being the best and 0.5 being the worst (from [43])

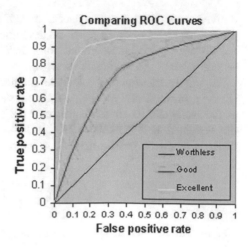

account any information about a particular new data vector (test data). An error is measured to estimate how well the new data fits into the model.

The models are in most cases *global models*, covering the whole problem space. Such models are for example: regression functions; some ANN models, and also— some support-vector machine (SVM) models, depending on the kernel function they use. These models are difficult to update on new data without using old data, previously used to derive the models. Creating a global model (function) that would be valid for the whole problem space is a difficult task, and in most cases—it is not necessary to solve.

Some global models may consist of many local models, that collectively cover the whole space and can be adjusted incrementally on new data. The output for a new vector is calculated based on the activation of one or several neighbouring local models. Such systems are the evolving connectionist systems (ECOS), for example—EFuNN and DENFIS (Chap. 2).

*Transductive modelling*

In contrast to the inductive learning and inference methods, transductive inference methods estimate the value of a potential model (function) only in a single point of the space (the new data vector) utilizing additional information related to this point [26]. This approach seems to be more appropriate for clinical and medical applications of learning systems, where the focus is not on the model, but on the individual patient. Each individual data vector (e.g.: a patient in the medical area; a future time moment for predicting a time series; or a target day for predicting a stock index) may need an individual, local model that best fits the new data, rather then—a global model. In the latter case the new data is matched into a model without taking into account any specific information about this data when creating the model.

Transductive inference is concerned with the estimation of a function in a single point of the space only. For every new input vector $x_i$ that needs to be processed for a prognostic task, the $N_i$ nearest neighbours, which form a sub-data set $D_i$, are derived from an existing data set D and, if necessary, generated from an existing model M. A new model Mi is dynamically created from these samples to approximate the function around point $x_i$. The system is then used to calculate the output value $y_i$ for this input vector $x_i$.

A simple transductive inference method is the k-nearest neighbour method (K-NN). In the K-NN method, the output value $y_i$ for a new vector $x_i$ is calculated as the average of the output values of the k nearest samples from the data set Di. In the weighted K-NN method (WKNN) the output $y_i$ is calculated based on the distance of the Ni nearest neighbour samples to $x_i$:

$$y_i = \frac{\sum_{j=1}^{Ni} w_j y_j}{\sum_{j=1}^{Ni} w_j} \tag{1.24}$$

where: $y_j$ is the output value for the sample $x_j$ from Di and $w_j$ are their weights measured as:

$$w_j = \frac{\max(d) - [d_j - \min(d)]}{\max(d)} \tag{1.25}$$

The vector d = [d1, d2, ... dNi] is defined as the distances between the new input vector $x_i$ and Ni nearest neighbours $x_j$, for j = 1 to Ni; max(d) and min(d) are the maximum and minimum values in d respectively. The weights $w_j$ have the values between min(d)/max(d) and 1; the sample with the minimum distance to the new input vector has the weight value of 1, and it has the value min(d)/max(d) in case of maximum distance.

Distance is usually measured as Euclidean distance:

$$\|\mathbf{x} - \mathbf{y}\| = \left[ \frac{1}{P} \sum_{j=1}^{P} |x_j - y_j|^2 \right]^{\frac{1}{2}} \tag{1.26}$$

Distance can be also measured as Pearson correlation distance, Hamming distance, cosine distance, etc. [27].

*WWKNN: Weighted examples, weighted variables K-NN* [28]

In the WKNN above the calculated output for a new input vector depends not only on the number of its neighboring vectors and their output values (class labels), as it is in the KNN method, but on the distance between these vectors and the new vector which is represented as a weight vector (W). It is assumed that all v input variables are used and the distance is measured in a v-dimensional Euclidean space with all variables having the same impact on the output variable.

But when the variables are ranked in terms of their discriminative power of class samples over the whole v-dimensional space, we can see that different variables have different importance to separate samples from different classes, therefore—a different impact on the performance of a classification model. If we measure the discriminative power of the same variables for a sub-space (local space) of the problem space, the variables may have a different ranking.

Using the ranking of the variables in terms of a discriminative power within the neighborhood of K vectors, when calculating the output for the new input vector, is the main idea behind the WWKNN algorithm [28], which includes one more weight vector to weigh the importance of the variables. The Euclidean distance dj between a new vector $x_i$ and a neighbouring one xj is calculated now as:

$$dj = SQR[SUM1 = 1 \text{ to } v(ci, l(xi, 1 - xj, 1))SQ2] \tag{1.27}$$

where: SQR denotes square root; SQ denotes square; SUM denotes summation function; ci, 1 is the coefficient weighing variable xl for in neighbourhood of xi. It can be calculated using a Signal-to-Noise Ratio (SNR) procedure that ranks each variable across all vectors in the neighborhood set Di of Ni vectors Ci = (ci, 1; ci, 2; ...; ci, v)

$$ci, l = Sl/SUM(Sl), \quad \text{for: } l = 1, 2, \ldots, v, \tag{1.28}$$

$$\text{where: } Sl = abs(Ml(\text{class } 1) - Ml(\text{class } 2))/(Stdl(\text{class } 1) + Stdl(\text{class } 2)) \tag{1.29}$$

Here Ml (class 1) and Stdl (class 1) are respectively the mean value and the standard deviation of variable xl for all vectors in Di that belong to class 1.

The new distance measure, that weighs all variables according to their importance as discriminating factors in the neighborhood area Di, is the new element in the WWKNN algorithm when compared to the WKNN.

Using the WWKNN algorithm, a "personalized" profile of the variable importance can be derived for any new input vector that represents a new piece of "personalised" knowledge. Weighting variables in personalized models is used in the TWNFI models (Transductive Weighted Neuro-Fuzzy Inference) in [29].

There are several open problems related to transductive learning and reasoning, e.g. how to choose the optimal number of vectors in a neighbourhood and the optimal number of variables, which for different new vectors may be different [30].

## 1.7.3 Model Validation

When a machine learning model is built based on a data set S, it needs to be validated in terms of its generalisation ability to produce good results on new, unseen data samples. There are several ways to validate a model:

- Train-test split of data: Splitting the data set S into two sets: Str for training, and Sts for testing the model;
- N-fold cross validation (e.g. 3, 5, 10): in this case the data set S is split randomly into k sub-sets S1, S2, ..., Sk and i = 1, 2, ... k times a model Mi is created on a

the data set S–Si and tested on the set Si; the mean accuracy across all k experiments is calculated.

- Leave-one-out cross validation (a partial case of the above method when the data set S is split N times, in each sub-set there is only one sample).

What concerns the whole task of feature selection, model creation and model validation, the above methods can be applied in two different ways:

- A "biased" way—features are selected from the whole set S using a filtering based method, and then a model is created and validated on the selected already features.
- An "un-biased" way—for every data subset Si in a cross validation procedure, first—features Fi are selected from the set S after set Si is removed from S (using some of the above discussed methods, e.g. SNR) and then—a model is created based on the feature set Fi; the model Mi is validated on Si using features Fi.

## 1.8  Brain-Inspired AI

*Artificial Intelligence (AI)* is part of the interdisciplinary information sciences area that develops and implements methods and systems that manifest cognitive behaviour [31–39].

Main features of AI are:

- learning,
- adaptation,
- generalisation,
- inductive and deductive reasoning,
- human-like communication.

Some more features are currently being developed:

- consciousness,
- self-assembly,
- self-reproduction,
- AI social networks.

Marvin Minsky (1961) [40] articulated the term *Artificial Intelligence* as computer systems that are able to perform: search, pattern recognition, learning, planning, inductive reasoning.

In [41] AI is defined as computer systems that exhibit human like intelligence. It is a group of science fields and technologies concerned with creating machines that take intelligent actions based on inputs. And also in [41] AI is defined as "… advanced digital technologies that enable machines to reproduce or surpass abilities that would require intelligence if humans were to perform them. This includes technologies that enable machines to learn and adapt, to sense and interact, to reason and plan, to optimise procedures and parameters, to operate autonomously, to be creative and to extract knowledge from large amounts of data...."

There is a trend in AI called *Artificial General Intelligence* (*AGI*) that considers machines to become able to perform any intellectual task that humans can do.

Another trend in AI is called *Technological Singularity*. This trend argues that machines will become super intelligent that they take over from humans and develop on their own, beyond which point the human societies can collapse in their present forms, which may ultimately lead to the perish of humanity.

**Stephen Hawking (b.1942–d.2018)** commented: "I believe there is no real difference between what can be achieved by a biological brain and what can be achieved by a computer. AI will be able to redesign itself at an ever-increasing rate. Humans, who are limited by slow biological evolution, couldn't compete and could be superseded by AI. AI could be either the best or the worst thing ever to happen to humanity..."

A new trend in AI is the *Brain-Inspired AI* (*BI-AI*), which is being developed and presented in this book. BI-AI systems use principles of deep learning in the human brain to reveal deep knowledge and to enable machines to manifest cognitive functions. BI-AI systems adopt structures and methods from the human brain to intelligently learn spatio-temporal data.

BI-AI systems have six distinctive features:

(1) They have their structures and functionality inspired by the human brain; they consist of spatially located neurons that create connections between them through deep learning in time-space by exchanging information—spikes. They are built of spiking neural networks (SNN), as explained in Chaps. 4–6 in the book.
(2) Being brain-inspired, BI-AI systems can achieve not only *deep learning*, but *deep knowledge* representation as well. *They are transparent.*
(3) They can manifest cognitive behaviour.
(4) They can be used for knowledge transfer between humans and machines as a foundation for the creation of symbiosis between humans and machines, ultimately leading to the integration of human intelligence and artificial intelligence (HI + AI) as discussed in the last chapter of the book.
(5) BI-AI systems are universal data learning machines, being superior than traditional machine learning techniques when dealing with time-space data.
(6) BI-AI systems can help understand-, protect-, and cure the human brain.

Box 1.1 elaborates further on the main features above and lists 20 features of BI-AI as presented and demonstrated in various chapters of the book. Some of them are in a preliminary stage of development and more can be expected in the future.

**Box 1.1.  Twenty structural, functional and cognitive features of BI-AI systems**

*Structural Features:*

1. The structure and organisation of a system follows the structure and organisation of the human brain, for example through using a 3D brain template.

2. Input data and information is encoded and processed in the system as spikes over time.

3. A system is built of spiking neurons and connections, forming SNN.

4. A system is scalable, from hundreds to billions of neurons and trillions of connections.

5. Inputs are mapped spatially into the 3D system structure.

6. Output information is also presented as spike sequences.

*Functional Features*

7. A system operates in a highly parallel mode, potentially all neurons operating in parallel.

8. A system can be implemented on various computer platforms, but more efficiently on neuromorphic highly parallel platforms and on quantum computers (if available).

9. Self-organised unsupervised, supervised and semi-supervised *deep learning* is performed using brain-inspired spike-time learning rules.

10.     The learned spatio-temporal patterns represent *deep knowledge*.

11.     A system operates in a fast, incremental and predictive learning mode.

12.     Different time scales of operation, e.g. nanoseconds, milliseconds, minutes, hours, days, millions of years (e.g. genetics), can be represented, possibly in their integration

13.     A system can process multimodal data from all levels of functionality per Fig. 1.1 (e.g. quantum; genetic;neuronal; ensembles of neurons; etc.), possibly in their integration.

*Cognitive features*

14.     A system can communicate with humans in a natural language.

15.     A system can make abstractions and discover new knowledge (e.g. rules) through self-observing its structure and functions.

16.     A system can process all kinds of sensory information that is processed by the human brain, including: visual-, auditory-, sensory-, olfactory-, gustatory, if necessary in their integration.

17.     A system can manifest both sub-conscious and conscious processing of stimuli.

18.     A system can recognise and express emotions and consciousness.

19.     Deep knowledge can be transferred between humans and machines using brain signals and other relevant information, e.g. visual, etc.

20.     BI-AI systems can form societies and communicate between each other and with humans achieving a constructive symbiosis between humans and machines.

We will argue and will demonstrate in this book that BI-AI systems, if properly developed and used, can bring a tremendous technological progress across all areas of human activities and sciences and technologies, such as:

- Early disease diagnosis and disease prevention (Chap. 18);
- Affective robots for homes and for elderly (Chaps. 8 and 14);
- Improved decision support and productivity (Chap. 20);
- Improved human intelligence and creativity (Chaps. 12, 13 and 22);
- Improved lives and longevity (Chaps. 17 and 18);
- Predicting and preventing hazardous events (Chap. 19);
- And many more.

Some of the above applications are developed and illustrated in the book.

## 1.9   Chapter Summary and Further Readings for Deeper Knowledge

This chapter discusses fundamentals of evolving processes in space and time and some of the challenges to model them and to acquire deep knowledge. All methods and concepts presented in this chapter are used in different chapters of the book as a fundamental information. More about this topic can be found in [17, 42].

With the large scale data collection across all science, technology and social areas, machine learning from data to create models and extract rules and knowledge became a necessity. This led to the establishment of artificial neural networks as major machine learning techniques that borrows some basic principle of information processing and learning from the brain (Chap. 2).

But, the human brain learns data in a *deep learning* mode and understands the evolving processes through the acquired *deep knowledge* (Chap. 3). How this could be used to create brain-inspired SNN systems is discussed in Chaps. 4–7 and how SNN can be used to create BI-AI application systems is presented in Chaps. 8–19. Chapters 20, 21 and 22 present some new directions of research in SNN and BI-AI.

Further recommended readings on specific topics can include:

- Aristoteles' epistemology [18, 19];
- Fuzzy logic [20, 21];
- Hidden Markov Models [25];
- Statistical Learning Theory [26, 27];
- Neuro-fuzzy systems [6, 28–30].

In [42] the author has expressed a view that in 20 years time or so AI will become common tools, perhaps as the spreadsheets are now, but scientists and technologists have to work hard in order to make this happened.

**Acknowledgements**  Some of the material in this chapter was first published by the author in [6, 9, 30].

# References

1. Stephen Hawking, *A Brief History of Time* (Bantam Books, NY, 1990)
2. A. Einstein, *Ideas and Opinions* (Wings Books, NY, 1954)
3. C. Darwin, *On the Origin of Species by Means of Natural Selection, or the Preservation of Favored Races in the Struggle for Life*, 1st edn. (John Murray, London, 1859), p. 502
4. C. Darwin, *On the Origin of Species by Means of Natural Selection, or the Preservation of Favored Races in the Struggle for Life*, 2nd edn. (John Murray, London, 1860)
5. J.H. Holland, *Adaptation in Natural and Artificial Systems*, 2nd edn. (MIT Press, Cambridge, MA, 1992) (1st edn., University of Michigan Press, 1975)
6. N.K. Kasabov, *Foundations of Neural Networks, Fuzzy Systems and Knowledge Engineering* (MIT Press, Cambridge, MA, USA, 1996)
7. G. Carpenter, S. Grossberg, *Adaptive Resonance Theory* (MIT Press, 2017)
8. R. Penrose, *The Emperors New Mind* (Oxford University Press, 1989)
9. N. Kasabov (ed.), *Springer Handbook of Bio-/Neuroinformatics* (Springer, 2014)
10. R. Dawkins, *The Selfish Gene* (Oxford University Press, 1989)
11. D. Hebb, *The Organization of Behavior* (Wiley, New York, 1949)
12. D.E. Goldberg, *Genetic Algorithms in Search, Optimization and Machine Learning* (Addison Wesley, 1989)
13. Abraham et al., Insect biochem. Mol. Biol. **23**(8), 905–912 (1993)
14. J. Gleick, *Chaos Making A New Science* (Viking Books, USA, 1987), p. 1987
15. O. Barndorff-Nielsen et al., *Chaos and Networks: Statistical and Probabilistic Aspects* (Chapman & Hall, London, New York, 1993), p. 1993
16. F.C. Hoppensteadt, Intermittent chaos, self organizing and learning from synchronous synaptic activity in model neuron networks. Proc. Natl. Acad. Sci. USA **86**, 2991–2995 (1989)
17. Protein Database, https://www.ncbi.nlm.nih.gov/protein
18. J. Ferguson, *Aristotle* (Twayne Publishers, New York, 1972)
19. J. De Groot, *Aristotle's Empiricism: Experience and Mechanics in the 4th Century BC* (Parmenides Publishing, 2014). ISBN 978-1-930972-83-4
20. L.A. Zadeh, Fuzzy sets. Inf. Control **3**(8), 338–353 (1965)
21. L.A. Zadeh, Fuzzy logic. IEEE Comput. **4**(21), 83–93 (1988)
22. B. Chandrasekaran, S. Mittal, Deep versus compiled knowledge approaches to diagnostic problem-solving. Int. J. Man Mach. Stud. **19**(2), 425–436 (1983). https://doi.org/10.1016/S0020-7373(83)80064-9
23. P.L. Rogers, G.A. Berg, J.V. Boettcher, C. Howard, L. Justice, K.D. Schenk, *Encyclopedia of Distance Learning*, vol. 4 (IGI Global, 2009)
24. A. Bennet, D. Bennet, *Knowledge-Based Development for Cities and Societies: Integrated Multi-Level Approaches* (IGI Global, 2010), https://doi.org/10.4018/978-1-61520-721-3.ch009
25. L.R. Rabiner, A tutorial on hidden Markov models and selected applications in speech recognition. Proc. IEEE **77**(2), 257–285 (1989)
26. V.N. Vapnik, *Statistical Learning Theory* (Wiley, New York, 1998), p. 1998
27. V. Cherkassky, F. Mulier, *Learning from Data: Concepts, Theory, and Methods* (Wiley, New York, 1998), p. 1998
28. N. Kasabov, Adaptation and Interaction in dynamical systems: modelling and rule discovery through evolving connectionist systems. Appl. Soft Comput. **6**(3), 307–322 (2006)
29. Q. Song, N. Kasabov, NFI: a neuro-fuzzy inference method for transductive reasoning. IEEE Trans. Fuzzy Syst. **13**(6), 799–808 (2005)
30. N. Kasabov (2007) *Evolving Connectionist Systems: The Knowledge Engineering Approach*, 2nd ed. (Springer, London, 2007)
31. D. Dennett, *From Bacteria to Bach and Back. The Evolution of Minds* (W. W. Norton & Co., 2017)

32. S. Russell, P. Norvig, *Artificial Intelligence: A Modern Approach*, 3rd edn. (Upper Saddle River, NJ, Prentice Hall, 2009)
33. N.J. Nilsson, *The Quest for Artificial Intelligence: A History of Ideas and Achievements* (Cambridge University Press, Cambridge, UK, 2010)
34. P. McCorduck, *Machines Who Think: A Personal Inquiry into the History and Prospects of Artificial Intelligence*, 2nd edn. (Natick, MA, 2004)
35. N. Bostrom, V.C. Millar, Future progress in artificial intelligence: a survey of expert opinion, in *Fundamental Issues of Artificial Intelligence*, ed. by V.C. Muller (Springer, Synthese Library, Berlin, 2016), p. 553
36. J.H. Andreae (ed.), *Man-Machine Studies*. ISSN 0110 1188, nos. UC-DSE/4 and 5 (1974)
37. J.H. Andreae: *Thinking with the Teachable Machine* (Academic Press, 1977)
38. R.S. Sutton, A.G. Barto, *Reinforcement Learning* (MIT Press, Barto, 1998)
39. A. Clark, *Surfing Uncertainty* (Oxford University Press, 2016)
40. M. Minski, Steps toward artificial intelligence. Proc. IRE **49**, 8–30 (1961)
41. The New Zealand AI Forum, http://aiforum.org.nz
42. The AUT AI Initiative, http://www.aut.ac.nz/aii
43. Wikipedia, http://www.wikipedia.org

# Chapter 2
# Artificial Neural Networks. Evolving Connectionist Systems

Classical artificial neural networks (ANN) were developed to learn from data. Evolving connectionist systems (ECOS) were further developed by the author and taken further by other researchers not only to learn in an adaptive, incremental way from data that measure evolving processes, but to extract rules and knowledge from the trained systems. Both methods were initially inspired by some principles of learning in the brain, but then they were developed mainly as machine learning and AI tools and techniques, with a wider scope of applications. Many of the architectures and learning methods of ANN and ECOS were used in the development of SNN, deep learning systems and brain-inspired AI discussed in other chapters of the book.

The chapter is organised in the following sections:

2.1. Classical Artificial Neural Networks: SOM, MLP, CNN, RNN.
2.2. Hybrid and knowledge-based ANN: Opening the "black box".
2.3. Evolving Connectionist Systems.
2.4. Evolving Fuzzy Neural Networks. EFuNN.
2.5. Dynamic Evolving Neuro Fuzzy Systems. DENFIS.
2.6. Other ECOS methods and systems.
2.7. Summary and further readings for deeper knowledge.

## 2.1 Classical Artificial Neural Networks: SOM, MLP, CNN, RNN

ANNs are computational models that mimic the nervous system in its main function of adaptive learning and generalization. ANNs are universal computational models. One of the most popular artificial neuron models is the McCulloch and Pitts neuron developed in 1943. It was used in early ANNs, such as Rosenblatt's Perceptron [1] and multilayer perceptron [2–5]. A simple example is given in Fig. 2.1.

© Springer-Verlag GmbH Germany, part of Springer Nature 2019
N. K. Kasabov, *Time-Space, Spiking Neural Networks and Brain-Inspired Artificial Intelligence*, Springer Series on Bio- and Neurosystems 7,
https://doi.org/10.1007/978-3-662-57715-8_2

**Fig. 2.1** A diagram of a
simple artificial neuron

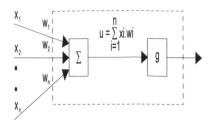

Various types of ANN architectures and learning algorithms have been developed, e.g.:

- Self-Organising maps (SOM) and unsupervised learning algorithms [6–8];
- Multilayer perceptrons (MLP) and back propagation supervised learning algorithm [3–5];
- Adaptive Resonance Theory (ART) [9];
- Recurrent ANN and reinforcement learning [10];
- Convolutional and deep learning ANN [11–13].

This section covers some classical models of ANN. They have also influenced the development of the brain-like spiking neural networks (SNN) and brain-inspired AI techniques presented in other chapters of the book.


### 2.1.1   Unsupervised Learning in Neural Networks.
###           Self-organising Maps (SOM)

Unsupervised learning is concerned with finding structures in the data. Techniques include:

- Clustering of data;
- Vector quantisation.

A basic technique to apply when finding structures in data is measuring distance (or similarity) between data vectors (data samples).

*Measuring distance (or similarity)* is a fundamental issue in all statistical and ANN learning methods. The following are some of the most used methods for measuring distance, illustrated on two, n-dimensional data vectors $\mathbf{x} = (x_1, x_2, \ldots, x_n)$ and $\mathbf{y} = (y_1, y_2, \ldots, y_n)$:

- Euclidean distance:

$$D(\mathbf{x}, \mathbf{y}) = \sqrt{\left[\left(\sum\nolimits_{i=1,\ldots,n} (x_i - y_i)^2\right)/n\right]} \qquad (2.1)$$

- Hamming distance (for binary vectors):

$$D(\mathbf{x}, \mathbf{y}) = \left( \sum_{i=1,\dots,n} |x_i - y_i| \right) / n \qquad (2.2)$$

where absolute values of the difference between the two vectors are used.
- Local fuzzy normalized distance [14–17]:

A local normalized fuzzy distance between two fuzzy membership vectors $\mathbf{x}_f$ and $\mathbf{y}_f$ that represent the membership degrees to which two real vector data $\mathbf{x}$ and $\mathbf{y}$ belong to pre-defined fuzzy membership functions is calculated as:

$$D(\mathbf{x_f}, \mathbf{y_f}) = \|\mathbf{x_f} - \mathbf{y_f}\| / \|\mathbf{x_f} + \mathbf{y_f}\|, \qquad (2.3)$$

where $\|\mathbf{x_f} - \mathbf{y_f}\|$ denotes the sum of all the absolute values of a vector that is obtained after vector subtraction (or summation in case of $\|\mathbf{x_f} - \mathbf{y_f}\|$) of two vectors $\mathbf{x_f}$ and $\mathbf{y_f}$ of fuzzy membership values; "/" denotes division.

- Cosine distance:

$$D = 1 - SUM(\sqrt{xiyi}/\sqrt{xi^2} \sqrt{yi^2}) \qquad (2.4)$$

- Correlation distance:

$$D = 1 - \sum_{i=1}^{n} (xi - \overline{xi})(yi - \overline{yi}) / \sum_{i=1}^{n} (xi - \overline{xi})^2 (yi - \overline{yi})^2 \qquad (2.5)$$

where $\overline{xi}$ is the mean value of the variable $x_i$.

Many unsupervised learning neural network methods are based on clustering of input data. Clustering is the process of defining how data are grouped together based on similarity.

*Clustering* results in the following outcomes:

- Cluster centres: these are the geometrical centres of the data grouped together; their number can be either pre-defined (batch-mode clustering), or not defined a priori but evolving;
- Membership values, defining for each data vector to what cluster it belongs to. This can be either a crisp value of 1 (the vector belongs to a cluster), or 0—it does not belong to a cluster (as it is in the k-means method), or a fuzzy value between 0 and 1 showing the level of belonging—in this case the clusters may overlap (fuzzy clustering).

*Self-organizing Maps—SOMs*

Here, the principles of the traditional SOMs are outlined first, and then some modifications that allow for dynamic, adaptive node creation, are presented.

Self-organizing maps belong to the vector quantisation methods where proto-types are found in a prototype (feature) space (map) of dimension l rather than in the input space of dimension d, l < d. In Kohonen's self-organizing feature map (SOM) [7, 8, 18, 19] the new space is a topological map of 1-, 2-, 3-, or more dimensions (Fig. 2.2).

The main principles of learning in SOM are as follows:

- Each output neuron specializes during the training procedure to react to similar input vectors from a group (cluster) of input data. This characteristic of SOM tends to be biologically plausible as some evidences show that the brain is organised into regions which correspond to similar sensory stimuli. A SOM is able to extract abstract information from multi-dimensional primary signals and to represent it as a location, in one-, two-, and three- etc. dimensional space.

- The neurons in the output layer are competitive ones. Lateral interaction between neighbouring neurons is introduced in such a way, that a neuron has a strong excitatory connection to itself, and less excitatory connections to its neighbouring neurons in a certain radius; beyond this area, a neuron either inhibits the activation of the other neurons by inhibitory connections, or does not influence it. One possible neighbouring rule that implements the described strategy is the so called *"Mexican hat"* rule. In general, this is "the winner-takes all" scheme, where only one neuron is the winner after an input vector was fed, and a competition between the output neurons has taken place. The fired neuron represents the class, the group (cluster), the label, or the feature to which the input vector belongs.

- SOMs transform or preserve similarity between input vectors from the input space into *topological closeness of neurons* in the output space represented as a topological map. Similar input vectors are represented by near points (neurons) in the output space. Example is given in Fig. 2.3.

The unsupervised algorithm for training a SOM, proposed by Teuvo Kohonen, is outlined in Box 2.1. After each input pattern is presented, the winner is established and the connection weights in its neighbourhood area $N_t$ increase, while the

**Fig. 2.2** Example of a simple SOM architecture of 2 input neurons and 2D output topological map (after [17, 20])

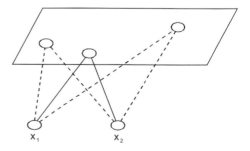

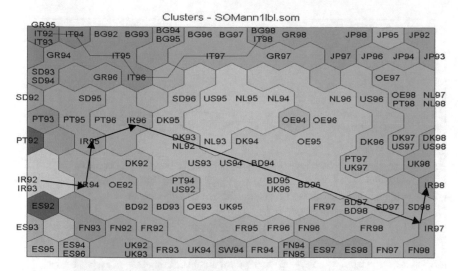

**Fig. 2.3** An example of SOM topological map trained on macroeconomic annual data of European countries. Countries with similar economic parameters are clustered together. The change in the economic development of Ireland is traced over years on the map (after [17])

connection weights outside the area are kept unchanged. $\alpha$ is a learning parameter. Training is done through a number of training iterations so that at each iteration the whole set of input data is propagated through the SOM and the connection weights are adjusted.

SOMs learn statistical features. The synaptic weight vectors tend to approximate the density function of the input vectors in an orderly fashion. Synaptic vectors $w_j$ converge exponentially to centres of groups of patterns and the nodes of the output map represent to a certain degree the distribution of the input data. The weight vectors are also called *reference vectors*, or reference codebook vectors. The whole weight vector space is called a *reference codebook*.

In SOM the topology order of the prototype nodes are pre-determined and the learning process is to "drag" (in terms of connection weights of the output nodes to the input variables) the ordered nodes onto the appropriate positions in the low dimensional feature map. As the original input manifold can be complicated and an inherent dimension larger than that of the feature map (usually set as 2 for visualization purpose), the dimension reduction in SOM may become inappropriate for complex data analysis tasks.

---

**Box 2.1 The Self-organsing Map (SOM) Training Algorithm**

K0.   Assign small random numbers to the initial weight vectors $\mathbf{w}j(t= 0)$, for every neuron j from the output map.

K1    Apply an input vector $\mathbf{x}$ at the consecutive time moment t.

K2    Calculate the distance $d_i$ (in n-dimensional space) between x and the weight vectors $\mathbf{w}j(t)$ of each neuron j. In Euclidean space this is calculated as follows:
$d_j = \mathrm{sqrt}((\Sigma((x_i - w_{ij})^2)))$.

K3    The neuron k which is closest to x is declared the winner. It becomes a centre of a neighbourhood area Nt.

K4    Change all the weight vectors within the neighbourhood area:
$\mathbf{w}_j(t + 1) = \mathbf{w}_j(t) + \alpha \cdot (\mathbf{x} - \mathbf{w}_i(t))$, if $j \in Nt$,
$\mathbf{w}_j(t + 1) = \mathbf{w}_j(t)$, if j is not from the area Nt of neighbours.

All of the steps from K1 to K4 are repeated for all the training instances. Nt and $\alpha$ decrease in time. The same training procedure is repeated again with the same training instances until convergence.

---

Some of the principles of SOM, such as topological mapping, are used and further developed in the evolving SOM (further section in this chapter) and to certain degree in the brain-inspired SNN as discussed in Chap. 6.

## 2.1.2   Supervised Learning in ANN. Multilayer Perceptron and the Back Propagation Algorithm

Connectionist systems for supervised learning learn from pairs of data $(\mathbf{x}, \mathbf{y})$, where the desired output vector $\mathbf{y}$ is known for an input vector $\mathbf{x}$.

If the model is incrementally adaptive, new data will be used to adapt the system's structure and function incrementally. If a system is trained incrementally, the generalization error of the system on the next new input vector (or vectors) from the input stream is called here *local incrementally adaptive generalization error*. The local incrementally adaptive generalization error at the moment *t*, for example, when the input vector is $\mathbf{x}(t)$, and the calculated by the system output vector is $\mathbf{y}(t)'$, is expressed as $\mathrm{Err}(t) = \|y(t) - y(t)'\|$.

The local incrementally adaptive root mean square error, and the local incrementally adaptive non-dimensional error index LNDEI(t) can be calculated at each time moment *t* as:

$$\text{LRMSE}(t) = \sqrt{\left(\sum\nolimits_{i=1,2\ldots,t}(\text{Err}(i)^2)/t\right)};\tag{2.6}$$

$$\text{LNDEI}(t) = \text{LRMSE}(t)/\text{std}(y(1):y(t)),\tag{2.7}$$

where std(y(1):y(t)) is the standard deviation of the output data points from time unit 1 to time unit t.

In a general case, the global generalization root mean square error RMSE and the non-dimensional error index are evaluated on a set of *p* new (future) test examples from the problem space as follows:

$$\text{RMSE} = \sqrt{\left(\sum\nolimits_{i=1,2,\ldots p}[(\mathbf{y_i} - \mathbf{y_i'})^2]/p\right)};\tag{2.8}$$

$$\text{NDEI} = \text{RMSE}/\text{std}(\mathbf{y_1}:\mathbf{y_p}),\tag{2.9}$$

where std($\mathbf{y_1}$: $\mathbf{y_p}$), is the standard deviation of the data from 1 to p in the test set.

After a system is evolved on a sufficiently large and representative part of the whole problem space Z, its global generalization error is expected to become satisfactorily small, similar to the off-line, batch mode learning error.

Multilayer perceptron (MLP) trained with a backpropagation algorithm (BP) use a global optimization function in both incrementally adaptive (pattern mode) training, and in a batch mode training [2, 4, 5]. The batch mode, off-line training of a MLP is a typical learning method. Figures 2.4 and 2.5 depict a typical MLP architecture and Box 2.2 depicts the batch mode backpropagation algorithm.

In the incremental, pattern learning mode of the backpropagation algorithm, after each training example is presented to the system and propagated through it, an error is calculated and then all connections are modified in a backward manner. This is one of the reasons for the phenomenon called catastrophic forgetting—if examples are presented only once, the model may adapt to them too much and "forget" previously learned examples, if the model is a global model. This phenomenon is

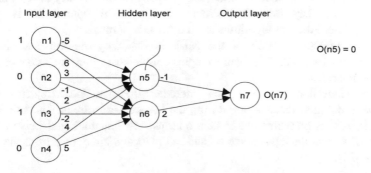

**Fig. 2.4** An example of a simple feedforward ANN (after [17])

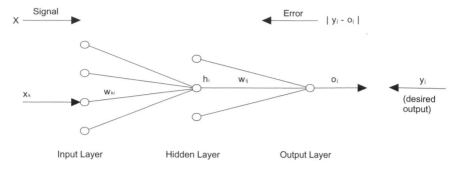

**Fig. 2.5** A schematic diagram used to explain the error backpropagation algorithm (after [17])

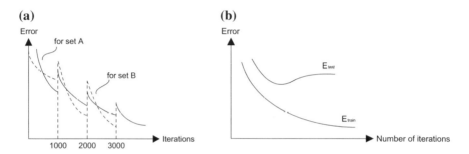

**Fig. 2.6** **a** Illustration of learning with catastrophic forgetting in MLP (after [17]). **b** Illustration of overfitting the data in MLP (after [17])

illustrated on Fig. 2.6a, where after training a MLP on a data set A it is trained on data set B and it 'forgets' a lot about data set A, etc.

In an incrementally adaptive learning mode, same or very similar examples from the past need to be presented many times again, in order for the system to properly learn new examples without forgetting them. The process of learning new examples and presenting previously used ones is called "rehearsal" training [21].

MLP can be trained in an incrementally adaptive mode, but they have limitations in this respect as they have a fixed structure and the weight optimisation is a global one if a gradient descent algorithm is used for this purpose.

A very attractive feature of the MLP is that they are universal function approximators (see [22, 23]) even though in some cases they may converge in a local minimum.

Some connectionist systems, that include MLP, use local objective (goal) function to optimise the structure during the learning process. In this case when a data pair $(\mathbf{x}, \mathbf{y})$ is presented, the system optimises its functioning always in a local vicinity of $\mathbf{x}$ from the input space X, and in the local vicinity of $\mathbf{y}$ from the output space Y [24].

---

**Box 2.2 The Backpropagation Algorithm**

Forward pass:

BF1. Apply an input vector **x** and its corresponding output vector **y** (the desired output).

BF2. Propagate forward the input signals through all the neurons in all the layers and calculate the output signals.

BF3. Calculate the $Err_j$ for every output neuron j as for example:
$Err_j = y_j - o_j$, where $y_j$ is the jth element of the desired output vector **y**.

Backward pass:

BB1. Adjust the weights between the intermediate neurons i and output neurons j according to the calculated error:
$\Delta wi_j(t+1) = lrate \cdot o_j(1 - o_j) \cdot Err_j \cdot o_i + momentum \cdot \Delta w_{ij}(t)$

BB2. Calculate the error $Err_i$ for neurons i in the intermediate layer:
$Err_i = \sum Err_j \cdot w_{ij}$.

BB3. Propagate the error back to the neurons k of lower level:
$\Delta w_{ki}(t+1) = lrate \cdot o_i(1 - o_i) \cdot Err_i \cdot x_k + momentum \cdot \Delta w_{ki}(t)$.

---

When a MLP is trained for too many iterations, its generalization ability (to recognize new data) may deteriorate which is known as *overfitting* (see Fig. 2.6b)

Principles of MLP and backpropagation algorithms are used and further developed for multiple layers as illustrated in Fig. 2.7, called deep ANN.

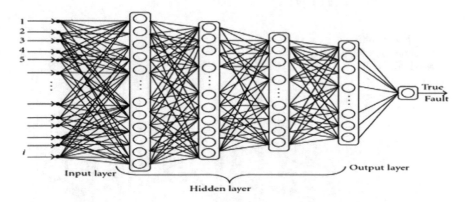

**Fig. 2.7** Multiple layers of MLP (also called deep ANN) as a simple example

### 2.1.3   Convolutional Neural Networks (CNN)

Fukushima proposed a biologically inspired MLP in which the first layer performs feature extraction from subspace of the image data and the other layers combine these features, similar to the visual cortex. He called these ANN Cognitron (1975) and Neocognitron (1980) [11, 12]—Fig. 2.8. This is perhaps the first deep NN structure that was inspired by the structure and functionality of the visual cortex.

The Neocognitron principles were further developed in a series of ANN called convolutional ANN (CNN) illustrated in Fig. 2.9a and multiple layer convolutional MLP—Fig. 2.9b.

CNN and deep ANN have been successfully developed for large scale image classification (e.g. 14 mln images in the ImageNet data set), recognizing spoken words from a large corpus of data (e.g. TIMIT), structuring large repositories of data in the IBM Watson question-answering systems, playing games such as Go with human masters, etc. [26–32].

The CNN and the deep ANN are excellent tools for vector, frame-based data (e.g. image recognition), but not much for spatio-temporal data that measure evolving processes, as these models can manifest catastrophic forgetting when trained on new data incrementally. There is no *time* of asynchronous events learned in the models and they are difficult to adapt to new data and change their structures. Even though this approach allows for deep learning of vector based data across many layers of neurons, it still lacks methods for deep knowledge representation in time-space as defined in Chap. 1.

Knowledge representation in an ANN was achieved in the knowledge–based ANN and more specially in the evolving connection systems (ECOS) as presented in the next sections. It was also achieved in evolving spiking neural networks (eSNN) in Chap. 5. Deep learning and deep knowledge representation in time-space as defined in Chap. 1 is achieved in the brain-inspired SNN as discussed in Chap. 6 and in other chapters of the book.

**Fig. 2.8** Fukushima's neocognitron CNN (after [12] and also [25])

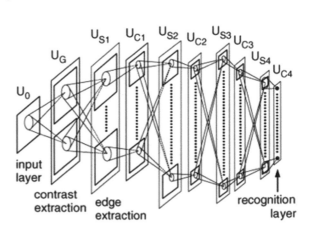

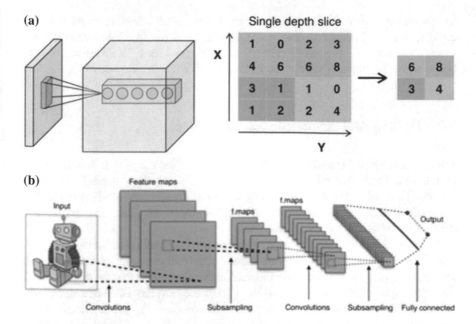

**Fig. 2.9** **a** Illustration of the principle of convolutional ANN (CNN) (after [25]). **b** Illustration of the principle of CNN and MLP with multiple layers (after [25])

## 2.1.4 Recurrent and LSTM ANN

Recurrent ANN (RNN) have feedforward, feedback and later connections [13, 26] —as illustrated with a simple example in Fig. 2.10.

As a further continuation of the RNN, the so called Long-Short Term Memory (LSTM), was developed [34, 35]. A LDTM RNN consists of units, each unit consisting of a cell, an input gate, an output gate and a forget gate. The cell is

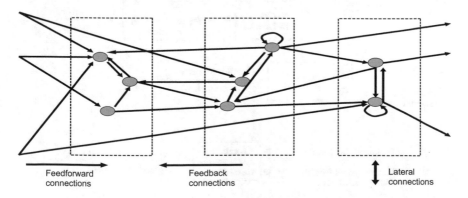

**Fig. 2.10** Recurrent ANN have feedforward, feedback and later connections (after [33])

responsible for remembering input data over time. Each of the three gates can be thought of as an artificial neuron, as in a multi-layer (or feedforward) neural network. They compute an activation based on a weighted sum. There are connections between these gates and the cell (see [25, 33–35] for more explanation).

## 2.2  Hybrid and Knowledge-Based ANN

Some of the ANN discussed above are considered 'black boxes' as it was difficult to interpret their internal structures and to articulate the essential knowledge learned. That led to the development of hybrid and rule based ANN that can both incorporate and extract essential information from the data and reveal new knowledge about the modelled processes.

In order to incorporate human knowledge into an intelligent system, an ANN module can be combined with a rule-based module in the same system. The rules can be fuzzy rules as a partial case [20]. An exemplar system is shown in Fig. 2.11, where, at a lower level, an ANN module predicts the next day value of a stock index and, at a higher level, a fuzzy reasoning module combines the predicted values with some macro-economic variables, using the following types of fuzzy rules [20]:

*IF <the stock value predicted by the ANN module is high>*

   *AND <the economic situation is good>*

*THEN <buy stock>*

Hybrid systems can also use crisp propositional rules, along with fuzzy rules [36]. The type of hybrid systems illustrated in from Fig. 2.11 are suitable to use when decision rules are available to integrate with a machine learning module that learns from incoming data.

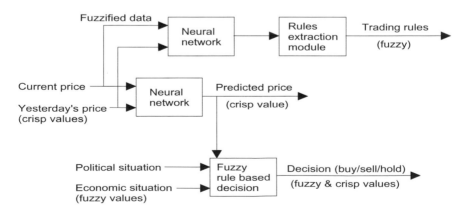

**Fig. 2.11** A hybrid ANN-fuzzy rule-based expert system for financial decision support [20]

Another group of ANN methods can be used not only to learn from data, but to extract rules from a trained ANN and/or also insert rules into an ANN as initialization procedure. These are the knowledge-based neural networks (KBNN).

*Types of Rules used in KBNN*

Different KBNNs are designed to represent different types of rules, some of them listed below:

(1) *Simple propositional rules* (e.g., IF x1 is A AND/OR x2 is B THEN y is C, where A, B and C are constants, variables, or symbols of true/false type) (see for example, [37–39]. As a partial case, interval rules can be used, for example: IF x1 is in the interval [x1min, x1max] AND x2 is in the interval [x2min, x2max] THEN y is in the interval [ymin, ymax], with Nr1 examples associated with this rule.

(2) *Propositional rules with certainty factors* (e.g., IF x1 is A (CF1) AND x2 is B (CF2) THEN y is C (CFc)), (see for example [40]).

(3) *Zadeh-Mamdani fuzzy rules* (e.g., IF x1 is A AND x2 is B THEN y is C, where A, B and C are fuzzy values represented by their membership functions) (see for example Fig. 2.12 and [41, 42]).

(4) *Takagi-Sugeno fuzzy rules* (for example, the following rule is a first order rule: IF x1 is A AND x2 is B THEN y is $a \cdot x1 + b \cdot x2 + c$, where A and B are fuzzy values and a, b and c are constants) ([43, 44]). More complex functions are possible to use in higher-order rules.

(5) *Fuzzy rules* with degrees of importance and certainty degrees (e.g.; IF x1 is A (DI1) AND x2 is B (DI2) THEN y is C (CFc), where DI1 and DI2 represent the importance of each of the condition elements for the rule output, and the CFc represents the strength of this rule) (see [20]).

(6) Fuzzy rules that represent associations of clusters of data from the problem space (e.g., Rule j: IF [an input vector x is in the input cluster defined by its centre (x1 is Aj, to a membership degree of MD1j, AND x2 is Bj, to a membership degree of MD2j) and by its radius Rj-in] THEN [y is in the output cluster defined by its centre (y is C, to a membership degree of MDc) and by its radius Rj-out, with Nex(j) examples represented by this rule]. These are the EFuNN rules discussed in a next section.

(7) *Temporal rules* (e.g., IF x1 is present at a time moment t1 (with a certainty degree and/or importance factor of DI1) AND x2 is present at a time moment t2 (with a certainty degree/importance factor DI2) THEN y is C (CFc)).

(8) *Temporal, recurrent rules* (e.g., IF x1 is A (DI1) AND x2 is B (DI2) AND y at the time moment $(t - k)$ is C THEN y at a time moment $(t + n)$ is D (CFc)).

(9) *Type-2 fuzzy rules*, that are fuzzy rules of the form of: IF x is A~ AND y is B~ THEN z is C~, where A~, B~, and C~ are type-2 fuzzy membership functions using intervals rather than single membership values [45].

The integration of ANN and fuzzy systems into one system attracted many researchers. The integration of fuzzy rules into a single neuron model and then into

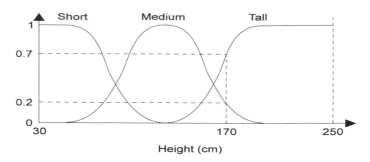

**Fig. 2.12** Example of fuzzy Gaussian membership functions that represent a variable Height

larger neural network structures, tightly coupling learning and fuzzy reasoning rules into connectionists structures, was initiated by Professor Takeshi Yamakawa and other Japanese scientists [46]. Many models of fuzzy neural networks are developed based on these principles [20, 47, 48]. These are adaptive neural networks for incremental learning and rule extraction: The neuro-fuzzy systems (no more the "black box curse"). As a general case, input and/or output variables can be non-fuzzy (crisp) or fuzzy. Example of fuzzy Gaussian membership functions is shown in Fig. 2.12.

Hybrid connections systems can incorporate fuzzy rules. They can also be used to extract fuzzy rules from already trained ANN called fuzzy neural networks as it is the case with the ECOS discussed in the next section. An example of a fuzzy rule is given below:

*IF Input 1 is High and Input 2 is Low THEN Output is Very High*

A typical example of hybrid ANN are ECOS, presented in the next section.

## 2.3  Evolving Connectionist Systems (ECOS)

### 2.3.1  *Principles of ECOS*

In the evolving connectionist systems (ECOS), introduced by the author, instead of training a fixed ANN through changing its connection weights, the connectionist structure and its functionality are evolving from incoming data, often in an on-line, one-pass learning mode and then it can be used to extract rules as knowledge representation [14–17, 49].

ECOS are modular connectionist based systems that evolve their structure and functionality in a continuous, self-organized, on-line, adaptive, interactive way from incoming information [14]. They can process both data and knowledge in a supervised and/or unsupervised way. ECOS learn local models from data through clustering of the data and associating a local output function for each cluster represented in a connectionist structure. They can learn incrementally single data items or chunks of data and also incrementally change their input features [15–17]. Elements of ECOS have been proposed as part of the classical neural network models, such as Self-Organizing Maps, Radial Basis Functions, Fuzzy ARTMap, growing neural gas, neuro-fuzzy systems, Resource Allocation Network (for a review see [17]). Other ECOS models, along with their applications, have been reported in [50, 51].

The principle of ECOS is based on *local learning*—neurons are allocated as centers of data clusters and the system creates local models in these clusters. Methods of fuzzy clustering, as means to create local knowledge-based systems, were developed by Bezdek, Yager, Filev and others [52, 53].

To summarize, the following are the main principles of ECOS as stated in [14]:

(1) Fast learning from large amount of data, e.g. using "one-pass" training, starting with little or no prior knowledge;
(2) Adaptation in real-time and in an on-line mode where new data is accommodated as it comes based on local learning;
(3) "Open", evolving structure, where new input variables (relevant to the task), new outputs (e.g. classes), new connections and neurons are added/evolved "on the fly";
(4) Both data learning and knowledge representation is facilitated in a comprehensive and flexible way, e.g., supervised learning, unsupervised learning, evolving clustering, "sleep" learning, forgetting/pruning, fuzzy rule insertion and extraction;
(5) Active interaction with other ECOSs and with the environment in a multi-modal fashion;
(6) Representing both space and time in their different scales, e.g., clusters of data, short- and long-term memory, age of data, forgetting, etc.;
(7) System's self-evaluation in terms of behavior, global error and success, and related knowledge representation.

## 2.3.2 Evolving Self-organising Maps

Several methods, such as: Dynamic Topology Representing Networks [54] and Evolving Self-organizing Maps (ESOM) [55] further developed the principles of SOM. These methods allow prototype nodes to evolve quickly in the original data space X, and at the same time acquire and keep a topology representation. The neighbourhood of the evolved nodes (neurons) is not pre-defined as it is in SOM. It is defined in an on-line mode according to the current distances between the nodes. These methods are free of the rigid topological constraints in SOM. They do not require searching for neighbourhood ranking as in the neural gas algorithm, thus improving the speed of learning.

Here, the ESOM method is explained in more detail.

Given an input vector $\mathbf{x}$, the activation on the $i$th node in ESOM is defined as:

$$a_i = e^{-\|x - w_i\|^2 / \varepsilon^2} \tag{2.10}$$

where $\varepsilon$ is a radial. Here ai can be regarded as a matching score for the i-th prototype vector $w_i$ onto the current input vector x. The closer they are, the bigger the matching score is.

The following on-line stochastic approximation of the error minimization function is used:

$$E_{app} = \sum_{i=1,n} a_i \|x - w_i\|^2 \tag{2.11}$$

where n is the current number of nodes in ESOM upon arrival of the input vector **x**.

To minimize the criterion function above, weight vectors are updated by applying a gradient descent algorithm. From Eq. (2.11) it follows:

$$\frac{\partial E_{app}}{\partial w_i} = a_i(w_i - x) + \|x - w_i\|^2 \partial a_i / \partial w_i \tag{2.12}$$

For the sake of simplicity, we assume that the change of the activation will be rather small each time when the weight vector is updated, so that $a_i$ can be treated as a constant. This leads to the following simplified weight-updating rule:

$$\Delta w_i = \gamma a_i(x - w_i), \quad for\ i = 1, 2, \ldots, n \tag{2.13}$$

Here $\gamma$ is a learning rate held as a small constant.

The likelihood of assigning the current input vector **x** onto the $i$th prototype $\mathbf{w}_i$ is defined as:

$$Pi(x, w_i) = a_i / \sum_{k=1,2,\ldots,n} (a_k) \tag{2.14}$$

*Evolving the Feature Map*

During on-line learning, the number of prototypes in the feature map is usually unknown. For a given data set the number of prototypes may be optimum at a certain time but later it may become inappropriate as when new samples are arriving the statistical characteristics of data may change. Hence it is highly desirable for the feature map to be dynamically adaptive to the incoming data.

The approach here is to start with a null map, and gradually allocate new prototype nodes when new data samples cannot be matched well onto existing prototypes. During learning, when old prototype nodes become inactive for a long time, they can be removed from the dynamic prototype map.

If for a new data vector **x** none of the prototype nodes are within a distance threshold, then a new node $\mathbf{w}_{new}$ is inserted representing exactly the poorly matched input vector $\mathbf{w}_{new} = \mathbf{x}$, resulting in a maximum activation of this node for **x**.

The ESOM learning algorithm is given in Box 2.3 (from [55]).

---

**Box 2.3 The ESOM Evolving Self-organised Maps Learning Algorithm**

*Step 1*: Input a new data vector **x**.

*Step 2*:  Find a set S of prototypes that are closer to **x** than a pre-defined threshold.

*Step 3*:  If S is null, go to step 4 (insertion), otherwise—calculate the activations $a_i$ of all nodes from S and go to step 5 (updating).

*Step 4 (insertion)*: Create a new node $\mathbf{w}_i$ for **x** and make a connection between this node and its two closest nodes (nearest neighbours) that will form a set S.

*Step 5 (updating)*: Modify all prototypes in S and re-calculate the connections s(i,j) between the winning node i (or the newly created one) and all the nodes j in the set S:

$$s(i,j) = a_i\,a_j/\max\{a_i,\ a_j\}$$

*Step 6*: After a certain number of input data are presented to the system, prune the weakest connections. If isolated nodes appear, prune them as well.

*Step 7*: go to step 1.

---

*Visualising the Feature Map*

Sammon projection or other dynamic visualisation techniques [17] can be used to visualize the evolving nodes of an ESOM at each time of incremental, evolving learning [55].

**Fig. 2.13** A schematic
diagram of a simple eMLP
[17, 51]

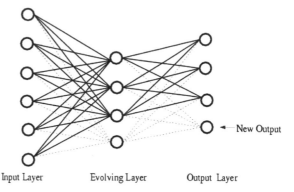

Input Layer          Evolving Layer          Output Layer

## 2.3.3 Evolving MLP

A simple evolving MLP method is called here eMLP and presented in Fig. 2.13 as a
simplified graphical representation [17, 51]. An eMLP consists of three layers of
neurons, the input layer, with linear transfer functions, an evolving layer, and an output
layer with a simple saturated linear activation function. It is a simplified version of the
evolving fuzzy neural network (EFuNN), presented later in this chapter.

The evolving layer is the layer that will grow and adapt itself to the incoming
data, and is the layer with which the learning algorithm is most concerned. The
meaning of the incoming connections, activation and forward propagation algorithms of the evolving layer all differ from those of classical connectionist systems.

If a linear activation function is used, the activation $A$ of an evolving layer node
$n$ is determined by Eq. (2.15)

$$A_n = 1 - D_n \qquad (2.15)$$

where $A_n$ is the activation of the node $n$ and; $D_n$ is the normalised distance between
the input vector and the incoming weight vector for that node.

Other activation functions, such as a radial basis function could be used. Thus,
examples which exactly match the exemplar stored within the neurons incoming
weights will result in an activation of 1 while examples that are entirely outside of
the exemplars region of input space will result in an activation of near 0.

The preferred form learning algorithm is based on accommodating, within the
evolving layer, new training examples by either modifying the connection weights
of the evolving layer nodes, or by adding a new node. The algorithm employed is
presented in Box 2.4 (from [17, 51]).

---

**Box 2.4 eMLP Learning Algorithm**

1. Propagate the input vector $I$ through the network

IF the maximum activation $a_{max}$ is less than a coefficient called sensitivity threshold $S_{thr}$ :

2. Add a node, ELSE

3. Evaluate the error between the calculated output vector $O_c$ and the desired output vector $O_d$

4. IF the error is greater than an error threshold $E_{thr}$ OR the desired output node is not the most highly activated,

5. Add a node, ELSE

6. Update the connections to the winning node in the evolving layer

7. Repeat the above procedure for each training vector

---

When a node is added, its incoming connection weight vector is set to the input vector $I$, and its outgoing weight vector is set to the desired output vector $O_d$.

The incoming weights to the winning node $j$ are modified according to Eq. (2.16), while the outgoing weights from node $j$ are modified according Eq. (2.17)

$$W_{i,j}(t+1) = W_{i,j}(t) + \eta_1 (I_i \times W_{i,j}(t)) \qquad (2.16)$$

where

| | |
|---|---|
| $W_{i,j}(t)$ | is the connection weight from input $i$ to $j$ at time $t$ |
| $W_{i,j}(t+1)$ | is the connection weight from input $i$ to $j$ at time $t+1$ |
| $\eta_1$ | is the learning rate one parameter |
| $I_i$ | is the $i$th component of the input vector $I$ |

$$W_{j,p}(t+1) = W_{j,p}(t) + \eta_2 (A_j \times E_p) \qquad (2.17)$$

where

| | |
|---|---|
| $W_{j,p}(t)$ | is the connection weight from $j$ to output $p$ at time $t$ |
| $W_{i,p}(t+1)$ | is the connection weight from $j$ to $p$ at time $t+1$ |
| $\eta_2$ | is the learning rate two parameter |
| $A_j$ | is the activation of $j$ |

$$E_p = O_{d(p)} - O_{c(p)} \qquad (2.18)$$

where $E_p$ is the error at $p$; $O_{d(p)}$ is the desired output at $p$; $O_{c(p)}$ is the calculated output at $p$.

The distance measure $D_n$ in Eq. (2.15) above is preferably calculated as the normalised Hamming distance, as shown in Eq. (2.19):

$$D_n = \frac{\sum_i^I |E_i - W_i|}{\sum_i^I |E_i + W_i|} \qquad (2.19)$$

where $I$ is the number of input nodes in the eMLP, $E$ is the input vector, $W$ is the input to evolving layer weight matrix.

Aggregation of nodes in the evolving layer can be employed to control the size of the evolving layer during the learning process. The principle of aggregation is to merge those nodes which are spatially, in terms of connection weights, close to each other. Aggregation can be applied for every (or after every $n$) training examples. It will generally improve the generalisation capability of EMLP. The aggregation algorithm is as follows:

FOR each rule node $r_j, j = 1: n$, where $n$ is the number of nodes in the evolving layer and $W1$ is the connection weights matrix between the input and evolving layers and $W2$ is the connection weights matrix between the evolving and output layers.

- find a subset $R$ of nodes in evolving layer for which the normalised Euclidean distances $D(W1_{rj}, W1_{ra})$ and $D(W2_{rj}, W2_{ra}) r_j, r_a \in R$ are below the thresholds $W_{thr}$.
- merge all the nodes from the subset $R$ into a new node $r_{new}$ and update $W1_{r_{new}}$ and $W2_{r_{new}}$ using the following formulae:

$$W1_{r_{new}} = \frac{\sum r_a \in R(W1_{r_a})}{m} \qquad (2.20)$$

$$W2_{r_{new}} = \frac{\sum r_a \in R(W2_{r_a})}{m} \qquad (2.21)$$

where $m$ denotes the number of nodes in the subset $R$.
- delete the nodes $r_a \in R$

Node aggregation is an important regularization. It is highly desirable in some application areas, such as speech or image recognition systems. In speech recognition, vocabulary of recognition systems needs to be customised to meet individual needs. This can be achieved by adding words to the existing recognition system or removing words from existing vocabulary.

eMLP are also suitable for on-line output space expansion because it uses local learning which tunes only the connection weights of the local node, so all the knowledge that has been captured in the nodes in the evolving layer will be local and only covering a "patch" of the input-output space. Thus, adding new class outputs or new input variables does not require re-training of the whole system on both the new and old data as it is required for traditional neural networks.

Based on the above theoretical considerations, the task here is to introduce an algorithm for on-line expansion and reduction of the output space in eMLP. As described above the eMLP is a three layer network with two layers of connections. Each node in the output layer represents a particular class in the problem domain. This local representation of nodes in the evolving layer enables eMLP to accommodate new classes or remove an already existing class from its output space.

In order to add a new node to the output layer, the structure of the existing eMLP first needs to be modified to encompass the new output node. This modification affects only the output layer and the connections between the output layer and the evolving layer. The graphical representation of this process is shown in Fig. 2.13. The connection weights between the new output in the output layer and the evolving layer are initialised to zero. In this manner the new output node is set by default to classify all previously seen classes as negative. Once the internal structure of the eMLP is modified to accommodate the new output class, the eMLP is further trained on the new data. As a result of the training process new nodes are created in the evolving layer to represent the new class.

The process of adding new output nodes to eMLP is carried out in a supervised manner. Thus, for a given input vector, a new output node will be added only if it is indicated that the given input vector is a new class. The output expansion algorithm is as follows:

FOR every new output class:

1. Insert a new node into the output layer;
2. FOR every node in the evolving layer $r_i, i = 1{:}n$, where $n$ is the number of nodes in the evolving layer, modify $W2$ the outgoing connection weights from the evolving to output layer by expanding $W2_{i,j}$ with set of zeros to reflect the zero output.

This is equivalent to allocating a part of the problem space for data that belong to new classes, without specifying where this part is in the problem space.

It is also possible to remove a class from an eMLP. It only affects the output and evolving layer of eMLP architecture:

FOR every output class $o$ to be removed,

1. Find set of nodes $S$ in the evolving layer which are committed to that output $o$
2. Modify $W1$ the incoming connection from input layer to evolving layer by deleting $S_i, i = 1{:}n$, where $n$ is the number of nodes in the set $S$ committed to output $o$
3. Modify $W2$ the outgoing connection weights from the evolving to output layer by deleting output node $o$.

The above algorithm is equivalent to dis-allocating a part of the problem space which had been allocated for the removed output class. In this manner, there will be no space allocated for the deleted output class in the problem space. In other words the network is unlearning a particular output class.

The eMLP is further studied and applied in [56, 57].

## 2.4   Evolving Fuzzy Neural Networks. EFuNN

Here the concept of ECOS is illustrated on two implementations: the evolving fuzzy neural network (EFuNN) [49] and the dynamic evolving neuro-fuzzy inference system (DENFIS) [16]. In ECOS, clusters of data are created based on similarity between data samples either in the input space (this is the case in some of the ECOS models, e.g., DENFIS), or in both the input and output space (this is the case, e.g., in the EFuNN models). Samples (examples) that have a distance to an existing node (cluster center, rule node) less than a certain threshold are allocated to the same cluster. Samples that do not fit into existing clusters form new clusters. Cluster centers are continuously adjusted according to new data samples, and new clusters are created incrementally. ECOS learn from data and automatically create or update a local fuzzy model/function, e.g.:

*IF  $<$ data is in a fuzzy cluster $C_i >$  THEN $<$ the model is $F_i >$*

where $F_i$ can be a fuzzy value, a logistic or linear regression function or ANN model [16, 17].

Generally speaking, fuzzy neural networks are connectionist structures that can be interpreted in terms of fuzzy rules [20, 46, 47, 58]. Fuzzy neural networks are NN, with all the NN characteristics of training, recall, adaptation, etc. while neuro-fuzzy inference systems are fuzzy rule based systems and their associated fuzzy inference mechanism that are implemented as neural networks for the purpose of learning and rule optimisation. The evolving fuzzy neural network (EFuNN) presented here is of the former type, while DENFIS systems are of the latter type. Some authors do not separate the two types that makes the transition from one to the other type more flexible and also broadens the interpretation and the application of each of these systems.

EFuNNs have a five-layer structure (Fig. 2.14). Here nodes and connections are created/connected as data examples are presented. An optional short-term memory layer can be used through a feedback connection from the rule (also called, case) node layer. The layer of feedback connections could be used if temporal relationships of input data are to be memorized structurally.

The input layer represents input variables. The second layer of nodes (fuzzy input neurons, or fuzzy inputs) represents fuzzy quantisation of each input variable space (similar to the factorizable RBF networks). For example, two fuzzy input neurons can be used to represent "small" and "large" fuzzy values. Different membership functions (MF) can be attached to these neurons.

The number and the type of MF can be dynamically modified. The task of the fuzzy input nodes is to transfer the input values into membership degrees to which

**Fig. 2.14** A simplified architecture of EFuNN

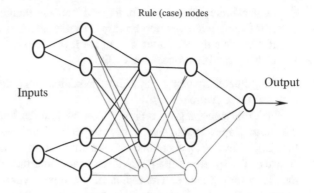

they belong to the corresponding MF. The layers that represent fuzzy MF are optional, as a non-fuzzy version of EFuNN can also be evolved with only three layers of neurons and two layers of connections as it is used in eMLP in the section above.

The third layer contains rule (case) nodes that evolve through supervised and/or unsupervised learning. The rule nodes represent prototypes (exemplars, clusters) of input-output data associations that can be graphically represented as associations of hyper-spheres from the fuzzy input and the fuzzy output spaces. Each rule node r is defined by two vectors of connection weights—W1(r) and W2(r), the latter being adjusted through supervised learning based on the output error, and the former being adjusted through unsupervised learning based on similarity measure within a local area of the problem space. A linear activation function, or a Gaussian function, is used for the neurons of this layer.

The fourth layer of neurons represents fuzzy quantization of the output variables, similar to the input fuzzy neuron representation. Here, a weighted sum input function and a saturated linear activation function is used for the neurons to calculate the membership degrees to which the output vector associated with the presented input vector belongs to each of the output MFs. The fifth layer represents the values of the output variables. Here a linear activation function is used to calculate the defuzzified values for the output variables.

A partial case of EFuNN would be a three layer network without the fuzzy input and the fuzzy output layers (e.g. eMLP, or an evolving simple RBF network). In this case a slightly modified versions of the algorithms described below are applied, mainly in terms of measuring Euclidean distance and using Gaussian activation functions.

The evolving learning in EFuNNs is based on either of the following two assumptions:

(1) No rule nodes exist prior to learning and all of them are created (generated) during the evolving process; or
(2) There is an initial set of rule nodes that are not connected to the input and output nodes and become connected through the learning (evolving) process. The latter case is more biologically plausible as most of the neurons in the

human brain exist before birth, and become connected through learning, but still there are areas of the brain where new neurons are created during learning if "surprisingly" different stimuli from previously seen are presented (see [59] as biological references for ECOS).

The EFuNN evolving algorithm presented next does not make a difference between these two cases.

Each rule node, e.g. $r_j$, represents an association between a hyper-sphere from the fuzzy input space and a hyper-sphere from the fuzzy output space (see Fig. 2.15), the $W1(r_j)$ connection weights representing the co-ordinates of the centre of the sphere in the fuzzy input space, and the $W2 (r_j)$—the co-ordinates in the fuzzy output space. The radius of the input hyper-sphere of a rule node $r_j$ is defined as $R_j = 1 - S_j$, where $S_j$ is the sensitivity threshold parameter defining the minimum activation of the rule node $r_j$ to a new input vector x from a new example $(\mathbf{x}, \mathbf{y})$ in order the example to be considered for association with this rule node.

The pair of fuzzy input-output data vectors $(\mathbf{x_f}, \mathbf{y_f})$ will be allocated to the rule node $r_j$ if $\mathbf{x_f}$ falls into the $r_j$ input receptive field (hyper-sphere), and $\mathbf{y_f}$ falls in the $r_j$ output reactive field hyper-sphere. This is ensured through two conditions, that a local normalised fuzzy difference between $x_f$ and $W1(r_j)$ is smaller than the radius $R_j$, and the normalised output error $Err = \|\mathbf{y} - \mathbf{y'}\|/Nout$ is smaller than an error threshold E. Nout is the number of the outputs and $\mathbf{y'}$ is the produced by EFuNN output. The error parameter E sets the error tolerance of the system.

**Fig. 2.15** EFuNN maps clusters from the input space to clusters in the output problem space, where the radii of the clusters can change during learning

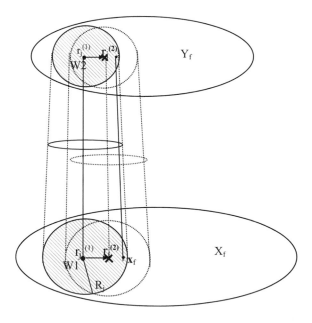

*Definition.* A local normalised fuzzy distance between two fuzzy membership vectors $\mathbf{d}_{1f}$ and $\mathbf{d}_{2f}$ that represent the membership degrees to which two real vector data $\mathbf{d}_1$ and $\mathbf{d}_2$ belong to pre-defined MFs, is calculated as:

$$D(d_{1f}, d_{2f}) = \|d_{1f} - d_{2f}\| / \|d_{1f} + d_{2f}\|, \tag{2.22}$$

where $\|x - y\|$ denotes the sum of all the absolute values of a vector that is obtained after vector subtraction (or summation in case of $\|x - y\|$) of two vectors x and y; "/" denotes division. For example, if $d_{1f} = (0, 0, 1, 0, 0, 0)$ and $d_{2f} = (0, 1, 0, 0, 0, 0)$, than $D(d_1, d_2) = (1 + 1)/2 = 1$ which is the maximum value for the local normalised fuzzy difference. In EFuNNs the local normalised fuzzy distance is used to measure the distance between a new input data vector and a rule node in the local vicinity of the rule node.

In RBF networks Gaussian radial basis functions are allocated to the nodes and used as activation functions to calculate the distance between the node and the input vectors.

Through the process of associating (learning) of new data points to a rule node $r_j$, the centres of this node hyper-spheres adjust in the fuzzy input space depending on the distance between the new input vector and the rule node through a learning rate $l_j$, and in the fuzzy output space depending on the output error through the Widrow-Hoff lest mean square (LMS) delta algorithm [60]. This adjustment can be represented mathematically by the change in the connection weights of the rule node $r_j$ from $W1(r_j^{(t)})$ and $W2(r_j^{(t)})$ to $W1(r_j^{(t+1)})$ and $W2(r_j^{(t+1)})$ respectively, employing the following vector operations:

$$
\begin{aligned}
W1\left(r_j^{(t+1)}\right) &= W1\left(r_j^{(t)}\right) + l_j \cdot \left(x_f - W1\left(r_j^{(t)}\right)\right) \\
W2\left(r_j^{(t+1)}\right) &= W2\left(r_j^{(t)}\right) + l_j \cdot (y_f - A2) \cdot A1\left(r_j^{(t)}\right)
\end{aligned}
\tag{2.23}
$$

where $A2 = f_2(W2 \cdot A1)$ is the activation vector of the fuzzy output neurons in the EFuNN structure when x is presented; $A1(r_j^{(t)}) = f_2(D (W1 (r_j^{(t)}), x_f))$ is the activation of the rule node $r_j^{(t)}$; a simple linear function can be used for $f_1$ and $f_2$, e.g. $A1(r_j^{(t)}) = 1 - D(W1 (r_j^{(t)}), x_f)$; $l_j$ is the current learning rate of the rule node $r_j$ calculated for example as $l_j = 1/Nex(r_j)$, where $Nex(r_j)$ is the number of examples currently associated with rule node $r_j$.

The statistical rationale behind this is that the more examples are currently associated with a rule node, the less it will "move" when a new example has to be accommodated by this rule node, i.e. the change in the rule node position is proportional to the number of already associated example to the new, single example.

**Fig. 2.16** Adjusting the rule
nodes during learning of 4
data points in EFuNN ([17])

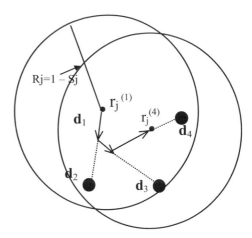

When a new example is associated with a rule node $r_j$ not only its location in the
input space changes, but also its receptive field expressed as its radius Rj, and its
sensitivity threshold Sj:

$$Rj^{(t+1)} = Rj^{(t)} + D\left(W1\left(r_j^{(t+1)}\right), W1\left(r_j^{(t)}\right)\right), \quad \text{respectively} \qquad (2.24)$$

$$Sj^{(t+1)} = Sj^{(t)} - D\left(W1\left(r_j^{(t+1)}\right), W1\left(r_j^{(t)}\right)\right) \qquad (2.25)$$

The learning process in the fuzzy input space is illustrated in Fig. 2.16 on four data
points $d_1$, $d_2$, $d_3$ and $d_4$. Figure 2.16 shows how the centre $r_j^{(1)}$ of the rule node $r_j$
adjusts (after learning each new data point) to its new positions $r_j^{(2)}$, $r_j^{(3)}$, $r_j^{(4)}$ when
one pass learning is applied. Figure 2.16 shows also how the rule node position
would move to new positions $r_j^{(2(2)}$, $r_j^{(3(2)}$, and $r_j^{(4(2)}$, if another pass of learning was
applied. If the two learning rate $l_1$ and $l_2$ have zero values, once established, the
centres of the rule nodes will not move.

The weight adjustment formulas (Eq. 2.23) define the standard EFuNN that has
the first part updated in an unsupervised mode, and the second part—in a supervised mode similar to the RBF networks. But here the formulas are applied once for
each example $(\mathbf{x}, \mathbf{y})$ in an incrementally adaptive mode, that is similar to the RAN
model [61] and its modifications. The standard supervised/unsupervised learning
EFuNN is denoted as EFuNN-s/u. In two other modifications of EFuNN, namely
double pass learning EFuNN (EFuNN-dp), and gradient descent learning EFuNN
(EFuNN-gd), slightly different update functions are used as explained in the next
section.

The learned temporal associations can be used to support the activation of rule nodes based on temporal pattern similarity. Here, temporal dependencies are learned through establishing structural links. These dependencies can be further investigated and enhanced through synaptic analysis (at the synaptic memory level) rather than through neuronal activation analysis (at the behavioural level). The ratio spatial similarity/temporal-correlation can be balanced for different applications through two parameters Ss and Tc such that the activation of a rule node r for a new data example $d_{new}$ is defined through the following vector operations:

$$A1(r) = \left| 1 - Ss \cdot D(W1(r), d_{newf}) + Tc \cdot W3\left(r_{max}^{(t-1)}, r\right) \right|_{[0,1]} \tag{2.26}$$

where $|.|[0,1]$ is a bounded operation in the interval $[0, 1]$; $D(W1(r), d_{newf})$ is the normalised local fuzzy distance value and $r_{max}^{(t-1)}$ is the winning neuron at the previous time moment. Here temporal connections can be given a higher importance in order to tolerate a higher distance. If $T_c = 0$, then the supervised learning in EFuNN is based on the above explained principles, so when a new data example $d = (x, y)$ is presented, the EFuNN either creates a new rule node $r_n$ to memorize the two input and output fuzzy vectors $W1(r_n) = x_f$ and $W2(r_n) = y_f$, or adjusts an existing rule node $r_j$.

After a certain time (when certain number of examples have been presented) some neurons and connections may be pruned or aggregated.

The supervised learning algorithms above allow for an EFuNN system to always evolve and learn when a new input-output pair of data becomes available. This is an *active learning mode*.

*EFuNN Sleep Learning Rules*

In another mode, *passive or sleep learning*, learning is performed when there is no input pattern presented. This may be necessary to apply after an initial learning has been performed. In this case existing connections, that store previously fed input patterns, are used as "echo" to reiterate the learning process. This type of learning may be applied in case of a short initial presentation of the data, when only small portion of data is learned in one-pass, incrementally adaptive mode, and then the training is refined through the sleep learning method when the system consolidates what it has learned before.

Sleep learning in EFuNN and in some other connectionist models is illustrated on several examples in [62].

*One-pass versus multiple-passes learning*

The best way to apply the above learning algorithms is to draw randomly examples from the problem space, propagate them through the EFuNN and tune the connection weights and the rule nodes, change and optimise the parameter values, etc., until the error becomes a desirably small one. In a fast learning mode, each example is presented only once to the system. If it is possible to present examples twice, or more times, the error may become smaller, but that depends on the parameter values of the EFuNN and on the statistical characteristics of the data.

The evolved EFuNN can perform *inference* when recalled on new input data. The EFuNN inference method consists of calculating the output activation value when a new input vector is applied. This is part of the EFuNN supervised learning method when only an input vector **x** is propagated through the EFuNN. If the new input vector falls in the receptive field of the winning rule node (the closest rule node to the input vector) *one-of-n* mode of inference is used that is based on the winning rule node activation (one rule inference). If the new input vector does not fall in the receptive filed of the closest to it rule node, than *m-of-n* mode is used, where *m* rule nodes (rules) are used in the EFuNN inference process, with an usual value of *m* being 3.

Different pruning rules can be applied for a successful *pruning* of unnecessary nodes and connections. One of them is given below:

IF $(Age(r_j) > OLD)$ AND (the total activation $TA(r_j)$ is less than a pruning parameter Pr times Age $(r_j)$) THEN prune rule node $r_j$,

where $Age(r_j)$ is calculated as the number of examples that have been presented to the EFuNN after $r_j$ had been fist created; OLD is a pre-defined age limit; Pr is a pruning parameter in the range [0,1], and the total activation $TA(r_j)$ is calculated as the number of examples for which $r_j$ has been the correct winning node (or among the m winning nodes in the m-of-n mode of operation).

The above pruning rule requires that the fuzzy concepts of OLD, HIGH, etc. are defined in advance. As a partial case, a fixed value can be used, e.g. a node is OLD if it has existed during the evolving of a FuNN from more than p examples. The pruning rule and the way the values for the pruning parameters are defined, depend on the application task.

One of the learning algorithms for EFuNN is shown in Box 2.5 [17].

**Box 2.5 EFuNN-s/u Learning Algorithm**

1. Set initial values for the system parameters: number of membership functions; initial sensitivity threshold (default $S = 0.9$); error threshold E; aggregation parameter Nagg—a number of consecutive examples after which an aggregation is performed (to be explained in a later section); pruning parameters OLD an Pr; a value for m (in m-of-n mode); thresholds $T_1$ and $T_2$ for rule extraction.

2. Set the first rule node to memorize the first example $(\mathbf{x}, \mathbf{y})$:

$$W1(r_0) = \mathbf{x_f}, \quad \text{and} \quad W2(r_0) = \mathbf{y_f}; \tag{2.27}$$

3. *Loop* over presentations of input-output pairs $(\mathbf{x}, \mathbf{y})$

{Evaluate the local normalized fuzzy distance D between $\mathbf{x_f}$ and the existing rule node connections W1 (formulae (2.22)).

Calculate the activation A1 of the rule node layer. Find the closest rule node $r_k$ (or the closest m rule nodes in case of m-of-n mode) to the fuzzy input vector $\mathbf{x_f}$.

*if* $A1(r_k) < S_k$ (sensitivity threshold for the node $r_k$), create a new rule node for $(\mathbf{x_f}, \mathbf{y_f})$

   *else*

Find the activation of the fuzzy output layer A2=W2.A1 and the output error

$$Err = || \mathbf{y} - \mathbf{y'} ||/Nout.$$

   *if* Err > E

   create a new  rule node to eMLPommodate the current example $(\mathbf{x_f}, \mathbf{y_f})$

      *else*

Update W1 $(r_k)$ and W2$(r_k)$ eMLPording to (2.23) (in case of *m-of-n* EFuNN update all the *m* rule nodes with the highest A1 activation).

Apply *aggregation* procedure of rule nodes after each group of N$_{agg}$ examples are
   presented.

   Update the values for the rule node r$_k$ parameters S$_k$, R$_k$, Age(r$_k$), TA (r$_k$).

   *Prune rule nodes* if necessary, as defined by pruning parameters.

   *Extract rules* from the rule nodes (as explained in a later sub-section)

   } End of the main loop.

At any time (phase) of the evolving (learning) process of an EFuNN fuzzy or
exact rules can be *inserted* and extracted. Insertion of fuzzy rules is achieved
through setting a new rule node r$_j$ for each new rule, such that the connection
weights W1(r$_j$) and W2(r$_j$) of the rule node represent this rule. For example, the
fuzzy rule (*IF x$_1$ is Small and x$_2$ is Small THEN y is Small*) can be inserted into an
EFuNN structure by setting the connections of a new rule node to the fuzzy condition nodes $x_1$-Small and $x_2$-Small and to the fuzzy output node y-Small to a value
of 1 each. The rest of the connections are set to a value of zero. Similarly, an exact
rule can be inserted into an EFuNN structure, e.g. *IF x$_1$ is 3.4 and x$_2$ is 6.7 THEN y
is 9.5*. Here the membership degrees to which the input values $x_1 = 3.4$ and
$x_2 = 6.7$, and the output value $y = 9.5$ belong to the corresponding fuzzy values are
calculated and attached to the corresponding connection weights.

Each rule node $r_j$ can be expressed as a fuzzy rule, for example:

Rule r$_j$: IF $x_1$ is Small 0.85 and $x_1$ is Medium 0.15 and $x_2$ is Small 0.7 and $x_2$ is
Medium 0.3 (Radius of the receptive field Rj = 0.1, maxRadiusj = 0.75) THEN y is
Small 0.2 and y is Large 0.8 (20 out of 175 examples associated with this rule),
where the numbers attached to the fuzzy labels denote the degree to which the
centres of the input and the output hyper-spheres belong to the respective MF. The
degrees associated to the condition elements are the connection weights from the
matrix W1. Only values that are greater than a threshold T1 are left in the rules as
the most important ones. The degrees associated with the conclusion part are the
connection weights from W2 that are greater than a threshold of T2. An example of
rules extracted from a bench-mark dynamic time series data is given in Sect. 3.5.
The two thresholds T1 and T2 are used to disregard the connections from W1 and
W2 that represent small and insignificant membership degrees (e.g., less than 0.1).

*Rule Node Aggregation in EFuNNs*

Another knowledge-based technique applied to EFuNNs is *rule node aggregation*.

For example, for the aggregation of three rule nodes r$_1$, r$_2$, and r$_3$ the following
two aggregation rules can be used to calculate the new aggregated rule node r$_{agg}$
W1 connections (the same formulas are used to calculate the W2 connections):

(a) as a geometrical centre of the three nodes:

$$W1(r_{agg}) = (W1(r_1) + W1(r_2) + W1(r_3))/3 \qquad (2.28)$$

(b) as a weighted statistical centre:

$$W2(r_{agg}) = (W2(r_1) \cdot Nex(r_1) + W2(r_2)Nex(r_2) + W2(r_3) \cdot Nex(r_3))/Nsum$$
$$(2.29)$$

where

$$Nex(r_{agg}) = Nsum = Nex(r_1) + Nex(r_2) + Nex(r_3);$$

$r_j$ is the rule node from the three nodes that ha a maximum distance from the new node $r_{agg}$ and Rj is its radius of the receptive field.

The three rule nodes will aggregate only if the radius of the aggregated node receptive field is less than a pre-defined maximum radius Rmax:

$$Rr_{agg} = D(W1(r_{agg}), W1(r_j)) + Rj \leq Rmax$$

In order for a given node $r_j$ to aggregate with other nodes, two subsets of nodes are formed—the subset of nodes $r_k$ that if activated to a degree of 1 will produce an output value $y'(r_k)$ that is different from $y'(r_j)$ in less than the error threshold E, and the subset of nodes that cause output values different from $y'(r_k)$ in more than E. The W2 connections define these subsets. Than all the rule nodes from the first subset that are closer to $r_j$ in the input space than the closest to $r_j$ node from the second subset in terms of W1 distance, get aggregated if the radius of the new node $r_{agg}$ is less than the pre-defined limit Rmax for a receptive field

Instead of aggregating all rule nodes that are close to a rule node $r_j$ than the closest node from the other class, it is possible to keep the closest to the other class node from the aggregation pool out of the aggregation procedure—as a separate node—a "guard", thus preventing a possible miss-classification of new data on the bordering area between the two classes.

Through node creation and their consecutive aggregation, an EFuNN system can adjust over time to changes in the data stream and at the same time—preserve its generalisation capabilities.

Through analysis of the weights W3 of an evolved EFuNN, *temporal correlation* between time consecutive exemplars can be expressed in terms of rules and conditional probabilities, e.g.:

$$IF\, r_1^{(t-1)} \quad THEN\, r_2^{(t)}(0.3) \qquad (2.30)$$

The meaning of the above rule is that some examples that belong to the rule (prototype) $r_2$ follow in time examples from the rule prototype $r_1$ with a relative conditional probability of 0.3.

## 2.5  Dynamic Evolving Neuro-fuzzy Inference Systems— DENFIS

The dynamic evolving neuro-fuzzy system, DENFIS, in its two modifications—for on-line—and for off-line learning, use Takagi-Sugeno type of fuzzy inference method [16, 17]. The inference used in DENFIS is performed on $m$ fuzzy rules as described below:

$$\begin{cases} \text{if } x_1 \text{ is } R_{11} \text{ and } x_2 \text{ is } R_{12} \text{ and } \dots \text{ and } x_q \text{ is } R_{1q}, & \text{then } y \text{ is } f_1\left(x_1, x_2, \dots, x_q\right) \\ \text{if } x_1 \text{ is } R_{21} \text{ and } x_2 \text{ is } R_{22} \text{ and } \dots \text{ and } x_q \text{ is } R_{2q}, & \text{then } y \text{ is } f_2\left(x_1, x_2, \dots, x_q\right) \\ \\ \text{if } x_1 \text{ is } R_{m1} \text{ and } x_2 \text{is } R_{m2} \text{ and } \dots \text{and } x_q \text{ is } R_{mq}, & \text{then } y \text{ is } f_m\left(x_1, x_2, \dots, x_q\right) \end{cases}$$

$$(2.31)$$

where "$x_j$ is $R_{ij}$", $i = 1, 2, \dots m$; $j = 1, 2, \dots q$, are $m \times q$ fuzzy propositions that form $m$ antecedents for $m$ fuzzy rules respectively; $x_j, j = 1, 2, \dots, q$, are antecedent variables defined over universes of discourse $X_j, j = 1, 2, \dots, q$, and $R_{ij}, i = 1, 2, \dots m$; $j = 1, 2, \dots, q$ are fuzzy sets defined by their fuzzy membership functions $\mu_{Rij}$: $X_j \rightarrow [0, 1]$, $i = 1, 2, \dots m$; $j = 1, 2, \dots, q$. In the consequent parts of the fuzzy rules, $y$ is the consequent variable, and crisp functions $f_i$, $i = 1, 2, \dots m$, are employed.

In the DENFIS on-line model, the first-order Takagi-Sugeno type fuzzy rules are employed and the linear functions in the consequence parts are created and updated through learning from data by using the linear least-square estimator (LSE).

If the consequent functions are crisp constants, i.e. $f_i(x_1, x_2, \dots, x_q) = C_i, i = 1, 2, \dots m$, we call such system a zero-order Takagi-Sugeno type fuzzy inference system. The system is called a first-order Takagi-Sugeno type fuzzy inference system if $f_i(x_1, x_2, \dots, x_q), i = 1, 2, \dots m$, are linear functions. If these functions are non-linear functions, it is called high-order Takagi-Sugeno fuzzy inference system.

For an input vector $x^0 = [x_1^0 \ x_2^0 \ \dots \ x_q^0]$, the result of inference, $y^0$ (the output of the system) is the weighted average of each rule's output indicated as follows:

$$y^0 = \frac{\sum_{i=1,m} \omega_i f_i\left(x_1^0, x_2^0, \dots, x_q^0\right)}{\sum_{i=1,m} \omega_i} \tag{2.32}$$

where: $\omega_i = \prod_{j=1}^{q} \mu_{R_{ij}}\left(x_j^0\right); \quad i = 1, 2, \dots, m; \quad j = 1, 2, \dots, q. \tag{2.33}$

In the DENFIS on-line model, the first-order Takagi-Sugeno type fuzzy rules are employed and a linear functions in the consequences can be created and updated by linear least-square estimator (LSE) on the learning data. Each of the linear functions can be expressed as follows:

$$y = \beta_0 + \beta_1 x_1 + \beta_2 x_2 + \cdots + \beta_q x_q. \tag{2.34}$$

For obtaining these functions a learning procedure is applied on a data set, which is composed of $p$ data pairs $\{([x_{i1}, x_{i2}, \ldots, x_{iq}], y_i), i = 1, 2, \ldots, p\}$. The least-square estimator (LSE) of $\boldsymbol{\beta} = [\beta_0\ \beta_1\ \beta_2 \ldots \beta_q]^T$, are calculated as the coefficients $b = [b_0\ b_1\ b_2 \ldots b_q]^T$, by applying the following formula:

$$b = \left(\mathbf{A}^T \mathbf{A}\right)^{-1} \mathbf{A}^T \mathbf{y} \tag{2.35}$$

where

$$\mathbf{A} = \begin{pmatrix} 1 & x_{11} & x_{22} & \cdots & x_{1q} \\ 1 & x_{21} & x_{22} & \cdots & x_{2q} \\ \cdot & \cdot & \cdot & \cdot & \cdot \\ \cdot & \cdot & \cdot & \cdot & \cdot \\ \cdot & \cdot & \cdot & \cdot & \cdot \\ 1 & x_{p1} & x_{p2} & \cdots & x_{pq} \end{pmatrix}$$

and $\mathbf{y} = [y_1\ y_2 \ldots, y_p]^T$.

A weighted least-square estimation method is used here as follows:

$$b_{\mathrm{w}} = \left(\mathbf{A}^T \mathbf{W} \mathbf{A}\right)^{-1} \mathbf{A}^T \mathbf{W} \mathbf{y}, \tag{2.36}$$

where

$$\mathbf{W} = \begin{pmatrix} w_1 & 0 & \cdots & 0 \\ 0 & w_2 & \cdots & 0 \\ \vdots & \vdots & \vdots & \vdots \\ 0 & \cdots & \cdots & w_p \end{pmatrix}$$

and $w_j$ is the distance between $j$-th example and the corresponding cluster centre, $j = 1, 2, \ldots p$.

We can rewrite the Eqs. (2.35) and (2.36) as follows:

$$\begin{cases} \mathbf{P} = \left(\mathbf{A}^T \mathbf{A}\right)^{-1}, \\ b = \mathbf{P} \mathbf{A}^T \mathbf{y}. \end{cases} \tag{2.37}$$

$$\begin{cases} \mathbf{P}_\mathrm{w} = \left(\mathbf{A}^T \mathbf{W} \mathbf{A}\right)^{-1}, \\ \boldsymbol{b}_\mathrm{w} = \mathbf{P}_\mathrm{w} \mathbf{A}^T \mathbf{W} \mathbf{y}, \end{cases} \tag{2.38}$$

Let the $k$th row vector of matrix A defined in Eq. (2.35) be $a_k^T = [1 \ x_{k1} \ x_{k2} \ \ldots \ x_{kq}]$ and the $k$th element of $y$ be $y_k$, then $b$ can be calculated iteratively as follows:

$$\begin{cases} b_{k+1} = b_k + \mathbf{P}_{k+1}\mathbf{a}_{k+1}\left(y_{k+1} - \mathbf{a}_{k+1}^T b_k\right), \\ \mathbf{P}_{k+1} = \mathbf{P}_k - \frac{\mathbf{P}_k \mathbf{a}_{k+1}\mathbf{a}_{k+1}^T \mathbf{P}_k}{1 + \mathbf{a}_{k+1}^T \mathbf{P}_k \mathbf{a}_{k+1}} \end{cases} \tag{2.39}$$

for k = n, n + 1, ... p − 1.

Here, the initial values of $\mathbf{P}_n$ and $b_n$ can be calculated directly from Eq. (2.39) with the use of the first $n$ data pairs from the learning data set.

Equation (2.39) is the formula of recursive LSE. In the DENFIS on-line model, we use a weighted recursive LSE with forgetting factor defined as the following equations:

$$\boldsymbol{b}_{k+1} = \boldsymbol{b}_k + w_{k+1}\mathbf{P}_{k+1}\mathbf{a}_{k+1}\left(y_{k+1} - \mathbf{a}_{k+1}^T \boldsymbol{b}_k\right),$$

$$\mathbf{P}_{k+1} = \frac{1}{\lambda}\left(\frac{\mathbf{P}_k}{} \quad \frac{w_{k+1}\mathbf{P}_k\mathbf{a}_{k+1}\mathbf{a}_{k+1}^T\mathbf{P}_k}{\lambda + \mathbf{a}_{k+1}^T \mathbf{P}_k \mathbf{a}_{k+1}}\right) \quad k = n, \ n+1, \ldots p - 1. \tag{2.40}$$

where $w$ is the weight defined in Eq. (2.36) and $\lambda$ is a forgetting factor with a typical value between 0.8 and 1.

In an on-line DENFIS model, the rules are created and updated at the same time with the input space partitioning using on-line evolving clustering method (ECM) and Eqs. (2.34) and (2.40). If no rule insertion is applied, the following steps are used for the creation of the first $m$ fuzzy rules and for the calculation of the initial values $\mathbf{P}$ and $b$ of the functions:

(1) Take the first $n_0$ learning data pairs from the learning data set.
(2) Implement clustering using ECM with these $n_0$ data to obtaining $m$ cluster centres.
(3) For every cluster centre $C_i$, find $p_i$ data points whose positions in the input space are closest to the centre, $i = 1, 2, \ldots, m$.
(4) For obtaining a fuzzy rule corresponding to a cluster centre, create the antecedents of the fuzzy rule using the position of the cluster centre and Eq. (2.34). Using Eq. (2.38) on $p_i$ data pairs calculate the values of $\mathbf{P}$ and $b$ of the consequent function. The distances between $p_i$ data points and the cluster centre are taken as the weights in Eq. (2.38).

In the above steps, $m$, $n_0$ and $p$ are the parameters of the DENFIS on-line learning model, and the value of $p_i$ should be greater than the number of input elements, $q$.

As new data pairs are presented to the system, new fuzzy rules may be created and some existing rules are updated. A new fuzzy rule is created if a new cluster centre is found by the ECM. The antecedent of the new fuzzy rule is formed by using Eq. (2.36) with the position of the cluster centre as a rule node. An existing fuzzy rule is found based on the rule node that is the closest to the new rule node; the consequence function of this rule is taken as the consequence function for the new fuzzy rule. For every data pair, several existing fuzzy rules are updated by using Eq. (2.40) if their rule nodes have distances to the data point in the input space that are not greater than $2 \times Dthr$ (the threshold value, a clustering parameter). The distances between these rule nodes and the data point in the input space are taken as the weights in Eq. (2.40). In addition to this, one of these rules may also be updated through changing its antecedent so that, if its rule node position is

**Fig. 2.17** Example of a fuzzy inference in DENFIS (after [15–17])

(a)  Fuzzy rule group 1 for a DENFIS (m = 3)

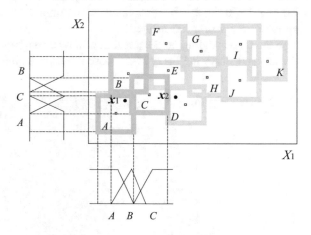

(b)  Fuzzy rule group 2 for a DENFIS (m = 3)

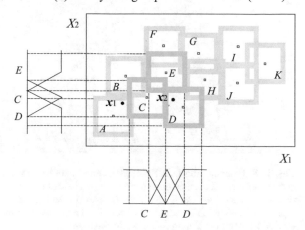

changed by the Evolving Clustering Method—ECM, the fuzzy rule will have a new antecedent calculated through Eq. (2.34).

*Takagi-Sugeno fuzzy inference in DENFIS*

The Takagi-Sugeno fuzzy inference system utilised in DENFIS is a dynamic inference. In addition to dynamically creating and updating fuzzy rules the DENFIS on-line model has some other major differences from the other inference systems.

First, for each input vector, the DENFIS model chooses m fuzzy rules from the whole fuzzy rule set for forming a current inference system. This operation depends on the position of the current input vector in the input space. In case of two input vectors that are very close to each other, especially in the DENFIS off-line model, the inference system may have the same fuzzy rule inference group. In the DENFIS on-line model, however, even if two input vectors are exactly the same, their corresponding inference systems may be different. It is due to the reason that these two input vectors are presented to the system at different time moments and the fuzzy rules used for the first input vector might have been updated before the second input vector has arrived.

Second, depending on the position of the current input vector in the input space, the antecedents of the fuzzy rules chosen to form an inference system for this input

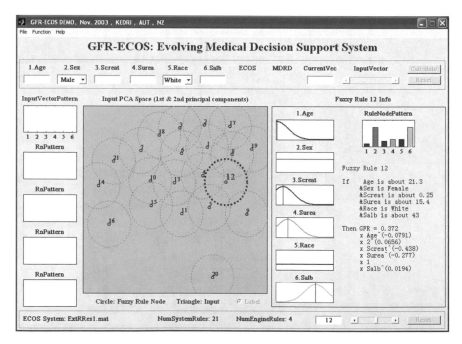

**Fig. 2.18** Example of a DENFIS application system using six input variables to train a DENFIS system for a medical decision support on a renal function GFR, where the evolved hidden nodes represent clusters of input data and the data in each of these clusters is approximated by a regression function [17]. Local fuzzy rule 12 is extracted that represents/approximates the data in cluster 12 using the shown membership functions

vector may vary. An example is illustrated in Fig. 2.17 where two different groups of fuzzy inference rules are formed depending on two input vectors $x_1$ and $x_2$ respectively in a 2D input space. We can see from this example that, for instance, the region $C$ has a linguistic meaning 'large', in the $X_1$ direction for Fig. 2.17a group, but for the group of rules from Fig. 2.17b it denotes a linguistic meaning of 'small' in the same direction of X1. The region $C$ is defined by different membership functions respectively in each of the two groups of rules.

Example of a DENFIS application system is given in Fig. 2.18. Six input variables are used to train a DENFIS system for a medical decision support on a renal function GFR, where the evolved hidden nodes represent clusters of input data and the data in each of these clusters is approximated by a regression function.


## 2.6   Other ECOS Methods and Systems

A special development of ECOS is *transductive reasoning and personalized modelling*. Instead of building a set of local models (i.e., prototypes) to cover the whole problem space and then use these models to classify/predict any new input vector, in transductive modelling for every new input vector a new model is created based on selected nearest neighbor vectors from the available data. Such ECOS models are the neuro-fuzzy inference system (NFI) and the transductive weighted NFI (TWNFI) [63]. In TWNFI, for every new input vector the neighborhood of closets data vectors is optimized using both the distance between the new vector and the neighboring ones and the weighted importance of the input variables, so that the error of the model is minimized in the neighborhood area [64].

In addition to the already presented methods of ECOS, following is a short summary list of other methods, systems and applications that use some of the principles of ECOS along with publications that reveal some more details of ECOS methods:

- Evolving Self-Organized Maps (ESOM) [65];
- Evolving Clustering Method (ECM) [66];
- Incremental feature learning in ECOS [67];
- On-line ECOS optimization [68];
- Assessment of EFuNN accuracy for pattern recognition using data with different statistical distributions [69];
- Recursive clustering based on a Gustafson–Kessel algorithm [70];
- Using a map-based encoding to evolve plastic neural networks [71];
- Evolving Takagi–Sugeno fuzzy model based on switching to neighboring models [72];
- A soft computing based approach for modeling of chaotic time series [73];
- Uninorm based evolving neural networks and approximation capabilities [74];
- Global, local and personalised modelling and profile discovery in Bioinformatics: An integrated approach [75];

- FLEXFIS: a robust incremental learning approach for evolving Takagi–Sugeno fuzzy models [76];
- Evolving fuzzy classifiers using different model architectures [77];
- RSPOP: Rough Set–Based Pseudo Outer-Product Fuzzy Rule Identification Algorithm [78];
- SOFMLS: online self-organizing fuzzy modified least-squares network [79];
- On-Line Sequential Extreme Learning Machine [80];
- Finding features for real-time premature ventricular contraction detection using a fuzzy neural network system [81];
- Evolving fuzzy rule-based classifiers [82];
- A novel generic Hebbian ordering-based fuzzy rule base reduction approach to Mamdani neuro-fuzzy system [83];
- Implementation of fuzzy cognitive maps based on fuzzy neural network and application in prediction of time series [84];
- Backpropagation to train an evolving radial basis function neural network [85];
- Smooth transition autoregressive models and fuzzy rule-based systems: Functional equivalence and consequences [86];
- Development of an adaptive neuro-fuzzy classifier using linguistic hedges [87];
- A meta-cognitive sequential learning algorithm for neuro-fuzzy inference system [88];
- Meta-cognitive RBF network and its projection based learning algorithm for classification problems [89];
- SaFIN: A self-adaptive fuzzy inference network [90];
- A sequential learning algorithm for meta-cognitive neuro-fuzzy inference system for classification problems [91];
- Architecture for development of adaptive on-line prediction models [92];
- Clustering and co-evolution to construct neural network ensembles: An experimental study [93];
- Algorithms for real-time clustering and generation of rules from data [94];
- SAKM: Self-adaptive kernel machine—A kernel-based algorithm for online clustering [95];
- A BCM theory of meta-plasticity for online self-reorganizing fuzzy-associative learning [96];
- Evolutionary strategies and genetic algorithms for dynamic parameter optimization of evolving fuzzy neural networks [97];
- Incremental leaning and model selection for radial basis function network through sleep learning [98];
- Interval-based evolving modeling [99];
- Evolving granular classification neural networks [100];
- Stability analysis for an online evolving neuro-fuzzy recurrent network [101];
- A TSK fuzzy inference algorithm for online identification [102];
- Design of experiments in neuro-fuzzy systems [103];
- EFuNNs ensembles construction using a clustering method and a co-evolutionary genetic algorithm [104];
- eT2FIS: An evolving type-2 neural fuzzy inference system [105];

- Designing radial basis function networks for classification using differential evolution [106];
- A meta-cognitive neuro-fuzzy inference system (McFIS) for sequential classification problems [107];
- An evolving fuzzy neural network based on the mapping of similarities [108];
- Incremental learning by heterogeneous bagging ensemble [109];
- Fuzzy associative conjuncted maps network [110];
- EFuNN ensembles construction using CONE with multi-objective GA [111];
- Risk analysis and discovery of evolving economic clusters in Europe [112];
- Adaptive time series prediction for financial applications [113];
- Adaptive speech recognition [114];
- and others [17].

## 2.7 Chapter Summary and Further Readings for Deeper Knowledge

The chapter presents foundations of artificial neural networks (ANN) and on one class of them—evolving connectionist systems. ECOS can not only be trained on data measuring evolving processes, but they can facilitate rule and knowledge extraction for a better understanding of these processes. Twenty four Centuries after Aristoteles, now the process of rule extraction and knowledge discovery from data can be automated. And not only that. The rules can be further adapted by incrementally training ECOS to accommodate new data and information about the problem in hand. These rules will no more be considered fixed and true for ever, but evolving as well.

Additional material to some of the sections can be found as follows:

- Neuro-fuzzy systems [20];
- ECOS [17, 59, 115];
- Fuzzy systems [116];
- Neural networks [116];
- ECOS development system NeuCom (www.theneucom.com).

The ANN and ECOS methods and systems presented above use predominantly the McCulloch and Pitts model of a neuron (Fig. 2.1). They have been efficiently used for wide range of applications as some of them listed above. Many of the principles of ANN and ECOS presented in this chapter have been further developed and used for the creation of SNN and evolving spiking neural networks (eSNN) correspondingly (Chaps. 4 and 5) and for brain-inspired SNN (Chap. 6). Overview of the development of ECOS, including eSNN, can be found in [115].

While the hybrid ANN and ECOS 'opened the black box" and provided means for rule and knowledge representation, these rules are "flat" rules, extracted from vector-based data. The ECOS methods were further developed into evolving SNN (eSNN) in Chap. 5. Instead of scalars as it is in ECOS, eSNN use information representation as spikes, the learning is based on times of spikes and fuzzy rules can be extracted.

Chapter 3 discusses how the brain learns from data as deep learning and how deep knowledge is represented. This is taken as inspiration for deep learning and deep knowledge representation in brain-inspired SNN, where deep spatio-temporal rules can be extracted. They are presented in Chap. 6 and used in other chapters.

**Acknowledgements** Some of the material in this chapter was first published by the author in [15–17, 20, 48, 49]. For the introduction of the principles of ECOS and of the ECOS models such as EFuNN and DENFIS, I have been inspired by the earlier work and by my discussions with pioneers, such as S. Amari, T. Kohonen, W. Freeman, L. Zadeh, T. Yamakawa, J. Taylor. I acknowledge the contribution of several of my co-authors of the early publications on ECOS methods, especially Qun Song, Mike Watts, Da Deng, Mathias Futschik, all Ph.D. students and postdoctoral fellows of mine at the University of Otago.

# References

1. F. Rosenblatt, The perceptron: a probabilistic model for information storage and organization in the brain. Psychol Rev **65**(1958), 386–402 (1958)
2. S. Amari, A theory of adaptive pattern classifiers. IEEE Trans. Electron. Comput. **EC-16**(3), 299–307 (1967)
3. S. Amari, Mathematical foundations of neurocomputing. Proc. IEEE **78**(9), 1443–1463 (1990)
4. D. Rumelhart, J. McLelland (eds.), *Parallel and Distributed Processing* (MIT Press, Cambridge, 1986)
5. P. Werbos, Backpropagation through time. Proc. IEEE **87**(10), 1990 (1990)
6. T. Kohonen, *Self-Organising Maps* (Springer, Berlin, 1992)
7. T. Kohonen, The self-organizing map. IEEE **78**(9), 1464–1480 (1990)
8. T. Kohonen, *Self-organizing Map* (Springer, New York, 1997). ISBN: 3-540-62017-6
9. G.A. Carpenter, S. Grossberg, *Adaptive Resonance Theory* (MIT Press, Cambridge, 1998)
10. R.S. Sutton, A.G. Barto, *Reinforcement Learning* (MIT Press, Cambridge, 1998)
11. K. Fukushima, Cognitron: a self-organizing multilayered neural network. Biol. Cybern. **20** (3–4), 121–136 (1975)
12. K. Fukushima, Neocognitron: a self-organizing neural network model for a mechanism of pattern recognition unaffected by shift in position. Biol. Cybern. **36**(1980), 193–202 (1980)
13. C.M. Bishop, *Pattern Recognition and Machine Learning* (Springer, Berlin, 2006)
14. N. Kasabov, Evolving fuzzy neural networks—algorithms, applications and biological motivation, in *Methodologies for the Conception, Design and Application of Soft Computing*, ed. by T. Yamakawa, G. Matsumoto (World Scientific, Singapore, 1998), pp. 271–274
15. N. Kasabov, *Evolving Connectionist Systems: Methods and Applications in Bioinformatics, Brain Study and Intelligent Machines* (SpringerVerlag, London, 2002)
16. N. Kasabov, Q. Song, DENFIS: dynamic evolving neural-fuzzy inference system and its application for time-series prediction. IEEE Trans. Fuzzy Syst. **10**(2), 144–154 (2002)
17. N. Kasabov, *Evolving Connectionist Systems: The Knowledge Engineering Approach*, 2nd ed. (Springer, London, 2007)
18. T. Kohonen, *Associative Memory—A System-Theoretical Approach* (Springer, Berlin, 1977)
19. T. Kohonen, Self-organized formation of topologically correct feature maps. Biol. Cybern. **43**(1), 59–69 (1982)

20. N. Kasabov, *Foundations of Neural Networks, Fuzzy Systems and Knowledge Engineering* (MIT Press, Cambridge, 1996), p. 1996
21. A.V. Robins, Consolidation in neural networks and the sleeping brain. Connect. Sci. **8** (1996), 259–275 (1996)
22. G. Cybenko, in *Continuous valued neural networks with two hidden layers are sufficient.* Technical Report, Department of Computer Science, Tufts University, 1988
23. S. Funahashi, C.J. Bruce, P.S. Goldman-Rakic, Mnemonic coding of visual space in the monkey's dorsolateral prefrontal cortex. J. Neurophysiol. **61**(1989), 331–349 (1989)
24. Y. Saad, Further analysis of minimum residual iteration. Numer. Linear Algebra Appl. **7**(2), 67–93 (2000)
25. Wikipedia, http://www.wikipedia.org
26. J. Schmidhuber, Deep learning in neural networks: An overview. Neural Networks **61**, 85–117 (2015)
27. Y. LeCun, Y. Bengio, G. Hinton, Deep learning. Nature **521**(7553), 436 (2015). https://doi.org/10.1038/nature14539
28. A. Krizhevsky, L. Sutskever, G.E. Hinton, *Image Net Classification with Deep Convolutional Neural Networks*, in Proceedings of Advances in Neural Information Processing Systems (2012), pp. 1097–1105
29. D. Ferrucci, E. Brown, J. Chu-Carroll, J. Fan, D. Gondek, A. Kalyanpur, A. Lally, W. Murdock, E. Nyberg, J. Prager, N. Schlaefer, Building Watson: an overview of the DeepQA project. AI Mag. **31**(3), 59–79 (2010). https://doi.org/10.2109/aimag.v31i3.2303
30. G. Hinton, L. Deng, D. Yu, G.E. Dahl, A.R. Mohamed, N. Jaitly, S.A. Senior, V. Vanhoucke, P. Nguyen, T. Sainath, B. Kingsbury, Deep neural networks for acoustic modeling in speech recognition: the shared views of four research groups. IEEE Signal Process. Mag. **29**(6), 82–97 (2012). https://doi.org/10.1109/msp.2012.2205597
31. L. Sutskever, O. Vinyals, Q.V. Le, Sequence to sequence learning with neural networks. Adv. Neural Inf. Process. Syst. 3104–3112 (2014)
32. D. Silver, A. Huang, C.J. Maddison, A. Guez, L. Sifre, G. Van Den Driessche, S. Dieleman, Mastering the game of Go with deep neural networks and tree search. Nature **529**(7587), 484–489 (2012). https://doi.org/10.1038/nature21961
33. L. Benuskova, N. Kasabov, *Computational Neurogenetic Modelling* (Springer, Berlin, 2007)
34. Sepp Hochreiter, Jürgen Schmidhuber, Long short-term memory. Neural Comput. **9**(8), 1735–1780 (1997). https://doi.org/10.1162/neco.1997.9.8.1735. (PMID9377276)
35. Felix A. Gers, Jürgen Schmidhuber, Fred Cummins, Learning to forget: continual prediction with LSTM. Neural Comput. **12**(10), 2451–2471 (2000)
36. N. Kasabov, S.I. Shishkov, A connectionist production system with partial match and its use for approximate reasoning. Connect. Sci. **5**(3–4) 275–305 (1993)
37. M. Feigenbaum, *Artificial Intelligence, a knowledge-based approach* (PWS-Kent, Boston, 1989)
38. S. Gallant, *Neural Network Learning and Expert Systems* (MIT Press, Bradford, 1993)
39. J. Hendler, L. Dickens, *Integrating Neural Network and Expert Reasoning: An Example*, in Proceeding of AISB Conference, ed. by L. Steels, B. Smith (Springer, New York, 1991), pp. 109–116
40. L. Fu, Integration of neural heuristic into knowledge-based inference. Connect. Sci. **1**(3), 1989 (1989)
41. L.A. Zadeh, Fuzzy sets. Inf. Control **8**(1988), 338–353 (1965)
42. E. Mamdani, Application on fuzzy logic to approximate reasoning using heuristic synthesis. IEEE Trans. Comput. **26**(12), 1182–1191 (1977)
43. T. Takagi, M. Sugeno, Fuzzy identification of systems and its applications to modelling and control. IEEE Trans. Syst. Man Cybern. **15**(1985), 116–132 (1985)

44. R. Jang, ANFIS: adaptive network based fuzzy inference system. IEEE Trans. Syst. Man Cybern. **23**(3), 665–685 (1993)
45. J. M. Mendel, *Uncertain Rule-Based Fuzzy Logic Systems: Introduction and New Directions* (Prentice-Hall, Upper-Saddle River, NJ, 2001)
46. T. Yamakawa, E. Uchino, T. Miki, H. Kusanagi, *A Neo Fuzzy Neuron and Its Application to System Identification and Prediction of the System Behavior*, in Proceedings of 2nd International Conference on Fuzzy Logic and Neural Networks, Japan, July 1992, pp. 477–483
47. T. Furuhashi, T. Hasegawa, S. Horikawa, Y. Uchikawa, *An Adaptive Fuzzy Controller Using Fuzzy Neural Networks*, in Proceedings of 5th International Fuzzy System Association World Congress, Korea, July 1993, pp. 769–772
48. N. Kasabov, J.S. Kim, M.J. Watts, A.R. Gray, FuNN/2—a fuzzy neural network architecture for adaptive learning and knowledge acquisition. Inf. Sci. Appl. **101**(3–4), 155–175 (1997)
49. N. Kasabov, Evolving fuzzy neural networks for on-line supervised/ unsupervised, knowledge–based learning. IEEE Trans. Syst. Man Cyber. Part B Cybern. **31**(6), 902–918 (2001)
50. M.E. Futschik, N. Kasabov, Fuzzy clustering in gene expression data analysis, in *Proceedings of the IEEE International Conference on Fuzzy Systems, USA* (2002), pp. 414–419
51. M.J. Watts, A decade of kasabov's evolving connectionist systems: a review. IEEE Trans. Syst. Man Cybern. Part C Appl. Rev. **39**(3), 253–269 (2009)
52. J.C. Bezdek, *Analysis of Fuzzy Information* (CRC Press, Boca Raton, 1987)
53. R.R. Yager, D.P. Filev, Generation of fuzzy rules by mountain clustering. J. Intell. Fuzzy Syst. Appl. Eng. Technol. **2**(3), 209–219 (1994)
54. J. Si, S. Lin, M.A. Vuong, Dynamic topology representing networks. Neural Network **13**, 617–627 (2000)
55. D. Deng, N. Kasabov, *ESOM: An Algorithm to Evolve Self-organizing Maps from Online Data Streams*, in Proceedings of IJCNN's 2000, Como, Italy, vol. VI (2000), pp. 3–8
56. M. Watts, N. Kasabov, Evolutionary computation for the optimization of evolving connectionist systems, in *Proceedings of WCCI'2002 (World Congress of Computational Intelligence), Hawaii* (IEEE Press, Washington, DC, 2002)
57. M.J. Watts, *Nominal-Scale Evolving Connectionist Systems*, in Proceedings of IEEE International Joint Conference on Neural Networks, Vancouver (IEEE Press, Washington, DC, 2006, pp. 4057-4061)
58. C.T. Lin, C.S.G. Lee, *Neuro Fuzzy Systems* (Prentice-Hall, Upper Saddle River, 1996)
59. N. Kasabov (ed.), *Springer Handbook of Bio-/Neuroinformatics* (Springer, Berlin, 2014)
60. B. Widrow, M.E. Hoff, Adaptive switching circuits. IPE WESCON Convention Rec. **4**, 96–104 (1960)
61. J. Platt, A resource allocating network for function interpolation. Neural Comput. **3**(1991), 213–225 (1991)
62. K. Yamauchi, J. Hayami, *Sleep Learning—An Incremental Learning System Inspired by Sleep Behavior*, in Proceedings of IEEE International Conference on Fuzzy Systems, Vancouver (IEEE Press, Piscataway, NJ, 2006), pp. 6295–6302
63. Q. Song, N. Kasabov, TWNFI—a transductive neuro-fuzzy inference system with weighted data normalization for personalized modelling. Neural Networks **19**(10), 1591–1596 (2006)
64. N. Kasabov, Y. Hu, Integrated optimisation method for personalised modelling and case study applications. Int. J. Funct. Inf. Personal. Med. **3**(3), 236–256 (2010)
65. D. Deng, N. Kasabov, On-line pattern analysis by evolving self-organizing maps. Neurocomputing **51**, 87–103 (2003)
66. Q. Song, N. Kasabov, NFI: a neuro-fuzzy inference method for transductive reasoning. IEEE Trans. Fuzzy Syst. **13**(6), 799–808 (2005)
67. S. Ozawa, S. Too, S. Abe, S. Pang, N. Kasabov, Incremental learning of feature space and classifier for online face recognition. Neural Networks 575–584 (2005)

68. Z. Chan, N. Kasabov, Evolutionary computation for on-line and off-line parameter tuning of evolving fuzzy neural networks. Int. J. Comput. Intell. Appl. **4**(3), 309–319 (2004)
69. R.M. de Moraes, *Assessment of EFuNN Accuracy for Pattern Recognition Using Data with Different Statistical Distributions*, in Proceedings of the 2nd Brazilian Congress on Fuzzy Systems, Brazil, November 2012, pp. 672–685
70. D. Dovžan, I. Škrjanc, Recursive clustering based on a Gustafson–Kessel algorithm. Evolv. Syst. **2**(1), 15–24 (2011)
71. P. Tonelli, J.B. Mouret, *Using a Map-Based Encoding to Evolve Plastic Neural Networks*, in Proceedings of the IEEE Workshop on Evolving and Adaptive Intelligent Systems, France, April 2011, pp. 9–16
72. A. Kalhor, B.N. Araabi, C. Lucas, Evolving Takagi–Sugeno fuzzy model based on switching to neighboring models. Appl. Soft Comput. **13**(2), 939–946 (2013)
73. J. Vajpai, J.B. Arun, *A Soft Computing Based Approach for Modeling of Chaotic Time Series*, in Proceedings of the 13th International Conference on Neural Information Processing, China, October 2006, pp. 505–512
74. F. Bordignon, F. Gomide, Uninorm based evolving neural networks and approximation capabilities. Neurocomputing **127**, 13–20 (2014)
75. N. Kasabov, Global, local and personalised modelling and profile discovery in bioinformatics: an integrated approach. Pattern Recogn. Lett. **28**(6), 673–685 (2007)
76. E.D. Lughofer, FLEXFIS: a robust incremental learning approach for evolving Takagi–Sugeno Fuzzy models. IEEE Trans. Fuzzy Syst. **16**(6), 1393–1410 (2008)
77. P. P. Angelov, E.D. Lughofer, X. Zhou, Evolving fuzzy classifiers using different model architectures. Fuzzy Sets Syst. **159**(23) (2008), 3160–3182
78. K. K. Ang, C. Quek, RSPOP: rough set-based pseudo outer-product fuzzy rule identification algorithm. Neural Comput. **17**(1), 205–243 (2005)
79. J. de Jesús Rubio, SOFMLS: online self-organizing fuzzy modified least-squares network. IEEE Trans. Fuzzy Syst. **17**(6), 1296–1309 (2009)
80. G.B. Huang, N.Y. Liang, H.J. Rong, On-line sequential extreme learning machine, in *Proceedings of the IASTED International Conference on Computational Intelligence, Canada*, July 2005, pp. 232–237
81. J.S. Lim, Finding features for real-time premature ventricular contraction detection using a fuzzy neural network system. IEEE Trans. Neural Networks **20**(3), 522–527 (2009)
82. P.P. Angelov, X. Zhou, F. Klawonn, Evolving fuzzy rule-based classifiers, in *Proceedings of the IEEE Symposium on Computational Intelligence in Image and Signal Processing, USA*, April 2007, pp. 220–225
83. F. Liu, C. Quek, G.S. Ng, A novel generic Hebbian ordering-based fuzzy rule base reduction approach to Mamdani neuro-fuzzy system. Neural Comput. **19**(6) (2007), 1656–1680
84. H. Song, C. Miao, W. Roel, Z. Shen, F. Catthoor, Implementation of fuzzy cognitive maps based on fuzzy neural network and application in prediction of time series. IEEE Trans. Fuzzy Syst. **18**(2) (2010), 233–250
85. J. de Jesús Rubio, D.M. Vázquez, J. Pacheco, Backpropagation to train an evolving radial basis function neural network. Evolv. Syst. **1**(3) (2010), 173–180
86. J.L. Aznarte, J.M. Benítez, J.L. Castro, Smooth transition autoregressive models and fuzzy rule-based systems: functional equivalence and consequences. Fuzzy Sets Syst. **158**(24) (2007), 2734–2745
87. B. Cetisli, Development of an adaptive neuro-fuzzy classifier using linguistic hedges. Expert Syst. Appl. **37**(8) (2010), 6093–6101
88. K. Subramanian, S. Suresh, A Meta-Cognitive Sequential Learning Algorithm for Neuro-Fuzzy Inference System. Applied Soft Comput. **12**(11) (2012), 3603–3614
89. G.S. Babu, S. Suresh, Meta-Cognitive RBF Network and Its Projection Based Learning Algorithm for Classification Problems. Applied Soft Comput. **13**(1) (2013), 654–666
90. S.W. Tung, C. Quek, C. Guan, SaFIN: a self-adaptive fuzzy inference network. IEEE Trans. Neural Networks **22**(12), 1928–1940 (2011)

91. S. Suresh, K. Subramanian, *A Sequential Learning Algorithm for Meta-cognitive Neuro-fuzzy Inference System for Classification Problems*, in Proceedings of the International Joint Conference on Neural Networks, USA, August 2011, pp. 2507–2512
92. P. Kadlec, B. Gabrys, Architecture for development of adaptive on-line prediction models. Memetic Comput. **1**(4), 241–269 (2009)
93. F.L. Minku, T.B. Ludemir, Clustering and co-evolution to construct neural network ensembles: an experimental study. Neural Networks **21**(9), 1363–1379 (2008)
94. D.P. Filev, P.P. Angelov, Algorithms for real-time clustering and generation of rules from data, in *Advances in Fuzzy Clustering and its Applications*, ed. by J. Valente di Oliveira, W. Pedrycz (Wiley, Chichester, 2007)
95. H. Amadou Boubacar, S. Lecoeuche, S. Maouche, SAKM: Self-adaptive kernel machine: a kernel-based algorithm for online clustering. Neural Networks **21**(9), 1287–1301 (2008)
96. J. Tan, C. Quek, A BCM theory of meta-plasticity for online self-reorganizing fuzzy-associative learning. IEEE Trans. Neural Networks **21**(6), 985–1003 (2010)
97. F.L. Minku, T.B. Ludemir, *Evolutionary Strategies and Genetic Algorithms for Dynamic Parameter Optimization of Evolving Fuzzy Neural Networks*, in Proceedings of the IEEE Congress on Evolutionary Computation, Scotland (2005), pp. 1951–1958
98. K. Yamauchi, J. Hayami, Incremental leaning and model selection for radial basis function network through sleep. IEICE Trans. Inf. Syst. **e90-d**(4), 722–735 (2007)
99. D.F. Leite, P. Costa, F. Gomide, *Interval-Based Evolving Modeling*, in Proceedings of the IEEE Workshop on Evolving and Self-developing Intelligent Systems, USA, March 2009, pp. 1–8
100. D.F. Leite, P. Costa, F. Gomide, Evolving granular neural networks from fuzzy data streams. Neural Networks **38**, 1–16 (2013)
101. J. de Jesús Rubio, Stability analysis for an online evolving neuro-fuzzy recurrent network, in *Evolving Intelligent Systems: Methodology and Applications*, ed. by P.P. Angelov, D. P. Filev, N.K. Kasabov (Wiley, Hoboken, 2010)
102. K. Kim, E.J. Whang, C.W. Park, E. Kim, M. Park, *A TSK Fuzzy Inference Algorithm for Online Identification*, Proceedings of the 2nd *International Conference on Fuzzy Systems and Knowledge Discovery, China*, August 2005, pp. 179–188
103. C. Zanchettin, L.L. Minku, T.B. Ludemir, Design of experiments in neuro-fuzzy systems. Int. J. Comput. Intell. Appl. **9**(2), 137–152 (2010)
104. F.L. Minku, T.B. Ludemir, *EFuNNs Ensembles Construction Using a Clustering Method and a Coevolutionary Genetic Algorithm*, in Proceedings of the IEEE Congress on Evolutionary Computation, Canada, July 2006, pp. 1399–1406
105. S.W. Tung, C. Quek, C. Guan, eT2FIS: an evolving type-2 neural fuzzy inference system. Inf. Sci. **220**, 124–148 (2013)
106. B.O'Hara, J. Perera, A. Brabazon, *Designing Radial Basis Function Networks for Classification Using Differential Evolution*, in Proceedings of the International Joint Conference on Neural Networks, Canada, July 2006, pp. 2932–2937
107. K. Subramanian, S. Sundaram, N. Sundararajan, A Metacognitive neuro-fuzzy inference system (McFIS) for sequential classification problems. IEEE Trans. Fuzzy Syst. **21**(6), 1080–1095 (2013)
108. J.A.M. Hernández, F.G. Castañeda, J.A.M. Cadenas, An evolving fuzzy neural network based on the mapping of similarities. IEEE Trans. Fuzzy Syst. **17**(6), 1379–1396 (2009)
109. Q.L. Zhao, Y.H. Jiang, M. Xu, *Incremental Learning by Heterogeneous Bagging Ensemble*, in Proceedings of the International Conference on Advanced Data Mining and Applications, China, November 2010, pp. 1–12
110. H. Goh, J.H. Lim, C. Quek, Fuzzy associative conjuncted maps network. IEEE Trans. Neural Networks **20**(8), 1302–1319 (2009)
111. F.L. Minku, T.B. Ludemir, *EFuNN Ensembles Construction Using CONE with Multi-objective GA*, in Proceedings of the 9th Brazilian Symposium on Neural Networks, Brazil, October 2006, pp. 48–53

112. N. Kasabov, Adaptation and interaction in dynamical systems: modelling and rule discovery through evolving connectionist systems. Appl. Soft Comput. **6**(3), 307–322 (2006)
113. H. Widiputra, R. Pears, N. Kasabov, Dynamic interaction network versus localized trends model for multiple time-series prediction. Cybern. Syst. **42**(2), 100–123 (2011)
114. A. Ghobakhlou, M. Watts, N. Kasabov, Adaptive speech recognition with evolving connectionist systems. Inf. Sci. **156**(2003), 71–83 (2003)
115. N. Kasabov, Evolving connectionist systems: From neuro-fuzzy-, to spiking—and neurogenetic, in *Springer Handbook of Computational Intelligence*, ed. by W. Kacprzyk, J. Pedrycz (Springer, Berlin, 2015), pp. 771–782
116. J. Kacprzyk (ed.), *Springer Handbook of Computational Intelligence* (Springer, Berlin, 2015)

# Part II
# The Human Brain

# Chapter 3
# Deep Learning and Deep Knowledge Representation in the Human Brain

Spiking neural networks (SNN) and the deep learning algorithms for them have been inspired by the structure, the organisation and the many aspects of deep learning and deep knowledge representation in the human brain. This chapter presents basic information about brain structures and functions and reveals some inner processes of deep learning and deep knowledge representation as inspiration for brain-inspired SNN (BI-SNN) and brain-inspired AI (BI-AI) presented in the next chapters. The presented here information is not intended for modeling the brain in its precise structural and functional complexity, but rather for: (1) Borrowing spatio-temporal information processing principles from the brain for the creation of brain-inspired SNN and brain-inspired AI as general spatio-temporal data machines for deep learning and deep knowledge representation in time-space; (2) Understanding brain data, when modeled with SNN, for a more accurate analysis and for a better understanding of the brain processes that generated the data.

The chapter has the following sections:

3.1. Time-space in the brain.
3.2. Learning and memory.
3.3. Neural representation of information.
3.4. Perception in the brain is always spatio/specro temporal.
3.5. Deep learning and deep knowledge representation in the brain.
3.6. Neurons and information transmission between neurons through synapses.
3.7. Measuring brain activities as spatio-temporal brain data (STBD).
3.8. Chapter summary and discussions.

## 3.1 Time-Space in the Brain

To use a metaphor, we can say here: T*ime is in our brain and the brain exists in Time*.

The brain (more than 80 bln neurons, 100 trillions of connections) has a complex spatial structure which has evolved over 200 mln or so years of evolution. It is

© Springer-Verlag GmbH Germany, part of Springer Nature 2019
N. K. Kasabov, *Time-Space, Spiking Neural Networks and Brain-Inspired Artificial Intelligence*, Springer Series on Bio- and Neurosystems 7, https://doi.org/10.1007/978-3-662-57715-8_3

the ultimate information processing machine. Three, mutually interacting, memory types learned in the brain are:

–  short term memory (neuronal membrane potential);
–  long term memory (synaptic weights);
–  genetic memory (genes in the nuclei of the neurons).

Spatio/spectro temporal evolving processes in the brain are manifested at different *time* scales, e.g.:

–  Nanoseconds—quantum processes;
–  Milliseconds—spiking activity of neurons;
–  Minutes—gene expressions;
–  Hours—learning in synapses;
–  Many years—evolution of genes.

More importantly, the brain learns as a *deep learning* mechanism, creating long neural network connections from external spatio-temporal data and from internal activities. These connections represent *deep knowledge*.

It is estimated that there are about $10^{11}$ to $10^{12}$ of neurons in the human brain [1]. Three quarters of neurons form a thick cerebral cortex that constitutes a heavily folded brain surface. Cerebral cortex is thought to be a seat of cognitive functions, like perception, imagery, memory, learning, thinking, etc. The cortex cooperates with evolutionary older subcortical nuclei that are located in the middle of the brain, in and around the so-called brain stem (Fig. 3.1).

Subcortical structures and nuclei are comprised for instance of basal ganglia, thalamus, hypothalamus, limbic system and dozens of other groups of neurons with more or less specific functions in operations of the whole brain. For example, the input from all sensory organs comes to the cortex pre-processed in thalamus.

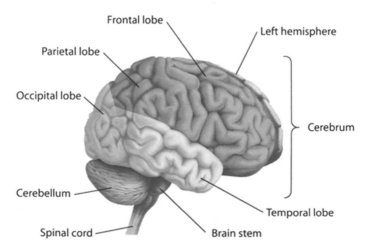

**Fig. 3.1**  Functional cortical areas of the brain (after [7])

Emotions and memory functions depend upon an intact limbic system. When one of its crucial parts, hippocampus, is lesioned, humans (and animals) loose their ability to store new events and form new memories. When a particular cortical area has been damaged, a particular cognitive deficit follows. However, all the brain parts, either cortical or subcortical, are directly or indirectly heavily interconnected, thus forming a huge recurrent neural network (in the terminology of artificial neural networks). Thus, we cannot speak of totally isolated neuroanatomic modules.

Figure 3.1, shows a schematic functional division of the human cerebral cortex and Fig. 3.2 shows both the cortical areas and the inner areas of the human brain. One third of the cortex is devoted to processing of visual information in the primary visual cortex and higher-order visual areas in the parietal cortex and in the infratemporal cortex. Association cortices take about one half of the whole cortical surface. In the parietal-temporal-occipital association cortex, sensory and language information are being associated. Memory and emotional information are associated in the limbic association cortex (internal and bottom portion of hemispheres). The prefrontal association cortex takes care of all associations, evaluation, planning ahead and attention. Language processing takes place within the temporal cortex, parietal-temporal-occipital association cortex, and frontal cortex.

At the border between the frontal and parietal lobes, there is a somatic sensory cortex, which processes touch and other somatosensory signals (temperature, pain, etc.) from the body surface and interior. In the front of it, there is a primary motor cortex, which issues signals for voluntary muscle movements including speech. These signals are preceded by the preparation and anticipation of movements that takes place in the premotor cortex. The plan of actions and their consequences, inclusion and exclusion of motor actions into and from the overall goal of an organism, are performed within the prefrontal association cortex. Subcortical basal

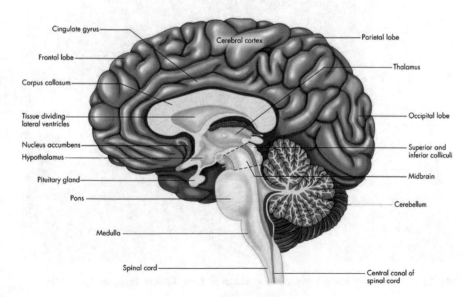

**Fig. 3.2** A view of the brain as both cortical and inner areas (after [71])

ganglia participate in preparation and tuning of motor outputs, in the sense of initiation and the extent of movements. Cerebellum executes routine automatized movements like walking, biking, driving, etc. We want to point out that there are far more anatomical and functional subdivisions within each of the mentioned areas.

Functions, or better, dominances of the right and left hemispheres in different cognitive functions are different [1] (Fig. 3.3). The dominant hemisphere (usually the left one) is specialized for language, logical reasoning, awareness of cognitive processes and awareness of the results of cognitive processes. Although the non-dominant hemisphere (usually the right one) is able to carry out cognitive tasks, it is not aware of them nor their results. It is specialized for emotional and holistic processing, intra- and extrapersonal representation of space. Its intactness is crucial for the awareness of the body integrity [2]. Lesion of the parietal cortex including the somatosensory cortex leads to the so-called anosognosia. The limbs and the body are intact but the cortical and mental representations become missing. Patients who have undergone a stroke to the right parietal lobe, neglect the left half of their body, in spite they can see it. It is not a consequence of the left hemi-paralysis. Mirror damage to the left parietal lobe does not lead to anosognosia. It seems that the right hemisphere is dominant in mental representations of intra- and extrapersonal space. In other words, subjective experience of the body self depends upon specific brain mechanisms, namely an integrity of primary and higher-order somatosensory cortical areas in the right hemisphere [2].

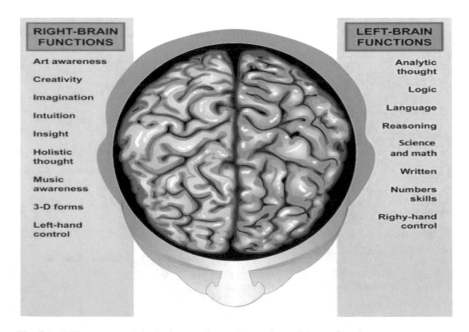

**RIGHT-BRAIN FUNCTIONS**

Art awareness

Creativity

Imagination

Intuition

Insight

Holistic thought

Music awareness

3-D forms

Left-hand control

**LEFT-BRAIN FUNCTIONS**

Analytic thought

Logic

Language

Reasoning

Science and math

Written

Numbers skills

Righy-hand control

**Fig. 3.3** Different part of the brain are allocated to perform different functions (after [71])

Several structural brain atlases have been created to support the study of the brain and to better structure brain data. Probably the first attempt was made by Korbinian Brodmann, who created a cytoarchitectonic map of the human brain, published in 1909. The map presents 47 distinctive areas of the cerebral cortex. Each Brodmann area (BA) is characterized by a distinct type of cells, but it also represents distinct structural area, distinct functional area (e.g. BA17 is the visual cortex), distinct molecular area (e.g. number of neurotransmitter channels) (see Fig. 3.4).

For many years, the standard 1998 Talairach Atlas of the human brain [3, 4] has served as the standard for reporting brain activation locations in the functional and structural brain mapping studies. They have created a co-planar 3D stereotaxic atlas of the human brain that can be used to study it from different subjects and collected using different methods. A software called Talairach Daemon (Fig. 3.5) is publicly available for download and can be used to calculate $(x, y, z)$ Talairach coordinates of any given point on the brain image together with its corresponding Brodmann area. By using this software, brain areas can be labelled accordingly in different visualization colours as depicted in Fig. 3.6.

While Talairach template is derived from the analysis of a single brain, another template which is referred to as Montreal Neurological Institute (MNI) coordinates is derived from the average of MRI data across individuals, for instance MNI152 and MNI1305 [5]. Another well-known brain template is proposed by the International Consortium for Brain Mapping (ICBM) and its few releases include ICBM452, ICBMChinese56, ICBM AD for Alzheimer Disease and ICBM MS for Multiple Sclerosis [6].

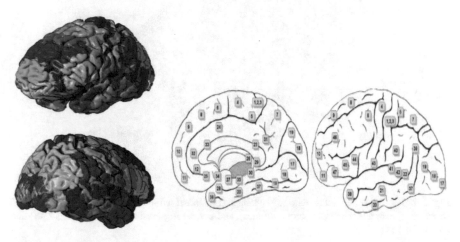

**Fig. 3.4** Brodmann areas (after [71])

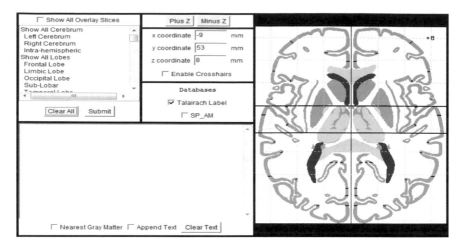

**Fig. 3.5** The Talairach Daemon Software for brain areas visualization (http://www.talairach.org/applet/)

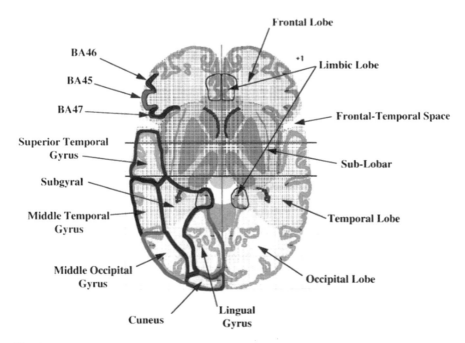

**Fig. 3.6** The Talairach atlas with lobe labels (illustrated with patterned colour fills), gyral structures (illustrated with bold colour outlines), and several Brodmann areas (illustrated with solid colour fills) [4]

## 3.2   Learning and Memory

Capability of learning and memory formation is one of the most important cognitive functions. Our identity largely depends upon what we have learned and what we can remember. We can divide the study of learning and memory into two levels:

1. The system level (where and when in term of space-time), which attempts to answer the question what brain parts and pathways the memory trace is stored in—the top-down approach, which will be the topic of this section, and
2. Molecular level (how?), which is devoted to investigation of the ways of coding and storage of information at the cellular and molecular level—the bottom-up approach, which will be introduced in Chaps. 15 and 16 and also in a later section in this chapter.

It has been long recognized that there is a short-term memory and a long-term memory. Short-term memory lasts for a few minutes and is also called the working memory. It occurs in the prefrontal cortex, although other parts of the cortex relevant to the memory content are activated too [8]. The learning process and the process of long-term memory formation can be divided into four stages: (Fig. 3.7).

1. Encoding. Attention focus and learning/entering of new information into the working memory. Finding associations with already stored memories.
2. Consolidation. The process of stabilization of new information, transformation into a long-term memory by means of learning/rehearsal.

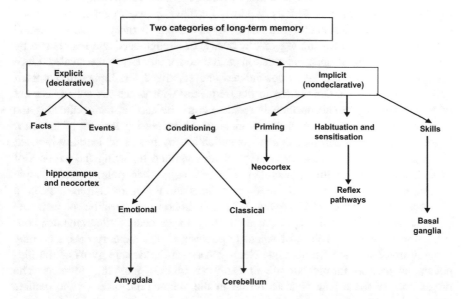

**Fig. 3.7** Different kinds of long-term memory fall under two general categories: explicit and implicit (from [7])

3. Storage. Long-term learning/storing of information in memory.
4. Recall. Retrieval of information from the working memory.

Based on clinical, imaging and animal studies we can divide long-term memory into two main categories that have different subtypes with different mechanisms and different localizations in the brain (Fig. 3.2). Explicit (declarative) memory is a memory of facts (semantic memory) and a memory of events (episodic memory). Recall from explicit memory requires conscious effort and stored items can be expressed in language. Hippocampus is a crucial but only a transitory stage in the explicit memory. How is the explicit memory formed? Information comes to brain through the sensory organs (visual, auditory, olfactory, tactile), and proceeds through subcortical sensory nuclei and sensory cortical areas into multimodal association areas, like for instance the parieto-temporo-occipital association cortex, limbic association cortex and the prefrontal association cortex. From there the information is relayed through parahippocampal cortex, perirhinal cortex and entorhinal cortex into the hippocampus. From hippocampus the information is relayed to subiculum from where it returns back to entorhinal cortex and all the way back to association cortical areas. Thus the brain circuit for the long-term explicit memory storage forms a re-entrant closed loop. According to experimental data, the "synaptic re-entry reinforcement" or SRR hypothesis and the corresponding computational model have been formulated and simulated [9, 10]. According to this hypothesis, after initial learning, reactivation of hippocampal memory traces repeatedly drives cortical learning. Thus, a memory trace (engram) is stored after many repetitions. Repeated reinforcement of synapses during the reactivation of memory traces could lead to a situation in which memory traces compete, such that the strengthening of one memory is always at the expense of others, which are either weakened or lost entirely. In other words, a single memory stored in a neural network is either lost (owing to synaptic decay) or strengthened and maintained by repeated rounds of synaptic potentiation each time the memory is reactivated. Once cortical connections are fully consolidated and stabilized, the hippocampus itself becomes dispensable. Differences in the frequency with which memory traces are either consciously or subconsciously recalled could be another factor affecting the selection of which memories are consolidated. An increasing amount of evidence suggests a role of sleep in memory consolidation by means of learning-induced correlations in the spontaneous activity of neurons and replaying the patterns of wake neural activities during sleep [11, 12]. Although others point out that people lacking REM sleep do not show memory deficits and that a major role of sleep in memory consolidation is unproven [13]. An interesting question is how the degradation of out-dated hippocampal memory traces occurs after memory consolidation is finished. The most recent hypothesis is that memory clearance may actually involve new-born neurons. Neurogenesis in the dentate gyrus of the hippocampus persists throughout life in many vertebrates, including humans. The progenitors of these new neurons reside in the subgranular layer of the dentate gyrus [14, 15].

The implicit or non-declarative memory serves to store the perceptual and motor skills and conditioned reactions. Recall of stored implicit information occurs without a conscious effort, automatically and the information is not expressed verbally. Basal ganglia and cerebellum are important for acquisition of motor habits and skills that are characterized by precise patterns of movements and fast automatic reactions. Cerebellum is the key structure for conditioning. Conditioned emotional reactions require amygdala in the limbic system. Nonassociative learning like habituation and sensitisation occur in primary sensory and reflex pathways. Priming is an increase in the speed or accuracy of a decision that occurs as a consequence of a prior exposure to some of the information in the decision context, without any intention or task related motivation, and occurs in neocortex.

Although implicit and explicit learning concern different memory contents, they share cellular and molecular mechanisms [7, 16]. These mechanisms will be one of the topics of the next chapter. Later we also introduce the genetics of learning and memory used also as inspiration for neurogenetic computational models as presented in Chap. 16.

## 3.3 Neural Representation of Information

The first principle of representation of information in the brain is redundancy. Redundancy means that every information (meant in any sense) is stored, transmitted and processed by a redundant number of neurons and synapses so that it does not become lost when neural networks undergo damage, for instance due to aging. When neural networks get damaged, their performance does not drop down to zero abruptly, like in a computer, but instead it degrades gracefully. Computer models of neural networks also confirm the idea that a degradation of performance with the loss of neurons and synapses is not linear but instead neural networks can withstand quite substantial damage, and still perform well. Next, the contemporary view on the nature of neural representation is such that information (in the sense of content or meaning) is represented by place in the cortex (or in general in the brain). However, this placing is a result of anatomical framework and shaping by input, i.e. by experience-dependent plasticity. For instance [7], a sound pattern for the word "apple" is represented in the auditory areas of the temporal cortex. It is represented as a spatial pattern of active versus inactive neurons. This neural representation is associated (connected) through synaptic weights with the neural representation of a visual image of apple in the parietal cortex, with the neural representation of an apple odor in the olfactory cortex, with memories on the grandma garden and facts about apples, being represented in some other areas of the cortex, etc. Neural representations (that is distributions or patterns of active neurons) within particular areas and their associations between areas appear as a result of learning (i.e. synaptic plasticity). Different objects are represented by means of different patterns or distributions of active neurons within cortical areas. Therefore we speak about the so-called distributed representations.

Current hypothesis states that recall from memory is an active process. Instead of passive processing of all electrical signals that arrive from hierarchically lower processing levels, cortical neural networks should be able to use fragments of activity patterns to fill in the gaps, and thus quickly re-create the whole neural representation. The filling-in process can be modelled by means of artificial neural networks, such as the Hopfield ANN (see [7]) (Fig. 3.8). Neural representations (patterns of activities) are stored in the matrix of synaptic weights through which neurons in the network are interconnected. The weight distribution storing a particular object representation is created due to an experience-dependent synaptic plasticity (learning). When a sufficiently large portion of this neural representation is activated from outside the network, few electric signals along all the synapses in the network quickly switch on the correct remaining neurons in the representation.

Neural representations in the sense of patterns of activity have a holistic character. Patterns of activity are being recalled (restored) as a whole. Thus, we can see a relation between the character of neural representations and gestalts. Gestalt psychology was developed at the beginning of the 20th century by Max Wertheimer, Kurt Koffka and Wolfgang Köhle in Germany. Gestalt psychology considers holistic mental gestalts (shapes, forms) to be the basic mental elements. For the gestalt to be stored and recalled, certain rules must be fulfilled, like the rules of proximity, good continuation, symmetry, etc. These rules have been experimentally verified.

To conclude, neural representations of objects are stored in the matrix of synaptic weights as a whole. We are not able to trace down a sequence of steps leading to the holistic percept. Synaptic weights implicitly bind together parts of the pattern.

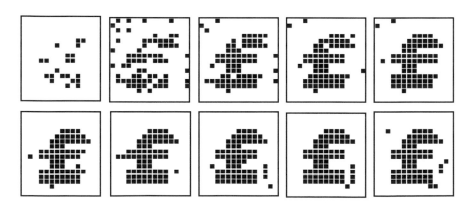

**Fig. 3.8** Illustration of spontaneous re-creation of neural representation after few input impulses (figure in the uppermost left corner). Black pixel represents a firing neuron while blank pixel represents a silent neuron. Between each pattern of activity from left to right (1 ms time frame), neurons in the network exchange only one impulse. Thus, basically after exchanging only two-three spikes, the memory pattern is re-created. Network can reverberate the restored memory pattern until a different external input arrives (from [7])

## 3.4 Perception in the Brain Is Always Spatio/ Spectro-temporal

Perception in the brain provides information for learning and development. The five senses of perception (visual auditory, tactile, gustatory and olfactory) send information to the brain as always spatio/spectro-temporal. Even a static picture is perceived in the retina as activity of cells that are activated differently in time and space for different colors and intensity of the pixels of the picture, the brighter pixels causing the first spikes to be sent to the visual cortex from the retina. This is also demonstrated in Chap. 9 when a SNN system is trained on fMRI data of a person seeing a picture.

Perception is accompanied by sensory awareness, and therefore we will describe the underlying neural processes. We will concentrate on visual perception and visual awareness since similar principles apply to all sensations. Neurons in different areas of the visual cortex respond to various elementary features, like oriented edges of light intensity (bars), binocular disparity, movement, color, etc. [1, 7]. Visual areas in the occipital, parietal and inferior temporal cortex, though reciprocally connected, are hierarchically organized. Results of processing at lower hierarchical levels are relayed to higher-order areas. Neurons in higher-order areas respond to various combinations of elementary features from lower-order areas. In primates, based on matching psychophysical and physiological data, three main visual systems, relatively independent but mutually heavily interconnected, have been identified: the "magno", "parvo" and the color system [7, 17]. The "magno" system is responsible for perception of movement, depth and space, and separation of objects. Several cues leading to the depth perception have been identified: stereopsy, depth from perspective, depth from mutual movement and occlusion, etc. The "parvo" system is responsible for shape recognition. For separation and recognition of objects, we use separation based on movement, separation from background, filling in borders, shape from shading, etc. The color system is responsible for color perception. With respect to cortical neurons belonging to these three systems, they possess different combinations and ranges of these four physiological properties: sensitivity to color (small/large), sensitivity to the light contrast (small/large), temporal resolution (small/large), spatial resolution (small/large). These are the so-called elementary features of visual objects. Elementary features belonging to one visual object activate different and spatially separated groups of neurons within the cerebral cortex.

Binding of spatially separated neurons coding for features belonging to one visual object could be performed by transient synchronization of firing of these neurons in time [18–20]. Similar synchronous oscillations of neurons were detected also in auditory, somatosensory, parietal, motor, and prefrontal cortices in the case of auditory, tactile and other perceptions, respectively [21]. Oscillations of neurons with frequencies around and above 40 Hz (known as gamma oscillations) have been detected in the cerebral cortex of humans, primates and other investigated mammals, in particular as a result of sensory stimulation. This synchronization occurs over relatively long distances (mm–cm), between different cortical areas, between cortex and thalamus, between the two hemispheres.

Synchronization means that neurons discharge (spike) with the same frequency and the same phase. This results in a distributed pattern of simultaneously firing neurons in space and time. Neural correlates of different objects can differ in: (a) which neurons are members of the pattern, (b) which is the particular frequency of their synchronization, and (c) which is the phase of their synchronization. Thus, transient synchronous gamma oscillations have been suggested as a possible candidate for the mechanism of binding many elementary features belonging to one object to one transient whole corresponding to a percept. Establishment of transient synchrony is based upon the underlying synaptic connectivity as a result of learning.

An experimental phenomenon strongly suggesting a one-to-one correspondence between transient synchronizations and perception is binocular rivalry. During binocular rivalry, each eye is constantly stimulated with a different pattern. Visual percept is neither an average of these two patterns nor their sum. Instead, a random alternation between the two percepts occurs as if they were competing with each other, hence the term binocular rivalry [22], discovered that neurons which respond to one or the other pattern are synchronized only during the corresponding percept. Thus, although the pattern is constantly stimulating an eye, cortical neurons get synchronized only when the pattern is perceived.

An important study of Rodriguez et al. in [23] has demonstrated that perception of faces in humans is accompanied by a transient ($\sim 180$ ms) synchronization of gamma activity in hierarchically highest visual areas in the parietal cortex and premotor areas in the frontal cortex (Fig. 3.9). Thus, transient synchronizations may accompany also other cognitive processes not only perception. Miltner et al. [24] indeed detected synchronization of gamma oscillations during an associative learning. Humans were supposed to learn to associate a visual stimulus with the tactile stimulus. A selective synchronization occurred between the visual cortex and that part of somatosensory cortex which represented the stimulated hand, during and after the learning. When people forgot the learned association, synchronization between these two stimuli, or rather between neural responses to these two stimuli, disappeared.

Currently, transient (100–200 ms) synchronous gamma oscillations are being studied as a promising candidate for the mechanism of binding many elementary features belonging to one object to one transient whole corresponding to a percept of that object [25, 26]. Such synchronized activity summates more effectively than nonsynchronized activity in the target cells at subsequent processing stages, and the activity can spread to a longer distances. If so, synchronization could increase the effect that a selected population of neurons has on other populations with great temporal specificity (in the range of milliseconds). There is also evidence that synchrony is important for inducing changes in synaptic efficacies and hence facilitate transfer of information into memory. Different objects in one scene may be associated with different phase-locked synchronous oscillations within the gamma frequency band. Thus, increased coherence between brain areas confined to a

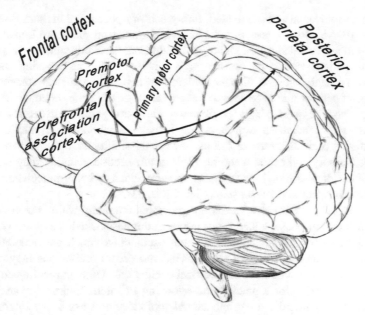

**Fig. 3.9** Corticocortical connections between the posterior parietal cortex and the main subdivisions of the frontal cortex. Illustrated areas showed increased coherence within the 40 Hz band in the Rodriguez et al.'s experiment on recognition of faces. When a human face was recognized, transient coherence occurred in the time window of 180 × 360 ms after the beginning of the picture presentation (from [7, 23])

narrow band around 40 Hz may denote a holistic perception of a complex stimulus. Based on experimental findings, crucial neural conditions for a conscious percept to be experienced are identified [19, 23, 27–29].

Generating sensory awareness involves the process of attention. Several areas in the prefrontal cortex are crucially involved in attention, namely areas 8Av (major connections with the visual system), 8Ad (major connections with the auditory system) and 8B (major connections with the limbic system) [8]. Attentional selection may depend on appropriate binding (coherence) of neuronal discharges in sensory areas in two simultaneously active directions: an attentional mechanism in prefrontal cortex could induce synchronous oscillations in selected neuronal populations (top-down interaction), and strongly synchronized cell assemblies could engage attentional areas into coherence (bottom-up interaction) [19]. Chapter 8 presents a SNN model to model attentional bias, which is manifested as brain activities when a person is reacting to non-targeted stimuli.

Another prefrontal areas activated during sensory perception include Brodmann areas 9, 10, 45, 46, 47 (see Fig. 3.10). These prefrontal areas are known to be involved in an extended action planning. In addition, these prefrontal areas plus the posterior parietal cortex are known to be involved in the working memory. Posterior parietal cortex is also known to be involved in mental imagery. For planning of actions it is necessary to keep track of at least one sequence of partial actions, hence the overlap between planning and memory mechanisms. It might be that sensory contents reach awareness only if they are bound to prefrontal areas via the posterior parietal cortex and thus have a possibility to become part of the working memory and action planning [25]. In turn, action planning may influence organization of attentional mechanisms and thus what is being perceived. Action planning can occur at a subconscious level [30, 31].

Coherences in the involved areas are generated internally within the cortex and although they are phase-locked, they are not stimulus locked. They are superimposed upon global thalamocortical gamma oscillations which are generated and maintained during cognitive tasks [32]. Thalamocortical oscillations may provide the basic oscillatory modulation of cortical oscillations. Other cortical mechanisms are then responsible for a precise phase-locking of internal cortical synchronous oscillations. In particular, these are lateral inhibitory and excitatory interactions, regularly bursting layer V pyramidal cells, and spike-timing dependent rapid

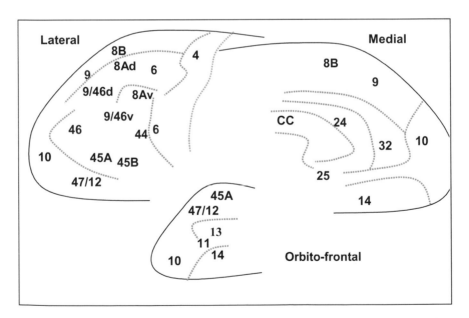

**Fig. 3.10** Human prefrontal cortex (after [7]). Lateral view (from outside), medial view (from inside) and the orbito-frontal view (from below) at the left hemispehere. The same divisions hold also for the right hemisphere. Numbers denote the corresponding Brodmann's areas. CC means Corpus Callosum

synaptic plasticity. In the latest, synapses and thus the inputs which do not drive the postsynaptic cell in synchrony are temporary weakened [33].

In [33], this is called ever changing semiglobal coherent activity, the *dynamic core*. The dynamic core corresponds to a large (semiglobal) continuous cluster of neuronal groups that are coherently active on a time scale of hundreds of milliseconds. Its participating neuronal groups are much more strongly interactive among themselves than with the rest of the brain. The dynamic core must also have an extremely high complexity as opposed to for instance convulsions. Each roughly 150 ms, a pattern of semiglobal activity must be selected within less than a second out of a very large, almost infinite, repertoire of options. Thus, the dynamic core changes in composition over time. As suggested by neuro-imaging, exact composition of the core varies significantly not only over time within one individual, but also vary significantly across individuals (Fig. 3.11).

According to [33], the dynamic core consists of a large number of distributed groups of neurons which enter the core temporarily based on their mutual coherence. Connecting groups of neurons into temporarily synchronized whole requires dense recurrent connections between brain areas, along which a reiterated re-entry of signals occurs. Neural reference space for any conscious state may be viewed as an abstract $N$-dimensional space, where each axis (dimension) stands for some participating group of neurons that code for (represent) a given aspect of the conscious experience. There can be hundreds of thousands of dimensions. The distance from the beginning of the axis represents the salience of that aspect. It may, for instance, correspond to the number of firing neurons within a given group. We

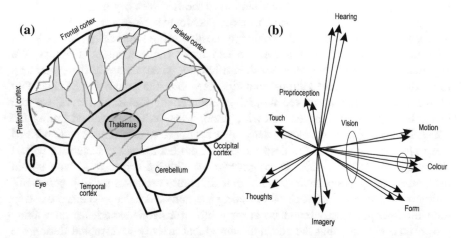

**Fig. 3.11 a** Illustration of the dynamic core, a changing coherent semiglobal spatio-temporal activity of the brain, which is supposed to be a neural correlate of consciousness. One configuration of the core lasts for about 150 ms. **b** Interpretation of the dynamic core as an $N$-dimensional neuronal reference space, where each axis (dimension) denotes some group of neurons which encodes (represents) a given aspect of the conscious experience. Each axis can be broken down into more elementary axes. There can be hundreds of thousands of spatio-temporal dimensions (from [7])

would like to point out the interesting similarity between this abstract $N$-dimensional neural space and the conceptual spaces introduced by [34].

What would be, in this theory, a neural basis for subconsciousness? The same group of neurons may at times be part of the dynamic core and underlie conscious experience, while at other times it may not be part of it and thus be involved in subconscious processing. In [27], have proposed that those active neurons which are not at the moment taking part in the semiglobal activity keep processing their inputs, and results of this processing may still affect behaviour.

We would like to mention also the explanation of neural correlate of qualia or the hard problem of consciousness, according to [33]. Qualia are specific qualities of subjective experiences, like redness, blueness, warmth, pain, and so on. According to the dynamic core hypothesis, pure redness would be represented by one particular state of the dynamic core that is by one and only one point in the $N$-dimensional neural space. This core state would certainly include large participation of neurons that code for the red colour and a small participation of neurons that code for other colours and for anything else. Coordinates of a point in the $N$-dimensional reference neural space are determined by activities of all neuronal groups that are at the moment part of the core. And these activities vary in time-space and across individuals. Thus, the subjective experience of redness will be different in different people and can be different for the same individual for instance in the morning and in the evening.

Sleep research has revealed that during sleep, humans normally go through two-three cycles of two sleep phases. One of these two phases is the so called REM sleep, according to the accompanying Rapid Eye Movements. EEG activity of the brain during the REM phase is very similar to the EEG activity of the awake brain during cognitive activity. Hence the term paradoxical sleep for the REM sleep phase, as it was not sleep at all. We dream mostly during REM sleep phases. When awakened during the REM phase, we can recall the content of a dream. We experience self-awareness when we dream but not when we are in the deep sleep [35]. Thus "I" is preserved during dreaming as well as the awake-like EEG activity of the brain. When awakened around at the end of the REM phase, we can remember that we dreamt, not knowing about what. When awakened during the non-REM sleep phase, we mostly deny any experience of dreaming. The non-REM sleep phase is also called the deep sleep, and the brain activity occurs in typical slow large regular waves. Recently, experiment with the spread of activity within neocortex during sleep have revealed that different cortical areas stop communicating over distance with each other during the non-REM sleep—a stage of sleep for which people mostly report no or very little conscious experience on waking [36]. Thus, it seems that the coherent semiglobal activity is disrupted during the non-REM sleep, and so is the conscious awareness [7]. Information consolidation in the brain during sleep was used as inspiration for the development of "sleep learning" algorithms for ECOS (Chap. 2). It can inspire the development of new algorithms for SNN as well.

## 3.5  Deep Learning and Deep Knowledge Representation in Time-Space in the Brain

As discussed above, the brain is a complex integrated spatio-temporal information processing machine. An animal or a human brain has a range of structural and functional areas that are spatially distributed in a constrained 3D space. When the brain processes information, either triggered by external stimuli, or by inner processes, such as visual-, auditory-, somatosensory-, olfactory-, control-, emotional-, environmental-, social, or all of these stimuli together, complex spatio—temporal pathways are activated and patterns are formed across the whole brain. For example, '...the language task involves transfer of stimulus information from the inner ear through the auditory nucleus in the thalamus to the primary auditory cortex (Brodmann's area 41), then to the higher-order auditory cortex (area 42), before it is relayed to the angular gyrus (area 39)...' [7, 37]. Many other studies of spatio-temporal pathways in the brain have been conducted, e.g. birdsong learning [38].

In principle, different 'levels' of spatio-temporal information processing can be observed in the brain, [37], all 'levels' acting in a concert. Spatio-temporal brain data (STBD) related to each of these 'levels' can be collected, but how do we integrate this information in a machine learning model?

Let us trace the visual brain processing in this experiment [7] (see Fig. 3.12). Projected image stimulates retina for 20 ms. In about 80 ms, neurons in the thalamic LGN (lateral geniculate nucleus) respond. Thalamic neurons activate neurons in the primary visual cortex (V1). Then, activation proceeds to and through higher-order visual areas, V2, V4 and IT. We speak about the so-called "WHAT" visual system, which is assumed to be responsible mainly for classification and recognition of objects. In the highest-order area of this system, i.e. the infratemporal (IT) cortex, activity appears after 150 ms since the picture onset (on average). It is thought that here, in the IT area, the classification process is completed [43]. If we divide 150 ms since the picture onset by the number of processing areas (i.e. retina, thalamus, V1, V2, V4), on average each of them has only 30 ms for processing of signals. The frontal areas, PFC, PMC and MC, are responsible for preparation and execution of motor response, for what they need only 100 ms. Divided by three, again we get about 30 ms for each area. Since each of the mentioned areas has further subtle subdivisions, each sub area can have only 10 ms to process signals and send them higher in the hierarchy of processing. At the same time, neurons in each area send signals up and down in the stream of hierarchical processing. Whether 10 or 30 ms, it is an extremely short time for processing in one single area. Cortical neurons, when naturally stimulated fire with frequencies of the order of 10–100 Hz. A neuron firing with an average frequency of 10 Hz (i.e., 10 impulses in 1000 ms), may fire the first spike in 100 ms from the beginning of stimulation. Thus, during the first 10–30 ms there will be no spikes from this neuron. Another neuron firing with the frequency of 100 Hz fires $1 \times 3$ spikes during the first

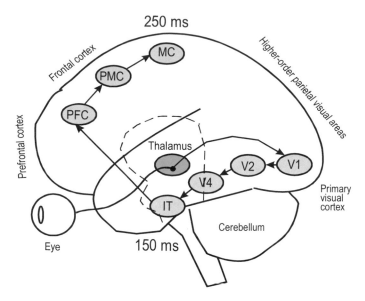

**Fig. 3.12** Deep spatio-temporal processing of visual stimuli in humans for image classification. Location of cortical areas: V1 = primary visual cortex, V2 = secondary visual cortex, V4 = quartiary visual cortex, IT = inferotemporal cortex, PFC = prefrontal cortex, PMC = premotor cortex, MC = motor cortex. The brain has learned through *deep learning* how to process visual stimuli, forming a *deep knowledge*, represented as connections between different spatially located parts of the brain, activated at different times (from [7]). We can represent the deep knowledge of classifying an image stimulus as a sequence of events (Ei), each of them consisting of a function Fi, that is activated as a location Si at a time Ti, and all of them connected as a piece of deep knowledge

$10 \times 30$ ms. In each of the above-mentioned areas, there are millions perhaps billions of neurons, then these neurons exchange only 1–3 spikes, and the result of this processing is sent higher to higher-order areas, and lower to the lower-order areas. Each neuron receives signals from say 10,000 other neurons and sends off signals to the next 10,000 neurons. Synaptic transmission delay in one synapse is about 1 ms. A neuron cannot wait 10,000 ms to receive signals from all its presynaptic neurons. Thus, the signals ought to come almost simultaneously, and not one after another.

Another complication in the neuronal processing of inputs is the fact that firing is a stochastic process. A good model for it is a Poisson stochastic process where the value of dispersion is equal to the value of the mean, thus the dispersion is large. Speaking about firing frequencies of 10 or 100 Hz, we mean average frequencies over relatively long time periods, let us say 500 ms (half of a second). Thus, a neuron firing with the average frequency of 100 Hz does not have to fire a single spike during the first 10–30 ms from the beginning of stimulation, and a neuron

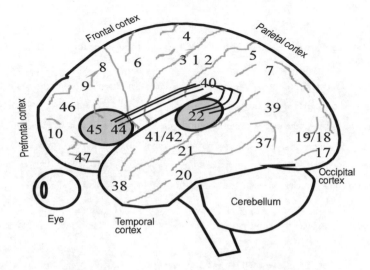

**Fig. 3.13** Deep spatio-temporal processing in the brain when dealing with words and language (from [7])

firing with the average frequency of 10 Hz may fire four spikes. Thus, to summarize, it is really a problem how neurons code information. So far, this problem has not been solved. In the following section we will introduce several current hypotheses.

Despite the very complex information processing in the brain, we can represent at abstract level the deep knowledge of classifying an image stimulus as a sequence of events (Ei), each of them consisting of a function Fi, that is activated as a location Si at a time Ti, and all of them connected as a piece of deep knowledge (as per the definitions of deep knowledge in Chap. 1).

Language processing during a simple task of repeating the word that has been heard is the Wernicke-Geschwind model [41] that is a deep spatio-temporal activation of brain areas as a result of deep learning beforehand. A language task involves many steps of processing as shown in Box 3.1 and Fig. 3.13, learned as deep learning and representing deep knowledge. The task involves different procedures (named as events Ei in Chap. 1), each event Ei consisting of a function Fi, spatial location Si and time of execution Ti, and all connected in a deep knowledge.

---

**Box 3.1. Deep knowledge learned and represented in time-space for a language task (see Fig. 3.13).**

---

1. Event E1: Transfer of information from the inner ear through the auditory nucleus in thalamus to the primary auditory cortex (Brodmann's area 41) (location S1) at time T1.

2. Event E2: Then to the higher-order auditory cortex (area 42) (location S2) at time T2;

3. Event 3: Then it is relayed to the angular gyrus (area 39) (location S3) at time T3. Angular gyrus is a specific region of the parietal-temporal-occipital association cortex, which is thought to be concerned with the association of incoming auditory, visual and tactile information.

4. Event 4: From here, the information is projected to Wernicke's area (area 22) (location S4) at time T4.

5. Event 5: Then, by means of the *arcuate fasciculus*, to Broca's area (44, 45), where the perception of language is translated into the grammatical structure of a phrase and where the memory for word articulation is stored (location S5) at time T5.

6. Event 6: This information about the sound pattern of the phrase is then relayed to the facial area of the motor cortex that controls articulation, so that the word can be spoken (location S6) at time T6.

---

Note: Times $T_i$ and locations $S_i$ of events $E_i$ can take either exact values or fuzzy values (e.g. around).

Similar pathway is involved in naming an object that has been visually recognized. This time, the input proceeds form retina and LGN (lateral geniculate nucleus) to the primary visual cortex, then to area 18, before it arrives to the angular gyrus, from where it is relayed by a particular component of arcuate fasciculus directly to Broca's area, bypassing Wernicke's area

The brain has learned through *deep learning* how to process visual stimuli, forming a *deep knowledge*, represented as connections between different spatially located parts of the brain, activated at different times.

The deep learning in the brain is achieved through creating connections between neurons in space and time. The patterns that are formed by these connection represent deep knowledge and enable people to perform different tasks. Similar to the discussed deep learning and deep knowledge representation in this section, in Chap. 8 we present methods for deep learning and deep knowledge representation from EEG data and in Chap. 10—from fMRI data.

# 3.6   Information and Signal Processing in Neurons and in the Brain

## 3.6.1   Information Coding

The brain consists of bullions of neurons and trillions of connections, and each neuron is a complex information processing machine, receiving thousands of signals from dendrites that receive signals from other neurons through synaptic connections. The neuron has just one output that emits spikes at certain times when the membrane of this neurons reaches a threshold (Fig. 3.14).

Information in the brain is represented and transferred as electrical potentials (spikes) under different encoding mechanisms as discussed below. Here we discuss some methods of coding information as spikes in the brain that have inspired methods for data coding in artificial SNN discussed in Chap. 4.

*Coding on Information Based on Spike Timing*

1. *Reverse correlation*. The first option is that the information about the salience of the object feature is encoded in the exact temporal structure of the output spike train. Let us say that two neurons fire three spikes within 30 ms. The first neuron fires a spike train with this temporal structure | || and the second neuron with this temporal structure | | |. By means of the techniques of reverse correlation, it is possible to calculate which stimulus exclusively causes which temporal pattern of which

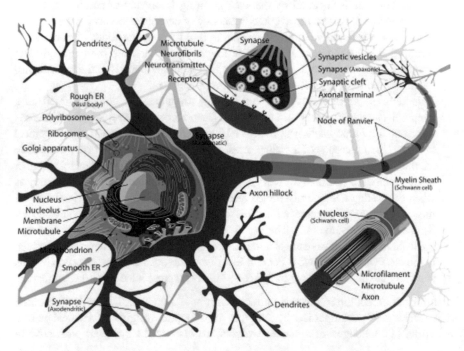

**Fig. 3.14**  A single neuron is a complex information processing machine (after [71])

neuron. The main proponents of this theory are Bialek and his co-workers who have made its successful verification in the fly visual system [42].

2. *Time to the first spike.* Let at time instant $t_0$ a stimulus arrives to the neural network. Neurons that fire the first (let us say in a window of 10 ms) carry the most important information about the stimulus features. The rest of neurons and the rest of impulses are ignored. This theory is favored by Thorpe [40, 43].

3. *Phase coding.* Information about the presence of a feature is encoded in the phase of neuron's impulses with respect to the reference background oscillation. Either they are in a phase lead or in a phase lag. The information can also depend on the magnitude of this phase lead (lag). This coding is preferred by people investigating hippocampus [44].

4. *Synchronization.* Populations of neurons that represent features belonging to one object can be bound together by synchronous firing in time. Such synchronization was discovered in the laboratory of W. Singer in the cat visual cortex to accompany percepts [22]. It was also detected in the human cortex during perception of meaningful stimuli (faces) [23].

Rate Coding

1. *Temporal average rate.* In this respect, works of an English physiologist Adrian from the 30-ties of the 20th century are being cited. Adrian found out that the average frequency of a neuron in the somatosensory cortex is directly proportional to the pressure applied to its touch receptor. Similar dependencies have been discovered in the auditory and visual cortices. That is, in the auditory cortex, the heard frequency is encoded by the average firing frequency of auditory neurons, and in the visual cortex, the average frequency of neurons encodes for the salience of its visual elementary feature. This coding is still being considered for stationary stimuli that last up to around 500 ms or longer, so that neurons have enough time to count (integrate) impulses over long time. Neurons that have the highest frequency signalize the presence of the relevant feature.

2. *Rate as a population average.* An average frequency is not calculated as a temporal average but rather as a population (spatio-temporal) average. One feature is represented by a population of many (10,000) neurons, for instance in one cortical column. Upon presence of a feature, most of them are activated. When we calculate the number of spikes in a 10 ms window of all these neurons and divide this number by the number of neurons, we will get approximately the same average frequency as when calculating a temporal average rate of any of these neurons (provided they all fire with the same average rate). This idea has been thoroughly investigated by Shadlen and Newsome [45]. They showed on concrete examples, that by means of population averaging we can get a reliable calculation of neuron's average rates even in the case when they have a Poisson-like distribution of output spikes. Populations that relay the highest number of spikes signalize the presence of the relevant feature.

Information coding in the brain affects learning. At present, it is widely accepted that learning is accompanied by changes of synaptic weights in cortical neural networks [1]. Changes of synaptic weights are also called *synaptic plasticity*. In

1949, the Canadian psychologist Donald Hebb formulated a universal rule for these changes: "When an axon of cell A excites cell B and repeatedly or persistently takes part in firing it, some growth process or metabolic change takes place in one or both cells so that A' s efficiency as one of the cells firing B is increased", which has been verified in many experiments and its mechanisms elucidated [46]. This learning principle is also known as "neurons that fire together, wire together". This principle is extended with the introduction of the time of spiking, leading to spike-time dependent learning rules used in the SNN models (see Chaps. 4, 5 and 6).

## *3.6.2   Molecular Basis of Information Processing*

In cerebral cortex and in hippocampus of humans and animals, learning takes place in excitatory synapses formed upon dendritic spines that use glutamate as their neurotransmitter. In the regime of learning, glutamate acts on specific postsynaptic receptors, the so-called NMDA receptors (*N*-methyl-D-aspartate). NMDA receptors are associated with ion channels for sodium and calcium (see Fig. 3.15). The influx of these ions into spines is proportional to the frequency of incoming presynaptic spikes. Calcium acts as a second messenger thus triggering a cascade of biochemical reactions which lead either to the long-term potentiation of synaptic

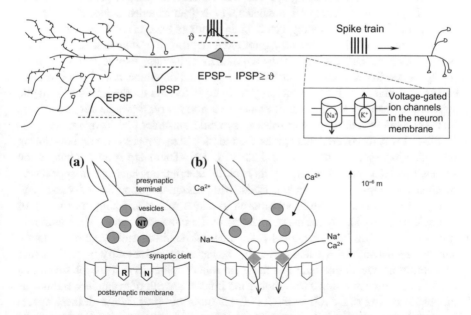

**Fig. 3.15** Scheme of synaptic transmission. **a** A synapse is ready to transmit a signal. **b** Transmission of electric signal (a spike) in a chemical synapse upon arrival of action potential into the terminal. NT = neurotransmitter, R = AMPA-receptor-gated ion channel for sodium, N = NMDA-receptor-gated ion channel for sodium and calcium (from [7])

weights (LTP) or to the long-term depression (weakening) of synaptic weights (LTD). In experimental animals, it has been recorded that these changes in synaptic weights can last for hours, days, even weeks and months, up to a year. Induction of such long-term synaptic changes involves transient changes in gene expression [47, 48].

A subcellular switch between LTD and LTP is the concentration of calcium within spines [49]. We speak about an LTD/LTP threshold. In turn, the intra-spine calcium concentration depends upon the intensity of synaptic stimulation that is upon the frequency of presynaptic spikes. That is, more presynaptic spikes means more glutamate within synaptic cleft. Release of glutamate must coincide with a sufficient depolarization of the postsynaptic membrane to remove the magnesium block of the NMDA receptor. The greater the depolarization, the more ions of calcium enter the spine. Postsynaptic depolarization is primarily achieved via AMPA (amino-methylisoxasole-propionic acid) receptors, however, recently a significant role of backpropagating postsynaptic spikes has been pointed out [50]. Calcium concentrations below or above the LTD/LTP threshold, switch on different enzymatic pathways that lead either to LTD or LTP, respectively. However, the current value of the LTD/LTP threshold (i.e. the properties of these two enzymatic pathways) can be influenced by levels of other neurotransmitters, an average previous activity of a neuron, and possibly other biochemical factors as well. This phenomenon is called metaplasticity, a plasticity of synaptic plasticity [51]. Dependence of the LTD/LTP threshold upon different postsynaptic factors is the subject of the Bienenstock, Cooper and Munro (BCM) theory of synaptic plasticity [52] (for a nice overview see for instance [53]). The BCM theory of synaptic plasticity has been successfully applied in computer simulations to explain experience-dependent changes in the normal and ultrastructrally altered brain cortex of experimental animals [54, 55].

Dendrites of cortical excitatory pyramidal neurons are abundant in tiny membrane extensions called spines. They are named so because they resemble in shape the spines on the rose stem. About 80% of all synaptic connections in the cerebral cortex are excitatory and vast majority of them is formed on the heads of synaptic spines. For many years the role of spines was a mystery. Nowadays it is accepted that they play several important roles in synaptic plasticity and learning.

First, it was discovered that spines change their size, shapes and numbers during the induction and maintenance of LTP [56, 57]. There are growth changes on spines, like spine head swelling, spine neck thickening, and increase in appearance of spines with mushroom-shaped heads. Morphological properties of spines and changes in their shape were first supposed to play a role in affecting the efficacy of synaptic transmission by means of changes in the input resistance [58]. Long, thin spines create a big input electrical resistance, while short, stubby spines create a smaller input resistance. Later, a role in sequestering and amplifying the calcium concentrations was suggested to be the main role of spines [59]. Through this role a mechanisms for saturation and stopping the infinite growth of synaptic weights was proposed, as well as the role in the LTP/LTD threshold [60]. While all these effects can take place, another important role for spines was suggested in the transport of new receptors into the spine head [61]. This model is based on our older hypothesis that the changes in efficacy of excitatory dendritic spine synapses can result from the fusion of transport vesicles carrying new membrane material with the postsynaptic membrane of spines [62]. Spacek and Harris indeed found structural evidence for exocytotic activity within spines in hippocampal CA1 pyramidal neurons [63]. Smooth vesicles of the diameter around 50 nm occurred in the cytoplasm of spine heads, adjacent to the spine plasma membrane, and fusing with the plasma membrane. In addition, Lledo et al. showed that inhibitors of membrane fusion blocked or strongly reduced LTP when introduced into CA1 pyramidal cells [64]. On the other hand, an increase in synaptic strength was elicited when membrane fusion was facilitated. In the CA1 region, LTP requires the activation of the NMDA glutamate receptors and a subsequent rise in postsynaptic calcium concentration. Besides other roles, $Ca^{2+}$ plays a crucial role in the final stage of vesicle fusion with the membrane, and the number of fused vesicles is proportional to $[Ca^{2+}]$ [65]. Since LTP in CA1 neurons is accompanied by appearance of AMPA subclass of glutamate receptors [66], it is reasonable to assume that vesicles can be a mean of their insertion. Indeed, Kharazia et al. [67] observed GluR1 (a subunit of AMPA receptors) containing vesicles associated with the cytoplasmic side of some GluR1-containing cortical synapses. Moreover, tetanic stimulation induces a rapid delivery of GluR1 into spines and this delivery requires activation of NMDA receptors [68].

Another effect of the vesicle fusion with the spine membrane would be the shaping and growth of the spine, which were observed during the induction and maintenance of LTP. However, prior to fusion the vesicles must get very close to the plasma membrane. The main mechanism for displacement of vesicles within axons and dendrites is the fast active transport with the speeds of 0.001–0.004 m/ms [69]. Fast transport depends on the direct interaction of transported vesicles with microtubules via the translocator kinesin-like molecules [69]. However, microtubules do not enter spines [63]. Thus, while the fast transport can bring vesicles close to the walls of dendritic shafts, another mechanism must come into play within spines themselves. The first natural candidate for this mechanism can be the diffusion of vesicles. However, we have shown that an electrophoretically driven, directed motion of negatively charged vesicles towards the spine head, evoked by the synapse stimulation itself can be ten times faster [61].

At a molecular level, different genes, that affect the activity of neuro-receptors and neuro-trasmitters, such as GABA, AMPA, NMDA are expressed differently in different parts of the brain that defines the functioning of these parts. An example is shown in Fig. 3.16. Information, related to the expression of genes in the brain can be used for neurogenetic modelling as discussed in Chap. 16.

## 3.7 Measuring Brain Activities as Spatio/ Spectro-temporal Data

### 3.7.1 General Notions

At present, a number of techniques is available to investigate where in the brain particular cognitive and other kinds of functions are based. In general, these methods are divided as being invasive or noninvasive. In medicine the term invasive relates to a technique in which the body is entered by puncture, incision or

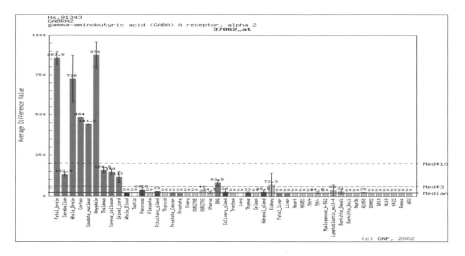

**Fig. 3.16** The expression of the GABRA2 gene causes the production of the GABA receptor in the synapses of neurons, and it is differently expressed in different spatially located parts of the brain (from Gene Expression Atlas, http://expression.gnf.org/cgi-bin/index.cgi) (see also [7])

other intrusion. Noninvasive means the opposite that is the technique that does not intrude into the body.

An invasive method of the brain study is the direct *stimulation*. Researchers perform electrical, magnetic or chemical stimulation of some neural circuit or part of it, and observe the consequences. Electrical stimulation is delivered through microelectrodes inserted into the brain. This type of research is done routinely on animals. It can be done on human subjects during the brain surgery when the skull has to be opened anyway and surgeons have to map the functions of the operated area and its surrounded parts. *Electrical stimulation of the brain* (ESB) can be also used to treat chronic tremors associated with Parkinson disease, chronic pain of patients suffering from back problems and other chronic injuries and illnesses. ESB is administered by passing an electrical current through a microelectrode implanted in the brain. With chemical stimulation, a particular chemical compound is administered into a chosen part of the brain that is supposed either to stimulate or inhibit neurons within it. The least invasive methods of the stimulation methods is magnetic stimulation, called the *Transcranial Magnetic Stimulation* (TMS). TMS and rTMS (repetitive TMS) are simply the applications of the principle of electromagnetic induction to get electric currents across the insulating tissues of the scalp and skull without the tissue damage. The electric current induced in the surface structure of the brain, the cortex, activates nerve cells in much the same way as if the currents were applied directly to the cortical surface. However, the path of this current is complex to model because the brain is a non-uniform conductor with an irregular shape. With stereotactic, MRI-based control (see below), the precision of targeting TMS can be as good as a few millimetres.

However, besides the invasiveness there are other problems with the methods of direct stimulation. Intensity of an artificial stimulation can be stronger or weaker than the level of spontaneous activity in the target circuit. Therefore artificial stimulation can engage more or respectively less of brain circuitry than is normally involved in the studied function. Thus, there are difficulties in determining which brain circuitries have been actually affected by the stimulation and thus which brain structures actually mediate the studied function. Further in this book we will consider only non-invasive measures of brain activity that are modelled with the use of SNN, even though these models can be also for modelling invasive brain data.

## 3.7.2  Electroencephalogram (EEG) Data

The oldest non-invasive method to measure electrical activity of the brain is the *electroencephalography* (EEG). An EEG is a recording of electrical signals from the brain made by attaching the surface electrodes to the subject's scalp (Fig. 3.17). These electrodes are located at exact locations of the scalp and measure corresponding activities as illustrated in Table 3.1. EEGs allow researchers to follow electrical potentials across the surface of the brain and observe changes over split

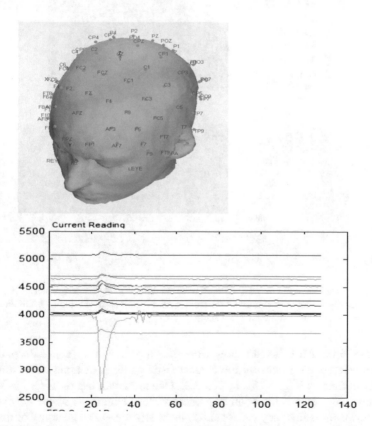

**Fig. 3.17** EEG signals taken from EEG electrodes spatially distributed on the scalp are spatio/spectro temporal data

**Table 3.1**  Anatomical locations of international 10–10 cortical projections

| Labels | Talairach coordinates | | | Gyri | | BA |
|---|---|---|---|---|---|---|
| | x avg (mm) | y avg (mm) | z avg (mm) | | | |
| FPl | −21.2 ± 4.7 | 66.9 ± 3.8 | 12.1 ± 6.6 | L FL | Superior frontal G | 10 |
| FPz | 1.4 ± 2.9 | 65.1 ± 5.6 | 11.3 ± 6.8 | M FL | Bilat. medial | 10 |
| FP2 | 24.3 ± 3.2 | 66.3 ± 3.5 | 12.5 ± 6.1 | R FL | Superior frontal G | 10 |
| AF7 | −41.7 ± 4.5 | 52.8 ± 5.4 | 11.3 ± 6.8 | L FL | Middle frontal G | 10 |
| AF3 | −32.7 ± 4.9 | 48.4 ± 6.7 | 32.8 ± 6.4 | L FL | Superior frontal G | 9 |
| AFz | 1.8 ± 3.8 | 54.8 ± 7.3 | 37.9 ± 8.6 | M FL | Bilat. medial | 9 |
| AF4 | 35.1 ± 3.9 | 50.1 ± 5.3 | 31.1 ± 7.5 | L FL | Superior frontal G | 9 |
| AF8 | 43.9 ± 3.3 | 52.7 ± 5.0 | 9.3 ± 6.5 | R FL | Middle frontal G | 10 |

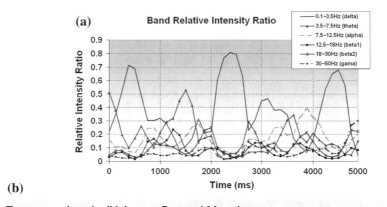

**(b)**

**Frequency band   (Hz)        General Meaning**

0.1-3.5 (delta)              Sleep or rest

3.5-7.5 (theta)              Learning, memory, sensory motor processing

7.5-12.5 (alpha)             Meditation, usually observed in the occipital lobe

12.5-30 (beta)               Active state, busy, or anxious thinking, concentration

30-100, (gamma)              Not known; consciousness usually 40

**Fig. 3.18  a** Different brain waves, characterised as different signal frequencies, could have different intensity at different times and different spatial locations in the brain. **b** Different brain waves are associated with different brain states

seconds of time. An EEG can show what state a person is in (e.g., asleep, awake, epileptic seizure, etc.) because the characteristic patterns of brainwaves differ for each of these states (Fig. 3.18a, b, Box 3.2). One important use of EEGs has been to show how long it takes the brain to process various stimuli. A major drawback of EEGs, however, is that they cannot show us the structures and anatomy of the brain and tell us which specific regions of the brain do what. In recent years, EEG has undergone technological advances that have increased its ability to read brain activity from the entire head from more than 512 sites simultaneously. The greatest advantage of EEG is that it can record changes in the brain activity almost instantaneously. On the other hand, the spatial resolution is poor, and thus should be combined with CT or MRI (see below). SNN methods for modeling and understanding EEG data, along with several applications, are presented in Chaps. 8, 9 and 14.

Box 3.2. EEG channels, corresponding Brodmann areas (BA) and functional/cognitive activity.

| EEG label | Main BA | Function |
| --- | --- | --- |
| AF3, AF4 | 9 | The frontal lobe contains most of the dopamine-sensitive neurons in the cerebral cortex. The dopamine system is associated with reward, attention, short-term memory tasks, planning, and motivation |
| F7, F8 | 45 | Together with BA 44, it comprises Broca's area, a region that is active in semantic tasks, such as semantic decision tasks (determining whether a word represents an abstract or a concrete entity) and generation tasks (generating a verb associated with a noun) |
| F3, F4 | 8 | Frontal cortex. The area is involved in the management of uncertainty. With increasing uncertainty there is increasing activation<br>An alternative interpretation is that this activation in frontal cortex encodes hope, a higher-order expectation positively correlated with uncertainty |
| FC5, FC6 | 6 | Premotor cortex and Supplementary Motor Cortex (Secondary Motor Cortex)— planning of complex, coordinated movements |
| T7 | 21 | Part of the temporal cortex .The region encompasses most of the lateral temporal cortex, a region believed to play a part in auditory processing and language. Language function is left lateralized in most individuals |
| T8 | 4 | Primary motor cortex of the human brain. It is located in the posterior portion of the frontal lobe |
| P7 | 37 | Part of the temporal lobe. The temporal lobe is involved in the retention of visual memories, processing sensory input, comprehending language, storing new memories, emotion, and deriving meaning |
| P8 | 19 | Parietal cortex; Visual areas designated as V3, V4, V5 (also known as the middle temporal area, or MT) and V6 (also known as dorsomedial area). BA 19 is the differentiation point of the two visual streams, of the 'what' and 'where' visual pathways. The dorsal region may contain motion-sensitive neurons, and ventral areas may be specialised for object recognition |
| O1, O2 | 18 | Occipital cortex—Primary visual cortex V1: Vision |

### 3.7.3 MEG

Related method to EEG, called *magnetoencephalography* (MEG) measures millisecond-long changes in magnetic fields created by the brain's electrical currents. MEG is a rare, complex and expensive neuroimaging technique. MEG machine uses a non-invasive, whole-head, e.g. 248-channel, *super-conducting-quantum-interference-device* (SQUID) to measure small magnetic signals reflecting changes in the electrical signals in the human brain. The incorporation of liquid helium creates the incredibly-cold conditions ($4.2°$ of Kelvin) necessary for the MEG's SQUIDS to be able to measure these brain magnetic fields that are billions of times weaker than the earth's magnetic force. Investigators use MEG to measure magnetic changes in the active, functioning brain in the speed of milliseconds. Besides its precision another advantage of MEG is that the biosignals it measures are not distorted by the body as in EEG. Used in conjunction with MRI or fMRI (see below), to relate the MEG sources to brain anatomical structures, researchers can localize brain activity and measure it in the same temporal dimension as the functioning brain itself. This allows investigators to measure, in real-time, the integration and activity of neuronal populations while either working on a task, or at rest. The brains of healthy subjects and those suffering from dysfunction or disease are imaged and analyzed.

### 3.7.4 CT and PET

The oldest among the noninvasive methods to study brain anatomy is *Computer Tomography* (CT). It is based on the classical X-ray principle. X-rays reflect the relative density of the tissue through which they pass. If a narrow X-ray beam is passed through the same point at many different angles, it is possible to construct a cross-sectional visual image of the brain. A 3D X-ray technique is called the CAT (Computerized Axial Tomography). CT is noninvasive and shows only the anatomical structure of the brain, not its function.

*Positron Emission Tomography* (PET) is used for studying the living brain activity. This noninvasive method involves an on-site use of a machine called cyclotron to label specific drugs or analogues of natural body compounds (such as glucose or oxygen) with small amounts of radioactivity. The labeled compound (a radiotracer) is then injected into the bloodstream which carries it into the brain. Radiotracers break down, giving off sub-atomic particles (positrons). By surrounding the subject's head with a detector array, it is possible to build up images of the brain showing different levels of radioactivity, and therefore, cortical activity. Thus, depending on whether we used glucose (oxygen) or some drug, PET can provide images of ongoing cortical or biochemical activity, respectively. Among the problems with this method are expense including the on-site cyclotrone and also technical parameters like the lack of temporal (40 s) and spatial (4 mm–1 cm) resolution. Usually the PET scan is combined either with CT or MRI to correlate the activity with brain anatomy.

*Single-Photon Emission Computed Tomography* (SPECT) uses gamma radioactive rays. Similar to PET, this noninvasive procedure also uses radiotracers and a scanner to record different levels of radioactivity over the brain. SPECT imaging is performed by using a gamma camera to acquire multiple images (also called projections) from multiple angles. A computer can then be used to apply a tomographic reconstruction algorithm to the multiple projections, yielding a 3D dataset (like in CT). Special SPECT tracers have long decay time, thus no on-site cyclotron is needed, which makes this method much less expensive than PET. However, the temporal and spatial resolution of brain activity is even smaller than in PET.

### 3.7.5  fMRI

*Magnetic Resonance Imaging* (MRI) uses the properties of magnetism instead of injecting the radioactive tracers into the bloodstream to reveal the anatomical structure of the brain. A large (and loud) cylindrical magnet creates a magnetic field around the subject's head. Detectors measure local magnetic fields caused by alignment of atoms in the brain with the externally applied magnetic field. The degree of alignment depends upon the structural properties of the scanned tissue. MRI provides a precise anatomical image of both surface and deep brain structures, and thus can be combined with PET. MRI images provide greater detail than CT images.

*Functional MRI* (fMRI) combines visualisation of brain anatomy with the dynamic image of brain activity into one comprehensive scan. This non-invasive technique measures the ratio of oxygenated to deoxygenate haemoglobin which have different magnetic properties. Active brain areas have higher levels of oxygenated haemoglobin than less active areas. An fMRI can produce images of brain activity as fast as every 100–500 ms with very precise spatial resolution of about $1 \times 2$ mm. Thus, fMRI provides both an anatomical and functional view of the brain and is very precise. FMRI is a technique for determining which parts of the brain are activated at what time by different types of brain activity, such as sight, speech, imagery, memory processes, etc. This brain mapping is achieved by setting up an advanced MRI scanner in a special way so that the increased blood flow to the activated areas of the brain shows up on fMRI scans.

fMRI imaging technique is non-invasive and radiation-free thus providing a safe environment to the subjects involved. The images are recorded in sequence either vertically or horizontally (Fig. 3.19), and over time, in a matrix of intensity values. They are captured in slices through the organs, generally in 8 or 16-bit (Fig. 3.18 right).

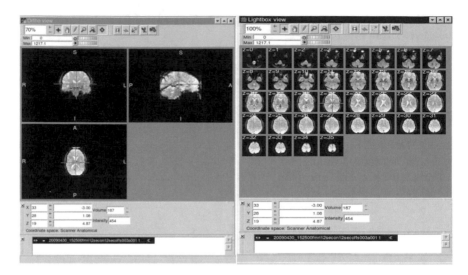

**Fig. 3.19** Brain images in vertical and horizontal slice: in sagittal, coronal and axial views (left). Slices of brain taken over time i.e. 32 images for a volume of brain (images are viewed using FSLView (FSLView, 2012) software (right) (after [75])

The images are constructed from two components—spatial/spectral (or spatio) and temporal. The first component is identified as the volume of a brain that can be further sub-divided into smaller 3D cuboids, known as voxels (volume element). In a typical fMRI study, a series of brain volumes are collected in quick succession and the value of BOLD response at all points in a 3D grid are recorded. A general 3D brain image typically contains 10,000 to 50,000 voxels, and each voxel consists of on the order of hundreds of thousands of neurons. Spatial image resolutions can be set either to have low or high resolution. Typical spatial resolution is 3 mm × 3 mm × 5 mm, corresponding to image dimensions in the order of 64 × 64 × 30 [70] and this resolution is relatively high compared to other imaging techniques.

The temporal component is acquired while scanning the whole volume of a brain that will take a few seconds to complete. In a single run of an experiment, 100 or more brain volumes are usually scanned and recorded for a single subject doing a particular sensorimotor or cognitive task. Temporal component depends on the time between acquisitions of each individual image, or the time of repetition (TR). In a typical experiment, TR ranges from 0.5 to 4.0 s and TR values in the range of 2 s are generally considered adequate [70].

The combination of this spatial and temporal information of the brain images will be the main concern investigated in this study. Chap. 10 presents SNN methods for modeling fMRI data with applications in cognitive studies, while Chap. 11 presents a method for the integration of fMRI and DTI (orientation) data.

Although a lot is known about the brain, issues about its functioning, representation and processing of information are still subjects of an intense research. The nature of brain dynamics is still unknown. Some researchers find evidence of chaos, whereas some are doubtful [72]. Main proponents of a chaotic dynamics, Freeman [73] and Tsuda [74], argue in favour of chaotic itinerancy based on EEG and other neurophysiological data. According to the picture of chaotic itinerancy, a complex system such as the (human) brain evolves by steps along a trajectory in the state space. Each step corresponds to a shift from one basin of attraction to another. Attractors represent classes for abstraction and generalization. Thus, the brain states evolve periodically through sequences of attractors. In a closed system the next attractor would be chosen solely by internal dynamics. In an open system, such as the brain, external inputs interfere with internal dynamics. Moreover, due to the changes induced by learning, trajectories continually change.

The self-organized criticality state can form the basis of the brain capacity to rapidly adjust to new external and internal stimuli. State changes resembling phase transitions occur continually everywhere in cortex at scales ranging from millimetres to $\sim 0.1$ m. Local neural activity can trigger a massive state change. However, several issues of caution should be pointed out. In spite the compelling evidence for self-organized criticality in the brain, the nature of the critical state is still unknown in neurobiological interpretation. It is high dimensional, noisy, non-Gaussian, and nonstationary [73]. Tremendous physical complexity of the brain arises also from the fact that it is not a homogenous tissue. Each part of the brain is morphologically different and has its own genetic profile as can be seen by analysis of large-scale human and mouse transcriptomes. Therefore the conditions for assessment of the type of dynamics are difficult to be met. Moreover brains are open systems driven by stochastic input. Thus it seems that the brain activity hardly can conform to the mathematical definitions of chaos. Whether the term chaotic itinerancy (or any other term from the chaotic vocabulary) is appropriate to describe state transitions in brain and cortex in particular remains open to challenge.

The complex spatio-temporal activity data from the brain still awaits explanation and proper modelling and this is what other chapters of the book present.

## 3.8   Chapter Summary and Further Readings for Deeper Knowledge

The presented in the chapter information is not intended for modeling the brain in its precise structural and functional complexity, but rather for: (1) Borrowing spatio-temporal information processing principles from the brain for the creation of brain-inspired SNN and brain-inspired AI as general spatio-temporal data machines for deep learning and deep knowledge representation in time-space; (2) Understanding brain data, when modeled with SNN, for a more accurate analysis and for a better understanding of the brain processes that generated the data. The chapter presents fundamentals of spatio-temporal information processing in the human brain and how that can be measured as data. Some of these principles are used in the rest of the book for the development of brain-inspired spiking neural networks (BI-SNN) as the main approach here to building brain-inspired artificial intelligence (BI-AI).

The aspects of *deep learning* and *deep knowledge representation* in the brain are especially important as these principles are used as inspiration for the BI-SNN and BI-AI systems (Chap. 6) also used in other chapters of the book.

Brain data such as EEG and fMRI have been modelled using evolving SNN (eSNN) and BI-SNN in Chaps. 8–11, 14 of the book.

More on the topic can be found in [7, 37]. Extended presentations on specific topics can be found in:

- Information processing in synapses (Chap. 36 in [75]);
- Understanding the brain via fMRI classification (Chap. 40 in [75]);
- Modelling vision with the neocognitron (Chap. 44 in [75]);
- Neurocomputational models of natural language (Chap. 48 in [75]);
- Integration of large-scale neuroinformatics (Chap. 50 in [75]).
- The brain and connectivity (Chap. 61 in [75]).
- The Allen brain atlas (Chap. 62 in [75]).

**Acknowledgements** Some of the text in this chapter is adopted from [7] and some figures are adopted from [7, 70]. I am highly indebted to Lubica Benuskova as my co-author of the Springer book [7], who contributed a great deal to the book and indirectly—to this chapter.

# References

1. E.R. Kandel, J.H. Schwartz, T.M. Jessell, *Principles of Neural Science*, 4th edn. (McGraw-Hill, New York, 2000), p. 2000
2. A.R. Damasio, *Descartes' Error* (Putnam's Sons, New York, 1994), p. 1994
3. J. Talairach, P. Tournoux, *Co-planar Stereotaxic Atlas of the Human Brain* (Thieme Medical Publishers, New York, 1988), p. 1988
4. J.L. Lancaster et al., Automated Talairach Atlas Labels for Functional Brain Mapping. Human Brain Mapp. **10**, 120–131 (2000)
5. G.A. Evans, H.L. Cromroy, R. Ochoa, The Tenuipalpidae of Honduras. Florida Entomologist **76**(1), 126–155 (1993)
6. A.W. Toga, P.M. Thompson, E.R. Sowell, Mapping brain maturation. Trends Neurosci. **2006** (29), 148–159 (2006)
7. L. Benuskova, N. Kasabov, *Computational Neurogenetic Modeling* (Springer, New York, 2007), p. 2007
8. A.C. Roberts, T.W. Robbins, L. Weikrantz, *The Prefrontal Cortex* (Oxford University Press, Oxford, 1998)
9. G.M. Wittenberg, M.R. Sullivan, J.Z. Tsien, Synaptic reentry reinforcement based network model for long-term memory consolidation. Hippocampus **12**, 637–647 (2002)
10. G.M. Wittenberg, J.Z. Tsien, An emerging molecular and cellular framework for memory processing by the hippocampus. Trends Neurosci. **25**(10), 501–505 (2002)
11. P. Maquet, The role of sleep in learning and memory. Science **2001**(294), 1048–1052 (2001)
12. R. Stickgold, J.A. Hobson, R. Fosse, M. Fosse, Sleep, learning, and dreams: off-line memory reprocessing. Science **2001**(294), 1052–1057 (2001)

13. D.J. Siegel, Memory: An overview with emphasis on the developmental, interpersonal, and neurobiological aspects. J. Am. Acad. Child Adolesc. Psychiatry **40**(9), 997–1011 (2001)
14. B. Seri, J.M. Garcia-Verdugo, B.S. McEwen, A. Alvarez-Buylla, Astrocytes give rise to new neurons in the adult mammalian hippocampus. J. Neurosci. **21**(18), 7153–7160 (2001)
15. R. Feng, C. Rampon, Y.-P. Tang, D. Shrom, J. Jin, M. Kyin, B. Sopher, G.M. Martin, S.-H. Kim, R.B. Langdon, S.S. Sisodia, J.Z. Tsien, Deficient neurogenesis in forebrain-specific *Presenilin-1* knockout mice is associated with reduced clearance of hippocampal memory traces. Neuron **32**, 911–926 (2001)
16. C.H. Bailey, E.R. Kandel, K. Si, The persistence of long-term memory: a molecular approach to self-sustaining changes in learning-induced synaptic growth. Neuron **44**, 49–57 (2004)
17. M. Livingstone, D. Hubel, Segregation of form, color, movement, and depth: anatomy, physiology, and perception. Science **240**, 740–749 (1988)
18. C.M. Gray, P. Konig, A.K. Engel, W. Singer, Oscillatory responses in cat visual cortex exhibit inter-columnar synchronization which reflects global stimulus properties. Nature **338**, 334–337 (1989)
19. W. Singer, Putative function of temporal correlations in neocortical processing, in *Large-Scale Neuronal Theories of the Brain*, ed. by K. Koch, J.L. Davis (The MIT Press, Cambridge, MA, 1994), pp. 201–239
20. P.R. Roelfsema, A.K. Engel, P. Konig, W. Singer, Visuomotor integration is associated with zero time-lag synchronization among cortical areas. Nature **385**, 157–161 (1997)
21. R.D. Traub, M.A. Whittington, I.M. Stanford, J.G.R. Jefferys, A mechanism for generation of long-range synchronous fast oscillations in the cortex. Nature **383**, 621–624 (1996)
22. P. Fries, P.R. Roelfsema, A.K. Engel, P. Konig, W. Singer, Synchronization of oscillatory responses in visual cortex correlates with perception in interocular rivalry. Proc. Natl. Acad. Sci. USA **94**, 12699–12704 (1997)
23. E. Rodriguez, N. George, J.-P. Lachaux, J. Martinerie, B. Renault, F.J. Varela, Perception´s shadow: long-range synchronization of human brain activity. Nature **397**, 434–436 (1999)
24. W.H.R. Miltner, C. Braun, M. Arnold, H. Witte, E. Taub, Coherence of gamma-band EEG activity as a basis for associative learning. Nature **397**, 434–436 (1999)
25. A.K. Engel, P. Fries, P. Konig, M. Brecht, W. Singer, Temporal binding, binocular rivarly, and consciousness. Conscious. Cogn. **8**, 128–151 (1999)
26. W. Singer, Neuronal synchrony: a versatile code for the definition of relations? Neuron **24**, 49–65 (1999)
27. C. Koch, F. Crick, Some further ideas regarding the neuronal basis of awareness, in *Large-Scale Neuronal Theories of the Brain*, ed. by C. Koch, J.L. Davis (MIT Press, Cambridge, MA, 1994), pp. 93–111
28. F. Crick, C. Koch, Are we aware of neural activity in primary visual cortex? Nature **375**, 121–123 (1995)
29. C. Koch, Towards the neuronal substrate of visual consciousness, in *Towards a Science of Consciousness: The First Tucson Discussions and Debates*, ed. by S.R. Hameroff, A.W. Kaszniak, A.C., Scott (The MIT Press, Cambridge, MA, 1996), pp. 247–258
30. B. Libet, Unconscious cerebral initiative and the role of conscious will in voluntary action. Behav. Brain Sci. **8**(8), 529–566 (1985)
31. B. Libet, Do we have free will? J. Conscious. Stud. **6**(8–9), 47–57 (1999)
32. U. Ribary, K. Ionnides, K.D. Singh, R. Hasson, J.P.R. Bolton, F. Lado, A. Mogilner, R. Llinas, Magnetic field tomography of coherent thalamocortical 40-Hz oscillations in humans. Proc. Natl. Acad. Sci. USA **88**, 11037–11401 (1991)
33. G.M. Edelman, G. Tononi, *Consciousness. How Matter Becomes Imagination* (Penguin Books, London, 2000), p. 2000
34. P. Gärdenfors, *Conceptual Spaces: The Geometry of Thought* (MIT Press, Cambridge, 2000)
35. R.R. Llinas, U. Ribary, Perception as an oneiric-like state modulated by senses, in *Large-Scale Neuronal Theories of the Brain*, ed. by C. Koch, J.L. Davis (The MIT Press, Cambridge, MA, 1994), pp. 111–125

36. M. Massimini, F. Ferrarelli, R. Huber, S.K. Esser, H. Singh, G. Tononi, Breakdown of cortical effective connectivity during sleep. Science **309**, 2228–2232 (2005)
37. N. Kasabov, *Evolving Connectionist Systems: The Knowledge Engineering Approach*, 2nd edn. (Springer, Berlin, 2007)
38. R.H. Hahnloser, C.Z. Wang, A. Nager, K. Naie, Spikes and bursts in two types of thalamic projection neurons differentially shape sleep patterns and auditory responses in a songbird. J. Neurosci. **28**, 5040–5052 (2008). [PubMed]
39. NeuCube. http://www.kedri.aut.ac.nz/neucube/
40. S. Thorpe, D. Fize, C. Marlot, Speed of processing in the human visual system. Nature **381**, 520–522 (1996)
41. R. Mayeux, E.R. Kandel, in Disorders of language: the aphasias, in *Principles of Neural Science*, vol. 1, 3rd edn., ed. by E.R. Kandel, J.H. Schwartz, T.M. Jessell (Appleton & Lange, Norwalk, 1991), pp. 839–851
42. F. Rieke, D. Warland, R. de Ruyter van Steveninck, W. Bialek, *Spikes—Exploring the Neural Code* (The MIT Press, Cambridge, MA, 1996)
43. S.J. Thorpe, M. Fabre-Thorpe, Seeking categories in the brain. Science **2001**(291), 260–262 (2001)
44. O. Jensen, Information transfer between rhytmically coupled networks: reading the hippocampal phase code. Neural Comput. **13**, 2743–2761 (2001)
45. M.N. Shadlen, W.T. Newsome, The variable discharge of cortical neurons: implications for connectivity, computation, and information coding. J. Neurosci. **18**, 3870–3896 (1998)
46. D. Hebb, *The Organization of Behavior* (Wiley, New York, 1949), p. 1949
47. M. Mayford, E.R. Kandel, Genetic approaches to memory storage. Trends Genet. **15**(11), 463–470 (1999)
48. W.C. Abraham, B. Logan, J.M. Greenwood, M. Dragunow, Induction and experience-dependent consolidation of stable long-term potentiation lasting months in the hippocampus. J. Neurosci. **22**(21), 9626–9634 (2002)
49. H.Z. Shouval, M.F. Bear, L.N. Cooper, A unified model of NMDA receptor-dependent bidirectional synaptic plasticity. Proc. Natl. Acad. Sci. USA **99**(16), 10831–10836 (2002)
50. H. Markram, J. Lübke, M. Frotscher, B. Sakmann, Regulation of synaptic efficacy by coincidence of postsynaptic APs and EPSPs. Science **275**(5297), 213–215 (1997)
51. W.C. Abraham, M.F. Bear, Metaplasticity: the plasticity of synaptic plasticity. Trends Neurosci. **19**(4), 126–130 (1996)
52. E. Bienenstock, L.N. Cooper, P. Munro, On the development of neuron selectivity: orientation specificity and binocular interaction in visual cortex. J. Neurosci. **1982**(2), 32–48 (1982)
53. P. Jedlicka, Synaptic plasticity, metaplasticity and the BCM theory. Bratislava Med. Lett. **103** (4–5), 137–144 (2002)
54. L. Benuskova, M.E. Diamond, F.F. Ebner, Dynamic synaptic modification threshold: computational model of experience-dependent plasticity in adult rat barrel cortex. Proc. Natl. Acad. Sci. USA **91**, 4791–4795 (1994)
55. L. Benuskova, M. Kanich, A. Krakovska, Piriform cortex model of EEG has random underlying dynamics, ed. by F. Rattay. Proceedings of World Congress on Neuroinformatics, vol. ARGESIM/ASIM-Verlag, Vienna, 2001
56. K.S. Lee, F. Schottler, M. Oliver, G. Lynch, Brief bursts of high-frequency stimulation produce two types of structural change in rat hippocampus. J. Neurophysiol. **44**(2), 247–258 (1980)
57. Y. Geinisman, L. deToledo-Morrell, F. Morrell, Induction of long-term potentiation is associated with an increase in the number of axospinous synapses with segmented postsynaptic densities. Brain Res. **566**, 77–88 (1991)
58. C. Koch, T. Poggio, A theoretical analysis of electrical properties of spines. Proc. Roy. Soc. Lond. B **218**, 455–477 (1983)
59. A. Zador, C. Koch, T. Brown, Biophysical model of a Hebbian synapse. Proc. Natl. Acad. Sci. USA **87**, 6718–6722 (1990)

60. J.I. Gold, M.F. Bear, A model of dendritic spine $Ca^{2+}$ concentration exploring possible bases for a sliding synaptic modification threshold. Proc. Natl. Acad. Sci. USA **91**, 3941–3945 (1994)
61. L. Benuskova, The intra-spine electric force can drive vesicles for fusion: a theoretical model for long-term potentiation. Neurosci. Lett. **280**(1), 17–20 (2000)
62. P. Fedor, L. Benuskova, H. Jakes, V. Majernik, An electrophoretic coupling mechanism between efficiency modification of spine synapses and their stimulation. Stud. Biophys. **92**, 141–146 (1982)
63. J. Spacek, K.M. Harris, Three-dimensional organization of smooth endoplasmatic reticulum in hippocampal CA1 dendrites and dendritic spines of the immature and mature rat. J. Neurosci. **17**, 190–204 (1997)
64. P.-M. Lledo, X. Zhang, T.C. Sudhof, R.C. Malenka, R.A. Nicoll, Postsynaptic membrane fusion and long-term potentiation. Science **1998**(279), 399–404 (1998)
65. T.C. Sudhof, The synaptic vesicle cycle: a cascade of protein-protein interactions. Nature **375**, 645–654 (1995)
66. D. Liao, N.A. Hessler, R. Malinow, Activation of postsynaptically silent synapses during pairing-induced LTP in CA1 region of hippocampal slice. Nature **375**, 400–404 (1995)
67. V.N. Kharazia, R.J. Wenthold, R.J. Weinberg, GluR1-immunopositive interneurons in rat neocortex. J. Comp. Neurol. **1996**(368), 399–412 (1996)
68. S.H. Shi, Y. Hayashi, R.S. Petralia, S.H. Zaman, R.J. Wenthold, K. Svoboda, R. Malinow, Rapid spine delivery and redistribution of AMPA receptors after synaptic NMDA receptor activation. Science **1999**(284), 1811–1816 (1999)
69. B.J. Schnapp, T.S. Reese, New developments in understanding rapid axonal transport. Trends Neurosci. **1986**(9), 155–162 (1986)
70. M.A. Lindquist, The statistical analysis of fMRI Data. Project Euclid **23**(4), 439–464 (2008)
71. Wikipedia. http://www.wikipedia.org
72. J. Theiler, On the evidence for low-dimensional chaos in an epileptic electroencephalogram. Phys. Lett. A **1995**(196), 335–341 (1995)
73. W.J. Freeman, Evidence from human scalp EEG of global chaotic itinerancy. Chaos **13**(3), 1–11 (2003)
74. I. Tsuda, Toward an interpretation of dynamic neural activity in terms of chaotic dynamicical systems. Behav. Brain Sci. **2001**(24), 793–847 (2001)
75. N. Kasabov (ed.), *Springer Handbook of Bio-/Neuroinformatics* (Springer, Berlin, 2014)

# Part III
# Spiking Neural Networks

# Chapter 4
# Methods of Spiking Neural Networks

Spiking neural networks (SNN) are biologically inspired ANN where information is represented as binary events (spikes), similar to the event potentials in the brain, and learning is also inspired by principles in the brain. SNN are also universal computational mechanisms [1]. These and many other reasons that are discussed in this chapter make SNN a preferred computational paradigm for modelling temporal and spatio-temporal data and for building brain-inspired AI. This chapter gives the background information for SNN that is further used in the rest of the book.

The chapter is organised in following sections:

4.1. Information representation as spikes. Spike encoding algorithms.
4.2. Spiking neuron models.
4.3. Methods for learning in SNN.
4.4. Spike pattern association neurons.
4.5. Why use SNN?
4.6. Chapter summary and further readings for deeper knowledge.

## 4.1 Information Representation as Spikes. Spike Encoding Algorithms

### 4.1.1 Rate Versus Spike Time Information Representation

The brain encodes external information into electrical pulses—spikes. The principle is that changes in the data in space and time are represented as binary events (spikes). This leads to several advantages in information processing:

- Compact information representation.
- Asynchronous data processing (not frame based or vector based).

© Springer-Verlag GmbH Germany, part of Springer Nature 2019
N. K. Kasabov, *Time-Space, Spiking Neural Networks and Brain-Inspired Artificial Intelligence*, Springer Series on Bio- and Neurosystems 7, https://doi.org/10.1007/978-3-662-57715-8_4

- Fast detection of changes in the environment—efficient for predictive modelling in a real time.
- Simple and fast processing (bits at times).
- Massively parallel, i.e. millions of neurons can exchange spike sequences in parallel.
- Energy efficient.

Already some devices convert real input data into spike sequences, such as Dynamic Vision Sensor (DVS) (https://inilabs.com; [2, 3]); Artificial cochlea (AER EAR) (https://inilabs.com).

Since data in SNN are communicated in terms of spikes and spike sequences, the methods used to encode real data into spikes are of a substantial step in creating spiking neural network systems.

The two main categories of neuronal information encoding schemes are rate code and pulse code that result in different spike characteristics. Their biological counterparts were discussed in Chap. 3.

*Rate Code*

Rate encoding or also known as firing rate is to encode a sequence of spikes based on the average number of spikes (or spikes count) over time i.e. how many spikes are emitted within a time encoding window. There are three different views of rate code, referring to different averaging methods: an average over time (single neuron, single run); or an average over several experiment repetitions (single neuron, repeated runs); or average over a populations of neuron (several neurons, single run) [4]. The measured spikes are those emitted within a specified time window that starts at stimulus onset and ends at stimulus termination.

The rate is calculated by dividing the number of spikes $(n_{sp})$ emitted in the duration $(T)$ with $T$, as presented in Eq. 4.1. However this encoding scheme was only suitable for stimulus which requires slow reaction of the organisms. This slow reaction was usually found in lab experiments, but not in many real biological brain functions. Real biological brain functions usually happen in much faster duration. In addition, any regularities found, may be considered as noise.

$$v = \frac{n_{sp}}{T} \tag{4.1}$$

The second view of rate code involves averaging the spikes over several experiment runs which are best suited for stationary and time-dependent stimulus. The same stimulus is repeated and the neurons' activity is recorded as spike density of Peri-Stimulus-Time Histogram (PSTH) [4]. As shown in Eq. 4.2, it is defined as $n_K(t; t + \Delta t)$ to be the total number of spikes in all runs; starting from stimulus sequence time, $t$; and $\Delta t$ is in the range of 1 or few milliseconds; divided with the number of repetitions $K$ which is then further divided with length of interval $\Delta t$. In a situation where a population of independent neurons receives the same stimulus, the mean firing rate is easier to record from a single neuron and average over $N$ repeated runs.

$$p(t) = \frac{1}{\Delta t} \frac{n_K(t; t + \Delta t)}{K} \tag{4.2}$$

The third view rooted from the notion of neurons population explained earlier which define rate code as the average of spikes over several neurons, i.e. neurons with the same characteristics and which respond to the same stimulus. As explained in [4] the rate $A(t)$ with units $s^{-1}$ is computed as in Eq. 4.3 where $N$ is the population size of neurons, $n_{act}(t; t + \Delta t)$ is total number of spikes emitted between $t$ and $t + \Delta t$ in the neuron population and $\Delta t$ is a small time interval.

$$A(t) = \frac{1}{\Delta t} \frac{n_{act}(t; t + \Delta t)}{N} = \frac{1}{\Delta t} \frac{\int_t^{t + \Delta t} \sum_j \sum_f \delta(t - t_j^{(f)}) dt}{N} \tag{4.3}$$

This approach solves the issue raised in the first approach, i.e. calculating the average in a single-neuron level; however it is questionable when it will be needed to calculate the average of spikes from a population of neurons with the same properties and connections. Nevertheless, the rate code is still practical in modelling spike activities in many brain areas and has been used in many successful experiments.

*Pulse Code (Time-based representation)*

Another encoding approach is based on the exact timing of spikes or better known as pulse or spike code. The idea of spike time in describing input stimulus has been the interest of many researchers such as in [5–9]. While rate-based representation defines spiking characteristics within a time interval, e.g. frequency, in time-based (temporal) representation information is encoded in the time of spikes. Every spike matters and its time—too! The two approaches are illustrated in Figs. 4.1 and 4.2.

## 4.1.2 Spike Encoding Algorithms

SNN use spike encoded information. This sub-section introduces some popular methods for encoding real value data (such as sound, speech, pixel image, video, temperature, seismic wave etc.) into spike sequences before they are processed (learned) in a SNN.

*Threshold-based encoding (or Temporal Contrast)*

A spike is generated only if a change in the input data occurs beyond a threshold— Fig. 4.3.

These algorithms belong to the class of temporal contrast encoding and decoding algorithms, examples are given in Box 4.1.

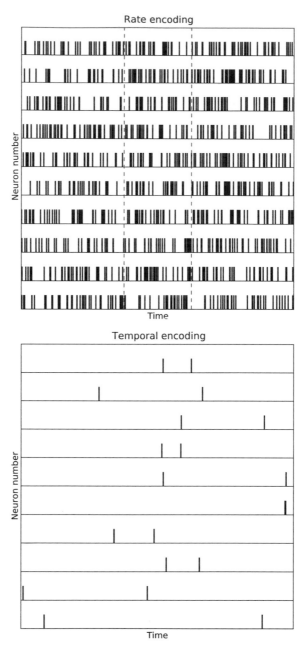

**Fig. 4.1** Rate base versus time based representation of information as spikes

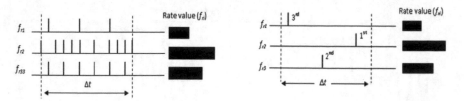

**Fig. 4.2** Representing the value of information (denoted as $f_n$ rate value) in a rate based versus time based representation as a hypothetical example

---

**Box 4.1. Examples of temporal contrast encoding and decoding algorithms**

---

**Temporal contrast encoding algorithm**

```
1: input: S, factor 2: output:B, threshold_TC
3: L ← length(S)
4: for t = 1 : L − 1 do
5:        diff ← |S(t + 1) − S(t)|
6: end for
7: threshold_TC ← mean(diff) + factor · std(diff)
8: diff ⇐ [0,diff]
9: for t = 1 : L do
10:        if diff(t) > threshold_TC then
11:            B(t) ← 1
12:        else if diff(t) < −threshold_TC then
13:            B(t) ← −1
14:      else
20:          B(t) ← 0
16:      end if
17: end for
```

---

**Temporal contrast decoding algorithm**

```
1: input: B, threshold_TC
2: output: Ŝ
3: Ŝ ← 0
4: L ← length(B)
5: for t = 2 : L do
6:        if Ŝ(t) > 0 then
7: Ŝ(t) ← Ŝ(t − 1) + threshold_TC
8: else if Ŝ(t) < 0 then
9:            Ŝ(t) ← Ŝ(t − 1) − threshold_TC
10:      else
11:          Ŝ(t) ← Ŝ(t − 1)
12:      end if
13: end for
```

---

## Rank Order Coding (ROC)

It is assume here that the first generated spike carries the most significant information and carries the most weight compared to the later spikes in the sequence [7, 8]. Based on this theory, there are two versions of encoding techniques:

– Rank Order Coding (ROC) [9];
– Population Rank Order Coding (POC) [10].

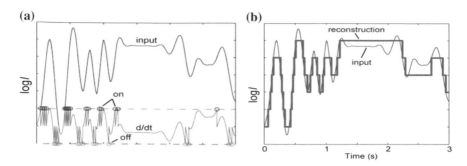

**Fig. 4.3** A graphical representation of threshold-based encoding method for a continuous signal into a spike sequence—**a** both positive and negative spikes represent changes of the intensity of the signal over time; **b** reconstructing the signal from the spike representation through a decoding procedure

In ROC, spikes are ordered according to their arrival, in which the first spike which arrives will be the first in the population, followed by the second spike and so on. To demonstrate this coding scheme, for neurons A, B, C, D and E in Fig. 4.4, the spikes are ranked as C > E > D > A > B.

*Population Rang Coding (POC)*

In contrast to ROC, POC is generated based on the firing time identified and calculated using intersection of sensitivity profiles such as Gaussian function [10]. In this scheme, a single input value $i$ is distributed into multiple input neurons each having overlapping receptive fields, represented as a continuous function, e.g. Gaussian (Fig. 4.5). Equation 4.4 is the Gaussian function used to calculate the firing time for the input neuron, where its centre $\mu_i$ and its width $\sigma$ are calculated in Eqs. 4.5 and 4.6 respectively. $[I_{min}, I_{max}]$ is the maximum and minimum range of input variable and $\beta$ (values between 1.0 and 2.0) controls the width of each Gaussian receptive field.

**Fig. 4.4** Ranking of spikes in a ROC encoding method

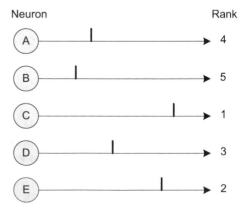

**Fig. 4.5** POC: The value of a single input variable is encoded into sequence of spikes by a population of 5 spiking neurons defined by their overlapping in the neighbourhood Gaussian receptive fields

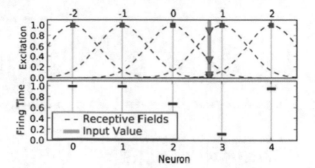

$$g(x) = \frac{1}{\sigma\sqrt{2\pi}} e^{-\frac{1}{2}\left(\frac{x-\mu}{\sigma}\right)^2} \tag{4.4}$$

$$\mu_i = I_{min} + \frac{2i - 3}{2} \cdot \frac{I_{max} - I_{min}}{M - 2} \tag{4.5}$$

$$\sigma = \frac{1}{\beta} \cdot \frac{I_{max} - I_{min}}{M - 2} \tag{4.6}$$

The method is illustrate in Figs. 4.5 and 4.6 [11, 12]. In Fig. 4.5 a value of a single variable causes a sequence of spikes to be emitted by a population of 5 neurons, the first spike being generated by the neuron to which receptive field the input value belongs to the highest membership degree and so on. In Fig. 4.6, six input neurons generate spikes based on the value of the input variables, the first spike (at time 0) being generated by neuron 3 which receives the highest input value.

Both ROC and POC have been successfully implemented in many experiments, for instance stroke classification and prediction [13], visual pattern recognition [14–16], feature and parameter optimization [12], string pattern recognition [17], audio

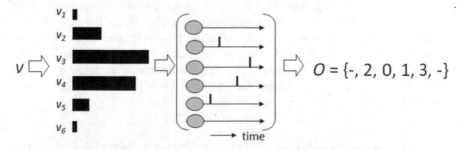

**Fig. 4.6** Six input neurons generate a sequence of spikes based on the membership value of 6 input variables to the receptive fields of the neurons, the higher the value, the earlier the spike is generated by the corresponding neuron

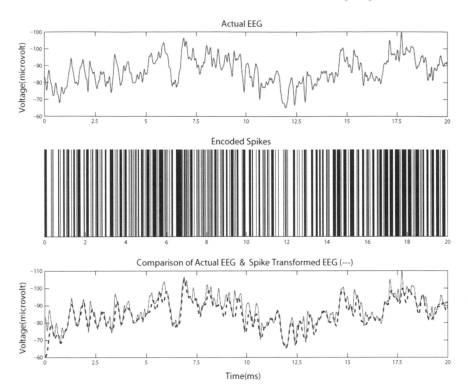

**Fig. 4.7** The top figure shows one channel of EEG signal for the duration of 20 ms. The middle figure is the spike representation of the above figure obtained using BSA. The bottom figure shows the actual channel EEG signal that has been superimposed with another signal (dashed lines) which represents the reconstructing BSA encoded spikes

recognition [18], text-independent speaker authentication [19]. Implementation based on Gaussian receptive fields was introduced in [20, 21].

*Ben's Spike Encoding Algorithm (BSA) [22]*

The key benefit of using BSA is that the frequency and amplitude features are smoother in comparison to another popular encoding algorithm—the HSA (Hough Spiker Algorithm), which is not discussed here. Moreover, due to the smoother threshold optimization curve, it is also less susceptible to changes in the filter and the threshold [21, 22]. Studies have shown that this method offers an improvement of 10–15 dB over the HSA spike encoding scheme. According to [22], a stimulus is estimated from a spike train by

$$S_{est} = (h \times x)(t) = \int\limits_{-\infty}^{+\infty} x(t-\tau)h(\tau)d\tau = \sum_{k=1}^{N} h(t - t_k) \qquad (4.7)$$

where, $t_k$ represents the neurons firing time, $h(t)$ denotes the linear filters impulse response and $x(t)$ is the spike of the neuron that can be calculated as

$$x(t) = \sum_{k=1}^{N} \delta(t - t_k) \tag{4.8}$$

For the example in Fig. 4.7, which is EEG data of one channel, the Finite Impulse Response (FIR) filter size is 20, and the BSA threshold is 0.955.

However, when the spike train $x(t)$ is applied with a discrete FIR filter, the Eq. (4.8) can be represented as

$$o(t) = (h \times x)(t) = \sum_{k=0}^{M} x(t - k)h(h) \tag{4.9}$$

where, $M$ refers to the number of filter taps. A more detailed explanation is given in [22].

The Ben's spike encoding and decoding algorithms are given in Box 4.2.

---

**Box 4.2. BSA encoding and decodinglgorithms**

---

**BSA encoding algorithm**

---

1: input: $S,filter,threshold_{BSA}$
2: output: $B$
3: $B \Leftarrow 0$
4: $L = length(S)$
5: $F = length(filter)$
6: **for** $t = 1 : (L - F + 1)$ **do**
7:      $e_1 \leftarrow 0$
8:      $e_2 \leftarrow 0$
9:      **for** $k = 1 : F$ **do**
10:         $e_1 += |S(t + k) - filter(k)|$
11:         $e_2 += |S(t + k - 1)|$
12:     **end for**
13:     **if** $e_1 \le (e_2 - threshold_{BSA})$ **then**
14:         $B(t) \leftarrow 1$
20:         **for** $k = 1 : F$ **do**
16:             $S(i + j - 1) -= filter(k)$
17:         **end for**
18:     **end if**
19: **end for**

---

**BSA decoding algorithm**

---

1: input: $B,filter$
2: output: $S$
3: $L = length(B)$
4: $F = length(filter)$
5: **for** t=1:L-F+1 **do**
6: **if** $B(t) == 1$ **then** 7: **for** $k = 1 : F$ **do**
8:             $S(t + k - 1) += filter(k)$
9:         **end for**
10:    **end if**
11: **end for**

Different spike encoding algorithms have distinct characteristics when representing input data. BSA is suitable for high frequency signals and because it is based on the Finite Impulse Response technique, the original signal can be recovered easily from the encoded spike train. Only positive (excitatory) spikes are generated by BSA, whereas all other techniques mentioned here can also generate negative (inhibitory) spikes. Temporal Contrast was originally implemented in hardware (Delbruck and Lichtsteiner 2007) [2, 3] in the artificial silicon retina. It represents significant changes in signal intensity over a given threshold, where the ON and OFF events are dependent on the sign of the changes. However if the changes of the signal intensity vary dramatically, it may not be possible to recover the original signal using the encoded spike train.

*Step Forward (SF) Encoding algorithm*

In [23] we proposed an improved spike encoding algorithm, SF (Step Forward encoding), to better represent the signal intensity. For a given signal $S(t)$ where ($t = 1, 2, ..., n$), we define a baseline $B(t)$ variation during time $t$ with $B(1) = S(1)$. If the incoming signal intensity $S(t1)$ exceeds the baseline $B(t1 - 1)$ plus a threshold defined as Th, then a positive spike is encoded at time $t1$, and $B(t1)$ is updated as $B(t1) = B(t1 - 1) + Th$; and if $S(t1) <= B(t1 - 1) - Th$, a negative spike is generated and $B(t1)$ is assigned as $B(t1) = B(t1 - 1) - Th$. In other situations, no spike is generated and $B(t1) = B(t1 - 1)$.

*Moving-Window (MW) Spike Encoding Algorithm,*

In another spike encoding algorithm also introduced in [23], called Moving-Window Spike Encoding Algorithm, the baseline $B(t)$ is defined as the mean of previous signal intensities within a time window T, thus this encoding algorithm can be robust to certain kinds of noise. Both SF and MW encoding algorithms result in a better recovery of the original encoded signals after decoding [24].

Before choosing a proper spike encoding algorithm, we need to figure out what information the spike trains shall carry for the original signals. After that, the underlying spike patterns in the spike trains will be better understood. Figure 4.8 shows spike trains generated by four different spike encoding algorithms with corresponding recovery signals. The blue (red) lines in (b), (c), (d), (e) are positive (negative) spikes, and the blue lines in (f), (g), (h), (i) are the original signals while the red dash lines are the signals reconstructed by corresponding spike trains. The threshold based (a temporal contrast) encoding is denoted as AER. The step-forward function (SF) and the MV (moving window) encoding algorithms are explained in details in [23]. A decision which encoding algorithm to chose for particular data and SNN models depends on the criterion for this. It can be that we want to have a more accurately recovered signals if and when decoded, or—a better classification results at the output of a SNN system as it is the case in the new method introduced in Chap. 21. In [24] a methodology for the selection and parameter optimisation of encoding algorithm is proposed.

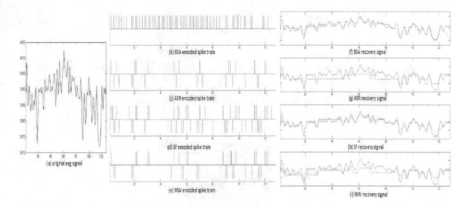

**Fig. 4.8** Spike trains generated by four different spike encoding algorithms with corresponding recovery signals after decoding. The blue (red) lines in (**b–e**) are positive(negative) spikes, and the blue lines in (**f–i**) are the original signals while the red dash lines are the signals reconstructed through decoding from corresponding spike trains. The threshold based encoding is denoted as AER (from [23])

## 4.2 Spiking Neuron Models

Several models of spiking neurons have been developed, some of them more biologically plausible and others—more computationally driven. This section presents them briefly.

### 4.2.1 Hodgkin-Huxley Model (HHM)

This model was introduced by Hodgkin and Huxley [25] who conducted the experiment on the giant axon of a squid. From the experiment, they have concluded that there are three ion channels in the neuron, which are Sodium (Na), Potassium (K) and leakage ($L$) channel with resistance. To calculate the total of ionic current $I_{ion}$, which is the sum of all participating channels; the formula in Eqs. 4.10 and 4.11 are used. In Eq. 4.10, $G_k$ represents all channels involved, $E_k$ represents the equilibrium potential and $V_m$ is the membrane potential. In addition, as elaborated by [4, 26] three gates of type "$m$" and one gate of type "$h$" are used to control the Sodium channel; and four gates of type "$n$" to control the Potassium channel. These gating variables are calculated using Eqs. 4.12, 4.13 and 4.14 where the transition rate for each gate from non-permissive to permissive states are represented by $\alpha_m(V)$, $\alpha_h(V)$ and $\alpha_n(V)$ and transition rate for each gate from permissive to non-permissive states are represented by $\beta_m(V)$, $\beta_h(V)$ and $\beta_n(V)$.

$$I_{ion} = \sum_k I_k = \sum_k G_k(V_m - E_k) \tag{4.10}$$

$$I_{ion} = G_{Na}m^3h(V_m - E_{Na}) + G_K n^4(V_m - E_K) + G_L(V_m - E_L) \tag{4.11}$$

$$\frac{m}{dt} = \alpha_m(V)(1 - m) - \beta_m(V)m \tag{4.12}$$

$$\frac{h}{dt} = \alpha_h(V)(1 - h) - \beta_h(V)h \tag{4.13}$$

$$\frac{n}{dt} = \alpha_n(V)(1 - n) - \beta_n(V)n \tag{4.14}$$

This neuron model only describes the channels and flow of ions in the neuron when generating spikes, which is far from the complex biological neuron, thus bearing several weaknesses as reviewed in [27] that include ignored events which may affect neuron's computation [28] and inaccurate prediction of the inactivation of the Sodium channel [29]. An electrical circuit that represents a simplified implementation of the HHM is given in Fig. 4.9. Despite HHM's limitation, it has become a fundamental and starting point for the development of many other simplified neuron models that will be discussed in the following sub sections.

### 4.2.2   Leaky Integrate-and-Fire Model (LIFM)

As compared from HHM that deals with ion channels and ion flows, LIFM view neuron as a leaky integrator, which will output a spike if the input voltage reaches a threshold and then reset to a resting state. Modelled by a differential equation, integrate-and-fire neuron which can be traced back from [30] is represented by a basic circuit that combine a capacitor $(C)$ and a resistor $(R)$ to produce current $(I(t))$. Equation 4.15 is the standard form of LIFM in which $u(t)$ is the membrane potential and $\tau_m = RC$ is the neuron's membrane time constant.

$$\tau_m \frac{du}{dt} = -u(t) + RI(t) \tag{4.15}$$

**Fig. 4.9**  An electrical circuit representing the Hodgkin-Huxley spiking neuron model [25]

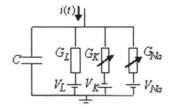

Spikes are described as events [4] indicated by a firing time $t^{(f)}$ defined by a threshold value (Eq. 4.16) and the potential will be reset to a new value $u_r < \vartheta$ (Eq. 4.17).

$$t^{(f)} : u\left(t^{(f)}\right) = \vartheta \tag{4.16}$$

$$\lim_{t \to t^{(f)}; t > t^{(f)}} u(t) = u_r \tag{4.17}$$

This model is viewed as the best-known instance of spiking neuron model because of its simplicity and low computational cost (see Fig. 4.10a, b).

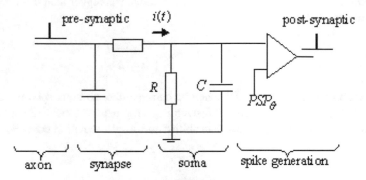

(a) Electrical circuit representing the LIF spiking neuron model

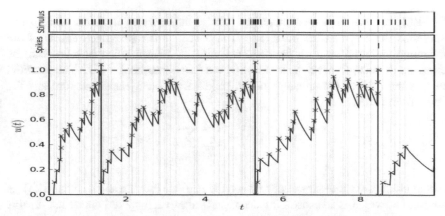

(b) The membrane potential of a LIF neuron accumulates input spikes as stimuli. When the

potential reaches a threshold, the neuron emits an output spike.

**Fig. 4.10 a** Electrical circuit representing the LIF spiking neuron model. **b** The membrane potential of a LIF neuron accumulates input spikes as stimuli. When the potential reaches a threshold, the neuron emits an output spike

### 4.2.3   Izhikevich Model (IM)

In IM [31], a simple spiking neuron is formulated by combining biologically plausibility of HHM and computational efficiency in LIF neurons. The model is defined as in Eq. 4.18 where $v$ is the membrane voltage, $u$ is a recovery variable used to adjust $v$, $I(t)$ is input currents, and $a$ and $b$ are adjustable parameters.

$$\frac{dv}{dt}(t) = 0.04v^2 + 5v + 140 - u + I(t) \tag{4.18}$$

$$\frac{du}{dt}(t) = a(bv - u) \tag{4.19}$$

A threshold value is set to 30 mV and if the voltage $v$ is bigger than this threshold, $v$ and $u$ are reset (Eq. 4.20).

$$if \ v \geq 30\,\text{mV}, \quad then \begin{cases} v = c \\ u = u + d \end{cases} \tag{4.20}$$

Based on parameter values of IM a neuron can manifest different spiking behaviour (Fig. 4.11). In [31] a comparison of several spiking neuron models is visualised in the dimensionality of biological plausibility and implementation cost (Fig. 4.12).

### 4.2.4   Spike Response Model (SRM)

In SRM, the neuron's membrane potential is summarized in terms of response kernel as described in Eq. 4.21 [32]. It is based on the integrated effects of the incoming spike arriving on the neuron $i$ with its neuron potential $u_i(t)$, and the emission of spike from the neuron if $u_i(t)$ reaches a threshold $\vartheta$ [33]. The potential $u_i(t)$ is the total of the influence of the spikes from pre-synaptic neurons and the spike from its own.

$$u_i(t) = \sum_{t_i^{(f)} \in \mathcal{F}_i} \eta_i\left(t - t_i^{(f)}\right) + \sum_{j \in \Gamma_i} \sum_{t_j^{(f)} \in \mathcal{F}_j} w_{ij} \in_{ij} \left(t - t_i^{(f)}\right) \tag{4.21}$$

Although it uses the same concept as the LIFM, the threshold $\vartheta$ in SRM is adjustable, which is increased (or decreased) after each spike occurrence. In this model, $t_i^{(f)}$ is the firing time of the last output spike, $\eta_i$ is a kernel function that describes spike emissions after action potential exceeds the threshold $\vartheta$ and its after-potential spikes, $\in_{ij}$ is a kernel function that describes the response of the post-synaptic neuron when receiving the spike from pre-synaptic neuron $j \in \Gamma_i$ and $w_{ij}$ is the response weight.

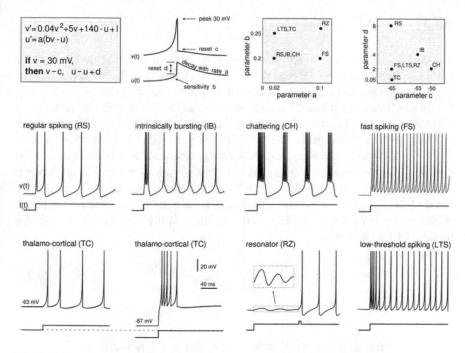

**Fig. 4.11** Based on parameter values of IM a neuron can manifest different spiking behaviour (after [31])

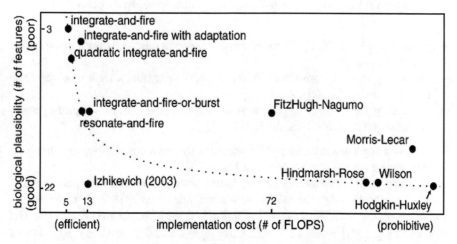

**Fig. 4.12** Comparison of spiking neuron models in the evaluation landscape of biological plausibility and implementation cost (adapted from [31])

### 4.2.5  Thorpe's Model (TM)

Inspiring from the integrate and fire capabilities of a neuron, TM defines that the first incoming spike carries the most information, because of the argument that the brain only can process one spike from each neuron at one particular processing step [7]. In this model the relation between stimulus saliency and spike relative timing plays a major role i.e. the first spike in the population is the most important in defining meaningful information. The membrane potential $u_i(t)$ is summarized as in Eq. 4.23 which will be reset to 0 after each spike emission. In the equation, $w_{ij}$ is the weight of the pre-synaptic neuron, *Mod* is modulation factor within the interval $[0, 1]$, and $order(j)$ is the spike rank of neuron $j$. The threshold $\vartheta = cu_{max}$ where $0 < c < 1$ and $u_{max}$ is maximum potential that a neuron can reach. Simulation software of this model, SpikeNET [34], has successfully simulated and modelled millions of LIF neurons.

$$u_i(t) = \begin{cases} 0 & \textit{if fired} \\ \sum w_{ji} Mod_i^{order(j)} & \textit{else} \end{cases} \tag{4.22}$$

### 4.2.6  Probabilistic and Stochastic Spiking Neuron Models

The probabilistic spiking neuron model (pSNM), introduced by the author [35], is a further extension of LIFM that includes three other probability parameters which are:

–  probability that a spike will arrive at post-synaptic neuron $n_i$ from pre-synaptic neuron $n_j$, $(p_{cj,i}(t))$;
–  probability that a synapse contributes to a spike potential after it receives spike from neuron $n_j$, $(p_{sj,i}(t))$;
–  probability that neuron $n_i$ generates an output spike if the total post-synaptic potential (PSP) reaches the threshold $(p_i(t))$.

A simplified representation of pSNM with one synaptic connection, together with the probability parameters is shown in Fig. 4.13.

The state of post-synaptic neuron $n_i$ is described as the total of inputs received from all $m$ synapses i.e. the post-synaptic potential $(PSP_i(t))$. The model is calculated using Eq. 4.23, where $e_j = 1$ if spike is emitted from neuron $n_j$ and $e_j = 0$ if otherwise; $g(p_{cj,i}(t)) = 1$ with a probability $p_{cj,i}(t)$, and 0 otherwise; $f(p_{sj,i}(t)) = 1$ if the synapse contributes to the potential with a probability $p_{sj,i}(t)$ and 0 otherwise; $w_{j,i}(t)$ is the connection weight; $t_0$ is the time of the last spike emitted by neuron $n_i$; and $\eta(t - t_0)$ is the decay.

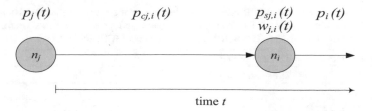

$$p_j(t) \qquad\qquad p_{cj,i}(t) \qquad\qquad\qquad \begin{array}{c} p_{sj,i}(t) \\ w_{j,i}(t) \end{array} \quad p_i(t)$$

$$n_j \longrightarrow n_i \longrightarrow$$

time $t$

**Fig. 4.13** Simplified representation of pSNM with all 3 probabilistic parameters and one synaptic connection

$$PSP_i(t) = \sum_{p=t_0}^{t} \sum_{j=1}^{m} e_j g(p_{cj,i}(t-p)) f(p_{sj,t}(t-p)) w_{j,t}(t) + \eta(t-t_0) \qquad (4.23)$$

If all probability parameters are equal to 1, the model is simplified to be similar to some well-known spiking neuron models, such as LIFM [4].

Stochastic neuronal models have some of their parameters change stochastically. The behaviour of such models is illustrated in Fig. 4.14 [35] on the cases of:

- Noisy Reset (NR);
- Step-wise Threshold (ST);
- Continuously changing Threshold (CT).

## 4.2.7 Probabilistic Neurogenetic Model of a Neuron

The activity of the spiking neuron models described so far has not been connected to the expression of genes and proteins in the neuron as parameters. Genes and

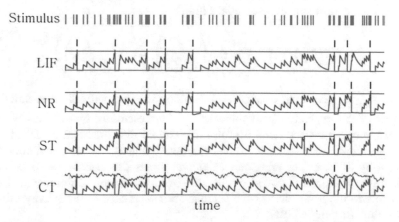

**Fig. 4.14** Spiking activities of several types of stochastic spiking neuron models as introduced in [35] when compared with the spiking behaviour of the deterministic LIF model: NR—noisy reset; ST—step-wise spiking threshold; CT—continuously changing stochastic threshold. All neuronal models receive the same spike sequence input stimulus (shown on the top of the figure) (after [35])

**Table 4.1** The dynamics in the increase and the decrease of the membrane potential of a spiking neuron depend very much on the expression of genes and proteins such as neuroreceptors (AMPAR, NMDAR, GABRA, GABARB) and ion channels (SCN, KCN, CLC)

| Neuronal parameters and related proteins | |
|---|---|
| Neuronal parameter amplitude and time constants of | Protein |
| Fast excitation PSP | AMPAR |
| Slow excitation PSP | NMDAR |
| Fast inhibition PSP | GABRA |
| Slow inhibition PSP | GABRB |
| Firing threshold | SCN, KCN, CLC |
| Late excitatory PSP through GABRA | PV |

proteins take a major role in the spiking activity of the neurons and this is implemented in a neurogenetic model of a spiking neuron, introduced in [36, 37]. The dynamics in the increase and the decrease of the membrane potential of a spiking neuron depend very much on the expression of genes and proteins such as neuroreceptors (AMPAR, NMDAR, GABRA, GABARB) and ion channels (SCN, KCN, CLC)—Table 4.1.

Using this information, the probabilistic model of a neuron pSNM from above has been extended to probabilistic neurogenetic model of a spiking neuron (PNGM) [36, 37]. As a partial case, when no probability parameters and no genetic parameters are used, the model is reduced to the LIFM.

In the PNGM four types of synapses for *fast excitation, fast inhibition, slow excitation, and slow inhibition* are used. The contribution of each one to the PSP of a neuron is defined by the level of expression of different genes/proteins along with the presented external stimuli. The model utilises known information about how proteins and genes affect spiking activities of a neuron. This information is used to calculate the contribution of each of the *four* different synapses $j$ connected to a neuron $i$ to its post synaptic potential PSPi(t):

$$\varepsilon_{ij}^{synapse}(s) = A^{synapse}\left(\exp\left(\frac{s}{\tau_{decay}^{synapse}}\right) - \exp\left(\frac{s}{\tau_{rise}^{synapse}}\right)\right) \qquad (4.24)$$

where: $\tau_{decay/rise}^{synapse}$ are time constants representing the rise and fall of an individual synaptic PSP; A is the PSP's amplitude; $\varepsilon_{ij}^{synapse}$ represents the type of activity of the synapse between neuron j and neuron i that can be measured and modelled separately for a fast excitation, fast inhibition, slow excitation, and slow inhibition (it is affected by different genes/proteins). External inputs can also be added to model background noise, background oscillations or environmental information. Genes that relate to the parameters of the neurons are also related to the activity of other genes, thus forming a GRN.

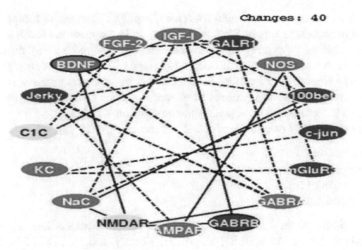

**Fig. 4.15** An example of a gene regulatory network model that is part of the functioning of a spiking neuron (from [36])

The PNGM can be further extended with the inclusion of other genes and proteins that regulate other functions of the neuronal cells and form gene regulatory network model with the genes and proteins from Table 4.1.

An example of a gene regulatory network model that is part of the functioning of a spiking neuron is shown in Fig. 4.15 [36, 37]. Such neurogenetic models can be used to model brain data such as related to AD [36–38]. More about computational neurogenetic modelling is presented in Chap. 16.

## 4.3 Methods for Learning in SNN

Learning in SNN relates to changes of the connection weights between two spiking neurons (Fig. 4.16). Several methods have been proposed so far, some of them presented in this section.

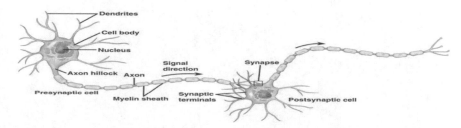

**Fig. 4.16** Learning in SNN relates to changes of the connection weights between two spiking neurons

As it is in biological neural networks (see Chap. 3), spike precise timing is one of the most important factors in SNN data coding and computation [1, 39] in order to generate efficient processing of information in the system. As explained earlier, information is represented and encoded into spikes which are very dependent on the exact firing timing, thus learning in SNN is a very complex process. In general, learning is defined as the process of parameter adaptation and learning rule is defined as the procedure of adjusting the connection weights.

Learning in SNN can be classified in several categories:

– Unsupervised;
– Supervised;
– Semi-supervised;
– Reinforcement,

which is similar to the learning in traditional neural networks (Chap. 2). The next sub-sections discuss the learning algorithms already designed for SNN, some of them reviewed in [40].

### 4.3.1  SpikeProp

Similar to backpropagation algorithm [41] designed for traditional ANN, SpikeProp [42] is designed to determine a set of desired firing times ($t_j^d$) of all output neurons, at the post-synaptic neurons for a given set of input pattern. This is achieved by applying an error function $E$, in the particular least mean squares error to minimize the error of squared difference between training output times $t_j$ and desired output times $t_j^d$. Nevertheless, two assumptions are mentioned: each neuron can fire only once in each processing step and the time course of the neuron's membrane potential after the firing is ignored. Weight $w_{ij}^k$ connecting pre-synaptic neuron and post-synaptic neuron is determined to minimize the error (Eq. 4.25) in which $\eta$ is the learning rate.

$$E = \frac{1}{2} \sum_j \left( t_j - t_j^d \right)^2 \tag{4.25}$$

$$\Delta w_{ij}^k = -\eta \frac{\partial E}{\partial w_{ij}^k}$$

## 4.3.2   *Spike-Time Dependent Plasticity (STDP)*

Another well-known learning paradigm inspired by the Hebbian learning principle is STDP [43–45] in which the synaptic weights are adjusted based on the temporal order of the incoming spike (pre-synaptic) and the output spike (post-synaptic). This synaptic weight adjustment determines synaptic potentiation known as long term potential (LTP) if the synaptic weight is increasing (positive change) and synaptic depression known as long-term depression (LTD) if the synaptic weight is decreasing (negative change). A particular connection is said to potentiate if a pre-synaptic spike arrives before a post-synaptic spike; and is said to depress if it arrives after a post-synaptic spike [45].

STDP is expressed in terms of STDP learning window $W\left(t_{pre} - t_{post}\right)$ in which the difference between arrival time of the pre-synaptic spike and the arrival time of post-synaptic spike will determine the synaptic weight (Eq. 4.26) (Fig. 4.17). In the equation, $\tau_+$ and $\tau_-$ refer to the pre-synaptic and post-synaptic time interval; and $A_+$ and $A_-$ refer to the maximum fraction of synaptic adjustment if $t_{pre} < t_{post}$ approaches to zero.

$$W\left(t_{pre} - t_{post}\right) = \begin{cases} A_+ \exp\left(\frac{t_{pre}-t_{post}}{\tau_+}\right) & if \ t_{pre} < t_{post}, \\ A_- \exp\left(-\frac{t_{pre}-t_{post}}{\tau_-}\right) & if \ t_{pre} > t_{post}, \end{cases} \quad (4.26)$$

Authors in [46] showed that using unsupervised learning with the STDP learning rule even a single spiking leaky integrate and fire neuron model (LIFM) can learn to react quickly on the onset of a spatio-temporal spiking pattern on 2000 synapses that the neuron has been presented with before, among them 1000 presenting noise. And the more often the LIFM 'sees' this pattern, the earlier the neuron recognises the onset of it (Fig. 4.18).

To summarise the principles of STDP:

- Hebbian form of plasticity in the form of long-term potentiation (LTP) and depression (LTD);
- Effect of synapses are strengthened or weakened based on the timing of pre-synaptic spikes and post-synaptic action potential.
- Pre-synaptic activity that precedes post-synaptic firing can induce LTP, reversing this temporal order causes LTD
- Through STDP, connected neurons learn consecutive temporal associations from data, forming chains of connections to represent patterns in the data.
- Several variations of the STDP exist.

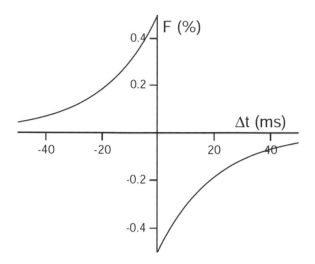

**Fig. 4.17** STDP is expressed in terms of STDP learning window $W\left(t_{pre} - t_{post}\right)$ in which the difference between arrival time of the pre-synaptic spike and the arrival time of post-synaptic spike will determine the synaptic weight; $\Delta t = tpre - tpost$

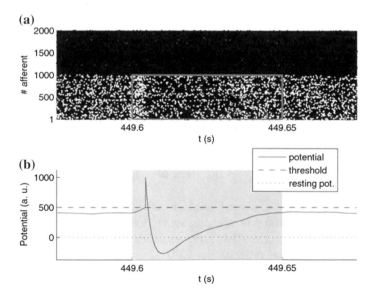

**Fig. 4.18** A single LIFM neuron can be trained with the STDP unsupervised learning rule to discriminate a repeating pattern of synchronised spiking on multiple synapses from noise [46]

### 4.3.3   Spike-Driven Synaptic Plasticity (SDSP)

SDSP, a variant of Spike Timing Dependent Plasticity (STDP), is a semi-supervised learning rule [47] that directs the change of the synaptic plasticity $V_{w0}$ of a synapse $w_0$ depending on the spike's time of the pre- and post-synaptic neurons. If a pre-synaptic spike arrives at the synaptic terminal while the post-synaptic neuron's membrane potential is higher than a given threshold value (i.e. normally shortly before a post-synaptic spike is emitted), the synaptic efficacy is increased (potentiation). However, when a pre-synaptic spike arrives at the synaptic terminal while the post-synaptic neuron's membrane potential is low (i.e. normally shortly after a spike is emitted), the synaptic efficacy is decreased (depression). Where $\Delta t_{spk}$ is the pre- and post-synaptic spike time window, this synaptic change can be expressed as:

$$\Delta V_{w0} = \begin{cases} \frac{I_{pot}\left(t_{post}\right)}{C_p} \Delta t_{spk}, & \text{if } t_{pre} < t_{post} \\ \frac{I_{dep}\left(t_{post}\right)}{C_d} \Delta t_{spk} & \text{if } t_{post} < t_{pre} \end{cases} \qquad (4.27)$$

SDSP introduces a dynamic 'drift' of the synaptic weights either to be 'up' or 'down', depending on the value of the weight itself [48]. If the weight is higher than the threshold value, then the weight is slowly driven (by the learning algorithm) to a fixed high value. On the contrary, the weight is slowly driven to a fixed low value if the weight is lower than the threshold value. These two values represent the two stable states and at the end of the learning process, the final weights can be encoded with 1 single bit [49].

### 4.3.4   Rank Order (RO) Learning Rule

In RO learning rule earlier coming spikes are considered more important (carry more information) and that is reflected in the learning rule, which increase the connection weights based on the order of spikes coming from different synapses (Eq. 4.28) (Fig. 4.19) [8]:

$$\Delta w_{ji} = m^{\text{order}(j)} \qquad (4.28)$$

where: m is a parameter called modulation factor.

The membrane potential PSP of a neuron is calculated as (Fig. 4.19):

$$u_i(t) = \begin{cases} 0 & \text{if fired} \\ \sum_{j|f(j)<t} w_{ji} m_i^{\text{order}(j)} & \text{else} \end{cases} \qquad (4.29)$$

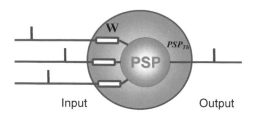 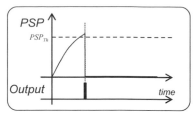

**Fig. 4.19** In RO learning rule earlier coming spikes are considered more important (carry more information) and that is reflected in the learning rule, which increase the connection weights based on the order of spikes coming from different synapses: The firing threshold of the neuron can be set to be a portion (e.g. C = 0.7) of the maximum PSP (Eq. 4.30) for a given input pattern, so that this neuron will emit a spike earlier than the pattern it is trained to spike on when the whole pattern is presented (Eq. 4.31)

$$PSP\,max(T) = SUM\left[\left(m^{order(j(t))}\right)w_{j,i}(t)\right], \quad for\ j = 1, 2, \ldots, k; \quad t = 1, 2, \ldots, T;$$

$$(4.30)$$

$$PSP_{Th} = C.PSPmax(T) \qquad\qquad (4.31)$$

The above ability of spiking neurons to be trained to spike early is essential feature of the brain. It has helped humans and animals to predict and to escape predators, for humans to play sport, e.g. predicting the trajectory of a ball and catching it, and for almost all other human activities. This feature is used in Chap. 18 for predicting individual occurrence of stroke one day ahead and in Chap. 19 for predicting probability of earthquakes few hours before the event.

### 4.3.5 Learning in Dynamic Synapses

A phenomenological model for modelling short-term dynamics of synapses has been proposed more than a decade ago by [50]. The model which is based on experimental data of biological synapses, suggests that the synaptic efficiency (weight) is a dynamic parameter that changes with every pre-synaptic spike due to two short-term synaptic plasticity processes: facilitation and depression. This inherent synaptic dynamics empower neural networks with a remarkable capability for carrying out computations on temporal patterns (i.e., time series) and spatio-temporal patterns. Maass and Sontag [51], in their theoretical analysis considering analogue input showed that with just a single hidden layer such networks can approximate a very rich class of non-linear filters. However there is a need for similar study in the presence of many inputs that carry sequences of spikes in a temporal relationship. It is suggested also that dynamic synapses work as memory buffers [52] due to the fact that a current spike is influenced by previous spikes.

Furthermore a SNN with dynamic synapses is showed to be able to induce a Finite State Machine mechanism [53]. A number of studies have utilized dynamic synapses in practical applications. One of the first practical application of dynamic synapses was speech recognition [54] and later-image filtering [55].

## 4.4 Spike Pattern Association Neurons and Neural Networks

### 4.4.1 Principles of Spike Pattern Association Learning. The SPAN Model

In this section, we present a supervised learning algorithm for SNN that enables a single neuron to learn spike pattern associations of input-output spike sequences at precise time of spikes. We refer to this learning neuron as SPAN for **S**pike **P**attern **A**ssociation **N**euron [56–59]. Using a SPAN neuron, one can build SNN to associate input to output temporal patterns of desired spike sequences.

In the SPAN learning algorithm, the input, output and desired spike trains are transformed into analogue signals by convolving the spikes with a kernel function. This transformation will simplify the computation of the error signal and, hence, allows the application of a gradient descent to optimize the synaptic weights.

In [58], the authors used such a signal transformation along with a Particle Swarm Optimizer in order to optimize the parameters of dynamic synapses enabling the network to learn a desired input/output mapping of spike trains. However, due to scalability issues when training big networks, learning algorithms based on evolutionary computation are less practical. Therefore, a gradient descent method was suggested in [59]. Preliminary experiments were conducted demonstrating the functioning of the algorithm. In this study, we present a comprehensive analysis of the SPAN method along with a theoretical investigation highlighting the relationship of SPAN to ReSuMe and Chronotron.

Similar to other supervised training algorithms, the synaptic weights of a neuron are adjusted iteratively in order to impose a desired input/output spike sequence mapping. We derive the proposed learning algorithm from the common Widrow-Hoff rule, also known as the Delta rule. For a synapse $i$, it is defined as:

$$\Delta w_i = \lambda x_i (y_d - y_a) = \lambda x_i \delta_i \qquad (4.32)$$

where $\lambda \in R$ is a real-valued positive learning rate, $x_i$ is the input transferred through synapse $i$, and $y_d$ and $y_a$ refer to the desired and the actual neural output, respectively. Note that $\delta_i = y_d - y_a$ is the difference or error between the desired and the actual output of the neuron.

This rule was introduced for traditional neural networks with linear neurons. For these models, the input and output corresponds to real-valued vectors. In SNN

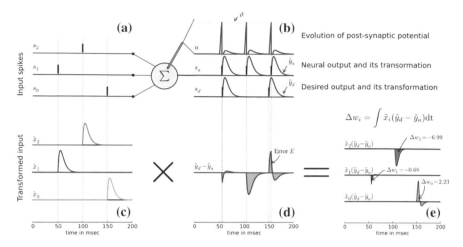

**Fig. 4.20** Illustration of the learning rule of SPAN (see text for detailed explanations of the figure) (from [56])

however, trains of spikes are passed between neurons rendering the Widrow-Hoff rule not applicable to SNN. More specifically, if $x_i$, $y_d$ and $y_a$ were considered as spike trains $s(t)$ in the form of

$$s(t) = \sum_f \delta(t - t^f) \tag{4.33}$$

where $t^f$ is the firing time of a spike and $\delta(\cdot)$ is the Dirac delta function $\delta(x) = 1$ if $x = 0$ and 0 otherwise, then the difference between two spike trains $y_d$ and $y_a$ does not define a suitable error landscape which can be minimized by a gradient descent. In this method, this issue is addressed by convolving each spike sequence with a kernel function $\kappa(t)$. This is similar to the binless distance metric used to compare spike trains [60]. A variety of kernel functions $\kappa(t)$ exist such as linear, (double) exponential, alpha and Gaussian kernels. In this study, we use an $\alpha$-kernel, $\alpha(t) = e\tau^{-1}te^{-t/\tau}H(t)$, however many other kernels appear suitable in this context. Using this kernel function, we can now perform the transformation of the spike sequences into analogue sequences (Fig. 4.20) and perform learning of the connection weights:

$$\Delta w_i(t) = \lambda \left(\frac{e}{2}\right)^2 \left[\sum_g \sum_f H\left(t - \max\{t_i^f, t_d^g\}\right)\left(t - t_d^g\right)\left(t - t_i^f\right)e^{-\frac{2t - t_i^f - t_d^g}{\tau}}\right.$$

$$\left. - \sum_h \sum_f H\left(t - \max\{t_i^f, t_d^g\}\right)\left(t - t_d^g\right)\left(t - t_i^f\right)e^{-\frac{2t - t_i^f - t_d^g}{\tau}}\right] \tag{4.34}$$

With a simple example, the behaviour of the presented learning rule can be demonstrated. Let us consider the case where the input, desired and actual spike trains have only a single spike at $t_i$, $t_d$, $t_a$, respectively and they satisfy $t_i \leq t_d \leq t_a$. Equation 4.34 then simplifies to:

$$\Delta w_i = \lambda \left(\frac{e}{2}\right)^2 \left[(t_d - t_i + \tau)e^{-\frac{t_d - t_i}{\tau}} - (t_a - t_i + \tau)e^{-\frac{t_a - t_i}{\tau}}\right] \tag{4.35}$$

And we note that:

$$\Delta w_i \begin{cases} > 0 & if \quad t_d < t_a \\ = 0 & if \quad t_d = t_a \\ < 0 & if \quad t_d > t_a \end{cases} \tag{4.36}$$

From Eq. 4.36 several observations can be made:

- if the actual spike occurs later than the desired spike ($t_d < t_a$), then the synaptic weight increases and so the output spike will be emitted earlier at a next input presentation (epoch);
- conversely, if the actual spike occurs earlier than the desired firing time ($t_a < t_d$), then the synaptic weight decreases and so the output spike will be emitted later;
- if the actual spike occurs exactly at the desired time ($t_a = t_d$), then the synaptic weight does not change;
- and the larger the difference between $t_a$ and $t_d$ is, the larger the size of synaptic weight change becomes.

Furthermore, we can observe that

- when $t_a \rightarrow \infty$, which means that no actual spike occurs, the synaptic weight increases to promote the emission of an output spike since $t_d < t_a$ holds,
- when $t_d \rightarrow \infty$, which means no output spike is desired, the synaptic weight decreases to promote a suppression of an output spike since $t_a < t_d$ holds.

These observations are intuitively valid and we can expect, by repeating these processes, that the learning rule drives the post-synaptic neuron to emit a spike at the desired time. Furthermore, we note that the smaller the value of $t_d - t_i$ or $t_a - t_i$ is, the larger the value of each term in the square brackets of Eq. 4.35 becomes. That means that only if the input spike at $t_i$ is temporally close to the desired or actual spike at $t_d$ or $t_a$, i.e. spike $t_i$ is the *cause* of spike $t_d$ or spike $t_a$, the corresponding synaptic weight $w_i$ changes significantly.

Weights are updated in an iterative process. In each iteration (or epoch), all input patterns are presented sequentially to the system. For each pattern the $\Delta w_i$ values are computed and accumulated. After the presentation of all patterns, the weights are updated to $w_i(e + 1) = w_i(e) + \Delta w_i$, where $e$ is the current epoch of the learning process.

We note that the algorithm is capable of training the weights of a single neural layer only. Related methods such as ReSuMe [61] and the Chronotron [62] exhibit similar restrictions. Therefore, a combination with the well-known Liquid State Machine (LSM) approach [63] was suggested in these studies. By transforming the input into a higher-dimensional space, the output of the LSM can potentially be mapped to any desired spike train.

Figure 4.20 illustrates the functioning of the proposed SPAN learning method. An output neuron is connected to three input neurons through three excitatory synapses with randomly initialized weights. For the sake of simplicity, each input sequence consists of a single spike only. However, the learning method can also deal with more than one spike per input neuron. The inputs spike trains $s_i$ are visualized in Fig. 4.20a. In this example, we intend to train the output neuron to emit two desired spikes at the pre-defined time $t_d^0$ and $t_d^1$.

Assume that, as shown in Fig. 4.20b, the presented stimulus causes the excitation of the output neuron resulting in the generation of three output spikes at times $t_a^0$, $t_a^1$ and $t_a^2$, respectively. Spike $t_a^0$ is temporally very close to the desired spike $t_d^0$; spike $t_a^1$ is undesired and should be suppressed by the learning method; and spike $t_a^2$ occurs slightly too late ($t_d^1 < t_a^2$). The evolution of the membrane potential $u$ ($t$) measured at the output neuron is shown in middle top diagram of the figure above the actual and the desired spike trains, cf. Fig. 4.20b.

The lower part in the figure (Fig. 4.20c–e) depicts a graphical illustration of Eq. 4.34. The input, actual and desired spikes trains are kernelized using the $\alpha$-kernel as defined in Eq. 4.35 (Fig. 4.20b, c). We define the area under the curve of the absolute difference $|y_d(t) - y_a(t)|$ as an error between actual and desired output.

Although this error is not used in the computation of the weight updates $\Delta w_i$, this metric is an informative measure of the achieved training status of the output neuron.

Figure 4.20e shows the weight updates $\Delta w_i$. We especially note the large decrease of weight $w_0$. The input spike train $s_0$ of the first input neuron causes an undesired spike at $t_a^1$ and lowering the corresponding synaptic efficacy potentially suppresses this behaviour. On the other hand, the synaptic weight $w_2$ is increased promoting the triggering of spike $t_a^2$ at an earlier time. Finally, weight $w_1$ remains almost unchanged since $t_a^1 \approx t_d^1$.

In an exemplified implementation, we employ the Leaky Integrate and-Fire (LIF) neuron which is one of the most widely used spiking neural models [4]. It is based on the idea of an electrical circuit containing a capacitor with capacitance $C$ and a resistor with resistance $R$, where both $C$ and $R$ are assumed to be constant. The dynamics of a neuron $i$ are then described by the following differential equation:

$$\tau_m = \frac{du_i}{dt} = -u_i(t) + RI_i^{syn}(t) \tag{4.37}$$

The constant $\tau_m = RC$ is called the membrane time constant of the neuron. Whenever the membrane potential $u_i$ crosses a threshold $\vartheta$ from below, the neuron fires a spike and its potential is reset to a reset potential $u_r$. Following [4], we define

$$t_i^{(f)} : u_i(t^{(f)}) = v, f \in \{0, \ldots, n-1\} \tag{4.38}$$

as the firing times of neuron $i$ where $n$ is the number of spikes emitted by neuron $i$. It is noteworthy that the shape of the spike itself is not explicitly described in the traditional LIF model. Only the firing times are considered to be relevant.

The synaptic current $I_i^{syn}$ of neuron $i$ is modeled using an $\alpha$-kernel:

$$I_i^{syn}(t) = \sum_j w_{ij} \sum_f \alpha(t - t_j^{(f)}) \tag{4.39}$$

where $w_{ij} \in R$ is the synaptic weight describing the strength of the connection between neuron $i$ and its pre-synaptic neuron $j$. The $\alpha$-kernel itself is defined as

$$\alpha(t) = e\tau_s^{-1} t e^{-t/\tau_s} H(t) \tag{4.40}$$

where $H(t)$ refers to the Heaviside function and parameter $\tau_s$ is the synaptic time constant.

### 4.4.2   Case Study Examples

The following case study examples are described in a reproducible way following [64]. In all experiments, the network architecture consists of single neuron driven by $n$ synapses. The input spike patterns stimulating the neuron are generated randomly. More specifically, each input spike train consists of a single spike generated randomly in the time interval (0, 200 ms). The simulation is performed using the NEST simulator [65]. We provide the setup details that are specific for a particular experiment in the individual sections below.

The purpose of the first experiment is to demonstrate the concept of the proposed learning method. The task is to learn a mapping from a random input spike pattern to specific target output spike train. This target consists of five spikes occurring at the times $t_d^0 = 33$, $t_d^1 = 66$, $t_d^2 = 99$, $t_d^3 = 132$ and $t_d^4 = 165$ ms. Initially, the synaptic weights are randomly generated. Over 100 epochs, we allow the output neuron to adjust its connection weights in order to produce the desired output spike train. The experiment is repeated for 100 runs each of them initialized with different random weights in order to guarantee statistical significance. In Fig. 4.21, the experimental setup of a typical run is illustrated. The left side of the diagram shows the network architecture as defined in the experimental setup above. The right side shows the desired target spike train (top) along with the produced spike trains by the output neuron over a number of learning epochs (bottom). We note that the after the learning process. Output spike trains in early epochs are very different from the desired target spike sequence. In later epochs the output spikes converge towards the desired sequence. We note that the neuron is able to reproduce the desired spike

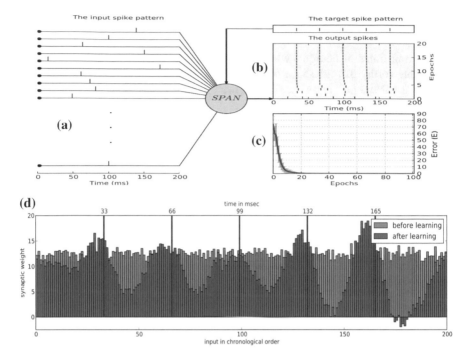

**Fig. 4.21 a** Learning spike pattern association with 400 input synapses. The neuron learns to map between spatiotemporal input pattern and output spike train; **b** the development of the output toward the target pattern for one of the trials; **c** the evolution of the error and the standard deviation; **d** the synaptic weights before and after training learn a desired input-output behaviour of the SPAN

output pattern very precisely in less than 20 learning epochs. Figure 4.21c shows the evolution of the average error over the performed 100 runs. We note the exponential decrease of the error. In 97% of all trials the target spike train could be reproduced in less than 30 epochs and even for the remaining three percent, the average temporal difference between learned and desired spike train was less than 0.2 ms.

The effect of the learning algorithm on the synaptic efficacy can be visualized by comparing the synaptic weights before and after the application of the learning process, cf. Fig. 4.21d. For the diagram, the neural inputs are chronologically sorted according to their spike firing times. A bar in the figure reflects the synaptic strength of a synapse that corresponds to a particular input.

In order to get an impression of the temporal causality of the weight changes, we overlay the plot with the desired firing times of the neuron (red vertical lines at 33, 66, 99, 132 and 166 ms). The figure presents the weight changes averaged over all 100 runs.

Due to the experimental setup, we observe a uniform distribution of the weights after the initialization of the algorithm. After the training over 100 epochs, the synapses that transfer input spikes which are temporally close to the desired target spikes are potentiated. On the other hand, synapses that transfer spike inputs at undesired times are inhibited. The sine-shaped form of the chronologically sorted synaptic efficacies is caused by the equidistant firing times of the spikes in the target sequence.

From this simple experiment, we conclude that the proposed learning method can be applied for a SPAN neuron to learn to associate a single multi-synaptic spiking input pattern to a desired spiking output sequence.

The previous experiment involved the learning of a single pattern only. In this experiment, we investigate the performance of SPAN when several input patterns have to be learned. Furthermore, we are interested in the behaviour of the method when the input stimuli are noisy which is important in the light of a real-world application. In the next sections, we investigate some more challenging learning scenarios for SPAN.

We construct an initial set of ten spike patterns each consisting of $n = 500$ input neurons that are allowed to emit a single spike only. With every presentation of an input pattern to the learning neuron, a noise is added to each spike in form of a jitter drawn from a Gaussian distribution. The strength of the jitter is controlled by the standard deviation of the Gaussian. In our experiments, we use different jitter strengths in order to investigate the impact of different noise levels on the learning performance of SPAN.

The neuron is trained for 400 epochs to emit a single spike at $t_d = 99$ ms in response to the input patterns. We call the output of the neuron successful, if the output sequence consists of a single spike only that occurs within the interval $[t_d - 5$ ms, $t_d + 5$ ms$]$. We define $P_s$ as the probability of a successful output. It is the ratio of the number of output spikes that match their desired spikes over all ten input patterns. We consider jitter strengths of 0, 5, 10, 15 and 20 ms. For each of them, an individual experiment is undertaken and repeated for 100 trials to guarantee statistical evidence.

Figure 4.22 presents the results of the experiment averaged over the 100 trials. The top row of diagrams show the obtained results for the noise-free case, i.e. a jitter strength of zero. On the left, the evolution of the error is presented. In the first few iterations of training, the neuron spikes arbitrary and the output does not match the desired target. We note that $P_s$ (depicted in the right top diagram) is low in the first few epochs of the training process. However, the output stabilizes quickly and $P_s$ increases rapidly indicating the neuron's ability to converge its output to the desired target spike.

In order to give an impression of the temporal difference between the obtained output spike and the target spike, we have computed the absolute difference $\Delta t = |t_d - t_a|$ for all successful output spikes. The evolution of $\Delta t$ is overlaid in the right top diagram. Clearly, the temporal difference is minimized quickly by SPAN's learning algorithm resulting in very precisely timed output spikes.

If noise is introduced to the presented input patterns, the difficulty of the learning task increases significantly. The diagrams in the middle row of Fig. 4.22 present the results for a jitter strength of 5 ms. As expected, the training error cannot become

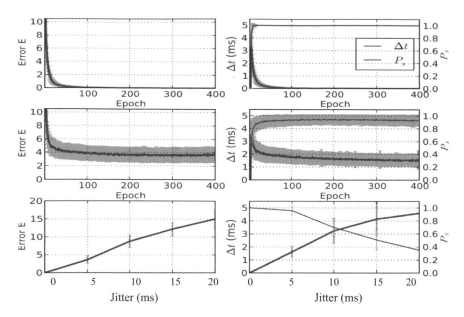

**Fig. 4.22** Learning multiple spike patterns using the SPAN learning rule. The top plots show the results when the patterns are learned without any noise applied. The diagrams below show the learning when jittered input patterns are used (jitter strength of 5 ms). A neuron is trained to fire a single spike at 99 ms. The success probability $P_s$ is computed in every epoch to indicate the number of times the output spikes matches the desired spike. The bottom diagrams show the final training error in dependence of the applied jitter strength

zero in this learning scenario. However, the evolution of the error indicates a certain convergence of the algorithm. Despite the noise, the training is very often successful. More than 90% of the output spikes fulfil the defined success criterion. The neuron is able to learn to fire within an average time shift $\Delta t = 2$ ms irrespective of the noise.

The performance of SPAN as a function of the jitter strength is depicted in the bottom plots of Fig. 4.22. For the diagrams we have used the neural outputs obtained during the last training epoch. Clearly, the error is proportional to the jitter strength. This relationship indicates a satisfying resistance of the SPAN rule to input pattern noise.

Even for large jitter strengths, the method is able to map around three out of ten pattern correctly, cf. right bottom diagram of the figure.

### 4.4.3 Memory Capacity of SPAN

An important issue related in the learning process is how much information the neuron can learn and memorize. We use the measure proposed in [66] to evaluate the memory capacity of SPAN. The memory capacity is described in term of the load factor which is defined as the ratio of the number of input patterns $p$ the neuron can classify correctly over the number of synapses $n$, i.e. $\frac{p}{n}$.

The $p$ input patterns are generated randomly, similar to the previous experiments, where each pattern consists of $n$ spike trains, each has a single spike at a random time instant. Subsequently, the patterns are assigned randomly to $c$ different classes, which is set to 5 in this experiment.

The task of the experiment is to train the neuron to classify the all patterns correctly in a maximum number of epochs of 500. The classification is performed by training the neuron to fire a single spike at a specified time instant $t_d^i$ when a pattern that belongs to class $i$ is presented at the input.

Thus, the class of the input pattern is identified by the time of the fired spike, $t_d^i$, which is set to 33, 66, 99, 132, or 165 ms to identify the five classes. The experiment is repeated on three network architectures having 200, 400 and 600 synapses. We report the success rate as a function of the number of the input patterns $p$. The success rate is the percentage of trials having the all input patterns classified successfully, also we report the average number of iterations required to achieve successful classification.

A pattern is decided as correctly classified if the 90% fired spike in response to that pattern is within 2 ms of the corresponding target spike. The learning rate is set to $\lambda = \frac{5c}{p}$ and the synaptic weights are initialized randomly using maximum synaptic weights of 5, 2.5 and 2 pA for the 200, 400 and 600 synapses respectively. These values were set based on trial and error.

Figure 4.23, shows the results of the experiment for the three cases of the synapses. From the figure, it is clear that increasing the number of synapses increases the number of patterns that can be remembered and classified correctly. However, more epochs and more computation time is required to adjust the synaptic weights. It is noted that after a certain number of input patterns, it becomes difficult for the neuron to recognize the patterns, hence, the success rate starts to drop. We consider the points where the success rate is 90% and above, which are indicated by the green diamond markers in Fig. 4.23. For these points, the value of $p$ is 15, 30, 35 with success rate of 96, 94, 90% respectively. Furthermore, the average number of epochs to achieve successful training is below 100. The load factor at these points is computed to be 0.075, 0.075 and 0.058 for the three cases of 200, 400 and 600 synapses respectively. To get a sense of these values, we have conducted an experiment to measure the memory capacity of ReSuMe learning rule at these points, i.e. with the same values of $p$ and $n$. For this experiment, a batch learning rule of ReSuMe was used (with a value of $a_R$ set to 0.025 and the learning rate set to 10). The obtained success rates of ReSuMe to learn to recognize the input patterns were 22, 10 and 52%. These values are lower than the success rates of SPAN, hence, ReSuMe has less memory capacity than SPAN.

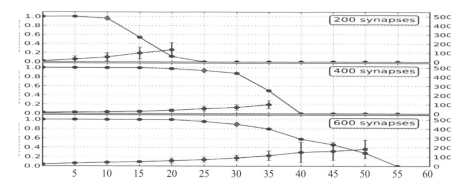

**Fig. 4.23** The memory capacity of SPAN with different number of synapses. The green diamond marker represents the maximum number of learned patterns for which the average number of successful training is above 90%

### 4.4.4  SPAN for Classification Problems

In this experiment a spatio-temporal classification task is performed. The objective is to learn to classify five classes of input spike patterns. The pattern for each class is given as a random input spike pattern that was created in a similar fashion as for the previous experiment. Fifteen copies for each of the five patterns are then generated by perturbing each pattern using a Gaussian jitter with a standard deviation of 3 ms resulting in a total of $15 \times 5 = 75$ samples in the training data set. Additionally, we create $25 \times 5 = 125$ testing samples using the same procedure. The output neuron is then trained to emit a single spike at a specific time for each class. Only the training set is used during training, while the testing set is used to determine the generalization ability of the trained neuron. The spike time of the output neuron encodes the class label of the presented input pattern. The neuron is trained to spike at the time instances 33, 66, 99, 132, and 165 ms respectively, each spike time corresponding to one of the five class labels. We allow 200 epochs for the learning method and we repeat the experiment in 30 independent runs. The number of synapses in this experiment was set to 200. For each run we chose a different set of random initial synaptic weights.

Figure 4.5a shows the evolution of the average error for each of the five classes. In the first few epochs, the value of the error oscillates and then starts to stabilize and decrease slowly. The learning error decreases for some classes faster than for others, e.g. class 3. We also note that the class reporting the highest error is class 1. This behaviour is expected and confirms a quite similar finding in [62]. In order to classify samples of class 1 correctly, the output neuron has to emit a very early spike at $t \approx 33$ ms. Consequently, the neuron needs to be stimulated by input spikes occurring at times before $t = 33$ ms. However, due to the random generation of the input data, only few input spikes occur before $t = 33$ ms. Most input spikes arrive after that time at the output neuron and therefore do not contribute to the correct classification of class 1 samples. The relationship between the accuracy and the output spike time was also noted in [62]. Future studies will further investigate this interesting observation.

In order to report the classification accuracy of the trained neuron, we define a simple error metric. We consider a pattern as correctly classified, if the neuron fires a single spike within $[t_d^f - 3\,\mathrm{ms}, t_d^f + 3\,\mathrm{ms}]$ of the desired spike time $t_d^f$. Any other output is considered as incorrect. It is noteworthy to mention that using this definition, an untrained neuron is very likely to produce incorrect outputs resulting in accuracies close to zero. Figure 4.24b shows the average classification error for each class in the training and testing phase. As mentioned above, for testing, the 125 unseen patterns of the test set are used. The neuron is able to learn to classify the 75 training patterns with an average accuracy of 94.8% across all classes. Once more, we note the comparatively poor classification performance of samples belonging to the first class. For the test patterns, the neuron is able to achieve average accuracy of 79.6% across all classes.

The experimental analysis presented in the previous section has demonstrated that, despite its algorithmic simplicity, the SPAN learning method can efficiently impose a desired input/output behaviour to a SNN. In this section, we compare the differences and the similarities between the proposed method and two related algorithms, the Chronotron [62] and the ReSuMe learning rule [61, 67].

Similar to SPAN, also the ReSuMe learning algorithm is derived from the Widrow-Hoff rule. ReSuMe interprets the Widrow-Hoff rule as a combination of an STDP and an anti-STDP process. With the introduction of an explicit learning window, the method emphasizes on the implementation of biologically plausible learning processes. The SPAN rule, on the other hand, follows a different idea. The sacrifice of biological realism allows the straightforward formulation of an efficient synaptic weight modification rule. By converting spike trains into analogue signals, the Widrow-Hoff rule can be directly applied to spiking neurons. Despite the fact

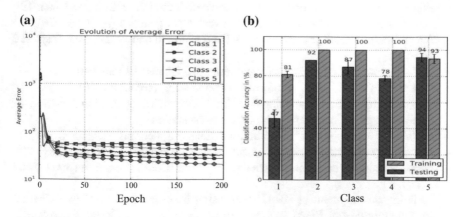

**Fig. 4.24**   **a** Evolution of the average errors obtained in 30 independent trails for each class of the training samples; **b** the average accuracies obtained in the training and testing phase

that the kernelization of spike trains was investigated in several studies before, we are not aware of any study that applies spike convolution in an algorithm for the learning of precisely timed spike train patterns. In [63, 67] kernel functions have been used to define spike train metrics and in [68] kernelized spike trains were studied in the context of classification problems using a nearest neighbour approach.

Although the biological plausibility of the SPAN learning method is at least questionable, a surprising observation can be made when the α-kernel is replaced by an exponential one.

In concept, the SPAN rule is also similar to the Chronotron E-learning rule [62]. Also in Chronotron the synaptic weights are modified according to a gradient descent in an error landscape. Its error function is based on the Victor & Purpura (VP) distance [69]. By finding a way to deal with the discontinuities of the VP metric, the Chronotron rule efficiently computes the error gradient and updates the weights accordingly. SPAN's error landscape, on the other hand, is based on a metric similar to the van Rossum metric [60] but with α kernels. This metric does not exhibit any discontinuities allowing the definition of a simple yet powerful learning rule. The differences and similarities of Chronotron and SPAN are discussed in [56, 57].

## 4.5   Why Use SNN?

SNN have some characteristics that make them superior in few aspects when compared with traditional machine learning techniques including the classical neural networks from Chap. 2, namely:

- Efficient modelling of temporal-, spatio-temporal or spectro-temporal data (Chaps. 8–22);
- Efficient modelling of processes that involve different time scales (see Chap. 19);
- Bridging higher level functions and "lower" level genetics (see Chap. 16);
- Integration of modalities, such as sound and vision in one system (see Chaps. 12 and 13);
- Predictive modelling and event prediction (see Chaps. 18 and 19);
- Fast and massively parallel information processing (see Chap. 20);
- Compact information processing (see Chap. 21 which is presenting a method for data compression based on spike-time representation);
- Scalable structures (from tens to billions of spiking neurons) (Chaps. 6 and 20);
- Low energy consumption if implemented on neuromorphic platforms (Chap. 20);
- Deep learning and deep knowledge representation in brain inspired SNN (Chap. 6);
- Enabling development of BI-AI when using brain inspired SNN (see Chap. 1 for the definition of BI-AI and their 20 features; and also Chaps. 6 and 22);

**Table 4.2**  A comparative analysis between SNN and other machine learning methods

| Method/features | Statistical methods (MLR, kNN, SVM) | ANN (e.g. MLP, CNN) | SNN |
|---|---|---|---|
| Information | Scalars | Scalars | Spike sequences |
| Data representation | Scalars, vectors | Scalars, vectors | Whole TSTD patterns |
| Learning | Statistical, limited | Hebbian rule | Spike-time dependent |
| Dealing with TSTD | Limited | Moderate | Excellent |
| Parallel computation | Limited | Moderate | Massive |
| Hardware support | Standard | VLSI | Neuromorphic VLSI |

TSTD is abbreviation of temporal or spatio-temporal data

Table 4.2 shows a brief comparative analysis between SNN and other statistical and machine learning techniques across various capabilities.

SNN opened the field of brain-inspired (cognitive, neuromorphic) computing. Dharmendra Modha, chief scientist of Brain-inspired Computing at IBM Research, commented "*The goal of brain-inspired computing is to deliver a scalable neural network substrate while approaching fundamental limits of time, space, and energy*".

## 4.6   Summary and Further Readings for Deeper Knowledge

This chapter gives some basic knowledge about SNN that is used in next chapters, where eSNN models are presented (Chap. 5) and brain-inspired SNN (Chap. 6).

More information on specific topics related to SNN can be found as follows:

- Selection and optimisation of spike encoding methods [24] and a software: http://www.kedri.aut.ac.nz/neucube/. Spiking Neuron Models [4];
- Spike-based strategies for rapid processing [15];
- Computational modelling with SNN (Chap. 37 in [38]);
- Brain-like information processing for spatio-temporal pattern recognition (Chap. 47 in [38]);
- Associative memory with SNN [32];
- SPAN [58, 59].

**Acknowledgements**  Along with traditional SNN models, this chapter presents original work by the author on probabilistic spiking neurons and neurogenetic spiking neurons, the latter being developed in collaboration with L. Benuskova [36]. The original work of spike-pattern association neurons (SPAN) was a team work by A. Mohemmed, S. Schliebs, S. Matsuda and the author [56, 57]. I acknowledge the contribution to the presentation of the spike encoding algorithms of Neelave Sengupta [70]. The software for the selection and parameter optimisation of spike encoding algorithm was developed by Balint Petro and available from: https://kedri.aut.ac.nz/R-and-D-Systems/neucube (Spiker).

# References

1. W. Maass, On the role of time and space in neural computation. Math. Found. Comput. Sci. **1998**, 72–83 (1998)
2. P. Lichtsteiner, T. Delbruck, A 64x64 aer logarithmic temporal derivative silicon retina. Res. Microelectron. Electron. PhD, **2**(1), 202–205 (2005). https://doi.org/10.1109/rme.2005. 1542972
3. T. Delbruck, jAER open source project (2007) http://jaer.wiki.sourceforge.net
4. W. Gerstner, W.M. Kistler, *Spiking Neuron Models: Single Neurons, Populations, Plasticity* (Cambridge University Press, Cambridge, 2002)
5. L.M. Optican, B.J. Richmond, Temporal encoding of two dimensional patterns by single units in promate inferior temporal cortex. III. Information theoretic analysis.pdf. J. Neurophysiol. **57**(1), 162–177 (1987)
6. R. Lestienne, Determination of the precision of spike timing in the visual cortex of anaesthetised cats. Biol. Cybern. **74**, 55–61 (1996). https://doi.org/10.1007/BF00199137
7. Z.F. Mainen, T.J. Sejnowski, Reliability of spike timing in neocortical neurons. Science **268** (5216), 1503–1506 (1995)
8. S. Thorpe, D. Fize, C. Marlot, Speed of processing in the human visual system. Nature **381** (6582), 520–522 (1996)
9. S. Thorpe, J. Gautrais, Rank order coding. Comput. Neurosci. Trends Res. **13**, 113–119
10. S.M. Bohte, H. La Poutre, J.N. Kok, unsupervised clustering with spiking neurons by sparse temporal coding and multilayer RBF networks. IEEE Trans. Neural Netw. **13**(2), 426–435 (2002)
11. H.N. Abdul Hamed, *Novel Integrated Methods of Evolving Spiking Neural Network and Particles Swarm Optimization* (Auckland University of Technology, 2012)
12. S. Schliebs, M. Defoin-Platel, N. Kasabov, Integrated feature and parameter optimization for an evolving spiking neural network. Adv. Neuro-Inf. 1229–1236 (2009). Retrieved from http://link.springer.com/chapter/10.1007/978-3-642-02490-0_149
13. N. Kasabov, V. Feigin, Z.-G. Hou, Y. Chen, L. Liang, R. Krishnamurthi et al., Evolving spiking neural networks for personalised modelling, classification and prediction of spatio-temporal patterns with a case study on stroke. Neurocomputing **134**, 269–279 (2014). https://doi.org/10.1016/j.neucom.2013.09.049
14. K. Dhoble, N. Nuntalid, G. Indiveri, N. Kasabov, Online spatio-temporal pattern recognition with evolving spiking neural networks utilising address event representation, rank order, and temporal spike learning, in *The 2012 International Joint Conference on Neural Networks (IJCNN)*, pp. 1–7. https://doi.org/10.1109/ijcnn.2012.6252439
15. S. Thorpe, A. Delorme, R. Van Rullen, Spike-based strategies for rapid processing. Neural Netw. Official J. Int. Neural Netw. Soc. **14**(6–7), 715–25 (2001). Retrieved from http://www. ncbi.nlm.nih.gov/pubmed/11665765
16. S.G. Wysoski, L. Benuskova, N. Kasabov, *On-Line Learning with Structural Adaptation in a Network of Spiking Neurons for Visual Pattern*, in Proceedings of International Conference on Artificial Neural Networks (Springer, Heidelberg, 2006), pp. 61–70

17. H.N. Abdul Hamed, N. Kasabov, Z. Michlovský, S.M. Shamsuddin, *String Pattern Recognition Using Evolving Spiking Neural Networks and Quantum Inspired Particle Swarm Optimization*, in Proceedings of International Conference on Neural Information Processing (Springer, Berlin, 2009), pp. 611–619

18. S.G. Wysoski, L. Benuskova, N. Kasabov, Fast and adaptive network of spiking neurons for multi-view visual pattern recognition. Neurocomputing **71**(13–15), 2563–2575 (2008)

19. S.G. Wysoski, L. Benuskova, N. Kasabov, *Spiking Neural Networks for Text-Independent Speaker Authentication, in Artificial Neural Networks–ICANN*, vol. 2 (Springer, Berlin, 2007), pp. 758–767

20. S.M. Bothe, H. La Poutré, J.N. Kok, Unsupervised clustering with spiking neurons by sparse temporal coding and multilayer RBF networks. IEEE Trans. Neural Netw. **13**(2002), 426–435 (2002)

21. J. Gautrais, S. Thorpe, Rate coding versus temporal order coding: a theoretical approach. BioSystems **48**(1998), 57–65 (1998)

22. B. Schrauwen, J. Van Campenhout, BSA, a fast and accurate spike train encoding scheme, in *Proceedings of the International Joint Conference on Neural Networks*, IEEE, vol. 4, pp. 2825–2830 (2003)

23. N. Kasabov, N. Scott, E. Tu, S. Marks, N. Sengupta, E. Capecci, M. Othman, M. Doborjeh, N. Murli, R. Hartono, J. Espinosa-Ramos, L. Zhou, F. Alvi, G. Wang, D. Taylor, V. Feigin, S. Gulyaev, M. Mahmoudh, Z.-G. Hou, J. Yang, Design methodology and selected applications of evolving spatio-temporal data machines in the NeuCube neuromorphic framework. Neural Netw. **78**, 1–14 (2016). http://dx.doi.org/10.1016/j.neunet.2015.09.011

24. B. Petro, N. Kasabov, R. Kiss, A methodology for selection and parameter optimisation of spike encoding algorithms, submitted, https://kedri.aut.ac.nz/R-and-D-Systems/neucube (Spiker)

25. A.L. Hodgkin, A.F. Huxley, A quantitative description of membrane current and its application to conduction and excitation in nerve. J. Physiol. **117**, 500–544 (1952)

26. M. Nelson, J. Rinzel, The Hodgkin-Huxley model. in *The book of Genesis*, ed. by J. M. Bower, D. Beeman (Springer, New York, 1995), pp. 27–51

27. C. Meunier, I. Segev, Playing the Devil's advocate : is the Hodgkin—Huxley model useful? Trends Neurosci. **25**(11), 558–563 (2002)

28. A.F. Strassberg, L.J DeFelice, Limitation of the Hodgkin-Huxley formalism: effects of single channel kinetics on transmembrane voltage dynamics. Neural Comput. **5**(6), 843–855 (1993) (MIT Press)

29. F. Bezanilla, C.M. Armstrong, Inactivation of the sodium channel. I. Sodium current experiments. J. General Physiol. **40**(5), 549–566 (1997)

30. L. Lapicque, Recherches quantitatives sur l'excitation electrique des nerfs traitee comme une polarization. Physiol. Pathol. Gen. **9**(1), 620–635 (1907). https://doi.org/10.1007/s00422-007-0189-6

31. E.M. Izhikevich, Simple model of spiking neurons. IEEE Trans. Neural Netw. **14**(2003), 1569–1572 (2003)

32. W. Gerstner, J.L. van Hemmen, Associative memory in a network of "spiking" neurons. Netw. Comput. Neural Syst. **3**(2), 139–164 (1992)

33. W. Gerstner, Spiking Neurons, in *Pulsed Neural Networks*, ed. by W. Maass, C.M. Bishop (MIT Press, Cambridge, 1998), pp. 3–54

34. A. Delorme, S.J. Thorpe, SpikeNET: an event-driven simulation package for modelling large networks of spiking neurons. Netw. Comput. Neural Syst. **14**(2003), 613–627 (2003)

35. N. Kasabov, To spike or not to spike: a probabilistic spiking neural model. Neural Netw. **23**(1), 16–19 (2010)

36. L. Benuskova, N. Kasabov, *Computational Neurogenetic Modelling* (Springer, New York, 2007)

37. N. Kasabov, N.R. Schliebs, H. Kojima, Probabilistic computational neurogenetic framework: from modelling cognitive systems to Alzheimer's disease. IEEE Trans. Auton. Mental Dev. **3**(4), 300–311 (2011)

38. N. Kasabov (ed.), Springer Handbook of Bio-/Neuroinformatics (Springer, Berlin, 2014)
39. S.M. Bohte, The evidence for neural information processing with precise spike-times: a survey. Nat. Comput. **3**(2), 195–206 (2004). https://doi.org/10.1023/b:naco.0000027755. 02868.60
40. A. Kasinski, F. Ponulak, Comparison of supervised learning methods for spike time. Int. J. Appl. Math. Comput. Sci. **16**(1), 101–113 (2006)
41. D.E. Rumelhart, G.E. Hinton, R.J. Williams, Learning representations by back-propagating errors. Nature 323(6088), 533–536 (1986)
42. S.M. Bohte, J.N. Kok, H. La Poutre, SpikeProp : Backpropagation for Networks of Spiking Neurons Error-Backpropagation in a Network of Spiking Neurons. ESANN (2000), pp. 419–424
43. C.C. Bell, V.Z. Han, Y. Sugawara, K. Grant, Synaptic plasticity in a cerebellum-like structure depends on temporal order. Nature **387**(1997), 278–281 (1997). https://doi.org/10.1038/387278a0
44. G. Bi, M. Poo, Synaptic modifications in cultured hippocampal neurons : dependence on spike timing, synaptic strength, and postsynaptic cell type. J. Neurosci. **18**(24), 10464–10472 (1998)
45. H. Markram, J. Lubke, M. Frotscher, B. Sakmann, Regulation of synaptic efficacy by coincidence of postsynaptic APs and EPSPs. Science **275**(January), 213–215 (1997)
46. T. Masquelier, R. Guyonneau, S.J. Thorpe, Spike timing dependent plasticity finds the start of repeating patterns in continuous spike trains. PLoS ONE **3**(1), e1377 (2008)
47. S. Fusi, M. Annunziato, D. Badoni, A. Salamon, D.J. Amit, Spike-driven synaptic plasticity: theory, simulation, VLSI implementation. Neural Comput. **12**(10), 2227–2258 (1999). https://doi.org/10.1162/089976600300014917
48. N. Kasabov, K. Dhoble, N. Nuntalid, G. Indiveri, Dynamic evolving spiking neural networks for on-line spatio- and spectro-temporal pattern recognition. Neural Netw. Official J. Int. Neural Netw. Soc. **41**(1995), 188–201 (2013). https://doi.org/10.1016/j.neunet.2012.11.014
49. S. Mitra, S. Fusi, G. Indiveri, Real-time classification of complex patterns using spike-based learning in neuromorphic VLSI. IEEE Trans. Biomed. Circuits Syst. **3**(1), 32–42 (2009). https://doi.org/10.1109/tbcas.2008.2005781
50. M. Tsodyks, K. Pawelzik, H. Markram, Neural networks with dynamic synapses. Neural Comput. **10**(4), 821–835 (1998)
51. W. Maass, E.D. Sontag, Neural systems as nonlinear filters. Neural Comput. **12**(8), 1743–1772 (2000)
52. W. Maass, T. Natschlager, H. Markram, Real-time computing without stable states: a new framework for neural computation based on perturbations. Neural Comput. **14**(11), 2531–2560 (2002)
53. T. Natschläger, W. Maass, Spiking neurons and the induction of finite state machines. Theor. Comput. Sci. Nat. Comput. **287**(1), pp. 251–265 (2002)
54. H. Namarvar, J.-S. Liaw, T. Berger, A new dynamic synapse neural network for speech recognition, in *2001 Proceedings of the International Joint Conference on Neural Networks, IJCNN '01* (2001)
55. N. Mehrtash, D. Jung, H. Klar, Image pre-processing with dynamic synapses. Neural Comput. Appl. **12**(33–41), 2003 (2003). https://doi.org/10.1007/s00521-030-0371-2
56. A. Mohemmed, S. Schliebs, S. Matsuda, N. Kasabov, Training spiking neural networks to associate spatio-temporal input-output spike patterns. Neurocomputing **107**, 3–10 (2013). https://doi.org/10.1016/j.neucom.2012.08.034
57. A. Mohemmed, S. Schliebs, S. Matsuda, N. Kasabov, SPAN: spike pattern association neuron for learning spatio-temporal sequences. Int. J. Neural Syst. **22**(4), 1–16 (2012)
58. A. Mohemmed, S. Schliebs, S. Matsuda, K. Dhoblea, N. Kasabov (2011), Optimization of spiking neural networks with dynamic synapses for spike sequence generation using PSO, in *International Joint Conference on Neural Networks*. IEEE Publishing, San Jose, California, USA (2011) (In Print)

59. A. Mohemmed, S. Schliebs, N. Kasabov, Method for training a spiking neuron to associate input output spike trains. In Engineering Applications of Neural Networks. Springer, Corfu, Greece (2011). (in Print)
60. M.C. van Rossum, A novel spike distance. Neural Comput. **13**(4), 751–763 (2001)
61. F. Ponulak, ReSuMe—new supervised learning method for spiking neural networks. Tech. report, Institute of Control and Information Engineering, Poznań University of Technology, Poznań, Poland (2005)
62. R.V. Florian, The chronotron: a neuron that learns to fire temporally-precise spike patterns. http://precedings.nature.com/documents/5190/version/1 (2010)
63. W. Maass, T. Natschläger, H. Markram, Realtime computing without stable states: a new framework for neural computation based on perturbations. Neural Comput. **14**(11), 2531–2560 (2002)
64. E. Nordlie, M.-O. Gewaltig, H.E. Plesser, Towards reproducible descriptions of neuronal network models. PLoS Comput. Biol. **5**(8), e1000456 (2009)
65. M.-O. Gewaltig, M. Diesmann, Nest (neural simulation tool). Scholarpedia **2**(4), 1430 (2007)
66. R. Gutig, H. Sompolinsky, The tempotron: a neuron that learns spike timing-based decisions. Nat. Neurosci. **9**(3), 420–428 (2006)
67. F. Ponulak, A. Kasiński, Supervised learning in spiking neural networks with ReSuMe: sequence learning, classification, and spike shifting. Neural Comput. **22**(2), 467–510 (2010). PMID: 19842989
68. B. Schrauwen, J.V. Campenhout, Linking nonbinned spike train kernels to several existing spike train metrics. Neurocomputing **70**(2007), 1247–1253 (2007)
69. J.D. Victor, K.P. Purpura, Metric-space analysis of spike trains: theory, algorithms and application. Netw. Comput. Neural Syst. **8**(2), 127–164 (1997)
70. N. Sengupta, Neuromorphic computational models for machine learning and pattern recognition from multi-modal time series data, PhD Thesis, Auckland University of Technology (2018)

# Chapter 5
# Evolving Spiking Neural Networks

Evolving SNN (eSNN) are a class of SNN and also a class of ECOS (Chap. 2) where spiking neurons are created (evolved) and merged in an incremental way to capture clusters and patterns from incoming data. This gives a new quality of the SNN systems to become adaptive, fast trained and to capture meaningful patterns from the data, turned into new knowledge, departing the "curse of the black box neural networks' and the "curse of catastrophic forgetting" as manifested by some traditional ANN models (Chap. 2). The inspiration comes from the brain as the brain always evolves its structure and functionality through continuous learning. It is always *evolving* and forming new knowledge.

The chapter is organised in the following sections:

5.1. Principles and methods of eSNN.
5.2. Convolutional eSNN.
5.3. Dynamic eSNN (deSNN).
5.4. Fuzzy rule extraction from eSNN.
5.5. Evolving SNN for reservoir computing.
5.6. Chapter summary and further readings for deeper knowledge.

## 5.1 Principles and Methods of Evolving SNN (ESNN)

The eSNN paradigm applies the ECOS principles (Chap. 2) to process spike-time information, namely:

(1) Fast learning from large amount of data, e.g. using mainly "one-pass" training, starting with little prior knowledge;
(2) Adaptation in real-time and in an on-line mode where new data is accommodated as it comes based on local learning;
(3) "Open", evolving structure, where new input variables (relevant to the task), new outputs (e.g. classes), new connections and neurons are added/evolved "on the fly";

© Springer-Verlag GmbH Germany, part of Springer Nature 2019
N. K. Kasabov, *Time-Space, Spiking Neural Networks and Brain-Inspired Artificial Intelligence*, Springer Series on Bio- and Neurosystems 7, https://doi.org/10.1007/978-3-662-57715-8_5

(4) Both data learning and knowledge representation is facilitated in a comprehensive and flexible way, e.g., supervised learning, unsupervised learning, evolving clustering, "sleep" learning, forgetting/pruning, fuzzy rule insertion and extraction;

(5) Active interaction with other systems and with the environment in a multi-modal fashion;

(6) Representing both space and time in their different scales, e.g., clusters of data, short- and long-term memory, forgetting, etc.;

(7) System's self-evaluation in terms of behavior, global error and success, and related knowledge representation.

There are several models of eSNN models.

The simplest eSNN model uses (see Chap. 4) (but not restricted to these limitations):

– Population spike coding algorithm (POC);
– Leaky integrate-and fire (LIF) model of a neuron;
– Rank-order (RO) learning rule.

This is schematically shown in Fig. 5.1a, b.

The RO learning motivation is based on the assumption that most important information of an input pattern is contained in earlier arriving spikes [1]. It establishes a priority of inputs based on the order of the spike arrival on the input synapses for a particular pattern. This is a phenomenon observed in biological systems as well as an important information processing concept for some spatio-temporal problems, such as computer vision and control. RO learning makes use of the information contained in the order of the input spikes (events). This method has two main advantages when used in SNN: (1) fast learning (as the order of the first incoming spikes is often sufficient information for recognising a pattern and for a fast decision making and only one pass propagation of the input pattern may be sufficient for the model to learn it); (2) asynchronous, data-driven processing. As a consequence, RO learning is most appropriate for AER input data streams as the address-events are conveyed into the SNN 'one by one', in the order of their happening [2, 3]. The RO coding for the eSNN structure from Fig. 5.1 is illustrated in Chap. 4.

An eSNN evolves its structure and functionality in an on-line manner, from incoming information. For every new input data vector, a new output neuron is dynamically allocated and connected to the input neurons (feature neurons). The neuron's connections are established using the RO rule for the output neuron to recognise this vector (frame, static pattern) or a similar one as a positive example. The weight vectors of the output neurons represent centres of clusters in the problem space and can be represented as fuzzy rules [4].

In some implementations neurons with similar weight vectors are merged based on Euclidean distance between them. That makes it possible to achieve a very fast learning (only one pass may be sufficient), both in a supervised and in an unsupervised mode [5]. When in an unsupervised mode, the evolved neurons represent a

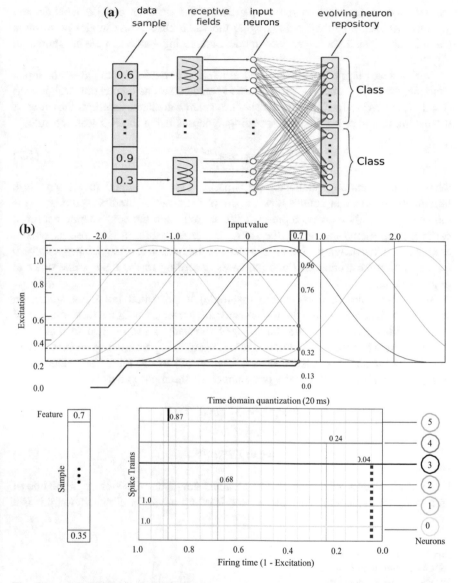

**Fig. 5.1 a** A principle diagram of the structure of an evolving SNN (eSNN) where the input variables are encoded into spike sequences though population coding (POC) and the output neuronal layer evolves with the presentation of every new data input vector, allowing also for merging of output neurons. **b** Example of how a random continuous value is encoded into spike trains using POC for a number of input neurons. The value of 0.7 activates 6 overlapping receptive fields, which excite 6 corresponding input neurons at different firing times

learned pattern (or a prototype of patterns). The neurons can be labelled and grouped according to their belonging to the same class if the model performs a classification task in a supervised mode of learning—an example is shown in Fig. 5.2.

During a *learning phase*, for each M-dimensional training input pattern (sample, example, vector) $P_i$ a new output neuron $i$ is created and its connection weights $w_{j,i}$ ($j = 1, 2, ..., M$) to the input (feature) neurons are calculated based on the order of the incoming spikes on the corresponding synapses using the RO learning rule:

$$w_{j,i} = \alpha \cdot mod^{order(j,i)} \tag{5.1}$$

where: $\alpha$ is a learning parameter (in a partial case it is equal to 1); *mod* is a modulation factor, that defines how important the order of the first spike is; $w_{j,i}$ is the synaptic weight between a pre-synaptic neuron $j$ and the postsynaptic neuron $i$; *order(j,i)* represents the order (the rank) of the *first* spike at synapse $j,i$ ranked among all spikes arriving from all synapses to the neuron $i$; *order(j,i)* has a value 0 for the first spike to neuron $i$ and increases according to the input spike order at other synapses.

While the input training pattern (example) is presented (all input spikes on different synapses, encoding the input vector are presented within a time window of T time units), the spiking threshold $Th_i$ of the neuron $i$ is defined to make this neuron spike when this or a similar pattern (example) is presented again in the recall mode. The threshold is calculated as a fraction (C) of the total $PSP_i$ (denoted as $PSP_{imax}$) accumulated during the presentation of the input pattern:

$$PSP_i^{max} = \sum_j mod^{order(j,i)} \tag{5.2}$$

$$\Theta = CPSP_i^{max} \tag{5.3}$$

If the weight vector of the evolved and trained new neuron is similar to the one of an already trained neuron (in a supervised learning mode for classification this is a

**Fig. 5.2** An example of an eSNN structure with n input variables and k output classes

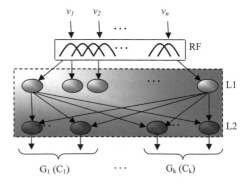

neuron from the same class pool), i.e. their similarity is above a certain threshold *Sim*, the new neuron will be merged with the most similar one, averaging the connection weights and the threshold of the two neurons [5, 6]. Otherwise, the new neuron will be added to the set of output neurons (or the corresponding class pool of neurons when a supervised learning for classification is performed). The similarity between the newly created neuron and a training neuron is computed as the inverse of the Euclidean distance between weight matrices of the two neurons. The merged neuron has weighted average weights and thresholds of the merging neurons.

While an individual output neuron represents a single input pattern, merged neurons represent clusters of patterns or prototypes in a transformed spatial—RO time-space. These clusters can be represented as fuzzy rules [4] that can be used to discover new knowledge about the problem under consideration.

The eSNN learning is adaptive, incremental, theoretically—'lifelong', so that the system can learn new patterns through creating new output neurons, connecting them to the input neurons, and possibly merging the most similar ones, following the ECOS principles from Chap. 2.

During the *recall phase*, when a new input vector is presented and encoded as input spikes, the spiking pattern is submitted to all created neurons during the learning phase. An output spike is generated by neuron $i$ at a time $l$ if the $PSP_i(l)$ becomes higher than its threshold $\Theta$. After the first neuron spikes, the PSP of all neurons are set to initial value (e.g. 0) to prepare the system for the next pattern for recall or learning.

The postsynaptic potential $PSP_i(l)$ of a neuron i at time $l$ is calculated as:

$$PSP_i(l) = \sum_{t=0,1,2,\dots,l} \sum_j e_j(t) \cdot \text{mod}^{order(j,i)} \qquad (5.4)$$

where: $e_j(t) = 1$ if there is a *first* spike at time $t$ on synapse j; *order* $(j,i)$ is the rank order of the first spike at synapse $j$ among all spikes to neuron $i$ for this recall pattern.

The parameter $C$, used to calculate the threshold of a neuron $i$, makes it possible for the neuron $i$ to emit an early output spike before the presentation of the whole learned pattern (lasting T time units) as the neuron was initially trained to respond. As a partial case $C = 1$.

During training of an eSNN, for each input vector the following steps are performed:

(a) Create (evolve) a new output spiking neuron and its connections;
(b) Propagate the input vector into the network calculating the PSP of the output neurons:

$$u_i(t) = \begin{cases} 0 & \text{if fired} \\ \sum\limits_{j|f(j)<t} w_{ji} m_i^{order\,(j)} & \text{else} \end{cases} \tag{5.5}$$

(c) Train the newly created neuron weights using RO learning based on the spike time arrival.
(d) IF similarity between a new and old neurons is greater than a Threshold THEN merge neurons, where N is the number of samples previously used to update the respective neuron.
(e) Update the corresponding threshold $\vartheta$ of the merged neuron.

The eSNN training algorithm is given in Box 5.1.

---

**Box 5.1.  The eSNN training algorithm**

1:    Initialize output neuron repository, $R = \{\}$

2:    Set eSNN parameters: $mod = [0,1], C = [0,1], sim = [0,1]$

3:    for $\forall$ input pattern $i$ that belongs to the same class do

4:    Encode input pattern into firing time of multiple pre-synaptic neurons $j$

5:    Create a new output neuron $i$ for this class and calculate the connection weights as $w_{ji} = mod^{order(j)}$

6:    Calculate $PSP_{max(i)} = Sum_j w_{ji} \times mod^{order(j)}$

7:    Calculate $PSP$ threshold value $\gamma_i = PSP_{max(i)} \times C$

8:    if The new neuron weight vector $\leq sim$ of trained output neuron weight vector in $R$ then

9:    Update the weight vector and threshold of the most similar neuron in the same output class group

10:   $w = \frac{w_{new} + w.N}{N+1}$

11:   $\gamma = \frac{\gamma_{new} + \gamma N}{N+1}$ where $N$ is the number of previous merges of the most similar neuron

13:   else

14:   Add the weight vector and threshold of the new neuron to the neuron repository $R$

15:   end if

16:   end for

17:   Repeat above for all input patterns of other output classes

---

The recall procedure can be performed using different recall algorithms implying different methods of comparing input patterns for recall with already learned patterns in the output neurons:

(a) The first one is described above. Spikes of the new input pattern are propagated as they arrive to all trained output neurons and the first one that spikes (its PSP is greater that its threshold) defines the output. The assumption is that the neuron (or several neurons in a k-nearest neighbor fashion) that best matches the input pattern will spike earlier based purely on the PSP (membrane potential). This type of eSNN is denoted as eSNNm.

(b) The second one implies a creation of a new output neuron for each recall pattern, in the same way as the output neurons were created during the learning phase, and then—comparing the connection weight vector of the new one to the already existing neurons using Euclidean distance. The closest output neuron (or neurons in the k-nearest neighbor fashion) in terms of synaptic connection weights is the 'winner'. This method uses the principle of transductive reasoning and nearest neighbour classification in the connection weight space. It compares spatially distributed synaptic weight vectors of a new neuron that captures a new input pattern and existing ones. We will denote this model as eSNNs.

The main advantage of the eSNN when compared with other supervised or unsupervised SNN models are:

- It is computationally inexpensive;
- It boosts the importance of the order in which input spikes arrive, thus making the eSNN suitable for a range of applications.
- It is one-pass, on-line learning method, where new data can be learned incrementally in a "life-long:" learning method, involving the merging/aggregating output neurons;
- It is knowledge-based, where the output neurons (after aggregation) represent prototypes in the data or cluster centres [6, 7];
- Allow for fuzzy rule extract ion as presented in a later section.

For a comprehensive study of eSNN see [5] and for a comprehensive review—[8].

The problem of the eSNN is that once a synaptic weight is calculated based on the first spike using the RO rule, it is fixed and does not change to reflect on other incoming spikes at the same synapse, i.e. there is no mechanism to deal with multiple spikes arriving at different times on the same synapse. The synapses are static. While the synapses capture some (long term) memory during the learning phase, they have limited abilities (only through the PSP growth) to capture short term memory during a whole spatio-temporal pattern presentation. Learning and recall of complex spatio-temporal patterns in an on-line mode would need not only fast initial set of connection weights, based on the first spikes, but also dynamic changes of these synapses during the pattern presentation.

An example of an eSNN is shown in Fig. 5.2 [5, 6, 8].

## 5.2 Convolutional ESNN (CeSNN)

Here, eSNN are used to build convolutional eSNN (CeSNN) [6]. The principles of convolutional NN was presented and illustrated in Chap. 2. Here CNN is created using the eSNN model and illustrated on image recognition problem as shown in Fig. 5.3.

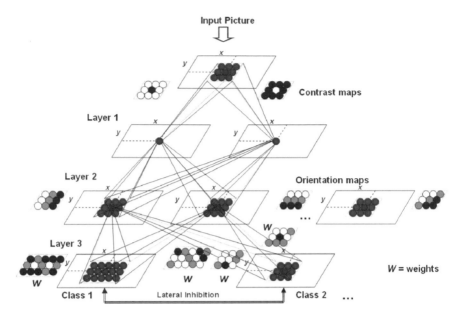

**Fig. 5.3** A convolutional eSNN for image pattern recognition [6]

The neural network is composed of 3 layers of integrate-and-fire neurons. The neurons have a latency of firing that depends upon the order of spikes received. Each neuron acts as a coincidence detection unit, where the postsynaptic potential for neuron $Ni$ at a time $t$ is calculated as:

$$PSP(i,t) = \sum \text{mod}^{order(j)} w_{j,i} \tag{5.6}$$

where mod $\in (0, 1)$ is the modulation factor, $j$ is the index for the incoming connection and $w_{j,i}$ is the corresponding synaptic weight.

Each layer is composed of neurons that are grouped in two-dimensional grids forming neuronal maps. Connections between layers are purely feed-forward and each neuron can spike at most once on a spike arrival in the input synapses. The first layer cells represent the ON and OFF cells of retina, basically enhancing the high contrast parts of a given image (high pass filter). The output values of the first layer are encoded to pulses in the time domain. High output values of the first layer are encoded as pulses with short time delays while long delays are given low output values. This technique called Rank Order Coding, already presented in a previous section and in Chap. 2 [9] here basically prioritizes the pixels with high contrast that consequently are processed first and have a higher impact on neurons' PSP.

Second layer is composed of eight orientation maps, each one selective to a different direction (0°, 45°, 90°, 135°, 180°, 225°, 270°, and 315°). It is important to notice that in the first two layers there is no learning, in such a way that the structure can be considered simply passive filters and time domain encoders (layers 1 and 2).

The theory of contrast cells and direction selective cells was first reported by Hubel and Wiesel. In their experiments they were able to distinguish some types of cells that have different neurobiological responses according to the pattern of light stimulus.

The third layer is where the learning takes place and where the main contribution of this work is presented. Maps in the third layer are to be trained to represent classes of inputs. In [9] the learning is performed off-line using the rule:

$$\Delta w_{j,i} = \frac{\text{mod}^{order(a_j)}}{N} \tag{5.7}$$

where $w_{j,i}$ is the weight between neuron $j$ of the 2nd layer and neuron $i$ of the 3rd layer, mod $\in (0,1)$ is the modulation factor, $order(a_j)$ is the order of arrival of spike from neuron $j$ to neuron $i$, and $N$ is the number of samples used for training a given class.

In this rule, there are two points to be highlighted: (a) the number of samples to be trained needs to be known a priori; and (b) after training, a map of a class will be selective to the average pattern.

In [6] a new approach is proposed for learning with structural adaptation, aiming to give more flexibility to the system in a scenario where the number of classes and/ or class instances is not known at the time the training starts. Thus, the output neuronal maps need to be created, updated or even deleted on-line, as the learning occurs. To implement such a system the learning rule needs to be independent of the total number of samples since the number of samples is not known when the learning starts.

The entire training procedure for the CeSNN from Fig. 5.3 follows steps described next and summarized in the flowchart of Fig. 5.4.

1. Propagate a sample $k$ of class $K$ for training into the layer 1 (retina) and layer 2 (direction selective cells—DSC);
2. Create a new map $Map_{C(k)}$ in layer 3 for sample $k$ and train the weights using the equation:

$$\Delta w_{j,i} = \text{mod}^{order(a_j)} \tag{5.8}$$

where $w_{j,i}$ is the weight between neuron $j$ of the layer 2 and neuron $i$ of the layer 3, mod $\in (0, 1)$ is the modulation factor, $order(a_j)$ is the order of arrival of spike from neuron $j$ to neuron $i$.

The postsynaptic threshold ($PSP_{threshold}$) of the neurons in the map is calculated as a proportion $c \in [0,1]$ of the maximum postsynaptic potential ($PSP$) created in a neuron of map $Map_{C(k)}$ with the propagation of the training sample into the updated weights, such that:

**Fig. 5.4** The training procedure for the CeSNN from Fig. 5.3

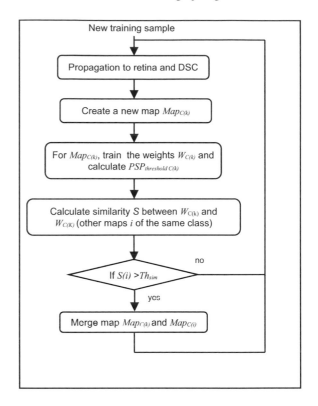

$$PSP_{threshold} = c \max(PSP) \qquad (5.9)$$

The constant of proportionality $c$ express how similar a pattern needs to be to trigger an output spike. Thus, $c$ is a parameter to be optimized in order to satisfy the requirements in terms of false acceptance rate (FAR) and false rejection rate (FRR).

3. Calculate the similarity between the newly created map $Map_{C(k)}$ and other maps belonging to the same class $Map_{C(K)}$. The similarity is computed as the inverse of the Euclidean distance between weight matrices.

4. If one of the existing maps for class $K$ has similarity greater than a chosen threshold $Th_{simC(K)} > 0$, merge the maps $Map_{C(k)}$ and $Map_{C(Ksimilar)}$ using arithmetic average as expressed in equation

$$W = \frac{W_{Map_{C(k)}} + N_{samples} W_{Map_{C(Ksimilar)}}}{1 + N_{samples}} \qquad (5.10)$$

where matrix W represents the weights of the merged map and $N_{samples}$ denotes the number of samples that have already being used to train the respective map. In similar fashion the $PSP_{threshold}$ is updated:

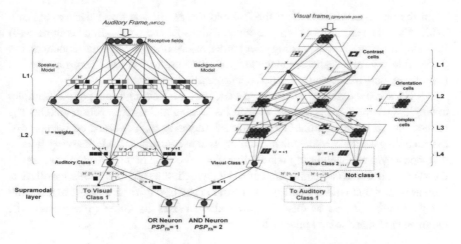

**Fig. 5.5** CeSNN for person identification using both speech and face image input data [6]

$$PSP_{threshold} = \frac{PSP_{MapC(k)} + N_{samples}PSP_{MapC(Ksimilar)}}{1 + N_{samples}} \qquad (5.11)$$

The convolutional eSNN first extract features from the input data through the convolutional layers. When input vectors are presented one after another, spikes are accumulated over time in the membrane potentials of the output neurons until an output neuron spikes identifying the class output. Then the membrane potentials are set to resting potential for the next input vectors to be received and classified. The convolutional eSNN for person authentication based on both speech and face data presented in Fig. 5.5 is described in details in Chap. 12 [6]. The CeSNN method presented here supports the creation of SNN that process spatio-temporal data. Space is represented as Maps and time is represented in the rate of the increase of the membrane potentials of the neurons in the last layer, as short term memory, before the neurons spike.

## 5.3 Dynamic Evolving SNN (DeSNN)

This method is first presented in [10]. The main disadvantage of the RO learning in eSNN is that the model adjusts the connection weight of each synapse once only (based on the rank of the first spike on this synapse), which may be appropriate for vector-based pattern recognition, but would not be efficient for complex TSTD (temporal or spatio-temporal data) where samples are not vectors, but whole temporal or spatio-temporal sequences/patterns of data. In the latter case the connection weights need to be further tuned based on the following spikes arriving on the same synapse over time as part of the whole input pattern and that is where the spike-time learning (e.g. STDP or SDSP) can be employed in order to implement dynamic synapses.

In the proposed deSNN both the RO and the SDSP learning rules are utilised. While the RO learning will set the initial values of the connection weights **w(0)** (utilising for example the existing event order information in the data sequence), the SDSP rule will adjust these connection weights based on further incoming spikes (events) as part of the same learned spatio-temporal pattern.

As in the eSNN, during a *learning phase*, for each training input pattern (sample, example, vector) $P_i$ a new output neuron $i$ is created and its connection weights $w_{j,i}$ to the input (feature) neurons are initially calculated as $w_{j,i}$ (0) based on the order of the incoming spikes on the corresponding synapses using the RO learning rule.

Once a synaptic weight $w_{j,i}$ is initialised, based on the first spike at the synapse j, the synapse becomes dynamic and adjusts its weight through the SDSP algorithm. It increases its value with a small positive value (positive drift parameter) at any time $t$ a new spike arrives at this synapse and decreases its value (a negative drift parameter) if there is no spike at this time.

$$\Delta w_{j,i}(t) = e_j(t) \cdot D \tag{5.12}$$

where: $e_j(t) = 1$ if there is a consecutive spike at synapse j at time t during the presentation of the learned pattern by the output neuron $i$ and $(-1)$ otherwise. In general, the drift parameter $D$ can be different for 'up' and 'down' drifts.

All dynamic synapses change their values in parallel for every time unit $t$ during a presentation of an input spatio-temporal pattern $P_i$ learned by an output neuron i, some of them going up and some—going down, so that all synapses (not a single one) of the neuron could collectively capture some temporal relationship of spike timing across the learned pattern.

While an input training pattern (example) is presented (all input spikes on different synapses, encoding the input vector are presented within a time window of T time units), the spiking threshold $Th_i$ of the neuron $i$ is defined to make this neuron spike when this or a similar pattern (example) is presented in the recall mode. The threshold is calculated as a fraction (C) of the total $PSP_i$ (denoted as $PSP_{imax}$) accumulated during the presentation of the whole input pattern:

$$PSP_{i\,max} = \sum_{t=1,2,\dots,T} \sum_{j=1,2,\dots,M} f_j(t) \cdot w_{j,i}(t) \tag{5.13}$$

$$Th_i = C.PSP_{i\,max} \tag{5.14}$$

*where:* T represents the time units in which the input pattern is presented; M is the number of the input synapses to neuron i; $f_j$ (t) = 1 if there is spike at time $t$ at synapse j for this learned input pattern, otherwise it is 0; $w_{j,i}$ (t) is the efficacy of the (dynamic) synapse between j and i neurons calculated at time t.

The resulted deSNN model after training will contain the following information:

- Number of input neurons M and output neurons N;
- Initial $\mathbf{w_i}$ (0) and final $\mathbf{w_i}$(T) vectors of connection weights and spiking threshold $Th_i$ for each of the output neurons i. The pairs [$\mathbf{w_i}$ (0), $\mathbf{w_i}$ (T)], i = 1,2, …,N would

capture collectively dynamics of the learning process for each spatio-temporal pattern and each output neuron (As a partial case only initial or final values of the connection weights can be considered or a weighted sum of them).
– deSNN parameters.

The overall deSNN training algorithm is presented in Box 5.2.

---

**Box 5.2. The deSNN Training Algorithm**

1  Set deSNN parameters* (including: Mod, C, Sim, and the SDSP parameters)

2  **For** every input spatio-temporal spiking pattern $P_i$ **Do**

   2a. Create a new output neuron i for this pattern and calculate the initial values of connection weights $w_i$ (0) using the RO learning formula (5.1).

   2b. Adjust the connection weights $w_i$ for consecutive spikes on the corresponding synapses using the SDSP learning rule (formula (5.8)).

   2c. Calculate $PSP_{imax}$ using formula (5.9).

   2d. Calculate the spiking threshold of the $i^{th}$ neuron using formula (5.10).

   2e. **(Optional) If** the new neuron weight vector $w_i$ is similar in its initial $w_i(0)$ an final $w_i(T)$ values after training to the weight vector of an already trained output neuron using Euclidean distance and a similarity threshold *Sim*, then merge the two neurons (as a partial case only initial or final values of the connection weights can be considered or a weighted sum of them)

   **Else**

   Add the new neuron to the output neurons repository.

   **End If**

   **End For** (Repeat for all input spatio-temporal patterns for learning)

   *: The performance of the deSNN depends on the optimal selection of its parameters as illustrated in the examples below.

---

Figure 5.6 illustrates the main idea of the deSNN learning algorithm. A single spatio-temporal pattern of four input spike trains is learned into a single output neuron. RO learning is applied to calculate the initial weights based on the order of the first spike on each synapse (shown in red). Then STDP (in this case—SDSP) rule is applied to dynamically tune these connection weights. The SDSP algorithm increases the assigned connection weight of a synapse which is receiving a following spike and at the same time depresses the synaptic connections of synapses that do not receive a spike at this time. Due to a bi-stability drift in the SDSP rule, once a weight reaches the defined High value (resulting in LTP) or Low value (resulting in LTD), this connection weight is fixed to this value for the rest of the training phase. The rate at which a weight reaches LTD or LTP depends upon the set parameter values.

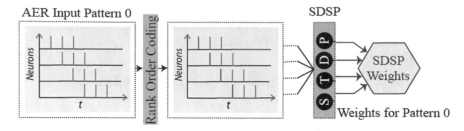

**Fig. 5.6** A simple example to illustrate the main principle of the deSNN learning algorithm [10]

For example, if input spikes arrive at times (0, 1, 2) ms on the first synapse, and are shifted by 1 ms for the other 3 synapses as shown in Fig. 5.6, the four initial connection weights $w_1, w_2, w_3, w_4$ to the output neuron will be calculated as: 1, 0.8, 0.64, 0.512 correspondingly, when the parameter mod = 0.8. If the SDSP High value is 0.6 and Low value is 0, the first three weights will be fixed to 0.6 and the fourth one will drift up 2 times. If the drift parameter is set to 0.00025, the final weight value of the fourth synapse will be 0.5125. After training both the initial and the final weights can be memorised.

So far, we have presented the learning phase of a deSNN model. In terms of *recall*, two types of deSNN are proposed that differ in the recall algorithms. They mainly correspond to the two types of eSNN from Sect. 5.2—eSNNs and eSNNm:

(a) deSNNm: After learning, only the initially created connection weights (with the use of the RO rule) are restored as long term memory in the synapses and the model. During recall on a new spatio-temporal pattern the SDSP rule is applied so that the initial synaptic weights are modified on a spike time basis according to the new pattern using formula (50) as it is during the SDSP learning phase. At every time moment t the PSP of all output neurons are calculated. The new input pattern is associated with the neuron $i$ if the $PSP_i(t)$ is above its threshold $Th_i$. The following formula is used:

$$PSP_i(t) = \sum_{l=1,2,...,t} \sum_{j=1,2,...,M} f_j(l).w_{j,i}(l) \qquad (5.15)$$

*where:* $t$ represents the current time unit during the presentation of the input pattern for recall; $M$ is the number of the input synapses to neuron $i$; $f_j(l) = 1$ if there is spike at time $l$ at synapse $j$ for this input pattern, otherwise it is 0; $w_{j,i}(l)$ is the efficacy of the dynamic synapse between j and i neurons at time $l$.

(b) deSNNs: This model corresponds to the eSNNs and is based on the comparison between the synaptic weights of a newly created neuron to represent the new spatio-temporal pattern for recall, and the connection weights of the created during training neurons. The new input pattern is associated with the closest output neuron based on the minimum distance between the weight vectors. As the synaptic weights are dynamic, the distance should be calculated in a different way than the distance measured in the eSNN possibly using both the initial $\mathbf{w}(0)$ and the final $\mathbf{w}(T)$ connection weigh vectors learned during training and recall. As a partial case, only the final weight vector $\mathbf{w}(T)$ can be used.

To summarise, deSNN combine:

- RO learning for weight initialisation based on the first spikes:
- Learning further input spikes at a synapse through a drift.

$$wj, i(t) = ej(t) \cdot \text{Drift} \qquad (5.16)$$

- A new output neuron may be added to a respective output repository for every new-input pattern.
- Two types of output neuron activation:

  - deSNNm (spiking is based on the membrane potential)
  - deSNNs (spiking is based on synaptic similarity between the newly created output neuron and the existing ones)

- Neurons may merge.

## 5.4  Fuzzy Rule Extraction from ESNN

One of the challenge with SNN is the difficulty to make sense out of the learned connections as they evolve from sequences of spikes. What do actually connections represent? Do they represent any meaningful patterns that can be interpreted for a better understanding of the data and the processes that generated it?

This section addresses this challenge by introducing a method for fuzzy rule extraction from eSNN.

### 5.4.1  Fuzzy Rule Extraction from ESNN

The rationale behind the presented here method for fuzzy rule extraction from eSNN is that input information is represented as spikes in time that account for the intensity of the input variables through population spike coding algorithm POC. The higher the membership degree of an input variable to a neuronal receptive filed in layer L1 of neurons (see Fig. 5.2), the earlier a spike is generated. The connection weights learned from these spikes in the L2 layer using ROC increase more the earlier spikes are generated in layer L1. So, in a sense the connection weights from layer L1 to L2 reflect on the values of the input variables and their membership degrees to the different receptive fields that can serve as fuzzy membership functions (see Chap. 2). This approach is presented here and illustrated on a simple example.

The eSNN structure used for the presentation of the method is shown in Fig. 5.7 illustrated only on one input variable. Each input variable $v_i$ is translated into trains of spikes and distributed to L1 neurons through delayed synaptic connections. The L2 neurons are created using an evolving learning algorithm and one-pass data propagation.

In this network, the input values of each input variable $v_i$ are encoded using a family of $m$ equally spaced overlapping Gaussian receptive fields [11, 12] (RF) and distributed to multiple L1 neurons through $m$ delayed synaptic connections. There is no delay at the centers of the Gaussian receptive fields and this delay increases towards the receptive field edges. Figure 5.8 shows an example of population encoding of two input variables $v_k$ and $v_l$, where $v_k < v_l$. Each input variable is encoded into a six-dimensional vector of spike times. The most stimulated L1 neuron fires first ($t_f = 0$) and the least stimulated neuron fires last ($t_f = t_{max}$). The values are deliberately chosen so that they cause a maximum excitation of two different L1 neurons, i.e. two earliest spikes appear at the $L1_{k3}$ and $L1_{l4}$ terminals at the same time, and two very similar spike patterns occur out of sequence by one L1 neuron. There is a relationship between which L1 neuron spikes first in a set of L1 neurons and the value encoded into this set of L1 neurons; when smaller values are encoded, the first neuron to spike tends towards the lower end of the observed set ($L1_{k1}$ and $L1_{l1}$ in this example) and when higher values are encoded the first neuron to spike tends towards the higher end of the observed set ($L1_{k6}$ and $L1_{l6}$ in this example).

The spikes are propagated from the excitatory L1 neurons to the computationally simple L2 integrate-and-fire neurons in a feed-forward manner. Each L2 neuron is a coincidence detection unit allowed to emit only one spike.

Each incoming spike from an L1 neuron influences the behaviour of the L2 neurons, changing their inner state (post synaptic potential or *PSP*). *PSP* of an L2 at time $t$ depends on the firing order $o_j$ of all its pre-synaptic neurons $L1_j$ as follows:

$$PSP_i(t) = \sum_j PSP_{ji} = \sum_j w_{ji} \times \text{mod}^{o_j} \qquad (5.17)$$

**Fig. 5.7** A simple structure of eSNN using only one input variable for the illustration of the method of fuzzy rule extraction

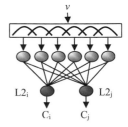

**Fig. 5.8** An example of population encoding of two input variables $v_k$ and $v_l$ ($v_k < v_l$) using six receptive fields ($m = 6$)

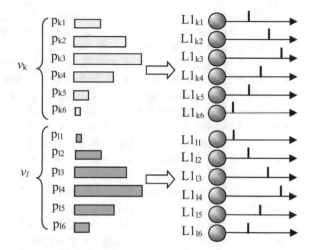

where $PSP_{ji}$ is the post-synaptic potential of neuron i, mod $\in (0, 1)$ is the modulation factor and $w_{ji}$ is the efficacy (weight) of the synapse connecting the L2$_i$ and an L1$_j$ neurons. The firing order $o_j$ depends on the timing of the spike emitted by L1$_j$ relative to the spikes emitted by other L1 neurons. A firing order $o_j = 0$ is assigned if neuron L1$_j$ spikes first amongst all L1 neurons. The L1 neuron that spikes second is assigned $o_j = 1$, and so on. The L1 neuron that spikes last is given $o_j = m - 1$.

An L2 neuron fires a post-synaptic spike when its $PSP$ reaches a certain threshold value $PSP_\theta$. This model does not include the refractory period, therefore, immediately after firing the post-synaptic spike, the neuron's PSP returns to 0.

The synapses between L1 and L2 neurons are dynamic; their values change over the timescale of training. During training, connections conveying earlier spikes are given a higher weight resulting in a greater strengthening of connections:

$$\Delta w_{ji} = \text{mod}^{o_j} \tag{5.18}$$

Hence, the synaptic connections conveying earlier spikes become more effective in causing the L2 neuron to fire in the future.

Each L1 neuron has a single connection to its input and one weighted synaptic connection to each of the L2 neurons denoted as a weight $w_{ji}$. The L2 layer is initially empty and all L2 neurons are built over time from the input data vectors by a fast evolving learning algorithm [5] that allows the learning of new input patterns without forgetting the older knowledge.

During training, a new neuron, L2$_n$, is created for a training pattern p$_i$, trained and its $PSP_{\theta n}$ is set to a proportion of its maximum post synaptic potential value $PSP_{nmax}$. Thus, L2 neurons have dynamic thresholds and their selectiveness can be controlled by adjusting their $PSP_\theta$.

Let the current pattern $p_i$ belong to a class $C_c$. The Euclidean distances between $L2_n$ to all pre-existing $L2_o \in \{L2|C = C_c\}$ are calculated and used as a similarity measure. If the similarity between $L2_n$ and $L2_o$ is below some threshold value $S_\theta$ ($S_\theta > 0$) then $L2_n$ is aggregated with $L2_o$. During aggregation the weights and threshold of $L2_n$ are averaged into the values of $L2_o$, and $L2_n$ is discarded. Thus, the network's structure continuously evolves through the creation and merging of neurons based on incoming data. As the L2 layer is being created clusters of L2 neurons are formed. The L2 neurons from a cluster $G_p$ are trained on the characteristics of only one class (positive examples). Hence, every $L2 \in G_p(C_p)$ learns to respond with earlier spikes when presented with input patterns from $C_p$.

Receptive fields not only increase the temporal distance between the input patterns, which improves the selectivity of the eSNN with rank order population coding, but it also opens the possibilities of using these networks for knowledge discovery. The eSNN builds a knowledge base that can be extracted during and after the learning process. These appear in the form of zero-order Takagi-Sugeno fuzzy rules (see Chap. 2—knowledge based ANN):

$$\text{IF}(v_1 \ is \ F_1) \ \text{AND} \ldots (v_n \ is \ F_n) \ \text{THEN} \ (y \ is \ C_k) \tag{5.19}$$

where $v_i$ is an input variable, $F_i$ is a linguistic value represented by its membership function such as SMALL, MEDIUM, LARGE etc., and $C_k$ represents a class label. The antecedent part in the rules is a composition, using the AND operators, of fuzzy conditions. The consequent part, the output, is a constant. The conditions, $v_i$ is $F_i$, are found through an analysis of the connections $w_{ji}$ between a pair of L1 and L2 neurons.

For example, the following two fuzzy rules can be extracted from a trained eSNN as explained below:

Rule 0:   IF ($v_1$ is SMALL) AND ($v_2$ is LARGE) AND ($v_3$ is LARGE)
               THEN Class 1 (e.g. water$_0$)
Rule 1:   IF ($v_1$ is LARGE) AND ($v_2$ is LARGE) AND ($v_3$ is SMALL)

$$\text{THEN Class 2 (e.g. water}_1). \tag{5.20}$$

If for example, a sample of the used case study data of water gives the readings small for sensor 1 and large for sensors 2 and 3, then this sample is water$_0$ in accordance with the rules stated above. For the sample to be water$_1$, the readings would be large for sensors 1 and 2 and small for sensor 3.

To better explain the idea, let us consider a simple network shown in Fig. 5.7 where two L2 neurons are created; a neuron $L2_i$ has been trained to recognize the class $C_i$ samples and a neuron $L2_j$ has been trained to recognize the class $C_j$ samples.

Let two one-dimensional vectors $v_i$ and $v_j$ belong to classes $C_i$ and $C_j$, respectively. It is postulated that knowledge about the relationship between two inputs, $v_i$ and $v_j$ (e.g. that $v_i < v_j$), is stored in the synaptic weights of the L2 neurons representing a long-term memory. The change in synaptic weight between L1 and L2 neurons in the eSNN depends on the firing time of the L1, i.e. $\Delta w_{ji} = f(t_j^f)$. As stated earlier, the synaptic weights associated with the connections which convey earlier spikes increase more than those which convey later spikes. Hence, based on the weight patterns it is possible to deduce the size of input values relative to other input values (small, medium, large) and the contribution of this values to the modeled output.

An example, where two L2 neurons have evolved is explained below. $L2_i$ has been trained to recognize the class $C_i$ samples and $L2_j$ has been trained to recognize the class $C_j$ samples. Assume that at time $t$, a spike arrives from $L1_n$ via two synaptic terminals $(w_{ni}, w_{nj})$. This spike excites the membrane potentials of both $L2_i$ and $L2_j$ as follows:

$$\Delta PSP_i = w_{ni} \times \text{mod}^{o_n} \tag{5.21}$$

$$\Delta PSP_j = w_{nj} \times \text{mod}^{o_n} \tag{5.22}$$

with the same convention as in (54) and where $o_n$ is in the range $[0, m - 1]$ and $m = 6$.

The difference between these two excitations is:

$$\begin{aligned} dPSP = \left| PSP_i - PSP_j \right| &= \left| \text{mod}^{o_n} \times (w_{ni} - w_{nj}) \right| \\ &= f(w_{ni} - w_{nj}) = f(d_w). \end{aligned} \tag{5.23}$$

Hence the difference in the excitation of the two L2 neurons is a function of the difference between their synaptic weights $(d_w)$. Thus, if two weights are identical, $d_w = 0$, both L2 neurons are equally excited by the incoming spike.

The $d_w$ values for the theoretical pattern of the $w_{ni}$ and $w_{nj}$ weights in Fig. 5.9 is shown in Fig. 5.10. It can be seen that the $L2_i$ neuron 'favors' lower input values, i.e. a lower input value will cause a bigger $\Delta PSP_i$. Thus, lower values cause $L2_i$ to spike before $L2_j$. The $L2_j$ neuron does the opposite; it spikes before $L2_i$ upon

**Fig. 5.9** Two patterns of connection weights between levels L1 and L2 neurons representing different input values of one input variable from class Ci and Cj samples

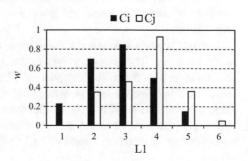

**Fig. 5.10** The difference of the connection weights $d_w$ ($d_w = w_{ni} - w_{nj}$) for a single input variable for the two class patterns from Fig. 5.9 shows a clear discrimination between class Ci and class Cj patterns that can be used and represented in fuzzy rules (5.24)

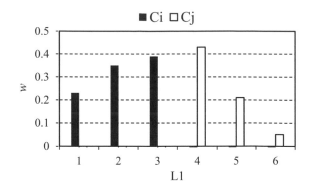

presentation of higher input values. Hence, the lower values $v$ will be classified as $C_i$ and the higher values as $C_j$ as shown below and in Figs. 5.9 and 5.10.

$$\text{Rule 0: IF } (v \text{ is SMALL}) \text{ THEN } C_i$$
$$\text{Rule 1: IF } (v \text{ is LARGE}) \text{ THEN } C_j$$

(5.24)

### 5.4.2 A Case Study of Fuzzy Rule Extraction from Water Tastant Sensory Data

The data samples were generated by an array of non-selective taste sensors based on conducting polymers that have been proven to be able to discriminate between basic tastants [13]. A data set D1 is comprised of 40 measured patterns from four different water types (two brands of mineral water, Milli-Q [14] water and distilled water), while a data set D2 contains 300 measured patterns from two brands of wine (Marcus James and Almadén Cabernet Sauvignon). Both sets are balanced, having an equal number of patterns of each type. The datasets were randomly shuffled and equally split into a training and testing set. To enable easier representation of the extracted rules only one L2 neuron per class has been evolved.

In each experiment, the eSNN model parameters (mod, C, Thr), including the number of receptive fields $m$, have been carefully chosen for accurate taste recognition.

Firstly, the knowledge accumulated by an eSNN about the mineral waters (water$_0$, water$_1$) is given. The average values of the mineral water patterns in the training set are shown in Fig. 5.11. The samples have been obtained using seven taste sensors; hence each pattern is a seven-dimensional vector. The average values of $v_6$ in both mineral waters are significantly smaller than averages of the other variables. The $v_6$ values are smaller than in any of the other variables; this variable $v_6$ is omitted for classification. The average values of all of the other variables show a considerable

**Fig. 5.11** Average values, $v_{avg}$, of the mineral waters' variables

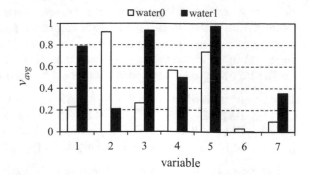

difference between the two mineral water types. For example, $v_1$ is much smaller in water$_0$ than in water$_1$, while $v_2$ is much smaller in water$_1$ than in water$_0$. This 'prior knowledge' is useful when the quality of knowledge representation is subsequently assessed.

Figure 5.12 shows the $d_w$ values of an eSNN trained using the water$_0$ and water$_1$ patterns. In this case eight receptive fields ($m = 8$) have been used to encode the input patterns. As a result, each $x_i$ feature pattern comprises of eight bars. However, if two weights are identical, $d_w = 0$, the corresponding bar will be missing in the feature pattern. This is observed in the pattern $x_4$. Furthermore, the feature patterns are influenced by Gaussian receptive fields. Gaussian receptive fields increase the sparseness in the original data, i.e., there is a smaller number of input neurons activated than without Gaussian receptive fields. As a result, only a subset of the connections convey spikes, others have weights equal to zero and their bars are missing from the pattern, as illustrated in $x_7$. Also, some weight may be almost identical, therefore the small $d_w$ will appear missing in the $d_w$ plots.

It is interesting to observe that in the sixth feature pattern, $x_6$, all $d_w$ values are zero because the weights are similar. As a result the contributions of the spikes through these terminals to the post-synaptic potential of L2 neurons are very similar. It can therefore be deduced that the values of the sixth input variable (with the rather small average values compared to other inputs) do not contribute to the

**Fig. 5.12** The $d_w$ values of an eSNN trained to distinguish between mineral water patterns ($m = 8$)

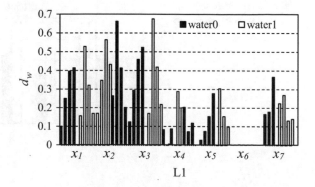

process of distinguishing between the mineral water samples. Also, the spikes coming from the $v_1$, $v_2$ and $v_3$ variables contribute more to the L2 neurons' *PSP*, than those coming from $v_4$, $v_5$ and $v_7$, indicating that these features are more important in the classification of the two mineral waters.

The following rules have been extracted for $water_1$ and $water_2$ samples:

Rule 0:  IF ($v_1$ is SMALL) AND ($v_2$ is LARGE) AND ($v_3$ is SMALL) AND ($v_4$ is LARGE) AND ($v_5$ is SMALL) AND ($v_7$ is SMALL) THEN $water_0$

Rule 1:  IF ($v_1$ is LARGE) AND ($v_2$ is SMALL) AND ($v_3$ is LARGE) AND ($v_4$ is SMALL) AND ($v_5$ is LARGE) AND ($v_7$ is LARGE) THEN $water_1$.

For example, the above rules state that $v_3$ is SMALL in $water_0$ (Rule 0) and LARGE in $water_1$ (Rule 1).

For the wine dataset D2 the following fuzzy rules have been extracted for the two brands of wine ($wine_0$, $wine_1$):

Rule 0:  IF ($v_1$ is SMALL) AND ($v_2$ is LARGE) AND ($v_3$ is SMALL) AND ($v_4$ is SMALL) AND ($v_5$ is LARGE) AND ($v_6$ is LARGE) AND ($v_7$ is SMALL) THEN $wine_0$

Rule 1:  IF ($v_1$ is LARGE) AND ($v_2$ is SMALL) AND ($v_3$ is SMALL) AND ($v_4$ is LARGE) AND ($v_5$ is LARGE) AND ($v_6$ is SMALL) AND ($v_7$ is LARGE) THEN $wine_1$.

It is noteworthy that all input variables contribute to the classification of the wines, including $v_6$ which seems irrelevant in the classification of the water types. We contribute this to the fact that the $v_6$ averages for both wines are similar (Fig. 5.13).

The two experiments described demonstrated knowledge discovery in a two-class problem scenario using eSNN. The knowledge extraction capability of the eSNN model on three-class and four-class problems has also been investigated. Firstly, the Milli-Q samples have been included in the experiment ($water_2$). The $water_2$ patterns have very small average values compared to the mineral water patterns (Fig. 5.14).

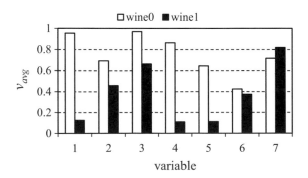

**Fig. 5.13** Average values, $v_{avg}$, of the wines' variables

**Fig. 5.14** Average values, $v_{avg}$, of the two mineral waters and Milli-Q patterns

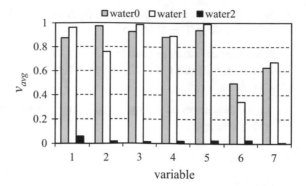

**Fig. 5.15** The $d_w$ values of an eSNN trained to distinguish between three waters ($m = 6$)

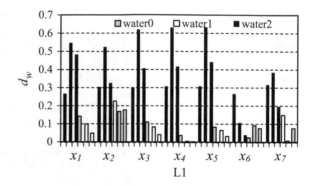

Six receptive fields have been used ($m = 6$) resulting in six bars within each feature pattern. Again one L2 neuron per water type have been evolved. Their distance weight $d_w$ values are shown in Fig. 5.15. It is interesting to see that while the average values of the water$_2$ samples are much smaller than the average values of water$_0$ and water$_1$, the weights of the L2 neuron that has learnt to recognize these samples are noticeably greater than the weights of the other neurons. This means that values of the water$_2$ patterns were encoded in such a way that they have been repeatedly enhancing the same set of connections. This results in the considerable observed strengthening of the connections.

Contrary to this, the mineral water samples have more evenly distributed their influences among a larger number of synaptic connections. This is due to the fact that the values in the water$_0$ and water$_1$ patterns are very similar and they compete for the same set of connections. This is also reflected in the extracted rules ('1' indicates SMALL, '2' indicates MEDIUM and '3' indicates LARGE):

Rule 0:   IF ($v_1$ is 2) AND ($v_2$ is 3) AND ($v_3$ is 2) AND ($v_4$ is 2) AND ($v_5$ is 2) AND ($v_6$ is 3) AND ($v_7$ is 3) THEN water$_0$

Rule 1:   IF ($v_1$ is 3) AND ($v_2$ is 2) AND ($v_3$ is 3) AND ($v_4$ is 3) AND ($v_5$ is 3) AND ($v_7$ is 2) THEN water$_1$

Rule 2:   IF ($v_1$ is 1) AND ($v_2$ is 1) AND ($v_3$ is 1) AND ($v_4$ is 1) AND ($v_5$ is 1) AND ($v_7$ is 3) THEN water$_2$.

The approach has also been tested using all four water types (i.e., two brands of mineral water, Milli-Q [14, 15] water and distilled water (water$_3$)) with following characteristics:

- a similarity between the water$_0$ and water$_1$ averages and between the water$_2$ and water$_3$ averages,
- a difference in magnitude between water$_2$ and water$_3$ averages and the water$_0$ and water$_1$ averages.

Again only one neuron per water type has been evolved and the fuzzy rules using three membership functions have been extracted as follows:

Rule 0:   IF ($v_1$ is 2) AND ($v_2$ is 3) AND ($v_3$ is 2) AND ($v_4$ is 2) AND ($v_5$ is 2) AND ($v_6$ is 2) THEN water$_0$

Rule 1:   IF ($v_1$ is 3) AND ($v_2$ is 2) AND ($v_3$ is 3) AND ($v_4$ is 3) AND ($v_5$ is 3) AND ($v_6$ is 2) AND ($v_7$ is 2) THEN water$_1$

Rule 2:   IF ($v_1$ is 1) AND ($v_2$ is 1) AND ($v_3$ is 1) AND ($v_4$ is 2) AND ($v_5$ is 2) AND ($v_7$ is 1) THEN water$_2$

Rule 3:   IF ($v_1$ is 2) AND ($v_2$ is 2) AND ($v_3$ is 2) AND ($v_4$ is 1) AND ($v_5$ is 1) AND ($v_6$ is 1) AND ($v_7$ is 2) THEN water$_3$.

The degree of granularity of the fuzzification has been increased from three to four where, for instance, '1' indicates VERY SMALL, '2' indicates SMALL, '3' indicates MEDIUM and '4' indicates LARGE:

Rule 0:   IF ($v_1$ is 3) AND ($v_2$ is 4) AND ($v_3$ is 3) AND ($v_4$ is 3) AND ($v_5$ is 3) AND ($v_6$ is 2) THEN water$_0$

Rule 1:   IF ($v_1$ is 4) AND ($v_2$ is 3) AND ($v_3$ is 4) AND ($v_4$ is 4) AND ($v_5$ is 4) AND ($v_6$ is 3) AND ($v_7$ is 3) THEN water$_1$

Rule 2:   IF ($v_1$ is 1) AND ($v_2$ is 1) AND ($v_3$ is 1) AND ($v_4$ is 2) AND ($v_5$ is 2) AND ($v_7$ is 1) THEN water$_2$

Rule 3:   IF ($v_1$ is 2) AND ($v_2$ is 2) AND ($v_3$ is 2) AND ($v_4$ is 1) AND ($v_5$ is 1) AND ($v_6$ is 1) AND ($v_7$ is 2) THEN water$_3$.

The resulting set of rules more accurately captures the data set characteristics. The high variable values that characterize the water$_0$ and water$_1$ patterns and their fine differences are described in Rule 0 and Rule 1, while the characteristics of the water$_2$ and water$_3$ values are embedded in Rule 2 and Rule 3.

The task of determining the number of L2 neurons is an important step in tuning the eSNN accuracy [15]. One L2 neuron per class might not be enough to achieve the best accuracy. As the number of neurons and rules increases it may become cumbersome to interpret these rules. This could become a problem if an eSNN is allowed to evolve in an open problem space without applying aggregation operator. The latter can be used to evolve prototype output neurons and to extract prototype rules of higher granularity.

## 5.5   Evolving SNN for Reservoir Computing

The eSNN principles presented and illustrated so far can be used to create evolving output classification or regression modules of reservoir types of SNN computational architectures which is the topic of this section.

### 5.5.1   Reservoir Architectures. Liquid State Machines (LSM)

Two major types of reservoir computing is liquid state machine (LSM) and echo state network [16–25]. LSM consists of many randomly interconnected recurrent neurons where each neuron receives input from other neurons in different times. This computational model is inspired from the idea of water ripples (output), which are generated after certain objects (input) are dropped on the still water. In an ideal setting, LSM that is constructed with a precise mathematical framework, promises universal computational power, for real-time computing on analogue function in continuous time. LSM is characterized as a model for adaptive computational system, which provides a method for employing randomly connected circuits, a theoretical context where various processors increase the computational power of a circuit and a method for multiplexing different computations within the same circuit [18].

During LSM simulation, the synaptic weights, neurons connectivity and neurons parameters are predefined and predetermined. Referring to Fig. 5.16, the continuous stream of input (e.g. trains of spikes) $u(t)$ is transmitted into liquid and will cause the neurons to respond and generate the liquid activity. The state of the liquid $x(t)$ that can be recorded at different time points is simply the current output of some operator or filter that maps input functions $u(t)$ onto function $x(t)$. This state is then passed to a readout function $f$ that will transform into output $v(t)$. In contrast to the classical LSM [18], where the Readout function is memory-less, here we propose that a trainable SNN readout classifier can be used, such as eSNN or deSNN.

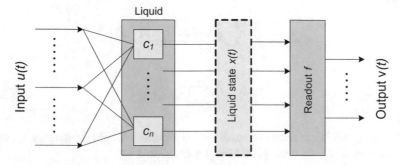

**Fig. 5.16** The LSM architecture consists of three main layers: input neuron layer, liquid state layer and readout function layer. eSNN and deSNN can be used as trainable classifiers or regressors in the Readout module

**Fig. 5.17** An example of a
spiking neuron connectivity
in a 3D reservoir of spiking
neurons using the small-world
connectivity rule

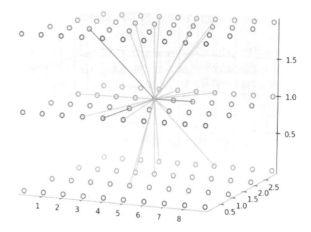

The connectivity in a LSM is usually pre-selected as a random connections or
fully connected neuronal structures. As a special case, a small world connectivity
(SWC) can be applied (Fig. 5.17), where neuron $a$ is connected to other neuron
$b$ with a probability $p_{a,b}$ that depends on the closeness of the two neurons. The
closer they are (the smaller the distance between them Da, b), the higher the
probability to connect them. This is a biologically plausible connectivity rule as the
brain is a small world connectivity system:

$$p_{a,b} = C \times e^{-D_{a,b}^2/\lambda^2} \tag{5.25}$$

An important characteristic of LSM is the separability of patterns that are activated
in the LSM.

*Separability* basically determines how well the liquid separates different classes
of input into different reservoir states. According to [19], separability can be
measured by dividing a set of states, $O(t)$, into $N$ subsets, $O_m(t)$ for every class. Here
$N$ denotes the total number of classes and $t$ represents the current iteration of the
reservoir. The inter-class distance, $C_d(t)$, and intra-class variance, $C_v(t)$, $C_d(t)$ can be
defined as:

$$C_d(t) = \sum_{m=1}^{N} \sum_{n=1}^{N} \frac{\|\mu(O_m(t)) - \mu(O_n(t))\|_2}{N^2} \tag{5.26}$$

$$C_v(t) = \frac{1}{N} \sum_{m=1}^{N} \rho(O_m(t)) \tag{5.27}$$

where, $\mu(O_m(t))$ is the center of mass for each class, $\rho(O_m(t))$ is average variance
from each state within class $m$ and can be calculated by

$$\mu(O_m(t)) = \frac{\sum_{O_n \in O_m(t)} O_n}{|O_m(t)|} \qquad (5.28)$$

$$\rho(O_m(t)) = \frac{\sum_{O_n \in O_m(t)} \|\mu(O_m(t)) - O_n\|_2}{|O_m(t)|} \qquad (5.29)$$

Therefore, based on Eqs. 5.27 and 5.28, the separability of the liquid (*Sep* Ψ) for a set of states $O(t)$ can be calculated as [19]:

$$Sep_\psi(O(t)) = \frac{C_d(t)}{C_v(t) + 1} \qquad (5.30)$$

Using different types of probabilistic neurons as discussed in Chap. 4, can result in different separability of patterns in the reservoir activated by different stimuli as shown in Fig. 5.18 and experimented in [20].

In regards to learning in the reservoir, there two types of reservoir architectures:

– SNN LSM, that involve no learning in the reservoir (the discussed so far above);
– SNN reservoir architectures that involve learning. Such architecture is NeuCbe discussed in the next chapter.

### 5.5.2 ESNN/DeSNN as Classification/Regression Systems for Reservoir Architectures

Some reservoir architectures involve learning in the output classification or regression module (the read-out function as per Fig. 5.17. This module is in principle capable of generating different responses from different dynamic reservoir input patterns (separation property) that also depends on the complexity of the

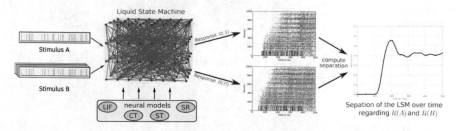

**Fig. 5.18** Experimental setup of the study of using different probabilistic models of spiking neurons in a LSM in regards of separability of patterns activated by different spiking stimuli sequences. In [20] a comparative analysis is performed between LSM when different types of spiking neuron models are used, such as LIF; CT (continuously changing spiking threshold); ST (step-wise threshold); SR (stochastic reset) (see Chap. 4)

reservoir structure. In addition, it also has 'approximation property' that depends on the adaptation ability of the readout function that can distinguish responses, their generalization and relation with the given output [21]. Any statistical analysis or classifier can be used to define the readout function. As the inputs can be in the form of continuous stream of data, such SNN architectures can be used to solve spatiotemporal problems such as in patterns recognition [22, 23], optimization [24] and classification problems [25].

Here we discuss how eSNN and deSNN can be used efficiently as output classification/regression modules for fast, evolving and meaningful supervised learning and pattern recognition of the patterns that are activated in the reservoir from input data.

The input-output mapping in the supervised learning can be imposed by means of error $E$ calculation, defined as

$$E = \frac{1}{2} \sum_j \left( t_j^{out} - t_j^{des} \right) \tag{5.31}$$

where, $E$ represents the difference between all network outputs $t_j^{out}$ and desired outputs $t_j^{des}$. The weights $w_{ij}^k$ is adjusted accordingly, so as to minimize the error E and can be expressed as

$$\Delta w_{ij}^k = -\eta \frac{dE}{dw_{ij}^k} \tag{5.32}$$

where, $\eta$ denotes the learning rate. However, in the process of implementing this model, there are several research questions that arise and need to be addressed. These research questions are mentioned in the following section.

Figure 5.19 shows a reservoir computational architecture where a deSNN model is used as a classifier of the spatio-temporal patterns in the reservoir.

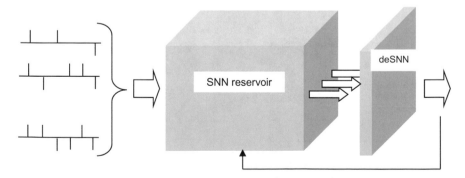

**Fig. 5.19** A reservoir computational architecture where a deSNN is used as a classifier of the dynamic patterns in the reservoir that are activated by different stimuli. A feedback from the deSNN output back to the reservoir can be established as well

There are several advantages of using eSNN/deSNN as classifiers/regressors for reservoir computing:

– one-line training on streaming data;
– adaptive training on new samples incrementally;
– allowing for new classes to be added incrementally;
– allowing for rules (fuzzy rules; prototype rules) to be extracted from the trained classifiers/regressors;
– allowing for associations to be analysed between the extracted rules in the output module and the patterns of activations in the reservoir.

The next chapter develops further the architecture from Fig. 5.19 and introduces a brain-like learning reservoir architecture that has an output functions eSNN/ deSNN.

## 5.6   Chapter Summary and Further Readings for Deeper Knowledge

This chapter presents the concept and some implementations of evolving SNN. The methods of evolving SNN add new properties to the concept of SNN, namely:

– Fast data processing, possibly one iteration of data presentation only;
– Adaptive learning, that addresses the "curse of catastrophic forgetting";
– Rule extraction, that addresses "the curse of NN black boxes";
– Integration of different modalities of data, all represented as spike sequences.

eSNN extend the principles of ECOS from Chap. 2 using spike information representation and spike-time learning. At the same time they keep all principles of ECOS that are knowledge based ANN. The rules extracted from eSNN represent 'flat' knowledge, even thou they can capture some temporal characteristics from data.

eSNN, including deSNN models are used as classifiers for brain-inspired SNN presented in Chap. 6, where deep learning and deep knowledge representation are facilitated.

Readings on specific topics of eSNN and deSNN can be found as follows:

– Rule extraction from eSNN [26];
– deSNN [10];
– Computational modelling with SNN (Chap. 37 in [27]);
– Brain-like information processing for spatio-temporal pattern recognition (Chap. 47 in [27]);
– Probabilistic SNN for reservoir computing [20];
– On the role of time and space in neural computation [28];
– Liquid State Machines [18];
– Overview of evolving connectionist systems [29];

- Basic information about SNN [30];
- Application of eSNN for Bioinformatics data (Chap. 15 in this book);
- Applications of eSNN for Biomedical data (Chap. 17 in this book);
- Applications of eSNN for financial data (Chap. 19 in this book);
- Applications of eSNN for environmental data (Chap. 19 in this book);
- Applications of deSNN for spatio-temporal data modelling and classification in NeuCube (Chap. 6 and also: http://www.kedri.aut.ac.nz/neucube/).

**Acknowledgements** Parts of the material in this chapter have been published previously as referenced in the corresponding sections. I acknowledge the contribution of my co-authors in these publications: L. Benuskova. S. Wysoski, S. Schliebs, S. Soltic, A. Mohemmed, K. Double, N. Nuntalid, G. Indiveri.

# References

1. S. Thorpe, D. Fize, C. Marlot, Speed of processing in the human visual system. Nature **381** (6582), 520–522 (1996)
2. P. Lichtsteiner, T. Delbruck, A 64 × 64 AER logarithmic temporal derivative silicon retina. Res. Microelectron. Electron. **2**(1), 202–205 (2005). https://doi.org/10.1109/rme.2005. 1542972
3. T. Delbruck, jAER open source project (2007). http://jaer.wiki.sourceforge.net
4. S. Soltic, N. Kasabov, Knowledge extract ion from evolving spiking neural networks with rank order population coding. Int. J. Neural Syst. **20**(6), 437–445 (2010)
5. N. Kasabov, *Evolving Connectionist Systems: The Knowledge Engineering Approach*, 2nd edn. (Springer, 2007) (1st edn., 2002)
6. S. Wysoski, L. Benuskova, N. Kasabov, Evolving spiking neural networks for audiovisual information processing. Neural Netw. **23**(7), 819–835 (2010)
7. S.M. Bohte, H. La Poutre, J.N. Kok, Unsupervised clustering with spiking neurons by sparse temporal coding and multilayer RBF networks. IEEE Trans. Neural Networks **13**(2), 426–435 (2002)
8. S. Schliebs, N. Kasabov, *Evolving Spiking Neural Networks: A Survey, Evolving Systems* (Springer, 2012)
9. S. Thorpe, J. Gautrais, Rank order coding. Comput. Neurosci. Trends Res. 113–119 (1998)
10. N. Kasabov, K. Dhoble, N. Nuntalid, G. Indiveri, Dynamic evolving spiking neural networks for on-line spatio- and spectro-temporal pattern recognition. Neural Netw. Off. J. Int. Neural Netw. Soc. **41**, 188–201 (2013). https://doi.org/10.1016/j.neunet.2012.11.014
11. P. Tiesinga, J. Fellous, T.J. Sejnowski, Regulation of spike timing in visual cortical circuits. Nat. Rev. Neurosci. **9**(2), 97–107 (2008)
12. E. Nichols, L.J. McDaid, N.H. Siddique, Case study on a self-organizing spiking neural network for robot navigation. Int. J. Neural Syst. **20**(6), 501–508 (2010). PMID: 21117272
13. A. Riul Jr., D.S. dos Santos Jr, K. Wohnrath, R. Di Tommazo, A.C.P.L.F. Carvalho, F. J. Fonseca, O.N. Oliveira Jr., D.M. Taylor, L.H.C. Mattoso, An artificial taste sensor: efficient combination of sensors made from Langmuir-Blodgett films of conducting polymers and a ruthenium complex and self-assembled films of an Azobenzene-containing polymer. Langmuir **18**(2002), 239–245 (2002)
14. Milli-Q, http://www.millipore.com/

15. S. Soltic, S.G. Wysoski, N. Kasabov, Evolving spiking neural networks for taste recognition, in *Proceedings of the International Joint Conference on Neural Networks, IJCNN 2008* (Hong Kong, 2008), pp. 2092–2098

16. W. Maass, T. Natschlager, H. Markram, Real-time computing without stable states: a new framework for neural computation based on perturbations. Neural Comput. **14**(11), 2531–2560 (2002)

17. H. Jaeger, H. Haas, Harnessing nonlinearity: predicting chaotic systems and saving energy in wireless communication. Science (New York, NY) **304**(5667), 78–80 (2004). https://doi.org/10.1126/science.1091277

18. W. Maass, *Liquid State Machines: Motivation, Theory, and Applications* (2010) (Chapter 1)

19. D. Norton, D. Ventura, Improving the separability of a reservoir facilitates learning transfer, in *Proceeding of the Seventh ACM Conference on Creativity and Cognition* (2009), pp. 339–340

20. S. Schliebs, A. Mohemmed, N. Kasabov, Are probabilistic spiking neural networks suitable for reservoir computing? in *International Joint Conference on Neural Networks* (San Jose, USA, 2011), pp. 3156–3163

21. B.J. Grzyb, E. Chinellato, G.M. Wojcik, W.A. Kaminski, Which Model to use for the liquid state machine? in *International Joint Conference on Neural Networks, 2009. IJCNN 2009.* IEEE (2009), pp. 1018–1024

22. E. Goodman, D. Ventura, Spatiotemporal pattern recognition via liquid state machines, in *IJCNN* (2006), pp. 3848–3853

23. S. Schliebs, N. Nuntalid, N. Kasabov, Towards spatio-temporal pattern recognition using evolving spiking neural networks. Neural Inf. Process. Theor. Alg. **6443**, 163–170 (2010). https://doi.org/10.1007/978-3-642-17537-4_21

24. Z. Yanduo, W. Kun, The application of liquid state machines in robot path planning. J. Comput. **4**(11), 1182–1186 (2009)

25. H. Ju, J. Xu, A.M.J. VanDongen, Classification of musical styles using liquid state machines, in The 2010 International Joint Conference on Neural Networks (IJCNN) (2010), pp. 1–7

26. S. Soltic, N.K. Kasabov, Knowledge extraction from evolving spiking neural networks with rank order population coding. Int. J. Neural Syst. **20**(6), 437–445 (2010)

27. N. Kasabov (ed.), *Springer Handbook of Bio-/Neuroinformatics* (Springer, 2014)

28. W. Maass, On the role of time and space in neural computation. Math. Found. Comput. Sci., 72–83 (1998)

29. N. Kasabov, Evolving connectionist systems: from neuro-fuzzy-, to spiking—and neurogenetic, in *Springer Handbook of Computational Intelligence*, ed. by J. Kacprzyk, W. Pedrycz (Springer, 2015), pp. 771–782

30. W. Gerstner, W.M. Kistler, *Spiking Neuron Models: Single Neurons, Populations, Plasticity* (Cambridge University Press, 2002)

# Chapter 6
# Brain-Inspired SNN for Deep Learning in Time-Space and Deep Knowledge Representation. NeuCube

This chapter introduces brain-inspired evolving SNN (BI-SNN) in which both the SNN architecture and learning are inspired by the structure, organisation and learning in the human brain. BI-SNN manifest deep learning from data and deep knowledge representation inspired by human brain as discussed in Chap. 3 of the book. In BI-SNN *data* is represented as spikes, *information* is represented as spatio-temporal spike patterns and *deep knowledge* is represented as patterns of connections that are subject to deep learning and can be interpreted by humans. One BI-SNN architecture introduced in the chapter is NeuCube. It is a open, evolving framework that is a set of algorithms allowing for the creation of SNN systems and BI-AI systems and also allowing for new algorithms to be developed in the future and explored as part of it.

The chapter is organised in the following sections:

6.1. Brain-Inspired SNN (BI-SNN). The NeuCube BI-SNN as a generic spatio-temporal data machine.
6.2. Deep learning in time-space and deep knowledge representation in NeuCube.
6.3. Modelling *Time* in NeuCube: the past, the present, the future and back in time.
6.4. A design methodology for application oriented spatio-temporal data machines.
6.5. Case studies for the design of classification and prediction spatio-temporal machines.
6.6. Chapter summary and further readings for deeper knowledge.

## 6.1 Brain Inspired SNN (BI-SNN). The BI-SNN NeuCube as a Generic Spatio-temporal Data Machine

### 6.1.1 A General Architecture of a BI-SNN

A general architecture of a BI-SNN architecture is shown in Fig. 6.1, while Fig. 6.2 shows a brain template that has been used to design the structure of the SNN reservoir, here named SNNcube.

© Springer-Verlag GmbH Germany, part of Springer Nature 2019
N. K. Kasabov, *Time-Space, Spiking Neural Networks and Brain-Inspired Artificial Intelligence*, Springer Series on Bio- and Neurosystems 7,
https://doi.org/10.1007/978-3-662-57715-8_6

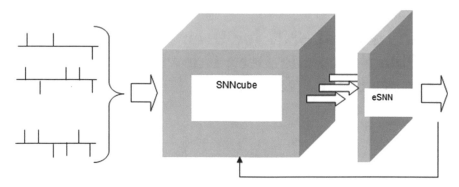

**Fig. 6.1** A general architecture of BI-SNN, where the SNN cube is structured as shown in Fig. 6.2

The main structural and functional principles of a BI-SNN are presented in Box 6.1.

---

**Box 6.1.   Main structural and functional characteristics of a BI-SNN**

---

1. Temporal inputs (features) are converted/encoded into spike trains (see Chap. 4).

2. Inputs are mapped spatially (brain-like) into a 3D SNNcube that consists of spiking neurons spatially organized in a topological 3D map. For modelling brain data the SNNcube is built with the use of a brain template (see Chap. 3) and for other types of data it is built to preserve the spatial information in the input data or to preserve the distance between data items.

3. Output classifier/regressor SNN is connected to neurons from the SNNcube (see Chap. 5).

4. SNNcube structure is organized as small world connectivity 3D structure of spiking neurons (see Chap. 5).

5. Unsupervised learning is performed in the SNNcube using spike-time learning rules, e.g. STDP (see Chap. 4).

6. Supervised learning is performed in the output SNN module, e.g. eSNN, deSNN, SPAN for classification or regression (see Chaps. 4 and 5).

7. Adaptive, *deep learning* of complex spatio-temporal patterns is performed in the SNNcube.

8. The BI-SNN operates in a fast, incremental learning mode.

9. The learned connectivity patterns in the SNNcube can be interpreted as *deep knowledge* representing deep spatio-temporal patterns in the data.

10. Learned connectivity patterns in the eSNN output module can be interpreted for rule extraction related to outputs (see Chap. 5).

---

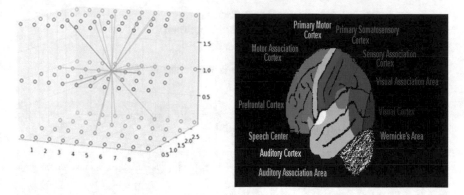

**Fig. 6.2** The SNNcube from Fig. 6.1 is organised as small world connectivity reservoir, which structure maps/resembles a brain template

## 6.1.2   The BI-SNN NeuCube as a Generic Spatio-temporal Data Machine

The general principles of the NeuCube BI-SNN architecture were presented in [1] and also in [2]. The NeuCube architecture is depicted in Fig. 6.3. It consists of the following functional modules:

- Input data encoding module;
- 3D SNN reservoir module (SNNcube);

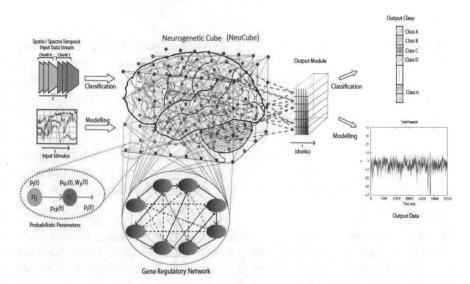

**Fig. 6.3** The BI-SNN NeuCube—a general architecture (from [1])

– Output function (classification) module;
– Gene regulatory network (GRN) module (optional);
– Parameter optimisation module (optional).

The NeuCube procedural meta-algorithm is presented in Box 6.2.

---

**Box 6.2. The NeuCube meta-procedural algorithm**

---

1. Encode input temporal or spatio-temporal data (TSTD) into spike sequences: continuous value input data is encoded into trains of spikes.
2. Construct and train in an unsupervised mode a   recurrent 3D SNNcube, to learn the spike encoded sequences that represent input data.
3. Construct and train in a supervised mode an evolving SNN classifier/regressor to learn to classify/predict different dynamic patterns of the SNNcube activities that represent different input patterns from SSTD that belong to different classes/output values;
4. Optimise the model through several iterations of steps (1) to (3) above for different parameter values until maximum accuracy is achieved. .
5. Recall and test the model on new data.

---

The above procedures and their corresponding modules in the NeuCube architecture from Fig. 6.3 are described further in this section.

*Input data encoding*

Continuous value input data need to be transformed into a train of spikes. Methods for spike-time encoding include [1, 3–6] (see Chap. 4)

– Threshold based encoding;
– Moving window encoding;
– Step-wise forward encoding;
– Ben Spike Algorithm (BSA) [6].

The transformed input data into spike series is entered (mapped) into spatially located neurons from the SNNcube. The mapping will depend on the problem in hand. If it is brain data, the mapping will utilise a brain template (Chap. 3).

*SNNcube training*

The SNNc is structured to spatially map data. A particular structure is small world connectivity [7]. In the brain, neurons in a structural or functional areas are more densely interconnected and the closer these areas are, the higher the connectivity between them [8–10].

The initial structure of the SNNcube is defined based on the available data and problem, but this structure can be evolving through the creation of new neurons and new connections based on the data using the ECOS principles [11] (Chap. 2). If new data do not sufficiently activate existing neurons, new neurons are created and allocated to match the data along with their new connections.

In one implementation, a SNNc can have a 3D structure connecting leaky-integrate and fire model (LIFM) spiking neurons with recurrent connections (for LIFM see Chap. 4). Input data is propagated through the SNNc and a method of *unsupervised learning* is applied, such as STDP. The neuronal connections are adapted through a spike-time learning, such as STDP (Chap. 4) and the SNNc learns to generate specific trajectories of spiking activities when a particular input pattern is entered. The SNNc accumulates temporal information of all input spike trains and transforms it into dynamic states that can be classified over time. The recurrent reservoir generates unique accumulated neuron spike time responses for different classes of input spike trains.

As an illustration, Fig. 6.4a illustrates the polychronisation process of neuronal activations during learning over time. Figure 6.4b shows the connectivity of a SNNc that are result of the learning procedure on illustrative spatio/spectro-temporal brain data, where the SNNc has 1471 neurons and the coordinates of these neurons correspond directly to the Talairach template coordinates with a resolution of 1 cm$^3$. It can be seen that as a result of training new connections have been created that represent *spatio-temporal interaction* between input variables captured in the SNNc from the data. The connectivity can be dynamically visualised for every new pattern submitted.

Figure 6.5 shows the activation level of the neurons in the SNNcube after unsupervised learning. The brighter the colour of a neuron, the higher its activation level is in terms of number of spikes emitted.

When spike sequence data is entered into the SNNcube and learning is applied, the spiking neurons start to spike. Figure 6.6 shows positive and negative spike emission histogram for all SNNcube 1471 neurons from Fig. 6.5.

Figure 6.7 shows a raster plot of the spikes entered into a SNNcube of 1471 neurons for one input spatio-temporal sample (pattern) of 14 inputs and 125 ms time.

Based on the input stream of spike sequences, clusters are created/emerged/evolved in the SNNcube during deep learning in the Cube. Two types of clusters are created based on:

– Neuronal connectivity (Fig. 6.8)
– Spiking activity as spike exchange between clusters (Fig. 6.9). The spikes emitted and transferred from a particular neuron (e.g. input neuron) to other neurons can be studied (Fig. 6.10).

**(a)**

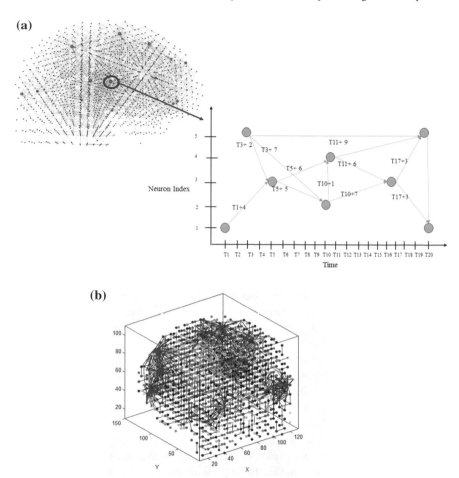

**(b)**

**Fig. 6.4** **a** An illustration of a polychronous SNN within the NeuCube framework activated during unsupervised learning (the figure was created by H. Bahrami). **b** A snapshot of the connections in a SNNcube after unsupervised learning on brain data. The connections represent spatio-temporal relationships between input data variables (EEG channels) over time and can be used for deep knowledge representation

*Evolving supervised learning for output classification/regression*

After a SNNc is trained on input data in an unsupervised mode, the same input data can be propagated again through the SNNc, pattern by pattern (each input pattern is spatio-temporal one over a time window), the state of the SNNc is measured for each pattern and an output classifier is trained to recognise this state in a predefined output class for this input pattern. For fast learning, we use evolving SNN classifiers (eSNN or deSNN) (Chap. 5). All neurons from the SNNc are connected to each of

**Fig. 6.5** Activation level of the neurons in the SNNcube after unsupervised learning. The brighter the colour of a neuron, the higher its activation level is in terms of number of spikes emitted

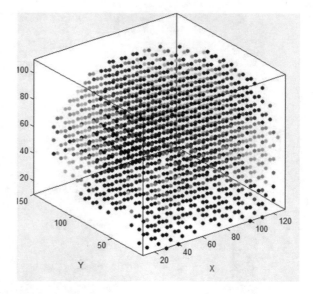

**Fig. 6.6** Positive and negative spike emission histogram for all SNNcube neurons

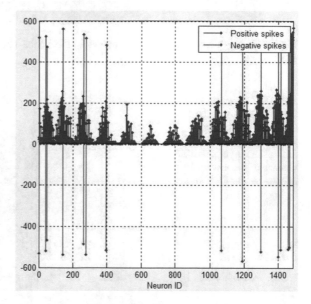

the evolved LIFM neurons of the eSNN classifier, but other connectivities can be established.

One of the originality of the NeuCube architecture is that it utilises the ability of the eSNN to learn to recognise complex spatio-temporal patterns generated in the SNNc before the whole input data pattern is entered. Different types of eSNN can be used as presented in Chap. 5, e.g. eSNN [11]; Dynamic eSNN (deSNN) [12], and also spike-pattern association SNN, e.g. SPAN [13] (Chap. 4).

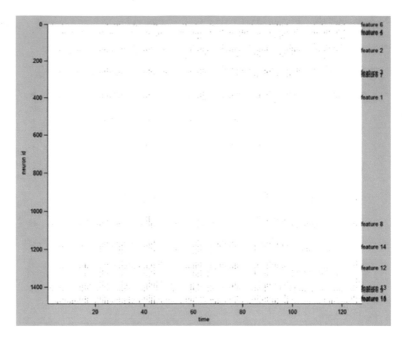

**Fig. 6.7** Raster plot of sample spike output by SNNcube (1471 neurons; 14 inputs/features; 125 ms time)

**Fig. 6.8** Clustering of the spiking neurons in the SNNcube by sowing significant connections from each of the input neurons/ features to the rest of the neurons in the SNNcube

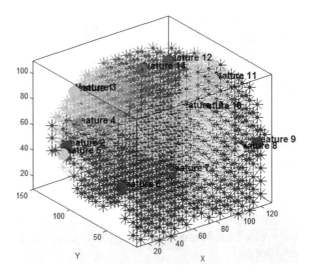

**Fig. 6.9** Clustering of the spiking neurons in the SNNcube by spike communication (number of spikes emitted by each of the input neurons/features and spread to other neurons in the SNNcube)

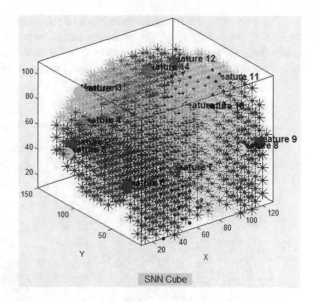

SNN Cube

The recall procedure, when new input data is presented for classification/ regression, can be performed using different recall algorithms applying different methods:

(a) A spike sequence that represents the response of the trained SNNc to new input data is propagated to all trained output neurons and the first one that spikes (its PSP is greater that its threshold) defines the output. The assumption is that the neuron that best matches the input pattern will spike earlier, based on the PSP threshold (membrane potential). This method is called eSNNm (deSNNm).

(b) The second method implies a creation of a new output neuron in the eSNN for each new input pattern from the SNNc, in the same way as the output neurons were created during the learning phase in the eSNN, and then—comparing the connection weight vector of the new one to the already existing neurons using Euclidean distance. The closest output neuron in terms of synaptic connection weights is the 'winner'. This method uses the principle of transductive reasoning and nearest neighbour classification in the connection weight space. It compares spatially distributed synaptic weight vectors of a new neuron that captures a new input pattern and existing ones. This method is called eSNNs (deSNNs) in Chap. 5.

The main advantage of using evolving SNN as classifiers/regressors, when compared with other supervised classification SNN models, is that eSNN are computationally inexpensive and boosts the importance of the order in which input spikes arrive, thus making the eSNN suitable for on-line learning and early prediction of temporal events.

To summarise, the main characteristics of the NeuCube BI-SNN architecture are shown in Box 6.3. NeuCube is an open, evolving framework that is a set of algorithms allowing for the creation of SNN systems and BI-AI systems and also allowing for new algorithms to be developed in the future and explored as part of it.

---

**Box 6.3. The main characteristics of the NeuCube BI-SNN architecture**

---

1. Whole input spatio-temporal patterns are entered and learned, rather than vector by vector.

2. Different temporal length of samples for training and recall is possible.

3. Chain-fire is applied after deep learning in the SNNcube, so that if only part of new input information is entered, the learned pattern in the SNNcube can be triggered leading to early and accurate prediction of an output.

4. Setting an early spike threshold in the classifier/regressor using the rank-order learning.

5. The system is responsive to changes in the input data through spike encoding.

6. Robust to noise through spike encoding.

7. Adaptable on new data.

8. Implementable on any software or hardware platforms.

9. Easily implementable on neuromorphic, highly parallel hardware platforms.

10. Fast, one pass learning of data.

11. The learned connectivity structures are interpretable for new information and deep knowledge discovery.

12. The activity of the SNN during training and recall can be used to extract spatio-temporal rules.

13. Tracing back in time the activity of the SNNcube during training or recall.

14. Incremental, "life-long" learning of streaming data.

15. Virtual Reality visualisation of the learned patterns in the SNNcube.

16. Allowing for different types of spiking neuron models to be experimented for different applications.

17. Allowing for different learning rules for spiking neurons to be experimented.

18. Allowing for building systems of multiple-SNNcubes.

19. Integrating multiple modality of data.

20. Data compression.

**Fig. 6.10** Tracing spiking
activity of a single spiking
neuron from the SNNcube

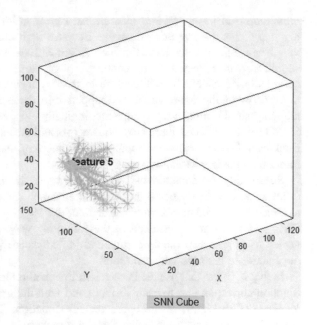

### 6.1.3   Mapping Input Temporal Variables into a 3D SNNcube Based on Graph Matching Optimisation Algorithm

This section introduces a method for mapping input temporal data into a NeuCube SNNcube, without having spatial data associated with it. Details of the presented here method can be found in [14].

Given a particular (spatio-) temporal data set, it is important to optimise the mapping of the data into the 3D SNNcube for an optimal learning of spatio-temporal patterns from this data and for an informative interpretation of the model that can lead to a better understanding of the temporal patterns in the data. For some spatio-temporal data, such as brain EEG, there is prior information about the location of each channel (input feature) and this information can be readily utilized for mapping the EEG temporal signal into the SNNcube [1, 15]. But for other common applications such as climate temporal data, we do not have such spatial mapping into the 3D SNNcube. And the way temporal data is mapped into the SNNcube would significantly impact on the results. Here we introduce a new method to map temporal input variables into the SNNcube for a better pattern recognition, a better and earlier event prediction and a better visualisation of the model to explain the data.

Suppose there are $s$ temporal samples in the data set, measured through $v$ temporal variables and the observed time length of each sample is $t$. We first choose randomly $v$ input neurons from the SNNcube. Then we map the variables into the SNNcube following the following principle: *input variables/features that, after the input data transformation into spike trains, represent highly correlated spike trains are mapped to nearby input neurons.* Because high correlation indicates that the variables are likely to be more time dependent with each other, and this relationship should also be reflected in the connectivity of the 3D SNNcube. Spatially close neurons in the SNNcube will capture in their connections more temporal interactions between the input variables mapped into these neurons. The principle of mapping similar input vectors into topologically close neurons is known from the SOM [16], but in SOM these are static vectors and the similarity is measured by the Euclidean distance. Now, we address the problem of mapping temporal sequences, rather than static vectors, into a 3D SNN.

Specifically, we construct two weighted graphs: the input neuron distance graph (NDG) and the time series/signal correlation graph (SCG). In the NDG, the input neurons' spatial 3D coordinates are denoted by $V_{NDG} = \{(x_i, y_i, z_i)|i = 1 \dots v\}$ and the graph edges are determined in the following way: each input neuron is connected to its $k$ nearest input neurons and the edges are weighted by the inverse of the Euclidean distance between them.

In the SCG, we first use the Parzen window method to estimate the spike density function corresponding to each variable and then the graph vertex set, denoted by $V_{SCG} = \{f_i|i = 1 \dots v\}$, is the spike density function. The graph edges are constructed in this way: each spike density function is connected to its $k$ highest correlated neighbours and the edges are weighted by the statistical correlation between the spike density functions of the input variables.

We adopt the graph matching technique, which is a powerful tool to solve mapping problems and has been widely used in computer vision and pattern recognition, to determine an optimal mapping between any two weighted graphs under the mapping rule. In our case, the two graphs are NDG and SCG. For these two graphs, we can compute their adjacency matrices, written as $A_n$ and $A_s$. The graph matching method is aimed to find out a permutation matrix $P$ that minimizes the following objective function:

$$\min_P ||A_n - PA_sP^T||_F^2 \qquad (6.1)$$

where $||\cdot||_F$ denotes the Frobenius matrix norm. Solving this problem exactly is known to be an NP hard problem due to its combinatorial optimization property. Many algorithms have been proposed to find an approximated solution.

Among these algorithms the Factor Graph Matching (FGM) algorithm [8] has been demonstrated to produce state-of-art results. So here we utilize the FGM algorithm to solve the SCG to NDG mapping problem with the following settings: suppose in NDG the sum of graph edge weights of an vertex, say vertex $i_{NDG} \in V_{NDG}$, to all other vertices is $d(i_{NDG})$, and, similarly, in SCG the sum of graph edge weights of vertex $i_{SCG} \in V_{SCG}$ to all other vertices is $c(i_{SCG})$, then the difference between $d(i_{NDG})$ and $c(i_{SCG})$ reflects the similarity of the positions of $i_{NDG}$ and $i_{SCG}$ in the corresponding graphs. So we define the vertex similarity as:

$$\exp\left(-|d(i_{NDG}) - c(i_{SCG})|^2/2\sigma_n^2\right); \quad i_{NDG}, i_{SCG} = 1\ldots v \tag{6.2}$$

and the edge similarity:

$$\exp\left(-\left|a_{ij}^{NDG} - a_{kl}^{SCG}\right|^2/2\sigma_e^2\right); \quad i,j,k,l = 1\ldots v \tag{6.3}$$

where: $a_{ij}^{NDG}, a_{kl}^{SCG}$ are graph edge weights in NDG and SCG, respectively; $\sigma_n$ and $\sigma_e$ are two parameters to control the affinity between neurons and edges, respectively.

Figures 6.11 and 6.12 show the input mapping result obtained by the proposed method for 2 exemplar temporal data.

The optimal input variable mapping makes it possible for early and accurate event prediction from temporal data. In many applications, such as pest population outbreak prevention, natural disaster warning and financial crisis forecasting, it is important to know the risk of event occurrence as early as possible in order to take some actions to prevent it or make adjustment in time, rather than waiting for the whole pattern of temporal data to be entered into the model. The main challenge in the early event prediction task is that the time length of the recall samples for prediction should be smaller than the time length of the training samples. This principle is shown in Fig. 6.13.

Traditional data modelling methods, such as SVM, kNN, and MLP, are no longer applicable for the early event prediction task, because they require the feature length of a prediction sample to be same as that of the training samples. Furthermore, it is also difficult for traditional methods to model both time and space components of the data because of the close interaction and interrelationship between the temporal variables in the (spatio-) temporal data.

In contrast, the proposed new mapping method would enable early event prediction as the connectivity of a trained SNNcube would reflect on the temporal relationship in the temporal data, so that if part of a new sample is presented this would fire a chain of activities in the SNNcube based on the established connections. This phenomenon is similar to the phenomenon of associative memories in Hopfield networks [17], but here we deal with temporal patterns rather than with static input vectors. Here, this is realized with use of the properties of the Leaky Integrate-and-Fire (LIF) neuronal model when the STDP learning rule is used [18]. An LIF neuron can learn *unsupervisedly* an arbitrary spatio-temporal pattern embedded in complex background spike trains and when its preferred spike sequence appears, the neuron can emit a spike very early at the start of the pattern. The utilized in the SNNcube chain-fire phenomenon was observed in zebra finches HVC area to control the precise temporal structure in birdsong [19], in which the neural activity is propagated in chain network to form the basic clock of the song rhythm. In the SNNcube we have also observed a similar chain-fire phenomenon while spike trains are presented to the network. These features endow the SNNcube with a powerful potential to encode complex spatio-temporal patterns contained in

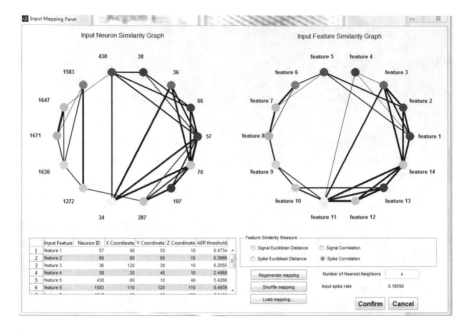

**Fig. 6.11** Matching results for an exemplar temporal data represented by 14 features. The left graph is the input NDG and the right graph is SCG. We can see that after matching, highly correlated features are mapped to nearby input neurons [14]

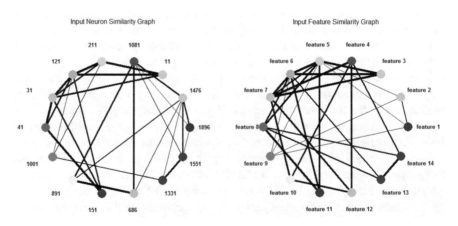

**Fig. 6.12** An input mapping result obtained by the proposed method for exemplar temporal data [14]

the input spike trains used for training and to respond early to the presence of a specific spatio-temporal pattern in a recall/prediction mode. Furthermore, our proposed mapping allows for uneven-length of samples to be mapped to even-length firing states of the neurons in the SNNcube.

**Fig. 6.13** A temporal data
model used for early event
prediction (the system is trained
on a whole input pattern, but
tested/recalled only on initial
part of a new pattern for early
event prediction)

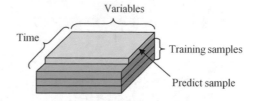

An optimal input variable mapping makes it possible for network structure
analysis and visualization and for a better data understanding. After it is trained, the
SNNcube has captured spatial and temporal relationships from the temporal data. It
is helpful to know how the neurons in the SNNcube are related to the input features
and what patterns have the SNNcube learned from the input signals. This information is important in order to understand the relationship between the model and
the temporal data set. We propose the following algorithm to unveil temporal data
patterns through analysing neuronal clusters in the SNNcube.

The neurons in the cube are indexed from 1 to $N$ according to the ascending
order of their $x$, $y$ and $z$ coordinates. We mark the input neurons as the information
source in the Cube and define a source matrix $F_{src} \in R^{N \times v}$ as follows: if neuron $i$ is
the input neuron to map variable $j$, then the entry $(i, j)$ of $F_{src}$ is 1, otherwise is 0.
The affinity matrix $A \in R^{N \times N}$ of the Cube is defined in the following way: the entry
$(i, j)$ of $A$ is the total spike amount transmitted between neuron $i$ and neuron $j$. Note
that more spike means stronger connections and more information shared between
the neurons. Then ratio of information for each neuron received from the input
information sources is computed as follows:

Step 1. Compute $S = D^{-1/2}AD^{-1/2}$, where $D$ is a diagonal matrix with
$D_{ii} = \sum_{j=1}^{N} A_{ij}$, $i = 1, 2, \ldots, N$

Step 2. Evaluate equation $\tilde{F} = I_{rate}S\tilde{F} + (I - I_{rate})F_{src}$ repeatedly until
convergence.

Step 3. Normalize $F = G^{-1}\tilde{F}$, where $G$ is a diagonal matrix with $G_{ii} = \sum_{j=1}^{C} \tilde{F}_{ij}$.

where $I$ is the identity matrix and $I_{rate}$ is a diagonal matrix defining the propagation
rates on different directions. In the first iteration $\tilde{F} = F_{src}$. The main principle
behind the algorithm is that information (or activation) is propagated in the network
and the propagation process is dominated by the network structure [20, 21]. During
iterations, the information is propagated from the source neurons to other neurons in
the reservoir, with respect to the connection strength between neurons and the
propagation factor matrix defined by $I_{rate}$.

The propagation factor matrix $I_{rate}$ controls the convergence of the propagation
algorithm and the amount of the information being propagated to other neurons in
the Cube. Here $I_{rate}$ is computed by $(I_{rate})_{ii} = \exp\left(-\bar{d}_i^2/2\sigma^2\right)$, where $\bar{d}_i$ is the mean

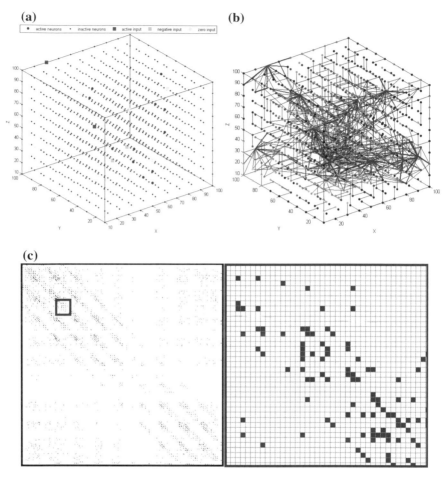

**Fig. 6.14** Snapshots from a dynamic visualisation of a $10 \times 10 \times 10$ SNNcube: **a** a SNNcube spiking state; **b** SNNcube connections with weights larger than 0.08; **c** Connection weight changes. Left: the whole weight matrix; right: magnified illustration of the activity in local clusters of neurons (red dots indicate decreasing and blue dots indicate increasing activities)

affinity value between a neuron and its 8 neighboring neurons, so that the information propagation between strongly connected neurons is large while the information propagated through weakly connected neurons is small.

Visualization of the neuron activities, the connection weight changes and the structure of the SNNcube are important for understanding the patterns in the temporal data and the processes that generated it. Since the SNNcube is a *white box*, we can visualise at each moment the spiking state of the neurons and their connection adjustments. Figure 6.14 shows a snapshot of a neuronal spiking state, the

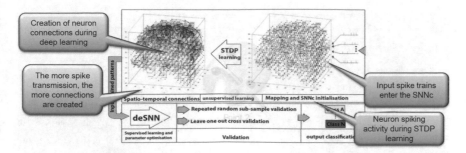

**Fig. 6.15** Deep learning in time-space in NeuCube

connections between neurons and the weight adjustment during SNNcube training. This is very different from traditional methods such SVM (Chap. 1) which have been used for same tasks but without offering facilities to trace the learning processes for the sake of data understanding.

## 6.2 Deep Learning in Time-Space and Deep Knowledge Representation in NeuCube

Learning in a NeuCube model is a two-phase process as it was described in the NeuCube framework above. The accuracy of a NeuCube model depends a great deal on the deep learning applied and on the SNNc learning parameters and the classifier/regressor parameters.

Spike trains are entered into the SNNcube and as result of the STDP or other spike-time learning rules, deep trajectories of connections are formed that represent different patterns of activity of the SNNcube. These patterns are then learned in a supervised way and classified in an output classifier (e.g. deSNN) as illustrated in Fig. 6.15.

This section presents different deep learning algorithms for NeuCube.

### 6.2.1 Deep Unsupervised Learning in Time-Space and Deep Knowledge Representation from Temporal or Spatio/ Spectro Temporal Data (TSTD)

The following is a procedure for deep unsupervised learning in the SNNcube.

---

**Box 6.4. A meta-algorithm for deep unsupervised learning and deep knowledge representation in NeuCube**

---

*1.      Initialisation of the SNN model:*

A model is pre-structured to map structural and functional areas of the modelled process presented by the temporal or spatio-temporal data. The SNN structure consists of spatially allocated spiking neurons, where the location of the neurons maps a spatial template of the problem space (e.g. brain template; geographic locations, etc.) if such information exists.

The input neurons are spatially allocated in this space to map the location of the input variables in the problem space. For temporal data for which spatial information of the input variables does not exist, the variables are mapped in the structure based on their temporal correlation—the more similar temporal variables are, the closer neurons they are mapped into [15]. The connections in the SNN are initialised using a small-world connectivity algorithm [1, 28, 35, 36].

*2.      Encoding of input data:*

Input data is encoded into spike sequences reflecting on the temporal changes in the data using some of the encoding algorithms, e.g. [1, 28, 35, 36].

*3.      Unsupervised learning in the SNN model:*

Unsupervised time-dependent learning is applied in the SNN model on the spike encoded input data. Different spike-time dependent learning rules can be used. The learning process changes connection weights between individual neurons based on the timing of their spiking activity. Through learning individual connections over time, whole areas (clusters) of spiking neurons, that correspond to input variables, connect between each other, forming deep patterns of connectivity of many consecutive clusters in a flexible way. The length of the temporal data and therefore — the patterns learned in the SNN model, are theoretically unlimited.

*4.      Deep knowledge representation) in the SNN model:*

A deep functional pattern is revealed as a sequence of spiking activity of clusters of neurons in the SNN model that represent active functional areas of the modelled process. Such patterns are defined by the learned structural patterns of connections. When same or similar input data is presented to a trained SNN model, the functional patterns are revealed as neuronal activity is propagated through the connectionist patterns. The depth of the obtained functional patterns is theoretically unlimited, depending on the resolution, e.g. from few to thousands and millions, if time is measured in milliseconds in the latter case.

The deep unsupervised learning in the SNNcube results in deep learned connectivity that can be interpreted as deep knowledge in the form as introduced in Chap. 1, linking events Ei and Ej. Events Ei and Ej for example are represented by corresponding functions Fi, Fj spatial locations Si, Sj, times Ti, Tj probabilities of the events to happen Pi, Pj and strength of the connection between the events Wi,j. All parameters of an event can be represented as crisp or as fuzzy values, e.g.:

- Location is around Si;
- Time is about Ti;
- Probability is about Pi (see about fuzzy probabilities in [22]);
- Strength is around; or strength is High;

A hypothetical example of deep knowledge represented as a *deep fuzzy rule* is given below:

*IF (event E1: function F1, location around S1, time about T1, probability about P1)*

  *AND (strength W1,2,)*
  *(event E2: function F2, location around S2, time about T2, probability about P2)*
  *AND (strength W2,3,)*
  *(event E3: function F3, location around S3, time about T3, probability about P3)*
  *AND ...*

*... ... ... ... ... ...*

  *(event En : function Fn, location around Sn, time about Tn, probability about Pn1)*
*THEN (Task event/task Q is executed)*

The fuzzy rule above allows for the event/task Q to be recognised even if only partial match of new data is entered and the rule applied. This is a brain-inspired principle, when we end up with crisp movements as a result of the activation of slightly different clusters of neurons at slightly different times in their sequence, as a reaction to certain crisp of fuzzy stimuli.

As a partial case, no fuzzy terms will be used, but crisp ones, e.g. the following *deep crisp rule*:

*IF (event E1: function F1, location S1, time T1)*

  *AND (strength W1,2,)*
  *(event E2: function F2, location S2, time T2)*
  *AND (strength W2,3,)*
  *(event E3: function F3, location S3, time T3)*
  *AND ...*

*... ... ... ... ... ...*

  *(event En : function Fn, location Sn, time Tn)*
*THEN (Task Q is executed)*

Crisp rules would be a case when activities of single neurons are measured in the brain at exact milliseconds time.

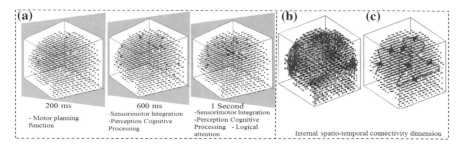

|  | 200 ms<br>- Motor planning<br>function | 600 ms<br>-Sensorimotor Integration<br>-Perception Cognitive<br>Processing | 1 Second<br>-Sensorimotor Integration<br>-Perception Cognitive<br>Processing  - Logical<br>attention | Internal spatio-temporal connectivity dimension |

**Fig. 6.16** (**a**) Illustration of a SNNcube that learns EEG data from 14 EEG channels when a person is moving a hand. (**b**) The connections of the trained SNNcube can be interpreted (see (**c**)) as deep knowledge showing four events E1,..., E4 that are executed at 4 large cortical locations in a sequence of 4 aggregated time segments (the figure is produced by Z. Gholami and M. Gholami) (from [23])

In Fig. 6.16 a SNNcube learns EEG data from 14 EEG channels when a person is moving a hand (Fig. 6.16a). The connections of the trained SNNcube (Fig. 6.16b) can be interpreted as deep knowledge showing four aggregated events E1,..., E4 that are executed at 4 different locations in a sequence of 4 time intervals. This example is also given in Chap. 8 with more detailed explanation (see also [23]).

## 6.2.2   Deep Supervised Learning in Time-Space

*Supervised learning for classification of learned patterns in a SNN model*:

When a SNN model is trained in an unsupervised mode on different temporal data, representing different classes, the SNN model learns different structural and functional patterns. When same data is propagated again through this SNN model, a classifier can be trained using the known labels, to learn to classify new input data that activate similar learned patters in the SNN model.

*Semi-supervised learning*:

The proposed approach allows for training a SNN on a large part of data (unlabelled) and training a classifier on a smaller part of the data (labelled), both data sets related to the same problem. This is how the brain learns too.

The dynamic spike patterns associated with each output class sample (prototype) can be analysed and deep spatio-temporal rules can be extracted, such as shown in Box 6.5 (Fig. 6.17).

---

Box 6.5. A hypothetical example of *deep knowledge representation* in a trained eSNN as a classifier for SNNcube

---

*IF (area (Xi,Yi,Zi) with a cluster radius Ri is activated at time about T1) AND*

   *(area (Xj,Yj,Zj) with a cluster radius Rj is activated at time about T2) AND*

   *(area (Xk,Yk,Zk) with a cluster radius Rk is activated at time about T3) AND*

   *(no other areas of the SNNcube are activated)*

*THEN  (The output class prototype is number 4 from class 1) (see Fig. 17).*

## 6.2.3   Deep Learning in Time-Space for Predictive Modelling in NeuCube. The EPUSSS Algorithm

One of the biggest challenges scientists are facing is making sense of complex dynamic patterns found in multimodal streaming data, 'hidden deep in time'. If such patterns can be interpreted then our ability to explain phenomena in nature, understand the mechanisms of human cognition, and to predict future events will be significantly improved. The current state-of-the-art of Artificial Intelligence (AI) is

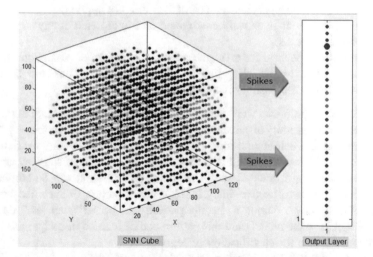

**Fig. 6.17** Colour representation of the time of activity of different clusters in the SNNcube that trigger the classification of prototype 4 from class 1 (in red). The brighter the colour, the earlier these neurons spike to activate a spike in the output neuron representing prototype 4 from class 1 (as a classification result). This can be interpreted as deep in time-space knowledge representation (see Box 6.5)

deep neural networks (DNN) (see Chap. 4). Despite their success in large-scale pattern recognition, they have severe constraints when learning from continuous streaming data. They have a fixed structure that cannot be changed over time; they do not capture patterns from data that include both time and space; they are slow to learn; and require processing of unnecessarily large amounts of data even though they may not be relevant to the outcome, and they are mainly applied on static data sets.

Inspired by the ability of the human brain to learn and predict long temporal sequences (e.g. music, texts, navigation pathways, etc.) here a computational model for *deep-in-time machine learning and predictive modelling of streaming data* is introduced. Data continually streams into a brain-like model at every time point (millisecond, day, etc.). As the patterns of causal relationships are learned, they are retained and modified in the evolving structure of the model. As more temporal data enters the model, the knowledge of the causal relationships becomes more deeply embedded. Our conjecture is that the deeper in time we go during training, the further we will be able to project into the future. If this is achieved, it will allow for forecasting and understanding events, hidden in time and space that occur in complex dynamical systems.

The model here is characterised by the following features:

(1) Patterns are learned from multimodal streaming data over time;
(2) Learning is fast, incremental, adaptive, and theoretically 'life-long';
(3) The model evolves its structure from data;
(4) Supervised and unsupervised modes of learning are applied;
(5) Learned patterns or rules can be extracted and interpreted at any stage during the learning process;
(6) The model can be used for early and accurate future event prediction including events that are hidden deep-in-time. Our intention is that the developed framework will successfully manifest all five features.

The goal of the proposed work is to create an integrated computational model of continuous deep learning of streaming data encoded as spike trains. As only a partial case, the data can be of a fixed time window, but without any restrictions in the length. A model will learn all data entered as a sequence of spike streams and will evolve meaningful internal spatio-temporal patterns that can be stored from time to time for a retrospective analysis. If an input pattern of specified length is entered, with a known output, a specific pattern will be activated (as a deep trajectory of spikes in the model) and this pattern can be learned in a supervised mode in an output module for classification of regression.

In Box 6.6 the EPUSSS learning principles are outlined.

---

**Box 6.6. The EPUSSS learning principles for Evolving Predictive Unsupervised/Supervised Spike Sequences**

---

1. Input variables are mapped in a 3D (2D) eSNN structure according to their 3D (2D) spatial coordinates or the similarity between the input data streams. Initially, all input/output neurons are either fully connected or their mutual connections evolve from the spike sequences using a modified STDP learning rule. These connections are called *meta synapses*. The rest of the connections are initialized using the small-world connectivity rule.

2. Learning from streaming data is as deep as necessary (life-long) as it is the learning in the brain.

3. The learning model is evolving, creating neurons and connections incrementally.

4. A model always learns from a time series data to predict the next values of the input variables, now also used as output variables in the next time point. The hypothesis is that if a model can predict next time-point values of a time series, it has learned the data well. In this approach to learn time series, we have both the input data (temporal variable values at time $t$ and the desired output data (the same variables at the time $(t + k)$ (where by default k = 1, but could be any future time window).

5. Predictive modelling is achieved through a supervised learning in an eSNN structure using error of prediction and a rule that changes the connection weights of the output neurons using this error. The principle is: only output neurons which have spiking error of 1 ( e.g. spike but should not spike) or —1 (do not spike, but should spike) will change their incoming connection weights at a needed depth of backpropagation. A global error is calculated across all output neurons Err(t) that is used for a control of the learning rate (*meta plasticity*).

6. A modified spike-time unsupervised learning in the eSNN is applied for the model to learn temporal associations between all neurons in the model. The model is learning spatio-temporal associations between input (output) variables (in a partial case same neurons can be used to represent both input and output temporal variables) and also across the whole model.

7. Deep learned patterns of spiking activities can be extracted from a model at any time of its training and for any desirable time window. These deep patterns of activities can be interpreted for a better understanding of data.

8. At any time of the model learning process, information 'exchange' between the input (output) variables can be extracted. This shows the one-to-one temporal association (temporal correlation) between changes in the variables that happen one after another. We call this *temporal regulatory graph* of the model variables.

9. The *temporal regulatory graph* can be represented as a set of temporal rules, so that the connection weights (meta synapses) from the input neurons Ii (i = 1,2,...,n) to an output

neuron/variables Oj(t) (j = 1,...,m) can be used to extract temporal association rules representing how Oj(t) depends  the input variables Ii(t —1), e.g.: high, moderate, low, depending on the value of the corresponding connection weight.

10. At any time of the model learning, important features can be extracted as the size of the cluster of connected neurons to each of the input variable.

11. Once a model has learned to predict a spike sequence one step ahead (k = 1) with a satis--factory error, longer term predictions (e.g. several steps ahead) can be modelled when the predicted output spikes at time (t + 1) are used as input spikes at (t + 1) to predict spikes at (t + 2) etc.

12. If an outcome information (e.g. class labels for classification or output scalar values for regression) is available related to certain input patterns, a classifier or regressor can be trained in a supervised learning mode. For this purpose, a whole input pattern is presented to the SNNcube and an output neuron is evolved and connection weights calculated. After training, classification (regression) training/validation error is calculated. The error can be used for:

(a) Modifying the connection weights of the classifier (regressor) to the SNNcube;

(b) Further training of the SNNcube on these particular whole input patterns for which output error is detected.

---

In Box 6.7 the above principles are implemented in an EPUSSS algorithm.

---

**Box 6.7. The EPUSSS meta-algorithm for deep Evolving, Predictive, Unsupervised and Supervised Spike Sequences learning**

---

1. Initialise a 3D SNN structure with sufficient number of neurons using a suitable initialisation of initial weights, e.g. small world connectivity within the structure (ordinary synapses), but fully connected input (output) neurons (meta synapses).

2. Map the input variables into the 3D SNN according to their similarity and define the output variables among them that will be used for supervised learning. As a general rule, all input variables can be used as output variables.

3. For i = 0 till end of the input spike stream DO:

   3a. Enter a spike vector at a time t;

   3b. Calculate the spiking activity of the whole SNN model for the next time (t + k); by default k = 1;

   3c. When the spike vector for time (t + 1) is available, calculate the error Err = (Predicted-Actual) spikes for each of the outputs and the total error as well.

   3d. Apply a Modified Perceptron learning rule to adjust the connection weights to each of the output neuron O using the error calculated for it as follows:

   – if neuron O was supposed to spike at (t + 1) and it did not, increase its incoming connection weights from the neurons connected to it that spiked at the time *t* and also of the incoming connection weights of the neurons connected to O that did not spike but have positive connection weights to the output neuron O;

   – if neuron O was not supposed to spike at (t + 1) and it did, decrease its incoming connection weights from the neurons connected to it that spiked at the time *t* and increase the incoming connection weights of the neurons connected to O that did not spike but have negative connection weights to the output neuron O;

   – if there is no error of the predicted spike of neuron O, do not apply learning.

   – The learning rate will depend on the global error (if the global error is large, then increase the learning rate; if no error, rate is 0).

   3e. The learning rule in 3d is applied several (e.g. 3–6) layers back which is a system parameter that can be optimized and depends on the inner relationship in the data.

   3f. (Optional) Apply a modified spike-time learning rule (e.g. modified STDP) to adjust all connection weights in the SNN model.

   3g. END

---

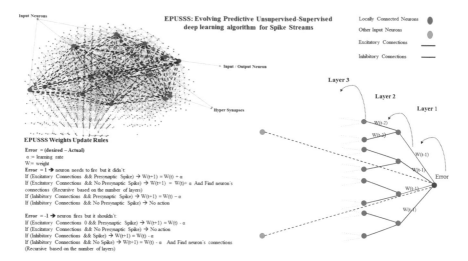

**Fig. 6.18** A graphical representation of the EPUSSS algorithm from Box 6.7 for deep in time-space learning (the figure was drawn by Helena Bahrami)

Figure 6.18 shows a graphical representation of the EPUSSS algorithm from Box 6.7 (the figure was created by H. Bahrami).

## 6.3    Modelling *Time* in NeuCube: The Past, the Present, the Future,... and Back to the Past

### 6.3.1    Event-Based Modelling. External Versus Internal Time. Past-, Present- and Future Time

The evolving processes modelled through spike encoding can also be viewed as *event based* processes. If an event happens (e.g. increase in the intensity of a pixel, or an earthquake), which we will call *external event*, then there will be a spike generated. A spike can be considered an *internal event*, representing changes in the model.

The time of the events measured in the data we call here *external time*. It can very from milliseconds (brain processes) to hundreds of light years (a light year is used to denote the space that the light can travel within a year, having a constant speed of 300,000 km/s).

The time of internal events (spikes) in a SNN model (called model internal time) is usually denoted as millisecond, in analogy with the spike times in the brain, but in a computational model this time is considered as a "time unit" and its duration depends on the implementation of the model.

In many cases, if a SNN model models EEG or EMG brain data measured in milliseconds, the spiking activity in the SNN can also be achieved in milliseconds, if a neuromorphic hardware or other super-computer platforms are used (see Chap. 20).

*Past data*, i.e. data measured in past time before training a model, is usually used to train a model and to capture meaningful patterns explaining the processes.

*Present* data, for which an output value may not exist, is used to recall a SNN model and to obtain a possible output. Once the real output for this data becomes known, these data items become past data and can be used to update the SNN model.

*Future* data can be obtained through predictive modelling, where after input data is entered, a chain of spikes is generated based on the learned connectivity from past data, and an output is calculated for events in a future time. Examples are the individual prediction of stroke (Chap. 18) and one hour ahead prediction of earthquake (Chap. 19).

### 6.3.2  Tracing Events Back in Time

Once a SNN model is created, the connections of the model are directed spatio-temporal associations between spiking activity of neurons that represent changes in the input data. The history of spike and connectivity creation during learning in a SNNcube can be traced back in time following the connectivity backwards. The spiking activity of the model can also be recorded from time to time, so it can be played back (Fig. 6.19).

## 6.4  A Design Methodology for Application Oriented Spatio-temporal Data Machines

Here we will discuss how the BI-SNN can be used to implement application-oriented spatio-temporal data machines for different tasks. These application systems will have features of BI-AI systems as defined in Box 6.8 (also in Box 1.1 of Chap. 1). The creation of BI-AI systems that manifest cognitive features are discussed in other chapters of the book, especially in Chaps. 8–14 and 22.

---

**Box 6.8. Twenty structural, functional and cognitive features of BI-AI systems**

---

*Structural Features*   :

1. The structure and organisation of a system follows the structure and organisation of the human brain through using a 3D brain template.

2. Input data and information is encoded and processed in the system as spikes over time.

3. A system is built of spiking neurons and connections, forming SNN.

4. A system is scalable, from hundreds to billions of neurons and trillions of connections.

5. Inputs are mapped spatially into the 3D system structure.

6. Output information is also presented as spike sequences.

*Functional Features*

7. A system operates in a highly parallel mode, potentially all neurons operating in parallel.

8. A system can be implemented on various computer platforms, but more efficiently on neuromorphic highly parallel platforms and on quantum computers (if available).

9. Self-organised unsupervised, supervised and semi-supervised deep learning is performed using brain-inspired spike-time learning rules.

10. The learned spatio-temporal patterns have a meaningful interpretation.

11. A system operates in a fast, incremental and predictive learning mode.

12. Different time scales of operation, e.g. nanoseconds, milliseconds, minutes, hours, days, millions of years (e.g. genetics), possibly in their integration.

13. A system can process multimodal data from all levels per Fig. 6.1 (e.g. quantum; genetic; neuronal; ensembles of neurons; etc.), possibly in their integration.

*Cognitive features*

14. A system can communicate with humans in a natural language.

15. A system can make abstractions and discover new knowledge (e.g. rules) through self-observing its structure and functions.

16. A system can process all kinds of sensory information that is processed by the human brain, including: visual-, auditory-, sensory-, olfactory-, gustatory, if necessary in their integration.

17. A system can manifest both sub-conscious and conscious processing of stimuli.

18. A system can recognise and express emotions and consciousness.

19. Knowledge can be transferred between humans and machines using brain signals and other relevant information, e.g. visual, etc.

20. BI-AI systems can form societies and communicate between each other and with humans achieving a constructive symbiosis between humans and machines.

---

**Fig. 6.19** Tracing processes back in time. As the model learns patterns of connections, these connections can be traced back and the time of the modelled evolving process can be reversed as a play back to understand better the data and the processes that generated it

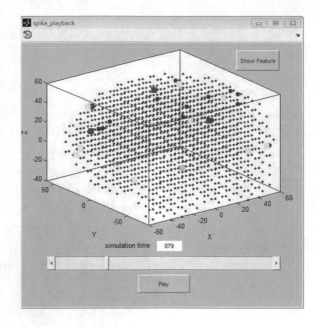

### 6.4.1   Design Methodology for Implementing Application Oriented Spatio-temporal Data Machines as BI-AI Systems in NeuCube

The BI-SNN NeuCube can be used to create BI-AI systems. For the design of an BI-AI system a number of research questions must be addressed. Here, we identify some of them:

1. Which input transformation function to use to encode the data as trains of spikes?
2. Which input variable mapping into the SNNc is used? Is there some a priori information we can use to spatially locate the input variables in the SNNc?
3. Which learning method to use in the SNNc?
4. Which output function is appropriate? Is it classification or regression?
5. How to visualize the developed BI-AI system for an improved understanding?
6. Which parameter optimisation method to apply?

The procedures for the creation of spatio-temporal data machines as BI-AI systems using a BI-SNN are listed in Box 6.9.

---

Box 6.9. A design methodology for spatio-temporal data machines as BI-AI systems using a BI-SNN

---

1. Input data transformation into spike sequences (see Chap. 4);

2. Mapping input variables into spiking neurons

3. Unsupervised learning spatio-temporal spike sequences in a scalable 3D SNN cube

4. Supervised learning and classification/regression of data;

5. Dynamic parameter optimisation

6. Evaluating the predictive modelling capacity of the system

7. Adaptation on new data, possibly in an on-line/real time mode

8. Model visualisation and interpretation for a better understanding of the data and the processes that generated it.

9. Implementation of a SNN model: von Neumann vs neuromorphic hardware versus quantum computing

All these procedures are explained in detail and illustrated in the following sub-sections.

## 6.4.2   Input Data Encoding

The input encoding module converts continuous data streams into discrete spike trains, suitable to be processed in the SNNcube. While in [1, 24–26] threshold-based encoding with fixed threshold was used, it is possible to apply a bi-directional thresholding of the signal gradient with respect to time, $d/dt$. The threshold is self-adaptive and is determined in the following way: for an input time series/signal $f(t)$, we calculate the mean $m$ and standard deviation $s$ of the gradient $df/dt$, then the threshold is set to $m + \alpha s$, where $\alpha$ is a parameter controlling the spiking rate after encoding. After this, we obtain a 'positive' spike train which encodes the segments in the time series with increasing signal and a negative spike train, which encodes the segments of decreasing signal.

There are different coding schemes for SNN, primarily rate (information as mean firing rates) or temporal (information as temporally significant) coding (see Chap. 4). In NeuCube, we use temporal coding to represent information. So far four different spike encoding algorithms have been integrated into the existing implementation of the NeuCube, namely the Ben's Spiker Algorithm (BSA), Temporal Contrast

(Threshold-based), Step-Forward Spike Encoding Algorithm (SF) and Moving-Window Spike Encoding Algorithm (MW) (see Chap. 4).

Different spike encoding algorithms have distinct characteristics when representing input data. BSA is suitable for high frequency signals and because it is based on the Finite Impulse Response technique, the original signal can be recovered easily from the encoded spike train.

Only positive (excitatory) spikes are generated by BSA, whereas all other techniques mentioned here can also generate negative (inhibitory) spikes. Temporal Contrast was originally implemented in hardware [27] in the artificial silicon retina. It represents significant changes in signal intensity over a given threshold, where the ON and OFF events are dependent on the sign of the changes. However if the changes of the signal intensity vary dramatically, it may not be possible to recover the original signal using the encoded spike train. In [24] we propose an improved spike encoding algorithm, SF, to better represent the signal intensity.

For a given signal S(t) where (t = 1, 2,…, n), we define a baseline B(t) variation during time t with $B(1) = S(1)$. If the incoming signal intensity S(t1) exceeds the baseline $B(t1 - 1)$ plus a threshold defined as Th, then a positive spike is encoded at time t1, and B(t1) is updated as $B(t1) = B(t1 - 1) + Th$; and if $S(t1) <= B (t1 - 1) - Th$, a negative spike is generated and B(t1) is assigned as $B(t1) = B (t1 - 1) - Th$. In other situations, no spike is generated and $B(t1) = B(t1 - 1)$.

As to the Moving-Window Spike Encoding Algorithm, the baseline B(t) is defined as the mean of previous signal intensities within a time window T, thus this encoding algorithm can be robust to certain kinds of noise.

Before choosing a proper spike encoding algorithm, we need to figure out what information the spike trains shall carry for the original signals. An example is given in Fig. 6.20. In [28] a methodology to select spike encoding algorithm and to optimise its

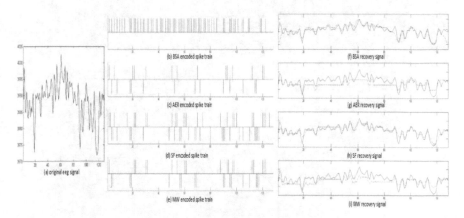

**Fig. 6.20** Illustration of the effect of using different methods for the encoding a continuous signal as part of the input data encoding (see also Chap. 4)

parameters is introduced and implemented as a software called Spiker (http://www.kedri.aut.ac.nz/neucube/).

### 6.4.3   Spatial Mapping of Input Variables

Mapping input variables into spatially located spiking neurons in the SNNc is a new approach towards modelling spatio-temporal introduced in [1] and is a unique feature of the NeuCube architecture and all systems developed with its use. The main principle is that if spatial information about the input variables is known it can help in (a) building more accurate models of the data collected through these variables and (b) a much better interpretation of the model and a better under-standing of the data can be achieved. This is very important for data such as brain data such as EEG (see [1, 29]) and for fMRI data [2], where patterns of interaction of brain signals can be learned and discovered. In some implementations we have used the Talairach brain template, mapped spatially into the SNNc. Another way of mapping, when there is no spatial information available for the input variables, is to measure the temporal similarity between the variables to map variables with similar patters into closer neurons in the SNNc [14]. This is the vector quantisation principle, where by 'vector' here we use time series, which do not necessarily have the same length (Fig. 6.21).

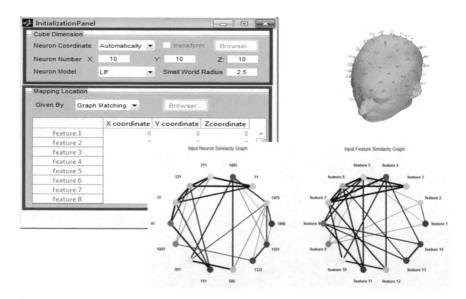

**Fig. 6.21** Spatial mapping of input variables can be done in two ways: mapping input variables using their 3D spatial coordinates, as exemplified with EEG data; mapping temporal variables that do not have spatial coordinates using a temporal similarity measure [14]

### 6.4.4   Unsupervised Training of the SNNcube

The NeuCube is trained in a two-stage learning process. The first stage is unsupervised learning that makes the SNNcube learn (spatio-) temporal relations from the input data by adjusting the connection weights in SNNcube. The second stage is supervised learning that aims at learning the class information associated with each training (spatio-) temporal sample.

The unsupervised learning stage is intended to encode 'hidden' (spatio-) temporal relationships from the input data into neuronal connection weights. According to the Hebbian learning rule, if the interaction between two neurons is persistent, then the connection between them will be strengthened. Specifically, we train the SNNcube using spike-timing dependent synaptic plasticity (STDP) learning rule [30]: if neuron $j$ fires before neuron $i$, then the connection weight from neuron $j$ to neuron $i$ will increase and, if the spikes are in a reverse order, the connection from neuron $i$ to neuron $j$ will decrease. This ensures that the time difference in the input spiking trains, which encode the temporal patterns in the original input signals, will be captured by the neuron firing state and the unsymmetrical connection weights in the reservoir.

The second training stage is to train an output classifier using class label information associated with the training temporal samples. The dynamic evolving Spike Neural Networks (deSNN) [31, 32] is used here as an output classifier, because deSNN is computationally efficient and emphasizes the importance of both first spikes arriving at the neuronal inputs (observed in biological systems [33]) and the following spikes (which in some stream data are more informative).

Once a NeuCube model is trained, all connection weights in the SNNcube and in the output classification layer are established. These connections and weights can change based on further training (adaptation), due to the evolvable characteristic of the architecture. For a given new temporal sample without any class label information, the trained NeuCube model can be used to predict the class label or an output value.

### 6.4.5   Supervised Training and Classification/Regression of Dynamic Spiking Patterns of the SNNcube in a SNN Classifier

Here we use an SNN for the output model of the type eSNN, deSNN (Chap. 5) or SPAN (Chap. 4). An eSNN or deSNN evolves its structure and functionality in an online manner, from incoming information. The training algorithms are given in Chaps. 4 and 5.

**(a)**

**(b)**

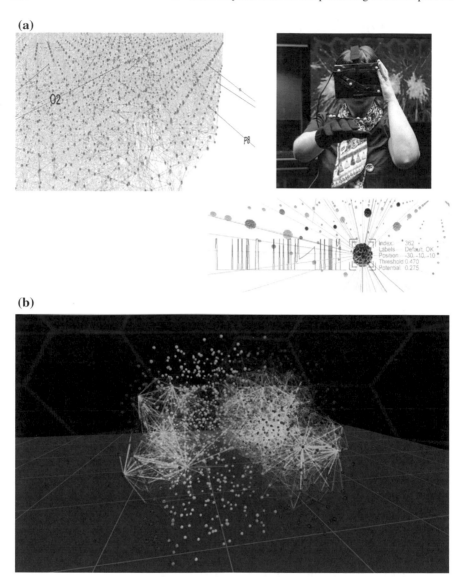

**Fig. 6.22** **a** 3D Visualisation of the SNNcube connectivity and spiking activity during training or recall (from [25]). **b** A VR visualisation of a NeuCube connectivity using the NeuVis system (the figure is created by Stefan Marks) (https://kedri.aut.ac.nz/R-and-D-Systems/virtual-reality)

### 6.4.6   3D Visualisation of the SNNcube

The 3d connectivity structure of the SNNcube during training or recall can be visualised using VR (Fig. 6.22a, b).

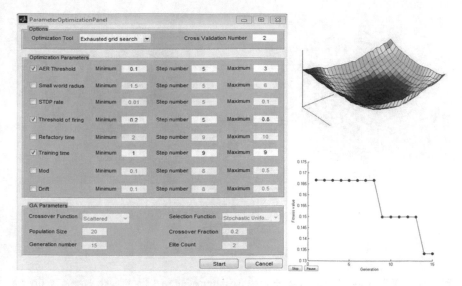

**Fig. 6.23** Using grid search to optimise some of the parameters of a NeuCube SNN model. The error produced by the model reduces by using just 15 generations of the GA optimisation of the parameters selected for optimisation as ticked in the above table

### 6.4.7   Optimisation of NeuCube Structure and Parameters

For an optimal performance of a NeuCube model its numerous parameters need to be optimised. Methods that can be used for the parameter optimisation include:

- Grid search (trying all combinations of the parameter values within a certain range);
- Genetic Algorithm (see Chap. 7);
- Particle Swarm Optimisation (PSO) algorithms (see Chap. 7);
- Quantum inspired Evolutionary Computation (QiEC) (see Chap. 7);
- Quantum Inspire PSO (see Chap. 7).

Using a grid search algorithm is illustrated in Fig. 6.23 to optimise some of the parameters of a NeuCube model. An optimum set of parameter values is found that minimises the classification error as an objective function after 15 iterations. SNN parameter optimisation is an open problem. Current research is directed towards "learning to learn" approach, i.e. a system will not only learn from data but will learn how to optimise its parameters as part of the learning process.

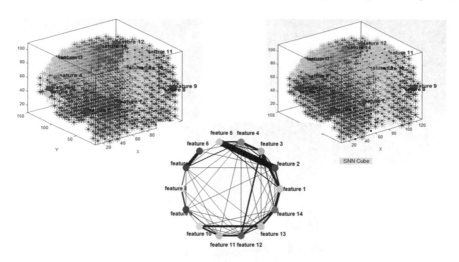

**Fig. 6.24** Different techniques can be used as information and knowledge extraction from a trained SNNcube: clustering according to connection weights; clustering according spiking activity; graph representation of variable (feature) cluster interaction in terms of dynamic information exchange. The latter can be used to extract *deep knowledge representation* linking the activities of different spatially located clusters of neurons over time

## 6.4.8  Model Interpretation, Rule Extraction, Deep in Time-Space Knowledge Representation

The NeuCube SNN model can be analysed for a better understanding the data and the processes that generated it. Different techniques can be used for this purpose, such as dynamic clustering (Fig. 6.24) [34]:

- Clustering according to connection weights;
- Clustering according to spiking activity;
- Graph representation of variable (feature) cluster interaction in terms of information exchange. This information can be used to extract deep knowledge representation linking the activities of different spatially located clusters of neurons over time and the intensity of all these interactions.

## 6.5  Case Studies of the Design and Implementation of Classification and Regression Spatio-temporal Data Machines

Two demo data sets, used for the illustration of the material in this section are available from: http://www.kedri.aut.ac.nz/neucube/. The development of the demo systems here are explained in terms of using the NeuCube development software from the same site.

### 6.5.1  A Case Study on the Design of a Classification Spatio-temporal Data Machine in NeuCube

The dataset used in this case study corresponds to a study participant moving their wrist either up or down, or holding their hand straight. This task was performed on a single subject and EEG data was sampled from 14 channels at a sampling rate of 128 Hz. 20 independent trials of 1 s duration were collected while the subject performed each movement task. The data set consists of the following files:

- Input sample files: Each sample file (sam1.csv, sam2.csv, sam3.csv ... sam60. csv in the example) contains data of one sample. Each sample corresponds to a data arranged in a matrix. The rows correspond to ordered time points, and the columns represent the features (in this case, EEG channels).
- Input target file: The target file stores the class label of each sample in a column, ordered in the same sequence as the numbers of the sample files.
- SNNcube coordinate file: This file describes the spatial coordinates of the neurons in the SNNcube. Every row in the SNNcube coordinate file contains the x, y, and z coordinates of a neuron. Talairach brain template is used, a template that makes it possible to represent any person's brain data in a standardised way. In this case we use 1471 spiking neurons in the SNNcube representing 1 cm$^3$ spatial resolution.
- Input coordinate file: This file describes the spatial location of the input neurons (features). Just like the SNNcube coordinate file, every row in the input coordinate file contains the x, y, and z coordinates of an input neuron.
- Feature name file: This file contains the names of the input features. In this example, it would contain a list of EEG channel names.

Figure 6.25 shows a set of parameters for unsupervised training of the SNNcube for the case study data and also the trained SNNcube connectivity. The SNNcube is structured according to the Talairach brain template with 1471 spiking neurons, each representing 1 cm$^3$ of brain area.

Figure 6.26 shows some parameters and results of the supervised training and classification of the case study data into 3 classes using deSNN classifier [12].

Analysis is performed to better understand the classification results (Fig. 6.27).

### 6.5.2  A Case Study on the Design a Regression/Prediction Spatio-temporal Data Machine in NeuCube

A demo dataset for regression analysis consists of 50 samples (http://www.kedri. aut.ac.nz/neucube/). Each sample consists of 100 timed sequences of daily closing prices for six different shares (Apple Inc., Google, Intel Corp, Microsoft, Yahoo,

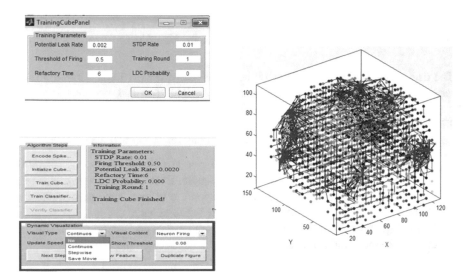

**Fig. 6.25** Parameters for unsupervised training of the SNNcube for the case study data and the trained SNNcube connectivity. The SNNcube is structured according to the Talairach brain template with 1471 spiking neurons, each representing 1 cm³ of brain area

and NASDAQ). The target values representing the closing price of NASDAQ at the next day are arranged in a column in the target file. For dataset like this financial dataset that does not have any natural spatial ordering, NeuCube automatically assigns spatial location based on a graph matching algorithm presented in Sect. 6.2.

In this experiment deSNN is used not as a classifier but as a regressor. This is achieved in the following way: each ouput neuron, corresponding to one input spatio-temporal or temporal sample is assigned a scallar output value from the targeted time series. The training algorithm is not changed. A k-nearest neighbour technique is used to calculate the output variable. Figure 6.28, shows the regression result produced by NeuCube on the demo regression dataset. The graph plots the true and predicted value of the validation samples. It also provides the Mean Squared Error (MSE) and Root Mean Squared Error (RMSE) as measures of performance on the validation set.

## 6.6   Chapter Summary and Further Readings for Deeper Knowledge

This chapter presents principles of brain-inspired SNN architectures exemplified by one of its implementations—NeuCube. Deep learning algorithms for NeuCube and a design methodology for using them to design of BI-AI systems are presented. The

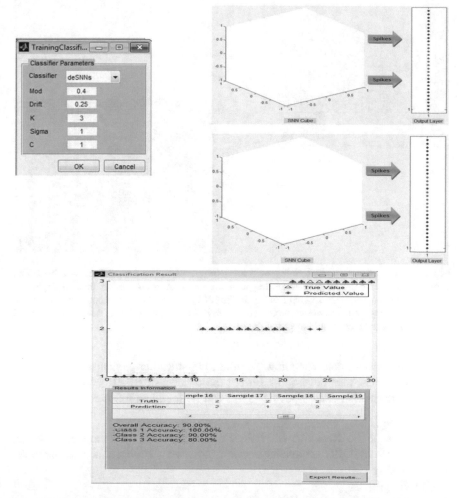

**Fig. 6.26** Parameters and results of supervised training and classification of the case study data into 3 classes using deSNN classifier [12]

chapter discusses how deep knowledge representation can be extracted from a trained SNNcube in an unsupervised mode and then—a trained classifier in a supervised mode. Two case studies are included here to illustrate this methodology and more are presented in the rest of the book. A limited version and an open source version (from 2019) of a NeuCube SNN development system are available from: http://www.kedri.acu.ac.nz/neucube/. NeuCube is an open, evolving framework that is a set of algorithms allowing for the creation of SNN systems and BI-AI systems and also allowing for new algorithms to be developed in the future and explored as part of it. I would like to encourage students and researchers to develop

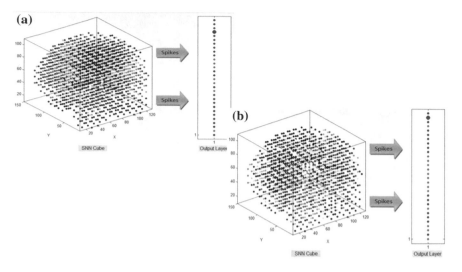

**Fig. 6.27 a** After supervised training of the eSNN classifier, the connections weights of the connections from the SNNcube to prototype 4 are clustered according to their strength (the brighter, the stronger); **b** the spiking neurons in the SNNcube are clustered according to their time of spiking activity related to output prototype 2 from class 1 (the brighter the neuron, the earlier it spikes for the activation of output prototype 2)

new algorithms for data encoding, learning, classification, visualisation and opti-misation of SNN, thus developing further the concept of BI-SNN and BI-AI.

More information about specific topics can be found as follows:

– NeuCube [1];
– Versions of the NeuCube SNN development system, including open source version (from 2019) are available from: http://www.kedri.acu.ac.nz/neucube/.
– eSNN [11];
– deSNN [12];
– design of BI-AI with BI-SNN [35];
– Computational modelling with SNN (Chap. 37 in [2]);
– Brain-like information processing for spatio-temporal pattern recognition (Chap. 47 in [2]);
– Overview of evolving connectionist systems (Chap. 40 in [36]).
– NeuVis: A VR visualization of NeuCube: https://kedri.aut.ac.nz/R-and-D-Systems/virtual-reality.
– A limited executable version and an open source version of the NeuCube development environment are available from http://www.kedri.aut.ac.nz/neucube/
– Applications of NeuCube for various types of problems and data modeling are available from: http://www.kedri.aut.ac.nz/R-and-D/;

**Fig. 6.28** Prediction results on the NASDAQ time series over 25 days, one day ahead, using 100 days back values of 6 stocks

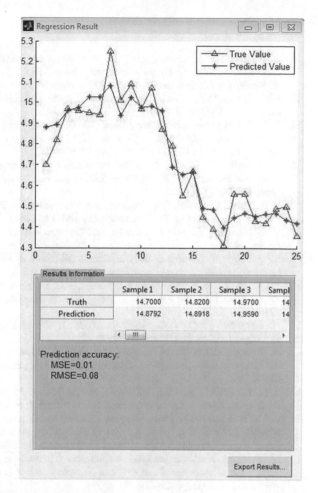

- Applications of NeuCube for spatio-temporal data modelling presented in other chapters of the book include: brain EEG (Chaps. 8, 9); brain fMRI (Chaps. 10, 11); brain fMRI + DTI (Chap. 11); Audio-visual (Chaps. 12, 13); Brain-computer interfaces (Chap. 14); Bioinformatics (Chap. 15); Neurogenetics (Chap. 16); Personalised modelling (Chaps. 17, 18).
- Ecological-, transport-, financial- and seismic data modeling with NeuCube are presented in Chap. 19.
- Implementing NeuCube on hardware platforms, including neuromorphic ones, is presented in Chap. 20.

**Acknowledgements**  Some parts of the material in this chapter have been published as referenced and cited in different sections of this chapter. I acknowledge the contribution of my co-authors in these publications Nathan Scott, Neelava Sengupta, Enmei Tu, Nelson and James from CASIA China, Maryam Gholami, Zohreh Gholami.

# References

1. N. Kasabov, NeuCube: a spiking neural network architecture for mapping, learning and understanding of spatio-temporal brain data. Neural Netw. **52**(2014), 62–76 (2014)
2. N. Kasabov (ed.), *Springer Handbook of Bio-/Neuroinformatics* (Springer, Berlin, 2014)
3. S.M. Bohte, The evidence for neural information processing with precise spike-times: a survey. Nat. Comput. **3** (2004)
4. T. Delbruck, jAER open source project (2007). http://jaer.wiki.sourceforge.net
5. P. Lichtsteiner, T. Delbruck, A 64 × 64 AER logarithmic temporal derivative silicon retina. Res. Microelectron. Electron. **2**, 202–205 (2005)
6. N. Nuntalid, K. Dhoble, N. Kasabov, in *EEG Classification with BSA Spike Encoding Algorithm and Evolving Probabilistic Spiking Neural Network*, LNCS, vol. 7062 (Springer, Heidelber, 2011), pp. 451–460
7. E. Bullmore, O. Sporns, Complex brain networks: graph theoretical analysis of structural and functional systems. Nat. Rev. Neurosci. **10**, 186–198 (2009)
8. V. Braitenberg, A. Schüz, *Statistics and Geometry of Neuronal Connectivity* (Springer, Berlin, 1998)
9. B. Hellweig, A quantitative analysis of the local connectivity between pyramidal neurons in layers 2/3 of the rat visual cortex. Biol. Cybern. **82**, 111–121 (2000)
10. Z.J. Chen, Y. He, P. Rosa-Neto, J. Germann, A.C. Evans, Revealing modular architecture of human brain structural networks by using cortical thickness from MRI. Cereb. Cortex **18**, 2374–2381 (2008)
11. N. Kasabov, *Evolving Connectionist Systems: The Knowledge Engineering Approach* (Springer, London, 2007) (first edition 2002)
12. N. Kasabov, K. Dhoble, N. Nuntalid, G. Indiveri, Dynamic evolving spiking neural networks for on-line spatio- and spectro-temporal pattern recognition. Neural Netw. **41**, 188–201 (2013)
13. A. Mohemmed, S. Schliebs, S. Matsuda, N. Kasabov, SPAN: spike pattern association neuron for learning spatio-temporal sequences. Int. J. Neural Syst. **22**(4), 1–16 (2012)
14. E. Tu, N. Kasabov, J. Yang, Mapping temporal variables into the NeuCube spiking neural network architecture for improved pattern recognition, predictive modelling and understanding of stream data. IEEE Trans. Neural Netw. Learn. Syst. **28**(6), 1305–1317 (2017)
15. N. Kasabov, NeuCube evospike architecture for spatio-temporal modelling and pattern recognition of brain signals, in *Artificial Neural Networks in Pattern Recognition* (Springer, Heidelberg, 2012), pp. 225–243
16. T. Kohonen, Self organising maps. Neural Comput. Appl. **7**, 273–286 (1998) (Springer)
17. J. Hopfield, Neural networks and physical systems with emergent collective computational abilities. Proc. Natl. Acad. Sci. U S A **79**(1982), 2554–2558 (1982)
18. T. Masquelier, R. Guyonneau, S.J. Thorpe, Spike timing dependent plasticity finds the start of repeating patterns in continuous spike trains. PLoS ONE **2008**(3), e1377 (2008)
19. Y. Ikegaya, G. Aaron, R. Cossart, D. Aronov, I. Lampl, D. Ferster et al., Synfire chains and cortical songs: temporal modules of cortical activity. Science **304**, 559–564 (2004)
20. J. Shrager, T. Hogg, B.A. Huberman, Observation of phase transitions in spreading activation networks. Science **236**(1987), 1092–1094 (1987)
21. D. Zhou, O. Bousquet, T.N. Lal, J. Weston, B. Schölkopf, Learning with local and global consistency. NIPS **2004**, 595–602 (2004)
22. N. Kasabov, *Foundations of Neural Networks, fuzzy Systems and Knowledge Engineering* (MIT Press, Cambridge, 1996)
23. Z. Doborjeh, N. Kasabov, M. Doborjeh, A. Sumich, Modelling Peri-Perceptual Brain Processes in a Deep Learning Spiking Neural Network Architecture. Nature, Scientific Reports **8**(8912) (2018). https://doi.org/10.1038/s41598-018-27169-8; https://www.nature.com/articles/s41598-018-27169-8

24. N. Kasabov, L. Zhou, M. Gholami Doborjeh, J. Yang, New algorithms for encoding, learning and classification of fMRI data in a spiking neural network architecture: a case on modelling and understanding of dynamic cognitive processes. IEEE Trans. Cogn. Dev. Syst. (2017). https://doi.org/10.1109/TCDS.2016.2636291
25. S. Marks, Immersive visualisation of 3-dimensional spiking neural networks. Evolving Syst. **8**, 193–201 (2017)
26. N. Kasabov, V. Feigin, Z.G.Y.C. Hou, L. Liang, R. Krishnamurthi et al., Evolving spiking neural network method and systems for fast spatio-temporal pattern learning and classification and for early event prediction with a case study on stroke. Neurocomputing **134**, 269–279 (2014)
27. T. Delbruck, P. Lichtsteiner, Fast sensory motor control based on event-based hybrid neuromorphic-procedural system, in *ISCAS* 2007, New Orleans, LA, pp. 845–848 (2007)
28. B. Petro, N. Kasabov, R. Kiss, Spiker: selection and optimisation of spike encoding methods for spiking neural networks, algorithms, (submitted). Software: http://www.kedri.aut.ac.nz/neucube/.
29. N. Kasabov, E. Capecci, Spiking neural network methodology for modelling, classification and understanding of EEG spatio-temporal data measuring cognitive processes. Inf. Sci. **294**, 565–575 (2015)
30. S. Song, K.D. Miller, L.F. Abbott, Competitive Hebbian learning through spike-timing-dependent synaptic plasticity. Nat. Neurosci. **3**(2000), 919–926 (2000)
31. K. Dhoble, N. Nuntalid, G. Indiveri, N. Kasabov, Online spatio-temporal pattern recognition with evolving spiking neural networks utilising address event representation, rank order, and temporal spike learning, in *The 2012 International Joint Conference on Neural Networks (IJCNN)*, IEEE, pp. 1–7 (2012)
32. J. Behrenbeck, Z. Tayeb, C. Bhiri, C. Richter, O. Rhodes, N. Kasabov, S. Furber, J. Conrad, Classification and Regression of Spatio-Temporal EMG Signals using NeuCube Spiking Neural Network and its implementation on SpiNNaker Neuromorphic Hardware. J. Neural Eng. (IOP Press, Article reference: JNE-102499) (2018). http://iopscience.iop.org/journal/1741-2552
33. S. Thorpe, J. Gautrais, Rank order coding, in *Computational Neuroscience* (Springer, New York, 1998), pp. 113–118
34. M.G. Doborjeh, N. Kasabov, Z.G. Doborjeh, Evolving, dynamic clustering of spatio/spectro-temporal data in 3D spiking neural network models and a case study on EEG data. Evolving Syst. 1–17 (2017)
35. N. Kasabov, N. Scott, E. Tu, S. Marks, N. Sengupta, E. Capecci, M. Othman, M. Doborjeh, N. Murli, R. Hartono, J. Espinosa-Ramos, L. Zhou, F. Alvi, G. Wang, D. Taylor, V. Feigin, S. Gulyaev, M. Mahmoudh, Z.-G. Hou, J. Yang, Design methodology and selected applications of evolving spatio-temporal data machines in the NeuCube neuromorphic framework. Neural Netw. **78**, 1–14 (2016). https://doi.org/10.1016/j.neunet.2015.09.011
36. N. Kasabov, Evolving connectionist systems: from neuro-fuzzy-, to spiking—and neurogenetic, in *Springer Handbook of Computational Intelligence*, ed. by J. Kacprzyk, W. Pedrycz (Springer, Berlin, 2015), pp. 771–782

# Chapter 7
# Evolutionary- and Quantum-Inspired Computation. Applications for SNN Optimisation

The chapter introduces the main principles and several algorithms of both evolutionary computation (EC) and its further development as quantum inspired evolutionary computation (QiEC). Evolution in nature is the slowest evolving process in time (takes thousands to millions of years for species to evolve through genetic reproduction), while quantum processes are the fastest (take about nano-seconds and pico-seconds in time).

The EC methods presented here include genetic algorithms and particle swarm optimization, while the QiEC methods include a versatile QiEC method and a quantum inspired particle swarm optimization method. The algorithms presented are for general use. They are also applied in the chapter for the optimization of evolving SNN structures and parameters. The results demonstrate that the QiEC methods lead to a faster and more accurate optimization. QiEC methods can be applied for the optimization of features and parameters of any other SNN and ANN models.

The chapter has the following sections:

7.1. Principles of Evolution in Nature and Methods of Evolutionary Computation.
7.2. Quantum Inspired Evolutionary Computation: Methods and Algorithms.
7.3. Quantum Inspired Evolutionary Computation for the Optimisation of SNN.
7.4. Quantum Inspired Particle Swarm Optimisation Algorithms.
7.5. Quantum Inspired Particle Swarm Algorithms for the Optimisation of SNN.
7.6. Chapter Summary and Further Readings for Deeper Knowledge.

© Springer-Verlag GmbH Germany, part of Springer Nature 2019
N. K. Kasabov, *Time-Space, Spiking Neural Networks and Brain-Inspired Artificial Intelligence*, Springer Series on Bio- and Neurosystems 7,
https://doi.org/10.1007/978-3-662-57715-8_7

## 7.1    Principles of Evolution and Methods of Evolutionary Computation

### 7.1.1    The Origin and the Evolution of Life

Evolution of species started after the origin of Life. The most obvious example of an evolutionary process is the evolution of Life. Life is defined in the Concise Oxford English Dictionary as "A state of functional activity and continual change peculiar to organized matter, and especially to the portion of it constituting an animal or plant before death; animate existence; being alive".

It is generally agreed that all Life today evolved by common descent from a single primitive lifeform. We do not know how this early form came about, but scientists suggest that it was a natural process which took place perhaps 3900 million years ago.

Charles Darwin suggested in 1871 that the original spark of life may have begun in a "...warm little pond, with all sorts of ammonia and phosphoric salts, lights, heat, electricity, etc. A protein compound was then chemically formed ready to undergo still more complex changes"....

*On the Origin of Species*, published on 24 November 1859, is a work of scientific literature by Charles Darwin which is considered to be the foundation of evolutionary biology. Darwin's book introduced the scientific theory that populations evolve over the course of generations through a process of natural selection. It presented a body of evidence that the diversity of life arose by common descent through a branching pattern of evolution. Darwin included evidence that he had gathered on the *Beagle* expedition in the 1830s and his subsequent findings from research, correspondence, and experimentation.

Nature's diversity of species is tremendous. How does mankind evolve in the enormous variety of variants—in other words, how does nature solve the optimisation problem of perfecting mankind? An answer to this question may be found in Charles Darwin's theory of evolution (1858).

**Charles Darwin (1809–1892)** developed a theory according to which evolution is concerned with the development of generations of populations of individuals governed by fitness criteria [1]. But this process is much more complex, as individuals, in addition to what nature has defined for them, develop in their own way —they learn and evolve during their lifetime.

Charles Darwin favoured the "Mendelian heredity" explanation that states that features are passed from generation to generation. In the early 1800's Jean-Baptiste Lamarck had expanded the view that changes in individuals over the course of their life were passed on to their progeny. This perspective was adopted by Herbert Spencer and became an established view along with the Darwin's theory of evolution.

Evolution is a process whereby populations are altered over time and may split into separate branches, hybridize together, or terminate by extinction. The evolutionary branching process may be depicted as a phylogenetic tree, and the place of each of the various organisms on the tree is based on a hypothesis about the sequence in which evolutionary branching events occurred.

In biology, **phylogenetics** is the study of evolutionary relationships among groups of organisms (e.g. species, populations), which are discovered through molecular sequencing data and morphological data matrices. Phylogenetic analyses have become essential to research on the evolutionary tree of life.

Natural evolution inspired the development of the theory of Evolutionary computation (EC). It is based on learning through evolution. It uses principles of the evolution theory, such as:

- Species adapt through genetic evolution (e.g. crossover and mutation of genes) in populations over generations.
- Genes are carrier of information: stability versus plasticity.
- A set of chromosomes define an individual.
- Survival of the fittest individuals within a population.

## 7.1.2  Methods of Evolutionary Computation (EC)

EC are stochastic search methods that mimic the behaviour of natural biological evolution. They differ from traditional optimization techniques in that they involve a search from a *population* of solutions, not from a single point, and carry this search over *generations*. So, EC methods are concerned with population-based search and optimisation of individual systems through generations of populations [2–5].

Several different types of evolutionary computation methods have been developed independently. These include:

- Genetic Programming (GP) which evolve programs [3];
- Evolutionary Programming (EP), which focuses on optimizing continuous functions without recombination [3];
- Evolutionary Strategies (ES), which focuses on optimizing continuous functions with recombination;
- Genetic Algorithms (GAs), which focuses on optimizing general combinatorial problems, the latter being the most popular technique [2, 4];
- Particle Swarm Intelligence [6];
- Firework EC algorithms [7].

EC has been applied so far to the optimisation of different structures and processes, one of them being the connectionist structures and connectionist learning processes [8, 9].

Methods of EC include in principle two stages:

1. A stage of creating new population of individuals.
2. A stage of development of the individual systems, so that a system develops and evolves through interaction with the environment, which is also based on the genetic material embodied in the system.

These stages are illustrated in Box 7.1.

---

**Box 7.1.** A typical EC meta- algorithm

---

1. Initialize population of possible solutions

2. WHILE a criterion for termination is not reached DO
    {
        2a. Crossover two specimens ("mother and father") and generate new individuals;
        2b. Select the most promising ones, according to a fitness function;
        2c. Development (if at all);
        2d. Possible mutation (rare) }
    }

---

The process of individual (internal) development has been ignored or neglected in many EC methods as insignificant from the point of view of the long process of generating hundreds of generations, each of them containing hundreds and thousands of individuals.

But my personal concern as an individual—and also as the author of the book—is that it matters to me not only how much I have contributed to the improvement of the genetic code of the population that is going to live, possibly, 2,000,000 years after me, but also how I can improve myself during my life time, and how I evolve as an individual in a particular environment, making the best out of my genetic material.

ECOS (including evolving SNNs and brain-inspired evolving SNNs) deal with the process of interactive off-line or on-line adaptive learning of a single system that evolves from incoming data. The system can either have its parameters (genes) predefined, or it can be self-optimised during the learning process starting from some initial values. But ECOS can also improve their performance and adapt better to a changing environment through evolution, i.e. through population-based improvement over generations of many ECOS models.

There are several ways in which EC and ECOS can be interlinked. For example, it is possible to use EC to optimise the parameters of an ECOS at a certain time of their operation, or to use the methods of ECOS for the development of the individual systems (individuals) as part of the global EC process.

Before we discuss methods for using EC for the optimisation of connectionist systems and evolving SNN and brain-inspired SNN in particular, I will present a short introduction to some of the most popular EC techniques—genetic algorithms (GA), evolutionary strategies (ES) and particle swarm optimisation.

### 7.1.3   Genetic Algorithms

Genetic algorithms (GA) were introduced for the first time in the work of John Holland in 1975. They were further developed by him and other researchers ([2–5]).

The most important terms used in GA are analogous to the terms used to explain the evolution processes. They are:

- gene—a basic unit that defines a certain characteristic (property) of an individual;
- chromosome—a string of genes; used to represent an individual or a possible solution to a problem in the solution space population—a collection of individuals;
- crossover (mating) operation—substrings of different individuals are taken and new strings (offspring) are produced mutation—random change of a gene in a chromosome;
- fitness (goodness) function—a criterion which evaluates how good each individual is;
- selection—a procedure of choosing a part of the population which will continue the process of searching for the best solution, while the other individuals "die".

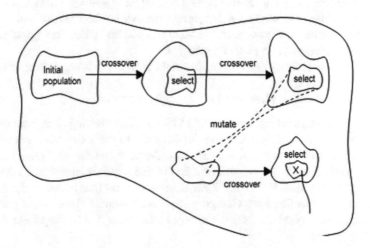

**Fig. 7.1** A schematic diagram of GA evolutionary process, starting with an initial population of solutions (represented as chromosomes) and ending with a good solution according to a set fitness function (optimisation criterion)

A simple genetic algorithm consists of steps shown in Fig. 7.1. The process over time has been "stretched" in space.

When using the GA method for a complex multi-option optimisation problem, there is no need for in-depth problem knowledge, nor is there a need for many data examples stored beforehand. What is needed here is merely a "fitness" or "goodness" criterion for the selection of the most promising individuals (they may be partial solutions to the problem). This criterion may require a mutation as well, which is a heuristic approach of "trial and error" type. This implies keeping (recording) the best solutions at each stages.

Many complex optimisation problems find their way to a solution through genetic algorithms. Such problems are, for example, the Travelling Salesman Problem (TSP)—finding the cheapest way to visit n towns without visiting a town twice; the Min Cut problem—cutting a graph with minimum links between the cut parts; adaptive control; applied physics problems; optimisation of the parameters of complex computational models; optimisation of neural network architectures [8] and finding fuzzy rules and membership functions [10].

The main issues in using genetic algorithms relate to the choice of genetic operations (crossover, selection, mutation). In case of the Travelling Salesman problem the crossover operation can be merging different parts of two possible roads ("mother" and "father" roads) until new usable roads are obtained. The criterion for the choice of the most prospective ones is minimum length (or cost).

A GA offers a great deal of parallelism as each branch of the search tree for a best individual can be utilised in parallel with the others. That allows for an easy realisation on parallel architectures. GAs are search heuristics for the "best" instance in the space of all possible instances. A GA model requires the specification of the following features:

- Encoding scheme, i.e. how to encode the problem in terms of the GA notation—what variables to choose as genes, how to construct the chromosomes, etc.
- Population size: how many possible solutions should be kept in a population for their performance to be further evaluated.
- Crossover operations—how to combine old individuals and produce new, more prospective ones.
- Mutation heuristic—when and how to apply mutation.

In short, the major characteristics of a GA are the following. They are heuristic methods for search and optimisation. In contrast to the exhaustive search algorithms, GA do not evaluate all variants in order to select the best one. Therefore they may not lead to the perfect solution, but to one which is closest to it taking into account the time limits. But nature itself is imperfect too (partly due to the fact that the criteria for perfection keep changing), and what seems to be close to perfection according to one "goodness" criterion may be far from it according to another.

*Selection, crossover and mutation operators*

The theory of GA and the other EC techniques includes different methods for selection of individuals from a population, different crossover techniques and different mutation techniques.

*Selection* is based on fitness that can employ several strategies. One of them is proportional fitness, i.e. "if A is twice as fit as B, A has twice the probability of being selected". This is implemented as roulette wheel selection and gives chances to individuals according to their fitness evaluation.

Other selection techniques include tournament selection (e.g. at every time of selection the roulette wheel is turned twice, and the individual with the highest fitness is selected), rank ordering, and so on [8]. Important feature of the selection procedure is that fitter individuals are more likely to be selected.

The selection procedure may also involve keeping the best individuals from previous generations (if this principle was used by Nature, Leonardo Da Vinci would still be alive, as he was one of the greatest artists ever, presumably having the best artistic genest). This operation is called *elitism*.

After the best individuals are selected from a population, a *crossover* operation is applied between these individuals. The crossover operator defines how individuals (e.g. "mother" and "father") exchange genes when creating the off-spring. Different crossover operations can be used, such as one-point crossover, two-points cross over, etc.

*Mutation* can be performed in several ways, e.g.:

- For a binary chromosome, just randomly "flip" a bit (a gene allele).
- For a more complex chromosome structure, randomly select a site, delete the structure associated with this site, and randomly create a new sub-structure.

Some EC methods just use mutation and no crossover ("asexual reproduction"). Normally, however, mutation is used to search in a "local search space", by allowing small changes in the genotype (and therefore hopefully in the phenotype) as it is in the evolutionary strategies (ES).

## 7.1.4  Evolutionary Strategies (ES)

Another EC technique is called Evolutionary Strategies (ES). These techniques use only one chromosome and a mutation operation, along with a fitness criterion, to navigate in the solution (chromosomal) space.

In the reproduction phase, the current population called the parent population is processed by a set of evolutionary operators to create a new population called the offspring population. The evolutionary operators include two main operators: mutation and recombination, both imitate the functions of their biological counterparts. Mutation causes independent perturbation to a parent to form an offspring and is used for diversifying the search. It is an asexual operator because it involves only one parent. In GA, mutation flips each binary bit of a parent string at a small, independent probability $p_m$ (which is typically in the range [0.001, 0.01]) to create an offspring. In ES, mutation is the addition of a zero-mean Gaussian random number to a parent individual to create the offspring. Let $s_{PA}$ and $s_{OF}$ denote the parent and offspring vector, they are related through the Gaussian mutation

$$s_{OF} = s_{PA} + z \quad z \sim N(0, s) \tag{7.1}$$

where $N(a, s)$ represents a normal (Gaussian) distribution with a mean $a$ and a covariance s and "$\sim$" denotes sampling from the corresponding distribution. ES uses the mutation as the main search operator.

The selection operator is probabilistic in GA and deterministic in ES. Many heuristic designs, like the Rank-based selection that assigns to the individuals a survival probability proportional (or exponentially proportional) to their ranking, have also been studied. The selected individuals then become the new generation of parents for reproduction. The entire evolutionary process iterates until some stopping criteria is met. The process is essentially a Markov Chain, i.e. the outcome of one generation depends only on the last. It has been shown that under certain design criteria of the evolutionary operators and selection operator, the average fitness of the population increases and the probability of discovering the global optimum tends towards unity. The search could however, be lengthy.

## 7.1.5 Particle Swarm Optimisation

In a GA optimisation procedure, a solution is found based on the best individual represented as a chromosome, where there is no communication between the individuals.

Particle Swarm Optimization (PSO), introduced by Kennedy and Eberhard (1995) [6] is motivated by social behaviour of organisms such as bird flocking, fish schooling, and swarm theory. In a PSO system, each particle is a candidate solution to the problem at hand. The particles in a swarm fly in multi-dimensional search space, to find an optimal or sub-optimal solution by competition as well as by cooperation among them. The system initially starts with a population of random solutions. Each potential solution, called a *particle*, is given a random position and velocity.

The particles have memory and each particle keeps track of its previous best position and the corresponding fitness. The previous best position is called the *pbest*. Thus, *pbest* is related only to a particular particle. The best value of all particles' *pbest* in the swarm is called the *gbest*. The basic concept of PSO lies in accelerating each particle towards its *pbest* and the *gbest* locations at each time step. This is illustrated in Fig. 7.2 for a two-dimensional space.

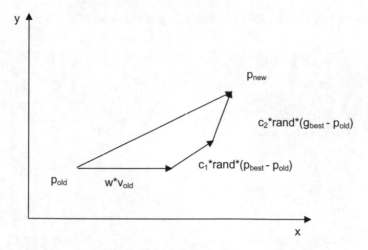

**Fig. 7.2** A graphical representation of the main idea of PSO using both local and global best solutions at the moment

PSO have been developed for continuous, discrete and binary problems. The representation of the individuals varies for the different problems. Binary Particle Swarm Optimization (BPSO) uses a vector of binary digits representation for the positions of the particles. The particle's velocity and position updates in BPSO are performed by the following equations:

$$v_{new} = w * v_{old} + c_1 * rand() * (p_{best} - p_{old}) + c_2 * rand() * (g_{best} - p_{old}) \quad (7.2)$$

$$p_{new} = \begin{cases} 0 & if\ r \geq s(v_{new}) \\ 1 & if\ r < s(v_{new}) \end{cases} \quad (7.3)$$

where:

$$s(v_{new}) = \frac{1}{1 + \exp(-v_{new})} \ and\ r \sim U(0, 1) \quad (7.4)$$

The velocities are still in the continuous space. In BPSO, the velocities are not considered as velocities in the standard PSO but are used to define probabilities that a bit flip will occur. The inertia parameter, $w$, is used to control the influence of the previous velocity on the new velocity. The term with $c_1$ corresponds to the cognitive acceleration component and helps in accelerating the particle towards the *pbest* position. The term with $c_2$ corresponds to the social acceleration component which helps in accelerating the particle towards the *gbest* position.

A simple version of a PSO procedure is given in Box 7.2.

---

**Box 7.2.** A pseudo code of a PSO algorithm

---

begin

        t ← 0 (time variable)

        1)  Initialize a population with random positions and velocities

        2)  Evaluate the fitness

        3)  Select the *pbest* and *gbest*

            while (termination condition is not met) do

            begin

                t ← t+1

        4)  Compute velocity and position updates

        5)  Determine the new fitness

        6)  Update the *pbest* and *gbest* if required

            end

end

---

## 7.1.6 Estimation of Distribution Algorithms (EDA)

In recent years another derivate of EC was proposed that uses a probabilistic model to guide its search. Let us consider a single gene in a chromosome. Such a gene can take different values according to a predefined alphabet. By associating a probability to each character of the alphabet a probability distribution for a gene can be defined. This distribution is usually unknown but can be estimated and iteratively learned by an algorithm. These algorithms represent a new class of EA and are called Estimation of Distribution Algorithms (EDA). An excellent survey about previous work in this field can be found in [11]. To generate new solutions in an EDA a sample from each distribution of each gene is drawn and formed to a complete chromosome.

By sampling a number of chromosomes a population of individuals is obtained which can be evaluated using a fitness criterion. A subset of appropriate individuals is selected and used to update the estimate of the current probability distribution. This process is iteratively repeated until a termination condition is met or all distributions have converged to some specific state.

A generic description of an EDA is illustrated in the Algorithm in Box 7.3.

---

**Box 7.3.** Estimation of Distribution Algorithm (EDA)

---

**Algorithm 1** Estimation of Distribution Algorithm (EDA)

1: $t \Leftarrow 0$
2: initialize the probabilistic model $\mathcal{P}(t)$
3: **while** not termination condition **do**
4:    sample $M$ new solutions from $\mathcal{P}(t)$ into $D(t)$
5:    evaluate the elements of $D(t)$
6:    select $L \leq M$ solutions from $D(t)$ into $D_s(t)$ using a selection method
7:    learn the probabilistic model $\mathcal{P}(t+1)$ from $D_s(t)$ and eventually from $\mathcal{P}(t)$
8:    $t \Leftarrow t+1$
9: **end while**

---

As stated in [12] one can divide EDA into three different classes based on the level of interactions between the variables (genes) that their models can represent. Here we are mainly interested in the first family of EDA that assumes independent variables and for which the probabilistic model is a simple vector of probabilities, such as population-based incremental learning (PBIL) [13], compact GA (cGA) [14] and univariate marginal distribution algorithm (UMDA) [11]. This family of EDA has been very useful to apply to complex optimisation problems, but some aspects of how they operate are still unclear [15]. A further development of EDA are the quantum-inspired algorithms as discussed further in this chapter.

## 7.1.7   Artificial Life Systems

The main characteristics of life are also main characteristics of a modelling paradigm called Artificial Live (ALife), namely:

(1) Self-organisation and adaptation
(2) Reproducibility
(3) Population/generation based.

A popular example of an ALife system is the so called Conways' Game of Life [16]: Each cell in a 2D grid can be in one of the two states either "on" (alive) or "off" (dead, unborn). Each cell has eight neighbours, adjacent across the sides and corners of the square.

Whether cells stay alive, die, or generate new cells depends upon how many of their eight possible neighbours are alive and is based on the following transition rule:

Rule S23/B3:    A live cell with two live neighbours, or any cell with three neigbhours, is alive at the next time step.

Example 1:      If a cell is *off* and has 3 living neighbours (out of 8), it will become alive in the next generation.

Example 2:      If a cell is *on* and has 2 or 3 living neighbours, it survives; otherwise, it dies in the next generation.

Example 3:      A cell with less than 2 neighbours will die of loneless and a cell with more then 3 neighbours will die of overcrowding.

In this interpretation, the cells (the individuals) never change the above rules and behave in this manner forever (until there is no individual left in the space). A more intelligent behaviour would be if the individuals change their rules of behaviour based on additional information they are able to collect. For example, if the whole population is likely to become extinct, than the individuals would create more offspring, and if the space became too crowded, the individual cells would not reproduce every time they are "forced" to reproduce by the current rule. In this case we are talking about emerging intelligence of the artificial life ensemble of individuals. Each individual in the Game of Life can be implemented as an ECOS that has connections with its neighbours and has three initial exact (or fuzzy) rules implemented, but at a later stage new rules can be learned.

## 7.2 Quantum Inspired Evolutionary Computation: Methods and Algorithms

### 7.2.1 Principles of Quantum Information Processing

Early 20th century experiments on particle and radiation physics showed that the subatomic size objects sometimes seemed to possess corpuscular and sometimes undulatory nature. Hence it was obligatory to rebuild the whole picture of microworld by developing a new kind of physics which would describe, explain, and predict the behaviour of object of very small size. The new physics launched in the first decades of 20th century was called quantum physics, and a more than a century after the beginning, its development work is still going on.

A most interesting feature of quantum physics is the principle of *superposition*. The machinery of classical physics allows constructions of new mixed states (which correspond to a probability distributions of pure states), and so does also quantum physics. Quantum physics allows also the construction of new pure states as superpositions of existing ones (for exact definitions of the terminology used here, we refer to [17–20]).

All information in the physical world is anyway represented by some physical system, and therefore also the nature of the information is affected by the nature of the physical world. It turns out that the information represented by quantum physical systems, *quantum information* differs from its classical counterpart in many notable parts, for example, it turns out that quantum information cannot be cloned arbitrarily [17]. As classical computing can be described as manipulating classical information, quantum computing is, in the same spirit, manipulation of quantum information. It is possible that the properties of quantum information help in resolving some computational tasks essentially more efficiently than classical information allows. In fact this was suggested already in [18], but a most interesting example was given in a very remarkable discovery where Shor [19] demonstrated that quantum computers would allow efficient integer factorization, a task assumed impossible for classical information processing. For a presentation of notable quantum algorithms, we refer to [20]. It is worth emphasizing here that the efficiency of quantum computing comes from ingenious use of superposition principle, not from the high "clock frequency" of quantum computers.

The mathematical machinery behind quantum physics is sometimes regarded as very involved, but its core is very straightforward. For the purposes of discrete information processing it is sufficient to consider only finite-level quantum systems with fixed set of physical observables, which will lead essentially to the following mathematical representation: A quantum system with $n$ perfectly distinguishable states is represented by using an $n$-dimensional vector space $H_n$ over complex numbers (Hilbert space). Any state of the system is an $n \times n$ complex matrix $\rho$ which is positive semidefinite, self-adjoint (meaning that $\rho = \rho^*$, where $\rho^*$ is the transpose of the complex conjugate of $\rho$, and has unit trace ($tr(\rho) = 1$, where $tr(\rho)$ stands for the sum of diagonal elements of $\rho$). Such a matrix is called a *density matrix*.

By fixed set of observables we mean that that we fix an orthonormal basis of $H_n$ to be the preferred basis, call the basis elements $\{|0\rangle, |1\rangle, \ldots, |n-1\rangle\}$, and say that the system represents one of the values $0, 1, \ldots n - 1$. For any basis element $|i\rangle$ we define a projection matrix $P_i$, which is a diagonal matrix with 1 as the $i$th and 0 as the other diagonal elements. So-called minimal interpretation of quantum physics then tells that $tr(P_i\rho)$ is the probability of seeing the quantum system in state $\rho$ to present value $i$. The minimal interpretation is hence an axiom relating the mathematical objects of quantum physics to statistical data obtained by observing quantum systems.

Quantum states $\rho$ evidently form a convex set, meaning that whenever $\rho_1$ and $\rho_2$ are density matrices, so is $p\rho_1 + (1 - p)\rho_2$ for any $p \in [0, 1]$. The extremals of the convex state set are called *pure states*, and it turns out that the pure states are characterized by the property $\rho^2 = \rho$. Pure states are hence projections onto one-dimensional subspaces and can therefore be presented by vectors instead of matrices. In so-called *vector state formalism* a pure state $\rho$ is hence replaced with a unit length-vector $\psi = \alpha_0|0\rangle + \alpha_1|1\rangle + \cdots + \alpha_{n-1}|n-1\rangle$, and the minimal interpretation turns into the following form: quantum system in pure state $\psi$ is seen to present value $i$ with probability $|\alpha_i|^2$ (mathematical details can be found in [13]).

The most well-known example of a quantum system is given when $n = 2$ and state is pure: The state of the system can then be written as $\psi = \alpha_0|0\rangle + \alpha_1|1\rangle$, where $|\alpha_0|^2 + |\alpha_1|^2 = 1$. Such a system is called a *quantum bit* or *qubit* in a pure state.

The way to represent joint quantum systems is via tensor product construction: If $\rho_1$ and $\rho_2$ are states of quantum systems 1 and 2, respectively, then $\rho_1 \otimes \rho_2$ is a state of the compound system. Especially, in the pure state formalism, this means that the basis vectors to represent multiple quantum bits ($n$) can be chosen as tensor products $|0\rangle \otimes |0\rangle \otimes \cdots \otimes |0\rangle$, etc., for which shorthand notations $|00\ldots00\rangle, |00\ldots01\rangle, |00\ldots10\rangle, |00\ldots11,\rangle\ldots, |11\ldots11\rangle$ are commonly used. Hence a general pure state of $n$ quantum bits can be written as

$$\alpha_0|00\ldots00\rangle + \alpha_1|00\ldots01\rangle + \alpha_2|00\ldots10\rangle + \alpha_3|00\ldots11\rangle + \alpha_{2^n-1}|11\ldots11\rangle \quad (7.5)$$

where $|\alpha_0|^2 + |\alpha_1|^2 + |\alpha_2|^2 + \cdots + |\alpha_{2^n-1}|^2 = 1$, and $|\alpha_i|^2$ stands for the probability that when observing state (line01), the bit string $b_1 b_2 \ldots b_n$ representing number $i$ is seen. Generally, a pure state in $H_n \otimes H_m$ is said to be *decomposable*, if it can be written as $|x\rangle \otimes |y\rangle$, where $|x\rangle \in H_n$ and $|y\rangle \in H_m$. State which is not decomposable, is *entangled*. Especially, a fully decomposable state (Eq. 7.1) can be written as

$$(\alpha_1|0\rangle + \beta_1|1\rangle) \otimes (\alpha_2|0\rangle + \beta_2|1\rangle) \otimes (\alpha_3|0\rangle + \beta_3|1\rangle) \otimes \cdots \otimes (\alpha_n|0\rangle + \beta_n|1\rangle) \quad (7.6)$$

where $|\alpha_i|^2 + |\beta_i|^2 = 1$ and $|\alpha_i|^2$ ($|\beta_i|^2$) is the probability of seeing 0 (1) when observing the $i$th quantum bit. Now comparison of Eqs. 7.5 and 7.6 shows that the former requires $2^n$ complex numbers, but the latter, fully decomposable state only $2n$ complex numbers. It follows that the decomposable state (Eq. 7.6) can be simulated by using classical computers in real time, meaning that the simulation time grows linearly with the number of qubits. The disadvantage of concentrating only on states of form (Eq. 7.6) is that those states do not fully exploit the features of quantum physics, but we may rather speak about *quantum-inspired* systems which make use of states of form (Eq. 7.6).

A (discrete) *time evolution* of a quantum system is depicted via completely positive mappings: A state $\rho_1$ transforms into $\rho_2 = V(\rho_1)$, where $V : L(H_n) \rightarrow L(H_n)$ is a completely positive mapping (see [21] for exact definitions). We say that the quantum system is *closed*, if its time evolution can be written as $V(\rho) = U\rho U^*$, where $U$ is a unitary mapping. It follows that for pure states, the time evolution can be written as $|x\rangle \rightarrow U|x\rangle$, where $U$ is unitary.

Quantum information principles, such as superposition, entanglement, interference, parallelism and other have been studied by a famous scientists, including Max Planck, Albert Einstein, Niels Bohr, W.Heisenberg. **Ernest Rutherford (1871–1937)** discovered that the atom is almost empty space, except a small space in it, where the total mass and energy of the atom is concentrated.

## 7.2.2    Principles of Quantum Inspired Evolutionary Algorithms (QEA)

A particular class of Quantum Evolutionary Algorithm (QEA) inspired by the concept of quantum principle was proposed by Han and Kim [20]. Since then, a lot of attention are drawn from researchers around the world on this technique with many advantages when compared to the classical EA. Inheriting from the basic EA concept, QEA is a population-based search method which simulates a biological evolution process and mechanism, such as selection, recombination, mutation and reproduction. Each individual in a population plays a role as a candidate solution and is evaluated by a fitness function to solve a given task. However, instead of using scalar values, information in QEA is represented in qubits so that, the value of a single qubit could be 0, 1, or a superposition of both. This probability presentation has a better characteristic of diversity than classical approaches. A single qubit which is the smallest information unit can be defined as $\begin{bmatrix} \alpha \\ \beta \end{bmatrix}$ which satisfies the probability fundamentals stating that $|\alpha|^2 + |\beta|^2 = 1$. A QEA individual is represented as a qubit vector $\begin{bmatrix} \alpha_1 & \alpha_2 & \cdots & \alpha_m \\ \beta_1 & \beta_2 & \cdots & \beta_m \end{bmatrix}$, where $\alpha$ and $\beta$ are complex numbers defining probabilities at which the corresponding state are likely to appear when a qubit collapses, for instance, when reading or measuring its value. QEA have been reported to successfully solve complex benchmark problems such as numerical [22], multiobjective optimisation [23] and several real world problems such as in [21, 24].

Many applications have been developed so far using the principles of QiEC, such as:

–   Specific algorithms with polynomial time complexity for NP-complete problems (e.g. factorising large numbers, [19]; cryptography)
–   Search algorithms [25], (having $O(N^{1/2})$ versus $O(N)$ complexity)
–   Quantum associative memories [26]
–   Quantum inspired evolutionary algorithms and neural networks [25, 27–30]
–   Algorithms for quantum computers even though such computers are still not available.

## 7.2.3    Quantum Inspired Evolutionary Algorithm (QiEA)

Defoin-Platel et al. [28] and Schliebs et al. [29] proposed an extended version of QEA. In [28] authors proposed a revisited description of QEA which I am going to summarize here. QiEA is a population-based search method. Its behaviour can be decomposed in three different and interacting levels see Fig. 7.3 and the explanation below.

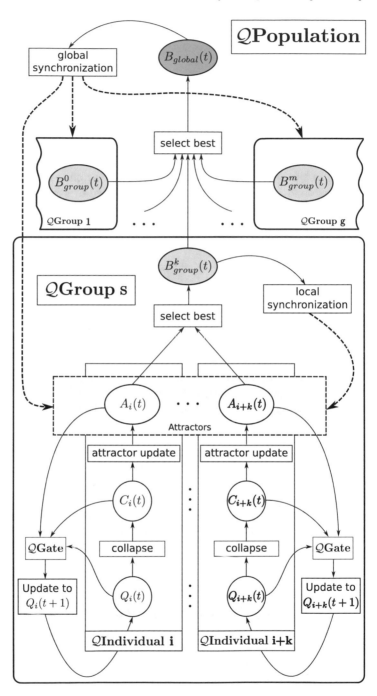

**Fig. 7.3** The QiEA block diagram (from [28])

*Quantum individuals*

The lowest level corresponds to *quantum individuals*.

A Qindividual $i$ at generation $t$ contains a Qbit string $Q_i(t)$ and two binary strings $C_i(t)$ and $A_i(t)$. More precisely $Q_i$ corresponds to a string of $N$ concatenated Qbits:

$$Q_i = Q_i^1 Q_i^2 \ldots Q_i^N = \begin{bmatrix} \alpha_i^1 & \alpha_i^2 & \cdots & \alpha_i^N \\ \beta_i^1 & \beta_i^2 & \cdots & \beta_i^N \end{bmatrix} \qquad (7.7)$$

For the purpose of fitness evaluation each $Q_i$ is first sampled (or collapsed) to form a binary individual $C_i$.

Each Qbit in $Q_i$ is sampled according to a probability defined by $\left|\beta_i^j\right|^2$. Thus $C_i$ represents a configuration in the search space which quality can be classically determined using a fitness function $f$. In the sense of EA, $Q_i$ is the genotype while $C_i$ is the phenotype of a given individual. In the sense of EDA, $Q_i$ defines a probabilistic model

$$P_i = \left[ \left|\beta_i^1\right|^2 \ldots \left|\beta_i^N\right|^2 \right] \qquad (7.8)$$

while $C_i$ is a realization of this model.

To each individual $i$ a solution $A_i$ is attached acting as an attractor for $Q_i$. Every generation $C_i$ and $A_i$ are compared in terms of both fitness and bit values. If $A_i$ is better than $C_i$ ($f(A_i) > f(C_i)$ assuming a maximization problem) and if their bit values differ, a quantum gate operator is applied on the corresponding Qbits of $Q_i$. Thus the probabilistic model $P_i$ defined by $Q_i$ is slightly moved toward the attractor $A_i$.

The update policy of an attractor $A_i$ can follow two distinctive strategies. In the original QEA [20] an *elitist* update strategy was used, in which the attractor $A_i$ is replaced by $C_i$ only if $C_i$ is better than $A_i$. In a *non-elitist* update strategy (firstly introduced in [28]) $C_i$ replaces $A_i$ at every generation. The choice of the update policy has great consequences for the algorithm and changes its behaviour completely. To emphasize the importance of the update rule the non-elitist version of QEA has been proposed as *Versatile QEA* (vQEA) [28] as the attractors are able to change every generation and therefore demonstrate a very high volatility. In the next section we give a more detailed explanation of the role of elitism.

In classical EA variation operators like crossover or mutation operations are used to explore the search space. The quantum analogue for these operators is called a quantum gate. In this study, the rotation gate is used to modify the Qbits. The $j$th Qbit at time $t$ of $Q_i$ is updated as follows:

$$\begin{bmatrix} \alpha_i^j(t+1) \\ \beta_i^j(t+1) \end{bmatrix} = \begin{bmatrix} \cos(\Delta\theta) & -\sin(\Delta\theta) \\ \sin(\Delta\theta) & \cos(\Delta\theta) \end{bmatrix} \begin{bmatrix} \alpha_i^j(t) \\ \beta_i^j(t) \end{bmatrix} \tag{7.9}$$

where the constant $\Delta\theta$ is a rotation angle designed in compliance with the application problem [30]. I note that the sign of $\Delta\theta$ determines the direction of rotation (clockwise for negative values). In this study the application of the rotation gate operator is limited in order to keep $\theta$ in the range $[0, \pi/2]$.

*Quantum Groups*

The second level corresponds to *quantum groups*. The population is divided into $g$ Qgroups each containing $k$ Qindividuals having the ability of synchronizing their attractors. For that purpose, the best attractor (in terms of fitness) of a group, noted $B_{group}$, is stored at every generation and is periodically distributed to the group attractors. This phase of local synchronization is controlled by the parameter $S_{local}$.

*Quantum Population*

The set of all $p = g \times k$ Qindividuals forms the *quantum population* and defines the topmost level of QEA. As for Qgroups, the individuals of a Qpopulation can synchronize their attractors. For that purpose, the best attractor (in terms of fitness) among all Qgroups, noted $B_{global}$, is stored every generation and is periodically distributed to the group attractors. This phase of global synchronization is controlled by the parameter $S_{global}$.

We can note that in the initial population all the Qbits are fixed with $|\alpha|^2 = |\beta|^2 = 1/2$ so that the two states "0" and "1" are equiprobable in collapsed individuals.

The QiEA is much faster to arrive at a global optimum than the exhaustive (grid) search and the GA methods as shown in Fig. 7.4, as it requires several magnitudes less number of iterations (evaluations).

## 7.2.4  Versatile QiEA (VQiEA)

In this section we present an improved version of QEA, called the Versatile Quantum-inspired Evolutionary Algorithm
(vQEA) avoiding the weaknesses reported above [28].

*Description of vQEA*

In order to prevent both the case of irreversible choice and the hitchhiking phenomenon, the strategy for updating attractors is modified. We introduce a new

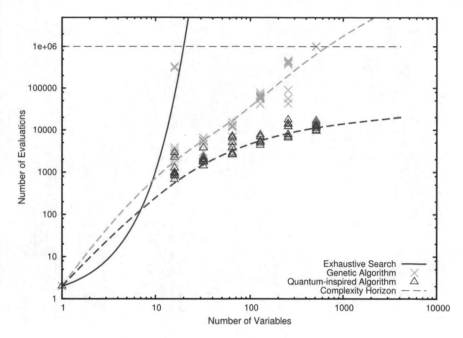

**Fig. 7.4** QiEA converge to an optimal solution in less number of generations (interactions, evaluations) and are applicable to a largere dimensionality space of variables for optimisation

parameter controlling this strategy based on *elitism*. In the classical QEA, the update procedure (called *attractor update* in Fig. 7.12) applies elitism such that an attractor Ai is replaced by Ci only if Ci is better. With vQEA this parameter is simply switched off. Therefore the attractors are replaced at every generation without considering their fitness and so they demonstrate a very high volatility. Moreover to ensure the convergence of vQEA, the global synchronization is also performed every generation in such way that all the attractors are identical and at generation t + 1 corresponds to the best solution found at generation t.

We note that with such a setting, the group size n and local synchronization parameters Slocal do not affect the algorithm anymore. With vQEA the information about the search space collected during evolution is not kept at the individual level but continuously renewed and shared among the whole population2. Nevertheless we think that the concept of group, which is similar to demes in classical EA, is interesting and we do not intend to remove it definitely. In this study, we avoid the tuning of n and Slocal and concentrate on the effects of removing elitism from QEA. Thus the simplified sequential procedure of vQEA is detailed in the Algorithm in Box 7.4. The sets of all the quantum individuals, collapsed individuals and attractors at generation t are noted Q(t), C(t) and A(t) respectively.

---

**Box 7.4.** The versatile quantum-inspired EA (vQEA)

---

**Algorithm.** The Versatile Quantum-inspired EA (vQEA)

1: $t \Leftarrow 0$

2: initialize Q(t)

3: **while** not termination condition **do**

4: make C(t) by observing the states of Q(t)

5: evaluate C(t)

6: store the best C(t) into Bglobal(t)

7: do global synchronization of A(t)

8: update Q(t) to Q(t + 1) using QGate

9: $t \Leftarrow t + 1$

10: **end while**

---

Figure 7.5 shows a hypothetical example of state convergence to local minima for a system described by a qbit register (chromosome) over 5 applications of a rotation quantum gate operator. The darker points represent system states described by the qubit vector that have a higher probability of occurrence.

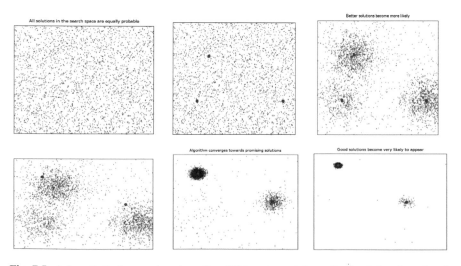

**Fig. 7.5** A hypothetical example, where the vQEA arrives at the optimal solution for a bench mark high dimensional problem in 5 iterations (generations), when compared to thousands iterations needed by traditional EC methods

## 7.2.5   *Extension of the VQiEA to Deal with Continuous Value Variables*

Since we want to consider also continuous search spaces now, we have to replace the Bernoulli distribution by a continuous one, such that it becomes possible to sample real values instead of discrete ones [29]. A number of approaches how to employ such distributions and how to model them have been studied in literature. Generally they are based on Gaussian distributions [31–35], histograms [31], or interval representations [36]. We consider a continuous EDA based on Gaussian distributions here.

For each dimension $j$ of the continuous search space and for each probabilistic model $i$, a random variable following a Gaussian distribution is evolved. Therefore the distribution is fully described by two parameters: The mean and the standard deviation $\sigma_i$. In each generation samples are drawn forming real-valued vectors, whose quality can be evaluated by the fitness measure. An update rule is then applied to update $\mu^{(j)}$ and $\sigma^{(j)}$ to move the search towards promising areas in the search space, making higher quality solutions more likely to be sampled in the next generation. We will first describe the basic structure of algorithm in detail, followed by the presentation of the chosen update rule.

The overall structure of the proposed extension is almost identical to vQEA. Like vQEA also the continuous version is a population-based search method [29]. Its behavior can be decomposed in three different interacting levels: Individual, group and population level.

*Individuals.* The lowest level corresponds to *individuals*. An individual $i$ at generation $t$ contains a probabilistic model $P_i(t)$ and two real-valued strings $C_i(t)$ and $A_i(t)$. More precisely $P_i$ corresponds to a string of $N$ pairs of values.

The pair $(\mu^{(j)},\ \sigma^{(j)})$ corresponds to the parameters of the distribution of the $j$th variable of the $i$th probabilistic model. Each variable in $P_i$ is sampled according to $\mu^{(j)}$ and $\sigma^{(j)}$, so that $C_i$ represents a configuration in the search space whose quality can be determined using a fitness function $f$. In most continuous optimization problems, the variables have a specific domain of definition. Without loss of generality we assume each $c^{(j)} \in C_i$ to be defined into the interval $[-1, 1]$. As a consequence, each $c^{(j)} \in C_i$ follows a *truncated normal distribution* in the range $[-1, 1]$. Truncated normals can be sampled using a simple numerical procedure and the technique is widely adopted in pseudo-random number generation, see e.g. [37] for an efficient implementation.

To each individual $i$ a solution $A_i$ is attached acting as an attractor for $P_i$. Every generation $C_i$ and $A_i$ are compared in terms of their fitness. If $A_i$ is better than $C_i$ (i.e. $f(A_i) > f(C_i)$ assuming a maximization problem), an update operation is applied on the corresponding model $P_i$. The update will move the mean values of the probabilistic model $P_i$ slightly towards the attractor $A_i$. The choice of a suitable model update operation is critical for the working of the algorithm. We will elaborate the details of the model update in a later section.

The update policy of an attractor $A_i$ can follow two distinctive strategies. In the original QEA [20], an *elitist* update strategy was used, in which the attractor $A_i$ is replaced by $C_i$ only if $C_i$ is better than $A_i$ in terms of fitness. In a *non-elitist* update strategy (firstly introduced in [28]) $C_i$ replaces $A_i$ at every generation. The choice of the update policy has great consequences for the algorithm and changes its behavior completely. To emphasize the importance of the update rule the non-elitist version of QEA has been proposed as *Versatile QEA* (vQEA) as the attractors are able to change every generation and therefore demonstrate a very high volatility. Since no experimental condition could be identified that favored the elitist attractor update policy, we will concentrate on the non-elitist version during the course of this paper.

*Groups.* The second level corresponds to *groups*. The population is divided into $g$ groups each containing $k$ individuals having the ability of synchronizing their attractors. For that purpose, the best attractor (in terms of fitness) of a group, noted *Bgroup*, is stored at every generation and is periodically distributed to the group attractors. This phase of local synchronization is controlled by the parameter *Slocal*.

*Population.* The set of all $p = g \times k$ individuals forms the *population* and defines the topmost level of the multi-model approach. As for the groups, the individuals of the population can synchronize their attractors, too. For that purpose, the best attractor (in terms of fitness) among all groups, noted *Bglobal*, is stored every generation and is periodically distributed to the group.

Figure 7.6 shows the update operation for a single Gaussian random variable. For each update the distance $d = a(t) - \mu(t)$ between the attractor $a(t)$ and the mean $\mu(t)$ of the Gaussian is computed at generation $t$. (a) If $d \geq \sigma(t)$ the attractor is considered distant. We interpret that situation by assuming that $\mu(t)$ does not represent a promising area in the search space. In this case the mean $\mu(t)$ is strongly shifted towards the attractor, while at the same time the standard deviation $\sigma(t)$ is increased to allow a wider search in the fitness landscape. (b) On the other hand, if the attractor is inside the boundaries defined by $\sigma(t)$, i.e. $d < \sigma(t)$, we assumed that $\mu(t)$ is already in a promising area of the search space. The algorithm starts to localize the search by shifting $\mu(t)$ only slightly towards the direction of the

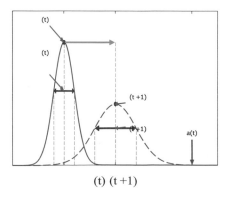
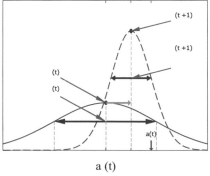

(t) (t +1)                                                     a (t)

**Fig. 7.6** Quanutm encoding of continuos value variables using Gaussion functions

attractor, while decreasing $\sigma(t)$ at the same time. This phase of global synchronization is controlled by the parameter *Sglobal*.

(a)  Update operation for distant attractors
(b)  Update operation for close attractors.

*Model Update*

The update of the probabilistic model is in particular interesting, since it governs how the search space is explored by the algorithm. Several continuous EDA have been proposed in literature ([31, 34, 35, 38, 39]), along with a number of different update rules, e.g. ([31, 32]). The common principle of all these continuous EDA is based on the sampling of a population. In vQEA (and thus also its extension) the situation is very different, since only a *single* solution (for each probabilistic model) is sampled in every iteration. Hence the model update can not rely on the density of a population, but has to use a single attractor instead to perform the desired update.

We formulate here an appropriate update rule for the probabilistic models. Updating the mean $\mu^{(j)}$ in the Gaussian variable $j$ appears to be straight forward. We adopt a mean shift towards the value of the current attractor $a^{(j)}$ at location $j$, which is quite similar to the mean update used in methods mentioned above. Depending on the distance $d = a^{(j)} - \mu^{(j)}$ a shift $\Delta\mu^{(j)}$ is computed at generation $t$ which is then used to perform the update:

$$\mu^{(j)}(t+1) = \mu^{(j)}(t) + \theta_\mu \Delta\mu^{(j)}(t) \tag{7.10}$$

In Eq. (7.10) a parameter $\theta_\mu$ is introduced, which we will refer to as the learning rate of the mean. We note that $\theta_\mu$ corresponds to the maximum mean shift in a single generation.

For the update of the standard deviation $\sigma^{(j)}$ we will exploit the idea that $\sigma^{(j)}$ should decrease whenever $\mu^{(j)}$ represents a "promising" area in the fitness landscape. We assume $\mu^{(j)}$ to be "fit" when $|d| < \sigma^{(j)}$. Thus, if the attractor $a^{(j)}$ is close to $\mu^{(j)}$ (within the boundaries defined by $\sigma^{(j)}$), the standard deviation $\sigma^{(j)}$ is decreased. It is noteworthy that solutions fulfilling this condition are more likely to be sampled, than other solutions, which means that on average $\sigma(j)$ will decrease. Attractors that are more distant to $\mu^{(j)}$ and thus $|d| \geq \sigma^{(j)}$, will cause an increase of $\sigma^{(j)}$, since it can be assumed that $\mu^{(j)}$ does not represent a promising area in the landscape. We define the standard deviation shift $\Delta\sigma^{(j)}$ at generation $t$ as:

$$\Delta\sigma^{(j)}(t) = \frac{1}{1 + e^{-10}(\sigma^{(j)}(t) - 0.5)} \tag{7.11}$$

and then use it to perform the update restrict the domain of $\sigma^{(j)}$ by defining upper and lower bounds, such that

$$\sigma_{\min} \leq \sigma^{(j)} \leq \sigma_{\max} \tag{7.12}$$

It is important to note, that the probabilistic update operator described above, is similar to the rotation gate used in QEA. As shown in [40] the size of an update step using the rotation gate depends on the convergence of the probabilistic model. This phenomenon was described as a form of deceleration of the algorithm before convergence. The sigmoid shape of the standard deviation update adopts a similar strategy, since also here the size of the shift $\Delta\sigma^{(j)}$ decreases with increasing convergence of the algorithm.

*Combined Search Spaces*

Many real-world problems require the exploration of combined search spaces: a binary and a continuous space. An example is the parallel evolution of the topology and the weight matrix of a neural network. Here the topology is encoded as a bit string, where "1" represents a present connection between two neurons and "0" encodes its absence. Another example is the wrapper based feature selection, where the presence/absence of a feature requires a binary search space, while appropriate configurations for the classification method may correspond to a continuous landscape.

It is now possible to employ vQEA on combined search spaces with two types of representation. Each representation uses its corresponding update operator to drive the probabilistic model towards promising areas in the search space. In every generation the models are sampled and then evaluated by a single fitness measure. The fitness evaluation uses the sampled binary and continuous solution part to determine the quality of the combined solution. According to the fitness of the obtained solution the models are updated. This extended vQEA allows us to enhance the original QiSNN.

We emphasize that the extended vQEA [29] is similar to a collaborative coevolutionary algorithm [41]. The evolution of the two representations proceeds more or less independently from each other. Both use their own solution representations and update operators and may explore their search space with different learning rates. Despite their independent evolution both representations share a single fitness function. The binary and continuous sub-solutions are the components of a combined solution, and both parts need to collaborate in order to maximize their fitness.

## 7.3 Quantum Inspired Evolutionary Computation for the Optimisation of SNN

### 7.3.1 *A Quantum-Inspired Representation of a SNN*

The approach to use QiEA for the optimization of SNN is based on the following principles [42–44]:

(a) *A quantum probabilistic representation of a spike:* A spike, at any time $t$, is both present (1) and not present (0), which is represented as a *qbit* defined by a probability density amplitude. When the spike is evaluated, it is either present or not present. To modify the probability amplitudes, a quantum gate operator is used, for example *the rotation gate*:

$$\begin{bmatrix} \alpha_i^j(t+1) \\ \beta_i^j(t+1) \end{bmatrix} = \begin{bmatrix} \cos(\Delta\theta) & -\sin(\Delta\theta) \\ \sin(\Delta\theta) & \cos(\Delta\theta) \end{bmatrix} \begin{bmatrix} \alpha_i^j(t) \\ \beta_i^j(t) \end{bmatrix} \tag{7.13}$$

More precisely, a spike arriving at a moment $t$ at each synapse $Sij$ connecting a neuron $Ni$ to a pre-synaptic neuron $Nj$, is represented as a qbit $Qij(t)$ with a probability to be in state "1" $\beta_{ij}(t)$ (probability for state "0" is $\alpha_{ij}(t)$). From the SNN architecture perspective this is equivalent to the existence (non-existence) of a connection Cij between neurons Nj and Ni.

(b) *A quantum probabilistic model of a spiking neuron:* A neuron $Ni$ is represented as a qbit vector, representing all m synaptic connections to this neuron:

$$\begin{bmatrix} \alpha & 1 & \alpha & 2 & \cdots & \alpha & m \\ \beta & 1 & \beta & 2 & \cdots & \beta & m \end{bmatrix} \tag{7.14}$$

At time $t$ each synapic qbit represents the probability for a spike to arrive at the neuron. The neuron are collapsed into spikes (or no spikes) and the cumulative input $u_i(t)$ to the neuron $Ni$ is calculated.

All input features $(x1, x2,...,xn)$, the eSNN parameters $(q1,q2,...,qs)$, the connections between the inputs and the neurons, including recurrent connections (C1, C1,..., Ck) and the probability of the neurons to spike $(p1,p2,...,pm)$ at time $(t)$ are represented in an integrated *qbit* register that is operated upon as a whole [42, 43].

This framework goes beyond the traditional "wrapper" mode for feature selection and modelling [45]. It was demonstrated that the vQEA is efficient for integrated feature and SNN parameter optimisation in a large dimensional space and also useful for extracting unique information from the modelled data [29]. All probability amplitudes together define a probability density $\psi$ of the state of a probabiltiy spiking neuron model (PSNM) (see Chap. 4) in a Hilbert space. This density will change if a quantum gate operator is applied according to an objective criterion (fitness function). This representation can be used for both tracing the learning process in an PSNM system or the reaction of the system to an input vector.

(c) *PSNN learning rules:* As the PSNM is an eSNN, in addition to the eSNN learning rules (Chap. 5) there are rules to change the probability density amplitudes of spiking activity of a neuron. The probability $\beta_{ij}(t)$ of a spike to arrive from neuron Nj to neuron Ni (the connection between the two be present) will change according to STDP rule, which is implemented using the quantum rotation gate. In a more detailed model, $\beta_{ij}(t)$ will depend on the strength and the frequency of the spikes, on the distance $Dij$ between neurons Nj and Ni, and on many other physical and chemical parameters that are ignored in this model but can be added if necessary.

(d) *The principle of feature superposition representation* [43, 44]: A vector of n qbits represents the probability of using each input variable $x1$, $x2$,...,$xn$ in the model at a time t. When the model computes, all features are "collapsed", where "0" represents that a variable is not used, and "1"—the variable is used.

- The principle of *feature superposition* [44]:
  At any time moment (t) a feature related to a given task is in a superposition of both *present* and *not present* states for a computational model, defined by the probability density amplitude. Before the model computes, a feature's state is collapsed into *present* or *not present*.
- Useful to capture patterns of interaction between features for a problem
- Integrates the environment with the model for a combined optimisation
- Useful to represent "*floating features*"
- The VQiEA performs much faster and more accurately than classical algorithms when evaluating combinations of interacting features, for a classification task.

Here the vQEA is applied for Evolving Spiking Neural Network (eSNN) optimisation (see Fig. 7.7). The result produces a faster convergence to the optimal solution with better accuracy when compared to traditional neural networks such as multilayer perceptrons and Naïve Bayesian Classifier (NBC).

SNN parameters optimised together with the features are: synaptic learning modulation factor; PSP threshold parameter; max # of output neurons per class [46].

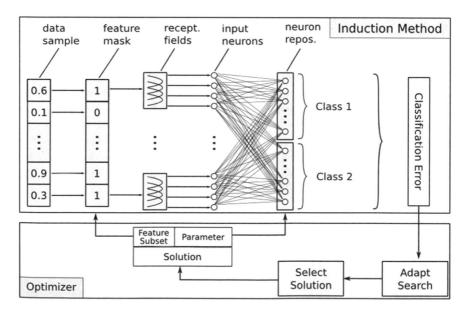

**Fig. 7.7** Integrated feature and eSNN parameter optimisation using quantum inspired evolutionary computation

## 7.3.2   Application of QiEA for the Optimisation of ESNN Classifier on Ecological Data

In [46] the original QiSNN framework was applied on an ecological modeling problem. Because of the promising results obtained from the benchmark studies before, we wanted to revisit the ecological data using the enhanced QiSNN for feature selection. For many invertebrate species little is known about their response to environmental variables over large spatial scales. That knowledge is important in order to predict the establishment of a species, that has the potential to cause great environmental harm. The usual approach to determine the importance of a range of environmental variables, that explain the global distribution of a species, is to train or fit a model to its known distribution using environmental parameters measured in areas where the species is present and where it is absent. In this study, meteorological data that comprised 68 monthly and seasonal temperature, rainfall and soil moisture variables for 206 global geographic sites were compiled from published records [47]. These variables were correlated to global locations where the Mediterranean fruit-fly (*Ceratitis capitata*), a serious invasive species and fruit pest, were recorded at the time of the study, as either present or absent. The dataset is balanced having equal number of samples for each of the two classes. Previous use of MLP on the data results in a classification accuracy of approximately 71% [48].

The experimental setup defined in [46] was kept mostly unchanged here to allow some comparison to previous results: Ten individuals are allowed to evolve in 4000 generations, statistical relevance is guaranteed by performing 30 independent runs and averaging the results. The additional parameters for the mean and standard deviation shift were set to $\theta_\mu = 0.1$ and $\theta_\sigma = 0.01$ respectively, the learning rate of the binary model was $\theta = \pi/100$. Figure 7.14 presents the results of the revisited experiment. Similar to the figures before the evolution of the average best feature subset is shown, where the color reflects how often a specific feature was selected at a certain generation. The comparison between NBC and the original QiSNN was discussed in great detail in [46], thus we will concentrate on the discussion of the performance of the two QiSNN only. Nevertheless the enhanced version reports greater consistency in the feature rejection. Also the enhanced QiSNN selected significantly less features than the original QiSNN. On average 14 features were selected using QiSNN, 9 in case of the enhanced QiSNN and 18 using NBC. Compared to the original QiSNN the enhanced version additionally rejected the following features: temp1, temp3, TAut2, TSpr1, Tannual, rain10, RSumR2, PEAnnual. The overall classification accuracy was similar between all tested algorithms.

From an ecological point of view the evolved feature subsets are coherent with the current knowledge in this area. Winter temperatures, autumn rainfall and the degree-days (DD5 and DD15) were particularly strong features.

Degree-days are the accumulated number of degrees of temperature above a threshold temperature (5° and 15° in this case) over time (in this dataset over the whole year). It would be expected that the latter two variables would be closely correlated. These results correspond to other analysis where more conventional statistical and machine learning methods were used to identify the contribution of environmental variables to *C. capitata* presence or absence [49]. While there is no indication from this analysis whether the features have a negative or positive effect on the distribution of the species, it is known that *C. capitata* is limited by the severity of temperatures in the winter and extremes of wet or dry conditions in the summer and autumn [50].

Figures 7.8, 7.9 and 7.10 show some experimental results on this case study, demonstrating integrated feature selection and model creation over 3000 generations of a QiEA, with a significant improvement of the accuracy of the eSNN model.

### 7.3.3 Integrative Computational Neuro Genetic Model (CNGM) Utilising Quantum-Inspired Representation

A schematic diagram of a CNGM is given in Fig. 7.11. The framework combines ESNN and a gene regulatory network (GRN) [51]. The qbit vector for optimization through the QEA is given in Fig. 7.12. In addition to the SNN parameters the CNGM has gene expression parameters g1,g2,…gl, each of them also represented

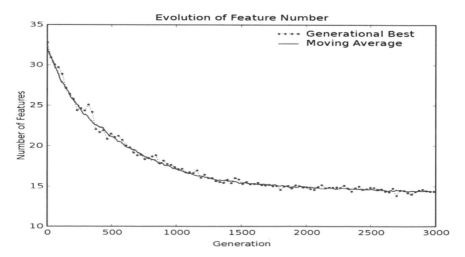

**Fig. 7.8** Evolution of the features on the climate data set using QiEA. The best accuracy model of eSNN is obtained for 15 features

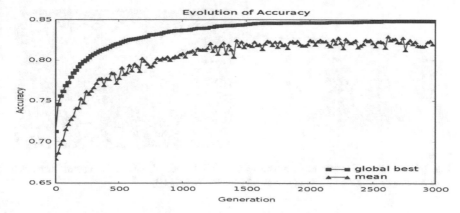

**Fig. 7.9** Evolution of classification accuracy of an eSNN classifier optimised with the use of QiEA on the climate data set after 3000 generations

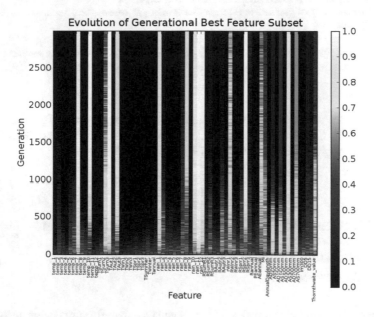

**Fig. 7.10** Evolution of features over 3000 generations for the case study problem using QiEA and an eSNN as a classifier

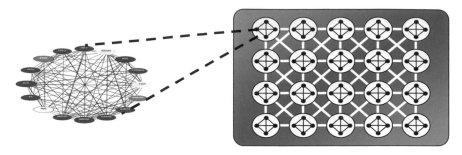

**Fig. 7.11** A schematic diagram of a neurogenetic SNN model. Each spiking neuron includes a gene regulatory network (GRN) model as parameters (after [51])

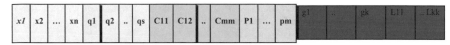

(b)  Input features    / NN p parameters /  NN connections/ Probability of  neurons spiking/Genes on/off  and their
connections

**Fig. 7.12** In addition to SNN parameters, a quantum chromosome of CNGM represents in a qbit register gene expression levels (g1,g2,…,gl) and the connections between the genes (L1,…,Lr) in the GRN

as a qbit with two states (state "1"—gene is expressed; state "0"—gene is not expressed" 0). Each link Li ($i = 1,2,…,r$) between two genes in the GRN is represented as a quantum bit with 3 states ("1" positive connection; "0"—no connection; "−1"—negative connection).

## 7.4  Quantum Inspired Particle Swarm Optimisation

### 7.4.1  *Quantum Inspired Particle Swarm Optimisation Algorithms*

Particle Swarm Optimisation (PSO) is a population based optimisation technique, developed by Eberhart and Kennedy in 1995 [6] as already presented in a previous section of this chapter. Individuals in PSO work together to solve a given problem, by responding to their own performance and the performance of the other particles in the swarm. Each particle has their own fitness value calculated during the optimisation process and the best fitness value achieved so far is stored and normally referred to as personal best or individual best (*pbest*). The overall best fitness value obtained by any particle in the population so far is called global best (*gbest*) which stores the best solution. Particle accelerates towards a new position by calculating the position velocity where the value of *pbest* and *gbest* would influence the direction of the particle in the next iteration. Equation 7.15 illustrates the velocity update and Eq. 7.16 is the calculation of the new particle position.

$$v_n = w * v_{n_{t-1}} + c_1 * rand() * (g_{best_n} - x_n) + c_2 * rand() * (p_{best_n} - x_n) \quad (7.15)$$

$$x_n = x_{n_{t-1}} + v_n \quad (7.16)$$

where value of the random number is between 0 and 1. $c_1$ and $c_2$ control the particle acceleration towards personal best or global best.

However, standard PSO is inadequate for a problem requiring a probability computation. Therefore, quantum principles have been embedded in PSO as a mechanism for the probability calculation and normally referred as Quantum inspired Particle Swarm Optimisation (QiPSO). QiPSO was first discussed in [30] and the main idea of QiPSO is to use the standard PSO function to update the particle position represented as a quantum angle ($\theta$). Quantum angle $\theta$ can be represented as $\begin{bmatrix} \cos(\theta) \\ \sin(\theta) \end{bmatrix}$, and it is equivalent to $\begin{bmatrix} \alpha \\ \beta \end{bmatrix}$ which satisfies the probability fundamental of $|\sin(\theta)|^2 + |\cos(\theta)|^2 = 1$. The velocity update formula in standard PSO is modified to get a new quantum angle which is translated to the new probability of the qubit by using Eq. 7.17.

$$\Delta\theta_n = w * \Delta\theta_{n_{t-1}} + c_1 * rand() * (\theta_{gbest_n} - \theta_n) + c_2 * rand() * (\theta_{pbest_n} - \theta_n)$$
$$(7.17)$$

Then, based on the new $\theta$ velocity, the new probability of $\alpha$ and $\beta$ is calculated using a rotation gate as follows:

$$\begin{bmatrix} \alpha \\ \beta \end{bmatrix} = \begin{bmatrix} \cos(\Delta\theta) & -\sin(\Delta\theta) \\ \sin(\Delta\theta) & \cos(\Delta\theta) \end{bmatrix} \begin{bmatrix} \alpha_{t-1} \\ \beta_{t-1} \end{bmatrix} \quad (7.18)$$

or by replacing the rotation gate with $\theta_t = \theta_{t-1} + \Delta\theta$ where $\theta$ is the new quantum angle of the quantum particle position.

### 7.4.2 Quantum Inspired Particle Swarm Optimisation Algorithm (QiPSO) for the Optimisation of ESNN

The algorithm for the method is presented in Box 7.5 [30].

---

**Box 7.5.** Integrated ESNN–QiPSO algorithm

---

**Algorithm** Integrated ESNN-QiPSO

---

1:  **for all** particle **do**
2:      **for all** ESNN parameters **do**
3:          **for all** qubit **do**
4:              initialise $\theta$
5:              get a collapsed state
6:          **end for**
7:          convert binary string to real value using Gray code
8:      **end for**
9:      **for all** feature qubits **do**
10:         initialise $\theta$
11:         get a collapsed state
12:     **end for**
13:     initialise fitness
14: **end for**
15: **while** not reaching maximum iteration **do**
16:     **for all** particle **do**
17:         get fitness from ESNN (accuracy of classification)
18:         **if** current fitness better than *pbest* fitness **then**
19:             assign current particle as *pbest*
20:             **if** current *pbest* fitness better than *gbest* fitness **then**
21:                 assign *pbest* as *gbest*
22:             **end if**
23:         **end if**
24:         **for all** ESNN parameters **do**
25:             **for all** qubits **do**
26:                 calculate velocity
27:                 apply rotation gate
28:                 get a collapsed state
29:             **end for**
30:             convert binary string to real value using Gray code
31:         **end for**
32:         **for all** feature qubits **do**
33:             calculate velocity
34:             apply rotation gate
35:             get a collapsed state
36:         **end for**
37:     **end for**
38: **end while**

### 7.4.3   Dynamic QiPSO

Here a method called Dynamic QiPSO (DQiPSO) is presented [52–54]. We apply this method as a model optimizer. This method further develops our previous QiPSO method [30], which may 'miss' to find optimal model parameter values when only using binary QiPSO. As the information is represented in the binary structure, the conversion from binary to real value may lead to inaccuracy especially if the number of the qubits selected to represent a parameter value is not sufficient. To overcome this problem, a combination of QiPSO and standard PSO is proposed where the QiPSO performs the probability calculation for feature selection task while standard PSO optimizes the parameters. This method not only effectively solves this problem but also eliminates one parameter which is the number of qubits to represent parameter values. The DQiPSO particle structure is depicted in Fig. 7.13.

Another element of enhancement from the previous method is the improvement in feature selection strategy. Standard PSO searching strategy is based on the random selection at the beginning of the process and each particle will update itself based on the best solution found subsequently. A major problem in this technique is having the chance of the relevant features not being selected at the beginning and affecting other particles in the entire process. This is due to each particle updating itself according to the particle without the relevant features. This problem does not only happen to the high dimensional problem, but also to the small dimensional problem. Therefore, a new strategy is proposed. Apart from the normal particle which updates itself based on the *pbest* and *gbest* information, where we referred as Update Particles, a new type of particles are added to the swarm, namely Random Particles and Filter Particles. Random particle randomly generates a new set of features and parameters in every iteration and increases the robustness of the search. While for the Filter Particles, it selects one feature at a time and feeds it to the networks and calculates the fitness value. This process is repeated to all features. After all features have been evaluated, the feature with fitness above threshold value will be considered as relevant. The threshold value is the average fitness value or

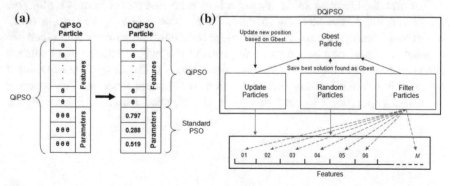

**Fig. 7.13  a** The proposed hybrid particle structure in DQIPSO [53, 54] and **b** DQIPSO feature selection strategy

can be manually adjusted. For the following iteration, features which are considered relevant will be randomly selected in order to find the best combination of relevant features. This strategy helps to solve unevaluated relevant features problem, reducing the search space and facilitating the optimizer in finding relevant features faster. Like other particles, if the Random Particles and Filter Particles are found to be the best solution, it will be stored as a *gbest* and Update Particles will update their position based on this new information. Some improvement in the update strategy is also proposed where *gbest* particle will only be replaced by new particles if the fitness is higher or has the same fitness value but with lower number of selected features. Due to the robust search space by DQiPSO, fewer particles are needed to perform the optimisation tasks which translates to faster processing time. The structure of this strategy is illustrated in Fig. 7.13.

## 7.4.4  Application of DQiPSO for Feature Selection and Model Optimisation

Using the well-known wrapper approach as described in [55–57], we introduce here a method for eSNN optimisation utilizing the DQiPSO method from above, first published in [52–54]. The QiPSO part of the DQiPSO method is used to optimise the features of the model while the PSO part of the DQiPSO method is used in a co-evolutionary mode to optimised the parameters of the eSNN. These parameters are: Modulation Factor (Mod); Proportion Factor (C) and Similarity (Sim) as described later and also in [44, 58]. All particles are initialized with random values and subsequently interacted with each other based on objective function—classification test accuracy. Since there are two components to be optimized, each particle is divided into two parts. The first part of each hybrid particle holds the feature mask where information is stored in a string of qubit which value 1 represents the feature selected, and value 0, otherwise. Another part holds parameters of eSNN. The proposed integrated framework is shown in Fig. 7.14.

The architecture of eSNN in this paper is based on the model from [44, 59]. This model consists of the encoding method of real value data to spike time, network model and learning method. Implementation for the information encoding methods is based on Population Encoding as proposed in [60] where a single input value is encoded to multiple input neurons (see also Chap. 4). The firing time of an input neuron $i$ is calculated using the intersection of Gaussian function. The centre is calculated using Eq. 7.19, and the width is computed using Eq. 7.20 with the

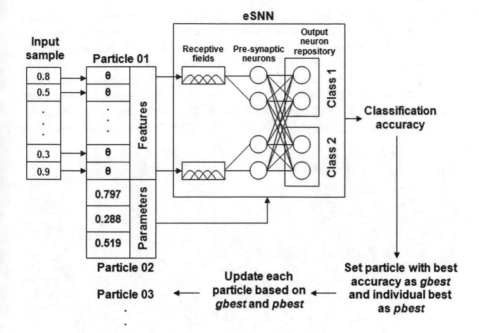

**Fig. 7.14** An integrated DQiPSO-eSNN framework for feature selection and model optimisation

variable interval of $[I_{min}, I_{max}]$. The parameter $\beta$ controls the width of each Gaussian receptive field.

$$\mu = I_{min} + (2 * i - 3)/2 * (I_{max} - I_{min})/(M - 2) \tag{7.19}$$

$$\sigma = 1/\beta(I_{max} - I_{min})/(M - 2) \text{ where } 1 \leq \beta \leq 2 \tag{7.20}$$

Thorpe's model [60] has been selected for the eSNN model because of its effectiveness and simplicity. The fundamental aspect of this model is that earlier spikes received by a neuron, have a stronger weight compared to later spikes. Once the neuron reaches a certain amount of spikes and the Post-Synaptic Potential (PSP) exceeds the threshold value, the neuron fires. Afterwards it becomes disabled. The neuron in this model can only fire once for a given stimulus. The computation of PSP of neuron $i$ is presented in Eq. 7.21,

$$N_i = \begin{cases} 0 & if \quad fired \\ \sum w_{ji} * Mod_i^{order(j)} & else \end{cases} \tag{7.21}$$

where $w_{ji}$ is the weight of pre-synaptic neuron $j$; $Mod_i$ is a parameter called modulation factor with interval of [0, 1] and $order^{(j)}$ represents the rank of a spike emitted by the pre-synaptic neuron. The $order^{(j)}$ starts with 0 if it spikes first among all pre-synaptic neuron and increases according to the firing time. For the eSNN's One-pass learning algorithm, each training sample creates a new output neuron. Trained threshold value and the weight pattern for the particular sample are stored in the neuron repository. However, if the weight pattern of the trained neuron is considered too similar with the neuron in the repository, the neuron will merge into the most similar one. The merging process involves modifying the weight pattern and the threshold to the average value. Otherwise, it will be added as a newly trained neuron to the repository. The major advantage of eSNN is the ability of the trained network to learn new samples incrementally, without the need for the SNN to be retrained on both old and new data. More details on eSNN can be found in Chap. 4 and also in [44] and [59].

The algorithm of an integrated ESNN and DQiPSO is given in Box 7.6.

---

**Box 7.6.** Integrated ESNN-DQiPSO

---

**Algorithm** Integrated ESNN-DQiPSO

---

1:  **for all** particles **do**

2:      initialise all ESNN parameters

3:      **for all** feature qubit **do**

4:          initialise $\theta$

5:          get a collapsed state

6:      **end for**

7:      initialise fitness

8:  **end for**

9:  **while** not reaching maximum iteration **do**

10:     **for all** particle **do**

11:         get fitness from ESNN (accuracy of the model)

12:         **if** (current fitness is better than *pbest* fitness) or ((current fitness $==$ *pbest*

                      fitness) and (feature selected is less than feature selected by *pbest*))

                  **then**

13:             assign current particle as *pbest*

14:             **if** (current *pbest* fitness is better than *gbest* fitness) or ((current *pbest* fitness $==$ *gbest* fitness) and (feature

                selected by *pbest* is less than feature selected by *gbest*)) **then**

15:                 assign *pbest* as *gbest*

16:             **end if**

17:         **end if**

18:         **for all** ESNN parameters **do**

19:             calculate velocity

20:             update parameters

21:         **end for**

22:         **for all** feature qubits **do**

23:             calculate velocity

24:             apply rotation gate

25:             get a collapsed state

26:         **end for**

27:     **end for**

28: **end while**

---

The DQiPSO method is applied here for the optimisation of eSNN trained on the two spirals benchmark classification problem. The problem is well known as a difficult non-linear separation problem introduced in [61]. In order to evaluate the performance in feature selection task, the two relevant data was copied with some noise added to the original data. These redundant dimensions are generated by adding a Gaussian noise using standard deviation of $\sigma = |p| * s$ with $|p|$ being the absolute value of vector $p$ while $s$ is a parameter controlling the noise strength to the original spiral points $p = (x, y)^T$. The noise increases linearly according to the distant from the spiral origin $(0, 0)^T$. Then the noise value is calculated as the $p_i$ centered Gaussian distributed random variable $N(p_i, \sigma^2)$. The information available in a feature decreases when stronger noise is applied Fig. 7.15. In addition to this, several irrelevant features with the random dimension value between [0,1] were also added into the dataset. The dataset in this experiment consisted of 20 features with 2 relevant features, 14 redundant features with the noise level $s$ from 0.2 to 0.8 and four random features. Detailed explanation of the data generation can be found in [29]. Then the features were arranged in a random order to simulate a real world problem scenario where relevant features are scattered in the dataset. 400 samples in two classes and equally distributed between classes was generated.

Based on our preliminary experiment, 20 eSNN receptive fields were chosen with the center is uniformly distributed between the maximum and minimum value of the data with the controlling parameter $\beta$ is 1.0. For the DQiPSO, 12 particles consisting of six Update Particles, four Filter Particles and two Random Particles were used. For the standard QiPSO, 20 particles were used. $c_1$ and $c_2$ were set to 0.05 which meant a balanced exploration between *gbest* and *pbest*. The inertia weight was set to $w = 2.0$. The dataset was applied to the proposed method and compared with our previous method [30] and eSNN using all features where all algorithms with parameter optimisation. 10 fold cross validation were used and the average result was computed in 300 iterations.

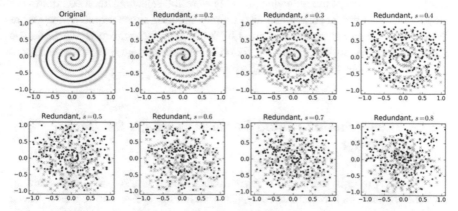

**Fig. 7.15** The two spirals benchmark data—the original data set and data sets with added various levels of noise using added redundant and randomly generated features to the original 2 ones

Figures 7.15 and 7.16 illustrate the selected features during the optimization process over 300 iterations, when the results from the use of the QiPSO method and the DQiPSO method are compared. Each model (particle) is trained and tested on the two spirals data using the 10 fold cross validation method to estimate the classification accuracy thus the fitness of the model. The lighter colour means more frequent corresponding features are selected and contradictory to the darker colour. The algorithm keeps eliminating irrelevant features in order to identify the most relevant features.

Figure 7.16 shows the results of the eSNN parameter optimisation. All eSNN parameters evolve steadily towards a certain optimal value, where the correct combination together with the selected relevant features lead to better classification accuracy. In terms of classification result, the average accuracy for DQiPSO is 93.4% with the highest single run accuracy found in this experiment is 97.2%. The result also shows that the average accuracy for DQiPSO at the beginning to the learning process is acceptably high, around 80%. This is due to the proposed DQiPSO particle structure that was able to select relevant features with nearly optimal parameter combination in the early stage of learning. For the QiPSO algorithm, the average accuracy is only 88.6%. While the QiPSO algorithm optimizes the eSNN parameters using all features, it delivers the worst result in this experiment 78.1% accuracy. In this case the algorithm entirely depends on the parameter optimisation which was inadequate to produce a satisfactory result. Overall, the proposed method demonstrates a satisfactory result in this experiment. For DQiPSO, the figure clearly shows the two relevant features which contained the most information, are constantly being selected in all runs. In contrast, the random value features are rejected during the learning process together with most of the redundant features. However, some redundant features with a noise level of 0.2, 0.3 and 0.4, are still occasionally selected. The reason is simply because these features contained some information that could be used to distinguish between classes, as we can see from Fig. 7.17. On the other hand, the QiPSO is generally able to select the relevant features in average of 7 times out of 10 runs. However, the ability to reject the irrelevant features was unsatisfactory. Most of the redundant features were still selected which contributed to the low classification accuracy and also more computing times required since more features have been selected. Since the QiPSO has no mechanism to stimulate the particle if there is no better solution found, therefore the algorithm may converge prematurely without the optimal results obtained.

## 7.5  Chapter Summary and Further Readings for Deeper Knowledge

The chapter presents methods of evolutionary computation (EC) and quantum inspired evolutionary computation (QiEC) including QiEA and QiPSO. These methods are generic methods, applicable to a wide range of problems and processes that can be optimised. In this chapter we show examples of their applications for the

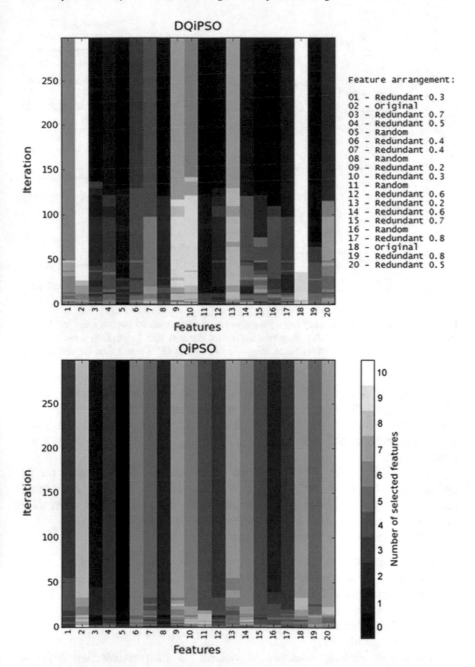

**Fig. 7.16** Evolution of feature selection is faster and more accurate when using DQiPSO versus using QiPSO algorithms for the two spiral problem with 18 additional noisy features added

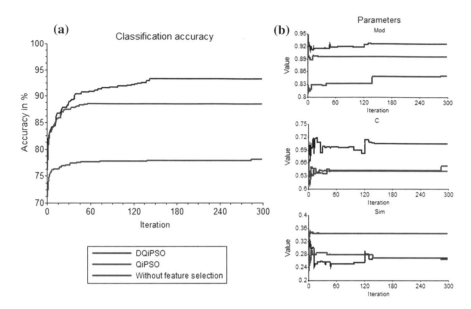

**Fig. 7.17** **a** Classification accuracy and **b** parameter optimisation result for the two spiral problem and with the added 18 noisy redundant features using eSNN classifier with 3 optimised parameters. The comparison shows a better accuracy when using DQiPSO versus QiPSO versus no features selected

optimization of features and parameters of SNN, and more specifically eSNN (Chap. 5). The QiEC methods result in a more efficient classification model design with optimal features (variables) selected and model parameters optimized rather than using standard EC techniques or without any optimisation of features and model parameters.

Applying the existing QiEC methods and developing new methods for the optimization of brain-like SNN, such as NeuCube is still a challenge for a future development. Chapter 22 discusses a potential future research direction where these methods are further integrated with molecular and brain inspired methods.

Further recommended readings include:

- Genetic Algorithm in Search, Optimization and Machine learning [2, 4, 5];
- Genetic Programming [3];
- Learning in ANN with EC [9];
- Learning in fuzzy systems through GA [10];
- Particle Swarm Optimization [6];
- Firework algorithms [7];
- Overview of evolving connectionist systems and their optimization ([62] and Chap. 40 in [63]).

**Acknowledgements** The chapter includes some materials published previously by the author in collaboration with colleagues. I acknowledge the contribution of the following colleagues in these publications: Stefan Schliebs, Haza Nuzly, Michael Defoin-Platel, Mike Watts. Some of the material is also taken from my previous books with Springer [44, 62, 64].

# References

1. C. Darwin, *On the Origin of Species by Means of Natural Selection, or the Preservation of 991 Favored Races in the Struggle for Life*, 1st edn. (John Murray, London, 1859). p. 502
2. D.E. Goldberg, *Genetic Algorithm in Search, Optimization and Machine learning* (Addison-Wesley Longman Publishing Co., Inc., Boston, MA, USA, 1989). ISBN 0201157675
3. J. Koza, *Genetic Programming of Computers by means of Natural Selection* (MIT Press, Cambridge, MA, USA, 1992). ISBN 0-262-11170-5
4. J.H. Holland, Genetic algorithm. Sci. Am. **1992**, 66–72 (1992)
5. J.H. Holland, *Emergence: From Chaos to Order* (Addison-Wesely, Redwood City, California, 1998). ISBN 0-201-14943-5
6. J. Kennedy, R. Eberhart, *Particle Swarm Optimization, IEEE*, in Proceedings of International Conference on Neural Networks (1995), pp. 1942–1948
7. Y. Tan (2018) *GPU-Based Parallel Implementation of Swarm Intelligence Algorithms* (Morgan Kaufmann, 2015)
8. D.B. Fogel, J.W. Atmar, Comparing genetic operators with Gaussian mutations in simulated evolutionary optimization. Biol. Cybem. **63**, 111–114 (1990)
9. X. Yao, A Review of evolutionary artificial neural networks. Int. J. Intell. Syst. **8**(4), 539–567 (1993)
10. T. Furuhashi, K. Nakaoka, Y. Uchikawa, *A New Approach to Genetic Based Machine Learning and an Efficient Finding of Fuzzy Rule*, in Proceedings of the 1994 IEEE/Nagoya-University World Wisepersons Workshop (WWW'94), Lecture Notes in Artificial Intelligence, ed by T. Furuhashi, vol. 1011 (Springer, 1994), pp. 173–189
11. H. Mühlenbein, G. Paass, *From Recombination of Genes to the Estimation of Distributions in Binary Parameters*, in Proceedings International Conference on Evolutionary Computation, Parallel Problem Solving From Nature—PPSN IV (1996), pp. 178–187
12. M. Pelikan, D. Goldberg, F. Lobo, A survey of optimization by building and using probabilistic model, 1999, IlliGAL, Tech. Rep. No. 99018 (1999)
13. C. Gonzalez, J. Lozano, P. Larranaga, Analyzing the PBIL algorithm by means of discrete dynamical systems. Complex Syst. **12**, 465–479 (2000)
14. G.R. Harik, F.G. Lobo, D.E. Goldberg, The compact genetic algorithm. IEEE Trans. Evol. Comput. **3**(4), 287–297 (1999)
15. A. Johnson, J.L. Shapiro, *The Importance of Selection Mechanisms in Distribution Estimation Algorithms*, in Proceedings 5th International Conference on Artificial Evolution AE01, Oct 2001, (2001), pp. 91–103
16. C. Adami, *Introduction to Artificial Life* (Springer, New York, 1998). ISBN 978-1-4612-7231-1
17. W.K. Wootters, W.H. Zurek, A Single Quantum Cannot Be Cloned. Nature **1982**(299), 802–803 (1982)
18. R.P. Feynman, Simulating physics with computers. Int. J. Theor. Phys. **21**(6/7), 467–488 (1982)
19. P.W. Shor, *Algorithms for Quantum Computation: Discrete Log and Factoring*, in Proceedings of the 35th Annual IEEE Symposium on Foundations of Computer Science—FOCS (1994), pp. 20–22
20. K.H. Han, J.H. Kim, Quantum-inspired evolutionary algorithm for class of combinatorial optimization. IEEE Trans. Evolut. Comput. **6**(6), 580–593 (2002)

21. J.S. Jang, K.H. Han, J.H. Kim, *Face Detection Using Quantum-Inspired Evolutionary Algorithm*, in Proceedings of the IEEE Congress on Evolutionary Computation (2004), pp. 2100–2106
22. A.V.A. da Cruz, M.M.B. Vellasco, M.A.C. Pacheco, *Quantum Inspired Evolutionary Algorithm for Numerical Optimization*, in Proceedings of the IEEE Congress on Evolutionary Computation (2006), pp. 2630–2637
23. H. Talbi, A. Draa, M. Batouche, A novel quantum inspired evaluation algorithm for multisource affine image registration. Int. Arab J. of Inf. Technol. **3**(1), 9–15 (2006)
24. G.K. Venayagamoorthy, G. Singhal, Quantum inspired evolutionary algorithms and binary particle swarm optimization for training MLP and SRN neural networks. J. Comput. Theor. Nanosci. **2**(4), 561–568 (2005)
25. L.K. Grover, *A Fast Quantum Mechanical Algorithm for Database Search*, in Proceedings 28th Annual ACM Symposium on the Theory of Computing (1996), pp. 212–129
26. D. Deutsch, Quantum computational networks. Proc. R. Soc. Lond. pp. 73–90 (1989)
27. A. Ezhov, D. Ventura, Quantum neural networks, in *Future Directions for Intelligent Systems* ed by N. Kasabov (Springer, Berlin, 2000)
28. M. Defoin-Platel, S. Schliebs, N. Kasabov, *A Versatile Quantum-Inspired Evolutionary Algorithm*, in Proceedings of the IEEE Congress on Evolutionary Computation (2007), pp. 423–430
29. S. Schliebs, M. Defoin-Platel, S. Worner, N. Kasabov, Integrated feature and parameter optimization for an evolving spiking neural network: exploring heterogeneous probabilistic models. Neural Netw. **22**(5–6), 623–632 (2009)
30. H.N.A. Hamed, N. Kasabov, S.M. Shamsuddin, *Integrated Feature Selection and Parameter Optimization for Evolving Spiking Neural Networks Using Quantum Inspired Particle Swarm Optimization*. In Proceedings of the International Conference of Soft Computing and Pattern Recognition (2009), pp. 695–698
31. B. Yuan, M. Gallagher, Playing in continuous spaces: some analysis and extension of population-based incremental learning, in The 2003 Congress on Evolutionary Computation. CEC'03, vol. 1, pp. 443–450
32. M. Gallagher, M. Frean, Population-based continuous optimization, probabilistic modelling and mean shift. Evol. Comput. **13**(1), 29–42 (2005)
33. M. Gallagher, M. Frean, T. Downs, *Real-Valued Evolutionary Optimization Using a Flexible Probability Density Estimator*, in Proceedings of the GECCO 1999 Genetic and Evolutionary Computation Conference (Morgan Kaufmann Publishers, 1999), pp. 840–846
34. P.A. Bosman, D. Thierens, Expanding from discrete to continuous estimation of distribution algorithms: the idea, in: In Parallel Problem Solving From Nature—PPSN VI (Springer, 2000), pp. 767–776
35. E. Mininno, F. Cupertino, D. Naso, Real-valued compact genetic algorithms for embedded microcontroller optimization. IEEE Trans. Evolut. Comput. **12**(2), 203–219 (2008)
36. I. Scrvet, L. Travée-Massuyès, D. Stern, Telephone network traffic overloading diagnosis and evolutionary computation techniques, in AE'97: Selected Papers from the Third European Conference on Artificial Evolution. Springer, London, UK, pp. 137–144 (1998)
37. J. Geweke, *Efficient Simulation from the Multivariate Normal and Student-t Distributions Subject to Linear Constraints and the Evaluation of Constraint Probabilities*, in: Computing Science and Statistics: Proceedings of the 23rd Symposium on the Interface. pp. 571–578 (1991)
38. M. Sebag, A. Ducoulombier, *Extending Population-Based Incremental Learning to Continuous Search Spaces*, in PPSN V: Proceedings of the 5th International Conference on Parallel Problem Solving from Nature (Springer, London, UK, 1998), pp. 418–427
39. H. Mühlenbein, T. Mahnig, A.O. Rodriguez, Schemata, distributions and graphical models in evolutionary optimization. J. Heuristics **5**(2), 215–247 (1999)
40. M. Defoin-Platel, S. Schliebs, N. Kasabov, Quantum-inspired evolutionary algorithm: a multimodel EDA, in Evolutionary Computation, IEEE Transactions onIn print (2009)
41. M.A. Pottcr, K.A.D. Jong, Cooperative coevolution: an architecture for evolving coadapted subcomponents. Evol. Comput. **8**, 1–29 (2000)

42. N. Kasabov, To spike or not to spike: a probabilistic spiking neural model. Neural Netw. **23** (1), 16–19 (2010)
43. N. Kasabov, Evolving intelligence in humans and machines: integrative evolving connectionist systems approach. IEEE Comput. Intell. Mag. **3**(3), 23–37 (2008)
44. N. Kasabov, *Evolving Connectionist Systems: The Knowledge Engineering Approach* (Springer, Berlin, 2007)
45. G. John, R. Kohavi, Wrappers for feature subset selection. Artif. Intell. **97**(1–2), 273–324 (1997)
46. S. Schliebs, M. Defoin-Platel, S. Worner, N. Kasabov, *Quantum-Inspired Feature and Parameter Optimization of Evolving Spiking Neural Networks with a Case Study from Ecological Modeling*, in Proceedings of the IJCNN 2009, Atlanta, 11–19 June (IEEE Press, 2009)
47. CABI, *Crop Protection Compendium, Global Module*, 5th edn. (2003)
48. M. Watts, S. Worner, Using MLP to determine abiotic factors influencing the establishment of insect pest species, in International Joint Conference on Neural Networks (IJCNN'06). IEEE, Vancouver, Canada, pp. 1840–1845 (2006)
49. S. Worner, G. Leday, T. Ikeda, *Uncertainty Analysis and Ensemble Selection of Statistical and Machine Learning Models that Predict Species Distribution*, in Ecological Informatics, Cancun, Mexico (2008)
50. M.T. Vera, R. Rodriguez, D.F. Segura, J.L. Cladera, R.W. Sutherst, Potential geographical distribution of the mediterranean fruit fly, ceratitis capitata (diptera:tephritidae) with emphasis on Argentina and australia. Environ. Entomol. **31**(6), 1009–1022 (2002)
51. L. Benuskova, N. Kasabov, Application of CNGM to learning and memory, in *Computational Neurogenetic Modeling, Topics in Biomedical Engineering*, International Book Series (Springer, Boston, MA, 2007)
52. N. Kasabov, H.N.A. Hamed, Quantum-inspired particle swarm optimisation for integrated feature and parameter optimisation of evolving spiking neural networks. Int. J. Artif. Intell. **7** (11), 114–124 (2011)
53. H.N.A. Hamed, N. Kasabov, S.M. Shamsuddin, Probabilistic evolving spiking neural network optimisation using dynamic quantum-inspired particle swarm optimisation. Aust. J. Intell. Inf. Process. Syst. **11**(1), 23–28 (2010)
54. H.N.A. Hamed, N. Kasabov, S.M. Shamsuddin, Dynamic quantum-inspired particle swarm optimization as feature and parameter optimizer for evolving spiking neural networks. Int. J. Mod. Optim. **2**(3), 2012 (2010). https://doi.org/10.7763/IJMO.2012.V2.108
55. G. John, R. Kohavi, K. Pfleger, *Irrelevant Features and the Subset Selection Problem*. In Proceedings of the 11th International Conference on Machine Learning (1994), pp. 121–129
56. G. John, R. Kohavi, Wrappers for feature subset selection. Artif. Intell. **97**(1–2), 273–324 (1997)
57. S.M. Bohte, J.N. Kok, H.L. Poutre, Error-backpropagation in temporally encoded networks of spiking neurons. Neurocomputing **48**(1) (2002)
58. S. Wysoski, L. Benuskova, N. Kasabov, *Text-Independent Speaker Authentication with Spiking Neural Networks*, in Proceedings of the ICANN 2007 (Springer, 2007)
59. S. Wysoski, L. Benuskova, N. Kasabov, fast and adaptive network of spiking neurons for multi-view visual pattern recognition. Neurocomputing **71**(13–15), 2563–2575 (2008)
60. S.J. Thorpe (1997), How can the human visual system process a natural scene in under 150 ms? Experiments and neural network models, in ESANN, 1997
61. K.J. Lang, M.J. Witbrock, *Learning to Tell Two Spirals Apart*, in Proceedings of the 1988 Connectionist Models Summer School SanMateo (Morgan Kauffmann, 1998), pp. 52–59
62. N. Kasabov (ed.) *Springer Handbook of Bio-/Neuroinformatics* (Springer, Berlin, 2014)
63. J. Kacprzyk, W. Pedrycz (eds.) *Springer Handbook of Computational Intelligence* (Springer, Berlin, 2015)
64. L. Benuskova, N. Kasabov, *Computational Neurogenetic Modelling* (Springer, New York, 2007)

# Part IV
# Deep Learning
# and Deep Knowledge
# Representation of Brain Data

# Chapter 8
# Deep Learning and Deep Knowledge Representation of EEG Data

This chapter presents general methods for deep learning and deep knowledge representation of EEG data in brain-inspired SNN (BI-SNN). These methods are applied to develop specific methods for EEG data analysis and for modelling brain cognitive functions, such as: performing cognitive tasks; emotion recognition from face expression; sub-conscious processing of stimuli; modelling attentional bias.

SNN are used here not to model the brain, but to model brain data. The chapter demonstrates that the BI-SNN can not only learn and classify brain EEG data much more accurately than traditional machine learning methods, but due to their organisation being a map of a brain template, they can represent deep knowledge from the EEG data in cortical brain areas. If trained on EEG data/knowledge when humans are performing cognitive functions, a BI-SNN can learn this knowledge and can possibly manifest it autonomously under certain conditions.

The chapter is organised in the following sections:

8.1. Time-space brain data. EEG data.
8.2. Deep learning and deep knowledge representation of EEG data in BI-SNN.
8.3. Deep learning, recognition and modelling of cognitive tasks.
8.4. Deep learning, recognition and expression of emotions in a BI-SNN.
8.5. Deep learning and modelling peri-perceptual processes.
8.6. Modelling attentional bias.
8.7. Chapter summary and further readings for deeper knowledge.

## 8.1 Time-Space Brain Data. EEG Data

### 8.1.1 Spatio-temporal Brain Data

Different types of spatio-temporal brain data (STBD) have been collected for many years at different 'levels' of information processing in the brain as discussed in Chap. 3. At the highest, cognitive level, the most common types are EEG, MEG,

© Springer-Verlag GmbH Germany, part of Springer Nature 2019
N. K. Kasabov, *Time-Space, Spiking Neural Networks and Brain-Inspired Artificial Intelligence*, Springer Series on Bio- and Neurosystems 7,
https://doi.org/10.1007/978-3-662-57715-8_8

fMRI, DTI, NIRS, single unit electrode data and others [1]. Electroencephalography (EEG) is the recording of electrical signals from the brain by attaching surface electrodes to the subject's scalp [2–4]. Magnetoencephalography (MEG) measures millisecond-long changes in magnetic fields created by the brain's electrical currents. MEG machines use a non-invasive, whole-head, e.g. 248-channel, *super-conducting-quantum-interference-device* (SQUID) to measure small magnetic signals reflecting changes in the electrical signals in the human brain. New methods for brain data collection are being developed and this area of research is likely to be further developed in the future.

Functional MRI (fMRI) combines visualisation of the brain anatomy with the dynamic image of brain activity into one comprehensive scan (e.g. [5–8]). This non-invasive technique measures the ratio of oxygenated to deoxygenate haemoglobin which have different magnetic properties. Active brain areas have higher levels of oxygenated haemoglobin than less active areas. An fMRI scan can produce images of brain activity at the time scale of seconds with precise spatial resolution of about 1–2 mm. Thus, fMRI provides both a 3D anatomical and functional view of the brain in the lower frequency spectrum. fMRI data modelling is discussed in Chaps. 10 and 11.

## 8.1.2  Brain Atlases

Over the last 30 years, neuroimaging techniques evolved a lot and have allowed neuroscientists to revisit the issue of mapping the human brain, such that a modern brain atlas is now expressed as a digital database that can capture the spatio-temporal distribution of a multitude of physiological and anatomical metrics.

Several structural brain atlases have been created to support the study of the brain and to better structure brain data. **Korbinian Brodmann (1868–1918)** was a German neurologist who segmented the cerebral cortex into 52 distinct regions from their cytoarchitectonic (histological) characteristics, known as Brodmann areas, published in 1909. The map presents 52 distinctive areas of the cerebral cortex. Each Brodmann area (BA) is characterized by a distinct type of cells, but it also represents distinct structural area, distinct functional area (e.g. BA17 is the visual cortex), distinct molecular area (e.g. number of neurotransmitter channels) [9]. EEG and fMRI data are often mapped into BA for a better interpretation of results [10].

An important contribution to the overall brain study and particularly to brain data analysis is the creation of a common coordinate system that can be used for a standardized study of brain data from different subjects and collected by different methods. Talairach and Tournoux (1988) [11] created a co-planar 3D stereotaxic atlas of the human brain. The Talairach coordinate space has its origin defined at the anterior commissure (AC), with $x$ and $y$ axes on a horizontal plan and $z$ axis on a vertical plane; in particular, the $y$ axis is defined by the line connecting the most superior of AC and the most inferior of the posterior commissure (PC); the $x$ axis is

**Fig. 8.1**  A view of brain
sections in terms of AC and
PC views

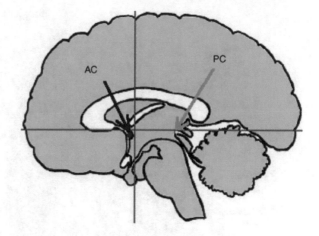

defined by the line that passes through the AC point and orthogonal to the AC-PC
line; whereas the $z$ axis is the line that passes through the interhemispheric fissure
and the AC point (Fig. 8.1.).

A software was also made available, called Talairach Daemon (www.talairach.
org) to calculate the Talairach coordinates (x, y, z) of any given point in a brain
image along with the corresponding BAs (Figs. 8.2,8.3) [12].

The Talairach atlas was generated from two series of sections from a single
60-year old female brain in 1967: one half was sectioned in the sagittal plane and
the other in the coronal plane. The transverse images in the atlas were manually
approximated from the information obtained in the sagittal and coronal planes. In
their work, Talairach and Tournou [11] identified the anatomical features from the
atlas and created a coordinate system related to anatomical landmarks.

The Talairach Atlas has been digitized and manually traced into a volume-
occupant hierarchy of anatomical regions. Hemispheres, lobes, lobules, gyri and
nuclei have been outlined and labelled. Grey matter, white matter and CSF regions
will also be defined. For cerebral cortex, all Brodmann areas have been traced and
expanded into 3-D volumes.

*Using brain templates to map brain data into a BI-SNN*

While the Talairach Atlas was derived from the analysis of a single brain, much
further development in stereotaxic mapping was achieved with the introduction of
the Montreal Neurological Institute (MNI) coordinates, based on averaged MRI
data across individuals, e.g. MNI152, MNI305 [13]. Mapping of standard brain
stereotaxic coordinates was further developed by the International Consortium for
Brain Mapping (ICBM) with the release of several brain map templates, such as:
ICBM452; ICBM Chinese56; ICBM AD (Alzheimer Disease); MS (multiple
sclerosis) and others [1]. Brain activity measurements, such as EEG and fMRI of
any subject can be represented in standard MNI coordinates. MNI coordinates can
be translated into Talairach coordinates and Brodmann Areas, and vice versa. The

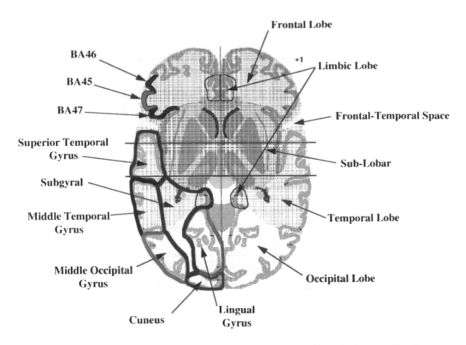

**Fig. 8.2** Structural and functional areas of the brain are annotated in a brain atlas. BA denotes a Brodmann area

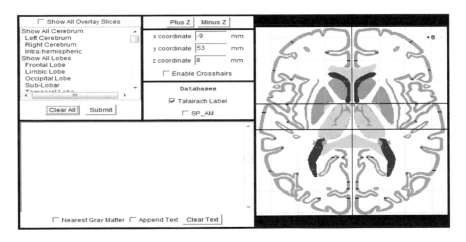

**Fig. 8.3** A snap shot of the Talairach software demon (http://www.talairach.org/daemon.html)

brain gene atlas, discussed further below, contains gene expression data collected from brain areas with identified MNI coordinates. MNI is a common standard now supported by many software systems, e.g. SPM [14].

At the lowest 'level' of information processing in the brain is the molecular information processing. Spatio-temporal activity in the brain depends on the internal brain structure, on the external stimuli and also very much on the dynamics at gene-protein level. This complex interaction is addressed through computational neurogenetic modelling [15]. The first issue is how to obtain gene data related to brain structures and functions. The Brain Atlas (www.brain-map.org) of the Allen Institute for Brain Science (www.alleninstitute.org) has shown that at least 82% of the human genes are expressed in the brain. For almost 1000 anatomical brain areas of two healthy subjects, 100 mln data points were collected that indicate gene expressions of several thousand genes and underlie the biochemistry of the sites [16]. This is in addition to the previously developed Mouse Brain Atlas.

The enormousness of brain data available and the complexity of the research questions that need answering through integrated models for brain data analysis are grand challenges for the areas of machine learning and information science in general as already pointed in some recent publications [17–20].

Accurate models of the brain have been developed (e.g. [1, 21, 22]). However, they cannot be used for machine learning and pattern recognition of spatio-temporal brain data (STBD) as their goal is to model the brain structurally and functionally and not learn and mine brain data. The human brain has evolved throughout more than 5mln years of human evolution with the more recent 10,000 years of human civilization. The accurate brain modelling may require modelling the principles of evolution in nature rather than just the brain as its product. Modelling the brain is a challenging task for many years to come (e.g. the EU Human Brain Project), but modelling and understanding brain data, that is available *now,* is a task that the neural network community needs to address *now.* And this is the goal perceived in this chapter and in the book as a whole.

## 8.1.3   EEG Data

Electroencephalography (EEG) is the recording of electrical signals from the brain by attaching surface electrodes to the subject's scalp [2–4]. These electrodes record brain waves which are electrical signals naturally produced by the brain. EEGs allow researchers to track electrical potentials across the surface of the brain and observe changes taking place over a few milliseconds. EEG data is spatio/spectro-temporal in the high frequency spectrum.

Electroencephalography (EEG) is the recording of electrical activity along the scalp. EEG measures voltage fluctuations resulting from ionic current flows within the neurons of the brain. This electrical signal can be measured by EEG data collected from the human scalp (Fig. 8.4).

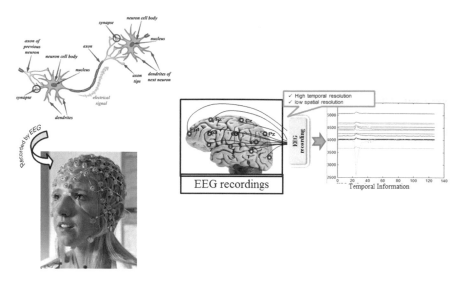

**Fig. 8.4**  EEG signals measure cortical spatio-temporal activity of a human brain along the scalp

*Properties of EEG signals*

- EEG provides high temporal resolution (sampling rates between 250 and 2000 Hz);
- Unable to provide a precise localisation of the neuron activation;
- electrodes record sums of activity from cortical sources (unclear spatial resolution).

Electroencephalography is the most known and old direct technique, indeed human EEG recordings started in 1920 with the German physiologist and psychiatrist Hans Berger.

The highest influence to EEG comes from electric activity of cerebral cortex due to its surface position.

Brain patterns form wave shapes that are commonly sinusoidal; they are usually measured from peak to peak and a normal range is between 0.5 and 100 µV in amplitude. The brain state of every single individual may make certain frequencies more dominant, however brain waves can be categorized (see Fig. 2.5) into five basic groups: beta (>13 Hz), alpha (8–13 Hz), theta (4–8 Hz), delta (0.5–4 Hz), gamma (>40 Hz), which characterize different brain states (Fig. 8.5).

In particular most of the signal comes from the neurons placed in the outermost layer of the cerebrum, the grey matter; they are the nearest to the EEG electrodes since the layer below them is made of white matter which creates a thickness that separates the other neurons more in the centre of the brain, from the head surface. This is important since the electric field decreases as a function of distance from the source and of the conductivity of the material that it goes through.

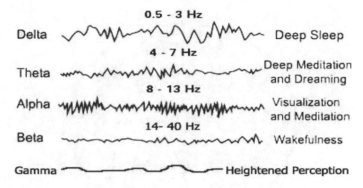

**Fig. 8.5** Different types of brain waves that can be detected from EEG signals

EEG, as MEG, can be used for many neurophysiological studies, one of the examples is tests on language functions, e.g. [15]: in a normal subject, during picture naming, visual and conceptual processes take place within the first 175 ms after stimulus presentation, followed by lexical retrieval (until 250 ms) and phonological encoding of the word form (250–450 ms); whereas, after neurological damage, different aspects of word-retrieval can be impaired (e.g., post-stroke anomia). Thus effects of damages and injuries can be outlined.

Other research and clinical applications of the EEG in humans and animals can be:

- locate areas of damage following head injury, stroke, tumour;
- monitor cognitive engagement;
- investigate epilepsy and locate seizure origin;
- test epilepsy drug effects;
- investigate sleep disorder and physiology.

The electroencephalogram (EEG) is defined as electrical activity of an alternating type recorded from the scalp surface after being picked up by metal electrodes and conductive media. In this work we will refer only to EEG measured from the head surface so without using depth probes; for this, the same procedure can be applied repeatedly to patients, normal adults, and children with virtually no risk or limitation.

EEG recording system consists of:

- electrodes with conductive media: they read the signal from the head surface
- amplifiers with filters: they bring the microvolt signals into the range where they can be digitalized accurately
- A/D converter: it changes signals from analogue to digital form
- recording device as personal computer (or other relevant device) which stores and displays obtained data.

We have said already that the EEG signals are scalp recordings of neuronal activity in the brain, in particular, they measure potential changes over time in basic electric

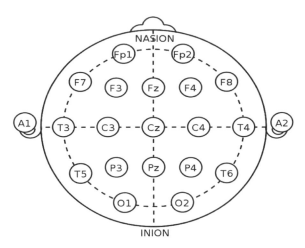

**Fig. 8.6** Labels for some EEG channels and their locations according to 10–20 electrode placement system

circuit conducting between signal (active) electrode and reference electrode. Moreover, an extra third electrode, called ground electrode, is needed for getting differential voltage by subtracting the same voltages showing at active and reference points. So the minimal configuration for monochannel EEG measurement consists of at least three electrodes: one active electrode, one (or two specially linked together) reference and one ground electrode. Nowadays there are also multi-channel configurations which can comprise 32, 64, 128 up to 256 active electrodes.

In 1958, International Federation in Electroencephalography and Clinical Neurophysiology adopted standardization for electrode placement called 10–20 electrode placement system which standardized physical placement and

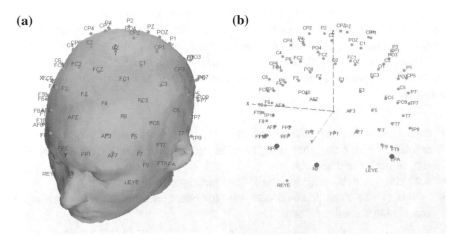

**Fig. 8.7** Overall location of the EEG channels according to the 10–20 system

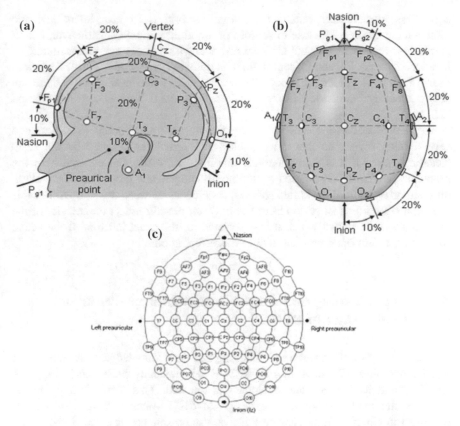

**Fig. 8.8   a** Precise location of the EEG electrodes according to the 10–20 system (after [70, 71]), **b** extended EEG electrode position nomenclature by American Electroencephalographic Society [70, 71]

designations of electrodes on the scalp (Figs. 8.6, 8.7 and 8.8). The head is divided into proportional distances from prominent skull landmarks (nasion, preauricular points, inion) to provide adequate coverage of all regions of the brain, so label 10–20 designates proportional distance in percents between ears and nose where points for electrodes are chosen. Following the same idea of proportions the 10–10 system has been developed too.

Electrode placements are labelled according adjacent brain areas: F (frontal), C (central), T (temporal), P (posterior), and O (occipital). The letters are accompanied by odd numbers at the left side of the head and with even numbers on the right side. Left and right side is considered by convention from point of view of a subject.

As it is known from tomography, different brain areas may be related to different functions of the brain, so each scalp electrode is located near certain brain centres, e.g. F7 is located near centres for rational activities, Fz near intentional and motivational centres, F8 close to sources of emotional impulses. Moreover, cortex electrodes around C3, C4, and Cz locations deal with sensory and motor functions; locations near P3, P4, and Pz contribute to activity of perception and differentiation; near T3 and T4 emotional processors are located, while near T5, T6 there are certain memory functions; primary visual areas can be found below points O1 and O2.

Between electrode and neuronal layers current penetrates through skin, skull and several other layers, so that a weak electrical signals is detected by the scalp electrodes; even if it is then massively amplified it is still characterized by some problems and limitations caused by the non-homogeneous properties of the skull, different orientation of the cortex sources, coherences between the sources. For this reason, the recording of scalp electrodes may not exactly reflect the activity of the particular area of cortex which it is associated to: the exact location of the active sources is still an open problem due to those limitations.

## 8.2   Deep Learning and Deep Knowledge Representation of EEG Data in BI-SNN

A methodology for deep learning, modelling and deep knowledge representation of STBD, including EEG data, in a BI-SNN is schematically presented in Figs. 8.9 and 8.10 and in Box 8.1 and explained in the text. This methodology follows precisely the methodology for the design of SNN systems using NeuCube as presented in Chap. 6. Here NeuCube is also used as an exemplar BI-SNN.

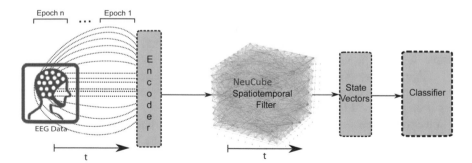

**Fig. 8.9**  A general scheme for using the NeuCube BI-SNN for EEG STBD modelling

**Box 8.1 A Methodology for Deep Learning, Modelling and Deep Knowledge Representation of EEG Data**

1. EEG input data transformation/encoding into spike sequences;
2. Spatial mapping of input EEG channels as variables into spiking neurons using a brain template, such as the Talairach template;
3. Unsupervised *deep learning* of the spatio-temporal spike sequences that encode the EEG signals in a scalable 3D SNN cube;
4. Supervised learning and classification/regression of data;
5. Dynamic parameter optimisation;
6. Evaluating the predictive modelling capacity of the system;
7. Adaptation on new data, possibly in an on-line/real time mode;
8. Model visualisation and connectivity and spiking activity pattern interpretation for a better understanding of the data and the brain processes that generated it;
9. Defining the *deep knowledge* represented in a trained SNNcube and output classifier as connectionist patterns;
10.     Implementation of a SNN model in a hardware/software platform, e.g. von Neumann versus neuromorphic hardware versus quantum computing.

It is important to highlight that the NeuCube model is a stochastic model (*i.e.* initial connection between the neurons of the reservoir are randomly generated) and therefore extremely sensitive to parameters' settings. Some of the major parameters that highly influence the model are:

- The spike encoding threshold of the encoding spike trains; a bi-directional threshold, which is applied to the signal's gradient according to the time. When a new data set is loaded, depending on the spike rate obtained, the threshold is determined by the model and by the used method for encoding (see Chap. 4).
- The connection distance between neurons of the network. According to the small world (SWO) connectivity principle, each neuron of a SNN reservoir is connected to its neighbouring neurons by a fixed distance multiplied by this parameter, the resulting value will be the one that establishes which neurons will be connected and which will not. By default this parameter is set to 0.15, which is a generic low value (see Chap. 6).

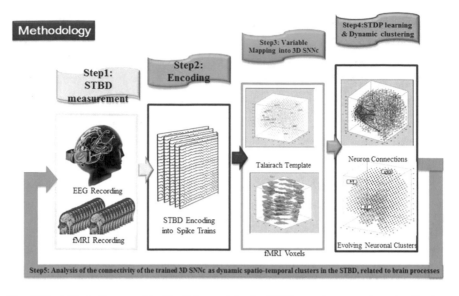

**Fig. 8.10** A block diagram of the NeuCube modules for STBD encoding, mapping, visualisation, learning and classification. The mapping module illustrates the allocation of EEG channels as input neurons and fMRI voxels in another SNNcube (the figure was created by M. Gholami)

- The threshold of firing, the refractory time and the potential leak rate of the LIF neurons (see Chap. 4). When a LIF neuron of the reservoir spikes its PSP increases gradually with every input spike according to the time, until it reaches an established threshold of firing. Then, an output spike is emitted and the membrane potential is reset to an initial state (refractory time). Between spikes, the membrane potential leaks, simulating biological membrane behaviour after which some equilibrium is not achieved after the diffusion of an ion.
- The STDP rate. According to the STDP learning rule (see [23] for more details), the firing activity of a particular neuron also cause its neighbour neurons to emit a spike. The membrane potential of this one will be increased with respect to its connection weight with the firing neuron. The membrane potential of each neuron has a small constant rate of leakage, which by default set as 0.01, identified as an appropriate value in most of the cases.
- The number of times that the NeuCube is trained.
- The variables *mod* and *drift* of the deSNN classifier. According to [24], every training sample is associated to an output neuron, which is connected to each and every other neuron of the reservoir. The initial connection weights of these output neurons are all set to zero. New connection weights are formed according to the rank-order (RO) learning rule. This are calculated depending of a

modulation factor (the variable *mod*) of the order of the incoming spikes. The new connection weights will then increase or decrease according to the number of spikes that follow the first one (the *drift* value). By default, these parameter values are 0.4 for the mod and 0.25 for the drift, also empirically defined.

A crucial step for the NeuCube is the optimization of these variables, which needs to be pursued in order to obtain desirable classification accuracy. Variables tuning can be achieved via grid search method, genetic algorithm or quantum-inspired evolutionary algorithm [25] (see also Chap. 7).

Three types of deep learning of EEG data is performed in NeuCube (also discussed in Chap. 6):

– Unsupervised learning in the SNNcube (Box 8.2);
– Supervised learning in the SNNcube and the output classifier (Box 8.3);
– Semi-supervised learning, combining the unsupervised and supervised learning (Box 8.3).

Figure 8.11 illustrates the evolution of neuronal connectivity and spiking activity in a SNNc of 1471 spiking neurons with Talairach- based coordinates [26] (a-b) the step-wise neuronal connectivity and the spiking patterns of the SNNc at two steps during a SNNc unsupervised learning (initial and final). The blue lines are positive (excitatory) connections, while the red lines are negative (inhibitory) connections. The brighter the colour of a neuron, the stronger its activity. Thickness of the lines identifies neuronal enhanced connectivity.

**(a)**                                              **(b)**

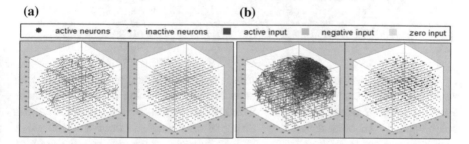

**Fig. 8.11** Dynamic visualisation of the evolution of neuronal connectivity and spiking activity in a SNNc of 1471 spiking neurons with Talairach- based coordinates [26] (**a**, **b**) the step-wise neuronal connectivity and the spiking patterns of the SNNc at two steps during a SNNc unsupervised learning (initial and final). The blue lines are positive (excitatory) connections, while the red lines are negative (inhibitory) connections. The brighter the colour of a neuron, the stronger its activity. Thickness of the lines identifies neuronal enhanced connectivity

**Box 8.2 Unsupervised Deep Learning in the NeuCube on STBD and Knowledge Representation**

*Initialisation of the SNN model:*

A model is pre-structured to map structural and functional areas of the modelled process presented by the temporal or spatio-temporal brain data. The SNN structure consists of spatially allocated spiking neurons, where the location of the neurons maps a spatial template of the problem space (e.g. brain template; if such information exists.

The input neurons are spatially allocated in this space to map the location of the input variables in the problem space. For temporal brain data for which spatial information of the input variables does not exist, the variables are mapped in the structure based on their temporal correlation—the more similar temporal variables are, the closer neurons they are mapped into.

The connections in the SNN are initialised using a small—world connectivity algorithm.

*Encoding of input data:*

Input data is encoded into spike sequences reflecting on the temporal changes in the brain data using some of the encoding algorithms (see Chap. 4)

*Unsupervised learning in the SNN model:*

Unsupervised time-dependent learning is applied in the SNN model on the spike encoded input data. Different spike-time dependent learning rules can be used. The learning process changes connection weights between individual neurons based on the timing of their spiking activity. Through learning individual connections over time, whole areas (clusters) of spiking neurons, that correspond to input variables, connect between each other, forming deep patterns of connectivity of many consecutive clusters in a flexible way. The length of the temporal data and therefore—the patterns learned in the SNN model, are theoretically unlimited.

*Obtaining deep knowledge representation as functional patterns in the SNN model:*

A deep functional pattern is revealed as a sequence of spiking activity of clusters of neurons in the SNN model that represent active functional brain areas of the modelled process. Such patterns are defined by the learned structural patterns of connections. When same or similar input data is presented to a trained SNN model, the functional patterns are revealed as neuronal activity is propagated through the connectionist patterns. The length of the obtained functional patterns is theoretically unlimited.

The learned connections can be observed, visualised and analysed for a deep knowledge representation and a better understanding of the data and the brain processes. The transparent structure of the SNNc and its spatial organisation that maps spatially the brain data allows tracing of the changes in the connections in a step-wise manner in response to the EEG spike input sequences. The evolution of SNNc connectivity throughout the learning process is illustrated in Fig. 8.11. What Fig. 8.11 shows is that starting with small random connections (initialised SNNc), the SNNc created new connections over time, reflecting the spatio-temporal relationships in the EEG data.

| **Box 8.3 Deep Supervised and Semi-supervised Learning in NeuCube** |
|---|
| *(b) Supervised learning for classification of learned patterns in a SNN model:* |
| When a SNN model is trained in an unsupervised mode on different temporal data, representing different classes, the SNN model learns different structural and functional patterns. When same data is propagated again through this SNN model, a classifier can be trained using the known labels, to learn to classify new input data that activate similar learned patters in the SNN model. |
| *(c) Semi-supervised learning:* |
| The proposed approach allows for training a SNN on a large part of data (unlabelled) and training a classifier on a smaller part of the data (labelled), both data sets related to the same problem. This is how the brain learns too. |

A BI-SNN can not only learn and classify brain EEG data, but due to their organisation being a map of a brain template, they can learn from the EEG data how humans perform motor-control or cognitive functions and then they can be used to represent these deep knowledge.

Figure 8.12 shows: (a) Three snapshots of spiking activity in a SNN model during training on one-second EEG data of wrist movement (time is in milliseconds); (b) The evolved connectivity in the SNN model after training (blue lines represent positive connection weights and red lines—negative connection weights); (c) A dynamic functional pattern is learned in the space of cortical brain functionality, representing deep procedural knowledge about how a human moves a hand.

The pattern activated in the trained SNNcube, shown in Fig. 8.12c, can be interpreted as deep knowledge (according to the definition in Chap. 1) represented as a sequence of events Ei, each defined by a function Fi, location of activity Si and time of activity Ti:

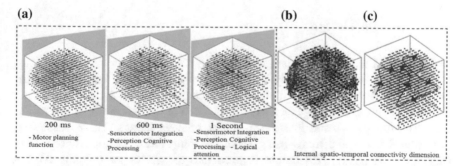

**(a)**                       **(b)**       **(c)**

| 200 ms | 600 ms | 1 Second |
|---|---|---|
| - Motor planning function | -Sensorimotor Integration -Perception Cognitive Processing | -Sensorimotor Integration -Perception Cognitive Processing  - Logical attention |

Internal spatio-temporal connectivity dimension

**Fig. 8.12  a** Three snapshots of spiking activity in a SNN model during training on one-second EEG data of wrist movement (time is in milliseconds); **b** the evolved connectivity in the SNN model after training (blue lines represent positive connection weights and red lines—negative connection weights); **c** a dynamic functional pattern is learned in the space of brain functionality that can be represented as deep knowledge as explained in the text

*IF (a person is moving a hand)*
*THEN (the following brain events are activated in space and time in a sequence):*
      *E1: Motor planning, in the Motor Planning functional brain area, time about 200 ms,*
      *AND*
      *E2: Sensorimotor integration, in the Sensorimotor integration brain area, time about 600 ms*
      *AND*
      *E3: Perception, in the Perception Cognitive brain area, time about 600 ms*
      *AND*
      *E4: Attention, in the Logical Attention brain area, time about 1 s.*

The resolution of the represented knowledge in time and space in Fig. 8.12 is hundreds of milliseconds and large brain areas. Knowledge at different time-space resolutions can be extracted from the SNN model, e.g. a single millisecond and small clusters of neurons (see for example [27], further presented in this chapter).

One can hypothesise that once a BI-SNN has learned some cognitive functions using the same structural and functional template as the human brain, this can be used as a "brain" of a machine that could manifest these functions on its own. This still needs an experimental proof.

## 8.3   Deep Learning, Recognition and Modelling of Cognitive Tasks

Details of the method presented here can be found in [28].

### 8.3.1   System Design

The scheme of the modelling is given in Fig. 8.13.

A NeuCube-based model is built that had a reservoir of 1471 spiking neurons. One of the great advantages of the NeuCube framework is that in many cases there is no need of pre-processing (such as normalization of the data, scaling, smoothing, etc.). The raw data is fed into the model as ordered vectors. These vectors are transformed into spike trains using spike encoding threshold for Address Even Representation (AER) [29] before mapped into the SNNc for training. AER is convenient when using continuous input data, such as EEG STBD, as this algorithm identifies just differences in consecutive values.

The input spike sequences are presented to the reservoir SNNc, which was implemented using leaky integrate and fire (LIF) neurons (see Chap. 4) [30–32]. The SNNc was trained using the spike timing dependant plasticity (STDP, [23]) learning rule. The STDP learning rule allows the spiking neurons of the SNNc to

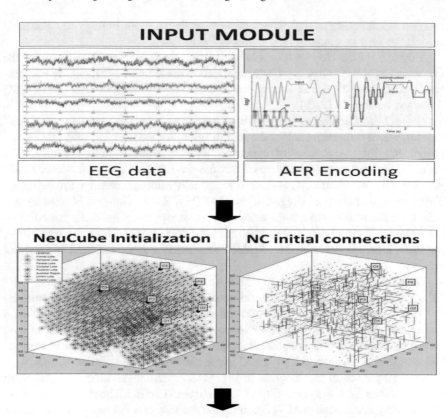

**Fig. 8.13** A graphical representation of the different steps from the proposed methodology applied for recognition and modelling of cognitive tasks using EEG data and NeuCube

learn consecutive temporal associations from data and therefore forming new connections in the architecture (*i.e.* the connection weights are changed during learning). This makes the NeuCube architecture useful for learning consecutive spatio-temporal patterns and therefore representing a more biologically plausible associative type of memory [33] as discussed in Chap. 6.

Although the SNNc can be evolving in size, for this research we explore the classification ability of the NeuCube architecture made up of 1471 spiking neurons, each representing the centre coordinates of a one cubic centimetre area from the 3D Talairach Atlas [11, 12, 34].

The 3D architecture of the SNNc is based on a "small-world organization" (SWO), which is fundamental for the initialization, for the learning processes of this model and for the process of capturing relevant patterns from the data. The encoding into spike sequences data from the six EEG recording device channels (C3, C4, P3, P4, O3 and O4) is entered to spatially allocated neurons following the Talairach coordinates as suggested in [34] (Fig. 8.14). Figure 8.14 also shows different areas of the SNNc that spatially represent regions of the brain according to the Talairach's template: frontal lobe, temporal lobe, parietal lobe, occipital lobe, posterior lobe, sub-lobar region, limbic lobe, anterior lobe, coloured in different colours.

We used the dynamic evolving SNN (deSNN [24]) algorithm to classify the EEG TSBD into the 5 brain states (classes). This classification method combines the rank-order learning rule [35] and the STDP [23] temporal learning for each output neuron to learn a whole spatio-temporal pattern using only one pass of data propagation. The classification results were evaluated using both repeated random sub-sampling validation (RRSSV) and leave-one-out cross-validation (LOOCV).

The last picture of the diagram in Fig. 8.13 represents another key advantage that NeuCube offers: the possibility of knowledge extraction. The state of the SNNc after training can be analysed. It can be observed that new connections are formed between the neurons that can be further interpreted in the context of different cognitive tasks.

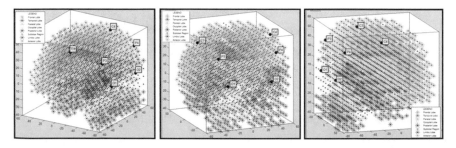

**Fig. 8.14** Different views of the SNNcube of 1471 neurons and the 6 input neurons for the case study EEG data and problem. Seven particular areas from the SNNcube that correspond spatially to brain regions according to the Talairach Atlas are also shown: in green—the frontal lobe; in magenta—the temporal lobe; in cyan—the parietal lobe; in yellow—the occipital lobe; in red—the posterior lobe; in orange—the sub-lobar region; in black—the limbic lobe; in light blue—the anterior lobe

## 8.3.2 Case Study Cognitive EEG Data

The data used for this study was recorded in an earlier experiment [36–38] and further studied in [39–41]. This data was collected from the cortex of seven healthy subjects (between 20 and 48 years old, six men and one woman, all right handed except for one subject) following five different scenarios, one resting task and four cognitive tests. A brain computer interface device was used to collect the information. The designed mental task scenarios consisted of: a "resting" task—a subject is relaxing, avoiding thoughts as much as possible (class 1); a "letter composition" task (class 2)—a subject is tasked with imagining writing a letter to someone without verbally expressing it; a "multiplication" task (class 3)—a subject is performing a non-simple two digit mental multiplication; a "counting" task (class 4)—a subject is visualising a blackboard on top of which numbers were sequentially being written; a "rotation" task (class 5)—a subject is mentally rotating on an axis a 3D geometric figure. Each recording session was carried out using six electrodes: C3, C4, P3, P4, O1, O2. Data was recorded for 10 s at 250 Hz, resulting in 2500 data points collected per session. Every task was repeated five times during a daily session. Some of the subject data was recorded on a one-day session, while other subjects repeated the five trial tasks for a second or third day session. The data of subject 4 was excluded from the experiment, as according to a previous study [41], the signal was repeatedly saturated or invalidated in several trials.

For our study, we resized each session dataset into two samples of 5 s each, 1250 data points per channel on every sample. Thus, for each of the 5 classes we had 10 samples of 1250 data points × 6 EEG channels, in total, we obtained 50 samples per subject and per session.

## 8.3.3 Experimental Results

In this study, we measured the classification accuracy of the NeuCube based model. Table 8.1 summarizes these results per subject and per session. Results are expressed as a percentage of accurately classified samples per class type and over all classes. Results obtained using RRSSV (50/50% train/test) are reported here.

As the data set was of a small size, it is not appropriate to draw any scientific conclusions about the mental tasks performance by different subjects, and that was not the goal of this study. We rather can conclude that it is feasible to consider the NeuCube-based method for further analysis and further experimental data modelling to become a widely used method for EEG data analysis related to mental tasks across applications. The results from this experiment still confirmed some expected phenomena:

**Table 8.1** Experimental results with the NeuCube-based model per subject and per session

| Samples | | Parameter setting | NeuCube with RRSSV | | | | | | |
|---|---|---|---|---|---|---|---|---|---|
| Subject and session | | Encoding threshold | Class 1 (%) | Class 2 (%) | Class 3 (%) | Class 4 (%) | Class 5 (%) | Average over all classes |
| 1 | Session 1 | 2.00 | 80 | 100 | 100 | 60 | 100 | 88 |
| | Session 2 | 1.99 | 100 | 100 | 100 | 100 | 100 | 100 |
| 2 | Session 1 | 1.30 | 100 | 100 | 100 | 100 | 20 | 84 |
| 3 | Session 1 | 4.07 | 80 | 40 | 80 | 40 | 40 | 56 |
| | Session 2 | 2.94 | 80 | 40 | 100 | 80 | 100 | 80 |
| 5 | Session 1 | 5.23 | 100 | 80 | 80 | 80 | 80 | 84 |
| | Session 2 | 2.95 | 20 | 60 | 100 | 100 | 80 | 72 |
| | Session 3 | 5.89 | 80 | 100 | 60 | 100 | 60 | 80 |
| 6 | Session 1 | 6.48 | 60 | 80 | 40 | 100 | 80 | 72 |
| | Session 2 | 5.57 | 80 | 100 | 100 | 80 | 60 | 84 |
| 7 | Session 1 | 1.70 | 60 | 100 | 100 | 100 | 80 | 88 |

Results reveal the classification accuracy percent obtained using RRSSV (per class type and as average over all classes)

**Table 8.2** NeuCube best results versus Nan-Ying Liang et al. [41] results

| Subjects | Session | NeuCube | | Nan-Ying Liang et al. [41] | |
|---|---|---|---|---|---|
| 1 | Session 2 | 100% | RRSSV | 86.70% | ELM with smoothing |
| 2 | Session 1 | 84% | RRSSV | 78.76% | SVM with smoothing |
| 3 | Session 2 | 80% | RRSSV | 64.60% | SVM with smoothing |
| 5 | Session 1 | 84% | RRSSV | 63.43% | SVM with smoothing |
| 6 | Session 2 | 84% | RRSSV | 69.47% | ELM with smoothing |
| 7 | Session 1 | 88% | RRSSV | 79.77% | SVM with smoothing |

- Subjects perform differently for different complex mental tasks (classes);
- Data for class 1 (relax) was the best classified across all subjects;
- The accuracy of classification increased with some manual parameter tuning showing that this is not the full potential of the NeuCube-based model and it still needs to be further optimised;
- Even dealing with very complex mental tasks, the classification accuracy was comparatively good (when compared to previously used classification models).

The above was confirmed as the results from the NeuCube based method were compared with the results obtained in previous experiments carried out the same data set [40, 41]. We obtained higher classification accuracy on the data per session, per subject and overall (see Table 8.2) when compared with other methods such as support vector machines (SVM) and extreme learning machines (ELM). When the SVM and the ELM methods were applied, the EEG data was first pre-processed (smoothed) and then—'compressed' into smaller number of input vectors, rather than treated as spatio-temporal stream data as it is in the NeuCube-model case.

In addition to the above, the NeuCube-based model has several other important advantages:

- It requires only one iteration data propagation for learning, while the classical methods of SVM and ELM require hundreds of iterations;
- The NeuCube-based model is adaptable to new data and new classes, while the other models are fixed and difficult to adapt on new data;
- The NeuCube-based model allows for a good interpretation of the data.

### 8.3.4 Model Interpretation

NeuCube constitutes a biologically inspired three-dimensional environment of SNN for on-line learning and recognition of spatio-temporal data. It takes into account data features, offering a better understanding of the information and the phenomena of study. This is illustrated in Fig. 8.15 which was obtained after the SNNc was trained with one of the data sets. From Fig. 8.15 we can notice that new connections are formed around the input neurons of the SNNc which were allocated to

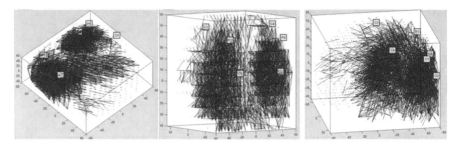

**Fig. 8.15** The SNNcube connectivity after training can be analysed and interpreted for a better understanding of the cognitive EEG data to identify differences between brain states representing different mental tasks and different subjects. The figure shows different views of the trained SNNcube for a better understanding of the learned connectivity

spatially map the EEG electrodes. Studying the picture, we could also deduce some implicit information, e.g. subjects where using actively their visual cortex (occipital lobes). Effectively, the subjects were performing each scenario with open eyes. Additionally, we can observe from the picture a high activity on the parietal lobe (integration of visual and other information).

The NeuCube model can be further trained incrementally on new data, including new classes, due to the capacity of the SNNcube to accommodate data in one pass learning and to the evolvability of the output classifier. The latter will generate a new output neuron for every new input pattern learned and will train it in one pass learning mode [24, 33, 42]. This ability of the NeuCube models will allow to trace the development of cognitive processes over time and to extract new information and knowledge about them.

## 8.4   Deep Learning, Recognition and Expression of Emotions in a BI-SNN

Human emotions are complex states of feelings that result in physical and psychological changes, which can be reflected by facial expressions, gestures, intonation in speech etc. Emotion models are necessary in the study of human emotions. A detailed description of the method and the experimental results presented in this section can be found in [43].

### 8.4.1   General Notions

Facial expression is a fundamental tool in human communication. Understanding the facial expression effects on a third person is of a crucial importance to develop a comprehensible communication. Neuropsychological studies reported that communications through facial expressions are highly related to the Mirror Neuron System (MNS). MNS principle has been introduced in 1990s by Rizzolatti when he discovered similar areas of the brain became activated when a monkey performed an action and when a monkey observed the same action performed by another [44]. The MNS in human was also confirmed by an experiment using functional magnetic resonance imaging (fMRI) data [45]. Different facial expressions of emotion have different effects on the human brain activity. The brain processes of perceiving an emotional facial expression and mimicking expression of the same emotion are spatio-temporal brain processes. The analysis of collected Spatio-Temporal Brain Data (STBD) related to these processes could reveal personal characteristics or abnormalities that would lead to a better understanding of the brain processes related to the MNS. This can be achieved only if the models created from the data can capture both spatio and temporal components from this data. Despite of the rich literature on the problem, such models still do not exist.

Here we study MNS phenomena using NeuCube due to the features of the NeuCube BI-SNN that have been demonstrated for mapping, learning, classification and visualization of STBD [26, 33, 46–49].

In this section, the NeuCube was used to model EEG data recorded during a facial expression task (both perceiving and mimicking) to investigate the cortical brain activity patterns elicited from 7 kinds of emotional faces (anger, contempt, disgust, fear, happiness, sadness, and surprise) in terms of similarity and differences. The models are analysed for a detail understanding on the problem.

### 8.4.2 Using a NeuCube Model for Emotion Recognition

The NeuCube architecture [33] was presented in Chap. 6 and it consists of: an input encoding module; a 3D recurrent SNN reservoir/cube (SNNc); an evolving SNN classifier. The encoding module converts continuous data streams into discrete spike trains. As one implementation, a Threshold Based Representation (TBR) algorithm is used for encoding. The NeuCube is trained in two learning stages. The first stage is unsupervised learning based on spike-timing-dependent synaptic plasticity (STDP) learning [23] in the SNNc. The STDP learning is applied to adjust the connection weights in the SNNc according to the spatiotemporal relations between input data variables. The second stage is a supervised learning that aims at learning the class information associated with each training sample. The dynamic evolving SNNs (deSNNs) [24] is employed as an output classifier.

In this study, the NeuCube is used for modelling, learning, and visualization of the case study EEG data corresponding to different facial expressions.

### 8.4.3   A Case Study of EEG Data for Emotion Recognition from Facial Expression

Eleven male Japanese participants, including 9 right-handed and 2 left-handed, aged between 22 and 25 years old (M = 23.2, SD = 1.2), participated in the case study of the facial expression task. As facial stimuli, JACFEE collection [50] was used, consisting of 56 colour photographs of 56 different individuals. Each individual illustrates one of the seven different emotions, i.e. anger, contempt, disgust, fear, happiness, sadness, surprise. The collection is equally divided into male and female populations (28 males, 28 female).

During the experiments, subjects were wearing EEG headset (Emotive EPOC+) which consists of 14 electrodes with the sampling rate of 128 Hz and the bandwidth is between 0.2 and 45 Hz.

The EEG data was recorded while the subjects were performing two different facial expression tasks. During the first presentation, subjects were instructed to perceive different facial expression images shown on a screen, and in the second presentation they were asked to mimic the facial expression images.

Each facial expression image was exposed for 5 s followed by randomly 5 to 10 s inter stimulus interval (ISI) as shown in Fig. 8.16.

### 8.4.4   Analysis of the Connectivity in a Trained SNNcube When a Person Is Perceiving Emotional Face and When a Person Is Expressing Such Emotions

A 3D brain-like SNNc is created to map the Talairach brain template of 1471 spiking neurons [11, 34]. The spatiotemporal data of EEG channels were encoded into spike trains and entered to the SNNc via 14 input neurons which spatial locations in the SNNc correspond to the 10–20 system location of the same channels on the sculp. The SNNc is initialized with the use of the "small world" connectivity [33].

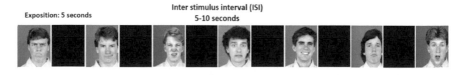

Exposition: 5 seconds                    Inter stimulus interval (ISI)
                                              5-10 seconds

**Fig. 8.16**  The facial expression-related task: the order of emotion expressions is: anger, contempt, disgust, fear, happiness, sad, and surprise. Each subject watched 56 images during an experiment

During the unsupervised STDP learning, the SNNc connectivity evolves with respect to the spike transmission between neurons. Stronger neuronal connection between two neurons means stronger information (spikes) exchanged between them. Figure 8.17 illustrates the trained SNNc with EEG data of perceiving and mimicking the 7 different facial expressions. It also shows the differences between the SNNc connectivity of perceiving *versus* mimicking, which was obtained after the two corresponding models were subtracted.

Figure 8.17 illustrates the experiment in this section as follows: (a) Exposing emotional facial expressions on a screen; (b) Connectivity of a SNNc trained on EEG data related to perceiving the facial expression images by a group of subjects; (c) Connectivity of a SNNc trained on EEG data related to mimicking the facial expressions; (d) Subtraction of the SNNc models from (a) and (b) to visualize, study and understand differences between perceiving and mimicking emotions.

It can be seen from Fig. 8.17 that when a SNNc was trained on the EEG data related to facial expressions of both perceiving and mimicking conditions, similar neuronal connections were evoked in the SNNc reflecting similar cortical activities.

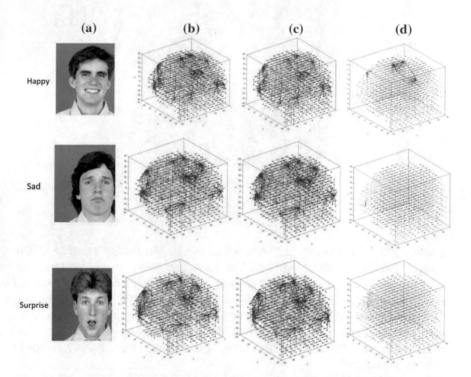

**Fig. 8.17** **a** Exposing emotional facial expressions on a screen; **b** Connectivity of a SNNc trained on EEG data related to perceiving the facial expression images by a group of subjects; **c** Connectivity of a SNNc trained on EEG data related to mimicking the facial expressions; **d** Subtraction of the SNNc models from (**b**) and (**c**) to visualize, study and understand differences between perceiving and mimicking emotions

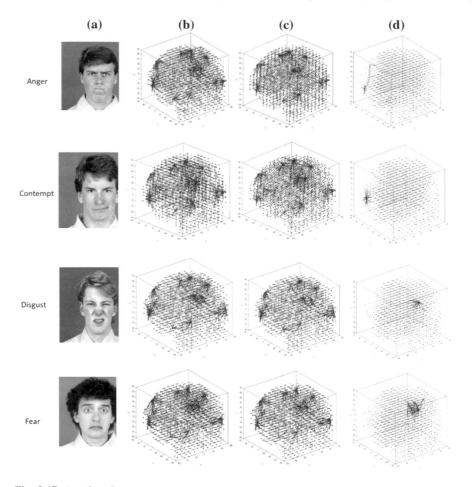

**Fig. 8.17** (continued)

Particularly, greater similarity can be observed in the right hemisphere of the SNNc for anger, contempt, sadness, surprise. This finding proves a neurological fact that this emotional information is usually processed across specific domains of the right hemisphere of the brain [51]. It also reflects the MNS principle in facial expression of emotion. Among the all presented emotional faces, some of them can be considered as dominant emotions if the brain activity patterns of perceiving and mimicking that emotions have a high level of similarity. This similarity is mostly observed for sadness.

Some differences between perceiving and mimicking emotions are also observed. It is seen from Fig. 8.18d that those neurons located around the T7 EEG channel represent the most differences between perceiving and mimicking facial expressions in anger, contempt, and less in sadness and surprise. We can also observe differences in the T6 area for fear, disgust and happy emotions.

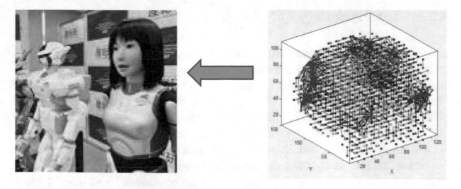

**Fig. 8.18** During the training of a SNNcube on brain signals that record human expression of different emotions the SNNcube forms distinctive connectivity and spiking activity patterns. Activation of such patterns through external stimuli or internal programme can trigger different emotional expression in an android (the figure on left is after [69])

Making use of the NeuCube SNN architecture allowed for the first time to discover the level of variation in the brain activity patterns against different facial expressions. We identified that role of mirror neurons can be dominant in sadness emotion when compared with the other emotions and that the biggest differences were recorded for fear and happiness in the T6 EEG channel area of the right hemisphere. This is only the first study in this respect. Further studies will require more subject data to be collected for a more models developed before the proposed method is used for cognitive studies and medical practice.

### 8.4.5  Can We Teach a Machine to Express Emotions?

During the training of a SNNcube on brain signals that record human expression of different emotions the SNNcube forms distinctive connectivity and spiking activity patterns. Activation of such patterns through external stimuli or internal programme can trigger different emotional expression in an android (Fig. 8.18).

## 8.5  Deep Learning and Modelling of Peri-perceptual Processes in BI-SNN

A full description of the method and experiments presented in this section can be found in [52].

## 8.5.1   The Psychology of Sub-conscious Brain Processes

Due to the vital role of emotional and unconscious processes in consumers' decision making, understanding the human brain activity and neural performance scope is of crucial importance to predict the human decisions, such as in the field of Neuromarketing for example. According to the studies that have been performed in neuromarketing field, most of the researches had widely considered the anterior area activities as the most important brain responses toward the marketing stimuli. In this study, we intended to make a different effort and check whether the pre-perceptual processes, such as Occipital and Parietal Lobes are also related to the preference and decision making against marketing stimuli (commercial brand logos). For this purpose, we used the NeuCube Spiking Neural Network (SNN) architecture for EEG-ERP data modelling, learning, classification, and visualization to reveal significant information about the consumers' brain processes. We analysed the EEG data recorded from 26 participants when they performed a cognitive task of familiar versus unfamiliar brand's logos. Tracing the NeuCube SNN-based model connectivity, enables us to find that the consumers' decision making is happening even before the consciousness. More importantly, it provides a better understanding of the EEG data source localization. In addition, the EEG data classification (familiar logo class vs. unfamiliar logo class) is done using the dynamic evolving SNN classifier of the NeuCube and the obtained accuracy is compared with traditional machine learning and analysis methods.

In the human behaviour study, subjects are not always honest. They may tell people what they think they want to hear, rather than what they really believe. In the context of market, neuromarketing is a new field that is developed through the collaboration of neuroscientific approaches and human behaviours to analyse subjects' reactions against certain stimuli to reveal the consumer preferences. The results of these trials can potentially predict differences in consumers' thought processes that might not necessarily be observable in their behaviour [53].

When human brain deals with market environments, external stimuli effects lead the brain functions to make the subject to choose the products. "Branding" plays a key role in the mechanism of preference. This directly influences the buying behaviour [54]. One of the methods which is typically employed as a mean of evaluating an individual's preference amongst products is *brand familiarity*. In the neuromarketing studies, brand familiarity and product preference has been correlated with neural activities [55–57]. Evidence has been found linking the medial prefrontal cortex with both brand familiarity and product preference [58].

Although post-perceptual involvement of brain regions has been excessively researched, influential features which appear before conscious and perception to the brands are not sufficiently presented. Also, in recent researches on neuromarketing field, although P300 component of ERPs has been excessively inquired, early components which happen before perceptual process have not assessed sufficiently. In this regards, we intended to discover whether the pre-perceptual areas of the brain, such as Occipital and Parietal Lobes are also related to the preference and decision making to marketing stimuli (commercial brand logos).

Recently, researchers in the field of marketing and advertising have sought the aid of neuroscience as a way to understand the foundations of the market behaviour, as the motivational aspects and decision making [59–62].On the same way, great attention has been given to methods of psychophysical and psychological scaling as a tool for quantitative study the consumers' behaviours related to choice and decision making [63, 64].

Since neuromarketing relies on the fact that above 70% of customers decisions are made at unconscious and sub-conscious level. *Therefore, many people cannot logically explain what was the reason of their decision making.*

Now a big challenge for neuroscientists, psychologists and economists is to find accurate brain data analysis models to study the STBD patterns and to precisely predict the consumers' decision making. In this study, we use the BI-SNN NeuCube [33] for EEG data modelling, learning, classification, and visualization. Through the interpretation of the model spiking activities and connectivity evoked by different mental tasks, the TSBD patterns will be more assessable. This investigation will demonstrate the differences of the marketing logo effects on the consumers' brain processes observed in the EEG data. By peering directly into brain data models, we can have a better interpretation of the brain processes that generated the data.

## 8.5.2   Experimental Setting and EEG Data Collection

Twenty-six right-handed volunteers were included in this study. There were 13 males with mean age of 24.40 and 13 females with mean age of 22.60. All had normal or corrected-to-normal vision, with no neurological abnormalities. All testing procedures were performed in "Hamrah clinic" of Tabriz [http://hamrahclinic.com/].

To address the studied issues, a cognitive task was designed. Before starting the experiment, in order to equalise the subjects' context and to increase their attention, a short story about choosing a drink brand was presented. The task was displayed and simultaneously the Event-Related Potentials (ERPs) was recorded.

Scalp potentials were recorded from 24 electrodes mounted in a custom elastic cap in configuration with the standard "10–20" location system. EEG data was recorded using an amplifier (Mitsar Instruments, Model 24 channels, EEG-202) with a band pass of 0.1–30 Hz, and digitized on-line at a sampling rate of 256 per second. All electrode impedances were kept below 5 k$\Omega$. Off-line ICA computerized artefact correction was used to eliminate trials during which detectable eye movements (>18), blinks or muscle potentials occurred. The resulting single-subject ERPs were then used to derive the group averaged waveforms for display and analyses.

This task is divided into three blocks. Each trial block began with the presentation of the target logo for 200 ms. As an initial instruction, participants are required to observe the logos on the screen and make a manual response to the target logo (water). Since the subjects asked to concentrate on the target logo, they are unconscious to the other logos (familiar and unfamiliar logos). Therefore, regardless of how the subjects like or dislike the logos, we can observe their preference to the familiar and unfamiliar logos while the subjects are conscious to the target. In this task, a total number of 8 logos of brands were used as the stimulus set. Although the logos contained verbal/lexical information, we did not consider this information and just concluded according to familiar and unfamiliar categories of the logos.

The presented images were a widely known or familiar brand (e.g., such as the coca cola) and unknown or unfamiliar brand (e.g., such as the Ayda cola) in two different categories: Four beverage brands and four beer brands.

The task was designed according to the *oddball paradigm* and it was divided into three blocks. Each trial block began with the presentation of the target logo (neutral stimulus) which remained on screen for 200 ms to remind that a manual response is required. The duration of each stimulus presentation was 200 ms, and the interval between the stimuli was randomly varied between 1300 and 1500 ms. the target logo was appeared 28 times in each block and each of the 8 none-target logos were presented 14 times, with the random order of presentation between the three blocks. A total number of 140 stimuli was presented (Fig. 8.19).

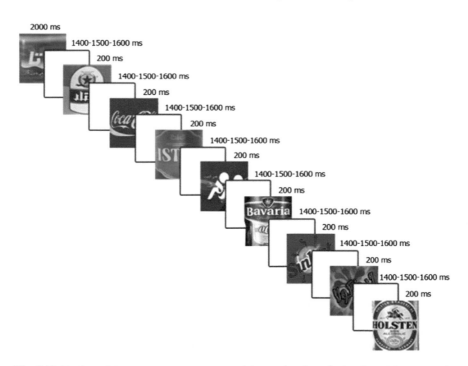

**Fig. 8.19** Design of the experiment: the presented logos, duration of stimuli, and time intervals (from [52])

As an initial instruction, participants had been asked to observe the logos on the screen and make a manual response to the target (water) as soon as they observe it.

### 8.5.3 The Design of a NeuCube Model

Making use of the proposed model, we aimed to investigate the patterns of electrical activity of neurons elicited during a cognitive task related to the drink brands. In contrast to the statistical analysis methods used so far, we present the potential of using SNN for analysing the complex dynamic brain activity during a mental task. In order to analyse the EEG data via NeuCube, the input EEG patterns were encoded into the spike trains using a Threshold-Based Technique and then entered into the spatially mapped spiking neural network cube (SNNc) via allocated input neurons.

In order to train the SNNc, Spike-Timing Dependent Plasticity (STDP) learning rule is used in unsupervised learning phase [23]. After the unsupervised learning process is finished, during the supervised learning phase, the input data is propagated again through the trained SNNc. Output neurons are evolved and trained using the same data that was used during the unsupervised learning. The classification results can be evaluated using random sub-sampling cross validation or leave one out cross validation. The NeuCube-based methodology proposed here for learning, classification and comparative analysis of EEG data is shown graphically in Fig. 8.20 where the resulted SNNcubes after training on familiar and non-familiar objects are presented for a comparative analysis.

We also conducted experiments when EEG data was divided into three classes (familiar, unfamiliar, and targeted stimuli) to train a SNNc. The trained SNNc is used to visualise different model connectivity generated during the mental activities. The SNNc connections reflect the brain processes corresponding to the subjects' preference (before making a decision) in unconscious level.

Figure 8.21a illustrates where 19 EEG channels are allocated in a 3D SNNc (where the spatial location of the input neurons are as same as their (x, y, z) coordinates in Talairach brain template). The initial connections between the neurons are created using small world connectivity rule [33]. After unsupervised training the SNNc with EEG patterns of Fam, Unfam, and target logos, new neural connections are created and evolved between the neurons.

Figure 8.21 illustrates that the neural connections are generated differently while the subjects are dealing with different mental activities. Asking subjects to observe the target stimulus (water) in the task, make them conscious to the target stimulus and unconscious to the non-target logos stimuli. Consequently, we can analyse the unconscious behaviour of the consumers impressed by the marketing stimuli (commercial logos).

By comparative analysis of the neural connections created for Fam *versus* unfam logos, we found that Fam logos generated more neural connections in left hemisphere and less connections in right hemisphere in compare with UnFam logos.

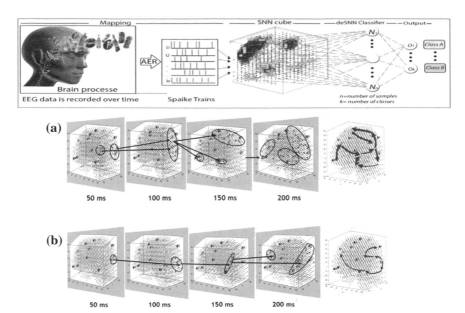

**Fig. 8.20** Top: NeuCube architecture with its main modules for EEG data mapping, learning, and classification; **a** spiking activity over 200 ms of a trained NeuCube model when a familiar object is presented and the trajectory of these activities as a deep knowledge representation; **b** spiking activity over 200 ms of a trained NeuCube model when a non-familiar object is presented and the trajectory of these activities as a deep knowledge representation (the figures are prepared by Z. Doborjeh and M. Doborjeh) (from [52])

The result of Neucube confirm the statistical analysis however Neucube enable us to have precise visualization in terms of predicting consumer preferences

The spike train of every EEG channel is entered as input data into a particular neuron of the SNNc, so that the (x, y, z) 3D location of this neuron maps the exact location of the EEG channel according to the Talairach template [11]. Each input neuron representing an EEG channel, which is a source of information that sends spikes to those neurons that are connected to it. After the unsupervised learning, the model information routes are captured between every input neuron $i$ and those neurons that received the spikes from $i$. The more the spikes transmitted between $i$ and the connected neurons, the thicker the connections captured between them.

Figure 8.22 illustrates the model information routes generated from 6 EEG channels, namely F7 and F8, O1 and O2, and P3, P4 while the subjects were dealing with familiar versus unfamiliar logos.

As it is clear from Fig. 8.22, more and greater activities are around the input neurons located in Occipital and parietal lobes when the subjects are seeing the familiar logos. These connections were created during 200 ms after stimulus presentation. It shows pre perceptual processes towards the marketing stimuli.

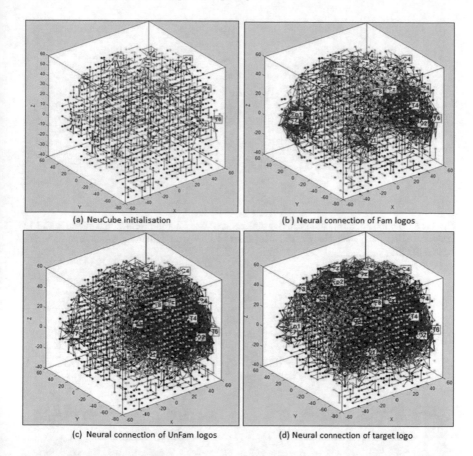

(a) NeuCube initialisation

(b) Neural connection of Fam logos

(c) Neural connection of UnFam logos

(d) Neural connection of target logo

**Fig. 8.21** Comparative visualization of the SNNcubes trained by EEG data against familiar, unfamiliar, and target brand logos. **a** The SNNcube mapped the Talairach brain template and the input EEG channels are allocated in their (x, y, z) coordinates. Initial connections between the input neurons are created based on the small world connectivity rule; **b–d** the neuron connections are created after the NeuCube unsupervised learning. The blue lines are positive (excitatory) connections, while the red lines are negative (inhibitory) connections. The brighter the colour of a neuron, the stronger its activity. Thickness of the lines identifies neuronal enhanced connectivity [52]

Recently, Event-related potential (ERP) studies have attempted to discover the processes that underlie conscious visual perception by contrasting ERPs produced by stimuli that are consciously perceived with those that are not. Figure 8.23 illustrates that different input neurons in a SNNcube that correspond to different EEG channels manifest different spiking activity that results in a different connectivity created.

Researchers suggests that the early parts of conscious processing can proceed independently of top-down attention, although top-down attention may modulate visual processing even before consciousness. Numerous studies used a variety of

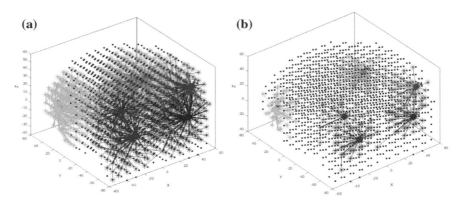

**Fig. 8.22** The model information activity clusters captured after the NeuCube learning with EEG data for 6 active EEG channels (F7, F8, O1, O2 and P3, P4) in familiar logos—(**a**) and unfamiliar logos—(**b**) [52]

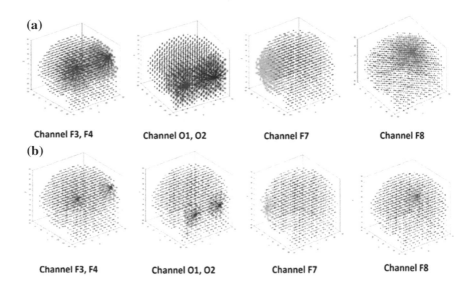

**Fig. 8.23** Different input neurons in a SNNcube that correspond to different EEG channels manifest different spiking activity that results in a different connectivity created: **a** for familiar stimuli; **b** for unfamiliar stimuli

methods for manipulating visual awareness, to find out pre perceptual process area in order to assess the consumers decision making. In this regards, we visualized the brain activity patterns generated by Familiar-related-stimuli *versus* unfamiliar-related stimuli with respect to the spatio-temporal relationships between the continuous EEG data streams, by using a BI-SNN Neucube.

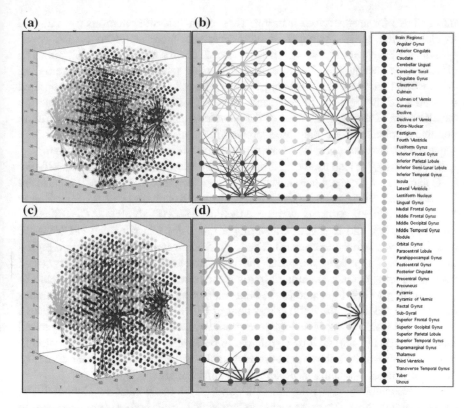

**Fig. 8.24** The SNNc is labelled by Talairach brain regions. The regions are captured in clusters of neurons by different colours. **a–b** 3D and 2D visualisation of the model information pathways created after the NeuCube unsupervised leaning by EEG data related to the familiar logos with the corresponding brain regions associated with these functional pathways. **c–d** 3D and 2D visualisation of the brain information rotes created a after the NeuCube unsupervised leaning by EEG data related to the unfamiliar logos with the corresponding brain regions associated with these functional pathways

Figure 8.24 shows a trained SNNcube with familiar versus non-familiar logos, where the spiking neurons are labelled by the Talairach brain areas to represent the interaction between different parts of the brain while the brain is dealing with marketing logos.

Through the comparative analysis of the brain information routes generated by familiar vs. unfamiliar logos, we can conclude that the brain functional pathways created in the left hemispheres are longer and more in familiar logos in comparison with the unfamiliar logos.

In order to learn and classify the EEG signal patterns recorded from subject while they were looking at Fam versus Unfam logos, the EEG data were entered into a SNNc for unsupervised learning. Then output classifier neurons were trained to classify the activity patterns of the SNNc that were activated by the corresponding EEG patterns into the pre-defined 2 classes of Fam and Unfam. The classification accuracy results were evaluated using repeated random rub-sampling cross validation (RRSV). In this experiment, the RRSV method has been applied with 50% of the data for training and 50% for testing. In order to optimise the classification accuracy, the values of the NeuCube parameters were altered through iterative applications of the three procedures: encoding the TSBD into spike sequences; SNNc unsupervised learning; supervised learning of the classifier.

In this study, we also used traditional machine learning techniques to classify the EEG data into two classes (subjects are looking at familiar brand logos, subjects are looking at unfamiliar brand logos). Multiple Linear Regression (MLR), Multi-Layer Perceptron (MLP), Support Vector Machine (SVM), Evolving Classification Function (ECF), and Evolving Clustering Method for Classification (ECMC) are used (http://www.theneucom.com). The classification accuracy results are reported in Table 8.3. The results show that the NeuCube was able to classify the brain activity patterns of Fam vs. Unfam logos more accurately than the other methods.

According to previous studies on neuromarketing field, most of the researches had widely considered the P300 component of ERP. Hence, we intended to make a different effort and check if the pre perceptual components such as P200 in parietal lobe and N100 in occipital lobe be relative to preference towards marketing stimuli. In our experiment by using the traditional methods we found significant effects of pre-perceptual components in preference against brands.

By demonstrating the role of N100 and P200 component in Occipital and parietal lobe in this research as a component which occur before comprehension in the brain, we acknowledged that the brain is able to process decisions much faster than what expected. Being meaningful, the changes on N100 and P200 amplitude of occipital and parietal lobes respectively can affect anterior areas widely. As a result, the amplitude of P300 and other late components in central and frontal lobes would definitely be affected by early components and it proves the significant role of pre-comprehensive attended regions in prediction of preference.

Our findings prove the BI-SNN, and more specifically—NeuCube potential to deal with both spatio and temporal content of the data without losing meaningful information. Since, statistical machine learning techniques are not able to deal with temporal data, we had to take the average over the temporal content of the EEG data. Therefore the classification accuracy results are less than the results obtained via NeuCube. We have found an increased amplitude in Occipital and Parietal area by traditional methods, however Neucube may tell what neurons have made contribution to this increase and more importantly can be used to obtain new findings that could not be obtained with the use of traditional statistical methods. This is also observed in other applications of NeuCube for STBD [64–66].

**Table 8.3** EEG data classification results using statistic method versus NeuCube model

| Method | MLR | MLP | SVM | ECF | ECMC | NC |
|---|---|---|---|---|---|---|
| Classification accuracy in % | 50.38 | 49.03 | 46.41 | 50.64 | 50.38 | 70.00 |
| Parameter setting | Not applicable for this method | Number of hidden units: 5<br>Number of training cycles: 600<br>Output value precision: 0.0001<br>Output function precision: 0.0001<br>Output activation function: linear<br>Optimisation: scg | SVM Kernel: polynomial degree, gamma, N/A: 1 | Maximum influence field: 1<br>Minimum influence field: 0.01<br>M of N: 3<br>Number of membership functions: 2<br>Number of epochs: 4 | Maximum influence field: 1<br>Minimum influence field: 0.01<br>M of N: 3 | AER threshold:0.575<br>Connection dist: 0.208<br>STDP rate: 0.010<br>Firing threshold: 0.500<br>Refractory time: 6.000<br>Train times: 1.000<br>deSNN mod: 0.400<br>deSNN drift: 0.250 |

## 8.6    Modelling Attentional Bias in BI-SNN

A full description of the method and experimental results in this section can be found in [67].

### 8.6.1   Attentional Bias

Attentional bias is a brain state when the brain gets activated when perceiving non-targeted object. This could be result of previous experience, preferences, and other causes. Attentionla bias can influence the decision making process in humans.

In this section the same data as in Sect. 8.5 is used, but in this case the EEG data will measure attention with a given targeted object as a stimulus.

### 8.6.2   Experimental Settings

Inspired by importance of the attentional bias principle in human choice behavior, we formed a NeuCube based SNN model for efficient recognition of attentional bias as influential factor in a consumer's preferences. The model was tested on a case study of EEG data collected from a group of moderate drinkers when they were presented by different drink product features.

Figure 8.25 illustrates the trained SNNcube models on the EEG data of several subjects when perceiving the target stimuli along with other stimuli, i.e. presentation of a target drink—water, and in between—different other drinks and brands.

### 8.6.3   Results

Our case study findings suggest that a product brand name may not significantly impress consumers by itself. However, when the name of a brand comes along with a context, such as design, color, alcoholic or non-alcoholic features, etc. it may direct the consumers attention to certain features and lead the consumers to choose a product. In this particular case study, we found that attentional bias towards alcoholic-related stimuli had stronger effects on the brain activity of the moderate drinkers as shown in the SNN connectivity in Fig. 8.26.

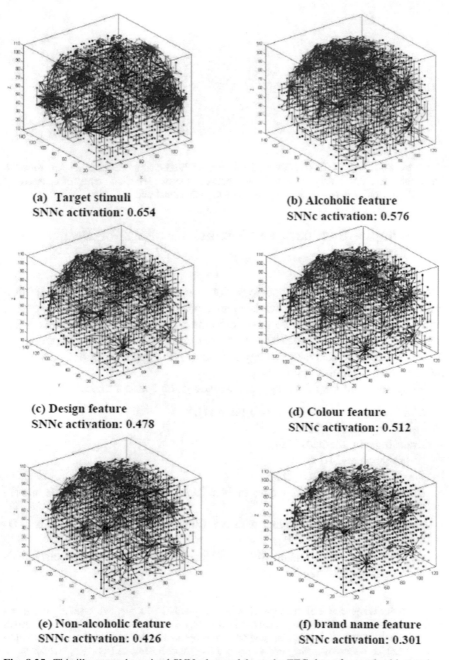

(a)  Target stimuli
SNNc activation: 0.654

(b) Alcoholic feature
SNNc activation: 0.576

(c) Design feature
SNNc activation: 0.478

(d) Colour feature
SNNc activation: 0.512

(e) Non-alcoholic feature
SNNc activation: 0.426

(f) brand name feature
SNNc activation: 0.301

**Fig. 8.25** This illustrates the trained SNNcube models on the EEG data of several subjects when perceiving the target stimuli along with other stimuli, i.e. presentation of a target drink—water, and in between—different other drinks and brands, such as alcoholic drinks

(a)                            (b)                            (c)

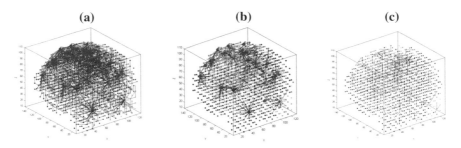

**Fig. 8.26** a A NeuCube based SNN model trained on EEG data of alcoholic-related stimuli; b non-alcoholic-related stimuli; c the difference between the two models in terms of connectivity (from [67]). Both features are non-targeted showing attentional bias

## 8.7    Chapter Summary and Further Readings for Deeper Knowledge

The chapter presents a methodology for deep learning, modelling and deep knowledge representation of EEG data and applies this methodology for the design of SNN for: learning cognitive tasks; emotion recognition; peri-perceptual information processing; attentional bias. These are only few illustrations for the applicability of the presented methodology and many other applications and studies can be developed using this approach.

We recommend the following further readings on specific topics:

– Multimodal atlases of the human brain [1];
– Talairach atlas [11, 12];
– Brain imaging methods [19];
– EEG mapping [34];
– The Blue Brain project [21];
– Demo on modelling EEG data in NeuCube: https://kedri.aut.ac.nz/R-and-D-Systems/neucube/eeg;
– Demo on EEG data modelling with NeuCube: https://kedri.aut.ac.nz/R-and-D-Systems/eeg-data-modelling;
– Neuromarketing demo using EEG and NeuCube: https://kedri.aut.ac.nz/R-and-D-Systems/neuromarketing.

**Acknowledgements** Some of the material in this chapter has been first published in journal and conference publications as referenced and cited in corresponding sections of the chapter and also in Springer book volumes [15, 68]. I acknowledge the contribution of my co-authors of these publications Lubica Benuskova, Maryam Doborjeh, Elisa Capecci, Zohreh Doborjeh, Nathan Scott, Alex Sumich. Most of the experiments in Sects. 8.4, 8.5 and 8.6 were conducted by Z. Doborjeh and M. Dorojeh, while experiments in Sect. 8.3 were conducted by E. Capecci.

# Appendix

(from [34]). Anatomical locations of international 10–10 EEG cortical projections into Talairach coordinates. Same coordinates are used in a SNNc of a NeuCube model.

| EEG chan. | Talairach coordinates | | | Gyri | | Brodmann area |
|---|---|---|---|---|---|---|
| | x av (mm) | y av (mm) | z av (mm) | | | |
| FP1 | −21.2 ± 4.7 | 66.9 ± 3.8 | 12.1 ± 6.6 | L FL | Superior frontal G | 10 |
| FPz | 1.4 ± 2.9 | 65.1 ± 5.6 | 11.3 ± 6.8 | M FL | Bilat. medial | 10 |
| FP2 | 24.3 ± 3.2 | 66.3 ± 3.5 | 12.5 ± 6.1 | R FL | Superior frontal G | 10 |
| AF7 | −41.7 ± 4.5 | 52.8 ± 5.4 | 11.3 ± 6.8 | L FL | Middle frontal G | 10 |
| AF3 | −32.7 ± 4.9 | 48.4 ± 6.7 | 32.8 ± 6.4 | L FL | Superior frontal G | 9 |
| AFz | 1.8 ± 3.8 | 54.8 ± 7.3 | 37.9 ± 8.6 | M FL | Bilat. medial | 9 |
| AF4 | 35.1 ± 3.9 | 50.1 ± 5.3 | 31.1 ± 7.5 | L FL | Superior frontal G | 9 |
| AF8 | 43.9 ± 3.3 | 52.7 ± 5.0 | 9.3 ± 6.5 | R FL | Middle frontal G | 10 |
| F7 | −52.1 ± 3.0 | 28.6 ± 6.4 | 3.8 ± 5.6 | L FL | Inferior frontal G | 45 |
| F5 | −51.4 ± 3.8 | 26.7 ± 7.2 | 24.7 ± 9.4 | L FL | Middle frontal G | 46 |
| F3 | −39.7 ± 5.0 | 25.3 ± 7.5 | 44.7 ± 7.9 | L FL | Middle frontal G | 8 |
| F1 | −22.1 ± 6.1 | 26.8 ± 7.2 | 54.9 ± 6.7 | L FL | Superior frontal G | 6 |
| Fz | 0.0 ± 6.4 | 26.8 ± 7.9 | 60.6 ± 6.5 | M FL | Bilat. medial | 6 |
| F2 | 23.6 ± 5.0 | 28.2 ± 7.4 | 55.6 ± 6.2 | R FL | Superior frontal G | 6 |
| F4 | 41.9 ± 4.8 | 27.5 ± 7.3 | 43.9 ± 7.6 | R FL | Middle frontal G | 8 |
| F6 | 52.9 ± 3.6 | 28.7 ± 7.2 | 25.2 ± 7.4 | R FL | Middle frontal G | 46 |
| F8 | 53.2 ± 2.8 | 28.4 ± 6.3 | 3.1 ± 6.9 | R FL | Inferior frontal G | 45 |
| FT9 | −53.8 ± 3.3 | −2.1 ± 6.0 | −29.1 ± 6.3 | L TL | Inferior temporal G | 20 |
| FT7 | −59.2 ± 3.1 | 3.4 ± 5.6 | −2.1 ± 7.5 | L TL | Superior temporal G | 22 |
| FC5 | −59.1 ± 3.7 | 3.0 ± 6.1 | 26.1 ± 5.8 | L FL | Precentral G | 6 |

(continued)

(continued)

| EEG chan. | Talairach coordinates | | | Gyri | | Brodmann area |
|---|---|---|---|---|---|---|
| | x av (mm) | y av (mm) | z av (mm) | | | |
| FC3 | −45.5 ± 5.5 | 2.4 ± 8.3 | 51.3 ± 6.2 | L FL | Middle frontal G | 6 |
| FC1 | −24.7 ± 5.7 | 0.3 ± 8.5 | 66.4 ± 4.6 | L FL | Superior frontal G | 6 |
| FCz | 1.0 ± 5.1 | 1.0 ± 8.4 | 72.8 ± 6.6 | M FL | Superior frontal G | 6 |
| FC2 | 26.1 ± 4.9 | 3.2 ± 9.0 | 66.0 ± 5.6 | R FL | Superior frontal G | 6 |
| FC4 | 47.5 ± 4.4 | 4.6 ± 7.6 | 49.7 ± 6.7 | R FL | Middle frontal G | 6 |
| FC6 | 60.5 ± 2.8 | 4.9 ± 7.3 | 25.5 ± 7.8 | R FL | Precentral G | 6 |
| FT8 | 60.2 ± 2.5 | 4.7 ± 5.1 | −2.8 ± 6.3 | L TL | Superior temporal G | 22 |
| FT10 | 55.0 ± 3.2 | −3.6 ± 5.6 | −31.0 ± 7.9 | R TL | Inferior temporal G | 20 |
| T7 | −65.8 ± 3.3 | −17.8 ± 6.8 | −2.9 ± 6.1 | L TL | Middle temporal G | 21 |
| C5 | −63.6 ± 3.3 | −18.9 ± 7.8 | 25.8 ± 5.8 | L PL | Postcentral G | 123 |
| C3 | −49.1 ± 5.5 | −20.7 ± 9.1 | 53.2 ± 6.1 | L PL | Postcentral G | 123 |
| C1 | −25.1 ± 5.6 | −22.5 ± 9.2 | 70.1 ± 5.3 | L FL | Precentral G | 4 |
| Cz | 0.8 ± 4.9 | −21.9 ± 9.4 | 77.4 ± 6.7 | M FL | Precentral G | 4 |
| C2 | 26.7 ± 5.3 | −20.9 ± 9.1 | 69.5 ± 5.2 | R FL | Precentral G | 4 |
| C4 | 50.3 ± 4.6 | −18.8 ± 8.3 | 53.0 ± 6.4 | R PL | Postcentral G | 123 |
| C6 | 65.2 ± 2.6 | −18.0 ± 7.1 | 26.4 ± 6.4 | R PL | Postcentral G | 123 |
| T8 | 67.4 ± 2.3 | −18.5 ± 6.9 | −3.4 ± 7.0 | R TL | Middle temporal G | 21 |
| TP7 | −63.6 ± 4.5 | −44.7 ± 7.2 | −4.0 ± 6.6 | L TL | Middle temporal G | 21 |
| CP5 | −61.8 ± 4.7 | −46.2 ± 8.0 | 22.5 ± 7.6 | L PL | Supramarginal G | 40 |
| CP3 | −46.9 ± 5.8 | −47.7 ± 9.3 | 49.7 ± 7.7 | L PL | Inferior parietal G | 40 |
| CP1 | −24.0 ± 6.4 | −49.1 ± 9.9 | 66.1 ± 8.0 | L PL | Postcentral G | 7 |
| CPz | 0.7 ± 4.9 | −47.9 ± 9.3 | 72.6 ± 7.7 | M PL | Postcentral G | 7 |
| CP2 | 25.8 ± 6.2 | −47.1 ± 9.2 | 66.0 ± 7.5 | R PL | Postcentral G | 7 |
| CP4 | 49.5 ± 5.9 | −45.5 ± 7.9 | 50.7 ± 7.1 | R PL | Inferior parietal G | 40 |
| CP6 | 62.9 ± 3.7 | −44.6 ± 6.8 | 24.4 ± 8.4 | R PL | Supramarginal G | 40 |
| TP8 | 64.6 ± 3.3 | −45.4 ± 6.6 | −3.7 ± 7.3 | R TL | Middle temporal G | 21 |
| P9 | −50.8 ± 4.7 | −51.3 ± 8.6 | −37.7 ± 8.3 | L TL | Tonsile | NP |

(continued)

(continued)

| EEG chan. | Talairach coordinates | | | Gyri | | Brodmann area |
|---|---|---|---|---|---|---|
| | x av (mm) | y av (mm) | z av (mm) | | | |
| P7 | −55.9 ± 4.5 | −64.8 ± 5.3 | 0.0 ± 9.3 | L TL | Inferior temporal G | 37 |
| P5 | −52.7 ± 5.0 | −67.1 ± 6.8 | 19.9 ± 10.4 | L TL | Middle temporal G | 39 |
| P3 | −41.4 ± 5.7 | −67.8 ± 8.4 | 42.4 ± 9.5 | L PL | Precuneus | 19 |
| P1 | −21.6 ± 5.8 | −71.3 ± 9.3 | 52.6 ± 10.1 | L PL | Precuneus | 7 |
| Pz | 0.7 ± 6.3 | −69.3 ± 8.4 | 56.9 ± 9.9 | M PL | Superior parietal L | 7 |
| P2 | 24.4 ± 6.3 | −69.9 ± 8.5 | 53.5 ± 9.4 | R PL | Precuneus | 7 |
| P4 | 44.2 ± 6.5 | −65.8 ± 8.1 | 42.7 ± 8.5 | R PL | Inferior parietal L | 7 |
| P6 | 54.4 ± 4.3 | −65.3 ± 6.0 | 20.2 ± 9.4 | R TL | Middle temporal G | 39 |
| P8 | 56.4 ± 3.7 | −64.4 ± 5.6 | 0.1 ± 8.5 | R TL | Inferior temporal G | 19 |
| P10 | 51.0 ± 3.5 | −53.9 ± 8.7 | −36.5 ± 10.0 | L OL | Tonsile | NP |
| PO7 | −44.0 ± 4.7 | −81.7 ± 4.9 | 1.6 ± 10.6 | R OL | Middle occipital G | 18 |
| PO3 | −33.3 ± 6.3 | −84.3 ± 5.7 | 26.5 ± 11.4 | R OL | Superior occipital G | 19 |
| POz | 0.0 ± 6.5 | −87.9 ± 6.9 | 33.5 ± 11.9 | M OL | Cuneus | 19 |
| PO4 | 35.2 ± 6.5 | −82.6 ± 6.4 | 26.1 ± 9.7 | R OL | Superior occipital G | 19 |
| PO8 | 43.3 ± 4.0 | −82.0 ± 5.5 | 0.7 ± 10.7 | R OL | Middle occipital G | 18 |
| O1 | −25.8 ± 6.3 | −93.3 ± 4.6 | 7.7 ± 12.3 | L OL | Middle occipital G | 18 |
| Oz | 0.3 ± 5.9 | −97.1 ± 5.2 | 8.7 ± 11.6 | M OL | Cuneus | 18 |
| O2 | 25.0 ± 5.7 | −95.2 ± 5.8 | 6.2 ± 11.4 | R OL | Middle occipital G | 18 |

# References

1. A. Toga, P. Thompson, S. Mori, K. Amunts, K. Zilles, Towards multimodal atlases of the human brain. Nat. Rev. Neurosci. **7**, 952–966 (2006)
2. E. Niedermeyer, F.H. L. Da Silva, *Electroencephalography: Basic Principles, Clinical Applications, and Related Fields* (Lippincott Williams & Wilkins, Philadelphia, 2005), 1309p
3. D.A. Craig, H.T. Nguyen, *Adaptive EEG Thought Pattern Classifier for Advanced Wheelchair Control*, in International Conference of the IEEE Engineering in Medical and Biology Society (2007), pp. 2544–2547

4. F. Lotte, M. Congedo, A. Lécuyer, F. Lamarche, B.A. Arnaldi, Review of classification algorithms for EEG-based brain-computer interfaces. J. Neural Eng. **4**(2), R1–R15 (2007)
5. R.C. deCharms, Application of real-time fMRI. Nat. Rev. Neurosci. **9**(9), 720–729 (2008)
6. T. Mitchel et al., Learning to decode cognitive states from brain images. Mach. Learn. **57**, 145–175 (2004)
7. K. Broderson et al., Generative embedding for model-based classification of fMRI Data. PLoS Comput. Biol. **7**(6), 1–19 (2011)
8. K. Broderson et al., Decoding the perception of pain from fMRI using multivariate pattern analysis. NeuroImage **63**, 1162–1170 (2012)
9. K. Zilles, K. Amunts, Centenary of Brodmann's map—conception and fate. Nat. Rev. Neurosci. **11**, 139–145 (2010)
10. S. Eickhoff et al., A new SPM toolbox for combining probabilistic cytoarchitectonic maps and functional imaging data. Neuroimage **25**, 1325–1335 (2005)
11. J. Talairach, P. Tournoux, *Co-planar Stereotaxic Atlas of The Human Brain. 3-Dimensional Proportional System: An Approach to Cerebral Imaging* (Thieme Medical Publishers, New York, 1988)
12. J. Lancaster et al., Automated Talairach Atlas labels for functional brain mapping. Hum. Brain Mapp. **10**, 120–131 (2000)
13. A.C. Evans, D.L. Collins, S.R. Mills, E.D. Brown, R.L. Kelly, T.M. Peters, *3D Statistical Neuroanatomical Models from 305 MRI Volumes*, in IEEE-Nuclear Science Symposium and Medical Imaging Conference (IEEE Press, 1993), pp. 1813–1817
14. J. Ashburner, Computational anatomy with the SPM software. Magn. Reson. Imaging **27**(8), 1163–1174 (2009)
15. L. Benuskova, N. Kasabov, *Computational Neuro-genetic Modelling* (Springer, New York, 2007), p. 290
16. M. Hawrylycz et al., An anatomically comprehensive atlas of the adult human brain transcriptome. Nature **489**, 391–399 (2012)
17. W. Gerstner, H. Sprekeler, G. Deco, Theory and simulation in neuroscience. Science **338**, 60–65 (2012)
18. C. Koch, R. Reid, Neuroscience: observation of the mind. Nature **483**(7390), 397–398 (2012)
19. J.B. Poline, R.A. Poldrack, Frontiers in brain imaging methods grand challenge. Front. Neurosci. **6**, 96 (2012). https://doi.org/10.3389/fnins.2012.00096
20. Van Essen et al., The human connectome project: a data acquisition perspective. NeuroImage **62**(4), 2222–2231 (2012)
21. H. Markram, The blue brain project. Nat. Rev. Neurosci. **7**, 153–160 (2006)
22. E.M. Izhikevich, G.M. Edelman, Large-scale model of mammalian thalamocortical systems. PNAS **105**, 3593–3598 (2008)
23. S. Song, K. Miller, L. Abbott et al., Competitive Hebbian learning through spike-timing-dependent synaptic plasticity. Nat. Neurosci. **3**, 919–926 (2000)
24. N. Kasabov, K. Dhoble, N. Nuntalid, G. Indiveri, Dynamic evolving spiking neural networks for on-line spatio- and spectro-temporal pattern recognition. Neural Networks **41**, 188–201 (2013)
25. M. Defoin-Platel, S. Schliebs, N. Kasabov, Quantum-inspired evolutionary algorithm: a multi-model EDA. IEEE Trans. Evol. Comput. **13**(6), 1218–1232 (2009)
26. M.G. Doborjeh, Y. Wang, N. Kasabov, R. Kydd, B. Russell, A spiking neural network methodology and system for learning and comparative analysis of EEG data from healthy versus addiction treated versus addiction not treated subjects. IEEE Trans. Biomed. Eng. **63** (9), 1830–1841 (2016)
27. M. Doborjeh, N. Kasabov, Z.G. Doborjeh, Evolving, dynamic clustering of spatio/ spectro-temporal data in 3D spiking neural network models and a case study on EEG data. Evolving Syst. 1–17 (2017). https://doi.org/10.1007/s12530-017-9178-8

28. N. Kasabov, E. Capecci, Spiking neural network methodology for modelling, classification and understanding of EEG spatio-temporal data measuring cognitive processes. Inf. Sci. **294**, 565–575 (2014). https://doi.org/10.1016/j.ins.2014.06.028

29. K. Dhoble, N. Nuntalid, G. Indiveri, N. Kasabov, *On-line Spatiotemporal Pattern Recognition with Evolving Spiking Neural Networks Utilizing Address Event Representation, Rank Oder and Temporal Spike Learning*, in *Proceedings of WCCI 2012* (IEEE Press, 2012), pp. 554–560

30. W. Gerstner, What's different with spiking neurons?, in *Plausible Neural Networks for Biological Modelling*, ed. by H. Mastebroek, H. Vos (Kluwer Academic Publishers, Dordrecht, 2001), pp. 23–48

31. G, Indiveri, B. Linares-Barranco, T. Hamilton, A. Van Schaik, R. Etienne-Cummings, T. Delbruck, S. Liu, P. Dudek, P. Hafliger, S. Renaud et al., Neuromorphic silicon neuron circuits. Front. Neurosci. **5** (2011)

32. N. Scott, N. Kasabov, G. Indiveri, *NeuCube Neuromorphic Framework for Spatio-Temporal Brain Data and Its Python Implementation*, in Proceedings of ICONIP 2013. LNCS, vol. 8228 (Springer, Berlin, 2013), pp.78-84

33. N. Kasabov, NeuCube: a spiking neural network architecture for mapping, learning and understanding of spatio-temporal brain data. Neural Networks **52**, 62–76 (2014)

34. L. Koessler, L. Maillard, A. Benhadid et al., Automated cortical projection of EEG sensors: anatomical correlation via the international 10–10 system. NeuroImage **46**, 64–72 (2009)

35. S. Thorpe, J. Gautrais, Rank order coding. Comput. Neurosci. Trends Res. **13**, 113–119 (1998)

36. C.W. Anderson, M. Kirby, in *Colorado State University—Brain Computer Interface Laboratory—1989 Keirn and Aunon*. Internet: www.cs.colostate.edu/eeg/main/data/1989_Keirn_and_Aunon (2003) [April 14, 2014]

37. Z. K. Keirn, in Alternative modes of communication between man and machine. Master's Thesis, Electrical Engineering Department, Purdue University, USA

38. Z.A. Keirn, J.I. Aunon, A new mode of communication between man and his surroundings. IEEE Trans. Biomed. Eng. **37**(12), 1209–1214 (1990)

39. C. Anderson, D. Peterson, Recent advances in EEG signal analysis and classification, in *Clinical Applications of Artificial Neural Networks*, ed. by R. Dybowski, V. Gant (Cambridge University Press, Cambridge, 2001), pp. 175–191

40. C. Anderson, Z. Sijercic, *Classification of EEG Signals from Four Subjects During Five Mental Tasks*, in *Solving Engineering Problems with Neural Networks: Proceedings of Conference on Engineering Applications in Neural Networks (EANN '96)*, ed. by A.B. Bulsari, S. Kallio, D. Tsaptsinos (Systems Engineering Association, PL 34, FIN-20111 Turku 11, Finland, 1996), pp. 407–414

41. L. Nan-Ying et al., Classification of mental tasks from EEG signals using extreme learning machine. Int. J. Neural Syst. **16**(01), 29–38 (2006)

42. N. Kasabov, *Evolving Connectionist Systems: The Knowledge Engineering Approach* (Springer, London, 2007) (first edition 2002)

43. H. Kawano, A. Seo, Z. Gholami, N. Kasabov, M.G. Doborjeh, *Analysis of Similarity and Differences in Brain Activities between Perception and Production of Facial Expressions Using EEG Data and the NeuCube Spiking Neural Network Architecture*, in ICONIP, Kyoto. LNCS (Springer, Bernin, 2016)

44. V. Gallese, L. Fadiga, L. Fogassi, G. Rizzolatti, Action recognition in the premotor cortex. Brain **119**, 593–609 (1996)

45. M. Iacoboni, R.P. Woods, M. Brass, H. Bekkering, J. C. Mazziotta, G. Rizzolatti, Cortical mechanisms of human imitation. Science **286**(5449), 2526–2528 (1999)

46. E. Tu, N. Kasabov, J. Yang, Mapping temporal variables into the NeuCube for improved pattern recognition, predictive modelling and understanding of stream data. IEEE Trans. Neural Netw. Learn. Syst. 1–13 (2016)

47. N. Kasabov, N. Scott, E. Tu, S. Marks, N. Sengupta, E. Capecci, Evolving spatio-temporal data machines based on the NeuCube neuromorphic framework: design methodology and selected applications. Neural Networks **78**(2016), 1–14 (2016)
48. M.G. Doborjeh, E. Capecci, N. Kasabov, Classification and Segmentation of fMRI Spatio-Temporal Brain Data with a Neucube Evolving Spiking Neural Network Model, in Analysis of Similarity and Differences in Brain Activities. IEEE SSCI, Orlando, USA (2014), pp. 227–228
49. M.G. Doborjeh, N. Kasabov, *Dynamic 3D Clustering of Spatio-temporal Brain Data in the NeuCube Spiking Neural Network Architecture on a Case Study of fMRI Data*, in Neural Information Processing, Istanbul (2015)
50. D. Matsumoto, P. Ekman, in Japanese, caucasian facial expressions of emotion (JACFEE) [Slides]. Intercultural and Emotion Research Laboratory, Department of Psychology, San Francisco State University, San Francisco (1988)
51. K.M. Alfano, C.R. Cimino, Alteration of expected hemispheric asymmetries: valence and arousal effects in neuropsychological models of emotion. Brain Cogn. **66**, 213–220 (2008)
52. Z. Doborjeh, N. Kasabov, M. Doborjeh, A. Sumich, Modelling peri-perceptual brain processes in a deep learning spiking neural network architecture. Nature, Scientific Reports **8**, 8912 (2018). https://10.1038/s41598-018-27169-8; https://www.nature.com/articles/s41598-018-27169-8
53. S. Venkataraman, S.D. Sarasvathy, N. Dew, W.R. Forster, Reflections on the 2010 AMR decade award: whither the promise? Moving forward with entrepreneurship as a science of the artificial. Acad. Manage. Rev. **37**, 21–33 (2012)
54. Z.O. Touhami, L. Benlafkih, M. Jiddane, Y. Cherrah, O. Malki, A. Benomar, Neuromarketing: where marketing and neuroscience meet. Afr. J. Bus. Manage. **5**(5), 1528–1532 (2011). https://doi.org/10.5897/AJBM10.729
55. S.M. McClure, J. Li, D. Tomlin, K.S. Cypert, L.M. Montague, P.R. Montague, Neural correlates of behavioural preference for culturally familiar drinks. Neuron **44**, 379–387 (2004)
56. M. Schaefer, H. Berens, H.J. Heinz, M. Rotte, Neural correlates of culturally familiar brands of car manufacturers. Neuroimage **31**(2), 861–865 (2006)
57. H. Walter, B. Abler, A. Ciaramidaro, S. Erk, Motivating forces of human actions. Neuroimaging reward and social interaction. Brain Res. Bull. **15**(5), 368–381 (2005)
58. A.R. Damasio, The somatic marker hypothesis and the possible functions of the prefrontal cortex. Philos. Trans. R. Soc. Lond. B Biol. Sci. **351**, 1413–1420 (1996)
59. N. Lee, A.J. Broderick, L. Chamberlain, What is neuromarketing? A discussion and agenda for future research. Int. J. Psychophysiol. **63**, 199–204 (2007)
60. A. Morin, Self-awareness part 1: definition, measures, effects, functions, and antecedents. J. Theor. Social Psychol. **5**(10), 807–823 (2011)
61. C. Oreja-Guevara, Neuromarketing. Neurologia Supl. **5**(1), 4–7 (2009)
62. C.S. Crandall, Psychophysical scaling of stressful life events. Psychol. Sci. **3**, 256–258 (1992)
63. D. Labbe, N. Pineau, N. Martin, Measuring consumer response to complex precision: scaling vs. categorization task. Food Quality Prefer. **23**(2), 134–137 (2012)
64. J. Hu, Z.G. Hou, Y.X. Chen, N. Kasabov, N. Scott, *EEG-Based Classification of Upper-limb ADL Using SNN for Active Robotic Rehabilation*, in IEEE RAS & EMBS International Conference on Biomedical Robotics and Biomechatronics (BooRob), Sao Paulo, Brazil (2014), 409–414
65. N. Kasabov, N. Scott, E. Tu, S. Marks, N. Sengupta, E. Capecci, M. Othman, M. Doborjeh, N. Murli, R. Hartono, J. Espinosa-Ramos, L. Zhou, F. Alvi, G. Wang, D. Taylor, V. Feigin, S. Gulyaev, M. Mahmoudh, Z.-G. Hou, J. Yang, Design methodology and selected applications of evolving spatio-temporal data machines in the NeuCube neuromorphic framework. Neural Netw. **78**, 1–14 (2016). http://dx.doi.org/10.1016/j.neunet.2015.09.011
66. E. Capecci, N. Doborjeh, N. Mammone, F. La Foresta, F.C. Morabito, N. Kasabov, *Longitudinal Study of Alzheimer's Disease Degeneration through EEG Data Analysis with a NeuCube Spiking Neural Network Model*, in Proceedings of WCCI—IJCNN 2016, Vancouver, 24–29 July 2016 (IEEE Press, 2016), pp. 1360–1366

67. Z.G. Doborjeh, M.G. Doborjeh, N. Kasabov, Attentional bias pattern recognition in spiking neural networks from spatio-temporal EEG data. Cognitive Comput. **10**, 35–48 (2018). https://doi.10.1007/s12559-017-9517-x
68. N.Kasabov (ed.), *Springer Handbook of Bio-Neuroinformatics* (Springer, Berlin, 2014)
69. Wikipedia: http://www.wikipedia.com
70. American Electroencephalographic Society, American electroencephalographic society guidelines for standard electrode position nomenclature. J. Clin. Neurophysiol. **8**(2), 200–202 (1991)
71. G.H. Klem, H.O. Lüders, H. Jasper, C. Elger, The ten-twenty electrode system of the International Federation. Electroencephalogram Clin. Neurophysiol. **52**, 3 (1999)

# Chapter 9
# Brain Disease Diagnosis and Prognosis Based on EEG Data

This chapter applies the methodology for learning and pattern recognition with BI-SNN, introduced in Chap. 8 on EEG data measuring changes in brain states due to a brain disease or treatment. While this approach can be widely used, data related to two well-spread brain abnormalities are used here: possible progression to Alzheimer's disease; response to treatment of drug addicts.

The chapter is organised in the following sections:

9.1. SNN for modelling EEG data to assess a potential progression from MCI to AD.
9.2. SNN for predictive modelling of response to treatment using EEG data.
9.3. Chapter summary and further readings for deeper knowledge.

## 9.1 SNN for Modelling EEG Data to Assess a Potential Progression from MCI to AD

This section presents a method and experimental results of using the BI-SNN NeuCube to learn brain EEG data from patients of mild cognitive impairment (MCI) and the brain signals of same patients, some of them developed AD. A comparative analysis shows a clear indication in brain activities that can be used to predict a possibility for a new patient with MCI if they can develop AD in the future. The material was first published in [1].

© Springer-Verlag GmbH Germany, part of Springer Nature 2019
N. K. Kasabov, *Time-Space, Spiking Neural Networks and Brain-Inspired
Artificial Intelligence*, Springer Series on Bio- and Neurosystems 7,
https://doi.org/10.1007/978-3-662-57715-8_9

## 9.1.1   Design of the Study and Data Collection

(1)  *EEG Data Description*

The EEG data was collected by the selected patients for the analysis. They underwent cognitive and clinical assessments, including mini-mental state examination (MMSE). Diagnosis of AD was made according to the criteria of the National Institute of Ageing-Alzheimer's Association. After diagnostic confirmation, patients were discriminated by gender, age, education, dementia onset, marital status and MMSE scores. All patients were under the influence of drug treatments such as cholinesterase inhibitors (ChEis), Memantine, anti-depressants, anti-psychotics and anti-epileptic drugs. The dosage of each drug administrated for the three-month period prior to the experiment was carefully monitored.

A total of seven patients were selected for the EEG data collection: three affected by AD and four diagnosed as suffering from MCI. They were all followed longitudinally for three months. During this period of time, the EEG data was recorded twice, at the beginning of the study and at the end of it, denoted as t0 and *t*1.

Before data collection, all patients and their caregivers went through a semi-structured interview, which included questions regarding the quality and duration of their sleep the night before the experiment along with the food consumed and the time it was consumed. Recordings are carried out using 19 EEG channel locations: Fpl, Fp2, 7, F3, Fz, F4, F8, TI, C3, Cz, C4, T4, *TS,* P3, Pz, P4, T6, 01, 02 and the *A*2 electrode was used as reference. These were placed according to the sites defined by the standard 0–20 international system. Data was recorded at a sampling ate of 1024 Hz for 5 min and a 50 Hz notch filter was applied during collection. Data was collected in the morning and under resting conditions, with subjects awake with their eyes closed and always under vigilant control. Data was collected in the region of Reggio di Callabria in the group of Prof. Morabito [1].

(2)  *Data pre-processing:* The EEG data was down-sampled o 256 Hz and processed using a 5 s sliding temporal window (i.e. one window includes 1280 EEG samples). The EEG signal was divided into rhythms of type $\delta$, $\theta$, $\alpha$ and $\beta$ by using a set of four band-pass filters implemented with the use of inverse Fast Fourier Transform (FFI'). The four EEG sub-bands were partitioned into *m* non-overlapping windows, here *m* depends of the length of the recording, which was minutes on average.

## 9.1.2   Design of a NeuCube Model

The same design procedure as introduced in Chap. 8 is used here. Only specific design issues are discussed for this task. The functional block diagram of the NeuCube BI-SNN is shown in Fig. 9.1, also presented in Chap. 6. The architecture consists of the following modules:

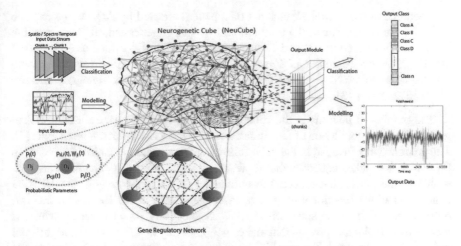

**Fig. 9.1** The NeuCube general functional architecture (from [6], also in Chap. 6)

(1) *Input Module and EEG Data Encoding:* The EEG data was first ordered as a sequence of real-value data. Every data sequence was transformed into a spike train using the threshold base representation *TBR_thr* algorithm (Chap. 4) [2]. This threshold was used to generate two types of spike sequences: a positive spike train corresponding to the signal increment, which is mapped to a specific input neuron in the SNNc; and a negative spike train, corresponding to the signal decline, which is mapped into another input neuron of the SNNc that is placed in the same position as the positive one. Algorithms that apply bi-directional threshold to transform vectors of consecutive values into trains of spikes, well suit EEG data as they identify only significant differences in the signal gradient (as demonstrated in Fig. 9.2). In the example shown in Fig. 9.2, 115 spikes were generated after applying the *TBR_thr* algorithm to the first 500 EEG data points

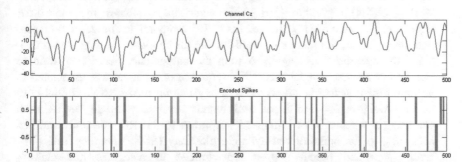

**Fig. 9.2** Example of encoding spatio-temporal EEG data into trains of spikes using the TBR_thr algorithm. The image shows the first 500 data points only of one EEG channel (the central Cz channel). The EEG signal (0–64 Hz) recorded from a patient affected by AD

recorded at the central $Cz$ channel of a patient affected by AD. As we can see from the figure, out of the total amount of spikes generated, 58 were positive spikes (identified as +1) and 57 were negative spikes (identified as −1).

(2) *The SNNcube Module and Unsupervised Learning:* The spike sequences were presented to the SNNc, which was implemented using leaky integrate-and-fire (LIP) neurons [3].

The number of neurons in the cube was set as 1471 neurons; each neuron represents 1 cm$^3$ of population of the human neural cells of the Talairach brain atlas [4]. The neurons were mapped in the cube following the standard mapping suggested in [5]. Thus, the spike sequences that represent the data from EEG channels are presented to the SNNc that reflects the number of input variables (e.g. the 19 EEG channels) and the functional brain areas associated with them. The SNNc was initialised according to the small-world (SW) connectivity [6, 7] instance, which is based on the biological process that makes neighbouring neural cells to be highly and strongly interconnected. Neurons' initial connection weights were calculated as the product of a random number [−0.1, +0.1] and the multiplicative inverse of the Euclidean distance d($i, j$) between pre-synaptic $i$ and a post-synaptic neuron $j$ (calculated according to their ($x, y, z$) coordinates). 20% of these weights are selected to be negative (inhibitory connection weights), as in the mammalian brain the number of GABAergic neurons found to be about 20–30% [8], while 80% were positive excitatory connection weights. The SNNc was trained in an unsupervised mode using the spike time dependent plasticity STOP) [9] learning rule, as it allows spiking neurons to learn consecutive temporal associations from the EEG data within and across EEG channels. By using this unsupervised learning rule, a connection between two neurons become stronger as their temporal order of activation persists and repeats with the time. After learning, the final connectivity and spiking activity generated in the network was analysed and interpreted for a better understanding of the data and the brain processes that generate it (as demonstrated in a later section). This makes the NeuCube useful for learning spatio-temporal patterns from the TBD.

(3) *Output Module far Supervised Learning:* The output classifier was trained via supervised learning method, using the dynamic evolving spiking neural network (deSNN) algorithm [10]. This algorithm combines the rank-order (RO) learning rule [11] with the STDP [9] temporal learning. In one pass data propagation, the same data used for the unsupervised raining was propagated through the SNNc again to train the output classifier. Every training sample that represents a labelled EEG sequence of a patient was associated to an output neuron that is connected to every neuron in the SNNc. Initial connection weights between input and output neurons were all set to zero. Connection weights were initialised according to the RO rule and modified according to the spike driven synaptic plasticity (SDSP) learning rule [12]. Every generated output neuron was trained to recognize and classify spatio-temporal spiking patterns of the SNNc triggered by a corresponding labelled input data sample (as demonstrated in a later section).

### 9.1.3   Classification Results

To investigate whether data collected during the two different sessions ($t_0$ and $t_1$) discriminates different stages of neural degeneration (from early MCI to advanced AD), we classified the data samples by using the entire EEG signal from 0–64 Hz. Data was divided into four classes: data collected at $t_0$ from subjects diagnosed as having MCI was labelled as class 1 (MCI $t_0$), while the data collected at $t_1$ from the same subjects was labelled as class 2 (MCI $t_1$); and data collected at $t_0$ from AD patients was labelled as class 3 (AD $t_0$), while the data collected as $t_1$ from the same patients was labelled as class 4 (AD $t_1$). In total we obtained 14 samples, two for each of the seven subjects, one at $t_0$ and one at $t_1$.

Even though, every subject underwent several minutes of data recording, we resized each samples to 42,240 data points for 19 EEG channels, as this was the size of the smallest sample available. A crucial step in obtaining desirable results from the NeuCube model is the optimisation of its numerous parameters. This can be achieved via grid search method, genetic algorithm, or quantum-inspired evolutionary algorithm [13, 14]. Therefore, unsupervised and supervised training, and validation are repeated changing the values of the parameters until the desired classification output is achieved. In this study, this was obtained via grid search method that evaluated the best combination of parameters that resulted in the highest classification accuracy. The optimised parameter values are:

- The $TBR_{Thr}$ for encoding algorithm was set at 0.5;
- The SW connectivity radius was set at 2.5;
- The threshold of firing e, the refractory time $r$ and the potential leak rate $l$ of the LIF neuron model were set at 0.5, 6 and 0.002 respectively;
- The STDP rate parameter $\alpha$ of the unsupervised learning algorithm was set at 0.01; rate was set at 0.001;
- The variables *mod* and *drift* of the deSNN classifier were set at 0.8 and 0.005 respectively [10].

In Table 9.1, we report the classification accuracy obtained using the above parameter values. The results, obtained after testing, are presented in the confusion table as the number of correctly classified samples *versus* the number of misclassified samples.

**Table 9.1** The NeuCube confusion table obtained by classifying EEG test data (50%) from 7 patients as a test subset into the four classes: MCIt$_0$, MCIt$_1$, AD $t_0$ and ADt$_1$. The correctly predicted classes are located in the diagonal of the table

| Confusion Table | | | | |
|---|---|---|---|---|
| | MCI $t_0$ | MCI $t_1$ | AD $t_0$ | AD $t_1$ |
| MCI $t_0$ | **2** | 1 | 0 | 0 |
| MCI $t_0$ | 0 | **1** | 0 | 0 |
| AD $t_0$ | 0 | 0 | **1** | 0 |
| AD $t_1$ | 0 | 0 | 0 | **1** |

As a result of training a NeuCube model to classify data from the four classes: $MCIt_0$, $MCIt_1$, $ADt_0$ and AD $t_1$, the testing results showed a perfect classification of three classes, but not MCI $t_1$. These results demonstrated the potential of the NeuCube to achieve high classification accuracy for the classes MCI $t_0$, AD $t_0$ and AD $t_1$, but also to indicate if some of the patients data from the MCI $t_1$ class is closer to the data from the MCI $t_0$ class or to the AD $t_0$ class, pointing to a possible development of the disease in the future. As reported in Table 9.1, one of the two subjects from the MCI $t_1$ class showed similar EEG patterns at $t_1$ as in $t_0$, indicating that his subject is not likely to develop AD in the near future. The four classes were in effect identifying different stages of neural degeneration (from early MCI to advanced AD). This is a good indication that a NeuCube model can be used in the future for predicting if MCI patients will develop AD. As the experiment was done on a small data set it is not possible to make any conclusions that would be clinically applicable at this stage.

### 9.1.4   Analysis of Functional Changes in Brain Activity from MCI to AD

Figure 9.3, shows the SNNc connectivity generated after unsupervised learning of the EEG signal (0–64 Hz) at $t_0$ and $t_1$ of a MCI subject who developed AD at time $t_1$.

The figures show significant decrease in neural activity from $t_0$ to $t_1$. The observed reduction in the model neural connectivity is compatible with neuronal changes associated with the advance of the disease. AD is a degenerative brain disorder that eventually destroys brain cells causing decline in cognitive activity and memory loss [15]. Using the NeuCube SNN-based visualization, we can obtain a better understanding and interpretation of the physiological brain ageing of AD patients.

More information can be extracted from the data by identifying relevant EEG sub-bands for AD to study the patient neural activity (Fig. 9.4).

## 9.2   SNN for Predictive Modelling of Response to Treatment Using EEG Data

This section presents a method for predicting response to treatment using deep learning of EEG data in a SNN model. EEG data measuring brain activities before treatment and after treatment are used to evaluate the response. A case study on drug addicts responding to methadone (MMT) is presented. Brain EEG data of three groups of subjects are recorded and analysed using a NeuCube model:

- Control (normal) subjects;
- Drug addicts subjects without treatment;
- Drug addicts who take treatment.

Some more details of the method along with experimental results can be found in [16].

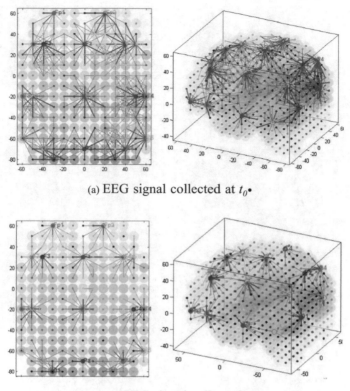

(a) EEG signal collected at $t_0$.

(b) EEG signal collected at $t_1$.

**Fig. 9.3** Connectivity generated after unsupervised learning of the SNNc was performed on the encoded EEG signal (0–64 Hz) of a MCI subject at time $t_0$ who developed AD at time $t_1$. The figure shows a xy-plane projection and the 3D SNNc

### 9.2.1    Conceptual Design

The experimental design for this case study is shown in Fig. 9.5 and explained further in this section.

### 9.2.2    The Case Study Problem Specification and Data Collection

*EEG Data Acquisition*

The EEG data was recorded via 26 cephalic sites: Fp1, Fp2, Fz, F3, F4, F7, F8, Cz, C3, C4, FC3, FCz, FC4, T3, T4, T5, T6, Pz, P3, P4, O1, O2, and Oz electrode sites (10–20 International System).

*The GO/NOGO Task*

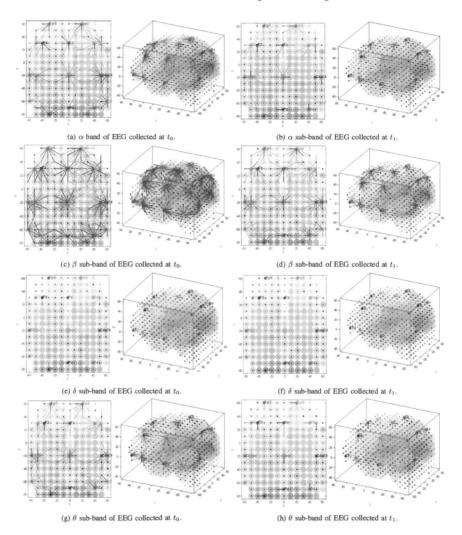

(a) $\alpha$ band of EEG collected at $t_0$.

(b) $\alpha$ sub-band of EEG collected at $t_1$.

(c) $\beta$ sub-band of EEG collected at $t_0$.

(d) $\beta$ sub-band of EEG collected at $t_1$.

(e) $\delta$ sub-band of EEG collected at $t_0$.

(f) $\delta$ sub-band of EEG collected at $t_1$.

(g) $\theta$ sub-band of EEG collected at $t_0$.

(h) $\theta$ sub-band of EEG collected at $t_1$.

**Fig. 9.4** Connectivity generated in the SNNc after unsupervised learning of the EEG data of a MCI subject at time $y_0$ who developed AD at time $t_1$ in $\alpha$, $\beta$, $\delta$ and $\theta$ sub-bands at $t_0$ and $t_1$. The figure shows the 2D (xy) plane and the 3D (x,y,z) SNNc. Significant reduction of connectivity is observed in the created NeuCube models from time $t_0$ to $t_1$ in the $\beta$ sub-band and less in the $\alpha$ and $\theta$ sub-bands, across the cortical areas

GO/NOGO task is a psychological test to measure a participant's capacity for response control and sustained attention. During the task, the participants were repeatedly presented with the word 'PRESS' (for 500 ms). The color of the word 'PRESS' was presented randomly in either red or green. Participants were instructed to respond by pressing a button with the index finger of both hands in response to the word that appeared in green (GO) and not respond to the word that appeared in red (NOGO).

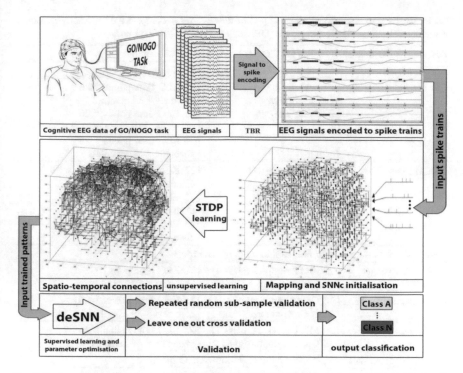

**Fig. 9.5** A conceptual diagram of using NeuCube BI-SNN for the analysis of EEG data in this case study

Participants were asked to complete the practice trial prior to the real test to ensure that they understood the task. At this stage, the word 'PRESS' was presented in the same color 6 times in a row. There were 28 sequences, 21 of which were presented in green and 7 in red, presented in a pseudo-random order, with an inter-stimulus interval of 1 s. The task duration was approximately 5 min. Speed and accuracy of response were stressed equally in the task instructions.

Input EEG Data Preparation for the SNN NeuCube Modelling

In this study, the EEG signal data of the MMT, OP, and control subjects were used as the input to the SNNc to demonstrate the differentiation between their brain activity patterns against the GO/NOGO task. For this purpose, we extracted several EEG sample files from the recorded EEG data and analyzed them separately using the NeuCube model during 3 experimental sessions.

The details of the data are presented in Table 9.2.

**Table 9.2** EEG data sets for the three experimental sessions to compare the brain activity patterns of the control (CO), MMT, and opiate (OP) subjects in a GO/NOGO task

| Session I: EEG data sample files for GO versus NOGO classification | | |
|---|---|---|
| Classifications | Samples per class | EEG sample file size |
| GO trials class | 21 control subjects 18 OP subjects 29 MMT subjects | 75 EEG time points * 26 channels * 21 samples 75 EEG time points * 26 channels * 18 samples 75 EEG time points * 26 channels * 29 samples |
| NOGO trials class | 21 control subjects 18 OP subjects 31 MMT subjects | 75 EEG time points * 26 channels * 21 samples 75 EEG time points * 26 channels * 18 samples 75 EEG time points * 26 channels * 31 samples |
| Session II: EEG data sample files captured during GO trials | | |
| MMT class versus CO class | 29 MMT samples (class 1) 21 control samples (class 2) | 75 EEG time points * 26 channels * 50 samples |
| OP class versus CO class | 18 Opiate samples (class 1) 21 control samples (class 2) | 75 EEG time points * 26 channels * 39 samples |
| MMT class versus OP class | 29 MMT samples (class 1) 18 Opiate samples (class 2) | 75 EEG time points * 26 channels * 47 samples |
| Session III: EEG data sample files captured during NOGO trials | | |
| MMT class versus CO class | 31 MMT samples (class 1) 21 control samples (class 2) | 75 EEG time points * 26 channels * 52 samples |
| OP class versus CO class | 18 OP samples (class 1) 21 control samples (class 2) | 75 EEG time points * 26 channels * 39 samples |
| MMT class versus OP class | 31 MMT samples (class 1) 18 OP samples (class 2) | 75 EEG time points * 26 channels * 49 samples |

## 9.2.3   Modelling the EEG Data in a NeuCube Model

The parameter values of the NeuCube models are presented in Table 9.3. During the learning process in SNNc, when a neuron $n_i$ fires at time t, neurons that are connected to $n_i$ will receive a spike from it and their potentials increase by synaptic weight of the entered spikes. However, the potentials of those neurons that do not receive the spike will leak. Hence, greater transmitted spikes between two neurons lead to stronger connectivity appears.

According to Fig. 9.6, control subjects exhibited a less excitation in NOGO trials when the response must be withheld in comparison with GO trials when the response is required. In contrast, excitations induced during the NOGO trials and were much greater than those induced during the GO trials in either MMT or OP subjects. These findings reflect the group differences on brain activity induced by the two competing response tendencies (GO versus NOGO), implicating deficits in

**Table 9.3** The optimal NeuCube parameters that resulted from a grid search to optimise the classification accuracy as an objective function

| Session | EEG sample files used in NeuCube classification | TBR$_{thr}$ | D$_{thr}$ | STDP rate |
|---------|---------------------------------------------------|-------------|-----------|-----------|
| I | Control subjects in GO versus control subjects in NOGO | 0.551 | 0.150 | 0.010 |
| | MMT subject in GO versus MMT subject in NOGO | 0.949 | 0.150 | 0.010 |
| | OP subjects in GO versus OP subjects in NOGO | 0.777 | 0.150 | 0.010 |
| II | MMT subject versus control subjects (GO task) | 0.463 | 0.225 | 0.014 |
| | Opiate subjects versus control subjects (GO task) | 0.450 | 0.075 | 0.013 |
| | MMT subject versus opiate subjects (GO task) | 0.669 | 0.208 | 0.008 |
| III | MMT subjects versus control subjects (NOGO task) | 0.532 | 0.225 | 0.006 |
| | Opiate subjects versus control subjects (NOGO task) | 0.468 | 0.175 | 0.005 |
| | MMT subjects versus OP subjects (NOGO task) | 0.886 | 0.225 | 0.014 |

inhibition to prevent the execution of the GO response in the subjects with history of opiate dependence no matter what their current treatment status. After the SNNc unsupervised training, neuronal connections with stronger weights reflect more spike transmissions between neurons' synapses. Therefore, the induced brain functional pathways that reveal the connection strength in SNNc, can be visualised. Here we generated and illustrated the pathways initiated from 5 EEG channels, namely C3, Fz, Cz, C4, and P4. These channels were chosen because of their strong involvement in the human response inhibition.

Figure 9.6 represents this information for the control, MMT, and OP subjects while they were responding to GO trials versus NOGO trials. The functional pathways of the control subjects (Fig. 9.7a-1) show that the spatio-temporal relationship was extensively observed in the neurons connected to the allocated input neuron for the Cz channel. By tracing the neuron connections that contain the most number of transmitted spikes, several functional pathways were traced for the Cz channel as a spike sender neuron. Figure 9.7b-1 illustrates the brain information pathways captured from the MMT subjects during the GO trials. The spike transition from the Cz was decreased in the MMT subjects in comparison with the Control subjects. On the other hand, the functional pathways generated by Fz channel were increased. Although the brain activity patterns of the Cz and Fz channels appeared differently in MMT and Control subjects, their brain functional pathways were comparable. In contrast, the brain functional pathways of the Opiate subjects were significantly different from either the Control or the MMT subjects indicated by the absence of functional pathways initiated from the Cz channel

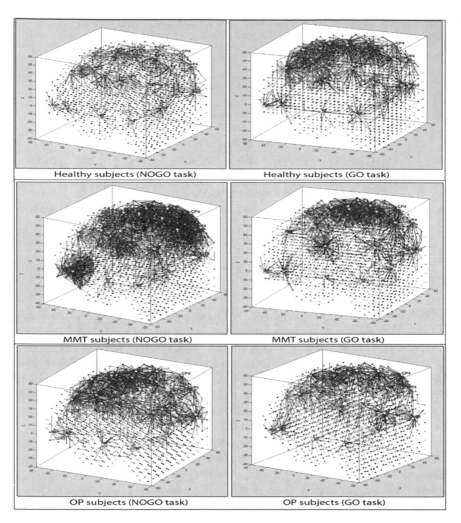

**Fig. 9.6** Illustration of the SNNc connectivity after the NeuCube learning with EEG data of 26 features (channels) for the experimental GO/NOGO task. The learnt connectivity of the SNNc is different for the control (healthy), MMT, and OP subjects related to the GO/NOGO task. The blue lines are positive (excitatory) connections, while the red lines are negative (inhibitory) connections. The brighter the colour of a neuron, the stronger its activity with neighbouring neurons. Thickness of the lines also identifies the neuron's enhanced connectivity. The 1471 neurons of the brain-like SNNc are spatially mapped according to the Talairach brain atlas [4]

(Fig. 9.7c-1). Consistent with previous studies [17–19], these findings indicate the possible abnormality of brain function associated with long term exposure to opioid type drugs. However, patients undertaking MMT for opiate addiction appeared less impaired than those current opiate users.

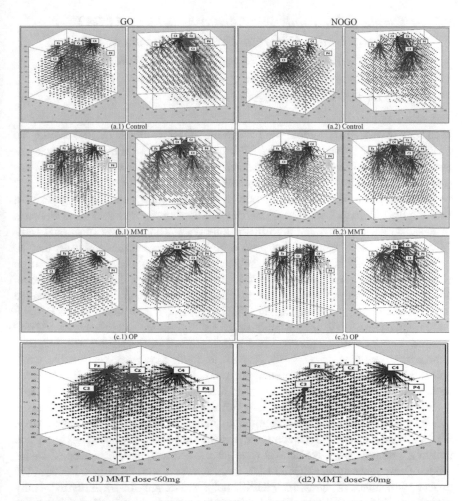

**Fig. 9.7** Functional pathways for Control, MMT, and OP subjects generated between 5 EEG channels (sender spike neurons) and the rest of the neurons inside the brain-like SNNc (receiver spike neurons) while doing GO trials versus NOGO trials. The big solid dots represent the input neurons and other neurons that are labelled with * sign are receiver spike neurons. The lines represent the connectivity between neurons. The unconnected dot means no spike arrived at that neuron. **a.1** the brain functional pathways of the control subjects in GO trials; **a.2** the brain functional pathways of the control subjects in NOGO trials; **b.1** the brain functional pathways of the MMT subjects in GO trials; **b.2** the brain functional pathways of the MMT subjects in NOGO trials; **c.1** the brain functional pathways of the OP subjects in GO trials; **c.2** the brain functional pathways of the OP subjects in NOGO trials; **d.1** the brain functional pathways of MMT group that received less than 60 mg methadone dose per day; **d.2** the brain functional pathways of MMT group that received more than 60 mg methadone dose per day

### 9.2.4 Comparative Analysis of Brain Activities of MMT Subjects Under Different Drug Doses Versus CO and OP Subjects. Modelling and Understanding the Information Exchange Between Brain Areas Measured Through EEG Channels

Members of the MMT group were receiving different doses of methadone. To examine the dose-related effects, the EEG patterns of the MMT subjects were categorized into two groups based on their current methadone dose: High dose (>60 mg/day) and low dose ($\leq 60$ mg/day). The EEG patterns of these two groups were learned in a SNNc and their functional pathways were visualized. Figure 9.7-d captures the differences between functional pathways generated by 5 EEG channels in MMT subjects on low and high methadone dose. The captured functional pathways of those MMT subjects that used a high dose were more similar to the OP group. On the other hand, the MMT subjects with less amount of methadone dose performed similar functional pathways to the control group. The NeuCube model allows also to perform modelling and understanding the information exchange between brain areas measured through EEG channels. This is illustrated in Fig. 9.8 and explained below. Figure 9.8 captures the spike communication between 26 EEG electrodes after NeuCube unsupervised learning. Each vertex represents a neuronal cluster corresponding to an EEG channel and the arcs represent relative spike amounts transmitted between different neuronal clusters. The wider the line between input neurons, the more spikes were transmitted between the corresponding clusters.

In Fig. 9.8a, by comparing two graphs obtained from control subjects in GO versus NOGO trials, it is clear that the spike communication was especially enhanced between neuronal clusters while the subjects were performing GO trials. Consequently, we can conclude that less spike interactions were manifested while subjects increased inhibition of responses during NOGO trials in comparison with GO trials. Perhaps, the green appurtenance of the word 'PRESS' helps to strength the visibility to healthy subjects and induces an enhanced activation in the central parietal and occipital areas, which probably encompasses the primary and secondary visual areas.

However, this trend is absent in either the MMT or opiate subjects. Furthermore, both the MMT and opiate subjects demonstrated increased spike communication in a wide range of areas, in particular, in the frontal, central, and temporal areas during the NOGO trials, implicating increased stimulation induced by NOGO stimuli in the areas related to attention, visual memory, and execution of voluntary movements. Our findings suggest anatomically and functionally different inhibition processes between people with history of opiate use and healthy control subjects. It is also noted that alternation of inhibition process are greater in the opiate users compared to the MMT subjects. For opiate subjects in NOGO trials, the majority of the wide lines were created between channel F4 and channels T6, P4, PZ, P3, T5, CP4, T4, C4, and CZ.

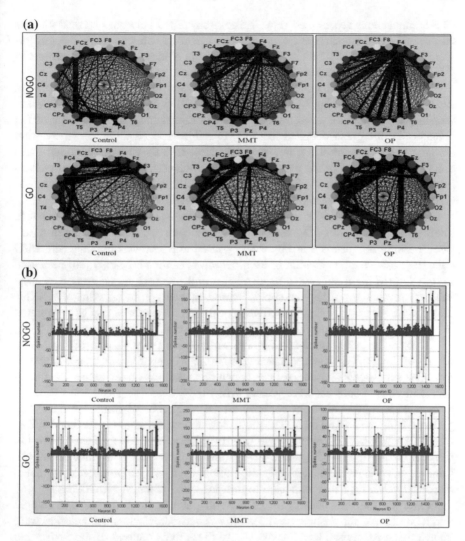

**Fig. 9.8  a** The total interaction between 26 neuronal clusters representing 26 EEG channels in terms of spike communication as a measure of information exchange between cortical brain areas. The thicker the line that connects two neurons that represent the corresponding electrodes, the more spikes are transmitted between corresponding clusters; **b** The number of spikes emitted by each neuron of the SNNc after SNNc unsupervised training with an exemplar EEG data recorded from Control, MMT, and Opiate subjects in GO versus NOGO trials. The blue lines are the positive spikes (excitatory) emitted by all neurons in the SNNc, while the red lines are negative spikes (inhibitory) emitted only by the input neurons representing the EEG channels. *EEG GO/ NOGO Pattern Classification Using the NeuCube Model*

These connections represented more spikes transmitted between neuronal clusters corresponding to the channel F4 and the other neuronal clusters while the subjects were undertaking NOGO trials. Consequently, the ability of the opiate subjects to inhibit their voluntary responses may be impaired during NOGO trials.

On the other hand, the interactions between these channels are not observed in the control subjects during NOGO trials. It means that there were not many spikes transmitted between the neuron clusters related to channel FZ and the other EEG channels. In the graph obtained from MMT subjects in NOGO trials, there were strong spike communications between FZ, CP4, and T4 clusters, although these connections were less in comparison with opiate subjects. The observed differences in spike communication implicate that the control and Opiate subjects performed differently while they were doing cognitive GO/NOGO tasks.

To achieve a better understanding of the spike occurrence and propagation inside the SNNc, the number of the spikes emitted by each neuron during the unsupervised training is illustrated in Fig. 9.8b. While the SNNc was training with EEG data, the post synaptic potential of each neuron $n_i$ at time t, $PSPi(t)$ [20] increased by the sum of the input spikes received from all pre-synaptic neurons. Once the PSPi (t) reaches the firing threshold, neuron $n_i$ emits a spike. After the SNNc unsupervised learning, temporal activities of the spiking neurons can be interpreted in terms of brain activities measured by the corresponding EEG channels.

An example of the number of spikes emitted by every neuron of the SNNc related to the EEG data is given in Fig. 9.8-b. By comparing the results obtained from GO versus NOGO trials, we can conclude that the average number of emitted spikes in control subjects were greater when they were doing GO trials in comparison with NOGO trials. In contrast for OP subjects, the emitted spikes were greater during the NOGO trials. The plots indicate that the number of emitted spikes of each neuron was less than 100 in control subjects and greater than 100 in OP subjects during the NOGO trials. These findings support our argument that OP subjects may experience difficulty in inhibiting their inappropriate automated responses when they were expected to not press the button in NOGO trials.

In order to learn and classify the EEG signal patterns, the EEG data was entered into a 3D SNNc for unsupervised learning. Then output classifier neurons were trained using supervised learning to classify the activity patterns of the SNNc into the pre-defined classes. The classification accuracy results were evaluated using repeated random rub-sampling cross validation (RRSV). In this experiment, the RRSV method was applied with 50% of the data for training and 50% for testing. In order to optimize classification accuracy, the values of the NeuCube parameters were altered through iterative applications of the NeuCube modules (1)–(3) as discussed in Sect. III.B.

In this experiment, the TBR threshold, Connection distance ($D_{thr}$), and STDP rate parameters were changed during 1000 optimization iterations and then the best accuracy was recorded. The parameters that generated the best classification accuracy are reported in Table 9.3. The firing threshold, the mod, and drift parameters were set to 0.05, 0.4, and 0.250 respectively.

## 9.2.5   *Analysis of Classification Results*

Classification accuracy results were compared with results obtained using traditional machine learning methods, as these methods are still being actively used in the literature for the purpose of classification of EEG data.

The methods we used for comparison are: Support Vector Machine (SVM); Multiple Linear Regression (MLR); Multi-Layer Perceptron (MLP); and Evolving Clustering Method (see www.theneucom.com). The classification accuracy results of the three experimental sessions for the three output classes of subjects are summarized in Table 9.4.

The classification accuracy results obtained in session I show that the control subjects took actions differently in GO trials versus NOGO trials. Therefore, their EEG spike trains were classified with a higher accuracy of 90.91% in comparison with MMT and opiate subjects. In session II, the classification accuracy of 85% in OP versus CO is higher than the accuracy of 77% in MMT versus CO, which means that the similarity between the EEG data of the MMT and control subjects was greater than the similarity between EEG data of the opiate and control subjects.

**Table 9.4** The EEG data classification accuracy results from three experimental sessions via RRSV method of the NeuCube. 50% of the data was used for training and 50% used for testing. The results of the traditional methods were obtained via leave one out cross validation (LOOCV)

| Control subjects (CO), Opiate subjects (OP), Accuracy is reported in % | | | | | | |
|---|---|---|---|---|---|---|
| Session | Classification | NewCube | SVM | MLP | MLR | ECMC |
| Session I: GO versus NO classification | Control subjects in GO versus NOGO | 90.91 | 50.55 | 48.50 | 50.38 | 29.71 |
| | MMT subjects in GO versus NOGO | 83.87 | 50.39 | 49.72 | 50.17 | 42.65 |
| | Opiate subjects in GO versus NOGO | 83.33 | 50.40 | 47.81 | 50.00 | 45.43 |
| Session II: OP, MMT, CO classification in GO | MMT subject versus control subjects (GO) | 77.00 | 47.12 | 45.36 | 49.86 | 50.47 |
| | Opiate subjects versus control subjects (GO) | 85.00 | 50.50 | 50.64 | 48.60 | 48.60 |
| | MMT subject versus opiate subjects | 79.00 | 47.90 | 45.22 | 50.53 | 49.98 |
| Session III: OP, MMT, CO classification in NOGO | MMT subjects versus control subjects | 85.00 | 49.13 | 48.62 | 50.49 | 50.15 |
| | Opiate subjects versus control subjects | 90.00 | 50.24 | 49.83 | 50.24 | 49.83 |
| | MMT subjects versus OP subjects | 88.00 | 46.57 | 50.51 | 50.00 | 48.71 |

Consequently, we can conclude that some of the MMT subjects respond to the methadone treatment and their brain activity patterns may be improved and become comparable to the CO subjects. Also, the classification accuracy of 100% in MMT versus OP demonstrates that all MMT subjects were classified correctly into the MMT class. In fact, this result indicates that the EEG data patterns of the MMT subjects are greatly different from opiate subjects.

The classification accuracy of 90% in OP versus CO is higher than the classification accuracy of 85% in MMT versus CO. These results show that the differences between the brain activity patterns of MMT and control groups can be minimum in contrast to OP group, and MMT group. It may represent that the MMT has a potential positive effect on brain function and contribute to functional recovery.

The experimental results demonstrated that:

(1) In all experiments, the NeuCube-based models obtained superior classification accuracy when compared with traditional machine learning methods.
(2) The brain activity patterns of healthy volunteers were significantly different from people with history of opiate dependence. The differences appeared less pronounced in people undertaking MMT compared to those current opiate users.
(3) The brain functional pathways of the healthy volunteers were greater and broader than either people undertaking MMT or those opiate users.
(4) The STBD patterns of people on low dose of methadone appeared more comparable to healthy volunteers compared to those on high dose of methadone.

## 9.3    Chapter Summary and Further Readings for Deeper Knowledge

The chapter presents some applications of the methodology for deep learning and modelling of EEG data and for the design of BI-AI systems from Chaps. 6 and 8, here for modelling EEG data related to brain diseases, on the case of AD and drug addiction. These are only two illustrations for the applicability of the introduced SNN methodology. Further applications and studies can be developed using this approach.

The following readings on specific and related topics can be recommended:

- Multimodal atlases of the human brain [21];
- Electroencephalography [22];
- Brain imaging [15];
- Alzheimer's Disease (Chaps. 51, 52 in [23] and also [24]);
- SNN for modelling EEG data of drug addicts under treatment [16–19];
- Demo on modelling EEG AD data in NeuCube: https://kedri.aut.ac.nz/R-and-D-Systems/neucube/eeg;

- NeuCube software development system (http://www.kedri.aut.ac.nz/neucube/);
- Methods and systems for measuring and modelling brain structures and functions, with applications [25–42].

**Acknowledgements**  Some of the material in this chapter has been first published in journal and conference publications as referenced and cited in corresponding sections and sub-sections and also in book volumes [2, 16, 23, 24]. I acknowledge the great contribution of my co-authors of these publications Maryam Doborjeh, Elisa Capecci, Zohreh Doborjeh, Nathan Scott, Carlo Morabito, Nadia Mammone, F. La Foresta, Grace Wang.

# References

1. E. Capecci, Z. Doborjeh, N. Mammone, F. La Foresta, F.C. Morabito, N. Kasabov, *Longitudinal Study of Alzheimer's Disease Degeneration through EEG Data Analysis with a NeuCube Spiking Neural Network Model*, in Proceedings WCCI—IJCNN, Vancouver (IEEE Press, 24–29 July 2016), pp. 1360–1366
2. N. Kasabov, N. Scott, E. Tu, S. Marks, N. Sengupta, E. Capecci, M. Othman, M. Doborjeh, N. Murli, R. Hartono, J. Espinosa-Ramos, L. Zhou, F. Alvi, G. Wang, D. Taylor, V. Feigin, S. Gulyaev, M. Mahmoudh, Z.-G. Hou, J. Yang, Design methodology and selected applications of evolving spatio-temporal data machines in the NeuCube neuromorphic framework. Neural Netw. **78**, 1–14 (2016). http://dx.doi.org/10.1016/j.neunet.2015.09.011
3. W. Gerstner, What's different with spiking neurons?, in *Plausible Neural Networks for Biological Modelling*, ed. by H. Mastebroek, H. Vos (Kluwer Academic Publishers, Dordrecht, 2001), pp. 23–48
4. J. Talairach, P. Tournoux, *Co-planar Stereotaxic Atlas of The Human Brain: 3-Dimensional Proportional System: An Approach to Cerebral Imaging* (Thieme Medical Publishers, New York, 1988)
5. L. Koessler, L. Maillard, A. Benhadid et al., Automated cortical projection of EEG sensors: Anatomical correlation via the international 10–10 system. NeuroImage **46**, 64–72 (2009)
6. N. Kasabov, NeuCube: a spiking neural network architecture for mapping, learning and understanding of spatio-temporal brain data. Neural Netw. **52**, 62–76 (2014)
7. J. Hu, Z.G. Hou, Y.X. Chen, N. Kasabov, N. Scott, *EEG-Based Classification of Upper-limb ADL Using SNN for Active Robotic Rehabilation*, in IEEE RAS & EMBS International Conference on Biomedical Robotics and Biomechatronics (BooRob) (Sao Paulo, Brazil, 2014), pp. 409–414
8. V. Capano, H.J. Herrmann, L. de Arcangelis, Optimal percentage of inhibitory synapses in multi-task learning. Sci. Rep. **5**, 9895 (2015)
9. S. Song, K. Miller, L. Abbott et al., Competitive Hebbian learning through spike-timing-dependent synaptic plasticity. Nat. Neurosci. **3**, 919–926 (2000)
10. N. Kasabov, K. Dhoble, N. Nuntalid, G. Indiveri, Dynamic evolving spiking neural networks for on-line spatio–and spectro-temporal pattern recognition. Neural Netw. **41**, 188–201 (2013)
11. S. Thorpe, J. Gautrais, Rank order coding. Computational Neuroscience: Trends Res. **13**, 113–119 (1998)
12. S. Fusi, Spike-driven synaptic plasticity for learning correlated patterns of asynchronous activity. Biol. Cybern **87**, 459–470 (2002)
13. M. Defoin-Platel, S. Schliebs, N. Kasabov, Quantum-inspired evolutionary algorithm: a multi-model EDA. IEEE Trans. Evol. Comput. **13**(6), 1218–1232 (2009)

14. M. Fiasce, M. Taisch, On the use of quantum-inspired optimization techniques for training spiking neural networks: a new method proposed, in *Advances in Neural Networks: Computational and Theoretical Issues* (Springer, 2015), pp. 359–368

15. J.B. Poline, R.A. Poldrack, Frontiers in brain imaging methods grand challenge. Front. Neurosci. **6**, 96 (2012). https://doi.org/10.3389/fnins.2012.00096

16. M.G. Doborjeh, N. Kasabov, Z. Doborjeh, SNN for Modelling Dynamic Brain Activities during a GO/NO_GO Task: A Case Study on Using EEG Data of Healthy Vs Addiction vs Treated Subjects. IEEE Trans. Biomed. Eng. **63**(9), 1830–1841 (2016)

17. G.Y. Wang et al., Changes in resting EEG following methadone treatment in opiate addicts. Clin. Neurophysiol. **126**(5), 943–950 (2015)

18. G. Y. Wang et al. Quantitative EEG and low-resolution electromagnetic tomography (LORETA) imaging of patients undergoing methadone treatment for opiate addiction. Clin. EEG Neurosci. (2015) https://doi.org/10.1177/1550059415586705

19. G.Y. Wang et al., Auditory event-related potentials in methadone substituted opiate users. J. Psychopharmacol. **29**(9), 983–995 (2015)

20. N. Kasabov, To spike or not to spike: a probabilistic spiking neuron model. Neural Netw. **23**(1), 16–19 (2010)

21. A. Toga, P. Thompson, S. Mori, K. Amunts, K. Zilles, Towards multimodal atlases of the human brain. Nat. Rev. Neurosci. **7**, 952–966 (2006)

22. E. Niedermeyer, F.H.L. Da Silva, *Electroencephalography: Basic Principles, Clinical Applications, and Related Fields* (Lippincott Williams & Wilkins, Philadelphia, 2005), p. 1309

23. N. Kasabov ed., *Springer Handbook of Bio-Neuroinfortics* (Springer, New York, 2014)

24. L. Benuskova, N. Kasabov, *Computaional Neurogenetc Modelling* (Springer, New York, 2007)

25. D.A. Craig, H.T. Nguyen, *Adaptive EEG Thought Pattern Classifier For Advanced Wheelchair Control*, in International Conference of the IEEE Engineering in Medical and Biology Society (2007), pp. 2544–2547

26. F. Lotte, M. Congedo, A. Lécuyer, F. Lamarche, B.A. Arnaldi, Review of classification algorithms for EEG-based brain-computer interfaces. J. Neural Eng. **4**(2), R1–R15 (2007)

27. R.C. deCharms, Application of real-time fMRI. Nat. Rev. Neurosci **9**(9), 720–729 (2008)

28. T. Mitchel et al., Learning to decode cognitive states from brain images. Mach. Learn. **57**, 145–175 (2004)

29. K. Broderson et al., Generative embedding for model-based classification of fMRI data. PLoS Comput. Biol. **7**(6), 1–19 (2011)

30. K. Broderson et al., Decoding the perception of pain from fMRI using multivariate pattern analysis. NeuroImage **63**, 1162–1170 (2012)

31. K. Zilles, K. Amunts, Centenary of Brodmann's map—conception and fate. Nat. Rev. Neurosci. **11**, 139–145 (2010)

32. S. Eickhoff et al., A new SPM toolbox for combining probabilistic cytoarchitectonic maps and functional imaging data. Neuroimage **25**, 1325–1335 (2005)

33. J. Lancaster et al., Automated talairach atlas labels for functional brain mapping. Hum. Brain Mapp. **10**, 120–131 (2000)

34. A.C. Evans, D.L. Collins, S.R. Mills, E.D. Brown, R.L. Kelly, T.M. Peters, *3D Statistical Neuroanatomical Models from 305 MRI Volumes*, in IEEE-Nuclear Science Symposium and Medical Imaging Conference (IEEE Press, 1993), pp. 1813–1817

35. J. Ashburner, Computational anatomy with the SPM software. Magn. Reson. Imaging **27**(8), 1163–1174 (2009)

36. L. Benuskova, Kasabov, *Computational Neuro-genetic Modelling* (Springer, New York, 2007), p. 290

37. M. Hawrylycz et al., An anatomically comprehensive atlas of the adult human brain transcriptome. Nature **489**, 391–399 (2012)

38. W. Gerstner, H. Sprekeler, G. Deco, Theory and simulation in neuroscience. Science **338**, 60–65 (2012)

39. C. Koch, R. Reid, Neuroscience: observation of the mind. Nature **483**(7390), 397–398 (2012)
40. Van Essen et al., The human connectome project: a data acquisition perspective. NeuroImage **62**(4), 2222–2231 (2012)
41. H. Markram, The blue brain project. Nat. Rev. Neurosci. **7**, 153–160 (2006)
42. E.M. Izhikevich, G.M. Edelman, Large-scale model of mammalian thalamocortical systems. PNAS **105**, 3593–3598 (2008)

# Chapter 10
# Deep Learning and Deep Knowledge Representation of fMRI Data

The chapter presents first background information about functional magnetic-resonance imaging (fMRI) and then introduces methods for deep learning and deep knowledge representation from fMRI data using brain-inspired SNN. These methods are applied to develop specific methods for fMRI data analysis related to cognitive processes.

The chapter is organised in the following sections:

10.1. Brain fMRI data and their analysis.
10.2. Deep learning and deep knowledge representation of fMRI data in BI-SNN.
10.3. Mapping, learning and classification of fMRI data in NeuCube on the case study of STAR/PLUS data.
10.4. Algorithms for modelling fMRI data that measure cognitive processes.
10.5. Chapter summary and further readings for deeper knowledge.

## 10.1  Brain fMRI Data and Their Analysis

### 10.1.1  What Are fMRI Data?

Functional MRI (fMRI) combines visualisation of the brain anatomy with the dynamic image of brain activity into one comprehensive scan [1–3]. This noninvasive technique measures the ratio of oxygenated to deoxygenated hemoglobin which have different magnetic properties. Active brain areas have higher levels of oxygenated hemoglobin than less active areas. An fMRI scan can produce images of brain activity at the time scale of seconds with precise spatial resolution of about 1–2 mm. Thus, fMRI provides both a 3D anatomical and functional view of the brain in the lower frequency spectrum.

fMRI can be used to visualize hemodynamic response in relation to neurons' activities in certain part of the brain [4]. This hemodynamic response is indicated by

© Springer-Verlag GmbH Germany, part of Springer Nature 2019
N. K. Kasabov, *Time-Space, Spiking Neural Networks and Brain-Inspired Artificial Intelligence*, Springer Series on Bio- and Neurosystems 7,
https://doi.org/10.1007/978-3-662-57715-8_10

the increasing amount of blood flow to that particular activated neurons area. The components of hemodynamic response include the changes in the oxyhemoglobin and deoxyhemoglobin concentration, in the cerebral blood volume (CBV) per unit of brain tissue and in the cerebral blood flow rate. There are different fMRI techniques that can capture the functional signals generated from the different components of hemodynamic response. One of the most common techniques is based on the concentration of oxyhemoglobin-deoxyhemoglobin component and it is known as blood-oxygen-level-dependent (BOLD) technique [5].

While MRI provides structural mapping of a brain, fMRI imaging technique combined with blood-oxygen-level-dependent (BOLD) technique [5], produced a better set of brain images, i.e. with excellent temporal as well as spatial information. In addition to structural mapping, fMRI generates functional mapping of the brain that actually takes advantage of iron in the blood-carrying-oxygen and blood-vessels-dilation physiological principle that occurs in activated regions. It is used to measure neural activity changes in the brain resulting from stimuli triggered externally or internally [6]. More precisely, fMRI measures the ratio of oxygenated haemoglobin to deoxygenated haemoglobin in the blood with respect to a control baseline, at many individual locations within the brain. It is widely believed that blood oxygen level is influenced by local neural activity, and hence this blood BOLD response is generally taken as an indicator of neural activity [2].

fMRI imaging technique is non-invasive and radiation-free thus providing a safe environment to the subjects involved. The images are recorded in sequence either vertically or horizontally (Fig. 10.1), and over time, in a matrix of intensity values. They are captured in slices of image data through the organs, generally in 8 or 16-bit (Fig. 10.1 right). There are a number of common formats in which the images are saved such as in DICOM, ANALYZE, NIFTI format or in raw voxel intensity values.

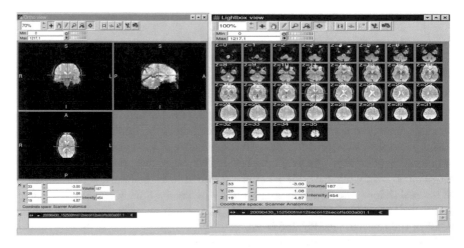

**Fig. 10.1** Brain images in vertical and horizontal slice: in sagittal, coronal and axial views (left). Slices of brain image data taken over time i.e. 32 images for a volume of brain (images are viewed using FSLView (FSLView, 2012) software (right)

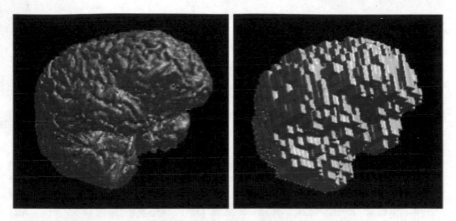

**Fig. 10.2** Surface renderings of 3D brain images. Small voxels (left, with 1 mm × 1 mm × 1.5 mm) versus large voxels (right, with 7 mm × 7 mm × 10 mm) (Smith 2004)

The images are constructed from two components—spatial/spectral (or spatio) and temporal. The first component is identified as the volume of a brain that can be further sub-divided into smaller 3D cuboids, known as voxels (volume element). In a typical fMRI study, a series of brain volumes are collected in quick succession and the value of BOLD response at all points in a 3D grid are recorded. A general 3D brain image typically contains 10,000–15,000 voxels, and each voxel consists of on the order of hundreds of thousands of neurons. Spatial image resolutions can be set either to have low or high resolution. As in Fig. 10.2 while high-resolution image provides more accurate information (e.g. voxels with dimensions of 1 mm × 1 mm × 1 mm) more CPU processing power is required and is not feasible at the moment. Typical spatial resolution is 3 mm × 3 mm × 5 mm, corresponding to image dimensions in the order of 64 × 64 × 30 [7] and still this resolution is relatively high compared to other imaging techniques.

The temporal component is acquired while scanning the whole volume of a brain that will take a few seconds to complete. In a single run of an experiment, 100 or more brain volumes are usually scanned and recorded for a single subject doing a particular sensorimotor or cognitive task. Temporal component depends on the time between acquisitions of each individual image, or the time of repetition (TR). In a typical experiment, TR ranges from 0.5 to 4.0 s and TR values in the range of 2 s are generally considered adequate [7].

The combination of this spatial and temporal information of the brain images will be the main concern investigated in this study.

## 10.1.2   Traditional Methods for fMRI Data Analysis

Choosing the best technique for fMRI data analysis is still a question that needs to be answered properly. There are many variables that have to be considered, for instance the weak signal of voxel of interest, the voxels being distributed among various spatial locations of the brain, different brain mapping of different brain sizes and spatially distributed noise. A popular analysis approach is pattern classification, where the brain patterns are observed to forecast the task being performed by the subject.

Naturally, brain activities are captured as fMRI data in a spatio-temporal format. In conducting the analysis, researchers often treat fMRI data classification either in univariate or multivariate; linear or non-linear; or as static or spatio-temporal approach.

In early years, the standard fMRI data analysis approach examines each brain voxel area in isolation (univariate) as static data as suggested in Statistical Parametric Mapping (SPM) [7, 8], which completely disregard the inherent spatio-temporal characteristics of fMRI data. Univariate approach processes fMRI voxels as independent individuals, thus no interaction and no relationship are measured among the voxels. This approach has been experimented with Gaussian Naïve Bayes method [2, 9]. However this approach neglects the collective information encoded by voxels patterns [10].

Multivariate analysis on the other hand, evaluates the correlation of brain patterns across the brain regions rather than examining them on a voxel-by-voxel basis. In [6] illustrated how multi-voxel patterns of activity can be used to distinguish between cognitive states when subjects were shown faces, houses and a variety of object categories. As different brain locations are triggered with the same (or different) stimulus, experiments should consider all relevant voxels instead of just considering a single particular voxel. This multi-voxel pattern analysis has been adapted by many researchers with various classifiers using either linear or non-linear classifiers: SVM [11–14], Gaussian Naïve Bayes (GNB) [15–17], neural networks without a hidden layer [18], non-linear SVM [13, 19] and neural networks with hidden layers [20]. All these studies only consider data at a single time or time interval.

Another fMRI classifier approach is whether they are linear or non-linear. While linear classifier divides the classes with a linear plane, non-linear classifier separates the classes using a more complex function [21]. Works related to fMRI on linear classifier includes [2, 12, 13, 22, 23]. Although this approach is more biased and less flexible than the non-linear classifier, several studies suggest that they could still generate accurate results [13, 23]. Non-linear classifier on the other hand has also produced good analysis [19, 24] although some other studies suggest that it produces the worst result [13, 25]. However, for robust classification, non-linear classifier requires larger training set [26].

In recent years, researchers are moving towards brain analysis that embed both spatial and temporal behaviour such as spatial-temporal SVM [27], Bayesian formulation [28] and Generalized Sparse Classifier (GSC) [29]. A research conducted by [30] selects a set of relevant voxels using General Linear Model (GLM) and then incorporates liquid state machine and Multi-Layer Perceptron (MLP). These researches focus on spatio-temporal classification, where multiple brain volumes within a trial are treated as a sample.

In conclusion, the study of fMRI characteristics and its relation with the behaviour of a classifier is still not well comprehended. Typically, fMRI datasets are ill-posed datasets that require massive computational power to process their voxels. In addition to this, the interaction of the classifier properties with BOLD signal properties of fMRI is still not well treated [30, 31].

## 10.1.3 Selecting Features from FMRI Data

In a typical fMRI experiment, a sequence of images related to the subject's brain activity every half seconds will be produced. The experiment usually consists of a set of trials and each trial produces many brain volumes over time. Each brain volume is comprised of voxels in the order of thousands and these voxels' intensities are the features to be classified. Learning this brain data poses many challenges especially in terms of the data being extremely sparse noisy data and high dimensional. This would cause over-fitting problem for the classifier. Hence it is necessary to apply feature selection methods to make learning tractable and to prevent over-fitting.

In selecting relevant features (voxels, or areas of voxels) that respond to a stimulus, it can either be done in a univariate or multivariate manner. Apart from the standard univariate approach multivariate pattern analysis approaches towards detection of ROI from fMRI data have been gaining a lot of attention recently. The advantage of multivariate method stem from the fact that even voxels with weak individual response may carry important cognitive information when analysed together.

Evolutionary feature selection is an algorithm that is based on evolutionary techniques (Chap. 7). This approach was proven effective in feature subset selection that detects which number and combination of individual voxels that carry information relevant to a stimulus [32]. These voxels are used as features in multiple linear regression (MLR) classifier and they proved that even the simple classification scheme can detect and distinguish relevant cortical information in noisy fMRI data. Although it considers voxels in multivariate way (analyse voxels collectively), voxels are only on a single volume and not tested on multiple volumes over time.

Another approach uses particle swarm optimization (PSO) based fMRI brain state classification algorithm, specifically designed to efficiently extract a subset of voxels optimal for classification task [33]. PSO is a stochastic optimization method [34] loosely based on the behaviour of swarming animals such as fish and birds (see Chap. 7). A number of particles, representing potential solution to the problem, are released in the search space of potential solutions. Each particle has a position and a velocity and is free to fly around the search space. And in the case of feature selection, this standard PSO is modified as proposed by [35] and not only achieves high performance scores but also identifies functionally relevant ROI [33].

In addition, methods that simultaneously select relevant voxels have been proposed which extend traditional classifiers by incorporating sparse regularization, which controls over fitting by encouraging zero weights to be assigned to irrelevant voxels [9, 36, 37]. And these works have been improved in the recently proposed Generalized Sparse Classifiers (GSC) [38] that permits more general penalties, such as spatial smoothness in addition to it being sparse and to be seamlessly integrated. Another improvement was Generalized Group Sparse Classifiers (GGSC) [39] that permits associations between features within predefined groups to be modelled.

## 10.2   Deep Learning and Deep Knowledge Representation of fMRI Data in NeuCube

This section presents generic methods for mapping, learning, modelling and understanding of fMRI data in a brain-inspired SNN architecture and illustrates the methods on modelling fMRI from cognitive tasks in the NeuCube SNN architecture from Chap. 6. Full description of the methods can be found in [40].

### 10.2.1   Why Use SNN for Modelling of fMRI Spatio-temporal Brain Data?

The brain processes input information in the form of spatio-temporal binary events called *spikes* [41–43]. SNN methods have been already developed and implemented as neuromorphic engineering systems, e.g.: neuromorphic hardware [44–46]; SNN for image and speech processing as trains of spikes [47–49]; unsupervised [50] and supervised learning and classification systems [51–53] etc.

Comparing to traditional neuronal networks, SNNs can integrate both spatio and temporal components of data which is important in modeling fMRI data. SNNs are considered the third generation of neural networks [54] and some of their remarkable features are: compact representation of space and time; fast data learning; time-based and frequency-based information representation; minimalistic information presentation; low energy consumption. Due to these reasons, SNN can be considered as suitable models for fMRI data. These features of the SNN are utilised in [55] for the creation of BI-SNN—NeuCube (see Chap. 6).

NeuCube is a BI-SNN model for learning, classification/regression, visualisation and interpretation of spatio-temporal data, initially proposed for brain data [55]. NeuCube consists of five main modules: data encoding and mapping; unsupervised learning in a SNNc; supervised learning and classification in eSNN; parameter optimisation; model visualisation and interpretation (see also Chap. 6). The size of the SNNc is scalable and controlled by three parameters: $n_x, n_y, n_z$ representing the neuron numbers along $x$, $y$ and $z$ directions. This cube can be used to map (x, y, z) coordinates of input variables, so that spatial information in the data is preserved. The SNNc is trained in an unsupervised mode on the spike sequences that represent the input spatio-temporal data. After this first phase of training, an eSNN output classifier is trained to learn the SNNc spatio-temporal activities that represent data patterns and their pre-defined classes. A dynamic evolving SNN (deSNN) can be used as an output classifier [51], but other classifiers can also be employed [52–54].

### 10.2.2   A Methodology for Deep Learning and Deep Knowledge Representation of fMRI Data in BI-SNN

This methodology includes several procedures as described below and schematically presented in Fig. 10.3.

The input data features (e.g., fMRI voxels) are spatially mapped into spatially allocated spiking neurons in a 3D SNNc according to the spatial location of these features as brain coordinates. A SNNc is created as a 3D SNN structure of a suitable size that maps spatially a brain template (such as Talairach [56], MNI [57]) or voxel coordinates of individual brain data. Then continuous value time series of voxel data that measure activity at a certain brain location is encoded into a spike train using Threshold-Based Representation method (TBR) or other methods (Chap. 4) [58]. The timing of the spikes corresponds to the time of the changes in the data. A spike time sequence, obtained after the encoding process, represents a new input information to the SNNc where the time unit maybe different from the real time of the data acquisition (machine computation time versus data acquisition time). The SNNc can be implemented using the popular leaky-integrate and fire neuronal model (LIFM or other SNN models [59]).

The neuronal post-synaptic potential (PSP), also called membrane potential u(t), increases with every input spike at a time t, multiplied by the synaptic efficacy (strength), until it reaches a threshold θ. After that, an output spike is emitted and the membrane potential is reset to an initial state. The membrane potential can have certain leakage between spikes, which is defined by a temporal parameter τ.

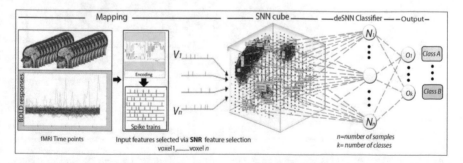

**Fig. 10.3** A schematic representation of the NeuCube-based methodology for fMRI data mapping, learning, visualisation, and classification (after [40])

The connectivity of the SNNc is initialized using the "small-world" connectivity rule (see also [55, 60, 61]). The small world connectivity rule is phenomenon observed in biological systems [62, 63]. Unsupervised learning is performed using Spike-Timing-Dependent Plasticity (STDP) learning rule [50] as one implementation. In this study, the unsupervised learning allows for the SNNc to evolve its connections so that they capture spatio-temporal associations between voxels representing consecutive spatio-temporal brain activities. For every input spatio-temporal fMRI sample, a trajectory of connections are formed in the SNNc. The length (the depth) of these trajectories depends on the spiking sequence representing the sample and the time of presentation.

Example is given below where the details of the case study data are presented in the following subsection.

Figure 10.4a presents three snapshots of deep learning of eight-second fMRI data in a NeuCube model when a subject is reading a negative sentence (time in seconds). Figure 10.4b captures the internal structural pattern, represented as spatio-temporal connectivity in the SNN model trained with eight-second fMRI data streams. The corresponding functional pattern is illustrated in Fig. 10.4c as a sequence of spiking activity of clusters of neurons in a trained NeuCube model representing deep knowledge. The internal functional dimensionally of the SNN model shows that while the subject was reading a negative sentence, the activated cognitive functions were initiated from the Spatial Visual Processing function. Then it was followed by the Executive functions, including decision making and working memory. From there, the Logical and Emotional Attention functions were involved. Finally, the Emotional Memory formation and Perception functions were evoked.

The trajectory of activities of spatially located brain areas over series of time intervals visualised in Fig. 10.4c can be represented as deep knowledge as a sequence of events Ei, each of them defined by a function Fi, location Si and time of execution Ti according to the definition of deep knowledge in Chap. 1 (see Box 10.1).

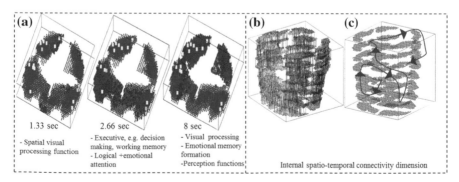

**Fig. 10.4  a** Three snapshots of *deep learning* of 8 sec of fMRI data in a NeuCube model when a subject is reading a negative sentence (time in seconds); **b** Internal structural pattern represented as spatio-temporal connectivity in the SNN model trained with 8 sec fMRI data stream; **c** A functional *deep knowledge* pattern represented as a sequence of spiking activity of clusters of spiking neurons in a trained model

---

**Box 10.1**  *Deep knowledge representation extracted from a trained NueCube model on fMRI STBD from a person reading a negative sentence*

---

IF (a person is reading a negative sentence)

THEN (the following events are triggered in space and time in a trained SNN model)

  E1: Vision, in the Spatial Visual Processing area, at time T1,

 AND E2: Decision making function, in the Decision making and working memory, at time T1,

 AND E3: Logical and Emotional Attention function, in the Attentional brain area, at time T3

 AND E4: Emotional functions, in the Emotional brain area, at time T4

 AND E5: Emotional memory formation function, in the Memory brain area, at time T5

 AND E6: Perception function, Perception brain area, at time T6.

---

Note: The times Ti and locations Si of the vents Ei can take either exact or fuzzy values (e.g. around, more or less, etc.)

---

An output classification module for supervised learning of spatio-temporal spike sequences, activated in the SNNc by the input data, is implemented using the deSNN classification algorithm [51]. During the supervised learning, output neurons are evolved and trained to recognize whole patterns of activities of the SNNc. A whole pattern of SNNc activity is defined as the spatio-temporal spiking activity of the SNNc during the time of the presentation of a whole input data sample labelled by a class label. The duration of the fMRI samples used can vary in time and number of voxels used. The use of eSNN allows for a further adaptation of the NeuCube model on new data in an incremental way without re-training the model on old data. The model can be further evolved, with new samples used for training and new classes introduced in an incremental way.

The output classification accuracy depends on the combination of NeuCube model parameter values. This combination can be optimized using different algorithms, such as: grid search (exhaustive search), genetic algorithm, and quantum inspired evolutionary algorithm [64] (Chap. 7). A number of default parameters are listed in Sect. 10.3.

The trained NeuCube model of fMRI data can be dynamically visualized in a 3D virtual reality space for the analysis of brain activities and for the discovery of new spatio-temporal causal relationships from the data [58].

The proposed here NeuCube-based methodology for mapping, learning, classification and knowledge representation of fMRI data is illustrated in the next sections on benchmark fMRI data sets.

## 10.3    Mapping, Learning and Classification of fMRI Data in NeuCube on the Case Study of STAR/PLUS Data

### 10.3.1    The STAR/PLUS Benchmark fMRI Data

The STAR/PLUS fMRI data set, originally collected by Marcel Just and his colleagues at Carnegie Mellon University's Center for Cognitive Brain Imaging (CCBI) [65, 66], was selected for the illustration of the proposed methodology. STAR/PLUS fMRI data sets consists of sequences of images from the whole brain volume captured every 500 ms during a cognitive task. For each subject conducting a picture versus sentence task, data from 40 trials has been collected, each trial starting by presenting a stimulus (picture or sentence) that remains on the screen for 4 s (8 brain images recorded). Then, a blank screen appears for another 4 s. After that, the next stimulus is presented for the next 4 s, etc. The fMRI data is spatially partitioned into 27 distinct regions of interest (ROIs), each corresponding to different number of voxels. From the STAR/PLUS fMRI data, two different subsets were extracted and used for two case studies illustrating our methodology. The first data set relates to modelling fMRI STBD when subjects are reading affirmative versus negative sentences. The second dataset relates to modelling fMRI STBD when a subject is seeing a picture versus reading a sentence. In order to analyze and classify voxel activity patterns generated by different stimuli type (picture/ sentence), the fMRI data is divided into two classes (1: a subject is seeing a picture; 2: a subject is reading a sentence). We will demonstrate in the next sections that using the proposed methodology we can not only classify these activities, but obtain a better understanding of their spatio-temporal manifestation in the brain.

To analyze the voxel activity patterns of the activated ROIs, either all voxels can be used and mapped in a SNNc model or a suitable subset of voxels can be selected. Different methods for feature selection can be used for the purpose. In our experiments we have used a standard statistical measure known as Signal-to-Noise Ratio (SNR) [52] via available online NeuCom platform [67].

For a two-class problem, a SNR index for a variable x is calculated as an absolute value of the difference between the mean value $M1x$ of the variable for class 1 and the mean $M2x$ of the variable for class 2, divided to the sum of the respective standard deviations. Figure 10.5 illustrates a set of selected voxels from the fMRI data for each of the two case studies, while Table 10.1 shows how many of these voxels belong to which ROI. We conclude from Table 10.1 (left column) that when a subject is making a decision about sentence polarity, more activated voxels are located on the Left Dorsolateral Prefrontal Cortex (LDLPFC), Left Temporal (LT), LOPER, and the Inferior Parietal Lobule (LIPL). Table 10.1 (right column) contains the selected voxels while the subject deals with picture/sentence stimuli. Calcarine (CALC) is more activated than the other parts of the brain.

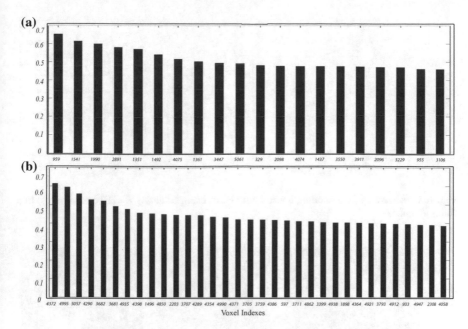

**Fig. 10.5** The SNR index (on the y-axis) of top voxels (on the x-axis) extracted from: **a** the affirmative versus negative sentence fMRI data set; **b** the picture vs sentence fMRI data set

**Table 10.1** Subset of voxels are selected via SNR (Signal-to-noise ratio) feature selection method from two fMRI data sets

| Activated brain regions in the affirmative versus Negative sentence task and the number of voxels selected in Fig. 10.5a that belong to each of these regions | Activated brain regions in the picture versus Sentence task and the number of voxels selected in Fig. 10.5b that belong to each of these regions |
|---|---|
| 'LT' (3), 'LOPER' (3), 'LIPL' (1), 'LDLPFC' (6), 'RT' (2), 'CALC' (1), 'LSGA' (1), 'RDLPFC' (1), 'RSGA' (1), 'RIT' (1) | 'CALC' (5), 'ROPER' (3), 'LT' (4), 'LOPER' (3), 'LSPL' (1), 'RIPS' (3), 'LPPREC' (1), 'RT' (4), 'LFEF' (1), 'LDLPFC' (3), 'RDLPFC' (1) 'LIPS' (2), 'RPPREC' (1), 'LIT' (1) |

## 10.3.2 fMRI Data Encoding, Mapping and Learning in a NeuCube SNN Model

### Model parameter setting

Threshold-Based Representation (TBR) method was applied on each voxel time series data to transfer the data into a sequence of spikes (see Chap. 4). If a voxel BOLD intensity value exceeds the TBR threshold, a spike occurs [58]. Figure 10.6 shows an example of 5 voxel time series.

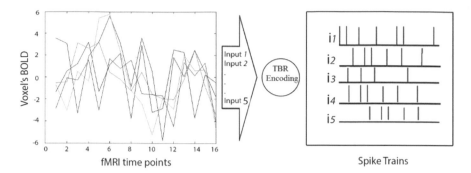

**Fig. 10.6** An example of encoding 5 voxel time series captured during 8 s (16 brain images) into trains of spikes

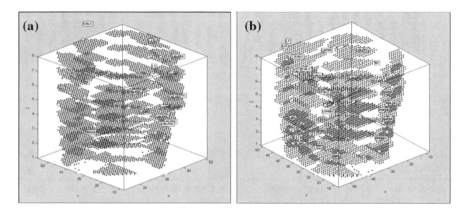

**Fig. 10.7** Direct mapping of fMRI voxels into a SNNc. The dimensions of the SNNc are defined by the maximum values of x, y, and z voxel coordinates, which in this case study equal to 51 × 56 × 8. In this dimensional space, 5062 voxels are mapped from the STAR/PLUS geometric voxel coordinates of a single person. The selected voxels in Fig. 10.5 for each case study problem are shown as input variables as circles, along with the ROI (as text in boxes) for: **a** affirmative versus negative sentence fMRI data; **b** picture versus sentence fMRI data

Here we have illustrated two types of voxel coordinate mappings in a 3D SNNc structure: (1) Direct mapping of individual fMRI voxel coordinates (Fig. 10.7.); (2) Mapping fMRI voxel coordinates first into a standard brain template, such as Talairach [56], and then mapping the Talairach coordinates into a 3D SNNc. This method is illustrated in Figs. 10.8 and 10.9 and explained below.

We have used the fMRI data of subject "*05680*" from the STAR/PLUS fMRI data. The fMRI data dimensions are defined by the maximum value of x, y, and z voxel coordinates, which equal in our case study data to 51 × 56 × 8 as can be seen in Fig. 10.7. Using these dimensions, 5062 voxel coordinates are recorded from the entire brain volume. We mapped all voxel coordinates into a SNNc so that the spiking neurons have the same 3D coordinates as the corresponding voxels.

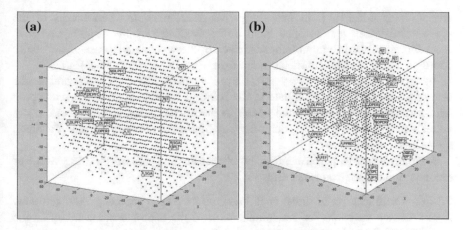

**Fig. 10.8** Mapping fMRI voxels into SNNc using the Talairach brain template. The 5062 voxel data of one subject were first mapped into 1471 Talairach template coordinates according to [55, 56, 69]. Then each template coordinate is mapped into a corresponding neuron from a SNNc. The selected top informative voxels in Fig. 10.5 for each case study problem are used as input variables and shown as circles along with the ROI (as text in boxes) for: **a** affirmative versus negative sentence fMRI data; **b** picture versus sentence fMRI data

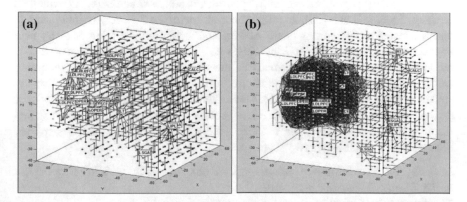

**Fig. 10.9** Voxels are mapped into SNNc using Talairach template: **a** Initial connections in a SNNc; **b** learned connections after STDP unsupervised learning using both affirmative and negative sentence fMRI samples when 20 input voxels selected as in Fig. 10.5. The dense areas of connectivity evolved in the SNNc can be analysed to understand the most active functional areas in the brain during these two tasks and how they interact dynamically

Figure 10.7a illustrates the spatial mapping of all fMRI voxels into a SNNc. 20 of these neurons are allocated and labelled to represent input features as per the selection in Fig. 10.5a. Figure 10.7b represents the same brain structure with different pre-selected voxels for the picture vs sentence data set for the same subject.

When we create a model of fMRI data collected from many subjects, we need to use a unifying structural brain template, such as the Talairach atlas [56], the MNI atlas [57] or other [68]. In this study, we transformed the coordinates of the pre-selected voxels and mapped them into a NeuCube of 1471 spiking neurons according the Talairach brain template. Each of these neurons represents the center coordinate of one cubic centimeter area from the 3D Talairach atlas [69].

In this experiment for every voxel from an fMRI data set, we calculate the nearest Talairach-based coordinate in the relevant Brodmann area. After mapping the coordinates of the preselected voxels to the Talairach-based coordinates, every voxel is mapped into a spiking neuron in the SNNc according to its new, Talairach transformed coordinates.

A NeuCube model performance is highly sensitive to parameter setting. Some of the most important parameters are:

$TBR_{thr}$: A self-adaptive bi-directional threshold for STBD encoding to spike trains.
$D_{thr}$: Distance threshold for the initialization of the neuronal connectivity in the used here small world connectivity rule.
$STDP$ learning rate ($\alpha$): A parameter used to modify the neuronal connections in a SNNc with respect to repetitively arrived spikes to the synapses. If a neuron $i$ fires before a neuron $j$ then its connection weight increases, otherwise it decreases with respect to the STDP learning rate ($\alpha$).
$(Th_o)$ : Threshold of firing for the neurons in the SNNcube.

$deSNN$ classifier parameters: These parameters are: mod and drift. As explained in Chap. 5 and [51] and [55], an output neuron is evolved for every training sample and connected to all neurons of the SNNc. The weight initialization of every new connection is based on the RO learning rule [70]. The weight is calculated as a modulation factor (the variable mod) to the power of the order of the incoming spikes. The initial connection weights are further modified to reflect the following spikes, using a drift parameter [51]. Once the structure of the NeuCube-model is defined, along with the method for data encoding and the method for voxel spatial mapping into a 3D SNNc, the model is trained and analyzed. These steps are illustrated in the following two sections.

*Learning and visualization of spatio-temporal connections in the SNNc with the use of the Talairach template mapping.*
*The case study of affirmative versus negative sentence*

Figure 10.9a shows the initial connections in the SNNc and the modified ones after the deep, unsupervised learning process using both affirmative and negative sentence fMRI samples. Our findings confirm studies that suggest that language comprehension, including a reading task, is processed in particular brain areas, such as Left Dorsolateral Prefrontal Cortex, Broca, and Wernicke [71]. Figure 10.9b shows more and stronger neuronal connections generated in the left hemisphere. These connections were established as a result of more spikes transferred between the neurons located in these areas, reflecting on the changes in the corresponding voxels in the fMRI data. Figure 10.10 shows connectivity after the SNNc was trained with only the affirmative or the negative sentence data, separately. The

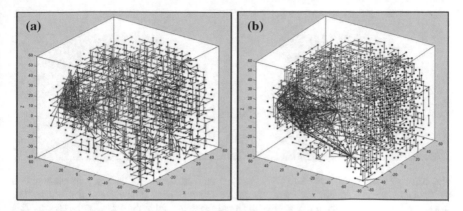

**Fig. 10.10** Voxels are mapped into SNNc using Talairach template: **a** Learned connections in a SNNc when only fMRI samples of affirmative sentences were used; **b** Learned connections in a SNNc when only fMRI samples of negative sentences were used. The initialisation is the same as in Fig. 10.9. The dense areas of connectivity evolved in the SNNc can be analysed to understand the difference between functional areas in the brain during each of the two tasks as dynamic interaction

observed connectivity from Fig. 10.10 confirms that the subject performs differently when reading an affirmative versus negative sentence and also suggests what the difference is in terms of brain spatio-temporal activity. In addition, we can observe that more and stronger connections are formed between neurons located in the left hemisphere (LDLPFC and LT) than in the right hemisphere (RDLPFC and RT) while the subject was reading a negative sentence. The connectivity is especially enhanced between the input neurons (i.e. the selected voxels) located in the LDLPFC and LOPER regions. Our interpretation of Fig. 10.10 is in line with the neuroscience literature, which reported that comprehension of negative sentences is cognitively different from affirmative sentences, involving different parts of the brain. Containing negative words, such as "not," in the middle of a sentence can make it more difficult to comprehend, due to their more complex syntactic and semantic structures. Therefore, this type of sentence engages more regions of the brain [72]. More detailed analysis on the connectivity related to the task can be performed by neuroscientists to answer different research questions. Another form of analysis of a trained SNNc is clustering of the neurons, that can performed with the use of the input variables (corresponding neurons) used as cluster centers. Each neuron in a trained SNNc belongs to the cluster from which center it has received most spikes as shown in Fig. 10.11. A spreading algorithm [73] was used to define these clusters. If there are more transmitted spikes between two neurons, there will be a stronger information route between them. Figure 10.11 shows the SNNc clusters after unsupervised training of a SNNc with the two fMRI time series separately. Figure 10.11a illustrates that there are not many functional pathways between LT region and the other parts of the brain while the subject is reading an

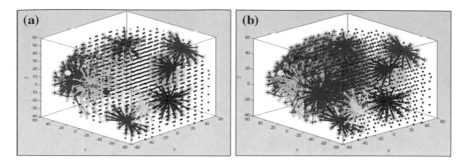

**Fig. 10.11** Voxels are mapped into SNNc using Talairach template: Clustering of neurons in a SNNc after unsupervised training for: **a** affirmative sentence data; **b** negative sentence data. The size of a cluster indicates the importance of the input feature/voxel at the centre of the cluster. This can be used for feature/voxel selection and marker identification for further studies

affirmative sentence. However, Fig. 10.11b shows that when a subject is reading a negative sentence, there is more interaction between neurons located in the left hemisphere. Therefore, more brain functional paths start from the input voxels located in the LT region (spike sender neuron) and continue up to the neurons located in the middle of the brain (spike receiver neurons).

*Learning and visualization of spatio-temporal connections in a trained SNNc using direct voxel mapping*

In order to visualise the neural connectivity and spiking activity inside a SNNc with 5062 spiking neurons for example (equal to the number of voxels in the STAR/PLUS fMRI data of an individual), we have loaded the whole fMRI voxel coordinates into the SNNc. Figure 10.12 shows the neuronal connections after unsupervised training of a SNNc with the use of two different data sets, related correspondingly to: affirmative sentence; negative sentence.

It is seen from this visualisation, that the exact locations of the fMRI voxels are mapped in the same 3D location of spatial located neurons. These neurons develop

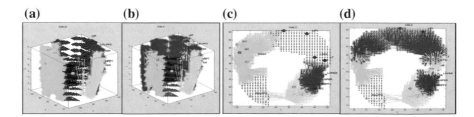

**Fig. 10.12** Voxels are directly mapped into a SNNc model: Clustering of the neurons in a trained SNNc with: **a** affirmative sentence data; **b** negative sentence data, along with their corresponding 2D projections shown in (**c**) and (**d**) correspondingly. The size of clusters indicates the importance of the feature voxel for the task and can be used for feature/voxel selection for further studies

their connections based on the temporal information in the fMRI data during the STDP learning. As seen, the neuronal connections in the SNNc here evolved differently during the unsupervised training of the SNNc with different fMRI data reflecting the different evoked cognitive functions in the brain.

### 10.3.3    Classification of the fMRI Data in a NeuCube-Based Model

While the SNNc learns fMRI data and creates spatiotemporal patterns of connectivity and spiking activity among spiking neurons as visualised in Fig. 10.13, the output classifier is to classify these patterns into pre-defined class labels [51, 55]. After completion of the unsupervised learning in the SNNc, input data is propagated again through the now trained SNNc in order to activate the learned patterns in the SNNc, so that a classifier can be trained to classify them. For every training sample, a new output neuron is evolved and connected to all neurons in the SNNc. Here we have used the deSNN classifier [51] (see Chap. 5). It is constructed and trained to learn and classify different trajectories of the SNNc spiking activities that represent different input patterns from the fMRI data that belong to different classes. As a result of the supervised learning in the classifier, once a new fMRI data sample of unknown class is entered, the classifier will classify this data into a known class, or will create a new class.

The deSNN classifier belongs to the class of evolving systems [52], so that it can incrementally add new samples and new classes with no need to retrain it with the old data and without manifesting the catastrophic forgetting phenomenon. The deSNN utilises a combination of rank-order learning [70] for the establishment of

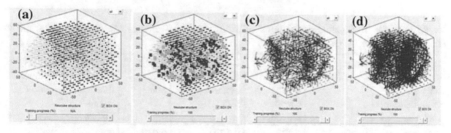

**Fig. 10.13** Visualization of fMRI data model and connectivity between neurons of eSNN: **a** no spiking activity yet, inactive neurons are in blue, fMRI data neurons are in yellow; **b** spiking activity: active neurons are represented in red, inactive neurons are represented in blue, positive input neurons are represented in magenta, negative input neurons are represented in cyan and zero input are represented in yellow; **c** neurons connectivity before training (SWC), positive connections are in blue and negative connections in red; **d** neurons connectivity after training

the initial weights of the synapses based on the order of the first arriving spike, and STDP-type learning for the tuning of these weights based on the following spikes arriving at the synapse.

*A NeuCube model parameter optimization and classification results for the benchmark data sets*

In a NeuCube fMRI model, the output classification accuracy depends on the parameter setting. In the experiments here, a grid search method was used, where for different combinations of parameter values (in our experiment 10,000), a model is created and its classification accuracy evaluated. Optimal parameter values of a model that are resulting in best classification accuracy are reported in Table 10.2.

Table 10.3 summarizes the fMRI data classification accuracy of the affirmative sentence class versus negative sentence class obtained using the NeuCube-based classification model. The results are compared with results obtained using traditional machine learning methods, as these methods are still being actively used in the literature for the purpose of classification of STBD. The methods used for comparison are: Support Vector Machine (SVM); Multiple Linear Regression (MLR); Multi-Layer Perceptron (MLP); Evolving Classification Function (ECF); Evolving Clustering Method (see www.theneucom.com). The already published classification result of the affirmative versus negative sentence fMRI data [74] is also reported. The NeuCube-based models achieved significantly better classification accuracy Table 10.3. In addition to a better classification results, visualization of the trained SNNc reveals new information about functional brain pathways.

In both experiments, the fMRI data was learned in the NeuCube models as whole spatio-temporal patterns. In contrast, the same fMRI data was learned in the other methods a vector-based, where vectors were formed through concatenating of temporal frames. No dynamic spatio-temporal fMRI patterns can be revealed while using these methods.

**Table 10.2** Optimal parameter settings of NeuCube-based models for different experiments (sessions) with the benchmark fMRI data

| Optimised parameters for the classification task | Experiment (Session) | TBR threshold for the encoding procedure | Connection distance (small-world radius) | STDP learning rate | SNN firing threshold in the SNNc | deSNN parameter —*mod* |
|---|---|---|---|---|---|---|
| Affirmative versus negative sentence data set | Session I | 3.327 | 0.128 | 0.010 | 0.5 | 0.4 |
| | Session II | 2.852 | 0.125 | 0.013 | 0.5 | 0.4 |
| | Session III | 2.0929 | 0.108 | 0.014 | 0.5 | 0.4 |

**Table 10.3** Classification accuracy of the affirmative sentence (class C1) versus negative sentence (class C2) data via a NeuCube model (50% of the data used for training and 50% used for testing as cross validation) and also traditional machine learning methods (obtained via NeuCom, www.theneucom.com), along with already published results [74]. The best classification accuracy among the tested methods is indicated in bold

| Method | Sessions and selected voxels for classification | C1 (affirm) (%) | C2 (negat) (%) | Total |
|---|---|---|---|---|
| NcuCube | *Session I*: 20 voxels selected from Table 10.1 (left column) | **80** | **100** | **90** |
| | *Session II*: 20 pre-selected voxels from RDLPFC region | **90** | **80** | **85** |
| | *Session III*: 20 pre-selected voxels from LDLPFC region | **90** | **80** | **85** |
| SVM results obtained in [9, 14] | *Session I*: classification based on the LDLPFC's voxels | 64 | 68 | 66 |
| | *Session II*: *classification based on the* RDLPFC's voxels | 65 | 69 | 67 |
| SVM | Parameter setting for traditional machine learning methods | 70 | 75 | 73 |
| | SVM Kernal: Polynomial, degree, gamma, N/A: 1 | | | |
| MLP | Number of hidden units = 180, Number of training cycles = 600, Output activation function—linear | 75 | 65 | 70 |
| ECF | Maximum field radius = 1, Minimum field radius = 0.01; M of N = 3; Number of membership functions = 2; Number of epochs = 4 | 55 | 70 | 63 |
| ECMC | Maximum field radius = 2; Minimum field radius = 0.01, M of N = 3 | 65 | 70 | 70 |
| MLR | Class performance variance = 0.26 | 65 | 60 | 63 |

The fMRI data sample file contains 40 samples (20 samples per class)

## 10.4 Algorithms for Modelling fMRI Data that Measure Cognitive Processes

This section presents a new method and algorithms for modelling cognitive tasks by deep learning of fMRI spatio-temporal brain data (STBD) in a SNN architecture. The method uses the same SNN NeuCube architecture as in Sects. 10.2 and 10.3 but different algorithms are introduced here for encoding the data, for learning and visualization of the models. Full description of the method can be found in [75].

## 10.4.1  Algorithm for Encoding Dynamic STBD into Spike Sequences

A continuous input brain data signal is encoded into a spike sequence so that the dynamics of the data is preserved. For a given STBD sequence $S(t)(t \in \{t_0, t_1, \ldots, t_L\})$, we first define the time $t_m$ when the signal reaches its minimum value:

$$t_m = arg \min_t S(t), \quad t \in \{t_0, t_1, \ldots, t_L\}. \tag{10.1}$$

The time period from $t_m$ to $t_L$ (the end time of the signal) is considered further and no spikes will be generated before time $t_m$. Based on the initial decrease in the signal, $t_m$ is set as the starting time point to capture the changes in the signal during a cognitive task. Let $B(t)$ denote the baseline for $S(t)$ at time $t(t \in [t_m, t_L])$ and $B(t_m) = S(t_m)$. If at a time moment $t_{i+1}(m \leq i < L)$, the signal value $S(t_{i+1})$ is above the previous baseline $B(t_i)$, we encode a spike at time $t_{i+1}$ and the baseline is updated as:

$$B(t_{i+1}) = \alpha S(t_{i+1}) + (1 - \alpha)B(t_i), \tag{10.2}$$

where $\alpha(\alpha \in [0, 1])$ is a parameter to control the signal's contribution to the increase of the baseline. Otherwise, if $S(t_{i+1})$ is below $B(t_i)$, then no spike is encoded at this time and the baseline is reset as $B(t_{i+1}) = S(t_{i+1})$. Successive spikes in the resulting spike sequence reflect the increase of the signal, whilst the absence of a spike means a decrease of the signal (Fig. 10.14a).

The proposed method accurately encodes the activation information of continuous temporal data into spike trains. This is important for the following interpretation of the trained SNNcube model, because it enables researchers to better understand brain processes that generate the data. This encoding is also robust to noise. Due to a minimum value threshold which is applied to changes in the signal value, small noise perturbations of the signal are not transformed into spikes. This transformation also accounts for the frequency of changes in the raw signal.

The timing of spikes corresponds with the time of change in the input data. The spike sequence is obtained after the encoding process which represents new input information to the SNN model, where the time unit maybe different from the real time of the data acquisition (machine computation time versus data acquisition time).

## 10.4.2  Connectivity Initialization and Deep Learning in a SNN Cube

After the STBD is encoded into spike trains, the next step is to train a SNNcube, where the spike sequences represent the input data. Input variables are mapped to

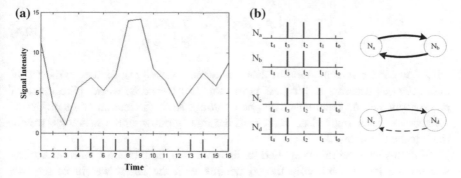

**Fig. 10.14 a** Spike sequence encoding for one signal. An example of one signal continuous values at 16 time points along with the encoded sequence of spikes (below); The successive spikes from time 4–9 represent the increase in the signal, while the absence of spikes from time 10–12 means a decrease in the signal; **b** connections established between two connected neurons after unsupervised learning in a SNNcube. Two examples of connection weights established through the proposed method for unsupervised learning between two connected neurons depending on the time of the pre- and post-synaptic spikes of the two neurons. The solid line is the final connection (a thicker line means a larger weight), while the dotted line is removed after learning because of its weaker connection weight. For example, spikes in neuron $N_c$ mostly precede those in neuron $N_d$, so the learned connection weight $w_{N_d N_c}$ is smaller, which will be removed after the unsupervised learning

corresponding spiking neurons in the 3D SNNcube with the same $(x, y, z)$ coordinates. The spike trains are then entered into the SNNcube as whole spatio-temporal patterns (samples) of many time units. A sample representing a labelled sequence of cognitive activity over a certain time period.

Before a learning rule is applied, the connections between spiking neurons in the SNNcube are initialized as follows:

Let $N_i$ denote the neighborhood of neuron $i$, defined as:

$$N_i = \{j : D_{ij} \leq T, i \neq j\}, \tag{10.3}$$

where $D_{ij}$ denotes the distance between neuron $i$ and neuron $j$, and T represents the maximum distance allowed for connections between two neurons (T is a parameter that is subject to optimization along with other model's parameters). For two neighboring neurons $i$ and $j$, bidirectional connections are created and connection weights are initialized to zero.

After initializing the connections, the input spike sequences are propagated through the SNNcube and the following learning rule is applied as introduced here: If neuron $i$ and $j$ are connected, and one spike from $i$ precedes that from $j$ within a certain time period, $w_{ij}$ will be increased and $w_{ji}$ left unchanged:

$$\Delta w_{ij} = \begin{cases} A_+ \exp\left(\frac{\Delta t}{\tau_+}\right) & if \ \Delta t \leq 0, \\ 0 & if \ \Delta t > 0, \end{cases} \tag{10.4}$$

where $\Delta w_{ij}$ is the synaptic modification (increment of weight); Similar to the STDP parameters as describe in [76], $\Delta t$ is the time difference between spike times of pre-synaptic neuron $i$ and post-synaptic neuron $j$. $A_+$ is the maximum quantities of synaptic modification; and $\tau_+$ represents the time window within which the weight modification is allowed.

After this learning rule is applied to the input data, both bidirectional connection weights are learned, but only the connection with the larger weight of the two bidirectional connections is retained as a final connection between the two neighboring neurons (Fig. 10.14b). This learning rule is spike time dependent, but different from the STDP rule [50, 76, 77] used in the NeuCube models developed so far for fMRI data [40, 55, 78, 79].

The weaker connection, of the two neuronal connections between neurons $i$ and $j$, is removed and the remaining connection represents a stronger, possible temporal relationship between the two neurons. The removed connection weights are all reset to zero to maintain symmetry of the equation and enable further adaptive training from new data. The trained SNNcube forms a deep architecture as whole spiking input sequences which are learned as chains of connections and spiking activities, regardless of the number of data points measured for every input variable. Unlike hand-crafted layers used in second-generation neural network models [80–84], or randomly connected neurons in the computing reservoir of a liquid state machines [42], the chains of directional connections established in the SNNcube represent long spatio-temporal relationships between the sources of the spike sequences (the input variables). Due to the scalable size of a SNNcube, the chains of connected neurons are not restricted in length during learning, which can be considered as unrestricted deep learning, in contrast to existing deep learning methods that use fixed number of layers. As we can see in the following sections, this learning also results in automatic feature extraction, i.e. the automatic selection of a smaller subset of marker input variables.

To analyze the spiking activity of a neuron $i$ in the SNNcube, we define an indicator called activation degree $D_i$:

$$D_i = \frac{\sum_j \left(w_{ij} + w_{ji}\right)}{number \ of \ neurons \ in \ N_i} \quad j \in N_i. \tag{10.5}$$

The parameter $D_i$ represents the averaged activation degree of neuron $i$ after a summation of all its inward and outward connection weights. A higher degree of activation of a spiking neuron, represents a greater likelihood that the corresponding loci in the brain are activation foci.

After training the SNNcube, neurons sharing similar spike patterns will have larger weighted connections. This allows us to analyze and understand for example a single subject's response to different stimuli and to compare the responses of different subjects to the same stimulus. A set of spiking neurons with the highest degree of activation representing a given class of stimuli or a cognitive state, will represent a *feature set* of *markers* for this class; thus the automatic selection of features as part of the internal deep learning process.

In the following section we illustrate the above model on two case study fMRI data related to cognitive tasks. The SNNcube's parameters used for the two case study experiments are set as: $\alpha = 0.5$; $\Lambda_+ = 0.1$; $\tau_+ = 1$.

### 10.4.3   Deep Knowledge Representation in a Trained SNN Model

Once the SNNcube is trained with spike sequences of encoded STBD, we can interpret both the connectivity and spiking activity of the model, aiming at new knowledge representation about brain functional connectivity and cognitive processes.

A deep chain of connections is learned for each input pattern (sample) in the SNNcube. When entering new input data, the fired chain of neurons and connections will indicate as to which of the previous learned patterns the new one belongs to. This can be used to classify STBD (as shown in the experimental results later in the paper) and for a better understanding of the spatio-temporal brain dynamics.

### 10.4.4   A Case Study Implementation on the STAR/PLUS Data

We randomly selected two subjects' data from the StarPlus fMRI data related to two cognitive tasks [85]. Our experiments were performed on two subject's data (ID = 05680 and ID = 04820). FMRI data comprised 25 brain regions of interest (ROIs) represented by 5062 and 5015 voxels respectively. For convenience, we will use the terms ID = 05 and ID = 04 to refer to the above subjects' fMRI data respectively.

The fMRI data was captured every 0.5 s (two fMRI volume snapshots per second) while the subjects performed reading a sentence or watching a picture perception tasks during 40 trials. We consider here the first 8-s of recorded data for each trial, during which a 4-s stimulus (picture or sentence) was presented, followed by a 4-s rest. The first 16 volumes of the fMRI data extracted from each trial fell into two classes: watching a picture (Class Pic) or watching a sentence (Class Sen).

As the brain volume has a one-to-one mapping with the SNNcube model, the value of a brain voxel in a brain activation map is defined as the corresponding neuron's activation degree in the SNNcube.

The results from applying the proposed method on fMRI data of subject ID 05 are illustrated in Fig. 10.15.

Brain activation maps for Class Pic and Class Sen were obtained after learning had taken place in the SNNcube (Fig. 10.15Aa). The neuron's activation degree of the SNNcube was averaged over 20 trials for each class. The voxels in red suggest they were more likely to be activation foci in a certain cognitive state, whilst the blue voxels were less likely to be active. The activation maps were normalized respectively within each class. These maps can be further interpreted, for example, it can be seen that when the subject was watching a sentence, the BOLD response in the Calcarine (CALC) region was much stronger than in other regions.

Neurological studies [71, 72], suggest that reading a sentence is more difficult to comprehend than seeing a picture. Therefore, it strongly engages specific regions of the brain along with the visual cortex. The CALC sulcus begins near the occipital lobe, where the primary visual cortex ($V1$) is concentrated, and passes through the splenium of the corpus callosum, where it is joined at the parieto-occipital sulcus. Our findings confirm that language comprehension, including a reading task, requires more concentration which involves more regions of the brain to act and consequently increases the amount of oxygenated blood required by neurons.

To detect voxel activation, a threshold $T_D$ for the neuron's degree of activation and a threshold $T_w$ for the neighboring neurons' connection weights were defined. The detection procedure is based on the following steps:

Step 1.  Find the activation foci in the SNNcube where activation degrees are above $T_D$.

Step 2.  Set the activation foci as an initial centers of the activation regions R.

Step 3.  Expand the activation regions R in the SNNcube, i.e. add a neuron outside R if it satisfies the condition that its connection weight with a certain neuron in R is higher than $T_w$.

Step 4.  Repeat Step 3 until no neurons outside R can be included in R. The neurons in R imply that corresponding voxels in the brain volume are the detected activation voxels.

Figure 10.15 shows that there are more activated neurons in the CALC region during Class Sen than Class Pic. When the subject was watching a picture, the right hemisphere was slightly more active than the left, but when the subject was reading a sentence, more ROIs in the left hemisphere were involved, including the Left Inferior Parietal lobe (LIPL), Left Superior Parietal Lobe (LSPL), and Left Temporal lobe (LT). Increased activation of the left cerebral hemisphere is proving

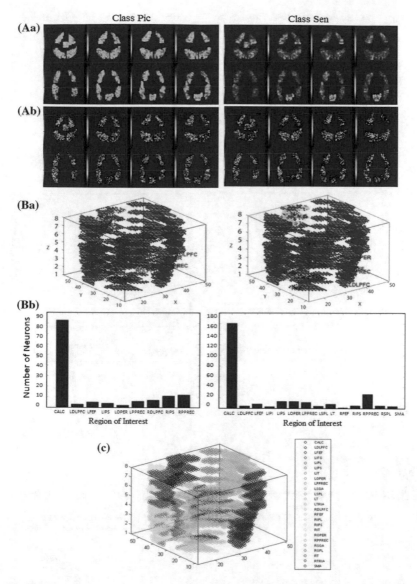

**Fig. 10.15** Brain activation detection and brain regions mapping in the SNNcube for subject ID 05. **Aa** 2D SNNcube activation maps for each class: watching a picture (Class Pic) or reading a sentence (Class Sen); **Ab** Probability map estimated by t-test for Class Pic (left) and Class Sen (right); **Ba** Locations of activation neurons in the averaged SNNcube; **Bb** Histogram of activated neurons with respect to different regions of interest (ROIs) for each class; **C** 25 ROIs were mapped into the SNNcube; **D** Averaged activation of the neurons in the SNNcube for individual trials for Class Pic and Class Sen. Abbreviations: CALC—calcarine; DLPFC—left dorsolateral prefrontal cortex; FEF—frontal eye fields; IFG—inferior frontal gyrus; IPL—left inferior parietal lobe; IPS —intraparietal sulcus; IT—inferior temporal gyrus; OPER—pars opercularis; PPREC—posterior precentral gyrus; SGA—supramarginal gyrus; SPL—superior parietal lobe; T—temporal lobe; TRIA—pars triangularis; SMA—supplementary motor area

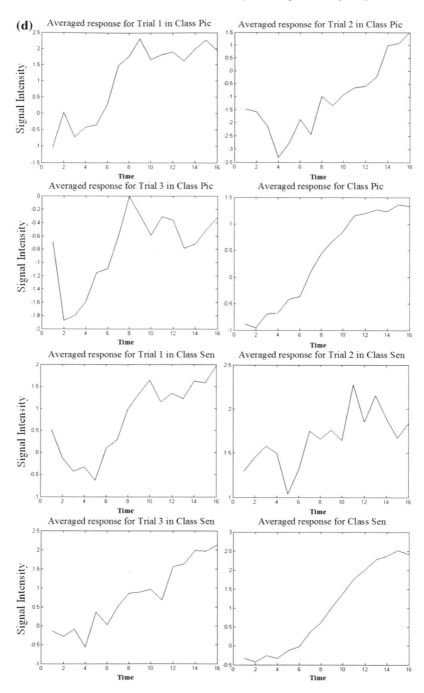

**Fig. 10.15** (continued)

to be a more important role for these areas during reading a sentence than during visual object processing. These activations evolved by transferring more spikes between the neurons located in these areas of the SNNcube, reflect more changes in the corresponding voxels' BOLD in the fMRI data.

Since we map voxels to spiking neurons, we can investigate how many activated voxels were involved in multiple brain activities. The percentage P of overlapped activation voxels is defined as follows:

$$P = \frac{R_{Pic} \cap R_{Sen}}{R_{Pic} \cup R_{Sen}}.$$   (10.6)

where $R_c$ denotes the activation voxels in Class c ($c \in \{Pic, Sen\}$). We obtained $P = 29.0\%$ for watching a picture and reading a sentence, indicating that a common part of the brain was engaged in both cognitive states.

Analysis of the spiking activity in the SNNcube confirms that BOLD responses differ across trials even of the same class, but the averaged BOLD response for each class corresponds to the hemodynamic response function (Fig. 10.15D). In this figure, the response of the activated voxels (shown in the histogram of activated neurons in Fig. 10.15B) is averaged over 16 fMRI time points and presented for 3 trails per class. We also presented the averaged response of all the trials per class.

To validate the extracted activated voxels, we conduct the t-tests of difference in mean responses of the activated voxels between the rest state and each class. The $p$-value for class Pic is 3.5622e−7, and 5.3622e−22 for class Sen. Thus, at significance level 99.5% the responses of such extracted activated voxels are significantly different from the rest state. We also compare the mean responses between class Pic and class Sen averaged over the extracted voxels, and it shows that the mean of the BOLD responses in class Sen is significantly larger than that in class Pic ($p$ = 8.0237e−8 using t-test).

During the SNNcube's learning process, the evolution of the neurons' activation degrees was also captured. The set of neurons with higher activation for one stimulus than another represents a *set of features* for this stimulus. To demonstrate this concept, we selected two sets of 500 neurons.

Figure 10.16 shows the evolution of neurons' activation degrees and the deep learning architecture formed in the SNNcube. (A) Neurons' activation degrees at three snapshots when the subject is watching a picture (one trial of Class Pic) or reading a sentence (one trial of Class Sen); The neurons' degrees are normalized at each snapshot for visualization purpose; (B) Locations of neurons with the top 500 activation degrees for Class Pic (upper row) and Class Sen (lower row). These neurons are used as spatio-temporal features for the classification of the two different brain activities; (C) Visualization of typical chains of connections for each class.

Figure 10.17 shows brain activation detection visualized in the SNNcube when trained on ID 04 fMRI data. (A) 2D SNNcube activation maps for class Pic and

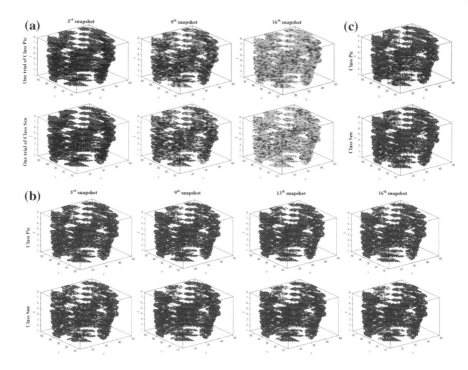

**Fig. 10.16** Evolution of neurons' activation degrees and the deep learning architecture formed in the SNNcube. **a** Neurons' activation degrees at three snapshots when the subject is watching a picture (one trial of Class Pic) or reading a sentence (one trial of Class Sen); The neurons' degrees are normalized at each snapshot for visualization purpose; **b** Locations of neurons with the top 500 activation degrees for Class Pic (upper row) and Class Sen (lower row). These neurons are used as spatio-temporal features for the classification of the two different brain activities; **c** Visualization of typical chains of connections for each class

class Sen; (B) Histogram of activated neurons with respect to different regions of interest (ROIs) for each class; (C) Locations of activation neurons in the averaged SNNcube; (D) Averaged activation of the neurons in the SNNcube for individual trials of Class Pic and Class Sen.

## 10.5  Chapter Summary and Further Readings for Deeper Knowledge

The chapter introduces methods for deep learning and classification of fMRI spatio-temporal brain data (STBD) and for modelling cognitive functions, such as:

– reading a sentence versus seeing a picture;
– reading a negative sentence versus reading a positive sentence.

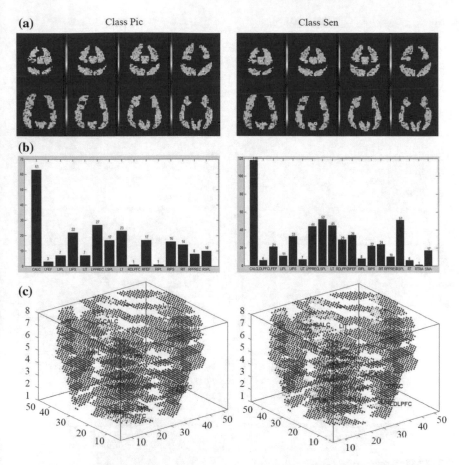

**Fig. 10.17** Brain activation detection is visualized in the SNNcube when trained on ID 04 fMRI data. **a** 2D SNNcube activation maps for class Pic and class Sen; **b** Histogram of activated neurons with respect to different regions of interest (ROIs) for each class; **c** Locations of activation neurons in the averaged SNNcube; **d** Averaged activation of the neurons in the SNNcube for individual trials of Class Pic and Class Sen

The methods presented here are further developed in Chap. 11, where the STDP unsupervised learning in the SNNcube now includes direction (orientation) data along with time-space data, illustrated there on fMRI + DTI data [86]. The new learning method is called orientation influenced STDP (oiSTDP).

Further readings will reveal more detailed information about specific topics, such as:

- fMRI data [1];
- Star/Plus data [65];
- NeuCube [55] and Chap. 6;
- Mapping and deep learning of fMRI data in NeuCube [40];

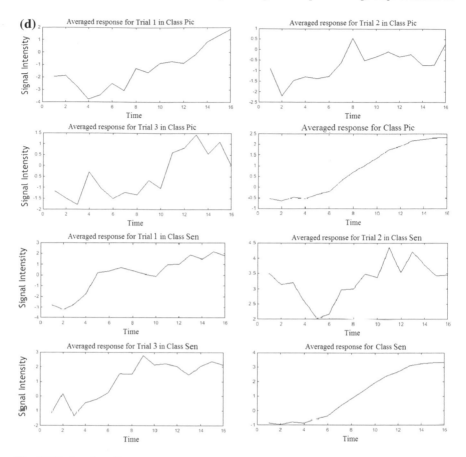

**Fig. 10.17** (continued)

- Algorithms for cognitive fMRI data in NeuCube [75];
- Understanding the brain via fMRI classification (Chap. 40 in [79]);
- Statistical methods for fMRI study (Chap. 38 in [79]).
- Demo on modelling fMRI data in NeuCube: https://kedri.aut.ac.nz/R-and-D-Systems/neucube/fmri
- Demo on fMRI data modelling with NeuCube: https://kedri.aut.ac.nz/R-and-D-Systems/fmri-data-modelling

**Acknowledgements**  Some of the material in this chapter is previously published as referenced in the corresponding sections. I would like to thank my co-authors of these publications Maryam Gholami, Lei Zhou, Norhanifah Murli, Jie Yang, Zohreh Gholami.

# References

1. R.C. DeCharms, Application of real-time fMRI. Nat. Rev. Neurosci. **9**, 720–729 (2008)
2. J.P. Mitchell, C.N. Macrae, M.R. Banaji, Encoding specific effects of social cognition on the neural correlates of subsequent memory. J. Neurosci. **24**(21), 4912–4917 (2004)
3. K.H. Brodersen, K. Wiech, E.I. Lomakina, C.S. Lin, J.M. Buhmann, U. Bingel, I. Tracey, Decoding the perception of pain from fMRI using multivariate pattern analysis. NeuroImage **63**(3), 1162–1170 (2012). https://doi.org/10.1016/j.neuroimage.2012.08.035
4. R.B. Buxton, K. Uludağ, D.J. Dubowitz, T.T. Liu, Modeling the hemodynamic response to brain activation. NeuroImage **23**(Suppl 1), S220–S233 (2004). https://doi.org/10.1016/j.neuroimage.2004.07.013
5. S. Ogawa, T.M. Lee, A.R. Kay, D.W. Tank, Brain magnetic resonance imaging with contrast dependent on blood oxygenation. Proc. Natl. Acad. Sci. **87**(24), 9868–9872 (1990)
6. J.V. Haxby, M.I. Gobbini, M.L. Furey, A. Ishai, J.L. Schouten, P. Pietrini, Distributed and overlapping representation of faces and objects in ventral temporal cortex. Science **293**(5539), 2425–2430 (2001)
7. M.A. Lindquist, The statistical analysis of fMRI data. Stat. Sci. **23**(4), 439–464 (2008). https://doi.org/10.1214/09-STS282
8. K.J. Friston, C.D. Frith, R.S. Frackowiak, R. Turner, Characterizing dynamic brain responses with fMRI: a multivariate approach. NeuroImage (1995). Retrieved from http://www.ncbi.nlm.nih.gov/pubmed/9343599
9. M.K. Carroll, G.A. Cecchi, I. Rish, R. Garg, A.R. Rao, Prediction and interpretation of distributed neural activity with sparse models. NeuroImage **44**(1), 112–122 (2009)
10. T. Schmah, R.S. Zemel, G.E. Hinton, S.L. Small, S.C. Strother, Comparing classification methods for longitudinal fMRI studies. Neural Comput. **22**(11), 2729–2762 (2010). https://doi.org/10.1162/NECO_a_00024
11. D.D. Cox, R.L. Savoy, Functional magnetic resonance imaging (fMRI) "brain reading": detecting and classifying distributed patterns of fMRI activity in human visual cortex. Neuroimage **19**(2), 261–270 (2003)
12. Y. Kamitani, F. Tong, Decoding the visual and subjective contents of the human brain. Nat. Neurosci. **8**(5), 679–685 (2005)
13. M. Misaki, Y. Kim, P.A. Bandettini, N. Kriegeskorte, Comparison of multivariate classifiers and response normalizations for pattern-information fMRI. NeuroImage **53**(1), 103–118 (2010). https://doi.org/10.1016/j.neuroimage.2010.05.051.Comparison
14. J. Mourão-Miranda, A.L.W. Bokde, C. Born, H. Hampel, M. Stetter, Classifying brain states and determining the discriminating activation patterns: support vector machine on functional MRI data. NeuroImage **28**(4), 980–995 (2005). https://doi.org/10.1016/j.neuroimage.2005.06.070
15. T.M. Mitchell, R. Hutchinson, M.A. Just, R.S. Niculescu, F. Pereira, X. Wang, in *Classifying Instantaneous Cognitive States from FMRI Data*. AMIA, Annual Symposium Proceedings/ AMIA Symposium, AMIA Symposium (2003), pp. 465–469. Retrieved from http://www.pubmedcentral.nih.gov/articlerender.fcgi?artid=1479944&tool=pmcentrez&rendertype=abstract
16. I. Rustandi, in *Classifying Multiple-Subject fMRI Data Using the Hierarchical Gaussian Naïve Bayes Classifier*. 13th Conference on Human Brain Mapping (2007a), pp. 4–5
17. I. Rustandi, in H*ierarchical Gaussian Naive Bayes Classifier for Multiple-Subject fMRI Data*. Submitted to AISTATS, (1), 2–4 (2007b)
18. S.M. Polyn, G.J. Detre, S. Takerkart, V.S. Natu, M.S. Benharrosh, B.D. Singer, J.D. Cohen, J.V. Haxby, K.A. Norman, A Matlab-based toolbox to facilitate multi-voxel pattern classification of fMRI data (2005)
19. Y. Fan, D. Shen, C. Davatzikos, in *Detecting Cognitive States from fMRI Images by Machine Learning and Multivariate Classification*. In Conference on Computer Vision and Pattern Recognition Workshop 2006 (IEEE, 2006), pp. 89–89. https://doi.org/10.1109/cvprw.2006.64

20. S.J. Hanson, T. Matsuka, J.V. Haxby, Combinatorial codes in ventral temporal lobe for object recognition: Haxby (2001) revisited: is there a "face" area? NeuroImage **23**(1), 156–166 (2004). https://doi.org/10.1016/j.neuroimage.2004.05.020

21. G. Yourganov, T. Schmah, N.W. Churchill, M.G. Berman, C.L. Grady, S.C. Strother, Pattern classification of fMRI data: applications for analysis of spatially distributed cortical networks. NeuroImage **96**, 117–132 (2014). https://doi.org/10.1016/j.neuroimage.2014.03.074

22. J.D. Haynes, G. Rees, Predicting the orientation of invisible stimuli from activity in human primary visual cortex. Nat. Neurosci. **8**(5), 686–691 (2005). https://doi.org/10.1038/nn1445

23. S. Ku, A. Gretton, J. Macke, N.K. Logothetis, Comparison of pattern recognition methods in classifying high-resolution BOLD signals obtained at high magnetic field in monkeys. Magn. Reson. Imaging **26**(7), 1007–1014 (2008). https://doi.org/10.1016/j.mri.2008.02.016

24. T. Schmah, G.E. Hinton, R.S. Zemel, S.L. Small, S. Strother, Generative versus discriminative training of RBMs for classification of fMRI images, in *Advances in Neural Information Processing Systems*, vol. 21, ed. by D. Koller, D. Schuurmans, Y. Bengio, L. Bottou (MIT Press, Cambridge, MA, 2009), pp. 1409–1416

25. S. LaConte, S. Strother, V. Cherkassky, J. Anderson, X. Hu, Support vector machines for temporal classification of block design fMRI data. NeuroImage **26**(2), 317–329 (2005). https://doi.org/10.1016/j.neuroimage.2005.01.048

26. N. Mørch, L. Hansen, S. Strother, C. Svarer, D. Rottenberg, B. Lautrup, in *Nonlinear vs. linear models in functional neuroimaging: Learning curves and generalization crossover*. Proceedings of the 15th international conference on information processing in medical imaging, volume 1230 of Lecture Notes in Computer Science (Springer, 1997) pp. 259–270

27. J. Mourão-Miranda, K.J. Friston, M. Brammer, Dynamic discrimination analysis: a spatial-temporal SVM. NeuroImage **36**(1), 88–99 (2007). https://doi.org/10.1016/j.neuroimage.2007.02.020

28. M.A.J. Van Gerven, B. Cseke, F.P. de Lange, T. Heskes, Efficient Bayesian multivariate fMRI analysis using a sparsifying spatio-temporal prior. NeuroImage **50**(1), 150–161 (2010). https://doi.org/10.1016/j.neuroimage.2009.11.064

29. B. Ng, R. Abugharbieh, in *Modeling Spatiotemporal Structure in fMRI Brain Decoding Using Generalized Sparse Classifiers*. 2011 International Workshop on Pattern Recognition in NeuroImaging (2011b), pp. 65–68. https://doi.org/10.1109/prni.2011.10

30. P. Avesani, H. Hazan, E. Koilis, L. Manevitz, D. Sona, in *Learning BOLD Response in fMRI by Reservoir Computing*. 2011 International Workshop on Pattern Recognition in NeuroImaging (2011), pp. 57–60. https://doi.org/10.1109/prni.2011.16

31. N. Kasabov, To spike or not to spike: a probabilistic spiking neuron model. Neural Netw. **23** (1), 16–19 (2010). https://doi.org/10.1016/j.neunet.2009.08.010

32. M. Åberg, L. Löken, J. Wessberg, in *An Evolutionary Approach to Multivariate Feature Selection for fMRI Pattern Analysis* (2008)

33. T. Niiniskorpi, M. Bj, J. Wessberg, in *Particle Swarm Feature Selection for fMRI Pattern Classification*. In BIOSIGNALS (2009), pp. 279–284

34. J. Kennedy, R. Eberhart, in *Particle Swarm Optimization*. Proceedings of ICNN'95— International Conference on Neural Networks, 4, 1942–1948. https://doi.org/10.1109/icnn. 1995.488968. C. Koch, R.C. Reid, in Observatories of the mind. Nature, **483**(22 March 2012), 397–398 (2012). https://doi.org/10.1038/483397a

35. X. Wang, J. Yang, X. Teng, W. Xia, R. Jensen, Feature selection based on rough sets and particle swarm optimization. Pattern Recogn. Lett. **28**(4), 459–471 (2007). https://doi.org/10. 1016/j.patrec.2006.09.003

36. S. Ryali, K. Supekar, D.A. Abrams, V. Menon, Sparse logistic regression for whole-brain classification of fMRI data. NeuroImage **51**(2), 752–764 (2010)

37. O. Yamashita, M. Sato, T. Yoshioka, F. Tong, Y. Kamitani, Sparse estimation automatically selects voxels relevant for the decoding of fMRI activity patterns. NeuroImage **42**(4), 1414–1429 (2008). https://doi.org/10.1016/j.neuroimage.2008.05.050

38. B. Ng, A. Vahdat, G. Hamarneh, R. Abugharbieh, *Generalized Sparse Classifiers for Decoding Cognitive States in fMRI*. Machine Learning in Medical Imaging (Springer, 2010), pp. 108–115

39. B. Ng, R. Abugharbieh, Generalized group sparse classifiers with application in fMRI brain decoding. Cvpr **2011**, 1065–1071 (2011). https://doi.org/10.1109/CVPR.2011.5995651

40. N. Kasabov, M. Doborjeh, Z. Doborjeh, Mapping, learning, visualization, classification, and understanding of fMRI data in the NeuCube evolving spatiotemporal data machine of spiking neural networks. IEEE Trans. Neural Netw. Learn. Syst. https://doi.org/10.1109/tnnls.2016.2612890, Manuscript Number: TNNLS-2016-P-6356, 2016

41. R. Brette et al., Simulation of networks of spiking neurons: a review of tools and strategies. J. Comput. Neurosci. **23**(3), 349–398 (2007)

42. E.M. Izhikevich, Polychronization: computation with spikes. Neural Comput. **18**(2), 245–282 (2006)

43. N. Scott, N. Kasabov, G. Indiveri, in *NeuCube Neuromorphic Framework for Spatio-Temporal Brain Data and Its Python Implementation*. Proc. ICONIP. Springer LNCS, vol 8228 (2013), pp. 78–84

44. S.B. Furber, F. Galluppi, S. Temple, L.A. Plana, The SpiNNaker project. Proc. IEEE **102**(5), 652–665 (2014)

45. P.A. Merolla et al., A million spiking-neuron integrated circuit with a scalable communication network and interface. Science **345**(6197), 668–673 (2014)

46. G. Indiveri et al., Neuromorphic silicon neuron circuits. Front. Neurosci. (2011) May 2011 [Online]. Available: http://dx.doi.org/10.3389/fnins.2011.00073

47. A. van Schaik, S.-C. Liu, AER EAR: a matched silicon cochlea pair with address event representation interface. Proc. IEEE Int. Symp. Circuits Syst. **5**, 4213–4216 (2005)

48. T. Delbruck. jAER, accessed on 15 Oct 2014. [Online] (2007). Available: http://sourceforge.net

49. P. Lichtsteiner, C. Posch, T. Delbruck, A dB using latency asynchronous temporal contrast vision sensor. IEEE J SolidState Circ. **43**(2), 566–576 (2008)

50. S. Song, K.D. Miller, L.F. Abbott, Competitive Hebbian learning through spike-timing-dependent synaptic plasticity. Nature Neurosci. **3**(9), 919–926 (2000)

51. N. Kasabov, K. Dhoble, N. Nuntalid, G. Indiveri, Dynamic evolving spiking neural networks for on-line spatio- and spectro-temporal pattern recognition. Neural Netw. **41**, 188–201 (2013)

52. N. Kasabov, *Evolving Connectionist Systems* (Springer, New York, NY, USA, 2007)

53. S.G. Wysoski, L. Benuskova, N. Kasabov, Evolving spiking neural networks for audiovisual information processing. Neural Netw. **23**(7), 819–835 (2010)

54. W. Maass, T. Natschläger, H. Markram, Real-time computing without stable states: a new framework for neural computation based on perturbations. Neural Comput. **14**(11), 2531–2560 (2002)

55. N.K. Kasabov, NeuCube: a spiking neural network architecture for mapping, learning and understanding of spatio temporal brain data. Neural Netw. **52**, 62–76 (2014)

56. J. Talairach, P. Tournoux, *Co-Planar Stereotaxic Atlas of the Human Brain: 3-Dimensional Proportional System: An Approach to Cerebral Imaging* (Thieme Medical Publishers, New York, NY, USA, 1998)

57. M. Brett, K. Christoff, R. Cusack, J. Lancaster, Using the Talairach atlas with the MNI template. NeuroImage **13**(6), 85 (2001)

58. N. Kasabov et al., Evolving spatio-temporal data machines based on the NeuCube neuromorphic framework: design methodology and selected applications. Neural Netw. **78**, 1–14 (2016). https://doi.org/10.1016/j.neunet.2015.09.011.2015

59. E.M. Izhikevich, Which model to use for cortical spiking neurons? IEEE Trans. Neural Netw. **15**(5), 1063–1070 (2004)

60. E. Tu et al., in *NeuCube(ST) for Spatio-Temporal Data Predictive Modeling with a Case Study on Ecological Data*, in Proceedings of International Joint Conference on Neural Networks (IJCNN), Beijing, China (2014), Jul 2014, pp. 638–645

61. E. Tu, N. Kasabov, J. Yang, Mapping temporal variables into the NeuCube for improved pattern recognition, predictive modeling, and understanding of stream data. IEEE Trans. Neural Netw. Learn. Syst., vol. **PP**(99), 1–13 (2016). https://doi.org/10.1109/tnnls.2016. 2536742.2016

62. E. Bullmore, O. Sporns, Complex brain networks: graph theoretical analysis of structural and functional systems. Nature Rev. Neurosci. **10**(3), 186–198 (2009)

63. V. Braitenberg, A. Schuz, *Cortex: Statistics and Geometry of Neuronal Connectivity* (Springer, Berlin, Germany, 1998)

64. S. Schliebs, N. Kasabov, Evolving spiking neural network—a survey. Evolving Syst. **4**(2), 87–98 (2013)

65. M. Just, StarPlus fMRI data, accessed on 13 Jul 2014. [Online] (2014). Available: http:// www.cs.cmu.edu/afs/cs.cmu.edu/project/theo-81/www

66. F. Pereira, (13 Feb 2002), E-print network, accessed on 13 Jul 2014 [Online]. Available: http://www.osti.gov/eprints/topicpages/documents/record/181/3791737.html

67. NEUCOM-KEDRI, Available: http://www.theneucom.com

68. K.A. Johnson, J.A. Becker, in *The Whole Brain Atlas.* Accessed on 16 Oct 2014 [Online]. Available: http://www.med.harvard.edu/AANLIB/home.html

69. L. Koessler et al., Automated cortical projection of EEG sensors: anatomical correlation via the international 10–10 system. NeuroImage **46**(1), 64–72 (2009)

70. S. Thorpe, J. Gautrais, in *Rank Order Coding.* Computational Neuroscience (Plenum Press, New York, NY, USA, 1998), pp. 113–118

71. M. Yuasa, K. Saito, N. Mukawa, Brain activity when reading sentences and emoticons: an fMRI study of verbal and nonverbal communication. Electron. Commun. Jpn. **94**(5), 17–24 (2011)

72. R.K. Christensen, Negative and affirmative sentences increase activation in different areas in the brain. J. Neurolinguist. **22**(1), 1–17 (2009)

73. D. Zhou, O. Bousquet, T.N. Lal, J. Weston, B. Schölkopf, Learning with local and global consistency. Proc. Adv. Neural Inf. Process. Syst. **16**, 321–328 (2004)

74. M. Behroozi, M.R. Daliri, RDLPFC area of the brain encodes sentence polarity: a study using fMRI. Brain Imag. Behav. **9**(2), 178–189 (2015)

75. N. Kasabov, L. Zhou, M. Gholami Doborjeh, J. Yang, New algorithms for encoding, learning and classification of fMRI data in a spiking neural network architecture: a case on modelling and understanding of dynamic cognitive processes. IEEE Trans. Cogn. Dev. Syst. (2017). https://doi.org/10.1109/tcds.2016.2636291

76. J. Sjöström, W. Gerstner, Spike-timing dependent plasticity. Front. Synaptic Neurosci. **5**(2), 35–44 (2010)

77. T. Masquelier, R. Guyonneau, S. J. Thorpe, Spike timing dependent plasticity finds the start of repeating patterns in continuous spike trains. PLoS ONE **3**(1), Art. no. e1377 (2008)

78. M.G. Doborjeh, E. Capecci, N. Kasabov, in Classification and Segmentation of fMRI Spatio-Temporal Brain Data with a NeuCube Evolving Spiking Neural Network Model. Proceedings of IEEE Symposium on Evolving and Autonomous Learning Systems (EALS), (Orlando, FL, USA, 2014), pp. 73–80

79. N. Kasabov (ed.), in Springer Handbook of Bio-/Neuroinformatics (Springer, 2014)

80. G.E. Hinton, R.R. Salakhutdinov, Reducing the dimensionality of data with neural networks. Science **313**(5786), 504–507 (2006)

81. G.E. Hinton, Learning multiple layers of representation. Trends Cogn. Sci. **11**(10), 428–434 (2007)

82. Y. Bengio, Learning deep architectures for AI. Found. Trends Mach. Learn. **2**(1), 1–127 (2009)

83. Y. LeCun, Y. Bengio, in *Convolutional Networks for Images, Speech, and Time Series.* The Handbook of Brain Theory and Neural Networks, vol 3361 (MIT Press, Cambridge, MA, USA, 1995), p. 1995

84. J. Schmidhuber, Deep learning in neural networks: an overview. Neural Netw. **61**, 85–117 (2015)
85. M. Just, StarPlus fMRI data. Accessed on 7 May 2016 [Online] (2016). Available: http://www.cs.cmu.edu/afs/cs.cmu.edu/project/theo-81/www/
86. N. Sengupta, C. McNabb, N. Kasabov, B. Russel, Integrating space, time and orientation in spiking neural networks: a case study on multimodal brain data modelling. IEEE Trans. Neural Netw. Learn. Syst. (2018)

# Chapter 11
# Integrating Time-Space and Orientation. A Case Study on fMRI + DTI Brain Data

This chapter introduces a new method for the integration of time-space data with additional and sometimes, a priory existing information, about the orientation (direction) of the spread of the temporal information. Examples of such data are many. A typical example is moving objects in the time-space, where measured direction can also be added for a better prediction of the movement of an object in time-space. Another example is integrating time-space brain data, such as fMRI with the orientation map of individual brain signals measured as DTI. The latter is developed in this chapter.

The chapter is organised in the following sections:

11.1. Introduction and Background Work.
11.2. A Personalised Modelling Architecture for fMRI and DTI Data Integration Based on the NeuCube BI-SNN.
11.3. Orientation-Influence Driven STDP (oiSTDP) Learning in SNN for the Integration of Time-Space and Direction, Illustrated on fMRI and DTI Data.
11.4. Experimental Results on Synthetic Data.
11.5. Using oiSTDP Learning for the Classification of Responding and Non-responding Schizophrenic Patients to Clozapine Monotherapy.
11.6. Summary and Further Readings for Deeper Knowledge.

## 11.1 Introduction and Background Work

In the recent past, non-invasive brain data collection techniques such as functional magnetic resonance imaging (fMRI), electroencephalography (EEG), diffusion tensor imaging (DTI) and others have made significant contributions to understanding various structural and functional properties of the human brain. Chapter 10 presented an introductory material and an extended bibliography about fMRI data (see also [1–30]).

© Springer-Verlag GmbH Germany, part of Springer Nature 2019
N. K. Kasabov, *Time-Space, Spiking Neural Networks and Brain-Inspired Artificial Intelligence*, Springer Series on Bio- and Neurosystems 7,
https://doi.org/10.1007/978-3-662-57715-8_11

There has been consistent development in data sampling technology over the past few years which has enabled simultaneous sampling of multiple modalities of brain data while a subject performs or does not perform a task. This provided an opportunity to perform pattern recognition using large quantities of such data. It is evident that each data modality provides a unique but limited perspective of the brain. For instance, fMRI measures neural activity indirectly by measuring changes in cerebral blood flow (the haemodynamic response) over time. Energy consumption increases in areas of the brain that are more active, leading to increases in blood flow to replace lost oxygen and glucose. This is a slow response, measured 6–10 s after the initial event of neuronal excitation. Though it provides poor temporal resolution, fMRI provides excellent spatial resolution, making it a useful tool for brain research. EEG, provides an outstanding temporal resolution (millisecond accuracy) at the expense of spatial resolution. EEG measures cortical electrical activity at the scalp surface and though the scalp does not impede electrical signals temporally, it causes spreading of the signal from its origin to a wider area, making source localisation much more complex. In the past, these data modalities were used independently for pattern recognition and overlooked the joint information present in the data [31]. Algorithms with the ability to extract and integrate relevant information from various data sources into a single model are crucial not only for the purpose of predictive modelling but also in terms of understanding the spatio-temporal relationships within the data.

In a clinical sense, pattern recognition algorithms can provide a novel and practical means to understand the differences between patients and healthy controls and predict individual patients' responses to treatment. Within psychiatric research, In particular, machine learning has gained considerable momentum as a useful tool for developing predictive models of treatment response. Incorporation of multiple imaging modalities into these algorithms could provide increased reliability, especially in disorders where clinical diagnosis does not necessarily guide treatment. In [32] recently applied machine learning algorithms to predict treatment response in late-life depression using a combination of clinical and imaging data. Comparing a number of algorithms, they determined that alternating decision trees could most accurately predict treatment outcome in this cohort using a combination of structural and functional connectivity data [32]. In [33] have used EEG data to predict response to selective serotonin reuptake inhibitors (SSRIs) in major depressive disorder and to clozapine in people with treatment-resistant schizophrenia [34]. In [35] also employed machine learning algorithms to predict response to clozapine, instead using a combination of clinical and pharmacogenetic data as input. Providing even more support for this approach, [36] employed machine learning techniques to predict treatment outcome in social anxiety disorder. Using task-based fMRI, they accounted for 40% of the variance in treatment response [36]. The challenge now is to create an algorithm that can incorporate brain data from different modalities.

One modality with potential for inclusion in such a model is DTI. DTI measures the net movement of water within a voxel. Water trapped within axons or dendrites is restricted to movement along the direction of those axons or dendrites, respectively; this can be measured using DTI, providing a map of neuronal tracts (white matter) within the brain. Decreased anisotropy within a voxel is often interpreted as a reduction in white matter integrity, implying that white matter has been damaged or is diminished in some way. However, a reduction in anisotropy may also be attributed to an increase in the number of crossing fibres, or in fact a reduction in the uniformity of fibre orientations. In terms of incorporation into a multimodal prediction model, DTI information might be utilised for its orientational implications rather than any interpretation regarding white matter integrity (Fig. 11.1).

Structural connectivity, as measured by DTI, has been demonstrated in several psychiatric disorders and has been shown to reflect functional dysconnectivity in some cases [37, 38]. In accordance with these theories, it would be appealing to incorporate dysconnectivity information into any algorithm that is designed to classify or predict outcomes in people with psychiatric disorders. This paper discusses the steps that we are undertaking to develop an algorithm that can incorporate orientational information from DTI along with the EEG/fMRI activity data.

A comprehensive review of the research in the direction of multi-modal brain data analysis (MBDA) is summarised in [31]. Some of the prominent work in MBDA includes integration of fMRI/EEG [39], fMRI/MEG [40] and fMRI/Gene expression [41]. In [31] has further classified MBDA into hypothesis-driven and data-driven approaches and argued the possibility of missing important connectivity links in the hypothesis-driven approach. The data-driven methods span across the domain of the combined blind source separation techniques such as Independent Component Analysis [42, 43] and its variants, multi-modal Cross-Correlation Analysis [44–46], Partial Least Squares [47], and others. A spiking neural network architecture

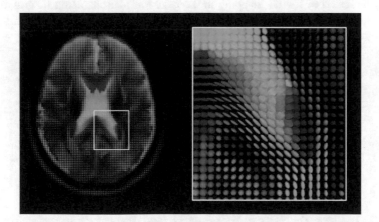

**Fig. 11.1** Orientational information from DTI image. Left image shows an axial slice of a single subject's DTI data, registered to structural and MNI standard space. The Right image shows a close-up of the right posterior corpus callosum. Directions corresponding to each colour are as follows: Red—left to right/right to left, green—anterior to posterior/posterior to anterior and blue —superior to inferior/inferior to superior [114]

NeuCube for incorporating spatio-temporal information was proposed in [48] and described in Chaps. 6 and 10 in this book. Within its vast range of applicability, this architecture has also been used in fMRI and brain data related studies [49]. We have used the NeuCube paradigm as the basis for the proposed pattern recognition architecture capable of integrating multidimensional information from multi-source data. Methods of SNN and some applications of NeuCube for fMRI and other time-space data along with their introductory information can be found in [48–82].

## 11.2   A Personalised Modelling Architecture for fMRI and DTI Data Integration Based on the NeuCube BI-SNN

As described in Chap. 6 NeuCube architecture [48] is a spiking neural network based pattern recognition system which is designed to represent and learn spatio-temporal associations in the data. The NeuCube system is designed as an extension of the liquid state machines (LSM) [83, 84] based reservoir computing paradigm. Figure 11.2 depicts the multi-stage pattern recognition process in NeuCube for spatio- and spectro-temporal data. The temporal information from a data source is passed through the encoding module to transform the real, continuous, and dynamic information into spike trains. In the unsupervised learning stage, the spike train is converted into a high dimensional space using a 3D cube of spiking neurons. In the final stage, a supervised linear discriminator uses the high-dimensional spike train to discriminate between patterns.

Some of the considerable departures of NeuCube from LSM as discussed in [48] are (1) Inclusion of spatial mapping in the reservoir neurons using natural dummy

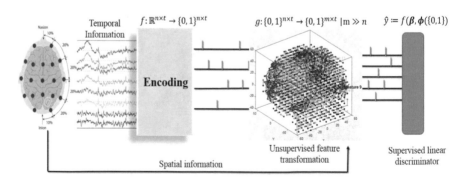

**Fig. 11.2** This figure depicts the multi-stage pattern recognition process in NeuCube for spatio- and spectro-temporal data. The temporal information from a data source is passed through the encoding module to transform the real, continuous, and dynamic information into spike trains. In the unsupervised learning stage, the spike train is converted into a high dimensional space using a 3D cube of spiking neurons. In the final stage, a supervised linear discriminator uses the high-dimensional spike train to discriminate between patterns

spatial 3D spatial coordinates. The spatial mapping of neurons takes its inspiration from Kohonen's self-organising maps (2) NeuCube learned reservoir has certain degree of visual interpretability due to the mapping and spatiotemporal learning in the reservoir.

The NeuCube architecture for personalised modelling on integrated data introduced here consists of three main modules:

(1) Temporal compression or encoding module (Chaps. 4 and 6): The data encoding layer transforms real continuous data $R^{n \times t}$ ($n$ is the number of features and $t$ represents time duration.) into spike trains $\{0,1\}^{n \times t}$. Numerous temporal encoding algorithms like BSA [84], temporal contrast, GAGamma [85] are proposed and used in an application specific manner. The data encoding module in NeuCube is a data compression system which has the unique property of compressing data in temporal dimension by representing useful events by spike-timings. In the temporal encoding scheme, the timings of the spike is considered to be useful rather than the quantity of the spike. This is much different from the traditional data compression algorithms like auto-encoder and PCA as the compression in the data is performed taking temporal dependencies in account. [86] describes the temporal encoding by spike-time representation in the light of data compression and information theory and compares the capabilities of different temporal encoding algorithms.

(2) Personalised SNNc learning module: The SNNc learning module refers to an unsupervised learning module in a three dimensional spatial grid of spiking and non-spiking neurons forming the liquid state. Each neuron inside the grid has a particular spatial location and resides within a neighbourhood of other neurons. This grid is known as the spiking neural network cube (SNNc) in the NeuCube architecture. The purpose of this layer is to transform the compressed spike representation of input data into a higher-dimensional space through unsupervised learning ($g: \{0,1\}^{n \times t} \to \{0,1\}^{m \times t} | m \gg n$) inside an SNNc, using a form of modified Hebbian based STDP learning [69]. It is imperative that the NeuCube SNNc is a spatially organised directed graph which is inspired from Kohonen's self-organising map [87] and the LSM [83]. However, the information that is represented in the SNNc map is distinctly different from the Kohonen's map. Contrary to the static information representation in the connection weights of the SOM, the SNNc presented in this article can capture the multidimensional information from static and dynamic data. Our SNNc approach differs from the other SNNc approaches in one other respect. All of the NeuCube system developed and used till now uses a single SNNc which is responsible for transforming the incoming data from a lower to higher dimensional space. As opposed to this, in the proposed architecture, we have used multiple SNNc's where each SNNc is responsible for transforming part of the incoming data. The outputs of the multiple SNNc are later merged in the supervised learning layer. The personalised SNNc based approach is unique and a departure from the 'one SNNc for all sample' approach of used in NeuCube systems earlier. The personalised SNNc architecture assumes the

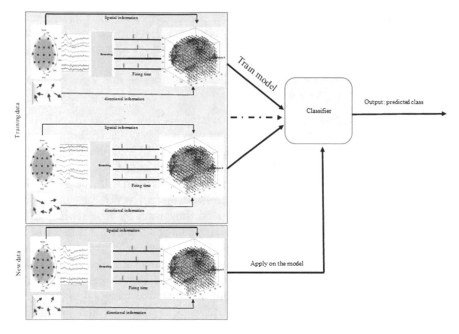

**Fig. 11.3** A Personalized modelling scheme using both fMRI and DTI data

existence of temporal relationships within a sample and not between samples, i.e. each sample maps its spatiotemporal relationship on its unique personal SNNc, which is used further for discrimination purpose. We will discuss the SNNc in detail in the later sections.

(3) Supervised learning module: This module uses the SNNc generated output spike sequences and/or the connection weights to learn a simple classifier or regressor [70]. K-NN based models [70] are the choice of supervised learning in almost all of the work done until now.

Figure 11.3 depicts the personalised modelling scheme using both fMRI and DTI data.

## 11.3   Orientation-Influence Driven STDP (oiSTDP) Learning in SNN for the Integration of Time-Space and Direction, Illustrated on fMRI and DTI Data

This section is dedicated for describing the learning algorithm for multidimensional information integration.

## 11.3.1   Architecture, Mapping and Initialization Scheme

The SNNc architecture consists of a spatially arranged (in three dimensions) set of neurons (computational units), partially connected together by synapses forming a directed acyclic incomplete graph. The network consists of two types of neuron. The neurons and synapses forms the vertices and the edges of the graph.

- Input neurons: The input neurons feed the input spike data to the SNNc. These neurons do not have any activations and do not perform any computations but rather act as an abstraction for pushing the data into the system. It is apparent that the input neurons does not have pre-synaptic connections i.e. to say an edge can only originate from such a neuron.
- Spiking neurons: The spiking neurons are leaky integrate and fire in nature and performs computations on input data (spikes). The details of the neuron model is described later. These neurons can act both as post and presynaptic (connection) neuron i.e. if we consider a pair of neuron connected by a directed edge (synapse), an edge can both originate and end at a spiking neuron.

The neurons in the SNNc are arranged spatially based on background knowledge about the problem or through different automated mapping algorithms [79] that transform some predefined similarity in the data to spatial Euclidean distance. The synaptic connectivity of the SNNc graph is created using the small world connectivity (SWC) algorithm [88, 89]. The SWC algorithm connects a neuron to its spatial neighbourhood (controlled by the hyperparameter radial distance $r_{swc}$) of neurons. However, as opposed to the random initialisation of the synaptic weights, we have initialised the synapses with a small constant weight of 0.05. We have not considered random initialisation as the unsupervised learning in the SNNc do not converge over time, rather the weight updates over time reflects levels of synchronicity in the input data.

## 11.3.2   Neuron Model

The activation of the spiking neurons present in the SNNc is modelled by the spike response model (SRM) which is a simplified realisation of the leaky integrate and fire (LIF) model. The SRM model generalises the differential equation based dynamics of the LIF model by replacing them with arbitrary kernels. Apart from being a powerful computational framework, SRM model captures the essential effects during spiking and has the advantages of an elegant mathematical formulation.

Figure 11.4 shows a typical connectivity configuration of a spiking neuron i. Neuron i typically has multi-input, multi-output configuration. The neuron pairs are connected by synapses represented by the synaptic strengths wi. Neuron i receives spikes from the pre-synaptic neurons and emits spikes when sufficiently stimulated.

**Fig. 11.4** A typical connectivity configuration of a spiking neuron i. Neuron i typically has multi-input, multi-output configuration. The neuron pairs are connected by synapses represented by the synaptic strengths w$_i$. Neuron i receives spikes from the pre-synaptic neurons and emits spikes when sufficiently stimulated (after [113])

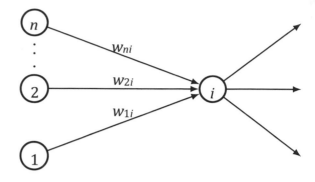

A spiking neuron receives (Fig. 11.4) spikes at different time instances from the pre-synaptic neurons and emit spikes when sufficiently stimulated. The activation state of a spiking neuron $i$ is described by the membrane potential $v_i$. In a non-stimulated state the membrane potential is said to be in the resting state $v_{rest} = 0$. The SRM model in our setup consists of multiple components and are described below:

(1) *Post-synaptic potential (PSP) kernel*: Firing of presynaptic neuron $j$ at time $t_j^f$, evokes a PSP in neuron $i$ and is modelled by the kernel response function $\varepsilon_0$ under the SRM paradigm.

$$\varepsilon_0 = \exp\left(-\frac{t - t_j^f}{\tau_m}\right)\mathcal{H}\left(t - t_j^f\right) \qquad (11.1)$$

where

$$H\left(t - t_j^f\right) = \begin{cases} 1, & \textit{if } t - t_j^f \geq 0 \\ 0, & \textit{Otherwise} \end{cases}$$

The PSP kernel is a function of $t - t_j^f$, representing the PSP trace over time generated by firing of neuron $j$ at time $t_j^f$. Figure 11.5 plots the PSP kernel as the function of $t - t^f$. The decay of the PSP kernel is given by the membrane time constant $\tau_m$ (Eq. 11.1). The choice of $\tau_m$ controls the velocity with which the impact of a pre synaptic spike decays. In our experiments, we have used a constant $\tau_m = 0.5$. This means the influence of a pre-synaptic spike diminishes from 1 to 0 within 5 discrete time intervals.

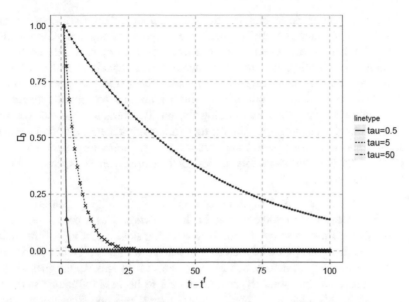

**Fig. 11.5** Post synaptic potential function $\in_0 (t - t^f)$. The curve decay with time is plotted for different values of the decay rate, given by the membrane time constant $\tau_m$ (after [113])

(2) *Temporal integration of PSP kernels and condition for spike emission*: It is imperative that the post synaptic potentials evoked by the pre-synaptic neurons needs some form of temporal integration to formalise the activation $v_i$. The overall contribution of the pre-synaptic spikes elicited by the presynaptic neurons $j$ at any time $t$ is given as part of Eq. 11.2 describing the SRM model:

$$v_i(t) = v_{rest} + \sum_{j \in T_i} w_{ji} \sum_{t_j^f \in F_j} \in_0 \left( t - t_j^f \right) \tag{11.2}$$

The inner sum adds up the PSP contributions due to the firings $t_j^f \in F_j$ of one pre-synaptic neuron. The outer sum adds up the PSP contributions of all the pre-synaptic neurons $j \in T_i$ connected to neuron $i$.

Equation 11.2 describes the membrane potential (activation state) $v_i$ of a spiking neuron $i$ can be calculated by adding the resting potential term and the temporal PSP sum. Each incoming spike perturbs the value of $v_i$ and if, after the summation of the inputs, the membrane potential reaches the threshold $v_{thr}$ then an output spike is generated. The firing time is given by the condition $v_i(t_i^f) > = v_{thr}$. After a neuron fires the neurons membrane potential is reset to $v_{rest}$.

(3) *Refractory period*: After emitting the spike a node enters a refractory period, when the membrane potential is unaffected by any incoming spike. In the SRM model the neuron behaviour in the refractory period depends only on the last firing moment leading to a short term memory like behaviour. In literature, the refractory period is modelled by absolute and relative refractory period. During the absolute refractory period, the neurons do not accumulate membrane potential and hence can't fire. During the relative refractory period it is relatively difficult but not impossible to fire. In our implementation we have used an absolute refractory period and not the relative refractory period for simplicity. The absolute refractory period of a neuron can be specified by the hyperparameter $\eta_{thr}$.

Figure 11.6 shows a plot of three simulations of a spiking neuron for 200 discrete time with random spike inputs. Each simulation uses a preset $v_{thr}$. At the beginning of the simulation the neuron is in a resting state $v_{t=0} = v_{rest}$. With arrival of spikes the membrane potential increases in a linear fashion and when sufficiently stimulated (sufficiency is determined by $v_{thr}$), the neuron spikes and goes back to the resting state. At this point the neuron is said to be in a refractory state. The neuron stays in this state for a predeterminer period $\eta_{thr}$ and then goes back to non-refractory state.

### 11.3.3  Unsupervised Weight Adaptation of Synapses

The unsupervised learning scheme is the most important aspect of our proposed architecture for integrating multidimensional information. In a neural network paradigm, learning is achieved through the synaptic strength updates of the network over time. Similar to our single neuron model approach, we believe, the learning behaviour of the SNNc can be explained using the learning model of a single spiking neuron. Considering the single neuron architecture, the unsupervised learning problem the problem of updating the $w_{ji}$'s by $\Delta w_{ji}(t)$ at any time $t$. In our recurrent neural network architecture in SNNc, our aim is to learn both dynamic influences from dynamic data (fMRI or DTI) and static orientational influence from static data (DTI).

(1) *Dynamic influence ($\varphi$) from fMRI or EEG and spiketime dependent plasticity:* In majority of the machine learning applications, models are trained on static data, where a sample is represented by a vector of numbers $\mathbf{x} = \{x_1, x_2, \ldots, x_n\}$, where each number represents the value of a feature. But in our case with fMRI or EEG data a sample is represented by the form of a matrix $\mathbf{X}_{seq} = \{\mathbf{x_1}, \mathbf{x_2}, \ldots, \mathbf{x_n}\}$, where $\mathbf{x_i} = \{x_i(1), x_i(2), \ldots x_i(t)\}$. This form of sample representation is unique not only in its two dimensional data representation but also in the ordering of the data in one of its dimensions. Learning from these type of data sequences in the machine learning domain is known as sequence learning and techniques like hidden Markov model and flavours of recurrent neural network

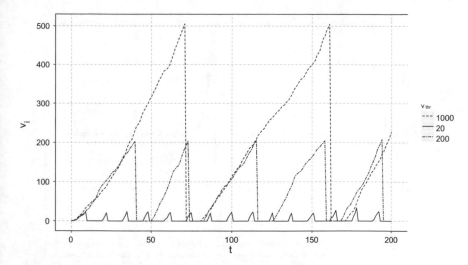

**Fig. 11.6** Plot of the membrane potentials ($v_i$) of a neuron i simulated over T = 200 time points using the SRM model. For the simulation, we connected 3 predecessor neurons to the neuron i. The spike data from the predecessor neurons are sampled randomly from uniform random distribution. The $\eta_{thr}$ for the neuron i was set to 10. Each of the three $v_i$ traces corresponds to a preset $v_{thr}$ mentioned in the label (after [113])

have shown promise in learning from such sequences. Here we describe an unsupervised sequence learning framework in NeuCube SNNc that uses data sequences as part of its learning. We call this the dynamic influence. The NeuCube SNNc being a recurrent spiking neural network architecture takes the data sequence in the form of spikes, i.e. $\mathbf{X} = \{0,1\}^{n \times t}$. The transformation of real continuous data to spike data is done by the encoding module as described earlier.

We model the dynamic influence of the spike-time data by the spike-time dependent plasticity learning rule. Spike time dependent plasticity (STDP) is a temporally asynchronous form of Hebbian learning ("neurons wire together, if they fire together") [90] induced by the temporal correlation of the spikes. This biological process in the brain adjusts the synaptic strength based on the relative timing of a neuron's input and the output spikes. With STDP, repeated pre-synaptic spike arrival just before post-synaptic spike leads to longterm potentiation (LTP) of the synapses establishing causal relationship, whereas repeated spike arrival after post-synaptic spikes leads to long-term depression (LTD) of the same synapse.

In [69, 91] the mathematical model of STDP learning is formalised as per Eq. 11.3 and Fig. 11.7. Symbols $j$ and $i$ are used to indicate pre and post synaptic neurons. In STDP learning, The dynamic influence $\varphi_{ji}$ is estimated using a learning window function $W(\cdot)$. The learning window takes the set of presynaptic firing times $\{t_j^1 \dots t_j^f\}$ and post-synaptic firing times $\{t_i^1 \dots t_i^g\}$ as input and calculates the LTP and LTD traces. Exponential decay functions are a popular choice for the

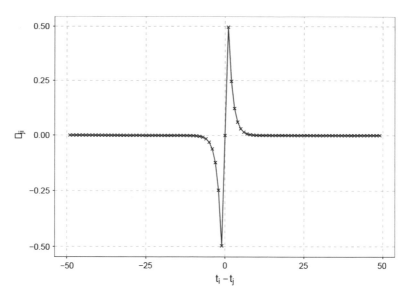

**Fig. 11.7** The STDP weight update as a function of the relative timing of the pre and post synaptic spikes (after [113])

learning window function and we use this learning window function for all our simulations. The $\kappa_+$ and $\kappa_-$ parameters control the maximum LTP and LTD update respectively and we have chosen $\kappa_- = \kappa_+ = 1$ to keep the bounds of dynamic influence between $[-1,1]$. From Eq. 11.3, it can be observed that the polarity of $(t_i^g - t_j^f)$ defines the polarity of $\varphi_{ji}$. This is a causal Hebbian relationship model where synapses are rewarded positively (strengthened) for causal firing ($i$ fires later than $j$ i.e. firing of $i$ is caused by firing of $j$) and penalised (weakened) for non causal firing. However, Eqs. 11.3 describes a batch update scheme and requires modification for on-line learning in the SNNc. [92] proposed a modified online STDP update rule. In the on-line setup, $\varphi$ is calculated every time neuron $i$ fires a spike or receives a spike from neuron $j$. Equation 11.4 formalises the dynamic influence update for on-line mode. The first term in the right hand side of Eq. 11.4 corresponds to the LTP update and is calculated when neuron $i$ fires a spike at time $t$. The second term is the LTD update and is calculated when neuron $i$ receives a spike from neuron $j$ at time $t$. Both the batch and on-line formalisations of STDP learning are extended from [92] which discusses the properties of the STDP learning model extensively.

Figure 11.7 shows the plot of the STDP learning function where the dynamic influence in quadrant I and III corresponds to LTP and LTD respectively.

$$\phi_{ij} := \sum_f \sum_g W\left(t_i^g - t_j^f\right)$$

$$W(s) := \begin{cases} k_+ \exp(-s), & \text{if } s > 0 \\ -k_- \exp(-s) & \text{if } s < 0 \end{cases} \tag{11.3}$$

$$\phi_{ij}(t) := \sum_f k_+ \exp\left(-\left(t - t_j^f\right)\right) - \sum_g k_- \exp(-(t - t_i^g))$$

It is evident from the discussion above that the STDP learning rule enhances or depletes the synaptic strength of the connections, based on the relative coincidence of the spikes. This behaviour mimics the ability of the biological neurons to encode information by detecting the occurrence of temporally close but spatially distributed input signals and thus incorporating spatio-temporal information in the model.

*DTI fibre tractography data and orientation influence:* Diffusion tensor weighed images are represented as a three dimensional image made of a set of spatially arranged glyphs. Each glyph/voxel (colour and orientation) in the image is characterised by a rotation invariant ellipsoid representing the properties of the molecular diffusion of water in that region. Due to the tensorial nature of the ellipsoid, the raw DTI voxel information is stored as a second order positive definite tensor:

$$D := \begin{pmatrix} D_{xx} & D_{xy} & D_{xz} \\ D_{xy} & D_{yy} & D_{yz} \\ D_{xz} & D_{yz} & D_{zz} \end{pmatrix} \tag{11.4}$$

The six unique elements of the tensor are the coefficients of the ellipsoid equation given by $D_{xx}x^2 + D_{yy}y^2 + D_{zz}z^2 + D_{yx}yx + D_{zx}zx + D_{xy}xy = 1$. The diffusion properties of the ellipsoid are characterised by the Eigen vectors and Eigen values of the tensor which correspond to the magnitudes and directions of the anisotropy. For example in the areas with isotropic diffusion, the shape of the ellipsoid will be nearly spherical with small anisotropy measure [93]. Fibre tractography is a very elegant method for delineating individual fibre tracts from diffusion images. In our work, we have used the DTI data in the form of orientation vectors representing mean orientation of the fibre tract at different voxel locations. The orientation vector of a sample DTI image is represented by a matrix $\mathbf{X_{or}} \in R^{n \times 3}$, where each feature is represented by a 3D vector describing the orientation of the fibre in the Cartesian coordinate system.

Here, we are establishing a learning rule that can accommodate both dynamic data influence, but also static orientation influence from the DTI data. The intuition behind the orientation influence can be explained again by a small SNNc architecture consisting of three neurons as shown in Fig. 11.8. The figure shows a single pre-synaptic neuron $j$ connected to two post synaptic neurons $i_1$ and $i_2$.

The important thing to note here is that the neurons in this diagram has spatial allocations. The location of the neurons are defined by the radial and the angular

**Fig. 11.8** Example of a
pre-synaptic neuron j
connected to two post
synaptic neurons $i_1$ and $i_2$.
Each neurons spatial location
is defined by the polar
coordinates (r,α) (after [113])

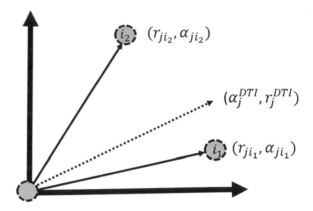

coordinate in the polar coordinate system. Now, we are interested in calculating the
orientation influence of neuron $j$ on neurons $\{i_1,i_2\}$. We call neuron $j$ as the pivot
neuron. Let the orientation vector of the pivot neuron (from DTI data) be represented by $(r_j,\alpha_j)$. The orientational influence of the pivot neuron on neurons $\{i_1,i_2\}$
are defined by their angular proximity pivot neuron's orientation vector. In that
way, as per our hypothesis, the pivot neuron wields a stronger angular influence on
the neurons as they lie closer angular proximity to the orientation vector of the pivot
neuron. Hence, the influence of neuron $j$ can be arranged as $i_1 > i_2$ due to the
angular proximity of $i_1$ and $j$ being greater than $i_2$ and $j$.

Even though we have used 2D vector space for explaining the intuition of
angular influence, our neurons in the SNNc reside in a 3D space. The intuition is
ofcourse extendible to a 3D vector space by adding another dimension in the
coordinate representation. In three dimensions the spherical coordinates of a point is
given by $(r,\alpha,\beta)$, where $r$ is the scalar distance of the point from the centre, $\alpha$ and $\beta$
are the elevation and azimuth angle from the centre. Gaussian radial basis function
(GRBF) is used to realise the elevation and azimuth orientational influences given
the elevation and azimuth data of the neurons. The elevation and azimuth orientational influences between pivot pre synaptic neuron $j$ and post-synaptic neuron
$i$ are:

$$\psi_{ji}^{\alpha} = e^{\dfrac{\left\| \alpha_{ji} - \alpha_j^{dti} \right\|^2}{2\sigma^2}} \tag{11.5}$$

$$\psi_{ji}^{\beta} = e^{\dfrac{\left\| \beta_{ji} - \beta_j^{dti} \right\|^2}{2\sigma^2}} \tag{11.6}$$

$$\psi_{ji} = \dfrac{\psi_{ji}^{\alpha} - \psi_{ji}^{\beta}}{2} \tag{11.7}$$

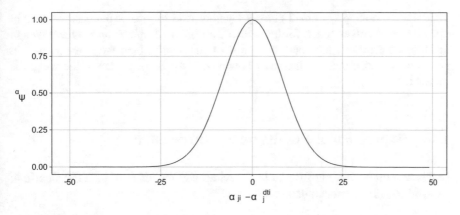

**Fig. 11.9** Plot of the elevation influence $\psi^\alpha$ as a function of the radial distance $\alpha_{ji} - \alpha_{dt}ij$ and $\sigma = 8$ (after [113])

GRBF exponentially decays the orientational influence as the Euclidean norm $\|\alpha_{ji} - \alpha_j\|$ and $\|\beta_{ji} - \beta_j\|$ increases. The variance hyperparameter $\sigma^2$ controls the speed with which the orientational influence decays with increasing radial distance (see Fig. 11.9). The overall orientational influence is calculated as the mean of the elevation and azimuth influence as shown in Eqs. (11.5–11.7).

Figure 11.10 shows the relationship of oiSTDP weight update $\Delta w$ with post and pre synaptic firing time difference $t_i - t_j$ and orientation distance $r_{ji}$. As the temporal difference between neuronal spikes decreases, the effect on weight updating

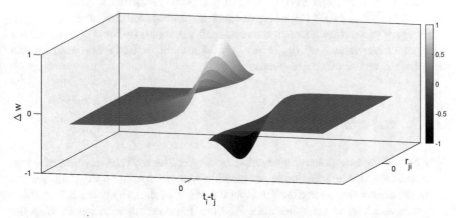

**Fig. 11.10** Graph showing the relationship of oiSTDP weight update $\Delta w$ with post and pre synaptic firing time difference $t_i - t_j$ and orientation distance $r_{ji}$. As the temporal difference between neuronal spikes decreases, the effect on weight updating increases, so that spikes timed closely together lead to greater increases in weight updating than spikes timed further apart. The order of spikes also affects weight updating. If neuron j fires before neuron i consistently, then the synaptic weight between them continues to increase; however, if the order switches, the weight is reduced (after [113])

increases, so that spikes timed closely together lead to greater increases in weight updating than spikes timed further apart. The order of spikes also affects weight updating. If neuron j fires before neuron i consistently, then the synaptic weight between them continues to increase; however, if the order switches, the weight is reduced.

## 11.4   Experimental Results on Synthetic Data

In this section we have analysed the behaviour and the effect of oiSTDP algorithm on synthetically generated datasets.

### 11.4.1   Data Description

In order to describe and evaluate the oiSTDP learning algorithm, we have used synthetically generated activity and orientation information. The input spike train $D_{seq}$ is of size $128 \times 14$, mimics a random sample of one second generated by a 14 channel EEG device with a sampling frequency of 128 Hz. All the experiments described here use an SNNc of 1485 neurons with sparse recurrent connections. The neurons in the SNNc reservoir are spatially distributed to mimic the shape of the brain [48]. The location of the input spike train in the reservoir is resolved as per the natural spatial ordering of EEG channels-AF3, F7, F3, FC5, T7, P7, O1, O2, P8, T8, FC6, F4 and F8. The SWC algorithm is used to initialise the SNNc network. We have used $r_{swc}= 0.02$ (Meaning connect neurons within 2% of the maximum distance) for connection generation and a small value of 0.05 for $W_{init}$. The default hyperparameter values of $(\eta_{thr} = 4, v_{thr} = 0.1, \kappa_- = \kappa_+ = 0.01)$ are used in the experiments, unless otherwise stated.

### 11.4.2   Experimental Results

(1) *Effect of the orientation information on SNNc*: The oiSTDP learning rule represents angular information in conjunction with the spatiotemporal information in the connection strengths. To show the effect of the orientation information, we have sampled the spike train $S_{in}$ from Poissons' distribution to keep the effect of spike synchronicity minimal in the SNNc map. Figure 11.11 shows qualitatively, the effect of the different orientation information on the final 3D SNNc map created by the oiSTDP learning algorithm. In first of the three experiments, all the SNNc neurons were directed in the $(\alpha = 0°, \beta = 0°)$ direction, i.e. parallel to the X axis and perpendicular to the Z axis. It is clearly visible from Figs. 11.11a–c that the strongest connections in the SNNc are

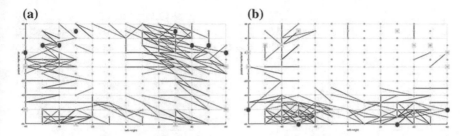

**Fig. 11.11** (a) Synchronous input spike train at locations AF3, F7, F3, FC5, FC6, F4, F8 (b) Synchronous input spike train at locations P7, O1, O2 and P8 and AF4. Comparison the effect of synchronous input spikes on the SNNc map generated by the orientation influence-driven STDP (oiSTDP) learning algorithm. Both figures are the top views (X Y plane) of the brain and the blue dots shows the synchronous input channels (after [113])

representative of the orientation information provided. The second and the third experiment uses ($\alpha = 45°$, $\beta = 45°$) and no orientation information respectively. It is evident from Fig. 11.11 that in absence of the temporal information (synchronicity), the angular information is described in the SNNc and as such, in simple cases, they are visually discriminatory. Since each of these learned SNNc is represented by a directed graph of weighed connections, connectomics analysis [94] can further be performed on the learned SNNc to extract new and useful knowledge in the spatiotemporal domain.

(2) *Effect of the spike synchronicity on SNNc*: The aim of this experiment is to show the effect of spike synchronicity, i.e. the effect of STDP learning on the SNNc map for different spatiotemporal patterns. According to the STDP learning rule, greater synchronicity leads to stronger connections through long term potentiation (LTP). To demonstrate the effect of the spatiotemporal synchronicity, we have created two samples of the input spike train. In the first sample, the spike sequences corresponding to the channels in the frontal lobe of the brain is kept the same (mimicking 100% synchronicity) and in the second sample, 100% spike synchronicity is kept at the occipital and parietal lobe. Figure 11.12 shows the comparison between the two SNNc maps created by the oiSTDP learning algorithm. The 'strongest connection' density is clearly more prominent in the frontal lobe in Fig. 11.12a due to the greater input spike synchronicity in that region. Similar clusters Fig. (11.12b) at the parietal and occipital lobe can be seen with when the second sample is used. Through these analysis we have demonstrated how different temporal patterns and the spatial arrangement of such patters affect the visual map of SNNc through the oiSTDP learning.

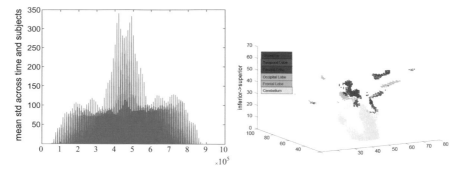

**Fig. 11.12** Voxel selection using absolute mean standard deviation: **a** A plot of voxel id vs. the absolute standard deviation (across time) of the **b** 3D MNI location of the selected voxels used for multimodal brain data voxel averaged over the subject modelling. The voxels are coloured by the anatomically defined regions in the MNI atlas

## 11.5   Using oiSTDP Learning for the Classification of Responding and Non-responding Schizophrenic Patients to Clozapine Monotherapy

### 11.5.1   Problem Specification and Data Preparation

This study was conducted as part of a large cross-sectional study investigating clozapine (CLZ) response in people with treatment-resistant schizophrenia (TRS) using EEG, MRI and genetic information (TRS study). The study was approved by the Health and Disability Ethics Committee and received locality approval from Auckland and Counties Manukau District Health Boards of New Zealand. CLZ is uniquely effective for treatment-resistant schizophrenia. However, many people still suffer from residual symptoms or do not respond at all (ultra-treatment resistant schizophrenia; UTRS) to CLZ.

In this study our aim was to build a model for discriminating CLZ monotherapy respondent and non-respondent individuals from multimodal fMRI and DTI brain data. It must be noted that the data used to build this model has been collected post CLZ treatment. For the purpose of our investigation, we used a subset of data collected from the TRS study with the intention of classifying subjects into groups with either TRS or UTRS using resting state fMRI and DTI data. A total of 25 subjects with no recorded head injury and aged from 18 to 45 years were chosen for the study. Fourteen subjects belonged to the TRS group and eleven from the UTRS group. Resting-state fMRI data was collected using a 3 T Siemens Skyra Magentom Scanner with TR = 3000 ms and TE = 30 ms for a duration of 8 min.

In the first stage, the fMRI and DTI data were preprocessed using standard preprocessing methods to control for head movement, registration and normalisation using FSL software [95]. Both fMRI and DTI data for each subject were registered to a subject specific structural image and normalised to the MNI-152

2 mm atlas [96, 97]. ICA-based Automatic Removal Of Motion Artifacts (ICA-AROMA) was used to remove motion artefacts from the fMRI data utilising FSLs FEAT output as input [98, 99]. The DTI data for each subject was further processed using the BEDPOSTX toolbox in FSL to generate the mean distribution of fibre orientation in vector form.

The second stage of data processing focused on selecting a set of voxels from the fMRI and DTI to be used to build the multimodal NeuCube model. As discussed before, since a major component of our model captures temporal variations in data and the noise reduction capabilities of SNN architectures through encoding [100], we hypothesise that the discriminatory information is hidden in the voxels with significant variation in the activity over time. Figure 11.13a shows the distribution of absolute mean standard deviation of all the voxels across time and subject. We selected a set of voxels with an absolute mean standard deviation of greater than 105 for our experiments. Figure 11.13b shows the 3D atlas locations of the selected voxels in the MNI coordinate system. The selected voxels are predominantly (>67%) located in the cerebellum of the brain. The second and third column of the ROI frequency table (see Table 11.1) also corresponds to the number and the percentage of voxels belonging to the different ROIs.

The final preprocessed dataset consists of dynamic fMRI trials $fMRI_{seq} \in R^{30 \times 238 \times 80}$ and static DTI orientation vector data $DTI_{stat} \in R^{30 \times 318 \times 3}$ of 30 subjects and 2318 voxels. Each fMRI voxel is sampled over 80 time points within a trial. The voxels of the DTI orientation data are represented by three dimensional vector signifying the primary orientation of the fibre tract at the voxel location.

Due to the multi-modular and rather flexible nature of the NeuCube eSTDM architecture, selecting baselines for comparison is a challenging task. In this work, we have used the NeuCube architecture as a combination of temporal feature compressor, spatial expander and classifier. The compressor and the SNNc module together is used for feature extraction in spatiotemporal domain. The classifier is then learned on the transformed feature representation of the data. Hence apart from proposing a spike-time based data representation our contribution lies in the feature extraction domain. Hence it is appropriate to compare our BSA + oiSTDP feature extraction method against other feature extraction methods in continuous data domain. We have compared the following feature extraction algorithms:

(1) Sparse autoencoder [101]: Autoencoders are shallow single hidden layered neural networks that can perform identity mapping of the input. The hidden layer of the autoencoder in that way learns non linear reduced dimensionality data representations. The sparse autoencoders impose sparsity regularisation constraint on the loss function to be optimised. In our experiments, we have used the fMRI data to learn a sparse autoencoder (with 1000 relu units hidden layer, L1 regularisation constraint of $10^{-5}$) that encodes the data into 1000 dimensional feature space using the python keras [102] API.

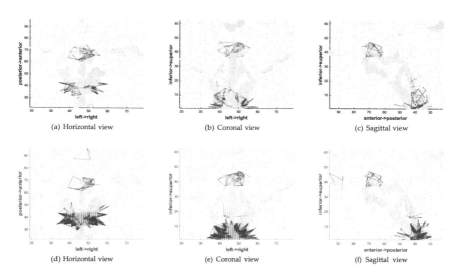

**Fig. 11.13** Visual comparison of the strongest connections (mean weight across subjects within a group) formed in the SNN model of the TRS (the top) and the UTRS group (bottom row). The yellow coloured cluster represents the input neurons and the green neurons are the computational spiking neurons

**Table 11.1** Frequency table of ROI's of the selected voxels

| ROI | # voxel | % |
|---|---|---|
| Frontal lobe | 177 | 7.64 |
| Insula | 16 | 0.69 |
| Temporal lobe | 138 | 5.95 |
| Cerebellum | 1557 | 67.17 |
| Occipital lobe | 25 | 1.08 |
| Parietal lobe | 134 | 5.78 |
| Thalamus | 187 | 8.07 |
| Caudate | 84 | 3.62 |

(2) Principle component analysis (PCA): PCA is a standard orthogonal linear feature transformation technique that transforms features into principle components. We have used scikit-learn API [103] to fit and transform the fMRI data to 1000 principle components.

(3) Independent component analysis (ICA): ICA is another statistical feature transformation technique, used to decompose feature space to statistically independent component space by maximising statistical independence of the estimated components. We have used scikit-learn [104] API's FastICA algorithm to fit and transform the fMRI data to 1000 independent components.

(4) Bernoulli restricted Boltzman's machine (RBM) [105]: Restricted Boltzmann machines (RBM) are unsupervised nonlinear feature learners based on a probabilistic model that has gained much popularity in the deep neural network domain. We have used the scikit-learn API to learn a Bernoulli RBM network with 1000 components using stochastic Maximum likelihood [104] learning.

### 11.5.2   Modelling and Experimental Results

Table 11.2 presents experimental result as comparisons. The rows of the table compares the methods for the classification task. (C) and (E) in the method names correspond to the custom and Euclidean distance function used as part of KNN respectively. The framework column specifies the role of each component in the method names. For example the proposed BSA + oiSTDP + KNN is a combination of Temporal feature compressor (TFC), Spatial expander (SE) and Classifier (C). The Performance of the binary classification task is measured by overall accuracy and Cohen's $\kappa$ statistic. The first 3 rows of the table compare the different NeuCube architectures. The BSA + oiSTDP + KNN (C) is the proposed architecture for fMRI and DTI integrated learning. The next two methods systematically removes (1) orientational influence from SNNc learning (STDP) and (2) the SNNc module to show the effect of inclusion of these artefacts on the performance. The best performance across the different methods is achieved by the proposed BSA + oiSTDP + KNN (C) architecture with overall accuracy of 72.4 ± 12.3% and Cohen's kappa of 0.44 ± 0.25. The classification accuracy increases by ≈8% and doubles the mean Cohen's $\kappa$ statistic when oiSTDP based SNNc learning is performed in the middle using fMRI and DTI data.

**Table 11.2** Comparative analysis of classification accuracy between the proposed method oiSTDP for integrated learning of fMRI and DTI data used in personalised SNN classifiers with other machine learning methods on the same fMRI classification data

| Method | Data | Temporal | Multi-dimensional | Accuracy (%) | Cohen's $\kappa$ |
|---|---|---|---|---|---|
| BSA + oiSTDP + KNN | fMRI + DTI | Yes | Yes | 72.3 ± 12.3 | 0.44 ± 0.25 |
| BSA + STDP + KNN | fMRI | Yes | No | 69.4 ± 13.9 | 0.38 ± 0.28 |
| BSA + KNN · | fMRI | No | No | 64.2 ± 12.4 | 0.22 ± 0.26 |
| Sparse autoencoder + KNN(E) | fMRI | No | No | 56.1 ± 7.2 | 0.01 ± 0.11 |
| PCA + KNN(E) | fMRI | No | No | 56.1 ± 11.3 | 0.13 ± 0.18 |
| ICA + KNN(E) | fMRI | No | No | 62.8 ± 12.3 | 0.26 ± 0.23 |
| RBM + KNN(E) | fMRI | No | No | 36.2 ± 4.9 | −0.23 ± 0.11 |
| LSTM | fMRI | Yes | No | 45.7 ± 9.6 | −0.15 ± 0.14 |
| GRU | fMRI | Yes | No | 45.2 ± 7.5 | −0.018 ± 0.13 |

Due to the non-temporal nature of the baseline feature compressors, the fMRI data for each subject is input to these feature extractors as a single vector(created by concatenating the temporal dimension) leading to a massive feature vector space. We have used KNN ($K = 1$) as classifier in the classification module to keep the comparisons as fair as possible. The disadvantage of the large feature space is quite imperative as it leads to potential over fitting of the data. We have avoided adding the DTI data to the already large feature space to avoid further over fitting. As the SNNc of NeuCube is a spiking recurrent neural network framework with temporal or sequential learning capabilities, we have also learned the binary classification task with other single hiddden layer recurrent neural network framework such as long short term memory (LSTM) [106] and gated recurrent units (GRU) [107]. Both LSTM and GRU networks were designed as shallow single hidden layered neural networks having 50 LSTM and GRU units. The networks were implemented in keras [102] API and learned by optimising binary crossentropy loss function using adaptive momentum optimiser. The results are shown in Table 11.2.

This study was conducted as part of a large cross-sectional study investigating clozapine (CLZ) response in people with treatment-resistant schizophrenia (TRS) using EEG, MRI and genetic information. CLZ is uniquely effective for treatment-resistant schizophrenia. However, many people still suffer from residual symptoms or do not respond at all (ultra-treatment resistant schizophrenia; UTRS) to CLZ. In this study, our aim was to build a predictive model for discriminating CLZ monotherapy respondent and non-respondent individuals using multimodal brain data.

For the purpose of our investigation, we used a subset of data (resting state fMRI and DTI data) collected from the TRS study with the intention of classifying subjects into groups with either TRS or UTRS. Both fMRI and DTI data for each subject were registered to a subject specific structural image and normalized to the MNI-152 2 mm atlas [96, 97].

As the fMRI data was collected during resting-state, the mean activity and deviation of activity from the voxels over time is negligible compared to task-driven fMRI data. Since a major component of our model is time dependent, we hypothesize that the discriminatory information is hidden in the voxels with significant variation in the activity over time. We selected a set of voxels with an absolute mean standard deviation of greater than 105. The final preprocessed dataset consists of one fMRI trial and one DTI trial of 2318 voxels per subject.

To create a personalized SNN model of the NeuCube, we used aiSTDP learning algorithm to train a set of 1000 computational spiking neurons, randomly scattered around the input neurons. The experimental results are reported after a grid based hyper-parameter search using the leave one out validation protocol. The best model achieved an overall cross validated accuracy of 72%. The area under the ROC curve for this model was 0.72. Evaluation of the confusion matrix showed equally distributed true positive/negative (UTRS: 73%, TRS: 71%) and false positive/negative (UTRS: 27%, TRS: 29%) rates.

We have further compared the classification performance of the model built on fMRI and DTI with models built using only fMRI through a number of pattern recognition algorithms (see Table 11.2). For modelling fMRI data, we have used three different algorithms. The personalized SNN + STDP method uses the canonical STDP to update the weights of the SNNc in the NeuCube architecture. The other two algorithms used are the standard machine learning (ML) algorithms like SVM and MLP. The proposed personalized SNN + aiSTDP outperformed the other algorithms, not only in the overall accuracy of the model but in the true positive and true negative metrics, which allows the model to be the most robust of all. Furthermore, we have individually scrutinized the connection weights for TRS and the UTRS groups, generated by the aiSTDP learning algorithm shows a comparison of the strongest mean connection weights of the TRS and the UTRS groups.

Figure 11.13 shows a visual comparison of the strongest connections (mean weight across subjects within a group) formed in the SNN model of the TRS (the top) and the UTRS group (bottom row). The yellow colored cluster represents the input neurons and the green neurons are the computational spiking neurons.

The majority of the strong connections are created in the lower cerebellum and thalamus. It has been shown that by connections via the thalamus, the cerebellum innervates with motor cortical, prefrontal and parietal lobes [108]. Following cerebellar damage, neurocognitive symptoms and a cognitive affective syndrome including blunted affect and inappropriate behaviour have been shown [109]. Our findings confirm the recent fMRI and PET studies that have demonstrated the involvement of cerebellum and thalamus in sensory discrimination [110], attention [111], and complex problem solving. All these functional modules are impaired in people with schizophrenia. Also a large density of strong connections is observed in the cerebellum region in the UTRS group compared to the TRS group. Similarly, larger number of strong connections are present in the thalamus region of the TRS as opposed to UTRS.

Table 11.3 shows a comparison of classification performance by different pattern recognition methods on the binary classification task described above.

**Table 11.3** Comparison of classification performance by different pattern recognition methods on the binary classification task

| Method | Data | Accuracy (%) | TP rate (%) | TN rate (%) |
|---|---|---|---|---|
| Personalized SNN + aiSTDP | fMRI + DTI | 72 | 73 | 71 |
| Personalized SNN + STDP | fMRI | 56 | 55 | 57 |
| SVM [23] | fMRI | 64 | 64 | 71 |
| autoMLP [24] | fMRI | 60 | 60 | 64.2 |

## 11.6    Chapter Summary and Further Readings
for Deeper Knowledge

The method presents the first attempt to integrate multiple modalities of information in a spiking neural network architecture. The novelty of this approach lies in the proposed personalised SNNc based architecture of NeuCube, and most importantly the proposed oiSTDP learning algorithm, which can integrate multiple dimensions of information including time, space, distance and orientation from data possessing heterogeneous spatial and temporal characteristics. The personalised modelling approach using multiple SNNc, on the other hand, negates any sequential bias of the samples on the generated model. Despite the assumptions on multimodal brain data, the proposed algorithm is not restricted to brain data and further possess the ability to handle any form of data that has some form of spatial, temporal and orientation information. Examples of such data include weather (change in temperature, wind movement, cloud movement etc.) and traffic data.

The experiments shown here were conducted to demonstrate the ability of the algorithm to capture discriminative joint information present in the data and represent this information within its connection strengths. The advantage of our formulation lies in its flexibility to include several dimensions of static and dynamic information. Successful integration of these data types in the present study provides a foundation on which more complex algorithms may be built. We have used the current design to incorporate DTI and fMRI from individuals initiating antipsychotic therapy to create a personalised classifier of treatment response in people with schizophrenia. Interrogation of the classification algorithm revealed increased network connectivity in the cerebellar region of the model, potentially implicating activity in this area of the brain as a biomarker of treatment response in schizophrenia. Inclusion of more participants and studies using specific task-based designs may expose other markers not currently identified in the literature and provide novel hypotheses regarding why some individuals respond to clozapine monotherapy while others do not.

Additional applications of the algorithm may include other disorders where treatment or clinical outcome is poorly understood. Application of a technique such as the one discussed in the current paper could increase our understanding of the mechanisms behind these outcomes, given that the proposed architecture of NeuCube provides a white box view of its classification decisions in an individual level.

The ability to incorporate data from multiple imaging modalities simultaneously could increase reliability of the model to predict treatment outcomes in future individuals. To date, studies have achieved high rates of accuracy in patient samples combining single imaging techniques alongside clinical and pharmacogenetic data [32, 33], though none have led to changes in clinical practice.

The presented in this chapter method for the integration of time-space and direction data is a generic method, applicable to various types of data across a large scale of tasks.

Further recommended readings include:

- fMRI [1];
- Modelling fMRI data in NeuCube [49, 112];
- A complete presentation of the proposed here method for fMRI + DTI modelling [113];
- Demo on modelling fMRI data in NeuCube: https://kedri.aut.ac.nz/R-and-D-Systems/neucube/fmri.

**Acknowledgements**  Some of the material in this chapter was first published in [113]. I would like to acknowledge the contribution of my co-authors Neelava Sengupta, C. McNabb and B. Russel and especially the first author of the paper Neel.

# References

1. R.C. DeCharms, Application of real-time fMRI. Nat. Rev. Neurosci. **9**, 720–729 (2008)
2. J.P. Mitchell, C.N. Macrae, M.R. Banaji, Encoding specific effects of social cognition on the neural correlates of subsequent memory. J. Neurosci. **24**(21), 4912–4917 (2004)
3. K.H. Brodersen, K. Wiech, E.I. Lomakina, C.S. Lin, J.M. Buhmann, U. Bingel, I. Tracey, Decoding the perception of pain from fMRI using multivariate pattern analysis. NeuroImage **63**(3), 1162–1170 (2012). https://doi.org/10.1016/j.neuroimage.2012.08.035
4. R.B. Buxton, K. Uludağ, D.J. Dubowitz, T.T. Liu, Modeling the hemodynamic response to brain activation. NeuroImage **23**(Suppl 1), S220–S233 (2004). https://doi.org/10.1016/j.neuroimage.2004.07.013
5. S. Ogawa, T.M. Lee, A.R. Kay, D.W. Tank, Brain magnetic resonance imaging with contrast dependent on blood oxygenation. Proc. Natl. Acad. Sci. **87**(24), 9868–9872 (1990)
6. J.V. Haxby, M.I. Gobbini, M.L. Furey, A. Ishai, J.L. Schouten, P. Pietrini, Distributed and overlapping representation of faces and objects in ventral temporal cortex. Science **293** (5539), 2425–2430 (2001)
7. M.A. Lindquist, The statistical analysis of fMRI data. Stat. Sci. **23**(4), 439–464 (2008). https://doi.org/10.1214/09-STS282
8. K.J. Friston, C.D. Frith, R.S. Frackowiak, R. Turner, Characterizing dynamic brain responses with fMRI: a multivariate approach. NeuroImage (1995). Retrieved from http://www.ncbi.nlm.nih.gov/pubmed/9343599
9. M.K. Carroll, G.A. Cecchi, I. Rish, R. Garg, A.R. Rao, Prediction and interpretation of distributed neural activity with sparse models. NeuroImage **44**(1), 112–122 (2009)
10. T. Schmah, R.S. Zemel, G.E. Hinton, S.L. Small, S.C. Strother, Comparing classification methods for longitudinal fMRI studies. Neural Comput. **22**(11), 2729–2762 (2010). https://doi.org/10.1162/NECO_a_00024
11. D.D. Cox, R.L. Savoy, Functional magnetic resonance imaging (fMRI) "brain reading": detecting and classifying distributed patterns of fMRI activity in human visual cortex. Neuroimage, **19**(2 Pt 1), 261–270 (2003)
12. Y. Kamitani, F. Tong, Decoding the visual and subjective contents of the human brain. Nat. Neurosci. **8**(5), 679–85 (2005)

13. M. Misaki, Y. Kim, P.A. Bandettini, N. Kriegeskorte, Comparions of multivariate classifiers and response normalizations for pattern-information fMRI. NeuroImage **53**(1), 103–118 (2010). https://doi.org/10.1016/j.neuroimage.2010.05.051.Comparison

14. J. Mourão-Miranda, A.L.W. Bokde, C. Born, H. Hampel, M. Stetter, Classifying brain states and determining the discriminating activation patterns: support vector machine on functional MRI data. NeuroImage **28**(4), 980–995 (2005). https://doi.org/10.1016/j.neuroimage.2005.06.070

15. T.M. Mitchell, R. Hutchinson, M.A. Just, R.S. Niculescu, F. Pereira, X. Wang, *Classifying Instantaneous Cognitive States from FMRI Data. AMIA*, in Annual Symposium Proceedings/ AMIA Symposium. AMIA Symposium (2003), pp. 465–469. Retrieved from http://www.pubmedcentral.nih.gov/articlerender.fcgi?artid=1479944&tool=pmcentrez&rendertype= abstract

16. I. Rustandi, *Classifying Multiple-Subject fMRI Data Using the Hierarchical Gaussian Naïve Bayes Classifier*, in 13th Conference on Human Brain Mapping (2007a), pp. 4–5

17. I. Rustandi, Hierarchical Gaussian Naive Bayes classifier for multiple-subject fMRI data. Submitted to AISTATS (1), 2–4 (2007b)

18. S.M. Polyn, G.J. Detre, S. Takerkart, V.S. Natu, M.S. Benharrosh, B.D. Singer, J.D. Cohen, J.V. Haxby, K.A. Norman, A Matlab-based toolbox to facilitate multi-voxel pattern classification of fMRI data, 2005

19. Y. Fan, D. Shen, C. Davatzikos, Detecting cognitive states from fMRI images by machine learning and multivariate classification, in Conference on Computer Vision and Pattern Recognition Workshop. IEEE (2006), pp. 89–89 https://doi.org/10.1109/cvprw.2006.64

20. S.J. Hanson, T. Matsuka, J.V. Haxby, Combinatorial codes in ventral temporal lobe for object recognition: Haxby (2001) revisited: is there a "face" area? NeuroImage **23**(1), 156–166 (2004). https://doi.org/10.1016/j.neuroimage.2004.05.020

21. G. Yourganov, T. Schmah, N.W. Churchill, M.G. Berman, C.L. Grady, S.C. Strother, Pattern classification of fMRI data: applications for analysis of spatially distributed cortical networks. NeuroImage **96**, 117–132 (2014). https://doi.org/10.1016/j.neuroimage.2014.03.074

22. J.D. Haynes, G. Rees, Predicting the orientation of invisible stimuli from activity in human primary visual cortex. Nat. Neurosci. **8**(5), 686–691 (2005). https://doi.org/10.1038/nn1445

23. S. Ku, A. Gretton, J. Macke, N.K. Logothetis, Comparison of pattern recognition methods in classifying high-resolution BOLD signals obtained at high magnetic field in monkeys. Magn. Reson. Imaging **26**(7), 1007–1014 (2008). https://doi.org/10.1016/j.mri.2008.02.016

24. T. Schmah, G.E. Hinton, R.S. Zemel, S.L. Small, S. Strother, Generative versus discriminative training of RBMs for classification of fMRI images, in *Advances in Neural Information Processing Systems, 21*, ed. by D. Koller, D. Schuurmans, Y. Bengio, L. Bottou (MIT Press, Cambridge, MA, 2009), pp. 1409–1416

25. S. LaConte, S. Strother, V. Cherkassky, J. Anderson, X. Hu, Support vector machines for temporal classification of block design fMRI data. NeuroImage **26**(2), 317–329 (2005). https://doi.org/10.1016/j.neuroimage.2005.01.048

26. N. Mørch, L. Hansen, S. Strother, C. Svarer, D. Rottenberg, B. Lautrup, *Nonlinear Versus Linear Models in Functional Neuroimaging: Learning Curves and Generalization Crossover*, in Proceedings of the 15th International Conference on Information Processing in Medical Imaging, vol. 1230 of Lecture Notes in Computer Science. Springer (1997), pp. 259–270

27. J. Mourão-Miranda, K.J. Friston, M. Brammer, Dynamic discrimination analysis: a spatial-temporal SVM. NeuroImage **36**(1), 88–99 (2007). https://doi.org/10.1016/j.neuroimage.2007.02.020

28. M.A.J. Van Gerven, B. Cseke, F.P. de Lange, T. Heskes, Efficient Bayesian multivariate fMRI analysis using a sparsifying spatio-temporal prior. NeuroImage **50**(1), 150–161 (2010). https://doi.org/10.1016/j.neuroimage.2009.11.064

29. B. Ng, R. Abugharbieh, *Modeling Spatiotemporal Structure in fMRI Brain Decoding Using Generalized Sparse Classifiers*, in 2011 International Workshop on Pattern Recognition in NeuroImaging (2011b), pp. 65–68. https://doi.org/10.1109/prni.2011.10

30. P. Avesani, H. Hazan, E. Koilis, L. Manevitz and D. Sona, *Learning BOLD Response in fMRI by Reservoir Computing*, in 2011 International Workshop on Pattern Recognition in NeuroImaging, (2011), pp. 57–60. https://doi.org/10.1109/prni.2011.16

31. J. Sui, T. Adalı, Q. Yu, J. Chen, V.D. Calhoun, A review of multivariate methods for multimodal fusion of brain imaging data. J. Neurosci. Methods, **204**(1), 68–81 (2012)

32. M.J. Patel, C. Andreescu, J.C. Price, K.L. Edelman, C.F. Reynolds, III, H.J. Aizenstein, Machine learning approaches for integrating clinical and imaging features in late-life depression classification and response prediction. Int. J. Geriatric Psychiatry **30**(10), 1056–1067 (2015)

33. A. Khodayari-Rostamabad, J.P. Reilly, G.M. Hasey, H. de Bruin, D.J. MacCrimmon, A machine learning approach using EEG data to predict response to ssri treatment for major depressive disorder. Clin. Neurophysiol. **124**(10), 1975–1985 (2013)

34. A. Khodayari-Rostamabad, G.M. Hasey, D.J. MacCrimmon, J.P. Reilly, H. de Bruin, A pilot study to determine whether machine learning methodologies using pre-treatment electroencephalography can predict the symptomatic response to clozapine therapy. Clin. Neurophysiol. **121**(12), 1998–2006 (2010)

35. C.-C. Lin et al., Artificial neural network prediction of clozapine response with combined pharmacogenetic and clinical data. Comput. Methods Programs Biomed. **91**(2), 91–99 (2008)

36. O. Doehrmann et al., Predicting treatment response in social anxiety disorder from functional magnetic resonance imaging. JAMA Psychiatry **70**(1), 87–97 (2013)

37. M.D. Greicius, K. Supekar, V. Menon, R.F. Dougherty, Restingstate functional connectivity reflects structural connectivity in the default mode network. Cereb. Cortex **19**(1), 72–78 (2009)

38. K.E. Stephan, K.J. Friston, C.D. Frith, Dysconnection in schizophrenia: from abnormal synaptic plasticity to failures of self-monitoring. Schizophrenia Bull. **35**(3), 509–527 (2009)

39. P.A. Valdes-Sosa et al., Model driven EEG/fMRI fusion of brain oscillations. Hum. Brain Mapping, **30**(9), 2701–2721 (2009)

40. S. M. Plis et al., Effective connectivity analysis of fMRI and MEG data collected under identical paradigms. Comput. Biol. Med. **41**(12), 1156–1165 (2011)

41. H. Yang, J. Liu, J. Sui, G. Pearlson, V.D. Calhoun, A hybrid machine learning method for fusing fMRI and genetic data: combining both improves classification of schizophrenia. Frontiers Hum. Neurosci. **4**(192), 3389 (2010)

42. V. D. Calhoun, J. Liu, T. Adalı, A review of group ICA for fMRI data and ICA for joint inference of imaging, genetic, and ERP data. NeuroImage, **45**(1), S163–S172 (2009)

43. S.J. Teipel, A.L. Bokde, T. Meindl, E. Amaro Jr, J. Soldner, M.F. Reiser, S.C. Herpertz, H. J. Möller, H. Hampel, White matter microstructure underlying default mode network connectivity in the human brain. NeuroImage, **49**(3), 2021–2032 (2010)

44. N.M. Correa, Y.O. Li, T. Adali, V.D. Calhoun, Canonical correlation analysis for feature-based fusion of biomedical imaging modalities and its application to detection of associative networks in schizophrenia. IEEE J. Sel. Topics Signal Process. **2**(6), 998–1007 (2008)

45. N.M. Correa, T. Eichele, T. Adalı, Y.-O. Li, V.D. Calhoun, Multiset canonical correlation analysis for the fusion of concurrent single trial ERP and functional MRI. NeuroImage, **50** (4), 1438–1445 (2010)

46. N. Correa, Y.-O. Li, T. Adalı, V.D. Calhoun, *Examining Associations Between fMRI and EEG Data Using Canonical Correlation Analysis*, in Proceedings 5th IEEE International Symposium on Biomedical Imaging, Nano Macro (ISBI), May 2008, pp. 1251–1254

47. K. Chen et al., Linking functional and structural brain images with multivariate network analyses: a novel application of the partial least square method. NeuroImage, **47**(2), 602–610 (2009)

48. N.K. Kasabov, NeuCube: a spiking neural network architecture for mapping, learning and understanding of spatio-temporal brain data. Neural Netw. **52**, 62–76 (2014)
49. N. Kasabov, M. Doborjeh, Z. Doborjeh, Mapping, learning, visualization, classification, and understanding of fMRI Data in the NeuCube evolving spatiotemporal data machine of spiking neural networks, in IEEE Transactions of Neural Networks and Learning Systems, https://doi.org/10.1109/tnnls.2016.2612890 Manuscript Number: TNNLS-2016-P-6356, 2016
50. N. Kasabov, To spike or not to spike: a probabilistic spiking neuron model. Neural Netw. **23** (1), 16–19 (2010). https://doi.org/10.1016/j.neunet.2009.08.010
51. M. Åberg, L. Löken, J. Wessberg, An evolutionary approach to multivariate feature selection for fMRI pattern analysis (2008)
52. T. Niiniskorpi, M. Bj, J. Wessberg, Particle swarm feature selection for fMRI pattern classification, in BIOSIGNALS (2009), pp. 279–284
53. J. Kennedy, R. Eberhart, *Particle Swarm Optimization*, in Proceedings of ICNN'95— International Conference on Neural Networks, vol. 4 (1995), pp. 1942–1948. https://doi.org/10.1109/icnn.1995.488968
54. C. Koch, R.C. Reid, Observatories of the mind. Nature **483**(22 March 2012), 397–398 (2012). https://doi.org/10.1038/483397a
55. X. Wang, J. Yang, X. Teng, W. Xia, R. Jensen, Feature selection based on rough sets and particle swarm optimization. Pattern Recogn. Lett. **28**(4), 459–471 (2007). https://doi.org/10.1016/j.patrec.2006.09.003
56. S. Ryali, K. Supekar, D.A. Abrams, V. Menon, Sparse logistic regression for whole-brain classification of fMRI data. NeuroImage **51**(2), 752–764 (2010)
57. O. Yamashita, M. Sato, T. Yoshioka, F. Tong, Y. Kamitani, Sparse estimation automatically selects voxels relevant for the decoding of fMRI activity patterns. NeuroImage **42**(4), 1414–1429 (2008). https://doi.org/10.1016/j.neuroimage.2008.05.050
58. B. Ng, A. Vahdat, G. Hamarneh, R. Abugharbieh, Generalized sparse classifiers for decoding cognitive states in fMRI, in *Machine Learning in Medical Imaging* (Springer, Berlin, 2010), pp. 108–115
59. B. Ng, R. Abugharbieh, Generalized group sparse classifiers with application in fMRI brain decoding. Cvpr **2011**, 1065–1071 (2011). https://doi.org/10.1109/CVPR.2011.5995651
60. R. Brette et al., Simulation of networks of spiking neurons: a review of tools and strategies. J. Comput. Neurosci. **23**(3), 349–398 (2007)
61. E.M. Izhikevich, Polychronization: computation with spikes. Neural Comput. **18**(2), 245–282 (2006)
62. N. Scott, N. Kasabov, G. Indiveri, *NeuCube Neuromorphic Framework for Spatio-Temporal Brain Data and Its Python Implementation*, in Proceedings ICONIP, vol. 8228 (Springer, LNCS) (2013), pp. 78–84
63. S.B. Furber, F. Galluppi, S. Temple, L.A. Plana, The SpiNNaker project. Proc. IEEE **102**(5), 652–665 (2014)
64. P. A. Merolla et al., A million spiking-neuron integrated circuit with a scalable communication network and interface. Science **345**(6197), 668–673 (2014)
65. G. Indiveri et al., Neuromorphic silicon neuron circuits, Frontiers Neurosci. May 2011. [Online]. Available: http://dx.doi.org/10.3389/fnins.2011.00073
66. A. van Schaik, S.-C. Liu, *AER EAR: A Matched Silicon Cochlea Pair with Address Event Representation Interface*, in Proceedings IEEE International Symposium Circuits System, vol. 5. May 2005, pp. 4213–4216
67. T. Delbruck, jAER. [Online]. Available: http://sourceforge.net. Accessed 15 Oct 2014
68. P. Lichtsteiner, C. Posch, T. Delbruck, A 128 × 128 120 dB 15 using latency asynchronous temporal contrast vision sensor. IEEE J. SolidState Circuits **43**(2), 566–576 (2008)
69. S. Song, K.D. Miller, L.F. Abbott, Competitive Hebbian learning through spike-timing-dependent synaptic plasticity. Nat. Neurosci. **3**(9), 919–926 (2000)

70. N. Kasabov, K. Dhoble, N. Nuntalid, G. Indiveri, Dynamic evolving spiking neural networks for on-line spatio- and spectro-temporal pattern recognition. Neural Netw. **41**, 188–201 (2013)
71. N. Kasabov, *Evolving Connectionist Systems* (Springer, New York, 2007), p. 2007
72. S.G. Wysoski, L. Benuskova, N. Kasabov, Evolving spiking neural networks for audiovisual information processing. Neural Netw. **23**(7), 819–835. (2010)
73. W. Maass, T. Natschläger, H. Markram, Real-time computing without stable states: a new framework for neural computation based on perturbations. Neural Comput. **14**(11), 2531–2560 (2002)
74. J. Talairach, P. Tournoux, *Co-Planar Stereotaxic Atlas of the Human Brain: 3-Dimensional Proportional System: An Approach to Cerebral Imaging* (Thieme Medical Publishers, New York, 1998), p. 1988
75. M. Brett, K. Christoff, R. Cusack, J. Lancaster, Using the Talairach atlas with the MNI template. NeuroImage **13**(6), 85 (2001)
76. N. Kasabov et al., Evolving spatio-temporal data machines based on the NeuCube neuromorphic framework: design methodology and selected applications. Neural Netw. **78**, 1–14 (2016). https://doi.org/10.1016/j.neunet.2015.09.011.2015
77. E.M. Izhikevich, Which model to use for cortical spiking neurons? IEEE Trans. Neural Netw. **15**(5), 1063–1070 (2004)
78. E. Tu et al., *NeuCube(ST) for Spatio-Temporal Data Predictive Modelling with a Case Study on Ecological Data*, in Proceedings International Joint Conference on Neural Networks (IJCNN), Beijing, China, July 2014, pp. 638–645
79. E. Tu, N. Kasabov, J. Yang, Mapping temporal variables into the NeuCube for improved pattern recognition, predictive modeling, and understanding of stream data. IEEE Trans. Neural Netw. Learn. Syst. **99**, 1–13 (2016). https://doi.org/10.1109/tnnls.2016.2536742. 2016
80. E. Bullmore, O. Sporns, Complex brain networks: graph theoretical analysis of structural and functional systems. Nature Rev. Neurosci. **10**(3), 186–198 (2009)
81. V. Braitenberg, A. Schuz, *Cortex: Statistics and Geometry of Neuronal Connectivity* (Springer, Berlin, 1998)
82. S. Schliebs, N. Kasabov, Evolving spiking neural network—a survey. Evolving Syst. **4**(2), 87–98 (2013)
83. M. Lukoševičius, H. Jaeger, Reservoir computing approaches to recurrent neural network training. Comput. Sci. Rev. **3**(3), 127–149 (2009)
84. B. Schrauwen, J. Van Campenhout, BSA, *a Fast and Accurate Spike Train Encoding Scheme*, in Proceedings International Joint Conference on Neural Networks (IJCNN), vol. 4. July 2003, pp. 2825–2830
85. N. Sengupta, N. Scott, N. Kasabov, *Framework for Knowledge Driven Optimisation Based Data Encoding for Brain Data Modelling Using Spiking Neural Network Architecture*, in Proceedings 5th International Conference on Fuzzy and Neural Computing (FANCCO), (2015), pp. 109–118
86. N. Sengupta, N. Kasabov, Spike-time encoding as a data compression technique for pattern recognition of temporal data. Inf. Sci. **406–407**, 133–145 (2017)
87. T. Kohonen, The self-organizing map. Proc. IEEE **78**(9), 1464–1480 (1990)
88. N. Kasabov et al., Evolving spiking neural networks for personalised modelling, classification and prediction of spatio-temporal patterns with a case study on stroke. Neurocomputing **134**, 269–279 (2014)
89. E. Tu et al., *NeuCube(ST) for Spatio-Temporal Data Predictive Modelling with a Case Study on Ecological Data*, in Proceedings International Joint Conference on Neural Networks (IJCNN), July 2014, pp. 638–645
90. D.O. Hebb, *The Organization of Behavior: A Neuropsychological Approach* (Wiley, Hoboken, NJ, USA, 1949), p. 1949
91. W. Gerstner, R. Kempter, J. L. van Hemmen, H. Wagner, A neuronal learning rule for sub-millisecond temporal coding. Nature **383**(6595), 76–78 (1996)

92. J. Sjöström, W. Gerstner, Spike-timing dependent plasticity. Front. Synaptic Neurosci. **5**(2), 35–44 (2010)
93. E. van Aart, N. Sepasian, A. Jalba, A. Vilanova, CUDA-accelerated geodesic ray-tracing for fiber tracking. J. Biomed. Imag. **2011**, (2011). https://doi.org/10.1155/2011/698908. Art. no. 6
94. M. Rubinov, O. Sporns, Complex network measures of brain connectivity: uses and interpretations. NeuroImage **52**(3), 1059–1069 (2010)
95. M. Jenkinson, C.F. Beckmann, T.E. Behrens, M.W. Woolrich, S.M. Smith, FSL. NeuroImage, **62**(2), 782–790 (2012)
96. V. Fonov et al., Unbiased average age-appropriate atlases for pediatric studies. NeuroImage **54**(1), 313–327 (2011)
97. V. Fonov, A.C. Evans, K. Botteron, C.R. Almli, R.C. McKinstry, D.L. Collins, Unbiased average age-appropriate atlases for pediatric studies. NeuroImage **54**(1), 313–327 (2011). [Online]. Available: http://www.sciencedirect.com/science/article/pii/S1053811910010062, https://doi.org/10.1016/j.neuroimage. 2010.07.033
98. R.H.R. Pruim, M. Mennes, J.K. Buitelaar, C.F. Beckmann, Evaluation of ICA-AROMA and alternative strategies for motion artefact removal in resting state fMRI. NeuroImage **112**, 278–287 (2015)
99. R.H.R. Pruim, M. Mennes, D. van Rooij, A. Llera, J.K. Buitelaar, C.F. Beckmann, ICA-AROMA: a robust ICA-based strategy for removing motion artifacts from fMRI data. NeuroImage **112**, 267–277 (2015)
100. N. Kasabov et al. (2016), Evolving spatio-temporal data machines based on the NeuCube neuromorphic framework: design methodology and selected applications. Neural Netw. **78**, 1–14 (2016)
101. A. Ng, Sparse autoencoder. CS294A Lect. Notes **72**, 1–19 (2011)
102. F. Chollet et al., Keras. [Online]. Available: https://github.com/fchollet/keras
103. F. Pedregosa et al., Scikit-learn: machine learning in Python, J. Mach. Learn. Res. **12**, 2825–2830 (2011)
104. G.E. Hinton, R.R. Salakhutdinov, Reducing the dimensionality of data with neural networks. Science **313**(5786), 504–507 (2006)
105. T. Tieleman, *Training Restricted Boltzmann Machines Using Approximations to the Likelihood Gradient*, in Proceedings 25th International Conference on Machine Learning (2008), pp. 1064–1071
106. S. Hochreiter, J. Schmidhuber, *LSTM Can Solve Hard Long Time Lag Problems*, in Proceedings Advances Neural Information Processing Systems (1997), pp. 473–479
107. K. Cho et al., Learning phrase representations using RNN encoder-decoder for statistical machine translation. [Online]. Available: https://arxiv.org/abs/1406.1078
108. F.A. Middleton, P.L. Strick, Anatomical evidence for cerebellar and basal ganglia involvement in higher cognitive function. Science **266**(5184), 458–461 (1994). [Online]. Available: http://science.sciencemag.org/content/266/5184/458, https://doi.org/10.1126/science.7939688
109. H. Baillieux, W. Verslegers, P. Paquier, P.P. De Deyn, P. Mariën, Cerebellar cognitive affective syndrome associated with topiramate. Clin. Neurol. Neurosurg. **110**(5), 496–499 (2008)
110. J.-H. Gao, L.M. Parsons, J.M. Bower, J. Xiong, J. Li, P.T. Fox, Cerebellum implicated in sensory acquisition and discrimination rather than motor control. Science **272**(5261), 545–547 (1996)
111. E. Courchesne, N.A. Akshoomoff, J. Townsend, O. Saitoh, A model system for the study of attention and the cerebellum: infantile autism. Electroencephalogr. Clin. Neurophysiol. Suppl. **44**, 315–325 (1995)
112. N. Kasabov, L. Zhou, M. Gholami Doborjeh, J. Yang, New algorithms for encoding, learning and classification of fMRI Data in a spiking neural network architecture: A case in modelling and understanding of dynamic cognitive processes. in IEEE Transaction on Cognitive and Developmental Systems, 2017. https://doi.org/10.1109/TCDS.2016.2636291

113. N. Sengupta, C. McNabb, N. Kasabov, B. Russel, Integrating space, time and orientation in spiking neural networks: a case study on multimodal brain data modelling, in IEEE Tr NNLS, 2018. https://ieeexplore.ieee.org/document/8291047/
114. Medical Image Computing, [Online]. Available: https://en.wikipedia.org/wiki/Medical_image_computing. Accessed 31 Jan 2018

# Part V
# SNN for Audio-Visual Data and Brain-Computer Interfaces

# Chapter 12
# Audio- and Visual Information Processing in the Brain and Its Modelling with Evolving SNN

This chapter presents first some background knowledge on how the human brain processes audio- and visual information. Then methods are presented for audio-, visual- and for the integrated audio and visual information processing using evolving spiking neural networks that include convolutional evolving spiking neural networks (CeSNN). Case studies are presented for person identification.

The chapter is organized in the following sections:

12.1. Audio and visual information processing in the human brain.
12.2. Modelling audio-, visual and audio-visual information processing with convolutional evolving spiking neural networks (CeSNN).
12.3. Case studies, experiments and results.
12.4. Summary and further readings for deeper knowledge.

## 12.1 Audio and Visual Information Processing in the Human Brain

In this section some basic facts about how the human brain processes audio and visual information are presented. Some more information can be found in [1, 2].

The human brain deals mainly with 5 sensory modalities: vision, hearing, touch, taste and smell. Each modality has different sensory receptors. After the receptors perform the stimulus transduction, the information is encoded through the excitation of neural action potentials. The information is encoded using average of pulses or time interval between pulses. This process seems to follow a common pattern for all sensory modalities, however there are still many unanswered questions regarding the way the information is encoded in the brain.

© Springer-Verlag GmbH Germany, part of Springer Nature 2019    431
N. K. Kasabov, *Time-Space, Spiking Neural Networks and Brain-Inspired Artificial Intelligence*, Springer Series on Bio- and Neurosystems 7,
https://doi.org/10.1007/978-3-662-57715-8_12

## 12.1.1  Audio Information Processing

The hearing apparatus of an individual transforms sounds and speech signals into brain signals. These brain signals travel further to other parts of the brain that model the (meaningful) acoustic space (the space of phones), the space of words, and the space of languages (see Fig. 12.1). The auditory system is adaptive, so new features can be included at a later stage and existing ones can be further tuned.

Precise modelling of hearing functions and the cochlea is an extremely difficult task, but not impossible to achieve [2]. A model of cochlea would be useful for both helping people with disabilities, and for the creation of speech recognition systems. Such systems would be able to learn and adapt as they work.

The ear is the front-end auditory apparatus in mammalians. The task of this hearing apparatus is to transform the environmental sounds into specific features and transmit them to the brain for further processing. The ear consists of three divisions: the outer ear, the middle ear and the inner ear—Fig. 12.2.

Figure 12.3 shows the human basilar membrane and the approximate position of the maximal displacement to tones of different frequencies. This corresponds to a filter bank of several channels, each tuned to a certain band of frequencies.

There are several models that have been developed to model auditory functions [2–9]. Very common are the Mel filter banks and the Mel scale cepstra coefficients [1]. For example, the centres of the first 26 Mel filter banks are the following frequencies [in hertz]: 86, 173, 256, 430, 516, 603, 689, 775, 947, 1033, 1130, 1392, 1550, 1723, 1981, 2325, 2670, 3015, 3445, 3962, 4565, 5254, 6029, 6997, 8010, 9216, 11,025. The first 20 Mel filter functions are shown in Fig. 12.4.

Other representations use a Gammatone function [10]. It is always challenging to improve the acoustic modelling functions and make them closer to the functioning of the biological organs, which is expected to lead to improved speech recognition systems.

The auditory system is particularly interesting because it allows not only to recognize sound but also to perform sound source location efficiently. Human ears are able to detect frequencies in the approximate range of 20–20,000 Hz. Each ear processes the incoming signals independently, which are later integrated considering signal's timing, amplitudes and frequencies—Fig. 12.5.

The narrow difference of time between incoming signals from the left and right ear results in a cue to location of signal origin.

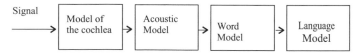

**Fig. 12.1** A typical model of spoken language processing

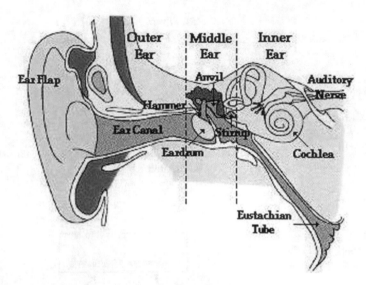

**Fig. 12.2** The human ear (from [34])

**Fig. 12.3** The human basilar membrane and the approximate position of the maximal displacement to tones of different frequencies. This corresponds to a filter bank of several channels, each tuned to a certain band of frequencies

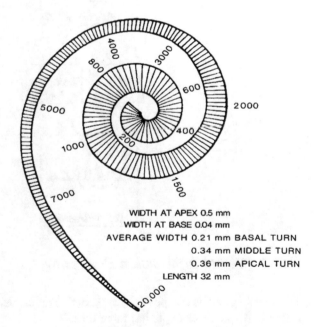

WIDTH AT APEX 0.5 mm
WIDTH AT BASE 0.04 mm
AVERAGE WIDTH 0.21 mm BASAL TURN
0.34 mm MIDDLE TURN
0.36 mm APICAL TURN
LENGTH 32 mm

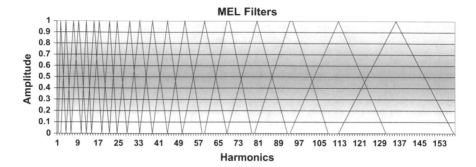

**Fig. 12.4** The first 20 Mel filter functions

**Fig. 12.5** A schematic diagram of integrated left- and right auditory information processing

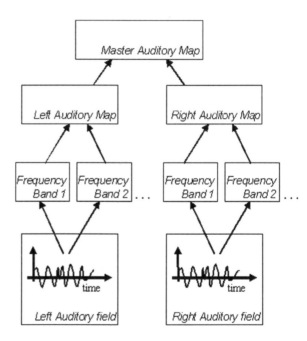

## 12.1.2   Visual Information Processing

The visual system is composed of eyes, optic nerves, many specialised areas of the cortex (the ape for example has more than 30).

The image on the retina is transmitted via the optic nerves to the first visual cortex (V1), which is situated in the posterior lobe of the brain. There the information is divided into two main streams, the "what" tract and the "where" tract.

The ventrical tract ("what") separates targets (objects and things) in the field of vision and identifies them. The tract traverses from the occipital lobe to the temporal lobe (behind the ears).

The dorsal tract ("where") is specialised in following the location and position of the objects in the surrounding space. The dorsal tract traverses from the back of the head to the top of the head.

The brain creates deep structures of neural networks when learns or perceives visual information—as shown in Fig. 12.6.

How and where is the information from the two tracts united to form one complete perception, is not completely known. On the subject of biological approaches for processing incoming information, Hubel and Wiesel received many awards for their description of the human visual system. Through neuro-physiological experiments, they were able to distinguish some types of cells that have different neurobiological responses according to the pattern of light stimulus. They identified the role that the retina has as a contrast filter as well as the existence of orientation selective cells in the primary visual cortex—Fig. 12.7. Their results have been widely implemented in biologically realistic image acquisition approaches.

The idea of contrast filters and orientation selective cells can be considered a feature selection method that finds a close correspondence with traditional ways of image processing, such as Gaussian and Gabor filters.

A Gaussian filter can be used for modelling ON/OFF states of receptive cells:

$$G(x,y) = e^{(\frac{x^2+y^2}{2\sigma^2})} \tag{12.1}$$

A Gabor filter can be used to model the states of orientation cells:

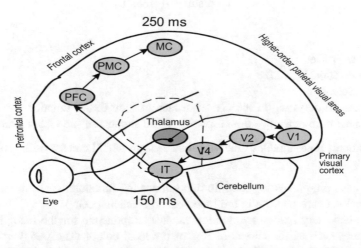

**Fig. 12.6** Deep serial processing of visual stimuli in humans for image classification. Location of cortical areas: V1 = primary visual cortex, V2 = secondary visual cortex, V4 = quartiary visual cortex, IT = inferotemporal cortex, PFC = prefrontal cortex, PMC = premotor cortex, MC = motor cortex [2, 35]

**Fig. 12.7** Contrast cells and direction selective cells

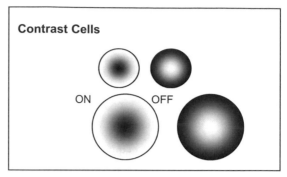

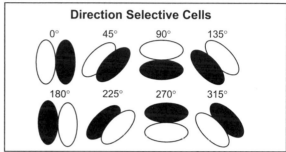

$$G(x, y) = e^{\left(\frac{x'^2 + \gamma^2 y'^2}{2\sigma^2}\right)} \cos\left(2\pi \frac{x'}{\lambda} + \phi\right)$$
$$x' = x\cos(\theta) + y\sin(\theta)$$  \hfill (12.2)
$$y' = -x\sin(\theta) + y\cos(\theta)$$

where:

$\varphi$ = phase offset
$\theta$ = orientation (0, 360)
$\lambda$ = wavelength
$\sigma$ = standard deviation of the Gaussian factor of the Gabor function
$\gamma$ = aspect ratio (specifies the ellipticity of the support of the Gabor function)

A computational model of the visual subsystem would consist of the following levels:

(a) A visual pre-processing module that mimics the functioning of the retina, the retinal network and the Lateral Geniculate Nucleus (LGN).
(b) An elementary feature recognition module, responsible for the recognition of features such as the curves of lips or the local colour. The peripheral visual areas of the human brain perform a similar task.

(c) A dynamic feature recognition module that detects dynamical changes of features in the visual input stream. In the human brain, the processing of visual motion is performed in the V5/MT area of the brain.
(d) An object recognition module that recognises elementary shapes and their parts. This task is performed by the infero-temporal (IT) area of the human brain.
(e) An object/configuration recognition module that recognises objects such as faces. This task is performed by the IT and parietal areas of the human brain.

## 12.1.3 Integrated Audio and Visual Information Processing

This subsection presents a method first published in [31].

How auditory and visual perception relate to each other in the brain is a fundamental question. A general diagram to represent these processes is shown in Fig. 12.8.

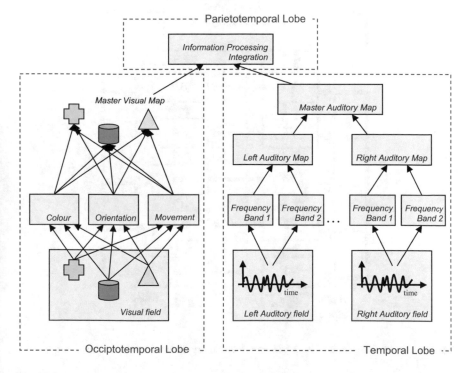

**Fig. 12.8** A schematic diagram of integrated audio-visual information processing in the brain

In the next section, the issue of integrating auditory and visual information in one information processing model is discussed. Such models may lead to better information processing and adaptation in the future intelligent systems.

In [31], a computational model called AVIS was proposed that models both the auditory and visual pathways in the brain but also accounts for the interaction between the different parts as shown in Fig. 12.8.

Below we describe the connectionist framework AVIS, which combines the principles from two preceding unimodal models. One model originates from multilingual adaptive speech processing [10] and the other from image processing using dynamic features [11, 12]. The global architecture of AVIS is illustrated in Fig. 12.9, and consists of three subsystems:

- an auditory subsystem;
- a visual subsystem;
- a higher-level conceptual subsystem.

Each of them is specified below, followed by a description of the modes of operation.

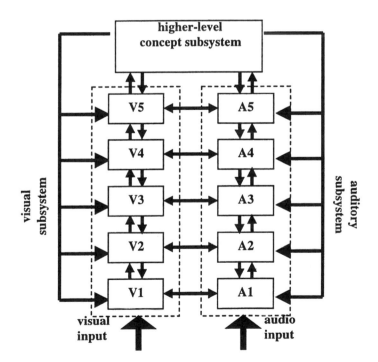

**Fig. 12.9** A block diagram of the framework for auditory visual information processing systems (AVIS) [31]

*The auditory subsystem*

The auditory subsystem consists of five modules. Below we give the main characterisations.

(a) The auditory pre-processing module transforms the auditory signal into frequency features, such as mel-scale coefficients. It accounts for time at a low level of synchronisation (i.e., milliseconds). Frequency, time and intensity features are spatially (tonotopically) represented as a sequence of vectors (i.e., a matrix). The functioning of the pre-processing module may be compared to the functioning of the cochlea.
(b) The elementary-sound recognition module is a basic building block of the subsystem. It is extendable so that new classes of sounds can be added during operation. A phoneme is adequately represented by a population activity pattern, i.e., an activity pattern distributed over a cluster of neurons. The position of the cluster centre can change through learning.
(c) The dynamic-sound recognition module accounts for the dynamical changes in the auditory information. The auditory cortex of the human brain functions analogously.
(d) The word-detection module attempts to identify the words. It uses a dictionary of pre-stored words. In the human brain the auditory detection of words is part of the cortical language areas [13].
(e) The language-structure detection module accounts for the order in which words are recognised. It uses linguistic knowledge, language knowledge, and domain knowledge as well as feedback from the higher-level conceptual sub-system.

The visual subsystem

The visual subsystem also consists of five modules. A characterisation follows below:

(a) The visual pre-processing module mimics the functioning of the retina, the retinal network, and the lateral geniculate nucleus (LGN).
(b) The elementary-feature recognition module is responsible for the recognition of features such as the curves of lips or the local colour. The peripheral visual areas of the human brain perform a similar task.
(c) The dynamic-feature recognition module detects dynamical changes of features in the visual input stream. In the human brain, the processing of visual motion is performed in area V5/MT.
(d) The object-recognition module recognises elementary shapes and their parts. This task is performed by the infero-temporal (IT) areas of the human brain.
(e) The object-configuration recognition module recognises configurations of objects such as faces. This task is performed by the IT and parietal areas of the human brain.

*The higher-level conceptual subsystem*

The higher-level conceptual subsystem takes its inputs from all modules of the lower-level subsystems and activates the clusters of neurons representing concepts (e.g., familiar persons) or meanings. The clusters of neurons are connected to the action part of the system (corresponding to the motor areas of the brain). In a person-identification task, the conceptual subsystem takes information from all the modules in the auditory and visual subsystems and makes a decision on the identity of the person observed.

## 12.2   Modelling Audio-, Visual and Audio-Visual Information Processing with Convolutional Evolving Spiking Neural Networks (CeSNN)

This section presents methods for audio-, for visual-, and for integrated audio-visual information processing using evolving SNN, and more specifically—convolutional eSNN. The methods were first proposed and published in [14, 15].

### 12.2.1   Issues with Modelling Audio-Visual Information with SNN

The integration of modalities for the purpose of pattern recognition often target tasks that cannot be solved by a single system or can be facilitated by using more than one source (generally where there is unimodal ambiguity, unimodal lack of data and/or correlation among modes). Many studies report considerable performance improvement [11, 16–18] as well as state that the use of modularity results in systems that are easy to understand and modify. In addition, modular approaches are well known for preventing modular damage, facilitating training and the inclusion of prior knowledge [16].

There are two classic issues when dealing with multimodal systems: how to perform the decomposition and recombination of modes:

- **Decomposition**: Decomposition can occur with modules and sub-modules, e.g. a visual can be decomposed into colour and shapes, which can be further decomposed into edges and borders, and so on. For the decomposition, the problems are not always well known and explicit as is the case with the visual and auditory modalities. In some cases, the decomposition can be done by automatically breaking down the problem based on intrinsic properties of the information provided [18].
- **Recombination**: The recombination of the modules can be cooperative (all modules contribute to the result), competitive (only the most reliable module is

responsible for the decision), sequential (the computation of one module depends on the output of the other), and supervised (one module is used to supervise the performance of others) [16, 17].

Sometimes in order to avoid the recombination process, systems perform the combination of information from different modalities before the recognition process is undertaken. One unique module is then used for recognition. While this approach is easier to design, often the unique module encounters difficulties during the learning process. Also in this configuration, the designer cannot include or extract explicitly any knowledge related to individual modalities during the recognition process.

The biologically inspired integration of modalities for pattern recognition uses the theory of spiking neural networks, where the individual modes and the integration procedure are implemented with spiking neurons. Each individual modality has its own network of spiking neurons. In general, the output layer of each modality is composed of neurons that authenticate/not authenticate a class they represent when output spikes are released.

The approach for integrating modalities consists of attaching a new layer onto the output of the individual modes. This layer (supramodal layer) represents the supramodal region and contains neurons that are sensitive to more than one modality [19]. In the implementation proposed here, the supramodal layer contains two spiking neurons for each class label. Each neuron representing a given class C in the supramodal layer has incoming excitatory connections from the output of class C neurons of each individual modality. The two neurons have the same dynamics, yet different thresholds for spike generation ($PSP_{Th}$). For one neuron, the $PSP_{Th}$ is set in such a way that an output spike is generated after receiving incoming spikes from any single modality (effectively it is a spike-based implementation of an OR gate). The other neuron has $PSP_{Th}$ set so that incoming spikes from all individual modalities are necessary to trigger an output spike (AND gate). AND neuron maximizes the accuracy and OR neuron maximizes the recall (see Fig. 12.10).

In addition to the supramodal layer, a simple way to perform crossmodal coupling of modalities is designed. The crossmodal coupling is set as follows: when output neurons of an individual modality emit spikes, the spikes not only excite the neurons in the supramodal layer, but also excite/inhibit other modalities that still have ongoing processes. Effectively the excitation/inhibition influences the decision on other modalities, biasing (making it easier/more difficult) the other modality to authenticate/not authenticate a pattern.

For the cross-modal coupling, different from the supra-modal layer connections that are only excitatory, both excitatory and inhibitory connections are implemented. With this configuration, the output of a given class C in one modality excites the class C neuronal maps in other modalities. In contrast, the output class $\hat{C}$ (not class C) in one modality has an inhibitory effect on class C neuronal maps in other modalities.

**Fig. 12.10** Integration of individual layers with a supra-modal layer and cross-modal connections. The individual and supra-modal layers are implemented using spiking neurons

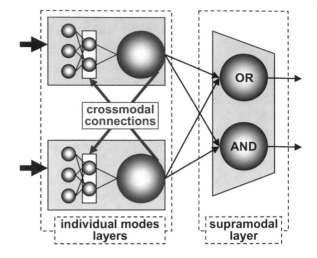

In the following section, the supra/cross modal concepts are applied to the case of audiovisual integration in a person authentication problem based on face and speech information. The implementation of the visual model follows the description given in Sect. 3.5 and the auditory model uses the architecture described in Sect. 4.3.3. A more detailed explanation of the implementation is also given.

## 12.2.2  Convolutional eSNN (CeSNN) for Modelling Visual Information

The visual system is modelled with a four-layer feed-forward network of spiking neurons, with the same configuration as described in Chap. 5 and in [15]. Figure 12.11 shows the network architecture, which combines opinions of being/ not being a desired face over several frames (multi-view face recognition). Basically, the network receives in its input several frames that are processed in a frame-by-frame manner. Neurons in the first layer (L1) represent the On and Off cells of the retina, enhancing the high contrast parts of a given image (high-pass filter). The second layer (L2) is composed of orientation maps for each frequency scale, each one being selective to different directions. They are implemented using Gabor filters in eight directions (0°, 45°, 90°, 135°, 180°, 225°, 270°, and 315°) and two frequency scales. Maps in the third layer are trained to be sensitive to complex visual patterns (faces in the case study evaluated here). In L3, neuronal maps are created or merged during learning in an adaptive online way.

L3 neurons receive crossmodal influences from other modalities. In other words, instead of L3 being composed of exclusively unimodal neurons sensitive to visual excitation, L3 has multisensory capabilities. L3 neurons are still mainly visual, but are also sensitive to stimuli from other modalities.

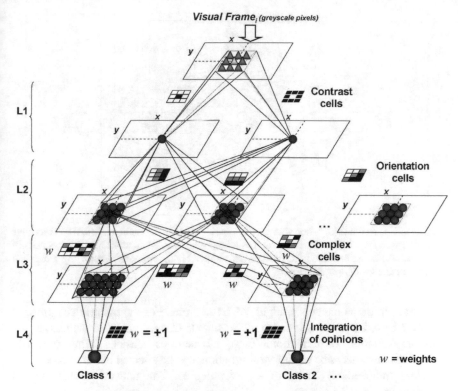

**Fig. 12.11** Evolving spiking neural network (eSNN) architecture for visual pattern recognition. Neurons in L1 and L2 are sensitive to image contrast and orientations, respectively. L3 has the complex cells, trained to respond to specific patterns. It is in L3 that crossmodal coupling occurs. L4 accumulate opinions from different input excitations over time [15]

Neurons in layer 4 (L4) accumulate opinions about being a certain class over several frames. If the opinions are able to trigger an L4 neuron to spike, the authentication is completed.

## 12.2.3 Convolutional eSNN (CeSNN) for Modelling Audio Information

The auditory system is modelled with a two-layer feed-forward network of spiking neurons [20]. Each speaker is represented by a set of prototype vectors that compute normalized similarity scores of MFCC (Mel Frequency Cepstrum Coefficients) considering speaker and background models. Prototypes of a given class are

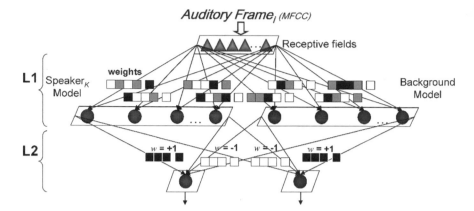

**Fig. 12.12** Speaker authentication with spiking neural networks. L1 neurons, with their respective connection weights, implement the prototypes of a given class. L1 neurons also are receivers of crossmodal excitation/inhibition. L2 neurons accumulate binary opinions about being a claimant over several frames of speech signals

memorized in the connection weights of L1 neurons. For the integrative approach described here, L1 neurons are also the recipients of crossmodal influences, in the form of excitation or inhibition. Thus, L1 neurons, besides being primarily responsible for processing auditory information, can be affected by other modalities (therefore multisensory units) to a lower degree. The network architecture is illustrated in Fig. 12.12.

There are two neurons in L2 for each speaker accumulating opinions over several frames of speech signals. One neuron is triggered if the speaker is authenticated and the other if the input excitation is more likely to be the background model.

### 12.2.4 *Convolutional eSNN (CeSNN) for Integrated Audio-Visual Information Processing*

The detailed audiovisual crossmodal integration architecture is shown in Fig. 12.13. The bottom part of Fig. 12.13 shows two neurons (OR and AND) representing the supramodal layer. Each spiking neuron in the supramodal layer operates in the same way as the neurons that compose the SNNs of individual modalities (fast integrate-and-fire neurons with modulation factor described in Chap. 5).

Even this simple configuration of the supramodal layer can have quite a complex behaviour that cannot be easily described in an analytical way. However, to facilitate the illustration of the integrative system, a particular case is described. The supramodal neurons are set with a modulation factor of mod = 1, and all the

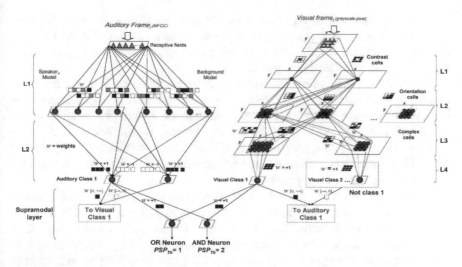

**Fig. 12.13** Integration of modalities using evolving SNNs. The supramodal layer integrates incoming sensory information from individual modalities and crossmodal connections enabling the influence of one modality upon the other

incoming excitatory connection weights ($W$) are set to 1. Thus, the $PSP_{Th}$ that implements the OR integration for two modalities is equal to 1. The neuron implementing the AND integration receives $PSP_{Th} = 2$. Notice that it is only possible to set these parameters deterministically because the neurons can spike only once during the entire simulation period (Fig. 12.13).

Once again, to facilitate the analysis, crossmodal influences between modalities are effectively modelled through the modification in the $PSP_{Th}$ of the crossmodal neurons, namely L3 neurons in the visual system and L1 neurons in the auditory system. Thus, instead of simulating crossmodal influences with spikes that will consequently excite/inhibit a neuron (increase/decrease neuron's PSP), which corresponds to the biological method, the crossmodal influence is implemented by increasing/decreasing the PSP threshold of the neurons. The effect in terms of network behaviour is the same, however it is found to be easier to parameterize the amount of crossmodal influence through the variation on the PSP thresholds. Thus, the strength of the crossmodal influences can be denoted with the following crossmodal parameters: $CM_{AVexc}$ (audio to video excitation), $CM_{AVinh}$ (audio to video inhibition), $CM_{VAexc}$ (video to audio excitation), $CM_{VAinh}$ (video to audio inhibition), which are implemented as a proportional change in the usual $PSP_{Th}$ values as:

$$PSP_{ThNew} = PSP_{ThOld}(1 + CM_{exc/inh}) \qquad (12.3)$$

where $CM_{exc/inh}$ is negative for crossmodal excitatory influence and positive for inhibitory influence.

In the simplest case, setting crossmodal coupling parameters to zero effectively means that each modality is processed separately, with a simple OR/AND fusion of opinions. Increasing the absolute value of crossmodal coupling parameters effectively increases the crossmodal influences.

Note that the definition of supramodal layer here is related only to the layer that effectively combines sensory information to make the final decision. It does not include all the areas where multisensory neurons are located. L3 neurons of the visual system and L1 neurons of the auditory system, despite being multisensory neurons, are considered a part of the individual pathways outside the supramodal layer. Thus, individual pathways could be more appropriately named as "mainly" visual and "mainly" auditory pathways.

Figure 12.14 illustrates the behaviour of the network over time. The dynamic behaviour of the integrated network is described as follows: each frame of the visual and auditory excitation (frames f1, f2,..., fN) are propagated through their corresponding individual architectures until the supramodal layer. Spikes of a given visual frame are propagated to L2 and L3 until a neuron belonging to a L3 map emits the first output spike, which is propagated to L4. L4 neurons accumulate opinions over several frames, whereas L1, L2 and L3 neurons are reset to their resting potential on a frame basis. The same occurs with auditory frames. Spikes are propagated to L1 neurons until a L1 neuron emits the first output spike, which is

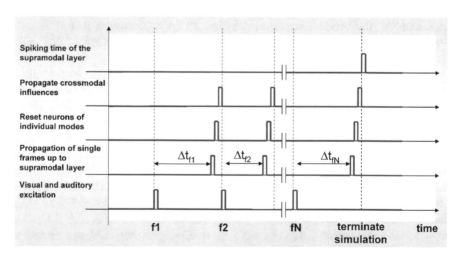

**Fig. 12.14** Typical behaviour of the integrated SNN architecture over time. The visual and auditory excitation (frames f1, f2,..., fN) are propagated through their corresponding individual architectures until the supramodal layer. Neurons of individual modalities are reset to their resting potential, namely L1, L2 and L3 neurons of the visual and L1 neurons of the auditory architecture. Crossmodal influences are propagated and a new frame is processed. The simulation is terminated when the supramodal layer spikes, both individual modes have released their opinions or there are no more frames to be processed

propagated to L2. L2 neurons accumulate opinions over several frames whereas auditory L1 neurons are reset to their resting potential before each frame is processed.

When auditory L2 neurons and/or visual L4 neurons release an output spike, the spikes are propagated to the supramodal layer. If there is no output spike in any visual L4 neuron and a visual L3 neuron has emitted a spike or there are no more spikes to be processed, the next visual frame can be propagated. In a similar fashion, if there is no output spike in any auditory L2 neuron and an auditory L1 neuron has emitted a spike or there are no more spikes to be processed, the next auditory frame can be propagated.

Visual L4 neurons and auditory L2 neurons retain their PSP levels that are accumulated over consecutive frames, until a class is recognized with an L4 neuron output spike or until there are no more frames to be processed. Crossmodal influences, if existent, are propagated synchronously before a new frame is processed. The crossmodal influence starts when one individual modality produces a result (output spike in a auditory L2 neuron or in a visual L4 neuron) and lasts until the processing is completed in all modalities.

In this model, the processing time for auditory and visual frames are considered the same, i.e., the supramodal layer receives synchronous information in a frame basis, although it is well known that auditory stimuli are processed faster than visual [19].

Note that when resetting the PSP in the visual L2 and L3 neurons and auditory L1 neurons in each frame, information about dynamic changes of the patterns are lost, i.e., the model does not keep track of the variations of a visual pattern nor how the pattern changes over time. Each visual frame is considered independently and the last layer of each individual modality effectively accumulates opinions about whether it is a trained pattern.

With respect to the processing speed, in principle, the crossmodal connections decrease the time required to authenticate true claimants and increase the time needed to reject false claims when compared with a purely AND integration. In other words, it speeds up the processing of correlated information from different modes because once an individual modality finishes its analysis and labels a pattern, it exerts excitatory influence on the neurons of other modalities with the same label. The bias effect towards the second modality facilitates its decision in case true information about the claimant is also provided, which causes a resultant decision to be achieved quickly. On the other hand, the time needed to reject false claimants increases. Should the first modality results in a negative opinion about the claimant, the crossmodal connections send inhibitory signals to the claimant's neurons on other modalities, making its authentication harder. If the claimant provides true information on the second modality, due to the negative opinion given by the first modality, the second modality will be more rigorous on the authentication process, which consequently affects the time required to release the overall result.

## 12.3   Case Studies, Experiments and Results

### 12.3.1   Data Sets

The integration of audiovisual modalities with a network of spiking neurons is evaluated with the VidTimit dataset [21], which contains video and audio recordings of 43 individuals. The test setup deals specifically with the audiovisual person authentication problem. A person is authenticated based on spoken phrases and the corresponding facial information as the utterances are recorded (faces are captured in frontal view).

The following items present the configuration details of each individual system as well as the parameters used on the integration mechanism:

- **Visual**: Face detection is accomplished with the Viola and Jones algorithm [22] implemented in the OpenCV library (Intel OpenCV, 2007). Faces are converted into greyscale, normalized in size (height = 60 × width = 40), convolved with an elliptical mask, and encoded into spikes using rank order coding [23, 24]. SNN does not require illumination normalization [24]. There are two scales of On/Off cells (4 L1 neuronal maps). In scale 1, the retina filters are implemented using a 3 × 3 Gaussian grid with $\sigma = 0.9$ and scale 2 uses a 5 × 5 grid with $\sigma = 1.5$. In L2, there are eight different directions in each frequency scale with a total of 16 neuronal maps. The directionally selective filters are implemented using Gabor functions with aspect ratio $\gamma = 0.5$ and phase offset $\varphi = \pi/2$. In scale 1 a 5 × 5 grid with a wavelength of $\lambda = 5$ and $\sigma = 2.5$ is used and in scale 2 a 7 × 7 grid with $\lambda$ and $\sigma$ set to 7 and 3.5, respectively. The modulation factor for the visual neurons was set to 0.995.
- **Auditory**: Speech signals are sampled at 16 kHz, and features extracted using standard MFCC with 19 MEL filter sub-bands ranging from 200 Hz to 7 kHz. Each MFCC is then encoded into spikes using rank order coding [23]. One receptive field neuron is used to represent each MFCC (19 input receptive fields). A specific background model is trained for each speaker model. For the sake of simplicity, the following procedure is applied: the background model of a speaker $i$ is trained using the same amount of utterances used to train the speaker model. The utterances are randomly chosen from the remaining training speakers. For the experiments, the number of neurons in the auditory L1 neuronal maps for the speaker and background model are defined a priori (50 neurons each). The modulation factor for auditory neurons is set to 0.9.
- **Integration**: The crossmodal parameters according to Eq. 12.3 are set as: $CM_{AVexc} = CM_{VAexc} = 0.1$ and $CM_{AVinh} = CM_{VAinh} = 0$. Results that do not take into account the crossmodal coupling are also presented, i.e., $CM_{AVexc} = CM_{VAexc} = CM_{AVinh} = CM_{VAinh} = 0$, which effectively correspond to AND or OR integration.

## 12.3.2 Experimental Results

The system is trained to authenticate 35 persons using six utterances from each individual. To train the visual part, only two frames from each individual are used, collected when uttering two distinct phrases from the same recording session were uttered.

The test uses two phrases (each phrase corresponding to one sample) recorded in two different sessions, therefore 35 users × 2 samples = 70 positive claims. Acting as impostors, the eight remaining users attempt to deceive each of the 35 users' models with two utterances, which give a total of 560 false claims.

The test is carried out frame-by-frame keeping the time correspondence between speech and visual frames. However, to speed up the computational simulations, the visual frames are downsampled. Five visual frames per second are used whereas the speech samples have a rate of 50 frames per second (Fig. 12.15). The downsampling of the visual frames does not affect the performance, as for a period lower than 200 ms no substantial differences between one facial posture and another can be noticed in the VidTimit dataset.

Figure 12.15 shows typical input streams to the SNN-based audiovisual person authentication system, where frames of detected faces are sampled at 200 ms (5 frames/s) and 19 MFCC extracted from the detected speech parts are processed every 20 ms (50 frames/s).

The supramodal layer and the crossmodal coupling are updated when an individual modality outputs a spike, which may occur once in every frame. Here, it is assumed that the processing time for one frame is the same, regardless of the modality, although it is well known that auditory stimuli are processed faster than visual (difference of approximately 40–60 ms [19].

For the speech mode, the number of opinions to validate a person is set proportionally to the size of a given utterance (20% of the total number of frames in an utterance is used). For the visual mode, the number of opinions to authenticate a person is set to two (two frames). Figure 12.16 shows the best performance obtained on each individual modality. While the best total error (TE) for the face

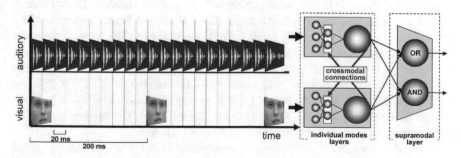

**Fig. 12.15** Frame-based integration of modalities (after [14])

**Fig. 12.16** Performance of individual modalities for different values of auditory (L1 $PSP_{Th}$) and visual parameters (L3 $PSP_{Th}$). Top: auditory system. Bottom: visual system. FAR is the false acceptance rate, FRR is the false rejection rate and TE is the total error (FAR + FRR) (after [14])

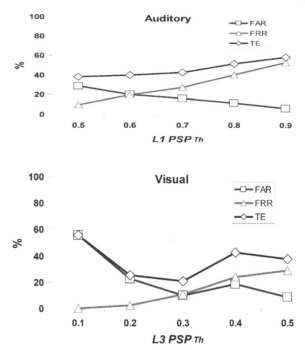

authentication is 21%, the auditory authentication is TE ≈ 38% (varying values of L1 $PSP_{Th}$ in the auditory system and L3 $PSP_{Th}$ in the visual system).

Figure 12.16 shows performance of individual modalities for different values of auditory (L1 $PSP_{Th}$) and visual parameters (L3 $PSP_{Th}$). Top: auditory system. Bottom: visual system. FAR is the false acceptance rate, FRR is the false rejection rate and TE is the total error (FAR + FRR).

Figure 12.17 shows the best performance of the system considering the type of integration held in the supramodal layer. First, the crossmodal coupling parameters are set to zero, simulating only the OR and AND integration of individual modalities done by the supramodal layer. Then, the crossmodal coupling is made active ("Crossmodal AND"), setting $CM_{AVexc} = CM_{VAexc} = 0.1$ and $CM_{AVinh} = CM_{VAinh} = 0$. The same parameters are used for individual modalities in this experiment, i.e., auditory parameters (L1 $PSP_{Th}$) and visual parameters (L3 $PSP_{Th}$) ranging from [0.5, 0.9] to [0.1, 0.5], respectively. The x-axis represents different combinations of L1 and L3 $PSP_{Th}$ ordered according to the performance.

Figure 12.18 shows the potential advantages of the integration module. When the system needs to operate with low FAR levels (below 10%), AND and "Crossmodal AND" provide lower FRR than any singular modality. When the system is required to operate with low FRR (below 10%), OR integration can be used instead, providing lower FAR for the same FRR levels.

In another scenario, the influence of crossmodal connections on the integrated system is evaluated. A subset of the VidTimit dataset is used for this purpose. The

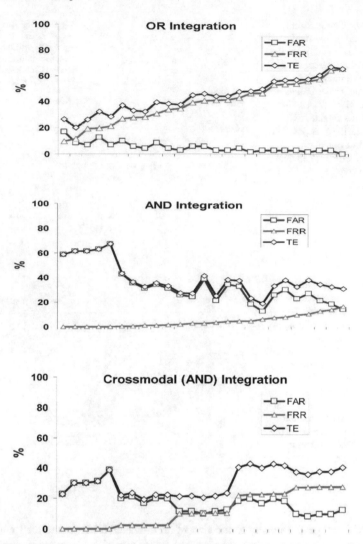

**Fig. 12.17** Performance of the OR and AND integration of modalities with a supramodal layer of spiking neurons (upper and middle graphs, respectively). The bottom graph, when excitatory crossmodal influences are activated "Crossmodal AND" (for auditory L1 $PSP_{Th}$ and L3 $PSP_{Th}$ ranging from [0.5, 0.9] to [0.1, 0.5], respectively)

setup for training is composed of six utterances from 10 individuals, whereas 13 individuals (10 that participated in the training stage and two completely unknown individuals) are used for testing. Each of the 10 individuals has 4 attempts at the test in a total of 40 positive claims. Acting as impostors, two individuals attempt to authenticate each of the 10 trained models four times, with a total of $2 \times 4 \times 10 = 80$ impostor attempts (false claims). Similar to the previous experiments, the authentication

**Fig. 12.18** Comparison between individual modes (auditory and visual) and the corresponding integration. Overall, the integration presents better performance than individual modes. OR, AND, "Crossmodal AND" alternate in the best position for different points of operation. EER is the equal error rate (where FAR = FRR)

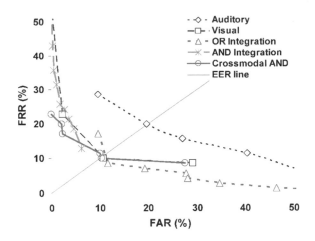

**Fig. 12.19** Performance of the network for different values of crossmodal excitation. There is a range of values of crossmodal influence for which the model gives similar performance, however, for all values, the integration presents better performance than individual modes and ANDs and OR configuration alternate as the best choices for different points of operation

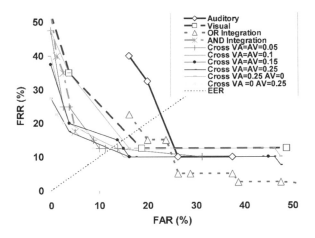

threshold is set proportionally to the size of an utterance (20% of the total number of frames needs to provide positive opinions) and only two visual frames are necessary to authenticate a person based on the face. Figure 12.19, shows the performance of the integrated network for different values of crossmodal excitation. From the graph it is not possible to detect the best crossmodal parameter values, which means that a range of parameter values can be used with the same result. However, once again is clear that OR integration works better for high FAR than any single modality, and AND integration works better for low FRR than any single modality.

## 12.4   Chapter Summary and Further Readings for Deeper Knowledge

This chapter presents some initial information about audio-, visual and integrated information processing in the human brain as inspiration for the presented SNN methods in this and in the next chapters. It also covers the integration of modalities for the purpose of audio-visual pattern recognition. Of particular interest was the compilation of biological findings that inspire the proposal of models to explain the way brains effectively process and integrate different sensory information. Through an evaluation of several models and theories describing brain activity, the focus is given to the understanding of two properties that can be useful in enhancing artificial pattern recognition tasks, in particular:

- the supramodal area, and;
- crossmodal connections between modalities.

The second part of the chapter describes a new simple way to integrate modalities using fast spiking neurons (See also [20]). In the new system, each individual modality utilizes specialized adaptive SNNs. The integration is done in a supramodal layer composed of multisensory neurons. In addition, one modality can influence another using a crossmodal mechanism.

The model also enables to set the strength of crossmodal connections individually for each pair of single modes. In biology, audiovisual crossmodal learning has been experimentally observed in [25]. In their experiments, after a training session with visual and auditory stimuli, when auditory stimuli alone were presented, areas of the visual cortex were also activated. In [26] the areas of neuronal changes (time-dependent plasticity) that may be related to the crossmodal operations are further investigated. However, there was no attempt to quantify or to define the rules for neuronal changes. In this respect, new neuronal models for exploring the mechanisms that govern such activities can underpin new discoveries. In the model proposed in this chapter, a proper training procedure for crossmodal connections can be explored and evaluated.

The proposed in this chapter model has several aspects that require further development, namely:

(a) the model cannot take into account some biological behaviours detected by psychological experiments, e.g., cannot cover familiarity decisions, semantic information, identity priming, and within and cross domain semantic priming [27–29].
(b) with respect to the implementation, the use of frames and the respective frame-by-frame synchronization seems to be very artificial, truncating the natural flow of information. In addition, at this stage, the difference in processing time in each modality [19] is ignored.

(c) the model can not emulate the mechanism that facilitates unimodal recognition when the training is done with more than one modality, behaviour which has been described in [30].

Under the pattern recognition perspective, the network was tested on the person authentication problem. Experiments clearly showed that the integration of modes enhances the performance in several points of operation of the system when the learning is done with the same training examples. For a comparative analysis, in [21], the integration of modalities is explored with the VidTimit dataset using a combination of mathematical and statistical methods. The auditory system alone, using MFCC features and GMM in a noise-free setup, reached TE (total error) = FAR (false acceptance rate) + FRR (false rejection rate) $\approx$ 22%. The visual system is reported to have TE $\approx$ 8% with features extracted using PCA (principal component analysis) and SVM (support vector machine) for classification. After testing several adaptive and non adaptive systems to perform integration, the best performance is obtained with a new approach that builds the decision boundaries for integration with consideration of how the distribution of opinions are likely to change under noisy conditions. The accuracy with the integration reached TE $\approx$ 6% involving 35 users for training and 8 users acting as impostors. To extract the best performance from the system and evaluate the crossmodal influence specifically on the pattern discrimination ability, an optimization mechanism needs to be incorporated. Similarly important is to explore different information coding schemes.

As pointed out in [15, 31] and in Chaps. 4–6, one of the promising properties of the computation with spiking neurons is that it enables the multi-criteria optimization of parameters according to accuracy, speed and energy efficiency. Since the integration is also based on spiking neurons, the optimization can be extended to cover the parameters used on integration as well (supramodal layer and crossmodal connection parameters).

Table 12.1 lists main features of the presented here convolutional eSNN for integrative audio-visual information processing.

The supramodal layer, as a first step, is implemented in this work with only two neurons. Two neurons were demonstrated to be sufficient to integrate incoming

**Table 12.1** Main features of the CeSNN for integrative audio-visual information processing

| Processing units | Spiking neurons are used as processing units in the individual and integrative information processing areas |
|---|---|
| Structure | The information of individual sensory modalities propagates with feed-forward connections into multiple layers composed of spiking neurons, representing the behaviour of various auditory and visual areas. Crossmodal connections and a supramodal layer integrate the systems (see Fig. 12.13) |
| Learning | Online evolving procedures enable the learning of external stimuli through synaptic plasticity and structural adaptation separately for each modality. Algorithms to train the strength of crossmodal connections and weights of the supramodal layer still need to be designed |

information from different modalities and to provide the system with complex dynamics that are difficult to evaluate analytically. In the simplest scenario, OR and AND integration has been simulated. Although a single neuronal unit can be interpreted as representing an entire ensemble of neurons, a more realistic implementation could be considered.

The underlying mechanisms that rule crossmodal activities remain the subject of further inquiry. The optimization of such connections and/or how to perform crossmodal learning is still an open field (a good introduction can be found in [25, 26]. The experiments presented in this chapter illustrate as a proof-of-concept how the crossmodal connections can be set up in a network of spiking neurons. Further evaluation, such as sensitivity analysis with respect to different performance criteria, and exploration of the best values of crossmodal influence (excitatory and inhibitory) still deserve special attention.

Further recommended readings include:

- AVIS [31];
- Human speech [6];
- Modelling vision with the neocognitron (Chap. 44 in [32]);
- Face recognition [27];
- NeuCube [33].

**Acknowledgements** Some of the material in this chapter was first published in [1, 2, 31, 14]. I acknowledge the contribution of my co-authors of these publications Simei Wysoski, Lubica Benuskova, Eric Postma and Jaap van den Herik.

# References

1. N. Kasabov, *Evolving Connectionist Systems: The Knowledge Engineering Approach* (Springer, Heidelberg, 2007)
2. L. Benuskova, N. Kasabov, *Computational Neurogenetic Modelling* (Springer, Heidelberg, 2007)
3. D.D. Greenwood, Critical bandwidth and the frequency coordinates of the basilar membrane. J. Acoust. Soc. Am. **33**(10) (1961). https://doi.org/10.1131/1.1908437
4. D. D. Greenwood, A cochlear frequency-position function for several species–29 years later. J. Acoust. Soc. Am. **87**(6), 2592–2605 (1990)
5. E. de Boer, H. R. de Jongh, On cochlear encoding: potentialities and limitations of the reverse-correlation technique. J. Acoust. Soc. Am. **63**(1), 115–135 (1978)
6. J.B. Allen, How do humans process and recognize speech? IEEE Trans. Speech Audio Process. **2**(4), 567–577 (1994)
7. E. Zwicker, Subdivision of the audible frequency range into critical bands (Frequenzgruppen). J. Acoust. Soc. Am. **33**, 248 (1961)
8. B.R. Glasberg, B.C. Moore, Derivation of auditory filter shapes from notched-noise data. Hear Res. **47**(1–2), 103–138 (1990)
9. T.J. Cole, J.A. Blendy, A.P. Monaghan, K. Krieglstein, A. Schmid, A. Aguzzi, G. Fantuzzi, E. Hummler, K. Unsicker, G. Schütz, Targeted disruption of the glucocorticoid receptor gene blocks adrenergic chromaffin cell development and severely retards lung maturation. Genes Dev. **9**(13), 1608–1621 (1995)

10. A.M. Aertsen, J.H. Olders, P.I. Johannesma, Spectro-temporal receptive fields of auditory neurons in the grassfrog. III. Analysis of the stimulus-event relation for natural stimuli. Biol. Cybern. **39**(3), 195–209 (1981)
11. E.O. Postma, H.J. van den Herik, P.T.W. Hudson, Image recognition by brains and machines, in *Brain-like Computing and Intelligent Information Systems*, ed. by S. Amari, N. Kasabov (Springer, Singapore, 1998), pp. 25–47
12. E.O. Postma, H.J. van den Herik, P.T.W. Hudson, SCAN: a scalable model of covert attention. Neural Netw. **10**, 993–1015 (1997)
13. K. Kim, N. Relkin, K.-M. Lee, J. Hirsch, Distinct cortical areas associated with native and second languages. Nature **388**, 171–174 (1997)
14. S. Wysoski, L. Benuskova, N. Kasabov, Evolving spiking neural networks for audio-visual information processing. Neural Netw. **23**(7), 819–835 (2010)
15. S. Wysoski, L. Benuskova, N. Kasabov, Fast and adaptive network of spiking neurons for multi-view visual pattern recognition. Neurocomputing **71**(13–15), 2563–2575 (2008)
16. A. Ross, A.K. Jain, Information fusion in biometrics. Pattern Recognit. Lett. **24**(13), 2115–2135 (2003)
17. C. Sanderson, K.K. Paliwal, Identity verification using speech and face information. Digital Signal Process. **14**, 449–480 (2004)
18. A. Sharkey, *Combining Artificial Neural Nets: Ensemble and Modular Multi-net systems* (Springer, Heidelberg, 1999)
19. B.E. Stein, M.A. Meredith, *The Merging of the Senses* (MIT Press, Cambridge, 1993)
20. S.G. Wysoski, L. Benuskova, N. Kasabov, Adaptive spiking neural networks for audiovisual pattern recognition, ICONIP. Lecture notes in computer science (Springer, 2007)
21. C. Sanderson, K.K. Paliwal, Identity verification using speech and face information. Digital Signal Process. **14**, 449–480 (2004)
22. P. Viola, M.J. Jones, Rapid object detection using a boosted cascade of simple features. Proceed. IEEE Comput. Soc. Conf. Comput. Vis. Pattern Recognit. **1**(2001), 511–518 (2001)
23. A. Delorme, L. Perrinet, S. Thorpe, Networks of integrate-and-fire neurons using rank order coding. Neurocomputing. 38–48 (2001)
24. A. Delorme, S. Thorpe, Face identification using one spike per neuron: resistance to image degradation. Neural Netw. **14**, 795–803 (2001)
25. A.R. McIntosh, R.E. Cabeza, N.J. Lobaugh, Analysis of neural interactions explains the activation of occipital cortex by an auditory stimulus. J. Neurophysiol. **80**(1998), 2790–2796 (1998)
26. D. Gonzalo, T. Shallice, R. Dolan, Time-dependent changes in learning audiovisual associations: a single-trial fMRI study. NeuroImage **11**, 243–255 (2000)
27. A.M. Burton, V. Bruce, R.A. Johnston, Understanding face recognition with an interactive activation model. B. J. Psychol. **81**, 361–380 (1990)
28. A.W. Ellis, A. Young, D.C. Hay, Modelling the recognition of faces and words, in *Modelling Cognition*, ed. by P.E. Morris (Wiley, London, 1987), p. 1987
29. H.D. Ellis, D.M. Jones, N. Mosdell, Intra- and inter-modal repetition priming of familiar faces and voices. B. J. Psycol. **88**, 143–156 (1997)
30. K. Kriegstein, A. von, Giraud, Implicit multisensory associations influence voice recognition. PLoS Biol. **4**(10), 1809–1820 (2006)
31. N. Kasabov, E. Postma, J. van den Herik, AVIS: a connectionist-based framework for integrated auditory and visual information processing. Inf. Sci. **133**, 137–148 (2000)
32. N.Kasabov (ed) (2014) Springer Handbook of Bio-/Neuroinformatics (Springer, Heidelberg, 2014)
33. N. Kasabov, NeuCube: a spiking neural network architecture for mapping, learning and understanding of spatio-temporal brain data. Neural Netw. **52**, 62–76 (2014). https://doi.org/10.1016/j.neunet.2014.01.006
34. Wikipedia: http://www.wikipedia.com
35. L. Benuskova, N. Kasabov, *Computational Neurogenetic Modelling* (Springer, Heidelberg, 2007)

# Chapter 13
# Deep Learning and Modelling of Audio-, Visual-, and Multimodal Audio-Visual Data in Brain-Inspired SNN

This chapter presents methods for audio-, visual- and for the integrated audio and visual information processing using brain-inspired SNN architectures such as NeuCube. Case studies are presented for short musical pieces recognition, fast moving object recognition, age-invariant face identification, moving digits recognition and other.

The chapter is organized in the following sections:

13.1. Deep learning of sound in brain-inspired SNN.
13.2. Deep learning of visual data in brain-inspired SNN architectures for fast moving object recognition and for gender recognition.
13.3 Retinotopic mapping and learning of dynamic visual information in a brain-like SNN architecture on the case of moving object recognition.
13.4. Chapter summary and further readings for deeper knowledge.

## 13.1 Deep Learning of Sound in Brain-Inspired SNN

### 13.1.1 Deep Learning of Audio Data in the Brain

The brain forms deep neuronal structures when perceives audio information as illustrated in Fig. 13.1. It is shown in Fig. 13.1 that hearing a word and repeating it requires a deep learning in the brain and recall of these pathways [1, 2]:

© Springer-Verlag GmbH Germany, part of Springer Nature 2019
N. K. Kasabov, *Time-Space, Spiking Neural Networks and Brain-Inspired Artificial Intelligence*, Springer Series on Bio- and Neurosystems 7,
https://doi.org/10.1007/978-3-662-57715-8_13

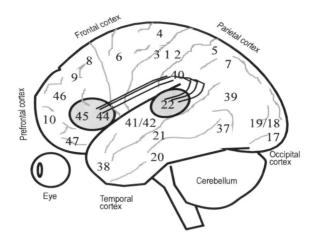

**Fig. 13.1** Hearing a word and repeating it requires a deep learning in the brain and recall of these pathways: Transfer of information from the inner ear through the auditory nucleus in thalamus to the primary auditory cortex (Brodmann's area 41); Then to the higher-order auditory cortex (area 42); Then it is relayed to the angular gyrus (area 39). Angular gyrus is a specific region of the parietal-temporal-occipital association cortex, which is thought to be concerned with the association of incoming auditory, visual and tactile information; From here, the information is projected to Wernicke's area (area 22); Then, by means of the *arcuate fasciculus*, to Broca's area (44, 45), where the perception of language is translated into the grammatical structure of a phrase and where the memory for word articulation is stored; This information about the sound pattern of the phrase is then relayed to the facial area of the motor cortex that controls articulation, so that the word can be spoken (from [2])

- Transfer of information from the inner ear through the auditory nucleus in thalamus to the primary auditory cortex (Brodmann's area 41);
- Then to the higher-order auditory cortex (area 42);
- Then it is relayed to the angular gyrus (area 39). Angular gyrus is a specific region of the parietal-temporal-occipital association cortex, which is thought to be concerned with the association of incoming auditory, visual and tactile information;
- From here, the information is projected to Wernicke's area (area 22);
- Then, by means of the *arcuate fasciculus*, to Broca's area (44, 45), where the perception of language is translated into the grammatical structure of a phrase and where the memory for word articulation is stored;
- This information about the sound pattern of the phrase is then relayed to the facial area of the motor cortex that controls articulation, so that the word can be spoken.

## 13.1.2 A BI-SNN Using Tonotopic and Stereo Mapping and Learning of Sound

In the computational model developed and experimented here a tonotopic mapping of audio data using a computational model of the cochlea is used when audio data is mapped into a SNN architecture as shown in Fig. 13.2. The NeuCube architecture and STDP learning in its SNNcube was explained in Chap. 6 and also shown in Fig. 13.3. Here this architecture uses different variable mapping and learning algorithm than the exemplified ones in Chap. 6 and other chapters of the book.

The NeuCube architecture from Fig. 13.3 and the tonotopic, stereo mapping of sound from Fig. 13.2 are used in the next section to develop a NeuCube-based model for musical pieces recognition.

## 13.1.3 Deep Learning and Recognition of Music

The human brain can learn and recall musical pieces by deep learning of dynamic, characteristic patterns of soundwaves through its auditory pathway and related brain areas.

How musical patterns evolve in the human brain? Music causes the emergence of patterns of activities in the human brain. This process is continuous, evolving,

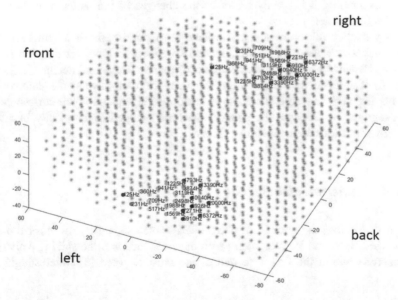

**Fig. 13.2** A brain-inspired *tonotopic, stereo* mapping of audio signals into a 3D SNN architecture (from [3])

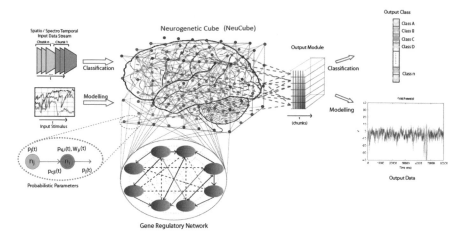

**Fig. 13.3** The NeuCube brain-inspired SNN architecture (see Chap. 6)

though in different pathways depending on the individual. Each musical piece is characterised by specific main frequencies (formants) and rules to change them over time. There is a large range of frequencies in Mozart's music, the greatest energy being in the spectrum of the brain activity (see Chap. 1)—Fig. 13.4. One can speculate that this fact may explain why the music of Mozart stimulates human creativity. But it is not the "static" picture of the frequencies that makes Mozart's music fascinating, it is the dynamics of the changes of the patterns of these frequencies over time.

Many studies of sound and music have been published where different techniques have been used [2, 4–24]. Figure 13.5 shows a flowchart diagram of our method. Here we use one implementation of the NeuCube architecture [25] (also Chap. 6). The advantage of using the Talairach brain template as the shape of our network is that we are able to enter the transformed signals into their corresponding brain regions; in this case the sound waves are mapped tonotopically into their respective locations in the auditory cortex.

## 13.1.4   Experimental Results

Figure 13.6 shows the three NeuCube models trained separately on musical pieces by Mozart, Bach and Vivaldi, along with the amplitude and the spiking activity of the neurons within the first 1 second of the musical piece [3]. Interestingly, we

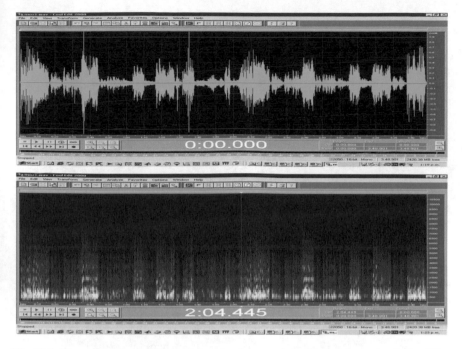

**Fig. 13.4** The wave form and the spectrum of a short segment of Mozart's music al piece "Eine kleine Nacht Musik" (see also Chap. 1)

could already observe notable differences in the spike trains created by the Cochlea module.

The differences in spiking activity were captured by the model and learned as connection weights by the network. Thus, our SNN model learns complex patters of frequency ranges of the different pieces of classical music and preserves the characteristics of the input signals. Table 13.1 shows classification results of the three types of music when a single NeuCube model was trained along with a classifier (deSNN [29]).

The experimental results confirm that using a BI-SNN with stereo and tonotopic mapping of sound is a promising approach that still needs to be explored further for

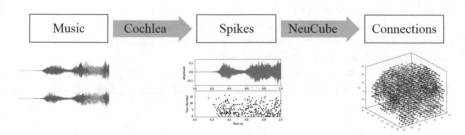

**Fig. 13.5** Method flowchart (the figure was created by Anne Wendt, [26])

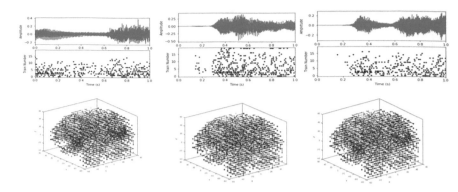

**Fig. 13.6** Tonotopic and stereo mapping of pieces of music by Mozart, Bach and Vivaldi as spectro-temporal data into a NeuCube model (adapted from [3, 26])

**Table 13.1** Confusion table of the classification of musical pieces by a NeuCube trained model (from [3])

|             | Mozart | Bach | Vivaldi |
|-------------|--------|------|---------|
| Predicted 1 | 171    | 3    | 1       |
| Predicted 2 | 9      | 176  | 1       |
| Predicted 3 | 0      | 1    | 178     |

wider applications including speech, music and multi-modal audio-visual data [2, 13–24].

## 13.2  Deep Learning and Recognition of Visual Data in a Brain-Inspired SNN for Fast Moving Object Recognition and for Gender Recognition

### 13.2.1  Two Approaches to Visual Information Processing

Visual information processing is a significant part of the current development in artificial intelligence [28, 30–37]. Two approaches to visual information processing are illustrated in Fig. 13.7. The first one is using a camera and processing the data frame by frame and it is called frame-based.

The second one is imitating the retina that encodes any changes in pixels into spikes and the rest of the pixels are not processed as not containing important for a task information, and it is called spike-based.

While both approaches have been widely used, we will use here the second approach – the spike-based one. Applications of this approach scan across:

- Surveillance systems;
- Cybersecurity;

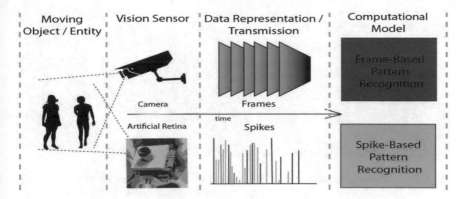

**Fig. 13.7** Two approaches to visual information processing. The first one is using a camera and processing the data pixel by pixel. The second one is using an artificial retina that encodes any changes in pixels into spikes and the rest of the pixels are not processed as not containing important for a task information

- Military applications;
- Autonomous vehicles.

## 13.2.2 Applications for Fast Moving Object Recognition

The method presented here is first published in [38], where more details can be found.

Moving object recognition is a challenging problem in computational intelligence. Fast moving object is considered as the one which could not easily be captured by conventional cameras in real time. The typical examples encompass fast moving cars, flying rockets, bouncing ping-pong balls, tennis balls, balancing pencils etc. It is impossible to recognize such moving objects without using a suitable algorithm and effective software system which are capable to learn and recognize patterns from complex Spatio- and Spectro-Temporal Data (SSTD).

Figure 13.8 shows a schematic diagram of a NeuCube-based system for fast moving object recognition along with the recognition accuracy when applied on experimental data.

The experimental datasets consist of four videos per group shown as Fig. 13.9 with the duration of 10 s, each video having four resolutions, namely, $1480 \times 720$, $640 \times 360$, $320 \times 180$ and $160 \times 90$ that makes four categories used for recognition. The video frame rate is 30 fps; therefore, we have 4800 samples in total for deep learning, each sample was encoded as an $10 \times 10$ array of events (spikes/no spikes) and transferred to the SNNcube for unsupervised learning, each frame generates 100 spikes. The data are propagated for deep learning in the SNN cube and the deSNN classifier. The accuracy of pattern classification by using the input

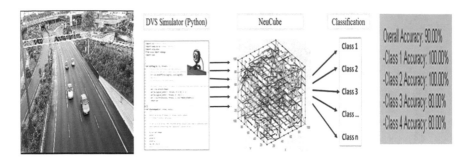

**Fig. 13.8** A schematic diagram of a NeuCube-based system for fast moving object recognition along with the recognition accuracy when applied on experimental data (from [38])

videos with its multiple resolutions is used to evaluate the performance of the proposed method.

We initialize NeuCube by using 1000 neurons to map 100 ($10 \times 10$) pixel addresses generated by our spike encoding simulator, every 100 pixels represent one video frame. The overall accuracy of classification is 97.92% (Fig. 13.10). For every input shown in Fig. 13.9, a new neuron is dynamically allocated in the deSNN and connected to the neurons in the SNNcube. In the NeuCube SNNcube, the more spikes are transmitted, the more connections are created.

### 13.2.3   Applications for Gender and Age Group Classification Based on Face Recognition

A full publication of the method presented here can be found in: [39]. We investigate age classification and gender recognition. The well-known FG-NET and MORPH Album 2 image gallery were used and antropometric features were extracted from landmark points on the face [38–41].

Aging is a slow process and its effects are visible only after a few years. But in spite of being slow, it remains a spatio-temporal phenomenon. The facial features of a person can be considered as a subspace and the aging over the years of this subspace is in turn a temporal process. It would be very useful to incorporate the temporal as well as spatial patterns in aging data as important components in classification. The raw data which has been used in this study is from (FG-NET) and MORPH image galleries. Age group classification and gender recognition have important applications for business managers and law enforcement agencies. In Human Computer Interaction (HCI) gender recognition can be used to make it more amenable to both genders. For example, it enables a computer to address a user by their correct title, Mr. or Mrs., as the case may be. Automatic gender recognition facilitates better interaction with humans as well as saves keystrokes in filling up forms.

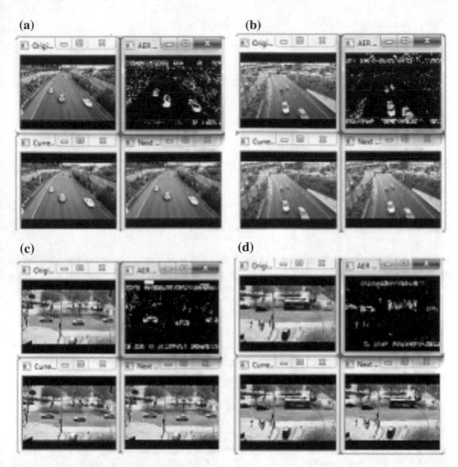

**Fig. 13.9** Video samples with cars moving fast to four directions, **a** moving up, **b** moving down, **c** moving left, **d** moving right (from [38])

Experiments were performed on the publicly available FG-NET (FG-NET 2002) and MORPH Album 2 (the largest publicly available face aging dataset) [40], both of which are used for bench- marking new methods. The lack of a large face aging database until recently limited research on age group classification. There are two desired attributes of a face aging database: (i) large number of subjects, and (ii) large number of face images per subject captured at many different ages. In addition, it is desired that these images should not have large variations in pose, expression, and illumination. The MORPH dataset has a large number of subjects while FGNET database has a large number of images. The MORPH dataset con- tains about 55,000 face images from 14,000 different people [40].

Two experiments are performed using the above data sets. Figure 13.11 shows: Top: A NeuCube model for facial mapping and recognition using NeuCube; Bottom: Application of the model for age-invariant person verification [42].

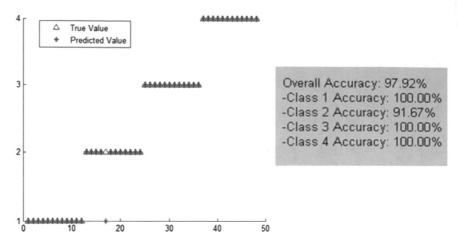

**Fig. 13.10** A classification accuracy of using the model from Fig. 13.8 for the classification of object movements in four classes (see the text above) (from [38])

Table 13.2 shows the results of age group classification using a Neu-Cube SNN model in comparison to traditional classifiers: SVM, MLP, NB.

The NeuCube SNN model results in better age classification results. More information can be obtained from [39] and also from: https://kedri.aut.ac.nz/R-and-D-Systems/age-invariant-face-recognition.

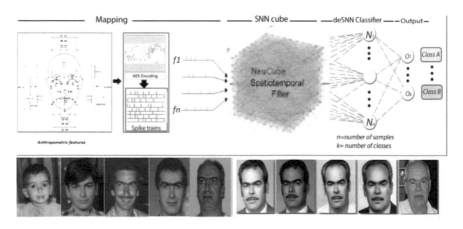

**Fig. 13.11** Top: A NeuCube model for facial mapping and recognition using NeuCube; Bottom: Application of the model for age-invariant person verification [39]

**Table 13.2** The Results of Age Group Classification using a Neu-Cube SNN model in comparison to traditional classifiers: SVM, MLP, NB

| Measure | NeuCube | MLP | Knn | NaiveBayes |
|---|---|---|---|---|
| Accuracy (%) | 95.00 | 82.30 | 89.03 | 67.60 |

## 13.3   Retinotopic Mapping and Learning of Dynamic Visual Information in a Brain-Like SNN Architecture on the Case Study of Moving Object Recognition

### 13.3.1   General Principles

The method presented here is published in full details in [43]. This sub-section introduces a new system for dynamic visual recognition that combines bio-inspired principles with a brain-like spiking neural network. The system takes data from a dynamic vision sensor (DVS) that simulates the functioning of the human retina by producing an address event output (spike trains) based on the movement of objects. The system then convolutes the spike trains and feeds them into a brain-like spiking neural network, called NeuCube, using the proposed here retinotopic mapping. Both the convolution algorithm and the mapping into the network mimic the structure and organization of retinal ganglion cells and the visual cortex. The method was tested on the benchmark MNIST dynamic vision sensor dataset and achieved an accuracy of 92.90%. Due to its bio-inspired and brain-like structure, analysing the connectivity of the spiking neural network also allows for a better understanding of the neural processes inside the visual cortex that underlie humans' ability to perform fast, accurate and energy efficient vision. The paper discusses advantages and limitations of the new method and concludes that it is worth exploring further on different datasets aiming for advances in dynamic computer vision with potential uses in self-driving cars, security systems, and robotics in general.

### 13.3.2   The Brain-Inspired SNN and the Proposed Retinotopic Mapping

The proposed here NeuCube-based SNN architecture incorporates several different principles of SNN and combines them into a single model for mapping, learning, and understanding of spatio-temporal data [25]. Signals are processed along successive stages as shown in Fig. 13.12.

Before going into detail about the learning algorithms used, we want to focus on the three-dimensional structure of NeuCube-based models and the bio-inspired way

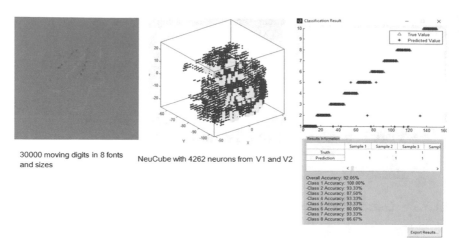

30000 moving digits in 8 fonts and sizes

NeuCube with 4262 neurons from V1 and V2

**Fig. 13.12** Brain-like dynamic vision data modelling and pattern recognition using NeuCube-based BI-SNN and retinotopic mapping of vision data (from [43])

we mapped the 148 input neurons into this structure. Our system uses a NeuCube initialized with 732 neurons, using the MNI coordinates of neurons from the primary visual cortex (V1, Brodman area 17), taken from the Atlas of the Human Brain (downloaded together with the xjView toolbox: http://www.alivelearn.net/xjview). The number of neurons is only bounded by computational limitations; it is possible to add further neurons from the secondary or third visual cortex or to represent the whole brain. Initial connections between the neurons are based on the "small-world" paradigm, where random connections within a pre-defined maximum distance of each neuron are formed. The mapping of the 148 input neurons into the 732 neurons of NeuCube mimics two important characteristics of the human visual cortex: cortical magnification and retinotopic mapping (Fig. 13.12).

Cortical magnification describes the overrepresentation of foveal signals inside the primary visual cortex. Although the fovea has a diameter of only 1.2 mm [44] and covers a very small fraction of the retina, its signals are processed by almost 50% of all neurons in V1. Therefore, we chose exactly 64 of our 148 input neurons to correspond to the central 64 DVS [45] pixels with a one-to-one relationship. This way, 50% of input neurons automatically correspond to the central pixels of the DVS, just like 50% of the primary visual cortex correspond to the central photoreceptors on the retina.

The second characteristic of the primary visual cortex that we adapted in our mapping is the preservation of spatial relationships between photoreceptors on the retina and their neural representation in the primary visual cortex, the so-called retinotopy. Signals from the top left of our visual field get mapped to the bottom right of V1 and vice versa. What humans see gets flipped upside down and mirrored, but things that appear next to each other in the visual field will still be represented next to each other in V1. Both the foveal as well as the peripheral ganglion cells follow this principle, although foveal signals get mapped into the posterior part and peripheral signals into the anterior part of V1 [44].

Figure 13.12 shows how the principle of retinotopy is applied to the mapping of the 148 input neurons to the 732 neurons of NeuCube.

### 13.3.3    Unsupervised and Supervised Learning of Dynamic Visual Patterns

Learning in the NeuCube-based model is performed in two stages: In the first step, unsupervised learning, for example using spike timing dependent plasticity (STDP, [46]), is performed to modify the initial connection weights. The SNN will learn to activate the same groups of spiking neurons when similar input stimuli are presented, and to change existing connections or create new ones that represent spatio-temporal patterns of the input data [25, 47]. NeuCube allows for a visualization of the learning process, and we discuss how the visualization can be used for a better understanding of the data and the neural processes.

In the second step, supervised learning is applied to the spiking neurons in the output classification module, where the same spike trains used for the unsupervised training are now propagated again through the trained SNN and output neurons are generated and trained to classify the spatio-temporal spiking pattern of the SNN into pre-defined classes [29, 47]. This kind of evolving classifier is computationally inexpensive and puts emphasis on the order in which input spikes arrive, making it suitable for on-line learning and early prediction of temporal events [25]. For a more detailed description of the NeuCube architecture see [25] and Chap. 6.

The methodology we propose for dynamic visual recognition consists of the following steps:

(1) Event-based video recording with DVS [45].
(2) Pooling and encoding of DVS output into spike trains for the input neurons of the SNN.
(3) Training NeuCube-based model on the spike data using unsupervised learning, e.g., STDP.
(4) Training of an output classifier in a supervised mode [29].
(5) Validating the classification results.
(6) Repeating steps (2)–(5) for different parameter values in order to optimize the classification performance. Record the model with the best performance.
(7) Visualizing the trained SNN and analyzing its connectivity and spiking activity for a better understanding of the data and the involved brain processes.

We present the application of this method on a benchmarking experiment with the MNIST-DVS dataset for spike-based dynamic visual recognition and go into further detail about the tuning of parameters and analysis of the SNN.

## 13.3.4   Design of an Experiment for the MNIST-DVS Benchmark Dataset

Based on the MNIST dataset of handwritten digits [42], which has been one of the most popular benchmarking datasets for image recognition for over 20 years, the MNIST-DVS dataset is part of the NE15-MNIST database (Neuromorphic Engineering 2015 on MNIST) [48–50]. NE15-MNIST consists of four subsets that all aim to provide a benchmark for spike-based visual recognition. While the *Poissonian* and the *FoCal* subset are generated from static MNIST images, the other two subsets are based on DVS recordings of the MNIST images. The MNIST-FLASH-DVS subset contains DVS recordings of MNIST digits that are flashed on a screen. Because we were interested in dynamic visual recognition of moving objects, we decided to work on the MNIST-DVS subset that consists of DVS recordings of MNIST digits that wiggle across a screen and thereby produce temporal contrast and DVS events on the digits' edges.

The MNIST-DVS dataset is available online [50]. It consists of 30,000 recordings of 10,000 original MNIST digits recorded at three different scales each (scale-4, scale-8, and scale-16). Each recording has a time length of about 2.5 s. The files are provided in the jAER format [45].

The only preprocessing we applied to the data was the removal of the 75 Hz timestamp harmonic. Stabilizing the video data would have been contrary to our intention to develop a system for dynamic visual recognition, and in fact preliminary experiments suggested that the system would perform better on the original unstabilized videos.

The pooling of the DVS spikes into 148 input spike trains (ganglion cells) for the SNN, remained the same throughout all experiments. Inside the spike encoding algorithm only those four thresholds were changed that determine how many pixels within the receptive field of a ganglion cell must fire within one time step to make the ganglion cell itself emit a spike. As a first step, we wanted to find out how the system would perform differently when these thresholds, and, thus, the average spike rate of the input data for the SNN, were changed. The ganglion cells' receptive fields decrease from the periphery towards the center. Starting from the periphery, ganglion cells in layer 1 integrate the signal of $32 \times 32 = 1.024$ DVS pixels, cells in layer 2 from $16 \times 16 = 256$ pixels, cells in layer 3 from $8 \times 8 = 64$ pixels, and cells in layer 4 from $4 \times 4 = 16$ pixels. Assigning the same percentage threshold to all four layers would result in very low or no activity in the peripheral ganglion cells, e.g., with a threshold of 10% it would take only two DVS events within the receptive field of a ganglion cell in layer 4 to trigger a spike, but 101 DVS events within the receptive field of a ganglion cell in layer 1. Especially with the MNIST-DVS dataset, where DVS events only occur at the edges of the moving digits and not in big blobs, this would make the peripheral ganglion cells redundant. On the other hand, increasing the thresholds too much from layer to layer towards the center would put more emphasis on the peripheral parts of the video than intended.

The mapping of the input spikes into the SNN NeuCube remained the same throughout all experiments. In all experiments NeuCube was initialized with 732 leaky integrate and fire neurons (LIF), representing the primary visual cortex. For future experiments with higher video resolutions and more input neurons, NeuCube can easily be extended to include neurons that represent the secondary and the third visual cortex. Initial connections are formed following "small-world" connectivity with random connections within a predefined maximum distance from each neuron. This maximum distance was set to 2.5 in all experiments.

Unsupervised learning using spike timing dependent plasticity (STDP, [46]) is performed first to learn spatio-temporal patterns by forming new connections between neurons. The output classifier we used is called the dynamic evolving SNN algorithm [29], and it combines rank-order learning [51] and STDP learning. The NeuCube architecture is a stochastic model and, therefore, sensitive to parameter settings. To find the best values for the major parameters that influence the system's performance, we applied a grid search method that tests the system on different combinations of parameters within a predefined range and used those parameter values that resulted in the best classification accuracy. For the firing threshold, the refractory time and the potential leak rate of the LIF neurons we used values of 0.5, 6, and 0.002, respectively. The STDP learning parameter was set to 0.01. The variables *Mod* and *Drift* of the deSNN classifier were set to 0.8 and 0.005. See Chap. 6 and [25] for a more detailed explanation of these parameters.

## 13.3.5   Experimental Results

To compare the system's performance, we performed 10-fold cross validation on 1000 videos (first 100 of each digit) for different video scales and average spike rates. Table 4 summarizes the results. As a general trend, with few exceptions, the classification accuracy increased together with the average spike rate of the input neurons. For all video scales, the classification accuracy also increased when the system was run on all 10,000 videos of a scale. The best classification results were achieved with all 10,000 videos of one scale, encoded with the highest possible spike rate (0% as spike encoding threshold for all four layers). Classification accuracies were 90.56, 92.03, and 86.09% for scale-4, scale-8, and scale-16, respectively. The best accuracy in a single run with 90% of randomly selected data samples for training and the remaining 10% for testing was 92.90% for 10,000 scale-8 videos with the highest possible spike rate. This result is comparable to previous results on the MNIST-DVS dataset, presented in Fig. 13.12.

The lower accuracies on the scale-4 and the scale-16 samples reflect the fact that in these videos, the MNIST digits fill out either the whole screen (scale-16) or only a very little region in the center (scale-4). For the scale-4 digits, the signals transmitted by ganglion cells from layers 1, 2, and 3 are mostly noise, and do not contain much information about the digits. In the scale-16 videos, there is almost no activity in the central region of the screen, and, thus, no information is transmitted by the 64 foveal ganglion cells. Since our method puts big emphasis on the center of the video (50% of the input neurons represent data from only the central 64 pixels), performance on the scale-16 videos is lower.

### 13.3.6   Model Interpretation for a Better Understanding of the Processes Inside the Visual Cortex

The main purpose of the above experiments, carried out on the MNIST-DVS dataset, is to confirm the system's classification performance on a benchmark dataset, and the wiggling digits do not represent a real-life scene. However, we want to show how the SNN after being trained can be analyzed, to see how its connectivity changes as a response to the data. A comparison with the connectivity after training the SNN on 1000 scale-16 videos shows that slightly less connections are formed between neurons processing foveal information, since the scale-16 videos contain less DVS events in the foveal region.

The proposed system achieves the highest classification performance on the benchmark MNIST-DVS dataset for spike-based dynamic visual recognition. Every part of the system, the DVS sensor, the algorithm for encoding the DVS output into spike trains, and the SNN NeuCube adapt features from the human visual system. This allows for future experiments where the same stimuli are presented to humans and the system, and brain processes visualized by neuroimaging methods can be compared to the network processes of the SNN. The parameters involved in the spike encoding algorithm can easily be tuned to mimic the original behavior of retinal ganglion cells in such an experiment. Furthermore, analyzing the learning processes inside the SNN can help to gain a better understanding of either the data or the neural processes inside the visual cortex.

Since so much is known about the human visual system, and we aimed to develop a biologically plausible, yet computationally feasible implementation, there are many details not included in our model. There already exist very advanced mathematical models for the function of retinal ganglion cells [52], and our spike encoding algorithm has by far not touched every detail of them. The receptive field of each ganglion cell, for example, is split into a center region and a surrounding region with opposite behavior towards light [53]. In so-called on-center cells the center region is stimulated, whereas the surrounding region is inhibited when exposed to light. So-called off-center cells exhibit converse behavior. Including the function of on- and off-center ganglion cells inside the spike encoding algorithm would highly increase the model's biological plausibility, but also increase its computational complexity.

The proposed system puts big emphasis on the central part of the videos in both the encoding of DVS events to spike trains and the representation inside the SNN. This is justified by analogous features of the *fovea centralis* in the center of the human retina, responsible for focused vision. However, an important characteristic of human vision is the very fast and simultaneous movement of both eyes, called saccades. Saccades help to scan a broader part of the visual field with the fovea and integrate this information into a detailed map [44]. Human eye movement is also controlled by the visual grasp reflex that directs the eyes towards salient events in the periphery of the visual field [54]. These mechanisms for eye movement could be implemented in the spike encoding algorithm by changing the coordinates for the pooling of DVS pixels for each time step, and thereby virtually moving the center of the visual field. However, this would require additional features to save the movement and integrate it into the SNN (see [55–58]).

### 13.3.7   Summary of the Proposed BI-SNN Retinotopic Mapping Method

Results on the MNIST-DVS dataset have shown that the system can exceed the classification performance of other methods for dynamic visual recognition. Furthermore, it is possible to dynamically visualize and analyze the activity inside the SNN.

Due to the promising benchmark results and the benefit of the visualization tools for in-depth understanding of the data and the network processes, we envisage further research using this approach. In particular, we suggest the exploration of new learning methods inside NeuCube and of different algorithms for the encoding of DVS data into spike trains. We also encourage the development of further benchmark datasets for spike-based visual recognition, e.g., spiking versions of the KTH and the Weizmann datasets of human actions [59, 60]. Since the NeuCube architecture is not bound to only consist of neurons representing the visual cortex, future directions can include the integration of our system for visual recognition inside a broader methodology for the processing of audio-visual data.

## 13.4   Chapter Summary and Further Readings for Deeper Knowledge

This chapter presents methods, systems and applications of BI-SNN for audio-, visual an integrated audio-visual information processing. There are several advantages of using BI-SNN for these tasks:

- Accounting of time and time changes in the data;
- Fast information processing in terms of spikes;
- Highly parallel information processing;
- Allowing for brain-like principles to be applied;
- Better accuracy of recognition of video data capturing moving objects.
  Further recommended readings include:

- Demo on face age recognition with NeuCube: https://kedri.aut.ac.nz/R-and-D-Systems/neucube/face-age-recognition.
- Age-invariant face recognition: https://kedri.aut.ac.nz/R-and-D-Systems/age-invariant-face-recognition
- Demo on fast moving object recognition using NeuCube: https://kedri.aut.ac.nz/R-and-D-Systems/fast-moving-object-recognition
- Audio-visual information processing in BI-SNN: https://kedri.aut.ac.nz/R-and-D-Systems/audio-visual-data-processing-and-concept-formation#auditory
- Digital vision: http://www.wisdom.weizmann.ac.il/~vision/SpaceTimeActions.html.
- Recognition of human actions: http://www.nada.kth.se/cvap/actions/.
- Actions as Space-Time Shapes: http://www.wisdom.weizmann.ac.il/~vision/SpaceTimeActions.html.
- MNIST-DVS and FLASH-MNIST-DVS Databases: http://www2.imse-cnm.csic.es/caviar/MNISTDVS.html.

**Acknowledgements**  The material in this chapter is partially published in several publications as referenced in the text. I acknowledge the contribution of my co-authors L. Paulin, A. Wendt, F. Alvi, W. Cui, W. Yan. L. Benuskova, G. Saraceno. The DVS (not used here as a hardware, but as a bench mark data set recorded with its use) is developed in the INI ETH/UZH by T. Delbruck and his team.

# References

1. N. Kasabov, *Evolving Connectionist Systems: The Knowledge Engineering Approach* (Springer, 2007)
2. L. Benuskova, N. Kasabov, *Computational Neurogenetic Modelling*, Topics in Biomedical Engineering. International Book Series, ISBN 978-0-387-48355-9
3. G. Saraceno, *Deep learning and memorizing of spectro-temporal data (music) in the spatio-temporal brain* (Master Thesis, University of Trento, 2017)
4. J.L. Eriksson, A.E.P. Villa, Artificial neural networks simulation of learning of auditory equivalence classes for vowels, in *International Joint Conference on Neural Networks, IJCNN*. (Vancouver, 2006), pp. 1453–1460
5. D.D. Greenwood, Critical bandwidth and the frequency coordinates of the basilar membrane. J. Acoust. Soc. Am. **33**(10), 1961 (1961). https://doi.org/10.1141/1.1908437
6. D.D. Greenwood, A cochlear frequency-position function for several species–29 years later. J. Acoust. Soc. Am. **87**(6), 2592–2605 (1990)
7. E. de Boer, H.R. de Jongh, On cochlear encoding: potentialities and limitations of the reverse-correlation technique. J. Acoust. Soc. Am. **63**(1), 115–135 (1978)
8. J.B. Allen, How do humans process and recognize speech? IEEE Trans. Speech Audio Process. **2**(4), 567–577 (1994)
9. E. Zwicker, Subdivision of the audible frequency range into critical bands (Frequenzgruppen). J. Acoust. Soc. Am. **33**(1961), 248 (1961)
10. B.R. Glasberg, B.C. Moore, Derivation of auditory filter shapes from notched-noise data. Hear Res. **47**(1–2), 103–138 (1990)

11. T.J. Cole, J.A. Blendy, A.P. Monaghan, K. Krieglstein, W. Schmid, A. Aguzzi, G. Fantuzzi, E. Hummler, K. Unsicker, G. Schütz, Targeted disruption of the glucocorticoid receptor gene blocks adrenergic chromaffin cell development and severely retards lung maturation. Genes Dev. 9(14), 1608–1621 (1995)

12. A.M. Aertsen, J.H. Olders, P.I. Johannesma, Spectro-temporal receptive fields of auditory neurons in the grassfrog. III. Analysis of the stimulus-event relation for natural stimuli. Biol. Cybern. 39(3), 195–209 (1981)

13. N. Kasabov, E. Postma, J. van den Herik, AVIS: a connectionist-based framework for integrated auditory and visual information processing. Inf. Sci. 143(2000), 147–148 (2000)

14. E.O. Postma, H.J. van den Herik, P.T.W. Hudson, Image recognition by brains and machines, in Brain-like Computing and Intelligent Information Systems, ed. by S. Amari, N. Kasabov (Springer, Singapore, 1998), pp. 25–47

15. E.O. Postma, H. J. van den Herik, P.T.W. Hudson, SCAN: a scalable model of covert attention. Neural Netw. 10, 993–1015 (1997)

16. K. Kim, N. Relkin, K.-M. Lee, J. Hirsch, Distinct cortical areas associated with native and second languages. Nature 388, 171–174 (1997)

17. S. Wysoski, L. Benuskova, N. Kasabov, Evolving spiking neural networks for audio-visual information processing. Neural Netw. 23(7), 819–835 (2010)

18. S. Wysoski, L. Benuskova, N. Kasabov, Fast and adaptive network of spiking neurons for multi-view visual pattern recognition. Neurocomputing 71(14–15), 2563–2575 (2008)

19. A. Ross, A.K. Jain, Information fusion in biometrics. Pattern Recognit. Lett. 24(14), 2115–2145 (2003)

20. C. Sanderson, K.K. Paliwal, Identity verification using speech and face information. Digital Signal Process. 14(2004), 449–480 (2004)

21. A. Sharkey, Combining Artificial Neural Nets: Ensemble and Modular Multi-net Systems (Springer, Heidelberg, 1999)

22. B.E. Stein, M.A. Meredith, The Merging of the Senses (MIT Press, Cambridge, 1993)

23. S.G. Wysoski, L. Benuskova, N. Kasabov, Adaptive spiking neural networks for audiovisual pattern recognition, ICONIP. Lecture notes in computer science (2007) (to appear)

24. C. Sanderson, K.K. Paliwal, Identity verification using speech and face information. Digital Signal Process. 14(2004), 449–480 (2004)

25. N. Kasabov, NeuCube: a spiking neural network architecture for mapping, learning and understanding of spatio-temporal brain data. Neural Netw. Off. J. Int. Neural Netw. Soc. 52, 62–76 (2014). https://doi.org/10.1016/j.neunet.2014.01.006

26. A. Wendt, G. Sraceno, L. Paulum, N. Kasabov, Audio-visual data processing and concept formation (internal report), https://kedri.aut.ac.nz/R-and-D-Systems/audio-visual-data-processing-and-concept-formation#auditory

27. C. Ge, N. Kasabov, Z. Liu, J. Yang, A spiking neural network model for obstacle avoidance in simulated prosthetic vision. Inf. Sci. 399(30–42), 2017 (2017)

28. A.R. McIntosh, R.E. Cabeza, N.J. Lobaugh, Analysis of neural interactions explains the activation of occipital cortex by an auditory stimulus. J. Neurophysiol. 80(1998), 2790–2796 (1998)

29. N. Kasabov, K. Dhoble, N. Nuntalid, G. Indiveri, Dynamic evolving spiking neural networks for on-line spatio- and spectro-temporal pattern recognition. Neural Netw. Off. J. Int. Neural Netw. Soc. 41, 188–201 (2014). https://doi.org/10.1016/j.neunet.2014.11.014

30. P. Viola, M.J. Jones, Rapid object detection using a boosted cascade of simple features. Proc. IEEE Comput. Soc. Conf. Comput. Vis. Pattern Recognit. 1(2001), 511–518 (2001)

31. A. Delorme, L. Perrinet, S. Thorpe, Networks of integrate-and-fire neurons using rank order coding. Neurocomputing 2001, 38–48 (2001)

32. A. Delorme, S. Thorpe, Face identification using one spike per neuron: resistance to image degradation. Neural Netw. 14(2001), 795–803 (2001)

33. D. Gonzalo, T. Shallice, R. Dolan, Time-dependent changes in learning audiovisual associations: a single-trial fMRI study. NeuroImage 11, 243–255 (2000)

34. A.M. Burton, V. Bruce, R.A. Johnston, Understanding face recognition with an interactive activation model. B. J. Psychol. **81**, 361–380 (1990)
35. A.W. Ellis, A. Young, D.C. Hay, Modelling the recognition of faces and words, in *Modelling Cognition*, ed. by P.E. Morris (Wiley, London, 1987), p. 1987
36. H.D. Ellis, D.M. Jones, N. Mosdell, Intra- and inter-modal repetition priming of familiar faces and voices. B. J. Psycol. **88**, 143–156 (1997)
37. K. Kriegstein, A. von, Giraud, Implicit multisensory associations influence voice recognition. PLoS Biol. **4**(10), 1809–1820 (2006)
38. W. Cui, W.Q. Yan, N. Kasabov, Deep learning with NeuCube for fats moving object recognition (KEDRI internal report), https://kedri.aut.ac.nz/R-and-D-Systems/fast-moving-object-recognition
39. F.B. Alvi, R. Pears, N. Kasabov, An evolving spatio-temporal approach for gender and age group classification with spiking neural networks. Evolv. Syst. (2017). https://doi.org/10.1007/s14530-017-9175-y
40. J.K. Ricanek, T. Tesafaye, Morph: a longitudinal image database of normal adult age-progression, in *7th International Conference on Automatic Face and Gesture Recognition. FGR 2006.* (IEEE, 2006), pp. 341–345
41. L.G. Farkas, *Anthropometry of the Head and Face* (Raven Press, New York, 1994)
42. Y. Lecun, L. Bottou, Y. Bengio, P. Haffner, Gradient-based learning applied to document recognition. Proc. IEEE **86**, 2278–2324 (1998). https://doi.org/10.1109/5.726791
43. L. Paulun, A. Abbott, N. Kasabov, A retinotopic spiking neural network system for accurate recognition of moving objects using NeuCube and dynamic vision sensors. Front. Comput. Neurosci. **12**, 42 (2018)
44. D. Purves, *Neuroscience* (Sinauer, Sunderland, MA, 2014)
45. T. Delbruck, Frame-free dynamic digital vision (University of Tokyo, 2008). L. Gorelick, M. Blank, E. Shechtman, Actions as Space-Time Shapes (2007). http://www.wisdom.weizmann.ac.il/∼vision/SpaceTimeActions.html. Accessed 29 Aug 2017
46. S. Song, K.D. Miller, L.F. Abbott, Competitive Hebbian learning through spike-timing-dependent synaptic plasticity. Nat. Neurosci. **3**, 919–926 (2000). https://doi.org/10.1038/78829
47. N. Kasabov, E. Capecci, Spiking neural network methodology for modelling, classification and understanding of EEG spatio-temporal data measuring cognitive processes. Inf. Sci. **294**, 565–575 (2015). https://doi.org/10.1016/j.ins.2014.06.028
48. T. Serrano-Gotarredona, B. Linares-Barranco, Poker-DVS and MNIST-DVS. Their history, how they were made, and other details. Front. Neurosci. **9**, 481 (2015). https://doi.org/10.3389/fnins.2015.00481
49. Q. Liu, G. Pineda-García, E. Stromatias, T. Serrano-Gotarredona, S.B. Furber, Benchmarking spike-based visual recognition: a dataset and evaluation. Front. Neurosci. **10**, 496 (2016). https://doi.org/10.3389/fnins.2016.00496
50. A. Yousefzadeh, T. Serrano-Gotarredona, B. Linares-Barranco, MNIST-DVS and FLASH-MNIST-DVS Databases (2015). http://www2.imse-cnm.csic.es/caviar/MNISTDVS.html. Accessed 21 Aug 2017
51. S. Thorpe, J. Gautrais, Rank order coding, in *Computational Neuroscience: Trends in Research, 1998*, ed. by J.M. Bower (Springer US, Boston, 1999), pp. 114–118
52. H. Wei, Y. Ren, A mathematical model of retinal ganglion cells and its applications in image representation. Neural Process. Lett. **38**, 205–226 (2014). https://doi.org/10.1007/s11063-014-9249-6
53. M. Nelson, J. Rinzel, The Hodgkin-Huxley model, in *The Book of GENESIS*, vol. 4, ed. by J. M. Bower, D. Beeman, (Springer, New York, 1995), pp. 27–51
54. S. Monsell, J. Driver, *Control of Cognitive Processes: Attention and Performance XVIII* (MIT Press, Cambridge, 2000)
55. J.A. Perez-Carrasco, C. Serrano, B. Acha, T. Serrano-Gotarredona, B. Linares-Barranco, spike-based convolutional network for real-time processing, in *Proceedings of 20th*

*International Conference on Pattern Recognition (ICPR)*, Istanbul, Turkey, 23–26 Aug 2010 (IEEE, Piscataway, NJ, 2010), pp. 3085–3088

56. O. Bichler, D. Querlioz, S.J. Thorpe, J.P. Bourgoin, C. Gamrat, Extraction of temporally correlated features from dynamic vision sensors with spike-timing-dependent plasticity. Neural Netw. Off. J. Int. Neural Netw. Soc. **32**, 339–348 (2014). https://doi.org/10.1016/j.neunet.2014.02.022

57. A. Jimenez-Fernandez, C. Lujan-Martinez, R. Paz-Vicente, A. Linares-Barranco, G. Jimenez, A. Civit, in *From Vision Sensor to Actuators, Spike Based Robot Control through Address-Event-Representation*. IWANN 2009: Bio-Inspired Systems: Computational and Ambient Intelligence (2009), pp. 797–804

58. F. Perez-Peña, A. Morgado-Estevez, A. Linares-Barranco, A. Jimenez-Fernandez, F. Gomez-Rodriguez, G. Jimenez-Moreno, et al., Neuro-inspired spike-based motion: from dynamic vision sensor to robot motor open-loop control through spike-VITE. Sensors **14**, 15805–15832 (2014). https://doi.org/10.3390/s141115805 (Basel, Switzerland)

59. I. Laptev, B. Caputo, Recognition of Human Actions (2005). http://www.nada.kth.se/cvap/actions/. Accessed 29 Aug 2017

60. L. Gorelick, M. Blank, E. Shechtman, Actions as Space-Time Shapes (2007). http://www.wisdom.weizmann.ac.il/~vision/SpaceTimeActions.html. Accessed 29 Aug 2017

61. C.A. Curcio, K.R. Sloan, R.E. Kalina, A.E. Hendrickson, Human photoreceptor topography. J. Comp. Neurol. **292**, 497–523 (1990). https://doi.org/10.1002/cne.902920402

# Chapter 14
# Brain-Computer Interfaces Using Brain-Inspired SNN

This chapter presents methods of BI-SNN for brain-computer interfaces (BCI). It introduces a new types of BCI, called brain-inspired BCI (BI-BCI). The BI-BCI can not only classify brain signals in a 'black-box' as the traditional BCI do, but they can create a model of the brain signals when a person is performing a task providing a neurofeedback, enabling a better understanding of the brain activities. This is also a step towards knowledge transfer from humans to machines. Applications for neuro-control, neurorehabilitation, cognitive games and others are discussed.

The chapter is organised in the following sections:

14.1. Brain computer interfaces (BCI).
14.2. A Framework for Brain-Inspired BCI (BI-BCI).
14.3. BI-BCI for motor execution and motor intention from EEG signals.
14.4. BI-BCI for neurorehabilitation with a neurofeedback and for neuro-prosthetics.
14.5. From BI-BCI to knowledge transfer between humans and machines.
14.6. Chapter summary and further readings for deeper knowledge.

## 14.1 Brain-Computer Interfaces

### 14.1.1 General Notions

The main idea behind BCIs is to record the brain's activity patterns (BAPs) when performing specific tasks that are associated with particular computer commands and then employ some powerful machine learning schemes to classify these patterns. When the user performs one of the tasks in real time, the classifier attempts to detect the associated command, which is then sent to the interface for execution, as indicated by [1]. Example is the P300 BCI, a communication tool used for spelling

© Springer-Verlag GmbH Germany, part of Springer Nature 2019                   479
N. K. Kasabov, *Time-Space, Spiking Neural Networks and Brain-Inspired
Artificial Intelligence*, Springer Series on Bio- and Neurosystems 7,
https://doi.org/10.1007/978-3-662-57715-8_14

purposes. This interface is controlled by the signals that are generated in the human brain as a result of visual stimulation.

The fundamentals of BCIs need to be clearly understood in order to achieve the study objectives. The general framework of a BCI is presented in Fig. 14.1. According to [2, 3], it comprises data acquisition, pre-processing, classification and biofeedback. These four steps are described in detail in the next section.

Figure 14.2 shows is a schematic representation of measuring brain signals when a person is performing a task.

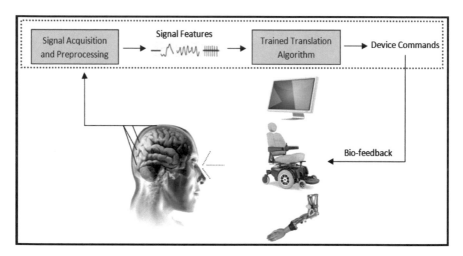

**Fig. 14.1** A general framework of a BCI system

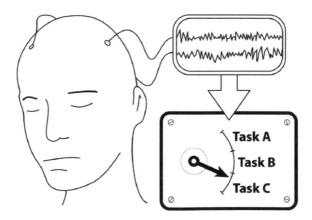

**Fig. 14.2** A schematic representation of measuring brain signals when a person is performing a task (from [52])

## 14.1.2  BCI Based on EEG

Electroencephalogram (EEG) data measure brain electrical potentials when a person is performing a task as discussed in Chaps. 3 and 8. Different frequency characteristics of EEG indicate different brain functions, at different time and space in the brain (see Table 14.1).

Figure 14.3, shows the electrodes' spatial positions and the channels' names in the 10–20 placement system.

## 14.1.3  Types and Applications of BCI

Mainly, there are two types of BCIs as reported in [4, 5] synchronous and asynchronous. A synchronous BCI is based on system initiation. Interaction is only

**Table 14.1** EEG bands and their properties [43]

| (see Chaps. 3 and 8) Signal | Frequency (Hz) | Shape | Properties |
|---|---|---|---|
| Delta | 1–3 | | This wave has high amplitude but low frequency. It is seen in young children normally, and also in adults when they are sleeping |
| Theta | 4–7 | | This signal is normally seen in young children, it could be as well generated in older children and adults in arousal or drowsiness. It is also associated with meditation, relaxation and creative status |
| Alpha | 8–13 | | This is the first type of wave discovered in the human brain. It has high amplitude. It emerges with eyes closing and relaxation, and attenuates with opening the eyes and mental exertion |
| Beta | 14–30 | | Beta wave can be also called sensorimotor rhythm, as it accrues when arms or hands idle. It could be associated with drugs and anxious thinking. It is generated from the frontal lobe, and is widely used for motor BCI applications. In the case of cortical damage this wave could be absent |
| Gamma | >30 | | This pattern is associated with alertness, working and motor movements |

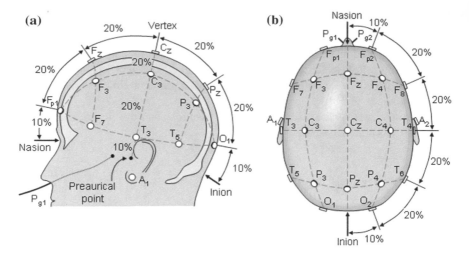

**Fig. 14.3** The electrodes' positions and the channels' names in the 10–20 international EEG replacement system [5] (Chaps. 3 and 8)

allowed in a fixed time window. Most synchronous interfaces count on event-related potentials that are generated by a stimulus, e.g., visual or auditory stimulus, produced in a known time frame. A good example is the P300 Speller. This system depends on the synchronisation between visual evocation and the brain activity patterns. This type is easier to design. Additionally, the classification is less affected by artifacts as a result of the windowing techniques.

In contrast, asynchronous interfaces depend on user initiation. They do not impose specific time frames for interaction and offer a more natural way for communication. However, designing and evaluating asynchronous systems is more complicated. To prevent accidental detection, the mental task must be unique. Appropriate control signals could be the sensor-motor rhythms as explained in [5].

BCI applications can be also divided into exogenous and endogenous interfaces [5]. Exogenous interfaces depend on external cues. Users training is not required since the control signal can be easily and quickly set up. Reasonable response can be achieved after a sufficient training, and good results can be achieved using a minimum number of channels, down to one. Nevertheless, this type may cause tiredness for some users while focusing their attention on the stimuli for long periods leading to significant decrease of the user performance. Contrary, endogenous interfaces are independent of any stimulation, thus, they are useful for users who are suffering from sensory organs damage. Despite that, user training is required and it is time consuming. Several months could be spent to reach a good performance, and still the speed is very low with an ITR of 3–35 bits/min. The study by [6] presents a good example of an endogenous interface.

When a targeted visual stimulus is presented to a subject, the visual cortex of the subjects is activated after 300 ms (considered to be the conscious perception of a

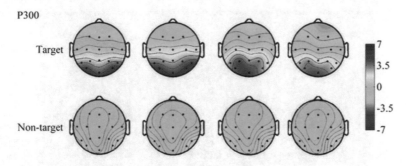

**Fig. 14.4** Scalp topographies recorded 300 ms after a target visual stimulation (top row) compared to the ones recorded 300 ms after a non-target stimulus [7]

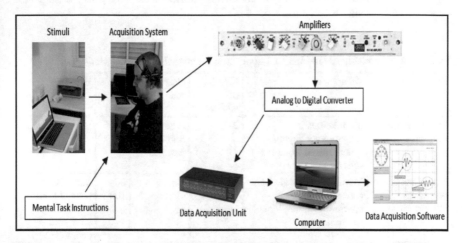

**Fig. 14.5** The measured EEG signals are amplified before they are used for BCI (from [8])

stimulus, in contrast to the peri-perceptual reaction as discussed in Chap. 8 [9]). Figure 14.4 shows scalp topographies recorded 300 ms after a target visual stimulation (top row) compared to the ones recorded 300 ms after a non-target stimulus [7].

The measured EEG signals are amplified before used for a BCI [8]—Fig. 14.5.

A typical use of a BCI is to control a device, such as a robot, or a symbol on a computer screen, using brain signals (Fig. 14.6).

Various brain control signals can be used for different applications (Table 14.2).

The next sections describes how BI-SNN, such as NeuCube, can be used to develop BCI, now called BI-BCI.

**Fig. 14.6** A typical use of a BCI is to control a device, such as a robot, or a symbol on a computer screen, using brain signals

**Table 14.2** Control signals used in BCI applications, and their main characteristics

| Signal | Phenomena | Number of choices | User training | ITR (bits/min) | Example |
|---|---|---|---|---|---|
| Sensor-motor rhythms | Modulations in sensorimotor rhythms synchronized to motor activities | 2–5 | Extensive training is required | 3–35 | BCI wheelchair |
| Visual evoked potentials (VEPs) [10] | Modulations in the visual cortex rhythms synchronized to a visual stimulus | High | No | 60–100 | VEP BCI to control a hand orthotic for paralyzed people [11] |
| P300 [12, 13] | Positive peaks in the brain waves due to infrequent visual, auditory or somatosensory stimuli. These peaks elicited about 300 ms after attending to an oddball stimulus among several frequent stimuli | High | No | 20–25 | P300 Speller [13] |
| Slow cortical potentials (SCPs) [14] (Hinterberger et al. 2004) | Slow voltage shifts in the brain waves correlated with increased/decreased neuronal activity | 2–4 | Extensive training is required | 5–12 | On-screen cursor control [14] |

## 14.2   A Framework for Brain-Inspired BCI (BI-BCI)

### 14.2.1   The NeuCube BI-SNN Architecture

Some general principles of the BI-SNN NeuCube architecture were presented in Chap. 6 and also in [15]. The NeuCube architecture is depicted in Fig. 14.7. It consists of the following functional modules:

- Input data encoding module;
- 3D SNN reservoir module (SNNc);
- Output function (classification) module;
- Gene regulatory network (GRN) module (Optional).

The process of creating a NeuCube model for a given task takes the following steps:

a. Encode input data into spike sequences: continuous value input information is encoded into trains of spikes;
b. Construct and train in an unsupervised mode a recurrent 3D SNN reservoir, SNNc, to learn the spike sequences that represent individual input patterns;
c. Construct and train in a supervised mode an evolving SNN classifier to learn to classify different dynamic patterns of the SNNc activities that represent different input patterns from SSTD that belong to different classes;

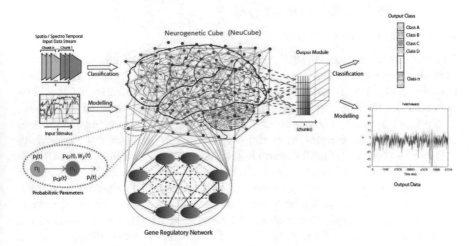

**Fig. 14.7** The NeuCube BI-SNN general architecture (Chap. 6 and also [15])

d. Optimise the model through several iterations of steps (a)–(c) above for different parameter values until maximum accuracy is achieved.
e. Recall the model on new data.

The above modules from (a) to (e) are described further in this section.

*Input data encoding module*

Continuous value input data can be transformed into spikes so that the current value of each input variable (e.g., pixel, EEG channel, fMRI voxel) is entered into a population of neurons that emit spikes based on how much the input value belongs to their receptive fields. This method is called population rank coding [16]—the higher the membership degree, the earlier a spike is generated.

Another method is the Threshold Based Encoding (TBE) method as demonstrated in the Silicon Retina [17, 18]. This is based on thresholding the difference between two consecutive values of the same input variable over time. This is suitable when the input data is a stream and only the changes in consecutive values can be processed.

In some specific applications, a method called Ben's Spike Algorithm (BSA) has been used for EEG data transformation into spike trains [19]. Methods for encoding input data into spike sequences are presented in Chap. 4. In [20] a method for selection and optimisation of the encoding algorithm is presented and implemented as software.

The transformed input data into spike series is entered (mapped) into spatially located neurons from the SNNc. The mapping will depend on the problem in hand. Here we enter brain data sequences to spatially located neurons in the SNNc that represent spatially brain areas where data is collected. Spike trains are continuously fed into the SNNc in their temporal order.

*The SNNcube module (SNNc)*

The SNNc is structured to spatially map brain areas for which time-space brain data (TSBD) or/and gene data is available. A neuronal SNNc structure can include known structural or functional connections between different areas of the brain represented in the data. Setting up a proper initial structural connectivity in a model, is important in order to learn properly spatio-temporal data, to capture functional connectivity and to interpret the model [21]. More specific structural connectivity data can be obtained using for example Diffusion Tensor Imaging (DTI) method.

Functional connectivity of the brain manifests the small-world organization across different time scales (e.g., seconds, milliseconds) [22]. Neurons in a structural or functional area of the brain are more densely interconnected and the closer these areas are, the higher the connectivity between them [23, 24]. Both structural connectivity, measured through MRI [25] and functional connectivity, measured through EEG and MEG [26] of the brain show small-world organization. This is the main reason for suggesting a type of small-world connectivity for a NeuCube initial structure, where clusters of neurons correspond to structural and functional areas related to the data [22]. If DTI data is available (see Chap. 11), this data can be used to preset some connections of the SNNc before the model is trained.

The initial structure of the SNNc is defined based on the available brain data and the problem, but this structure can be evolving through the creation of new neurons and new connections based on the using the ECOS principles [27]. If new data do not sufficiently activate existing neurons, new neurons are created and allocated to match the data along with their new connections.

In a current implementation, the SNNc has a 3D structure connecting leaky-integrate and fire model (LIFM) spiking neurons with recurrent connections. The input data is propagated through the SNNc and a method of *unsupervised learning* is applied, such as STDP. The neuronal connections are adapted and the SNNc learns to generate specific trajectories of spiking activities when a particular input pattern is entered. On Fig. 14.7 a special class of LIFM is shown—the probabilistic neuronal model that has probability parameters attached to the connections, the synapses and the output of the spiking neuron. The SNNc accumulates temporal information of all input spike trains and transforms it into dynamic states that can be classified over time. The recurrent reservoir generates unique accumulated neuron spike time responses for different classes of input spike trains.

Figure 14.8 shows a snapshot of the connections in the SNNcube after unsupervised learning. The connections represent spatio-temporal relationships between input data variables that encode brain areas of the source data over time. The SNNc has 1471 neurons and the coordinates of these neurons correspond directly to the Tailarach template coordinates with a resolution of 1 cm³. It can be seen that as a result of training new connections have been created that represent spatio-temporal interaction between input variables captured in the SNNc from the data. The connectivity can be dynamically visualised for every new pattern submitted. For example, the connectivity of the right part of the SNNc in Fig. 14.8 is larger as the SNNc has learned movement of the left hand of a subject, controlled by the right hemisphere of the brain.

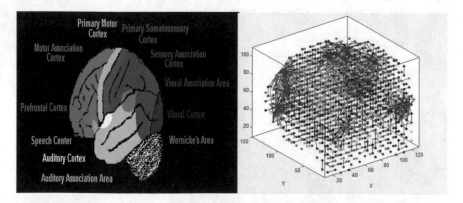

**Fig. 14.8** A snapshot of the connnections in the SNNcube after unsupervised learning. The connections represent spatio-temporal relationships between input data variables that encode brain areas of the source data over time. The connectivity of the right part of the SNNc is larger as the SNNc has learned movement of the left hand of a subject, controlled by the right hemisphere of the brain

## 14.2.2    A Brain-Inspired Framework for BCI (BI-BCI) with Neurofeedback

Using the ability of the NeuCube BI-SNN to learn as discussed in Chap. 6 and in other chapters of the book, here we demonstrate that NeuCube can be used to build a BI-BCI that also provides a neurofeedback showing which parts of the brain are active in time-space (Fig. 14.8).

Figure 14.9 shows a schematic diagram of using NeuCube for BI-BCI with neuro-feedback showing which parts of the brain are active in time-space.

The activity of a trained SNNc on EEG brain data related to certain tasks, can be classified in output classification module of the NeuCube BI-SNN architecture providing the control signal to an activator device to perform the task—Fig. 14.10 (see also Fig. 14.9).

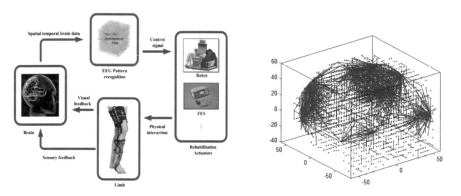

**Fig. 14.9** A schematic diagram of using NeuCube for BI-BCI with neuro-feedback showing which parts of the brain are active in time-space

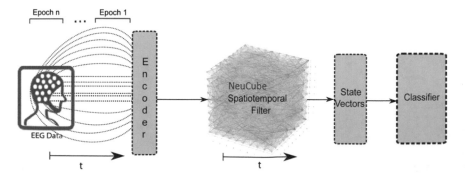

**Fig. 14.10** The activity of a trained SNNc on EEG brain data related to certain tasks, can be classified in output classification module of the NeuCube BI-SNN architecture providing the control signal to an activator device to perform the task—(see also Fig. 14.9)

Several BCI application systems, built with the use of the above methodology, are presented in the next sections.

## 14.3   BI-BCI for Detecting Motor Execution and Motor Intention from EEG Signals

The method an experimental results presented in this section are first published in [28].

### 14.3.1   Introduction

A focal neurological insult that causes changes to cerebral blood flow, such as in a stroke, can result in mild to severe motor dysfunctions on the contralateral side of the body. Although some spontaneous recovery usually occurs in the first 6 months after stroke only about 14% of people with stroke recover normal use of the upper limb [29]. The driver of functional recovery after stroke is neural plasticity, the propensity of synapses and neuronal circuits to change in response to experience and demand [30–34]. Whilst it is known that frequency and intensity of intervention following stroke is important high intensity rehabilitation is resource-limited. In order to deliver interventions at a high enough intensity and frequency for neural plasticity we need to develop devices that can assist with rehabilitation without the concentrated input of rehabilitation professionals.

Motor imagery (MI), or the mental rehearsal of a movement, is an approach used by rehabilitation professionals to encourage motor practice in the absence of sufficient muscle activity [33–35]. MI is thought to activate similar cortical networks as activated in a real movement, including activation of the primary motor cortex, premotor cortex, supplementary motor area and parietal cortices [36, 37]. Recent evidence suggests that although there are common cortical networks in real and imagined movement (frontal and parietal sensorimotor cortices) there are also important differences, with ventral areas being activated in imagined movement, but not in real movement. These specific additional activations in the extreme/external capsule may represent an additional cognitive demand of imagery based tasks.

Recovery of movement control is greater after motor execution training than after MI training alone. Interestingly the combination of MI training with even passive movement generates greater recovery than MI alone [38]. Combining motor imagery with functional electrical muscle stimulation, via Brain Computer Interface (BCI) devices, may result in greater neural plasticity and recovery than motor imagery alone, or motor imagery combined with passive movement. The additional feedback to the brain provided by executing a movement may enhance plasticity and reduce the cognitive demand of motor imagery. Many people following stroke or other neurological disorder have some residual muscle activity but fail to recruit enough motor units at an appropriate speed and pattern, to generate sufficient force to complete the desired movement [39, 40]. A BCI device in which motor imagery triggers an appropriate signal to a functional electrical stimulation system would facilitate the practice of real movements and potentially result in greater neural plasticity and functional recovery.

EEG records brain signals through electrodes on the scalp and is the most widely used method for recording brain data used in BCI devices. EEG is non-invasive and has good temporal and spatial resolution. However, EEG systems have been criticized because of the time consuming and complex training period for the potential user [41]. One advantage of the NeuCube framework is that intensive training of the user is not required as NeuCube classifies naturally elicited cortical activity, rather than a specific component of the EEG signal, such as the P300 wave, the production of which has to be learned by the user. In addition, the NeuCube is capable of learning in an on-line fashion, training as it is used.

As an example here we are investigating the feasibility of using NeuCube with EEG data to develop a functional electrical stimulation BCI system that is able to assist in the rehabilitation of complex upper limb movements. Two methods of use are under consideration, firstly for people who have no voluntary activity in a limb who would drive the device using MI, and secondly for people who have some residual activity in their muscles that, in addition to using MI, may augment the device with their own muscle activity. To do this it is important to establish a high degree of accuracy of classification of movement intention and movement execution to ensure that the appropriate electrical stimulation output is then provided. One of the challenges to any BCI system is the extent to which it accurately classifies the input signal.

In [41] real movement, consisting of a pinch grip to a specified force level, compared to a resting state, was used. Data were collected using functional Near Infrared Spectrometry (fNIRS) combined with other physiological data, such as blood pressure and respiratory information. Using hidden Markov Model's (HMM's) as the classifier framework accuracies ranging between 79.6 and 98.8% over 2 classes were achieved. Using fNIRS in a trial of MI [42] investigated the classification accuracy of a simple imagined tap of the thumb on a keyboard versus a complex multi-digit tapping sequence. Linear discriminant analysis (LDA) was used in combination with careful selection of the best performing data channel (out of 3 possible channels) and best 4 features for each participant. The study in [42] reported classification accuracies in a 2-class model (simple imagined movement or complex imagined movement) of between 70.8 and 91.7%. A Sparse Common Spatial Pattern (SCSP) optimization technique that reduced EEG channels by disregarding noisy channels and channels thought to be irrelevant was reported in [44], however this approach results in a loss of data that could be informative.

We were interested in determining if it was feasible to use the NeuCube framework as a driver of BCI devices. As a first step we wanted to determine if the NeuCube was at least equivalent in classifying movement tasks as other commonly used methods. As proof-of concept we designed a study that required NeuCube to classify imagined and real movements in two different directions and at rest (wrist flexion, extension or rest). The general hypothesis is that NeuCube using EEG data can correctly identify brain patterns corresponding to specific movements. Previous work from our lab in association with research collaborators has indicated the potential of NeuCube to identify different EEG patterns relating to different imagined movements from a commercially available 14 channel EEG headset. In this trial imagined wrist extension, rest and wrist flexion achieved accuracy in 1 individual of 88, 83 and 71% respectively [45].

The specific hypothesis for this study was that the NeuCube would accurately classify both single joint real and imagined movements of the hand into one of three classes, flexion, extension or rest. This paradigm built on the earlier work in [45] by increasing the complexity of the task by requiring the NeuCube to distinguish three conditions, two different muscle contraction patterns (flexion or extensor muscle activity) or rest [45]. A secondary hypothesis was that the NeuCube would perform better than other classification methods, including Multiple Linear Regression (MLR), Support Vector Machine (SVM), Multilayer Perceptron (MLP) and Evolving Clustering Method (ECM) [46], along with offering other advantages such as adaptability to new data on-line and interpretability of results.

## 14.3.2   Design of an Experimental BI-BCI System

Three healthy volunteers from our laboratory group participated in the study. None had any history of neurological disorders and all were right handed.

All measures were taken in a quiet room with participants seated in a dining chair. The task consisted of either performing the specified movements or imagining the movements, or remaining at rest. All tasks were completed with eyes closed to reduce visual and blink related artifacts. The movement execution task involved the participant resting, flexing the wrist or extending the wrist. The starting position was from mid-pronation with the forearm resting on the person's lap. The movement intention task involved the participant imagining or performing the movements as described above. Participants were required to imagine or perform each movement in 2 s and to repeat that 10 times.

A low-cost commercially available wireless Emotiv Epoc EEG Neuroheadset was used to record EEG data. The Epoc records from 14 channels based on International 10–20 locations (AF3, F7, F3, FC5, T7, P7, O1, O2, P8, T8, FC6, F4, F8, AF4). Two additional electrodes (P3, P4) were used as reference. Data were digitized at 128 Hz sampling rate and sent to the computer via Bluetooth. An important factor was that no filtering was applied to the data, either online or offline.

The data was separated into classes denoting each task. Each set of ten samples was then evenly divided into a training (seen) and a testing (unseen) set. The data was then converted into trains of spikes (one train per channel, 14 in total) with the

TBE algorithm, utilizing a spiking threshold of 6. No other data processing was applied.

### 14.3.3  Classification Results

Each training sample was presented to the NeuCube once, entered as 14 input streams of EEG continuous data collected at the msec time unit [47]. The spiking activity of every neuron was recorded over the time of the sample, and these presented to the deSNNs classifier. The deSNNs was initialized with a Mod of 0.9 and drift factor of 0.25 (empirically established values for this dataset) (see Chap. 5 for details of the deSNN). The synaptic weights for both the NeuCube and the deSNNs were then fixed at their final (learned) values for the validation phase. The unseen data samples were presented in the same way, and the predicted classes recorded. The predicted classes were then compared to the actual classes of those samples.

The NeuCube-based model described above was compared to some popular machine learning methods: MLR, SVM, MLP and ECM. The SVM method uses a Polynomial kernel with a rank 1; the MLP uses 30 hidden nodes with 1000 iterations for training. The ECM [48] uses m = 3; Rmax = 1; Rmin = 0.1. Data for these methods is averaged at 8 ms intervals and a single input vector is formed for each session, as is general practice.

Figure 14.11 shows the connectome of the trained NeuCube. Blue lines show strong excitatory connections between two neurons, and red strong inhibitory. Table 14.3 shows results of the comparative study—classification accuracy is expressed as percentage for real and imagined movements.

The classification accuracy of the NeuCube was on average 76%, with individual accuracies ranging from 70 to 85%. There was a consistent rate of recognition between the real and the imagined movement. In terms of the comparison with other classification approaches, it is clear from the results shown in Table 14.3 that the NeuCube performed significantly better than the other machine learning techniques with the highest average accuracy over all subjects and samples, whilst the closest competitor was SVM with the second highest average accuracy of 62%. MLR was the poorest performing, with an average accuracy of 50.5%, or just over the chance threshold.

### 14.3.4  Analysis of the Results

This was a feasibility study to investigate the potential of using NeuCube in BCI based rehabilitation devices. In considering the classification accuracies, which ranged from 70 to 85%, it is important to consider three factors. Firstly, the data were collected in an unshielded room using a commercially available gaming EEG headset, resulting in an EEG signal with relatively high signal to noise ratio.

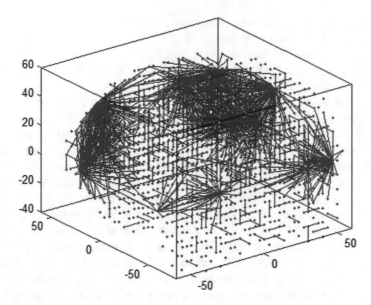

**Fig. 14.11** Connectome of the trained NeuCube. Blue lines show strong excitatory connections between two neurons, and red strong inhibitory [28]

**Table 14.3** Results of the comparative study; accuracy expressed as percentage for real and imagined movements [28]

| Subject/session | MLR | SVM | MLP | ECM | NeuCube |
|---|---|---|---|---|---|
| 1/Real | 55 | 69 | 62 | 76 | 80 |
| 1/Imagined | 63 | 68 | 58 | 58 | 80 |
| 2/Real | 55 | 55 | 45 | 52 | 67 |
| 2/Imagined | 42 | 63 | 63 | 79 | 85 |
| 3/Real | 41 | 65 | 41 | 45 | 73 |
| 3/Imagined | 53 | 53 | 63 | 53 | 70 |
| Average (appr.) | 52 | 62 | 55 | 61 | 76 |

Secondly, there was no processing or feature extraction performed on the data prior to classification, the raw, noisy, EEG data was used as the input. Thirdly, all comparative methods in this study, excepting NeuCube, were validated using Leave-One-Out (all but one sample used for training), while the NeuCube was validated with a more disadvantageous 50/50 (half used for training, half for testing) split. The accuracy of the NeuCube was still significantly higher than the other techniques and would likely rise when trained with leave-one-out paradigms.

Bearing these three factors in mind the classification accuracies obtained using NeuCube are in a similar range to those reported in other research and demonstrates that NeuCube is capable of accurately classifying noisy and relatively low-quality data. In addition, unlike many other approaches NeuCube does not require a lengthy feature extraction process, instead using all the raw data for classification, thus utilizing a rich data set that does not lose any potentially useful data.

We chose to use a relatively cheap and accessible EEG headset because two major factors that prevent the adoption of high technology interventions into rehabilitation practice are cost and complexity. EEG systems commonly used in research and clinical situations are expensive and unlikely to be widely available to rehabilitation specialists.

An advantage of the NeuCube is that it provides feedback and allows for interpretation of results and understanding of the data and the brain processes that generated it. This is illustrated in Fig. 14.11 where the connectivity of a trained SNNc is shown for further analysis. The SNNc and the deSNN classifier have evolvable structures, i.e., a NeuCube model can be trained further on more data and recalled on new data not necessarily of the same size and feature dimensionality. This allows in theory for a NeuCube to be partially trained on highly accurate data captured in a controlled manner with medical devices, and then further trained and adapted to the particular subject with a cheaper, less accurate device such as the Emotiv. This will increase potential uses in clinical and rehabilitation applications.

The large number of parameters that need to be optimized for every experiment to achieve the best results limits the current NeuCube. The results presented in this study are obtained through manual parameter optimization. To mitigate this, adaptive and evolutionary techniques (including the GRN discussed prior and quantum-inspired optimization) are being developed for this system, so that parameter selection is automated in a desirable way.

The results of this study support the premise that NeuCube is feasible to use in BCI based rehabilitation devices. Additionally, the ability of the NeuCube to both spatially and temporally represent brain data and provide visualization of the data could be useful in future applications. Observing changes in neural representation and spike timing throughout rehabilitation interventions could provide valuable information on human learning and adaptation to advance rehabilitation interventions.

## 14.4   BI-BCI for Neurorehabilitation with a Neurofeedback and for Neuro-prosthetics

### 14.4.1   General Notions

In every six seconds, someone in the world becomes physically disabled due to a stroke. To improve the quality of life of these stroke survivors, Neurorehabilitation aims at rebuilding the affected motor functions through regular exercises. This intends to strengthen the remaining neural connections by utilizing the brain's ability to build new neural pathways.

**Fig. 14.12** A diagram of the functional blocks in a neuro-rehabilitation device based on a trained NeuCube SNN on brain data that provides neurofeedback

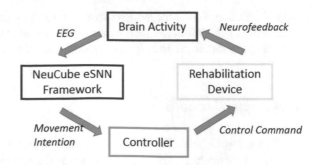

Figure 14.12 presents a diagram of the functional blocks in a neuro-rehabilitation device based on a trained NeuCube SNN on brain data that provides neurofeedback.

Figure 14.13 depicts a basic overview of this approach which facilitates a brain state-based classification of EEG signals using SNN. The module encloses a Finite State Machine which acts as a finite memory to the model and a biologically plausible NeuCube SNN architecture to decode state transitions over the time. The module follows the cue based (synchronous) BCI paradigm. While the subject is performing the task, EEG signals are recorded and classified. This classification

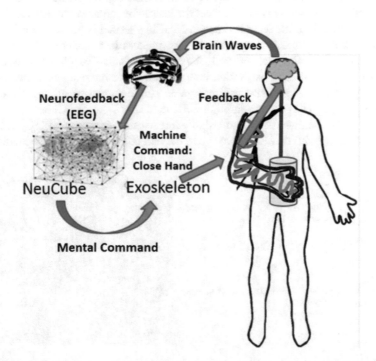

**Fig. 14.13** Basic functional flow of BCI based neurorehabilitation through NeuCube SNN architecture (the figure is drawn by K. Kumarasinghe)

output is used to control the rehabilitation robot arm. This approach enables the user to control the rehabilitation robots through their own thoughts and intentions and provides neurofeedback to help user to improve their brain functions [49].

### 14.4.2   Applications

In [49] the NeuCube BCI architecture was extended with the use finite automata, to control a robotic hand. The general functional diagram from Fig. 14.12 now is implemented for a control of a robotic hand (Figs. 14.14 and 14.15). The system demonstrated a higher accuracy of detecting human intention when compared to traditional machine learning methods.

In line with development of the NeuCube-based Neurorehabilitation, two cognitive games (called Grasp and NeuroRehab [50]) and one portable BCI have been developed as discussed below. The concept of Cognitive game does not only give a "fun" factor to the patient, but also trains them with the functionality of the product. These applications were developed for patients who have no voluntary muscular movements. The patients are trained with an imaginary task, which involves them to imagine moving a part of their body or imagining a series of relatively complex muscle movements. The patient is equipped with EEG cap on the scalp followed by the instruction on what to imagine, so that the instructor can record the neural activity of the brain. Based on the recorded data, a NeuCube model is trained, which can be used to control the objects. Once the training process is completed, the instructor performs an online classification with a new EEG data. The classified output is converted into a control signal, which intern controls the movements in the game.

For example, Fig. 14.16a is the Grasp game virtual environment, where a user is trained on how to hold a glass through EEG data using the NeuCube.

Figure 14.16b shows the NeuroRehab game virtual environment [50] as a two class problem, the aim of which is to move the ball either left or right depending on

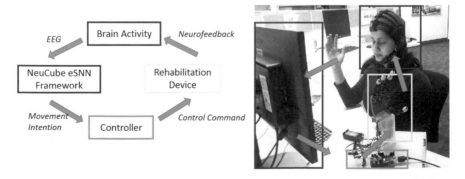

**Fig. 14.14** The general functional diagram from Fig. 14.12 now is implemented for a control of a robotic hand (from [49])

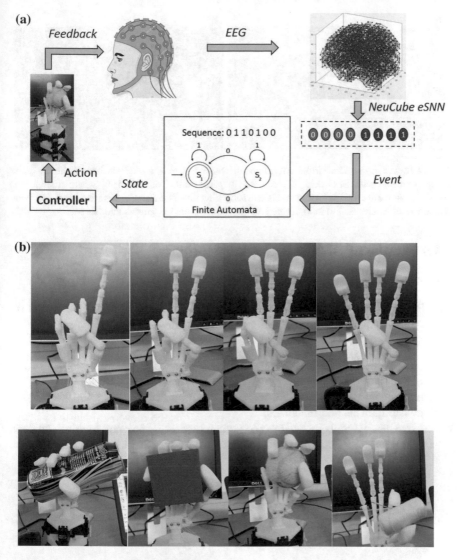

**Fig. 14.15  a** A brain-like motor controlling framework for prosthetic control using automata theory, cognitive computing & NeuCube evolving spiking neural network architecture (from [49]). **b** The BCI can use EEG signals to control fingers of a prosthetic hand (from [49])

the thought patterns of the patient. The subject can get the overview of how the NeuCube-based SNN connections are being formed while he/she is trying to move an object. Our preliminary studies [29] showed that when compared to standard machine learning algorithms, this approach allows to obtain higher pattern recognition accuracy, a better adaptability to new incoming data and a better interpretation of the models.

**(a)**                          **(b)**

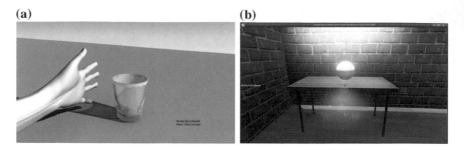

**Fig. 14.16** **a** Grasp game virtual environment, where a user is trained on how to hold a glass using NeuCube with EEG data. **b** NeuroRehab game virtual environment, where a subject is trained to move the ball left or right. If a wrong direction is chosen, a negative mark is given. These exercises are used to help the patients to improve their cognitive abilities

## 14.5 From BI-BCI to Knowledge Transfer Between Humans and Machines

The BI-BCI presented in this chapter are based on a trained NcuCube model on brain signals when humans are performing tasks. This can be further explored for knowledge transfer from humans to machines as graphically presented in Fig. 14.17. The motivations for this are the following:

- BI-BCI has the same brain template as the human brain, thus allowing for the development of new methods for an exchange of information and knowledge between humans and machines;
- A NeuCube BI-SNN, as well as the human brain, can integrate multimodal data, including EEG (Chap. 8), fMRI (Chap. 9), audio-visual (Chap. 13), tactile, etc.

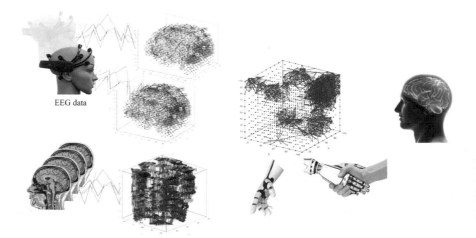

EEG data

**Fig. 14.17** From BI-BCI to knowledge transfer between humans and machines (the figure was drawn by Maryam Gholami)

That makes it possible to use all or most of the modalities for learning and only some of them for recall (e.g. visual information only);

- The brain signals used to train the system are representing procedural knowledge of how the human is performing a task;
- The learned connectivity in the system represents deep knowledge that mimics the human knowledge both in space and in time (see Chap. 6).

This topic is further discussed in Chap. 22. We have to note that while a human can control a machine with their brain signals, the other way is not possible, or is it?

## 14.6   Chapter Summary and Further Readings for Deeper Knowledge

The chapter presents methods and systems for using SNN for BCI with the emphasis on the development of brain-inspired BCI (BI-BCI) based on brain-inspired SNN. The BI-BCI not only classify brain signals into one of the commands passed for execution to an actuator device, but they provide a neurofeedback in terms of visualisation of brain activities in time-space, that reflect in the connectivity of the model. This is contrast with the traditional BCI models that comprise a black box.

Further readings on topics related to BCI can be found in several chapters in [51], such as:

- EEG signal processing for BCI (Chap. 46 in [51]);
- sEMG analysis for recognition and rehabilitation actions (Chap. 56 in [51]);
- Brain-like robotics (Chap. 57 in [51]);
- Development learning for user activities (Chap. 58 in [51]);
- NeuCube BCI demo: https://kedri.aut.ac.nz/R-and-D-Systems/brain-computer-interfaces-bci.

**Acknowledgements**   Some of the material in this chapter was first published as referenced in the text. I acknowledge the contribution of my co-authors of these publications Denise Taylor, Elisa Capecci, Kaushalya Kumarasinghe, Stefan Schliebs, Mahonri Owen, Nelson and James from Prof Hou's group of the China Academy of Sciences Institute of Automation in Beijing.

## References

1. L. Davlea, B. Teodorescu, Modular brain computer interface based on steady state visually evoked potentials (SSVEP). Paper presented on the E-health and bioengineering conference (EHB), Iasi, Romania. Retrieved from IEEE database (2011)
2. I. Sugiarto, I.H. Putro, Application of distributed system in neuroscience: a case study of BCI framework. The 1st international seminar on science and technology 2009 (ISSTEC 2009), Universitas Islam Indonesia, Yogyakarta, 2009

3. S.G. Mason, G.E. Birch, A general framework for brain-computer interface design. Neural Syst. Rehabil. Eng. **11**(1), 70–85 (2003). https://doi.org/10.1109/TNSRE.2003.810426

4. D. Plass-Oude Bos, H. Gürkök, B. Van de Laar, F. Nijboer, A. Nijholt, User experience evaluation in BCI: mind the gap! Int. J. Bioelectromagn. **13**(1), 48–49 (2011)

5. L. Fernando, N. Alonso, J. Gomez-Gil, Brain computer interfaces, a review. Sensors **12**(2), 1211–1264 (2012)

6. L.R. Hochberg, M.D. Serruya, G.M. Friehs, J.A. Mukand, M. Saleh, A.H. Caplan, A. Branner, D. Chen, R.D. Penn, J.P. Donoghue, Neuronal ensemble control of prosthetic devices by a human with tetraplegia. Nat. J. **442**, 164–171 (2006)

7. Y. Zhang, A novel BCI based on ERP components sensitive to configural processing of human faces. J. Neural Eng. **9**(1). https://doi.org/10.1088/1741-2560/9/2/026018 (2012)

8. Emotiv: https://www.emotiv.com

9. Z. Doborjeh, N. Kasabov, M. Doborjeh, A. Sumich, Modelling peri-perceptual brain processes in a deep learning spiking neural network architecture. Nature, Scientific reports **8**, 8912 (2018)

10. D. Sperber, F. Clement, C. Heintz, O. Mascaro, H. Mercier, G. Origgi, D. Wilson, Epistemic vigilance. Mind Lang. **25**(4), 359–393 (2010)

11. D. Ortner, D. Grabher, M. Hermann, E. Kremmer, S. Hofer, C. Heufler, The adaptor protein Bam32 in human dendritic cells participates in the regulation of MHC class I-induced CD8$^+$ T cell activation. J. Immunol. **187**(8), 3972–3978. https://doi.org/10.4049/jimmunol.1003072. (2011, epub 19 Sept 2011)

12. E.M. Mugler, C.A. Ruf, S. Halder, M. Bensch, A. Kubler, Design and implementation of a P300-based brain-computer interface for controlling an internet browser. IEEE Trans. Neural Syst. Rehabil. Eng. **18**(1), 599–609 (2010)

13. A. Furdea, S. Halder, D.J. Krusienski, D. Bross, F. Nijboer, N. Birbaumer, A. Kübler, An auditory oddball (P300) spelling system for brain-computer interfaces. J. Psychophysiol. **46**, 617–625 (2009)

14. T. Hinterberger, S. Schmidt, N. Neumann, J. Mellinger, B. Blankertz, G. Curio, N. Birbaumer, Brain-computer communication and slow cortical potentials. Biomed. Eng. **51**(1), 1011–1018 (2004)

15. N. Kasabov, NeuCube evospike architecture for spatio-temporal modelling and pattern recognition of brain signals, in *ANNPR* N. Mana, F. Schwenker, E. Trentin ed. by LNAI, vol. 7477 (Springer, Berlin, 2012), pp.225–243

16. S.M. Bothe, The evidence for neural information processing with precise spike-times: a survey. Neural Comput. **3**(2), 1–13 (2004)

17. T. Delbruck, jAER open source project (2007), http://jaer.wiki.sourceforge.net

18. D.A. Lichtenstein, G.A. Mezière, Relevance of lung ultrasound in the diagnosis of acute respiratory failure: the BLUE protocol. Chest **134**(1), 117–125. https://doi.org/10.1378/chest. 07-2800. (2008, epub 10 Apr 2008)

19. N. Nuntalid, K. Dhoble, N. Kasabov, in *EEG Classification with BSA Spike Encoding Algorithm and Evolving Probabilistic Spiking Neural Network*, LNCS, vol. 7062 (Springer, Berlin, 2011), pp. 451–460

20. B. Petro, N. Kasabov, R. Kiss, Selection and optimisation of spike encoding methods for spiking neural networks, algorithms, submitted; http://www.kedri.aut.ac.nz/neucube/

21. C.J. Honey, R. Kötter, M. Breakspear, O. Sporns, Network structure of cerebral cortex shapes functional connectivity on multiple time scales. Proc. Natl. Acad. Sci. **104**, 10240–10245 (2007)

22. E. Bullmore, O. Sporns, Complex brain networks: graph theoretical analysis of structural and functional systems. Nat. Rev. Neurosci. **10**, 186–198 (2009)

23. V. Braitenberg, A. Schüz, *Statistics and Geometry of Neuronal Connectivity* (Springer, Berlin, 1998)

24. B. Hellwig, A quantitative analysis of the local connectivity between pyramidal neurons in layers 2/3 of the rat visual cortex. Biol. Cybern. **2**, 111–121 (2000). https://doi.org/10.1007/PL00007964

25. Z.J. Chen, Y. He, P. Rosa-Neto, J. Germann, A.C. Evans, Revealing modular architecture of human brain structural networks by using cortical thickness from MRI. Cereb. Cortex **18**, 2374–2381 (2008)
26. C.J. Stam, Functional connectivity patterns of human magnetoencephalographic recordings: a small-world network? Neurosci. Lett. **355**, 25–28 (2004)
27. N. Kasabov, *Evolving Connectionist Systems: The Knowledge Engineering Approach* (Springer, London, 2007) (first edition, 2002)
28. D. Taylor, N. Scott, N. Kasabov, E. Tu, E. Capecci, N. Saywell, Y. Chen, J. Hu, Z.-G. Hou, *Detecting Motor Execution and Motor Intention from EEG Signals* (IEEE WCCI, Beijing, 2014)
29. A. Mohemmed, S. Schliebs, S. Matsuda, N. Kasabov, SPAN: spike pattern association neuron for learning spatio-temporal spike patterns. Int. J. Neural Syst. **22**(4), 1250012 (2012)
30. K. Kong, K. Chua, J. Lee, Recovery of upper limb dexterity in patients more than 1 year after stroke: frequency, clinical correlates and predictors. NeuroRehabilitation **28**(2), 105–111 (2011)
31. J. Kleim, T. Jones, Principles of experience-dependent neural plasticity: implications for rehabilitation after brain damage. J. Speech Hear. Res. **51**, S225–S239 (2008)
32. K. Fox, Experience-dependent plasticity mechanisms for neural rehabilitation in somatosensory cortex. Philosophical Trans. R. Soc. Lond. Series B, Biol. Sci. **364**(1515), 369–381 (2009)
33. A.L. Kerr, S.Y. Cheng, T.A. Jones, Experience-dependent neural plasticity in the adult damaged brain. J. Commun. Disord. **44**(5), 538–548 (2011)
34. M. Jeannerod, Neural simulation of action: a unifying mechanism for motor cognition. NeuroImage **14**(1), S103–S109 (2001)
35. G. Rizzolatti, L. Fogassi, V. Gallese, Neurophysiological mechanisms underlying the understanding and imitation of actions. Nat. Rev. Neurosci. **2**, 661–670 (2001)
36. M. Jeannerod, The representing brain: neural correlates of motor intention and imagery. Behav. Brain Res. **17**, 187–245 (1994)
37. L. Fadiga et al., Motor facilitation during action observation: a magnetic stimulation study. J. Neurophysiol. **73**(6), 2608–2611 (1995)
38. R. Grush, The emulation theory of representation: motor control, imagery, and perception. Behav. Brain Sci. **27**, 377–396 (2004)
39. S.J. Page, P. Levine, S. Sisto, M.V. Johnston, A randomized efficacy and feasibility study of imagery in acute stroke. Clin. Rehabil. **15**, 233–240 (2001)
40. V. Gray, C.L. Rice, S.J. Garland, Factors that influence muscle weakness following stroke and their clinical implications: a critical review. Physiother. Can. **64**(4), 415–411 (2012)
41. S-H. Chang, G.E. Francisco, P. Zhou, W.Z. Rymer, S. Li, Spasticity, weakness, force variability, and sustained spontaneous motor unit discharges of resting spastic-paretic biceps brachii muscles in chronic stroke. Muscle Nerve **48**(1), 85–92 (2013)
42. R. Zimmermann, L. Marchal-Crespo, J. Edelmann, O. Lambercy, M.C. Fluet, R. Riener, R. Gassert, Detection of motor execution using a hybrid fNIRS-biosignal BCI: a feasibility study. J. Neuroeng. Rehabil. **10**, 4 (2013)
43. NeuCube BCI demo; https://kedri.aut.ac.nz/R-and-D-Systems/brain-computer-interfaces-bci
44. L. Hopler, M. Wolf, Single-trial classification of motor imagery differing in task complexity: a functional near-infrared spectroscopy study. J. Neuroeng. Rehabil. **8**, 34 (2011)
45. M. Arvaneh, G. Cuntai, A. Kai Keng, Q. Chai, Optimizing the channel selection and classification accuracy in EEG-based BCI. IEEE Trans. Biomed. Eng. **59**(6), 1865–1873 (2011)
46. Y. Chen, J. Hu, N. Kasabov, Z. Hou, L. Cheng, in *NeuCubeRehab: A Pilot Study For EEG Classification in Rehabilitation Practice Based on Spiking Neural Networks.* Proceedings of the International Conference on Neural Information Processing, Daegu, Korea (2013)
47. L. Fernando, N. Alonso, J. Gomez-Gil, Brain computer interfaces, a review. Sensors **12**(2), 1211–1264 (2012)

48. N. Kasabov, Q. Song, in *ECM, A Novel On-line, Evolving Clustering Method and Its Applications*. Proceedings of the 5th Biannual Conference Artificial Neural Network Expert System (ANNES 2001) (2001), pp. 87–92

49. K. Kumarasinghe, M. Owen, N. Kasabov, D. Taylor, C.K. Au, FaNeuRobot: A Brain-Like Motor Controlling Framework for Prosthetic Control Using Automata Theory, in *Cognitive Computing & NeuCube Evolving Spiking Neural Network Architecture*, IEEE Robotics, Conference, Sydney, May 2018

50. A. Gollahalli, Masters thesis, Auckland University of Technology, (2017)

51. N. Kasabov (ed.), *Springer Handbook of Bio-/Neuroinformatics* (Springer, Berlin, 2014)

52. Wikipedia: www.wikipedia.com

53. N. Kasabov, L. Liang, R. Krishnamurthi, V. Feigin, M. Othman, Z.-G. Hou, P. Parmar, Evolving spiking neural networks for personalised modelling of spatio-temporal data and early prediction of events: a case study on stroke. Neurocomputing **134**, 269–279 (2014)

# Part VI
# SNN in Bio- and Neuroinformatics

# Chapter 15
# Computational Modelling and Pattern Recognition in Bioinformatics

This chapter explores the ability of SNN to capture changes in Bioinformatics data for predicting events or classifying biological states from DNA, gene and protein data. It starts with a bioinformatics primer.

The chapter is organised in the following sections:

15.1. Bioinformatics primer.
15.2. Biological databases. Modelling bioinformatics data.
15.3. Gene and protein interaction networks and the system biology approach.
15.4  Brain-inspired SNN architectures for deep learning of gene expression time series data and for the extraction of Gene Interaction Networks (GIN).
15.5. Chapter summary and further readings for deeper knowledge.

## 15.1  Bioinformatics Primer

This section is an introduction to both biology and computational modelling of biological data. Part of the material is published in [1, 2].

### 15.1.1  General Notions

Bioinformatics brings together several disciplines—molecular biology, genetics, microbiology, mathematics, chemistry and bio-chemistry, physics, and of course—informatics. Many processes in biology, as it was discussed in Chap. 1, are dynamically evolving and their modelling requires evolving methods and systems. In Bioinformatics new data is being made available with a tremendous speed that would require the models to be continuously adaptive. Knowledge-based modelling, that includes rule and knowledge discovery, is a crucial requirement. All

© Springer-Verlag GmbH Germany, part of Springer Nature 2019
N. K. Kasabov, *Time-Space, Spiking Neural Networks and Brain-Inspired Artificial Intelligence*, Springer Series on Bio- and Neurosystems 7,
https://doi.org/10.1007/978-3-662-57715-8_15

these issues make the evolving connectionist methods and systems needed for problem solving across areas of bioinformatics, from DNA sequence analysis, through gene expression data analysis, through protein analysis, and finally to modelling genetic networks and entire cells as a *system biology approach*. That will help to discover genetic profiles and to understand better diseases that do not have a cure so far, to better understand what the human body is made of and how it works in its complexity at its different levels of organisation.

## 15.1.2   DNA, RNA and Proteins. The Central Dogma of Molecular Biology and the Evolution of Life.

Nature evolves in time. The most obvious example of an evolving process is life. Life is defined in the Concise Oxford English Dictionary as follows: *A state of functional activity and continual change peculiar to organized matter, and especially to the portion of it constituting an animal or plant before death; animate existence; being alive.* The carrier of life over generations is the DNA.

The DNA (Dioxyribonucleic Acid) is a chemical chain, present in the nucleus of each cell of an organism, and it consists of ordered in a double helix pairs of small chemical molecules (bases, nucleotides) which are: Adenine (A), Cytosine (C), Guanidine (G), and Thymidine (T), linked together by dioxyribose sugar phosphate nucleic acid backbone.

A DNA molecule is organised as a double-helix structure, where A is connected to T molecules and C to G. Many disease are due to small changes in the DNA code, called Single Nucleotide Polymorphism (SNPs) as illustrated in Fig. 15.1, where instead of C-G link, there is a A-T link.

**Fig. 15.1**  A single nucleotide polymorphism

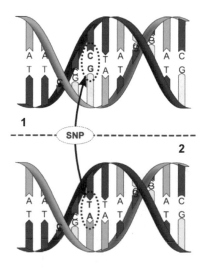

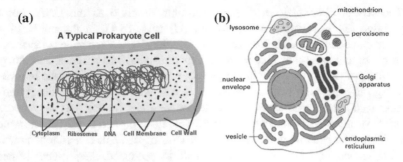

**Fig. 15.2** **a** Bacteria (prokaryotic cell) **b** typical eukaryotic cell

The DNA contains millions of nucleotide base pairs, but only 5% or so is used for the production of proteins, and these are the segments from the DNA that contain genes. Each gene is a sequence of base pairs that is used in the cell to produce proteins. Genes have length of hundreds to thousands of bases.

In simple organisms, bacteria (prokaryotic organisms) DNA is transcribed directly into mRNA that consists of genes that contain only codons (no intron segmens)—Fig. 15.2a. The translation of the genes into proteins is initiated by proteins called ribosomes that bind to the beginning of the gene (ribosome binding site) and translate the sequence until reaching the termination area of the gene. Finding ribosome binding sites in bacteria would reveal how the bacteria would act and what proteins will be produced.

In higher organisms (that contain a nucleous in the cell) the DNA is first transcribed into a pre-mRNA that contains all the regions from the DNA that contain genes. The pre-RNA is then transcribed into many sequences of functional mRNAs through a splicing process, so that the intron segments are deleted from the genes and only the exon segments that account for proteins are extracted. The functional mRNA is now ready to be translated into proteins (Fig. 15.2b).

The *central dogma of the molecular biology* (see Fig. 15.3) states that the DNA is transcribed into RNA, which is translated into proteins, which process is continuous in time until the organism is alive [2].

**Fig. 15.3** The central dogma of the molecular biology states that the DNA is transcribed into RNA, which is translated into proteins, which process is continuous in time until the organism is alive

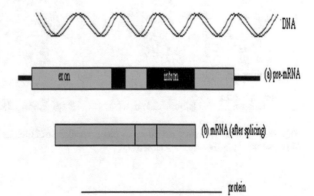

The RNA (ribonucleic acid) has a similar structure as the DNA, but here Thymidine (T) is substituted by Uridine (U) (Fig. 15.4). In the pre-RNA only segments that contain genes are extracted from the DNA. Each gene consists of two types of segments—exons, that are segments translated into proteins, and introns—segments that do not take part in the protein production. Removing the introns and ordering only the exon parts of the genes in a sequence is called splicing and this process results in the production of a messenger RNA (or mRNA) sequences.

mRNAs are directly translated into proteins. Each protein consists of a sequence of amino-acids, each of them defined as a base triplet, called a codon. From one DNA sequence there are many copies of mRNA produced, the presence of certain gene in all of them defines the level of the gene expression in the cell and can

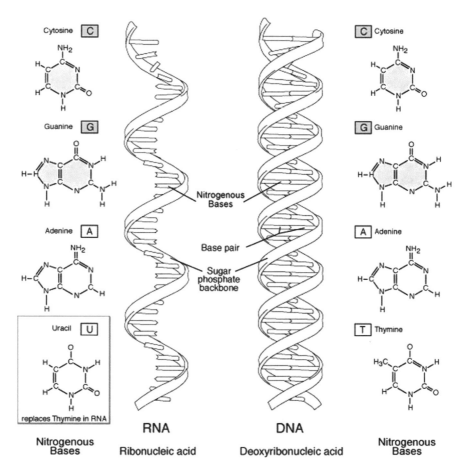

**Fig. 15.4** The RNA (ribonucleic acid) has a similar structure as the DNA, but here Thymidine (T) is substituted by Uridine (Uracil) (U)

indicate what and how much of the corresponding protein will be produced in the cell.

The above description of the central dogma of the molecular biology is very much a simplified one, but that would help to understand the rationale behind using connectionist and other information models in bioinformatics [2].

Genes are complex chemical structures and they cause dynamic transformation of one substance into another during the whole life of an individual, as well as the life of the human population over many generations [3–6]. When genes are "in action", the dynamics of the processes in which a single gene is involved are very complex, as this gene interacts with many other genes, proteins, and is influenced by many environmental and developmental factors.

The whole process of DNA transcription, gene translation, and protein production is continuous and it *evolves* over time. Proteins have 3D structures that unfold over time governed by physical and chemical laws. Proteins make some genes to express and may suppress the expression of other genes. The genes in an individual may mutate, change slightly their code, and may therefore express differently at a next time. So, genes may change, mutate, evolve in a life time of a living organism.

Only 2–5% of the human genome (the DNA) contains useful information what concerns the production of proteins. The number of genes contained in the human genome is about 40,000. Only the gene segments are transcribed into RNA sequences and then translated into proteins. The transcription is achieved through special proteins, enzymes called RNA polymerase, that bind to certain parts of the DNA (promoter regions) and start 'reading' and storing in a mRNA sequence each gene code. Analysis of a DNA sequence and identifying promoter regions is a difficult task. If it is achieved, it may make possible to predict, from a DNA information, how this organism will develop, or alternatively—what an organism looked like in retrospect. The promoter recognition process is part of a complex process of gene regulatory network activity, where genes interact between each other over time, defining the destiny of the whole cell.

RNA molecules are emerging as central "players" controlling not only the production of proteins from messenger RNAs, but also regulating many essential gene expression and signalling pathways. Mouse cDNA sequencing project FANTOM in Japan showed that non-coding RNAs constitute at least a third of the total number of transcribed mammalian genes [7]. In fact, about 98% of RNA produced in the eukaryotic cell is non-coding [8], produced from introns of protein-coding genes, non-protein-coding genes and even from intergenic regions, and it is now estimated that half of the transcripts in human cells are non-coding and functional.

These non-coding transcripts are thus not junk, but could have many crucial roles in the Central Dogma of Molecular Biology. The most recently discovered, rapidly expanding group of non-coding RNAs is microRNAs, which are known to have exploded in number during the emergence of vertebrates in evolution (see [9], also containing a good review on non-coding RNAs and their evolution). They are already known to function in lower eukaryotes in regulation of cell and tissue development, cell growth and apoptosis and many metabolic pathways, with similar likely roles in vertebrates [8].

MicroRNAs are encoded by long precursor RNAs, commonly several hundred basepairs long, which typically form fold-back structures resembling a straight hairpin with occasional bubbles and short branches. The length and the conservation of these long transcribed RNAs makes it possible to discover and classify by sequence similarity search method to discover and classify many phylogenetically related microRNAs in the Arabidopsis genome [10]. Such analysis has established that most plant microRNA genes have evolved by inverted duplication of target gene sequences. The mechanism of their evolution in mammals is less clear.

Lack of conserved microRNA sequences or microRNA targets between animals and plants suggests that plant microRNAs evolved after the split of the plant lineage from mammalian precursor organisms. This means that the information about plant microRNAs does not help to identify or classify most mammalian microRNAs. Also, in mammalian genomes the foldback structures are much shorter, down to only about 80 basepairs, making sequence similarity search a less effective method for finding and clustering remotely related microRNA precursors.

*Evolutionary processes* imply the development of generations of populations of individuals where crossover, mutation, selection of individuals, based on fitness (survival) criteria are applied in addition to the developmental (learning) processes of each individual (see Chap. 7).

A biological system evolves its structure and functionality through both, life-long learning of an individual, and evolution of populations of many such individuals, i.e. an individual is part of a population and is a result of evolution of many generations of populations, as well as a result of its own developmental, of its life-long learning process.

Same genes in the genotype of millions of individuals may be expressed differently in different individuals, and within an individual—in different cells of their body. The expression of these genes is a dynamic process depending not only on the types of the genes, but on the interaction between the genes, and the interaction of the individual with the environment (the Nurture versus Nature issue).

Several principles are useful to take into account from evolutionary biology:

– Evolution preserves or purges genes.
– Evolution is a non-random accumulation of random changes.
– New genes cause the creation of new proteins.
– Genes are passed on through *evolution: generations* of populations and selection processes (e.g. natural selection).

There are different ways of interpreting the DNA information, some of them scientific, some of them—artistic (see [3]):

- DNA as a source of information and cells as information processing machines [4];
- DNA and the cells as stochastic systems (processes are non-linear and dynamic, chaotic in a mathematical sense);
- DNA as a source of energy;
- DNA as a language;
- DNA as music;
- DNA as a definition of Life [1, 2].

Proteins provide the majority of the structural and functional components of a cell. The area of molecular biology that deals with all aspects of proteins is called proteomics. So far about 30,000 proteins have been identified and labelled, but this is considered to be a small part of the total set of proteins that keep our cells alive.

The mRNA is translated by ribosomes into proteins. A protein is a sequence of amino-acids, each of them defined by a group of 3 nucleotides (codons). There are 20 amino acids all together, denoted by letters (A, C-H, I, K-N, P-T, V, W, Y). The codons of each of the amino acids are given in Table 15.1, so that the column represents the first base in the triplet, the row—the second base, and the last column —the last base.

The length of a protein in number of amino-acids, is from tens to several thousands.

Each protein is characterized by some characteristics, for example [4, 11]:

- Structure;
- Function;
- Charge;
- Acidity;

**Table 15.1** The codons of each of the 20 aminoacids. The column represents the first base in the triplet, the row —the second base, and the last column—the last base

|   | U | C | A | G |   |
|---|---|---|---|---|---|
| U | Phe | Ser | Tyr | Cys | U |
|   | Phe | Ser | Tyr | Cys | C |
|   | Leu | Ser | – | – | A |
|   | Leu | Ser | – | Trp | G |
| C | Leu | Pro | His | Arg | U |
|   | Leu | Pro | His | Arg | C |
|   | Leu | Pro | Gln | Arg | A |
|   | Leu | Pro | Gln | Arg | G |
| A | Ile | Thr | Asn | Ser | U |
|   | Ile | Thr | Asn | Ser | C |
|   | Ile | Thr | Lys | Arg | A |
|   | Met | Thr | Lys | Arg | G |
| G | Val | Ala | Asp | Gly | U |
|   | Val | Ala | Asp | Gly | C |
|   | Val | Ala | Glu | Gly | A |
|   | Val | Ala | Glu | Gly | G |

– Hydrophilicity;
– Molecular weight.

An initiation codon defines the start position of a gene in a mRNA where the translation of the mRNA into protein begins. A stop codon defines the end position.

Proteins with a high similarity are called homologous. Homologous that have identical functions are called orthologues. Similar proteins that have different functions are called paralogues.

Proteins have complex structures that include:

– Primary structure (a linear sequence of the amino-acids)—see for example Fig. 15.5.
– Secondary structure (3D, defining functionality). An example of a 3D representation of a protein is given in Fig. 15.6.
– Tertiary structure (high level folding and energy minimisation packing of the protein).
– Quaternary structure (interaction between two or more protein molecules).

More information about molecular biology can be found in [2].

### 15.1.3  Phylogenetics

Evolution is a process whereby populations are altered over time and may split into separate branches, hybridize together, or terminate by extinction. The evolutionary branching process may be depicted as a phylogenetic tree, and the place of each of the various organisms on the tree is based on a hypothesis about the sequence in which evolutionary branching events occurred.

In biology, **phylogenetic** is the study of evolutionary relationships among groups of organisms (e.g. species, populations), which are discovered through molecular sequencing data and morphological data matrices. Phylogenetic analyses have become essential to research on the evolutionary tree of life.

**Fig. 15.5** A primary structure of a protein—a linear sequence of the amino-acids

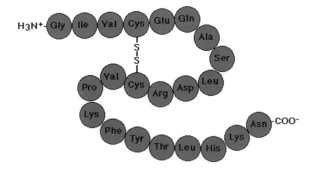

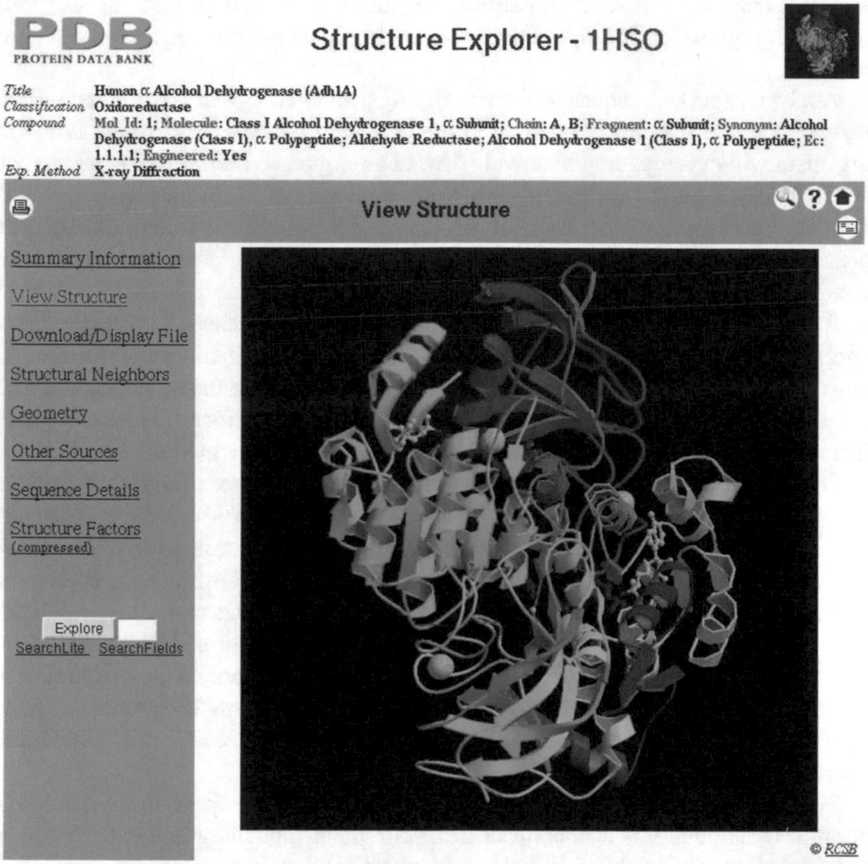

## Structure Explorer - 1HSO

*Title* **Human α Alcohol Dehydrogenase (Adh1A)**
*Classification* **Oxidoreductase**
*Compound* **Mol_Id: 1; Molecule: Class I Alcohol Dehydrogenase 1, α Subunit; Chain: A, B; Fragment: α Subunit; Synonym: Alcohol Dehydrogenase (Class I), α Polypeptide; Aldehyde Reductase; Alcohol Dehydrogenase 1 (Class I), α Polypeptide; Ec: 1.1.1.1; Engineered: Yes**
*Exp. Method* **X-ray Diffraction**

**View Structure**

Summary Information

View Structure

Download/Display File

Structural Neighbors

Geometry

Other Sources

Sequence Details

Structure Factors
(compressed)

Explore
SearchLite SearchFields

© RCSB

**Fig. 15.6** An example of a secondary structure (3D, defining functionality) of a protein obtained with the use of the PDB data set, maintained by the National Centre for Biological Information (NCBI) of the National Institute for Health (NIH) of the USA

## 15.1.4 The Challenges of Molecular Data Analysis

As it was mentioned previously, only 2–5% of the human genome (the DNA) contains information what concerns the production of proteins [2]. The number of genes contained in the human genome is about 40,000 [5]. Only the gene segments are transcribed into RNA sequences. The transcription is achieved through special proteins, enzymes called RNA polymerase, that bind to certain parts of the DNA (promoter regions) and start 'reading' and storing in an mRNA sequence each gene code.

Analysis of a DNA sequence and identifying promoter regions is a difficult task. If it is achieved, it may make possible to predict, from a DNA information, how this organism will develop, or alternatively—what an organism looked like in retrospect.

Analysis of gene expression data from microarrays is discussed in the next section. Here, some typical tasks of DNA and RNA sequence pattern analysis are presented, namely ribosome binding site identification, and splice junction recognition.

Recognizing patterns from DNA, or from mRNA sequences is a way of recognizing genes in these sequences and of predicting proteins in silico (in a computer). For this purpose, usually a "window" is moved along the sequence and data from this window is submitted to a neural network classifier (identifier) which identifies if one of the known patterns is contained in this window.

Only the gene segments are transcribed into RNA sequences and then translated into proteins as pointed out before. The transcription is achieved through special proteins, enzymes called RNA polymerase, that bind to certain parts of the DNA (promoter regions) and start 'reading' and storing in a mRNA sequence each gene code. Analysis of a DNA sequence and identifying promoter regions is a difficult task. If it is achieved, it may make possible to predict, from a DNA information, how this organism will develop, or alternatively—what an organism looked like in retrospect. The promoter recognition process is part of a complex process of gene regulatory network activity, where genes interact between each other over time, defining the destiny of the whole cell.

Finding the splice junctions that separate the introns from the exons in a pre-mRNA structure, is a difficult task for computer modelling and pattern recognition, that once solved would help understand what proteins would be produced from a certain mRNA sequences. This task is called splice junction recognition. Figure 15.7 shows the process of splicing primary RNA into mRNA.

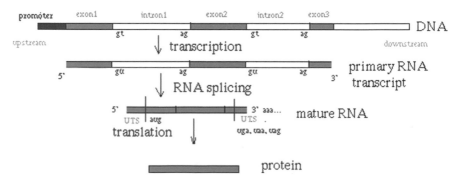

**Fig. 15.7** The process of splicing primary RNA into mRNA

But even having recognized the splice junctions in a pre-mRNA, it is extremely difficult to predict which genes will really become active, i.e. will be translated into proteins, and how much active they will be—how much protein will be produced. That is why gene expression technologies (e.g. microarrays) have been introduced, to measure the expression of the genes in mRNAs. The level of a gene expression would suggest how much protein of this type will be produced in the cell, but again this will only be an approximation.

RNA molecules are emerging as central "players" controlling not only the production of proteins from messenger RNAs, but also regulating many essential gene expression and signalling pathways. In fact, about 98% of RNA produced in the eukaryotic cell is non-coding [8], produced from introns of protein-coding genes, non-protein-coding genes and even from intergenic regions, and it is now estimated that half of the transcripts in human cells are non-coding and functional.

These non-coding transcripts are thus not junk, but could have many crucial roles in the Central Dogma of Molecular Biology. The most recently discovered, rapidly expanding group of non-coding RNAs is microRNAs, which are known to have exploded in number during the emergence of vertebrates in evolution (see [9], also containing a good review on non-coding RNAs and their evolution). They are already known to function in lower eukaryotes in regulation of cell and tissue development, cell growth and apoptosis and many metabolic pathways, with similar likely roles in vertebrates [8].

MicroRNAs are encoded by long precursor RNAs, commonly several hundred basepairs long, which typically form fold-back structures resembling a straight hairpin with occasional bubbles and short branches. The length and the conservation of these long transcribed RNAs makes it possible to discover and classify by sequence similarity search method to discover and classify many phylogenetically related microRNAs in the Arabidopsis genome [10]. Such analysis has established that most plant microRNA genes have evolved by inverted duplication of target gene sequences. The mechanism of their evolution in mammals is less clear.

There are many aspects of molecular biology that make their analysis and modelling difficult, such as:

- Abundance of genome data, RNA data, protein data and metabolic pathway data is now available (see http://www.ncbi.nlm.nih.gov) and this is just the beginning of computational modeling in Bioinformatics.
- Complex interactions:

  - between proteins, genes, DNA code
  - between the genome and the environment
  - much yet to to be discovered

- Stability and repetitiveness: Genes are *relatively* stable carriers of information.
- Many sources of uncertainty:

- Alternative splicing
- Mutation in genes caused by: ionising radiation (e.g. X-rays); chemical contamination, replication errors, viruses that insert genes into host cells, aging processes, etc.
- Mutated genes express differently and cause the production of different proteins.

- It is extremely difficult to model dynamic, evolving biological processes.

At the same time, researchers are always finding ways to address the above difficulties, addressing important problems, both small and large scale, such as:

- Discovering patterns (features) from DNA and RNA sequences (e.g. genes, promoters, RBS binding sites, splice junctions)
- Analysis of gene expression data and predicting protein abundance
- Discovering of gene networks—genes that are co-regulated over time
- Protein discovery and protein function analysis
- Predicting the development of an organism from its DNA code
- Reconstructing life from DNA
- Modeling the full development (metabolic processes) of a cell
- Genetic modification
- Medical decision support systems
- Precision medicine (personalised modelling)
- Disease diagnostic systems
- Treatment outcome prediction
- Artificial Life
- Synthetic food
- Protecting and preserving Life on the planet Earth
- Life beyond the planet Earth
- Life and Death.

Next sections offer more information about how bioinformatics data can be modelled and useful patterns and new knowledge discovered.

## 15.2 Biological Databases. Computational Modelling of Bioinformatics Data

### 15.2.1 Biological Databases

Databases are essential for bioinformatics research and applications. Many databases exist, covering various information types, for example:

- Gene Banks;
- DNA and protein sequences;
- molecular structures;

– phenotypes and biodiversity.

Databases may contain:

– empirical data (obtained directly from experiments);
– predicted data (obtained from analysis);
– both.

They may be specific to a particular organism, pathway or molecule of interest. Alternatively, they can incorporate data compiled from multiple other databases.

The biological databases vary in their format, access mechanism, and whether they are public or not.

Some of commonly used databases are listed below [2]:

- Biological sequence analysis: Genbank, UniProt.
- Protein Families and Motif Finding: InterPro, Pfam.
- Next Generation Sequencing: Sequence Read Archive.
- GRN Analysis: Metabolic Pathway Databases (KEGG, BioCyc), Interaction Analysis Databases, Functional Networks.
- Design of synthetic genetic circuits: GenoCAD.

A comprehensive description of biological databases can be found in [2].

## 15.2.2 General Information About Bioinformatics Data Modelling

Many statistical and machine learning methods have been used to analyse molecular data (see [2] and also Chap. 2). Artificial neural networks (ANN) and evolving connectionist systems (ECOS), described in Chaps. 2 and 4, have been widely used for pattern recognition from DNA and RNA data. General schemes of using ANN and eSNN for pattern identification from bioinformatics data are given in Figs. 15.8 and 15.9 respectively.

Many connectionist models have been developed for identifying patterns in a sequence of RNA or DNA [4, 6]. Most of them deal with a static data set and use multiplayer perceptrons MLP, or self-organising maps SOMs (Chap. 2).

In many cases, however, there is a continuous flow of data that is being made available for a particular pattern recognition task. New labelled data need to be added to existing classifier systems for a better classification performance on future unlabelled data. This can be done with the use of the evolving models and systems.

Several case studies are used here to illustrate the application of evolving systems for sequence DNA and RNA data analysis. Figure 15.9 illustrates the use of eSNN, where output nodes evolve to capture prototypes of input data. More about eSNN can be found in Chap. 4.

Several sequence similarity and RNA folding based methods have been developed to find novel microRNAs. Simple BLAST similarity search identified e.g. orthologues of let-7 microRNA in several species [11]. Another approach has been

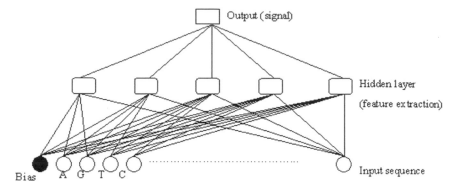

**Fig. 15.8** A general scheme of using MLP neural networks for sequence pattern identification from DNA data (Chap. 2)

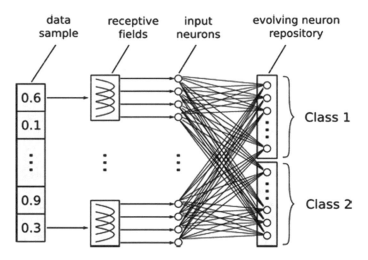

**Fig. 15.9** The structure of eSNN evolves output nodes to capture prototypes of input data in an incremental way (Chap. 5)

screening by RNA fold prediction algorithms (best known are Mfold and RNAfold) to look for stem-loop structure candidates having a characteristically low deltaG value indicating strong hybridization of the folded molecule, followed by further screening by sequence conservation between genomes of related species. Software systems called MIRseeker [12] and MIRscan [13] have been used in this fashion for fruitfly (*Drosophila*) and human microRNA discovery, respectively. Most recently, about a thousand candidate conserved human microRNAs have been found by phylogenetic conservation based search strategy [14]. This method is based on careful multiple alignment of many different sequences from closely related primate species to find accurate conservation at single nucleotide resolution.

The problem with all these approaches is that they require extensive sequence data and laborious sequence comparisons between many genomes as one key filtering step. Also, finding species-specific, recently evolved microRNAs by these methods is difficult, as well as evaluating the phylogenetic distance of remotely related genes which have diverged too much in sequence.

One tenet of microRNA is that the two-dimensional (2D) structure of many microRNAs (and non-coding RNAs in general) can give additional information which is useful for their discovery and classification, even with data from within only one species. This is analogous to protein three-dimensional (3D) structure analysis showing often functional and/or evolutionary similarities between proteins that cannot easily be seen by sequence similarity methods alone.

Protein 3D structural comparisons are based on accurate protein crystallization data on atomic coordinates of amino acids in the polypeptide macromolecule chain. Unfortunately, such molecular structure data is scanty for RNAs in general, and for microRNA precursors in particular. Also, RNA folding simulation in 3D is still a difficult computational problem, just as the traditional grand challenge of deducing protein folding ab initio from the amino acid sequence alone. Prediction of RNA folding in 2D is more advanced, and reasonably accurate algorithms are available, which can simulate the putative most likely and thermodynamically most stable structures of self-hybridizing RNA molecules. Many of such structures have been also verified by various experimental methods in the laboratory, corroborating the general accuracy of these folding algorithms.

## 15.2.3 Gene Expression Data Modelling and Profiling

One of the contemporary directions while searching for efficient drugs for many illnesses, such as cancer or HIV, is the creation of gene profiles of these diseases and subsequently finding targets for treatment through gene expression regulation. A gene profile is a pattern of expression of a number of genes that is typical for all, or for some of the known samples of a particular disease.

A disease profile would look like:

IF (gene g1 is highly expressed) AND (gene g37 is low expressed) AND (gene 134 is very highly expressed) THEN most probably this is cancer type C (123 out of available 130 samples have this profile).

Having such profiles for a particular disease makes it possible to set early diagnostic test, so a sample can be taken from a patient, the data related to the sample is processed, and a profile is obtained. This profile can be matched against existing gene profiles and based on similarity, it can be predicted with certain probability if the patient is in an early phase of a disease or he/she is at risk of developing the disease in the future with certain probability.

Microarray equipment is used widely at present to evaluate the level of gene expression in a tissue, or in a living cell [15]. Each point (pixel, cell) in a microarray represents the level of expression of a single gene. Five principal steps in the microarray technology are: tissue collection; RNA extraction; microarray gene expression calculation; scanning and image processing; data analysis—Fig. 15.10.

Techniques for analysis of DNA and RNA and disease profiling using evolving connectionist systems are published in [1, 2]—see Fig. 15.11.

The recent advent of cDNA microarray and gene chip technologies means that it is now possible to simultaneously interrogate thousands of genes in tumours. The potential applications of this technology are numerous and include identifying markers for classification, diagnosis, disease outcome prediction, therapeutic responsiveness, and target identification. Microarray analysis might not identify unique markers (e.g. a single gene) of clinical utility for a disease because of the heterogeneity of the disease, but a prediction of the biological state of disease is likely to be more sensitive by identifying clusters of gene expression (profiles) [16, 17].

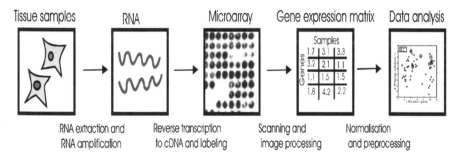

**Fig. 15.10** Five principal steps in the microarray technology are: tissue collection; RNA extraction; microarray gene expression calculation; scanning and image processing; data analysis (from [1])

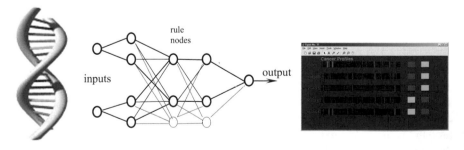

**Fig. 15.11** From DNA and RNA to disease profiling using evolving connectionist systems (from [1])

## 15.2.4   Clustering of Time Series Gene Expression Data

Each gene in a cell may express differently over time. And this makes the gene expression analysis based on static data ("one shot") not a very reliable mechanism. Measuring the expression rate of each gene over time gives the gene a temporal profile of its expression level. Genes can be grouped together according to their similarity of temporal expression profiles.

One of the main purposes for cluster analysis of time-course gene expression data is to infer the function of novel genes by grouping them with genes of well-known functionality. This is based on the observation that genes which show similar activity patterns over time (co-expressed genes) are often functionally related and are controlled by the same mechanisms of regulation (co-regulated genes). The gene clusters generated by cluster analysis often relate to certain functions, e.g. DNA replication, or protein synthesis. If a novel gene of unknown function falls into such a cluster, it is likely that this gene serves the same function as the other members of this cluster. This 'guilt-by-association' method makes it possible to assign functions to a large number of novel genes by finding groups of co-expressed genes across a microarray experiment [18].

Different clustering algorithms have been applied to the analysis of time-course gene expression data: k-means, SOM and hierarchical clustering, to name just a few [18]. They all assign genes to clusters based on the similarity of their activity patterns. Genes with similar activity patterns should be grouped together, while genes with different activation patterns should be placed in distinct clusters. The cluster methods used so far have been restricted to a one-to-one mapping: one gene belongs to exactly one cluster. While this principle seems reasonable in many fields of cluster analysis, it might be too limited for the study of microarray time-course gene expression data. Genes can participate in different genetic networks and are frequently coordinated by a variety of regulatory mechanisms. For the analysis of microarray data, we may therefore expect that single genes can belong to several clusters.

Several researchers have noted that genes were frequently highly correlated with multiple classes and that the definition of clear boarders between gene expression clusters seemed often arbitrary [19]. This is a strong motivation to use fuzzy clustering in order to assign single objects to several clusters. In fuzzy clustering each sample from a population can belong to several clusters to a membership degree, all membership degrees for this sample adding up to 1 [17–21] (Chaps. 1 and 2).

Another reason for applying fuzzy clustering is the large noise component in microarray data due to biological and experimental factors. The activity of genes can show large variations under minor changes of the experimental conditions.

Numerous steps in the experimental procedure contribute to additional noise and bias. A usual procedure to reduce the noise in microarray data is setting a threshold for a minimum variance of the abundance of a gene. Genes below this threshold are excluded from further analysis. However, the exact value of the threshold remains arbitrary due to the lack of an established error model and the use of filtering as pre-processing.

Since we usually have little information about the data structure in advance, a crucial step in the cluster analysis is selection of the number of clusters. Finding the 'correct' number of clusters addresses the issue of cluster validity. This has turned out to be a rather difficult problem, as it depends on the definition of a cluster. Without prior information, a common method is the comparison of partitions resulting from different numbers of clusters. For assessing the validity of the partitions, several cluster validity functionals have been introduced [20]. These functionals should reach an optimum if the correct number of clusters is chosen. When using evolving clustering techniques the number of the clusters does not need to be defined a priori.

Two fuzzy clustering techniques are the batch mode fuzzy C-means clustering (FCM) and an evolving clustering through evolving self-organised maps (ESOM) (scc Chap. 2).

In the FCM clustering experiment the (for more details see [21]) the fuzzification parameter m [20] is an important parameter for the cluster analysis. For the randomised data set, FCM clustering formed clusters only if m was chosen smaller than 1.15. Higher values of m led to uniform membership values in the partition matrix. This can be regarded as an advantage of FCM over hard clustering, which always forms clusters independently of the existence of any structure in the data. An appropriate choice for a lower threshold for m can therefore be set, if no cluster artefacts are formed in randomised data. An upper threshold for m is reached if FCM does not indicate any cluster in the original data. This threshold depends mainly on the compactness of the clusters. The cluster analysis with FCM showed that hyper-spherical distribution are more stable for increasing m than hyper-ellipsoid distributions. This maybe expected since FCM clustering with Euclidean norm favours spherical clusters.

The functional F that evaluates the "goodness" of the clustering [20] reached its maximum for 4 clusters for values of m < 1.35, while for larger m, F showed a monotonic decrease in number of clusters.

In another experiment, an evolving self-organising map ESOM (see Chap. 2) is evolved from the yeast gene temporal profiles used as input vectors. The number of clusters did not need to be specified in advance—Fig. 15.12 (from [1]).

It can be seen from Fig. 15.12, that clusters 72 and 70 are represented on the ESOM as neighbouring nodes. The ESOM on the figure is plotted as a 2D PCA projection. Cluster 72 has 43 members (genes that have similar temporal profiles), cluster 70–61 members, and cluster 5—only 3 genes as cluster members.

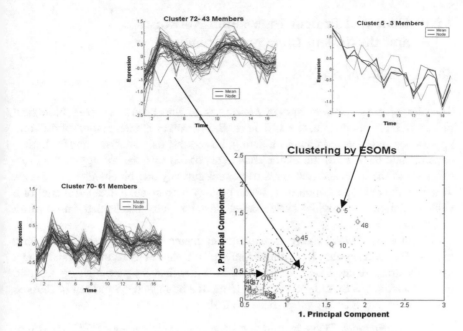

**Fig. 15.12** Using evolving self-organising maps (ESOM) (Chap. 2) for clustering gene expression time series data (from [1])

New clusters are created in an on-line mode if the distance between existing clusters and the new data vectors representing also time series with a fixed length are above a chosen threshold.

## 15.2.5   Protein Data Modelling and Structure Prediction

One task that has been explored in the literature, is predicting the secondary structure from the primary one. Segments of a protein can have different shapes in their secondary structure, which is defined by many factors, one of them being the amino-acid sequence itself. The main types of shape are:

– Helix
– Sheet
– Coil (loop).

Qian and Sejnowski [23] investigated the use of MLP for the task of predicting the secondary structure based on available labelled data, also used in the following experiment.

## 15.3 Gene and Protein Interaction Networks and the System Biology Approach

### 15.3.1 General Notions

The aim of *computational system biology* is to understand complex biological objects in their entirety, i.e. at a *system level*. It involves the integration of different approaches and tools: computer modeling, large-scale data analysis, and biological experimentation. One of the major challenges of the systems biology is the identification of the logic and dynamics of gene-regulatory and biochemical networks. The most feasible application of systems biology is to create a detailed model of a cell regulation to provide system-level insights into mechanism-based drug discovery.

*System–level understanding* is a recurrent theme in biology and has a long history. The term "system-level understanding" is a shift of focus in understanding a system's structure and dynamics in a whole, rather than the particular objects and their interactions. System-level understanding of a biological system can be derived from insight into four key properties [24, 25]:

1. *System structures.* These include the gene regulatory network (GRN) and biochemical pathways. They can also include the mechanisms of modulation the physical properties of intracellular and multi cellular structures by interactions.
2. *System dynamics.* System behavior over time under various conditions can be understood by identifying essential mechanisms underlying specific behaviors and through various approaches depending on the systems nature: metabolic analysis (finding a basis of elementary flux modes that describe the dominant reaction pathways within the network), sensitivity analysis (the study of how the variation in the output of a model can be apportioned, qualitatively or quantitatively, to different sources of variation), dynamic analysis methods such as phase portrait (geometry of the trajectories of the system in state space) and bifurcation analysis (bifurcation analysis traces time-varying change(s) in the state of the system in a multidimensional space where each dimension represents a particular system parameter (concentration of the biochemical factor involved, rate of reactions/interactions, etc.). As parameters varied, changes may occur in the qualitative structure of the solutions for certain parameter values. These changes are called bifurcations and the parameter values are called bifurcation values).
3. *The control method.* Mechanisms that systematically control the state of the cell can be modulated to change system behavior and optimize potential therapeutic effect targets of the treatment.
4. *The design method.* Strategies to modify and construct biological systems having desired properties can be devised based on definite design principles and simulations, instead of blind trial-and-error.

As it was mentioned above, in reality analysis of system dynamics and understanding the system structure are overlapping processes. In some cases analysis of the system dynamics can give useful predictions in system structure (new interactions, additional member of system). Different methods can be used to study the dynamical properties of the system:

- Analysis of steady-states allows finding the systems states when there are no dynamical changes in system components.
- Stability and sensitivity analyses provide insights into how systems behavior changes when stimuli and rate constants are modified to reflect dynamic behavior.
- Bifurcation analysis, in which a dynamic simulator is coupled with analysis tools, can provide a detailed illustration of dynamic behavior.

The choice of the analytical methods depends on availability of the data that can be incorporated in into the model and the nature of the model. It is important to know the main properties of the complex system under investigation, such as robustness.

Robustness is a central issue in all complex systems and it is very essential for understanding of the biological object functioning at the system level. Robust systems exhibit the following phenomenological properties:

- *Adaptation*, which denotes the ability to cope with environmental changes.
- *Parameter insensitivity*, which indicates a system's relative insensitivity (to a certain extent) to specific kinetic parameters.
- *Graceful degradation*, which reflects the characteristic slow degradation of a system's functions after damage, rather than catastrophic failure.

All the above features are present in some AI methods and techniques and make them very suitable to modelling complex biological systems [2]. Revealing all these characteristics of a complex living system helps choosing an appropriate method for their modelling, and also constitutes an inspiration for the development of new AI methods that possess these features.

Modelling living cells in silico (in a computer) has many implications; one of them is testing new drugs through simulation rather than on patients. According to recent statistics, human trials fail for 70–75% of the drugs that enter them.

Modelling living cells in silico (in a computer) has many implications, one of them is testing new drugs through simulation rather than on patients. According to [26] human trials fail for 70–75% of the drugs that enter them.

Computer modelling of processes in living cells is an extremely difficult task. There are several reasons for that, among them are:

- The processes in a cell are dynamic and depend on many variables some of them related to a changing environment.
- The processes of DNA transcription, and protein translation are not fully understood.

Several cell models have been created and experimented, among them:

- The Virtual Cell model [27];
- The e-cell model and the self-survival model [28];
- A mathematical model of a cell cycle [29].

A starting point to dynamic modelling of a cell would be dynamic modelling of a single gene regulation process. In [30] the following methods for single gene regulation modelling are discussed, that take into account different aspects of the processes (chemical reactions, physical chemistry, kinetic changes of states, thermodynamics):

- Boolean models, based on Boolean logic (true/false logic);
- Differential equation models;
- Stochastic models;
- Hybrid Boolean/differential equation models;
- Hybrid differential equations/stochastic models;
- Neural network models;
- Hybrid connectionist-statistical models.

The next step in dynamic cell modelling would be to try and model the regulation of more genes, hopefully a large set of genes (see [31]). Patterns of collective regulation of genes are observed in the above reference, such as chaotic attractors. Mutual information/entropy of clusters of genes can be evaluated.

A general, hypothetical evolving model of a cell is outlined below. It encompasses the *system biology approach*. It is based on the following principles:

1. The model incorporates all the initial information such as analytical formulas, databases, rules of behaviour.
2. In a dynamic way, the model adjusts, adapts over time during its operation.
3. The model makes use of all current information and knowledge at different stages of its operation (e.g., transcription, translation).
4. The model takes as inputs data from a living cell and models its development over time. New data from the living cell is supplied if such is available over time.
5. The model runs until it is stopped, or the cell has died.


## 15.3.2   Gene Regulatory Network Modelling

Modelling processes in a cell includes finding the genetic networks (the network of interaction and connections between genes, each connection defining if a gene is causing another one to become active, or to be suppresses). The reverse engineering approach is used for this task [32]. It consists of the following: Gene expression data is taken from a cell (or a cell line) at consecutive time moments. Based on

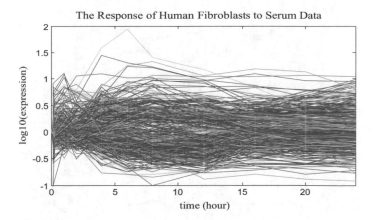

**Fig. 15.13** The time course data of the expression of genes in the Human fibroblast response to serum benchmark data [33]

these data a logical gene network is derived. For example, it is known that clustering of genes with similar expression patterns will suggest that these genes are involved in same regulatory processes.

Modelling *gene regulatory networks (GRN)* is the task of creating a dynamic interaction network between genes that defines the next time expression of genes based on their previous levels of expression.

Models of GRN, derived from gene expression RNA data, have been developed with the use of different mathematical and computational methods, such as: statistical correlation techniques; evolutionary computation; ANN; differential equations, both ordinary and partial; Boolean models; kinetic models; State-based models and others.

An example of GRN extraction from data is presented in [33] where the human response to fibroblast serum data is used (Fig. 15.13) and a GRN is extracted from it (Fig. 15.14).

### 15.3.3  Protein Interaction Networks

Proteins interact between each other in time-space, forming structures that define important fucntions for an organism. Figure 15.15 shows an example of network-based analysis using UniHI [2]. For proteins of interest such as p53, interaction partners can be queried and visualized. The derived networks can be subsequently filtered based on evidence (e.g., number of PubMed references reporting the interaction) or based on gene expression data. All networks can be readily inspected for enrichment in biological processes using an integrated tool.

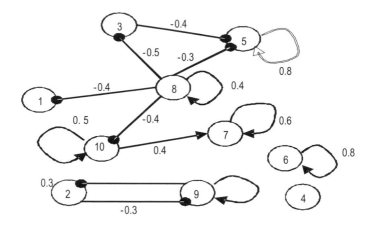

**Fig. 15.14** A GRN obtained with the use of the method from [33] on the data from Fig. 15.13, where 10 clusters of gene expression values over time are derived, each cluster represented as a node in the GRN and their interaction in time—as weighted arcs

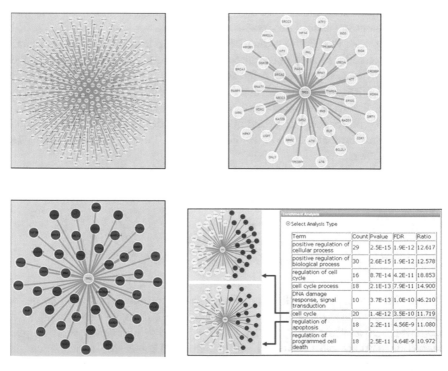

**Fig. 15.15** Example of network-based analysis using UniHI. For proteins of interest such as p53, interaction partners can be queried and visualized. The derived networks can be subsequently filtered based on evidence (e.g., number of PubMed references reporting the interaction) or based on gene expression data. All networks can be readily inspected for enrichment in biological processes using an integrated tool [2]

## 15.4 Brain-Inspired SNN Architectures for Deep Learning of Gene Expression Time Series Data and for the Extraction of Gene Regulatory Networks

This chapter presents a novel method for modelling gene interaction networks using a bran-inspired SNN exemplified by NeuCube (Chap. 6). The method is first published in [34] and applied on cancer data in [35] and on Ebola vaccine data in [36].

### 15.4.1 General Notions

Genes are the blueprint for proteins synthesis, which altered behaviour have been implicated in several pathologies, and that is why they have attracted researchers' attention for several years [37–39]. By identifying which genes are turned on in a particular cell, in which amount and when, can help us uncover the process behind the cell life, processes and understanding of how diseases work [40]. In the era of system biology, the most popular techniques used to quantify transcriptome data are: DNA hybridization-based microarrays and next-generation sequencing (RNA-Seq) technologies [41, 42]. Even though RNA-seq has emerged as the method of choice for measuring transcriptome of organisms [43], DNA microarrays are still widely used for targeting the identification of already known common allele variants, and public databases make this data available to the entire community. The transcriptome data provided can be considered representative of the entire community, as genome's sequence similarity over individuals is really high (about 99.9%), with the same genes in the same location [44]. People are unique by phenotype; however, every human being shares the same genetic blueprint [40]. While the measurement and analysis of steady-state microarray data are routine, the analysis of time-series gene expression profiling are of growing interest, as they can shed some light on the complex relation- ships between genes and how they work together over time, to determine adaptive phenotype and transcription factors that are associated with certain diseases. This type of data, time-series data, is difficult to analyse and using traditional statistical and artificial intelligent techniques may incur in a loss of information. Novel artificial intelligence techniques, such a spiking neural networks (SNN) [45–48] have emerged as the method of choice, as they have successfully demonstrated their ability to learn and extract meaningful patterns from time-series data by using biologically inspired net of neurons [49, 50]. Neurons are computational units that process binary information as spikes that encode significant changes "hidden" in the raw data, and learn how these changes of temporal activity interacts with each other over time. By extracting meaningful patterns from raw time-series data, we can understand mechanisms involved in the regulatory expression of genes. In our study, we use time-series microarray data available from the National Centre for Biotechnology Information (NCBI) database to analyse the feasibility of an SNN system for gene expression data modelling, classification and interpretation. We want to understand the genetic information available and analyse the interaction between genes over time to identify the clinical implication behind this process. We achieve this by integrating SNN techniques in a computational model able to derive the interaction between genes over time as a gene interaction network (GIN). This is a novel approach that could lead to biologically realistic models and to a better understanding of the phenomenon of study.

## 15.4.2  A SNN Based Methodology for Gene Expression Time Series Data Modelling and Extracting GRN

This method was fully published in [34].

To model and analyse the dynamic behaviour of a complex biological process, such as the interaction between thousands of genes over time, we have designed a novel system based on the NeuCube SNN architecture (Chap. 6) [51–55].

Figures 15.16 and 15.17 show the idea of using the NeuCube SNN system for time- series gene expression data analysis. This SNN system consists of the following modules:

- Input information encoding module, where the transcriptomics data is first modelled and then encoded into trains of spikes, each spike representing a change in the gene expression over time;
- 3D SNN cube (SNNc) module (the Cube), where the encoded time-series data is entered and learned;
- Output module for data classification;
- GIN module for knowledge discovery.

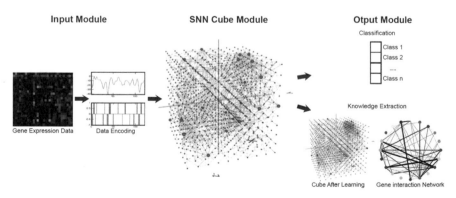

**Fig. 15.16** A graphical representation of the novel SNN system designed for time–series gene expression data analysis (from [34]) (the figure was created by E. Capecci)

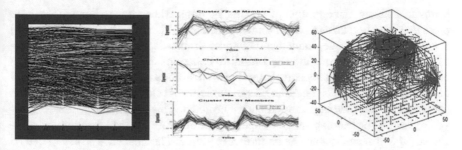

**Fig. 15.17** Time series gene expression data are first clustered and then used to train a NeuCube model

In the following paragraphs, this procedure is explained in detail.

*Input Module and Gene Expression Data Encoding*

The gene expression data is first ordered as a sequence of real-value data vectors. Every data vector is transformed into a spike train using a number of encoding algorithms, such as the adaptive threshold-based (ATB) encoding algorithm (Chap. 4) and time series mapping algorithm proposed in [56]. The encoding method uses a self- adaptive bi-directional threshold to discretise the signal gradient with respect to the time. The results are a positive spike train that encodes the point in which the signal increase and a negative spike which encode the point in which the signal de- crease. This was calculated for every vector of the time-series data. As a result, each spike train carries the information about the temporal regulation of the expression of a certain gene over time that was in principle "hidden" in the raw data.

*The SNN Cube Module and Unsupervised Learning*

The encoded spike trains are fed into a cube of, *e.g.* $10 \times 10 \times 10$, leaky integrate and-fire (LIF) neurons. Input features are mapped as proposed in [56] (computed according to equation in Appendix B), as this method optimises the mapping of the input variables using a graph matching algorithm. Optimising this process improved the results greatly, as the mapping influenced the learning, and therefore the understanding of the time-series patterns created by the data. The spiking LIF neurons of the Cube are connected using a small-world (SW) connectivity rule. These connections have weight values assigned and calculated as per Chaps. 4 and 5. Excitatory connection weights represent 80%, while inhibitory connection weights represent 20% of the total. Setting up a proper initial structural connectivity in a network of artificial spiking neurons is important, as it allows the SNN model to properly learn from the data and capture functional connectivity information from it. This preserves the spatio-temporal relationships within the data, which is a significant source of information generally overlooked by other techniques. Unsupervised learning of the Cube is performed to modify the initially set connection weights. The SNNc learns to activate same groups of spiking neurons when similar input *stimuli* are presented, also known as a polychronization effect [57]. Hebbian learning rules [58] are applied at this stage, implementing spike-timing-dependent plasticity (STDP)-like algorithms [59] (computed according to equations in Appendix D). The STDP protocol describes the connection between two neurons as stronger as their activation persists and repeats. The neurons become able to develop new connections in the network that can then be analysed and interpreted. This makes the proposed SNN system useful for learning spatio-temporal patterns from the data, forming an associative type of memory that can be further explored. The final cube structure can be visualised after unsupervised training for greater knowledge extraction that cannot be achieved with purely statistical or mathematical techniques.

*Output Module for Supervised Learning and Data Classification*

During supervised learning, the same data used for the unsupervised training is propagated again through the trained cube, and output neurons are generated and trained to classify the data patterns into output spike sequences (the classes). To learn and classify spiking patterns from the Cube, we use the dynamic evolving spiking neural network (deSNN) classifier, which allows simple class-based discrimination [60] (Chap. 5). The spiking activity generated during learning can be visualised and used as a bio-feedback.

## 15.4.3 Extracting GRN from a Trained Model

The learned spiking activity and connectivity generated by the time-series gene expression data can be analysed by means of a GRN (or also called in [34] gene interaction network or GIN). Each one of the input gene features is used as a source of information to define the centre of a cluster of neurons. Each unlabelled neuron is assigned to a cluster with which it exchanges the highest number of spikes. This is calculated, during the STDP learning process. The more spikes transmitted between clusters, the greater is the spatio-temporal interaction between genes over time. The obtained GIN represents a connectivity graph, where nodes represent gene variables, and the lines and their thickness define the amount of temporal information exchanged between genes. In other words, the interaction between input variables is captured in the GIN in terms of their changes in time. The information showed in the GIN is used to analyse the complex temporal patterns "hidden" in data, which can be used for the development of new methodologies of gene expression data modelling and understanding.

### 15.4.4 A Case Study Experimental Modelling of Gene Expression Time Series Data

This study makes use of the information provided by transcriptome data to classify and analyse the interaction of different genes over time as a GRN. For our experimental case study, the raw gene expression data was first modelled and encoded into trains of spikes. Then, patterns of temporal activity generated during learning were classified, to establish the effectiveness of the SNN system designed to separate different time-series gene expression data into different classes. Then, the learned patterns of temporal activity generated in the Cube were analysed in terms of spiking activity and connection weight changes and new information was extracted as a GRN to study the interaction of genes expressed over time.

Figure 15.18 shows and example of encoding the case study data, the training and the extraction of GRN for each of the two classes of data.

As a case study for our problem, gene expression data collected during elicitation of allergic contact dermatitis (ACD) over time has been used. Data was obtained from the publicly available Gene Expression Omnibus (GEO) repository of functional genomics data (NCBI GEO [61, 62] accession GSE6281 [63]). Data was collected from a control group, and a group of people that had shown an inflammatory response to a nickel patch test. Expression profiling was analysed by hybridization of high density oligonucleotide arrays obtained after skin biopsies of 7 nickel-allergic female subjects, and 5 non-allergic female controls. Biopsies were taken over four time periods: 0, 7, 48 and 96 h. The control group did not show eczema at any time-point, while nickel allergic patients reacted with eczema at 48 and 96 h only. Samples were analysed using microarray technology that measured

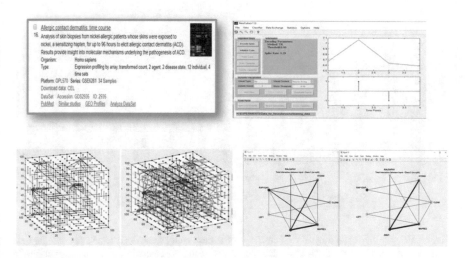

**Fig. 15.18** The process of encoding a case study data, the training and the extraction of GRN for each of the two classes of data [34]

transcript levels in biopsies. Each microarray contains thousands of data points, an enormous quantity of data. Although, time-series microarray data are really short and the number of gene expressed is huge compared with the short-time course data available. To adequately describe the distribution of the data, we need to reduce the number of features to approach the curse of dimensionality phenomenon, such as the Hughes phenomenon [64–66]. Thus, extracting the relevant number of genes that interact with each other over time is the first problem that we need to solve for developing a computational model of GIN. We need to evaluate how relevant the original variables or features of the model are, when using classification accuracy as an objective function to the problem. To achieve this, we have used the popular signal-to-noise ratio (SNR), as described in Chap. 1 and [67], to evaluate how important a variable is to discriminate samples belonging to different classes. SNR is a filtering method that selects and ranks features in advance, before the model is created. For the experiments, data was ordered according to the time of collection. We obtained a total of eight samples, four of each class (control patients or patients affected by ACD). Missing data was handled using linear interpolation method [68]. Then, SNR method was applied.

Figure 15.19 gives the SNR ranking of the features. As a result, our sample data consisted of a set of four time-series data per seven features selected out of the original 54675 variables. These corresponded to genes CLDN6, H72868, RALGAPA1, RAP1GAP, LEF1, ZMIZ1 and MAPRE3. In the encoding module, the ordered vectors of real-valued data were then converted into trains of spikes using the ATB encoding algorithm.

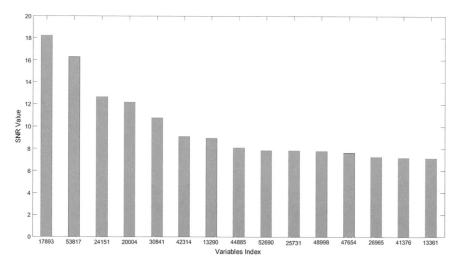

**Fig. 15.19** The SNR ranking of the features for the case study data. As a result, our sample data consisted of a set of four time-series data per seven features selected out of the original 54675 variables. These corresponded to genes CLDN6, H72868, RALGAPA1, RAP1GAP, LEF1, ZMIZ1 and MAPRE3 [34]

**Table 15.2** Classification accuracy percentage obtained after training the SNN system with the gene expression data available

| Measure | Overall accuracy % |
|---|---|
| Mean | 73 |
| Standard deviation | 4.5 |

To investigate whether the system was able to correctly discriminate the two different groups of people, we classified the data into two classes: class one—control—and class two—patients that shown ACD. In total, we had 8 samples, four for each class, four time steps of data collection (0, 7, 48 and 96 h), and 7 features/genes extracted after SNR dimensionality reduction technique was applied. Optimisation of the numerous parameters of the SNN system was performed via a grid search method that evaluated ten combinations of parameters and selected the one with the highest classification accuracy. This process was evaluated using the Monte-Carlo cross validation method. More specifically, we randomly selected a cube and used the leave-one-out cross validation (LOOCV) method to select the training and testing set. This process was repeated for 30 times, generating each time a random new cube for training and testing. LOOCV method has been chosen, as it best assesses the variables importance over a random set of entities, it best tackles the low-number-of-predictors versus high-number-of-dimensions problem, and it is the best method to validate a model, when only a small data set is available. The optimised parameter values obtained were (see Chap. 6):

- The *ATB* encoding algorithm was set at 0.01;
- The SW connectivity radius was set at 2.5;
- The parameters of the LIF neuron model were set at 0.5 (threshold of firing), 6 (the refractory time) and 0.002 (the potential leak rate);
- The STDP rate parameter of the unsupervised learning algorithm was set at 0.002;
- The variables *mod* and *drift* of the deSNN classifier were set at 0.8 and 0.005 respectively.

Table 15.2 reports the classification accuracy percentage obtained with this combination of parameters. These results demonstrated the capability of the SNN system to discriminate between the two classes, even when trained with such a small data set. This is a good indication that the proposed SNN system can be used for classifying time-series gene expression data.

### 15.4.5    Extracting GRN Form a Trained Model and Analysis of the GRN for New Knowledge Discovery

After the optimisation procedure, we retained the best SNN model parameters, the initial connectivity and temporal mapping of the input features. Network analysis was carried out by training the entire time series data available for each class separately. This process was iterated 100 times per class, to allow the system to learn from the short-time series data. First, the input features were clustered with other neurons of the Cube according to their connection weight value (see Chap. 6). Consequently, the neural activity generated per class revealed seven clusters of genes (Fig. 15.20 left). The proportion of activity generated in the entire 3D cube is illustrated in the pie chart (Fig. 15.20 right). As shown in Fig. 15.20, there are three major clusters obtained for each class. The major component for the two classes is

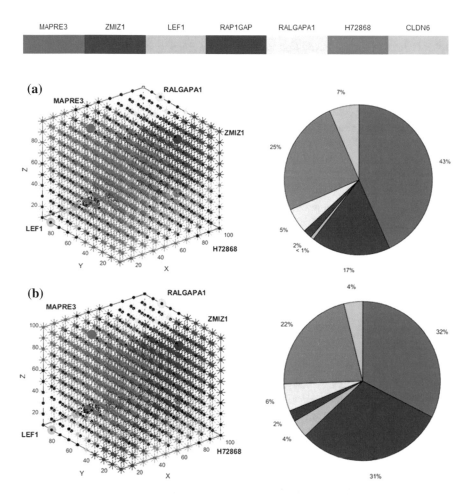

**Fig. 15.20** Left: the 3D SNN cube shows the clusters formed by CLDN6, H72868, RALGAPA1, RAP1GAP, LEF1, ZMIZ1 and MAPRE3 with the other neurons of the cube according to their connection weights value. Right: the pie chart illustrates the numerical proportion of the clustered activity expressed as a percentage with respect to the entire cube: **a** class one—control; **b** class two —patients that shown ACD. Every gene is shown in a different colour, as indicated in the bar at the top of the figure [34]

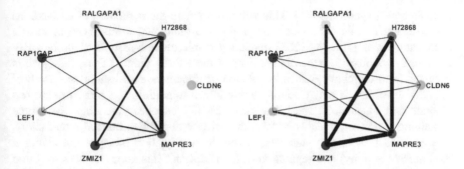

**Fig. 15.21** GRNs of the seven input features (genes CLDN6, H72868, RALGAPA1, RAP1GAP, LEF1, ZMIZ1 and MAPRE3) obtained per class. Left: class one—control. Right: class two—patients that shown ACD. The seven genes are indicated in different colours corresponding to each of the respective clusters. The stronger the interaction between genes, the thicker is the line that connects the node [34]

given by the cluster of gene MAPRE3 (43% for class one and 32% for class two); followed by H72868 for class one (25%) and ZMIZ1 for class two (31%); and ZMIZ1 for class one (17%), and H72868 for class two (22%). The total interaction between the seven clusters was calculate per class and visualised in a GIN (Fig. 15.21, left class one, right class two). Each cluster of genes is indicated by a different colour. The stronger the interaction between clusters, the thicker is the line that connects the genes. The two GINs obtained for each class show significance differences between subjects that are sensitised to nickel versus control subjects.

Some of the main points that we can extrapolate from the GINs are:

- First of all, we can appreciate that there is no temporal interactions exchanged between CLDN6 gene and other genes of the network for class one; while, CLDN6 gene shows significant temporal association with the four other genes of the network for class two.
- In comparison with GIN of class one, the GIN of class two shows stronger connections between MAPRE3 (the major component for the two classes, in term of neural activity exchanged as total connection weight) and ZMIZ1 (the second major cluster of temporal activity exchanged for class two).
- The expression of ZMIZ1 gene has shown a stronger temporal correlation with RALGAPA1 gene for class two compared with class one.

These observations appear to be consistent with the finding in the literature [69]. Claudin-6, is part of the claudin family, the most important component of tight junction strands. These are barrier between the cells of an epithelium that controls the paracellular flow of molecules in the intercellular space [70], thus is expression cause up-regulation of epithelial cells in patients showing skin eczema. Also, it is well known that nickel is a ubiquitous trace element and present also in most of the dietary items. Thus, food is a major source of nickel exposure for the population and can provoke not only dermatitis, but also gastrointestinal symptoms similar to irritable bowel syndrome [71]. This will also explain the regulation of CLDN6, as the detection of this gene is a marker for intestinal- and diffuse-type of gastric adenocarcinomas [70]. MAPRE3 encodes for microtubule-associated protein RP/EB Family Member 3, which is member of the RP/EB family. Thus, this gene is involved in microtubule growth, regulation of dynamics, stabilisation and anchoring at centrosomes. It may also play a role in cell migration. It is mostly expressed in brain and muscle. On the other hand, ZMIZ1 encodes zinc finger MIZ-type containing 1 protein, which regulates the activity of various transcription factors. The encoded protein may also play a role in sumoylation (i.e. post-translational modification activity) and regulation of transcription. This gene can be also found expressed, in thymus, small intestine, colon and peripheral blood leukocytes [69]. RALGAPA1 encode for a major subunit required for the heterodimerization activity of RALGTPase activating protein, the RalGTPase Activating Protein Catalytic Alpha Subunit 1. This gene is overexpressed in liver and esophagus and its expression is also revealed in nervous system, liver, skin and lymph node tissue [69]. These findings show that as a consequence of the assimilation of nickel, patients affected by ACD are to express genes that are revealed in skin, mussels and in the human digestive system, and that this expression is related with cell proliferation and immune response.

### 15.4.6  Discussions on the Method

The analysis of patterns generated by time-series gene expression data constitutes a major goal for the area of bioinformatics and system biology. In this study, we have modelled short-time transcriptome data and revealed meaningful patterns and new knowledge from the genes expressed over time by means of novel SNN system that creates a GIN model. For the case study data of allergy reaction against nickel, we have found that CLDN6 was the feature that scored the highest SNR value and its cluster revealed temporal association with other genes for class two only (positive reaction). Still, we have found that the exchange of temporal activity has been dominated by MAPRE3 and ZMIZ1 clusters. Also, H72868 cluster has shown relevant temporal association and interaction with other genes. According to the information available regarding the nature of these genes, it seems plausible that patients affected by ACD express genes that are revealed in skin, mussels and in the human digestive system, and that these genes are also related to cell proliferation and immune response. All of this, appear to be the consequence of the assimilation of nickel in the body and the increased of eczematous reactions over time. We can conclude that, SNNs are the method of choice when dealing with big transcriptomics data, as they extract only relevant patterns of temporal activity from the gene expression data; moreover, the clinical *focus* of this study demonstrated the model ability to constitute a valuable tool that can support experts in the area of bioinformatics and system biology in understanding and studying the interaction between genes expressed over time. Future work includes:

- Implementation of different feature selection methods for temporal gene expression data, such as the one proposed in [72];
- Studying the possible implication of directionality between temporal interaction of genes in the network;
- Comparative analysis of the results obtained with the proposed method with the ones obtained by [63] on the same data set;
- Further study on the application of the method for cancer time series data analysis [35];
- Further study on the application of the method for the analysis of gene expression data measuring response to treatment, such as to Ebola vaccine [36].

## 15.5 Chapter Summary and Further Readings

This chapter first introduces the basic knowledge about molecular biology, before presenting methods for modelling bioinformatics data such as gene expression using evolving spiking neural networks and BI-SNN. The latter are applied for deep learning and modelling of gene expression time series data, thus enabling the discovery of gene interaction networks (GIN) to better understand the biological processes. Analysis of genes related to brain functions is covered in Chap. 16 under the title—Computational neurogenetic modelling.

More about biological background and computational modelling in bioinformatics can be found in various chapters of [2], including:

- Understanding information processes in biological systems (Chaps. 2–7, in [2]);
- Molecular biology, genomics and proteomics (Chaps. 8–11 in [2]);
- Biological databases (Chap. 26 in [2]);
- Ontologies in bioinformatics (Chap. 27 in [2]);
- Path finding in biological networks (Chap. 19 in [2]);
- Inferring transcription networks from data (Chap. 20 in [2]);
- Inferring genetic networks with recurrent neural network model using differential evolution (Chap. 22 in [2]);
- Structural pattern discovery in protein-protein interaction networks (Chap. 23 in [2]);
- Cancer stem cells (Chap. 28 in [2]);
- Epigenetics (Chap. 29 in [2]);
- Autoimmune diseases (Chap. 30 in [2]);
- Nutrigenomics (Chap. 31 in [2]);
- Nanomedicine (Chap. 32 in [2]);
- Personalised medicine (Chap. 33 in [2]);
- Health informatics (Chap. 34 in [2]);
- Ecological informatics (Chap. 35 in [2]);
- Modelling bioinformatics data with BI-SNN: https://kedri.aut.ac.nz/R-and-D-Systems/bioinformatics-data-modelling-and-analysis.

**Acknowledgements** The introductory biological material in Sects. 15.1 and 15.2 are mainly covered in [1, 2] and Sect. 15.4—in [34]. I acknowledge the contribution of my co-authors of the referenced in this chapter publications Elisa Capecci, Jack Dray, Lucien Koefoed, Mattias Futschik, Mike Watts, Vinita Jansari, Dimitar Dimitrov, Josafath Espinosa.

# References

1. N. Kasabov, *Evolving Connectionist Systems: The Knowledge Engineering Approach* (Springer, London, 2007). (1st edn 2002)
2. N. Kasabov (ed.), *Springer Handbook of Bio-/Neuroinformatics* (Springer, Berlin, 2014)
3. D. Hofstadter, *Godel, Escher, Bach: An Eternal Golden Braid* (Basic Books, New York, 1979)
4. P. Baldi, S. Brunak, *Bioinformatics—A Machine Learning Approach* (MIT Press, Cambridge, 1998, 2001)
5. T. Friend, Genome projects complete sequence. *USA Today*, 23 June 2000
6. L. Fu, An expert network for DNA sequence analysis. IEEE Intell. Syst. Appl **14**(1), 65–71 (1999)
7. Y. Okazaki et al., Analysis of the mouse transcriptome based on functional annotation of 60,770 full-length cDNAs. Nature **420**(6915), 563–573 (2002)
8. J.S. Mattick, I.V. Makunin, Small regulatory RNAs in mammals. Hum Mol Genet. **14**(Spec No 1), R121–R132 (2005)
9. A.F. Bompfünewerer, et al., Evolutionary patterns of non-coding RNAs. Theory Biosci. **123**, 301–369 (2005)
10. J. Allen et al., Discrimination of modes of action of antifungal substances by use of metabolic footprinting. Appl. Environ. Microbiol. **70**(10), 6157–6165 (2004)
11. A.E. Pasquinelli, B.J. Reinhart, F. Slack, M.Q. Martindale, M.I. Kuroda, B. Maller, D.C. Hayward, E.E. Ball, B. Degnan, P. Müller, J. Spring, A. Srinivasan, M. Fishman M, Finnerty, J. Corbo, M. Levine, P. Leahy, E. Davidson, G. Ruvkun, Conservation of the sequence and temporal expression of let-7 heterochronic regulatory RNA. Nature **408**(6808), 86–89 (2000)
12. C.S. L. Lai, D. Gerrelli, A.P. Monaco, S.E. Fisher, A.J. Copp, FOXP2 expression during brain development coincides with adult sites of pathology in a severe speech and language disorder. Brain **126**, 2455–2462 (2003). https://doi.org/10.1093/brain/awg247
13. F.L. Lim et al., Mcm1p-induced DNA bending regulates the formation of ternary transcription factor complexes. Mol. Cell. Biol. **23**(2), 450–461 (2003)
14. E. Berezikov, R.H. Plasterk, Camels and zebrafish, viruses and cancer: a microRNA update. Hum. Mol. Genet. **14**, 183–190 (2005)
15. M. Schena (ed.), *Microarray Biochip Technology* (Eaton Publishing, Natick, MA, 2000)
16. M. Futschik, A. Jeffs, S. Pattison, N. Kasabov, M. Sullivan, A. Merrie, A. Reeve, Gene expression profiling of metastatic and non-metastatic colorectal cancer cell-lines. Genome Lett. **1**(1), 1–9 (2002)
17. M. Futschik, M. Sullivan, A. Reeve, N. Kasabov, Prediction of clinical behaviour and treatment of cancers. Appl. Bioinform. **3**, 553–558 (2003)
18. J.L. DeRisi, V.R. Iyer, P.O. Brown, Exploring the metabolic and genetic control of gene expression on a genomic scale. Science **278**(5338), 680–686 (1997)
19. H. Chu, C. Parras, K. White, F. Jimenez, Formation and specification of ventral neuroblasts is controlled by vnd in Drosophila neurogenesis. Genes Dev. **12**(22), 3613–3624 (1998)
20. N. Pal, J.C. Bezdek, On cluster validity for the fuzzy c-means model. IEEE Trans. Fuzzy Syst. 370–379 (1995)
21. M. Futschik, N. Kasabov, Fuzzy clustering in gene expression data analysis. In *Proceedings of the World Congress of Computational Intelligence WCCI'2002, Hawaii*, May 2002. IEEE Press

22. S. Brown, S. Holtzman, T. Kaufman, R. Denell, Characterization of the tribolium deformed ortholog and its ability to directly regulate deformed target genes in the rescue of a Drosophila deformed null mutant. Dev. Genes. Evol. **209**(7), 389–398 (1999)

23. N. Qian, T.J. Sejnowski, Predicting the secondary structure of globular protein using neural network models. J. Mol. Biol. **202**, 065–884 (1988)

24. D.S. Dimitrov, I.A. Sidorov, N.K. Kasabov, Computational biology, in *Handbook of Theoretical and Computational Nanotechnology*, vol. 1, ed. by M. Rieth, W. Sommers (American Scientific Publisher, 2004)

25. N. Kasabov, I.A. Sidorov, D.S. Dimitrov, Computational intelligence, bioinformatics and computational biology: a brief overview of methods, problems and perspectives. J. Comput. Theor. Nanosci. **2**(4), 473–491 (2005)

26. A. Zaks, Annuities under random rates of interest. Insur. Math. Econ. **28**, 1–11 (2001)

27. J. Schaff, L.M. Loew, The virtual cell., in *Pacific Symposium on Biocomputing* (1999), pp. 228–239

28. M. Tomita, Whole-cell simulation: a grand challenge of the 21st century. Trends Biotechnol. **19**(6), 205–210 (2001)

29. K.W. Kohn, D.S. Dimitrov, Mathematical Models of Cell Cycles, Computer Modelling and Simulation of Complex Biological Systems (1999)

30. M.A. Gibson, E. Mjolsness, Modelling the activity of single genes, in *Computational Modelling of Genetic and Biochemical Networks*, ed. by J.M. Bower, H. Bolouri (MIT Press, Cambridge, 2001), pp. 3–48

31. R. Somogyi, S. Fuhrman, X. Wen, Genetic network inference in computational models and applications to large-scale gene expression data, in *Computational Modelling of Genetic and Biochemical Networks*, ed. by J.M. Bower, H. Bolouri (MIT Press, Cambridge, 2001), pp. 120–157

32. P. D'haeseleer, S. Liang, R. Somogyi, Genetic network inference; from co-expression clustering to reverse engineering. Bioinformatics **16**(8), 707–726 (2000)

33. S.Z. Chan, N. Kasabov, L. Collins, A hybrid genetic algorithm and expectation maximization method for global gene trajectory clustering. J. Bioinform. Comput. Biol. **3**(5), 1227–1242 (2005)

34. E. Capecci, J.L. Lobo, I. Lana, J.I. Espinosa-Ramos, N. Kasabov, *Modelling Gene Interaction Networks from Time-Series Gene Expression Data using Evolving Spiking Neural Networks, Evolving Systems* (Springer, Berlin, 2018)

35. J. Dray, E. Capecci, N. Kasabov, Spiking neural networks for cancer gene expression time series modelling and analysis, in Proc. ICONIP, Springer, 2018

36. L. Koefoed, E. Capecci, V. Jansari, N. Kasabov, Analysis of gene expression data of Ebola vaccine using spiking neural networks, in Proc. IJCNN, 2018

37. C. Kuma, M. Mann, Bioinformatics analysis of mass spectrometry-based proteomics data sets. FEBS Lett. **583**(11), 1703–1712 (2009)

38. M. Pertea, S.L. Salzberg, Between a chicken and a grape: estimating the number of human genes. Genome Biol. **11**(5), 206 (2010)

39. I. Ezkurdia, D. Juan, J.M. Rodriguez, A. Frankish, M. Diekhans, J. Harrow, J. Vazquez, A. Valencia, M.L. Tress, The shrinking human protein coding complement: are there now fewer than 20,000 genes? ArXiv e-prints, 2013, 1312.7111 (2013)

40. E.H. Shen, C.C. Overly, A.R. Jones, The allen human brain atlas: comprehensive gene expression mapping of the human brain. Trends Neurosci. **35**(12), 711–714 (2012)

41. S. Panda, T.K. Sato, G.M. Hampton, J.B. Hogenesch, An array of insights: application of dna chip technology in the study of cell biology. Trends Cell Biol. **13**(3), 151–156 (2003)

42. X. Wang, M. Wu, Z. Li, C. Chan, Short time-series microarray analysis: methods and challenges. BMC Syst. Biol. **2**(1), 58 (2008)

43. A. Mortazavi, B.A. Williams, K. McCue, L. Schaeffer, B. Wold, Mapping and quantifying mammalian transcriptomes by rna-seq. Nat. Meth. **5**(7), 621–628 (2008)

44. L. Feuk, A.R. Carson, S.W. Scherer, Structural variation in the human genome. Nat. Rev. Genet. **7**(2), 85–97 (2006)

45. W. Maass, Networks of spiking neurons: the third generation of neural network models. Neural Networks **10**(9), 1659–1671 (1997)

46. W. Gerstner, Time structure of the activity in neural network models. Phys. Rev. E **51**(1), 738 (1995)

47. W. Gerstner, *Plausible Neural Networks for Biological Modelling*. What's different with spiking neurons? (Kluwer Academic Publishers, Dordrecht, 2001), p. 2345

48. W. Gerstner, H. Sprekeler, G. Deco, in Theory and simulation in neuroscience. Science **338** (6103), 60–65

49. S. Ghosh-Dastidar, H. Adeli, Improved spiking neural networks for eeg classification and epilepsy and seizure detection. Integr. Comput.-Aided Eng. **14**(3), 187–212 (2007)

50. N. Kasabov, E. Capecci, Spiking neural network methodology for modelling, recognition and understanding of eeg spatio-temporal data measuring cognitive processes during mental tasks. Inf. Sci. **294**, 565–575 (2015)

51. N. Kasabov, Neucube evospike architecture for spatio-temporal modelling and pattern recognition of brain signals, in *Artificial Neural Networks in Pattern Recognitioned*, vol. 7477, ed. by N. Mana, F. Schwenker, E. Trentin. Lecture Notes in Computer Science (Springer, Berlin, 2012), pp. 225–243

52. Y. Chen, J. Hu, N. Kasabov, Z.-G. Hou, L. Cheng, Neucuberehab: a pilot study for eeg classification in rehabilitation practice based on spiking neural networks. Neural Inf. Process. **8228**(2013), 70–77 (2013)

53. N. Kasabov, Neucube: a spiking neural network architecture for mapping, learning and understanding of spatio-temporal brain data. Neural Networks **52**(2014), 62–76 (2014)

54. E. Tu, N. Kasabov, M. Othman, Y. Li, S. Worner, J. Yang, Z. Jia, Neucube(st) for spatio-temporal data predictive modelling with a case study on ecological data, in *2014 International Joint Conference on Neural Networks (IJCNN)* (2014), pp. 638–645. https://doi.org/10.1109/ijcnn.2014.6889717

55. N. Kasabov, N.M. Scott, E. Tu, S. Marks, N. Sengupta, E. Capecci, M. Othman, M.G. Doborjeh, N. Murli, R. Hartono et al., Evolving spatio-temporal data machines based on the neucube neuromorphic framework: design methodology and selected applications. Neural Networks **78**(2016), 1–14 (2016)

56. E. Tu, N. Kasabov, J. Yang, Mapping temporal variables into the neucube for improved pattern recognition, predictive modeling, and understanding of stream data. IEEE Trans. Neural Networks Learn. Syst. **28**(6), 1305–1317 (2017)

57. E.M. Izhikevich, Polychronization: computation with spikes. Neural Comput. **18**(2), 245–282 (2006)

58. D.O. Hebb, *The Organization of Behavior: A Neuropsychological Approach* (Wiley, New York, 1949)

59. Song, K.D. Miller, L.F. Abbott, Competitive hebbian learning through spike-timing-dependent synaptic plasticity. Nat. Neurosci. **3**(9), 919–926 (2000)

60. N. Kasabov, K. Dhoble, N. Nuntalid, G. Indiveri, Dynamic evolving spiking neural networks for on-line spatio- and spectro-temporal pattern recognition. Neural Networks **41**(2013), 188–201 (2013)

61. R. Edgar, M. Domrachev, A.E. Lash, Gene expression omnibus: Ncbi gene expression and hybridization array data repository. Nucleic Acids Res. **30**(1), 207–210 (2002)

62. T. Barrett, S.E. Wilhite, P. Ledoux, C. Evangelista, I.F. Kim, M. Tomashevsky, K.A. Marshall, K.H. Phillippy, P.M. Sherman, M. Holko et al., Ncbi geo: archive for functional genomics data sets - update. Nucleic Acids Res. **41**(D1), 991–995 (2012)

63. M.B. Pedersen, L. Skov, T. Menn´e, J.D. Johansen, J. Olsen, Gene expression time course in the human skin during elicitation of allergic contact dermatitis. J. Invest. Dermatol. **127**(11), 2585–2595 (2007)

64. G. Hughes, On the mean accuracy of statistical pattern recognizers. IEEE Trans. Inf. Theory **14**(1), 55–63 (1968)

65. E. Keogh, A. Mueen, in *Curse of Dimensionality*, ed. by C. Sammut, G.I. Webb (Springer, Berlin, 2010), pp. 257–258

66. M.C. Alonso, J.A. Malpica, A.M. de Agirre, Consequences of the hughes phenomenon on some classification techniques, in *ASPRS 2011 Annual Conference, Milwaukee, Wisconsin*, May 2011, pp. 1–5

67. NeuCom, http://www.theneucom.com

68. The MathWorks Inc., Interpolation Methods (R2017b Documentation). https://au.mathworks.com/help/curvefit/interpolation-methods.html

69. Weizmann Institute of Science, GeneCards—Human Gene Database—GCID: GC16M003014, GC02P02693, GC10P079068. http://www.genecards.org/

70. E. Rendo´n-Huerta, F. Teresa, G.M. Teresa, G.-S. Xochitl, A.F. Georgina, Z.-Z. Veronica, L. F. Montan˜o, Distribution and expression pattern of claudins 6, 7, and 9 in diffuse-and intestinal-type gastric adenocarcinomas. J. Gastrointest. Cancer **41**(1), 52–59 (2010)

71. A. Rizzi, E. Nucera, L. Laterza, E. Gaetani, V. Valenza, G.M. Corbo, R. Inchingolo, A. Buonomo, D. Schiavino, A. Gasbarrini, Irritable bowel syndrome and nickel allergy: what is the role of the low nickel diet? J. Neurogastroenterol. Motility **23**(1), 101 (2017)

72. M. Radovic, M. Ghalwash, N. Filipovic, Z. Obradovic, Minimum redundancy maximum relevance feature selection approach for temporal gene expression data. BMC Bioinform. **18** (1), 9 (2017)

# Chapter 16
# Computational Neuro-genetic Modelling

While Chap. 15 gave the basics of molecular biology and some methods for modelling bioinformatics data, computational neurogenetic modelling (CNGM) takes inspiration from neuro-genetics and develops neural network models that include gene information in their structure and functionality, similar to the biological neural networks, that have genes in the nucleus of each neuron, that not only affect but also cause the spiking activity of the neurons. Here we will consider genes as parameters that affect the functioning of the spiking neurons and the SNN as a whole. CNGM is a new science direction with promising applications, some of them discussed in the chapter.

The chapter is organised in the following sections:

16.1. Computational neurogenetics.
16.2. Probabilistic neurogenetic model of a spiking neuron.
16.3. CNGM architectures.
16.4. Applications of CNGM.
16.5. Life, death and CNGM.
16.5. Chapter Summary and further readings for deeper knowledge.

## 16.1 Computational Neurogenetics

### 16.1.1 General Notions

Integrative computational neurogenetic modelling (CNGM) is concerned with the development of computational models that integrate brain structural and functional information with brain-related genetic information. In this chapter we will discuss brain-inspired CNGM based on spiking neural network computational models. These models can be successfully applied for mapping, learning and understanding of spatio-temporal and neurogenetic brain data related to the same problem of

© Springer-Verlag GmbH Germany, part of Springer Nature 2019
N. K. Kasabov, *Time-Space, Spiking Neural Networks and Brain-Inspired Artificial Intelligence*, Springer Series on Bio- and Neurosystems 7,
https://doi.org/10.1007/978-3-662-57715-8_16

interest. Data, such as EEG, fMRI, MEG, gene/protein expression data, and other, can be modelled in a CNGM in their integration and spatio-temporal interaction.

The brain is a complex spatio-temporal neurogenetic information processing machine that processes information at different functional levels (Fig. 16.1). Spatio-temporal activity in the brain depends on the internal brain structure, on the external stimuli and also very much on the dynamics at gene-protein level. Methods for measuring activity of the brain, such as EEG, fMRI, MEG, PET, DTI have been widely used, some of them presented in this encyclopedic volume. CNGM go one step further—integrating these data with relevant genetic data for a better understanding of the brain.

Genes are both the result of the evolution of species and the functioning of an individual brain during a life time. Different genes express as different mRNA, microRNA and proteins in different areas of the brain and are involved in all information processes, from spiking activity, to perception, decision making and emotions. Functional connectivity develops in parallel with structural connectivity during brain maturation where a growth-elimination process (synapses are created and eliminated) depends on gene expression and environment. For example, postsynaptic AMPA-type glutamate receptors (AMPARs) mediate most fast excitatory synaptic transmissions and are crucial for many aspects of brain functioning, including learning, memory and cognition [1]. In [2] performed weighted gene co-expression network analysis to define modules of co-expressed genes and identified 29 such modules, associated with distinct spatio-temporal expression patterns and biological processes in the brain. The genes in each module form a gene regulatory network (GRN).

The spiking activity of a neuron may act as a feedback and affect the expression of genes. As pointed out in [3] on a time scale of minutes and hours the function of neurons may cause changes in the expression of hundreds of genes transcribed into mRNAs and also in microRNAs. This links together the short-term, the long-term and the genetic memories of the neurons representing the global memory of the whole neuronal system.

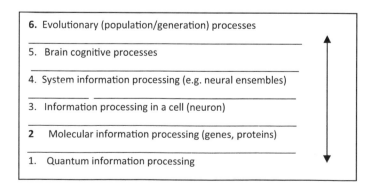

**Fig. 16.1** Different 'levels' of information processing in the brain [11]

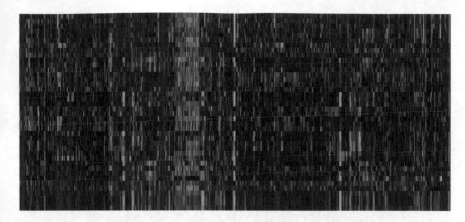

**Fig. 16.2** From the brain explorer: The expression level of several genes (on the vertical axis) in different areas of the brain (horizontal axis): ABAT A_23_P152505, ABAT A_24_P330684, ABAT CUST_52_PI416408490, ALDH5A1 A_24_P115007, ALDH5A1 A_24_P923353, ALDH5A1 A_24_P3761, AR A_23_P113111, AR CUST_16755_PI416261804, AR CUST_85_PI416408490, ARC A_23_P365738, ARC CUST_11672_PI416261804, ARC CUST_86_PI416408490, ARHGEF10 A_23_P216282, ARHGEF10 A_24_P283535, ARHGEF10 CUST_) (from www.brain-map.org) (http://www.alleninstitute.org)

The anatomically comprehensive atlas of the adult human brain *trascriptome* (www.brain-map.org) is a rich repository of bran-gene data that will definitely trigger new directions for research and computational modelling of neurogenetic data [4, 5, 12, 13]. Example of a gene expression brain map is given in Fig. 16.2.

Gene expression is clearly distinguished between structural and functional areas of the brain. Specific genes define specific functions of different sections of the brain. Specific genes relate to specific types of neurons and types of connections. For example, the gene expression level of genes related to dopamine-signalling (e.g. DRD5-DRD1, COMPT, MAOB, DDC, TH, etc.) is higher in areas of a normal subject brain that consist of neurons with larger number of dopamine regulated ion channels. These areas relate to dopamine–driven cognition, emotion and addiction. Such areas are: hippocampus, striatum, hypothalamus, amygdala, and pons. If these areas are activated normally it means that there is a sufficient dopamine signalling. In a diseased brain a non-activated area may suggest lack of dopamine.

In [6], for example, demonstrated that changes in the neurotransmitter receptor densities for important neurotransmitters coincide mostly with the Brodmann cytoarchitectonic borders. These neurotransmitter receptors are: $\alpha 1$, noradrenergic $\alpha 1$ receptor; $\alpha 2A$ noradrenergic $\alpha 2A$ receptor; AMPAR, $\alpha$-amino-3-hydroxy-5-methyl-4-isoxazole propionic acid receptor; GABAB, GABA ($\gamma$-aminobutyric acid)-ergic GABAB receptor; M2, cholinergic muscarinic M2 receptor; M3 cholinergic muscarinic M3 receptor; NMDAR, N-methyl-d-aspartate receptor.

Figure 16.3 shows that the expression of the GABRA2 receptor is different in different parts of the brain.

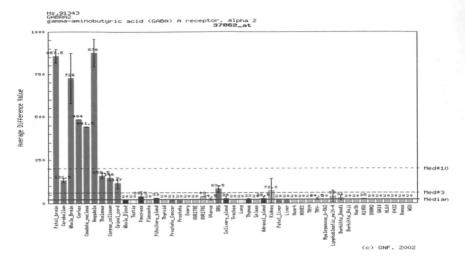

**Fig. 16.3** The expression of the GABRA2 receptor is differently expressed in different parts of the brain (from [13])

The enormousness of brain data available and the complexity of the research questions that need answering through integrated models for brain data analysis are grand challenges for the areas of machine learning and information science in general [7–10]. And this is where CNGM can help.

CNGM, as defined and proposed in [11–13], is based on computational models of spiking neurons, linked as spiking neural networks (SNN). A single spiking neuron model can integrate gene and spiking information related to spiking activity of the neuron, while a SNN can represent a pattern of brain activity. Something more, a probabilistic neurogenetic spiking neuron model can incorporate quantum information processing characteristics as discussed in the next section.

## 16.2   Probabilistic Neurogenetic Model (PNGM) of a Spiking Neuron

### 16.2.1   The PNGM of a Spiking Neuron

Several spiking neuronal models have been proposed so far (e.g. [14–17]) as also presented in Chap. 4. In this section the LIFM has been extended to probabilistic neurogenetic model (PNGM) [4, 13, 18, 19]—Fig. 16.4. As a partial case, when no probability parameters and no genetic parameters are used, the model is reduced to the LIFM.

In the PNGM four types of synapses for *fast excitation, fast inhibition, slow excitation, and slow inhibition* arc used. The contribution of each one to the PSP of

**Fig. 16.4** A diagram of the probabilistic neurogenetic model of a spiking neuron [19]

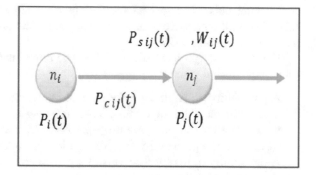

a neuron is defined by the level of expression of different genes/proteins along with the presented external stimuli. The model utilises known information about how proteins and genes affect spiking activities of a neuron. Table 16.1 shows what proteins affect each of the four types of synapses. Neuronal action potential parameters and related proteins and ion channels in the computational neuro-genetic model of a spiking neuron: AMPAR—(amino-methylisoxazole-propionic acid) AMPA receptor; NMDR—(N-methyl-D-aspartate acid) NMDA receptor; GABA$_A$R—(gamma-aminobutyric acid) GABA$_A$ receptor, GABA$_B$R—GABA$_B$ receptor; SCN—sodium voltage-gated channel, KCN—kalium (potassium) voltage-gated channel; CLC—chloride channel (from [13])

This information is used to calculate the contribution of each of the *four* different synapses $j$ connected to a neuron $i$ to its post synaptic potential PSPi(t):

$$\varepsilon_{ij}^{synapse}(s) = A^{synapse}\left(\exp\left(-\frac{s}{\tau_{delay}^{synapse}}\right) - \exp\left(-\frac{s}{\tau_{rise}^{synapse}}\right)\right) \qquad (16.1)$$

where: $\tau_{decay/rise}^{synapse}$ are time constants representing the rise and fall of an individual synaptic PSP; A is the PSP's amplitude; $\varepsilon_{ij}^{synapse}$ represents the type of activity of the synapse between neuron j and neuron i that can be measured and modelled separately for a fast excitation, fast inhibition, slow excitation, and slow inhibition (it is affected by different genes/proteins). External inputs can also be added to model

**Table 16.1** Different genes and proteins affect spiking activity (the PSP) of a neuron

| Different types of action potential of a spiking neuron | Related neurotransmitters and ion channels |
|---|---|
| Fast excitation PSP | AMPAR |
| Slow excitation PSP | NMDAR |
| Fast inhibition PSP | GABA$_A$R |
| Slow inhibition PSP Modulation of PSP | GABA$_B$R mGluR |
| Firing threshold | Ion channels SCN, KCN, CLC |

background noise, background oscillations or environmental information. Genes that relate to the parameters of the neurons are also related to the activity of other genes, thus forming a GRN.

The PNGM is a probabilistic model. In addition to the connection weights $w_{j,i}(t)$, three probabilistic parameters are defined:

- A probability $p_{cj,i}(t)$ that a spike emitted by neuron $n_j$ will reach neuron $n_i$ at a time moment $t$ through the connection between $n_j$ and $n_i$. If $p_{cj,i}(t) = 0$, no connection and no spike propagation exist between neurons $n_j$ and $n_i$. If $p_{cj,i}(t) = 1$ the probability for propagation of spikes is 100%.
- A probability $p_{sj,i}(t)$ for the synapse $s_{j,i}$ to contribute to the PSPi(t) after it has received a spike from neuron $n_j$.
- A probability $p_i(t)$ for the neuron $n_i$ to emit an output spike at time $t$ once the total $PSP_i$ (t) has reached a value above the PSP threshold (a noisy threshold).

The total PSPi(t) of the spiking neuron $n_i$ is now calculated using the following formula:

$$PSP_i(t) = \sum \left( \sum e_j f_1 \Big( p_{cj,i}(t-p) \Big) f_2 \Big( p_{sj,i}(t-p) \Big) w_{j,i}(t) + \eta(t-t_0) \right)$$
$$p = t_0, .., t \quad j = 1, .., m$$

(16.2)

where: $e_j$ is 1, if a spike has been emitted from neuron $n_j$, and 0 otherwise; $f_1(p_{cj,i}(t))$ is 1 with a probability $p_{cji}(t)$, and 0 otherwise; $f_2(p_{sj,i}(t))$ is 1 with a probability $p_{sj,i}(t)$, and 0 otherwise; $t_0$ is the time of the last spike emitted by $n_i$; $\eta(t-t_0)$ is an additional term representing decay in the $PSP_i$. As a special case, when all or some of the probability parameters are fixed to "1", the above probabilistic model will be simplified and will resemble the LIFM.

The probabilistic parameters of the PNGM of a neuron have also their biological analogues and are controlled by specific genes [20]. For example, the probability of a synapse to contribute to the post-synaptic potential after it has received a spike from a pre-synaptic neuron may be affected by different proteins, e.g.: proteins that affect the transmitter release mechanism from the pre-synaptic terminal such as the SNARE proteins (Syntaxin, Synaptobrevin II, SNAP-25), SM proteins (Munc18–1), the sensor Synaptotagmin, and Complexin, and also the proteins such as PSD-95 and Transmembrane AMPA receptor regulatory proteins (TARPs) in the postsynaptic site. The probability for a neuron to emit an output spike at the time when the PSP has reached a value above the threshold may be affected by different proteins, e.g.: density of the sodium channels in the membrane of the triggering zone. The time decay parameter in a LIFM may be affected by different genes and proteins depending on the type of the neuron. Such proteins are: transporters in the pre-synaptic membrane, the glial cells and the enzymes, which uptake and break down the neurotransmitters in the synaptic cleft (BAX, BAD, DP5); metabotropic GABAB Receptors; KCNK family proteins that are responsible for the leak conductance of the resting membrane potential.

## 16.2.2   Using the PNGM of a Neuron to Build SNN

The PNGM of a neuron can be used to build SNN of different types: simple one-layer feed-forward network; recurrent network; reservoir type network; 3D cube that maps the structure of the brain in the NeuCube architecture [4].

Different learning rules for SNN can apply for learning in a SNN. The STDP learning rule (Spike Timing Dependant Plasticity) [21] utilizes Hebbian plasticity [22] in the form of long-term potentiation (LTP) and depression (LTD). Efficacy of synapses is strengthened or weakened based on the timing of post-synaptic action potential in relation to the pre-synaptic spike. If the difference in the spike time between the pre-synaptic and post-synaptic neurons is negative (pre-synaptic neuron spikes first) then the connection weight between the two neurons increases, otherwise it decreases. Connected neurons, trained with STDP learning rule, learn consecutive temporal associations from data. New connections can be generated based on activity of consecutively spiking neurons. Pre-synaptic activity that precedes post-synaptic firing can induce long-term potentiation (LTP), reversing this temporal order causes long-term depression (LTD).

The PNGM and the STDP learning methods can be used to develop new types of eSNN models for spatio-temporal pattern recognition, extending SPAN [23, 24]; deSNN [25]; reservoir eSNN [26, 27]; ReSuMe [28]; Chronotron [29]; Tempotron [30] (see Chaps. 4 and 5). The dynamic eSNN (deSNN) [25] combines rank-order and temporal (e.g. STDP) learning rules as presented in Chap. 5. The initial values of synaptic weights are set according to the rank-order learning assuming the first incoming spikes are more important than the rest. The weights are further modified to accommodate following spikes activated by the same stimulus, with the use of a temporal learning rule–STDP.

When the PNGM of a neuron is used to build an eSNN or a deSNN, rank-order learning rule [31] uses important information from the input spike trains, namely the rank of the first incoming spikes on the neuronal synapses. It establishes a priority of inputs (synapses) based on the order of the spike arrival on these synapses for a particular pattern. This is a phenomenon observed in biological systems. The rank-order learning has several advantages when used in SNN, mainly: fast, one-pass learning (as it uses the extra information of the order of the incoming spikes) and asynchronous data entry (synaptic inputs are accumulated into the neuronal membrane potential in an asynchronous way). The postsynaptic potential of a neuron $i$ at a time $t$ is calculated as:

$$PSP(i,t) = \sum mod^{order(j)} w_{j,i} \qquad (16.3)$$

where: $mod$ is a modulation factor, that has a value between 0 and 1; $j$ is the index for the incoming spike at synapse $j,i$ and $w_{j,i}$ is the corresponding synaptic weight; $order(j)$ represents the order (the rank) of the spike at the synapse $j,i$ among all spikes arriving from all $m$ synapses to the neuron $i$. The $order(j)$ has a value 0 for the first spike and increases according to the input spike order. An output spike is generated by neuron $i$ if the *PSP* $(i,t)$ becomes higher than a threshold $PSP_{Th}(i)$. During the training process, for each training input pattern there is a new output neuron created and its connection weights are calculated based on the order of the incoming spikes:

$$\Delta w_{j,i}(t) = \mathrm{mod}^{\mathrm{order}(j,i(t))} \tag{16.4}$$

## 16.3  Computational Neurogenetic Modelling (CNGM) Architectures

### 16.3.1  CNGM Architectures

Figure 16.5 shows a general diagram of a CNGM system that consists of several modules hierarchically connected [20]:

– low molecular level modelling modules;

  – GRN modules;
  – SNN;
  – high level of SNN activity analysis module.

More about this architecture can be found in [20].

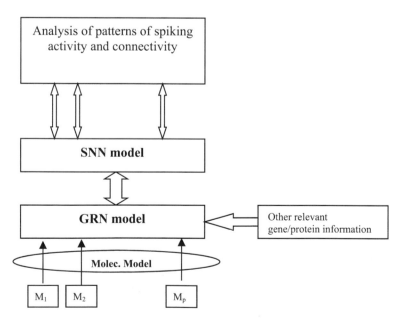

**Fig. 16.5** A general diagram of a CNGM. It consists of several modules hierarchically connected: a low molecular level modelling modules; GRN modules; SNN; high level of SNN activity analysis module [20]

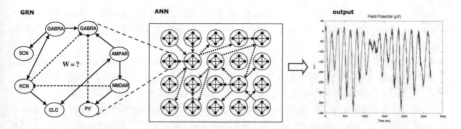

**Fig. 16.6** A CNGM architecture from [13]. It consists of a SNN, where each spiking neuron is governed by a gene regulatory network (GRN) of genes or proteins. More about this architecture can be found in [13]

Another architecture of CNGM is shown in Fig. 16.6. It consists of a SNN, where each spiking neuron is governed by a gene regulatory network (GRN) of genes or proteins. More about this architecture can be found in [13]. It is used as an optional implementation of the SNNcube in the NeuCube architecture (Chap. 6).

### 16.3.2 The NeuCube Architecture as a CNGM

A CNGM architecture, that was also presented in Chap. 6, is NeuCube (Fig. 16.7) [4]. It consists of the following functional modules:

- Input data encoding module;
- 3D SNN reservoir module (SNNr) also denoted as SNNcube or SNNc;
- Output function (classification) module;
- Gene regulatory network (GRN) module;
- Optimisation module.

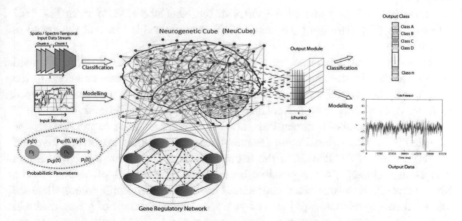

**Fig. 16.7** NeuCube as a CNGM architecture that uses a PNGM of a neuron and GRN of genes controlling the spiking activity of the SNNcube

Below we describe the functioning of the NeuCube as a CNGM architecture, which is complementary to the functional description of NeuCube presented in Chap. 6.

Continuous value input data is transformed into spikes (e.g. pixel, EEG channel, fMRI voxel) using population rank coding [32], Threshold Based Encoding (TBE) [33] or other methods, e.g. Ben's Spike Algorithm (BSA) [34].

The transformed input data is entered (mapped) into spatially located neurons from the SNNr. Brain data sequences are mapped to spatially located neurons in the SNNr that represent spatially brain areas where data is collected. Spike trains are continuously fed into the SNNr in their temporal order. The SNNr is structured to spatially map brain areas for which TSBD or/and gene data is available using some of the brain templates, such as Talairach [35], MNI [36] or other. A neuronal SNNr structure can include known structural or functional connections between different areas of the brain represented in the data. Setting up a proper initial structural connectivity in a model, is important in order to learn properly spatio-temporal data, to capture functional connectivity and to interpret the model [37]. More specific structural connectivity data can be obtained using for example Diffusion Tensor Imaging (DTI) method (Chap. 11).

The input time-space data is propagated through the SNNr and a method of *unsupervised learning* is applied, such as STDP [21]. The neuronal connections are adapted and the SNNr learns to generate specific trajectories of spiking activities when a particular input pattern is entered. The SNNr accumulates temporal information of all input spike trains and transforms it into dynamic states that can be classified over time. The recurrent reservoir generates unique neuronal spike time responses for different classes of input spike trains which effect is called polychronization [15].

After the SNNr is trained in an unsupervised model, the same input data is propagated again through the SNNr and an output classifier is trained to recognize the patterns of activity of the SNNr in a predefined output class for this input pattern.

The expression of genes in the GRN in the NeuCube CNGM from Fig. 16.7 affect the spiking activity of the whole SNN as explained below and illustrated in Fig. 16.8.

Since the NeuCube structure maps brain structural areas through standard stereotaxic coordinates (e.g. MNI, Talairach, etc.) gene data can be added to the NeuCube architecture if such data is available. Gene expression data can be mapped to neurons and areas from a NeuCube as a fifth dimension, in addition to the 3 spatial and one temporal dimensions, so that a vector of gene expression can be allocated for every neuronal group. Some of these genes would be directly involved in the function of the PNGM of the neurons. Neurons can share same expression vectors in a cluster. This is possible because spatial locations of neurons in the SNNr correspond to stereotaxic coordinates of the brain [5]. Furthermore, there are known chemical relationships between genes, or between groups of genes related to same brain function, forming gene regulatory networks (GRN) [38]. Therefore, the fifth dimension in a SNNr can be represented as a GRN.

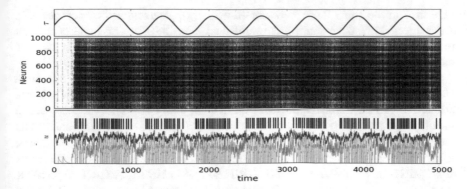

**Fig. 16.8** A single gene expression level over time can affect the pattern of activity of a whole SNNr of 1000 neurons. The gene controls the τ parameter of all 1000 LIF neurons over a period of five seconds and changes in this parameter affect the spiking activity of the neurons

The relationship between genes and spiking neuronal activities can be explored through varying gene expression levels and performing simulations, especially when both gene and brain data are available for some special cognitive or abnormal brain states. Some data related to expression of genes in different areas of the brain under different conditions are available from the Brain Atlas of the Allen Brain Institute.

This is illustrated with a simple example shown in Fig. 16.8. The response of the 1000 spiking neurons to changes of a neuronal parameter due to changes in the expression of a single gene is shown as a raster plot of spike activity. A black point in this diagram indicates a spike of a specific neuron at a specific time in the simulation (the x axis). The bottom diagram presents the evolution of the membrane potential of a single neuron from the network (green curve) along with its firing continuous noisy threshold (red curve). Output spikes of the neuron are indicated as black vertical lines in the same diagram (from [4]).

## 16.4    Applications of CNGM

### 16.4.1    Modelling Brain Diseases

Based on prior information and available data, different CNGM models can be created for the study of various brain states, conditions and diseases ([13] Genes and Diseases) such as: epilepsy; schizophrenia; mental retardation; brain aging; Parkinson disease; clinical depression; stroke; AD [4, 20, 38, 39]. Once learned in a CNGM model, the already known two-way links between spiking activity of the brain and gene transcription and translation, can be potentially used to evaluate gene mutations and the effects of drugs.

One of the most studied brain disease is Alzheimer's Disease (AD). Gene expression data at a molecular level from both healthy and AD patients have been published in the Brain Atlas (www.brain-map.org). Interactions between genes in AD subjects have been studied and published (e.g. [38]). An example of a GRN related to AD is given in Fig. 16.9. Atlases of structural and functional pathways data of both healthy and AD subjects have also been made available [40].

A possible scenario of studying AD through neurogenetic modelling in a NeuCube model will involve the GRIN2B gene. It has been found that subjects affected by AD have a deficit of NMDAR subunit, with GRIN2B level decreased in the hippocampus. A GRN of NMDAR will be constructed as the synthesis of this receptor is possible only due to the simultaneous expression of different genes, which are responsible for the subunits that form the macromolecule. Such genes are: GRIN1-1a, GRIN2A, GRIN2B, GRIN2D and GRIN3A. A GRN of AMPAR genes can also be developed and the two GRNs connected in a NeuCube SNNcube.

A similar approach can be applied for modelling data for Parkinson's disease, multiple sclerosis, stroke and other brain diseases for which both molecular and TSBD is available.

### 16.4.2  CNGM for Cognitive Robotics and Emotional Computing

Building artificial cognitive systems (e.g. robots, AI agents) that are able to communicate with humans in a human-like way has been a goal for computer scientists for decades now. Cognition is closely related with emotions. Basic emotions are happiness, sadness, anger, fear, disgust, surprise, but other human emotions play role in cognition as well (pride, shame, regret, etc.). Some primitive emotional robots or simulation systems have already been developed (e.g. see [41]). The area of affective computing, where some elements of emotions are modelled in a computer system, is growing [42].

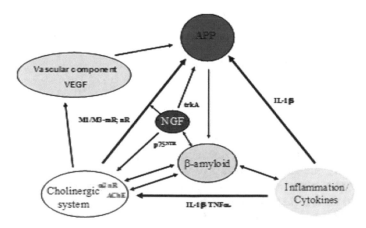

**Fig. 16.9**  Genes related to AD form a GRN that can be modelled as part of a CNGM [38]

A CNGM would make it possible to model cognition-emotion brain states that could further enable the creation of human-like cognitive systems. That would require understanding relevant brain processes as different levels of information processing. For example, it is known that human emotions depend on the expression and the dynamic interaction of neuromodulators (serotonin, dopamine, noradrenalin and acetylcholine) and some other relevant genes and proteins (e.g., 5-HTTLRP, DRD4, DAT), that are functionally linked to the spiking activity of the neurons in certain areas of the brain. They have wide ranging effects on brain functions. For example, Noradrenaline is important to arousal and attention mechanisms. Acetylcholine has a key role in encoding memory function. Dopamine is related to aspects of learning and reward seeking behaviour and may signal probable appetitive outcome, whereas serotonin may affect behaviour with probable aversive outcome. Modifying gene and protein expression levels of genes used in a particular CNGM would affect the learning and pattern recognition properties of that model. For example, the modification could cause connections and functional pathways to become stronger or weaker, which could be observed and further interpreted in terms of cognitive and emotional states.

## 16.5   Life, Death and CNGM

Life and death are two states of an organism. Life expires when certain molecular and brain functions stop. But when exactly they stop, would they be reversible in a time window and can CNGM help discover the mystery of that? Some biological facts about life and death are listed below.

– Apoptosis (from ancient Greek ἀπόπτωσισ "falling off") is a process of programmed cell death that occurs in multicellular organisms. Between 50 and 70 billion cells die each day due to apoptosis in the average human adult. In contrast to necrosis, which is a form of traumatic cell death that results from acute cellular injury, apoptosis is a highly regulated and controlled process that confers advantages during an organism's lifecycle. Because apoptosis cannot stop once it has begun, it is a highly regulated process. Apoptosis can be initiated through one of two pathways in the intrinsic pathway the cell kills itself because it senses cell stress, while in the extrinsic pathway the cell kills itself because of signals from other cells Excessive apoptosis causes atrophy, whereas an insufficient amount results in uncontrolled cell proliferation, such as cancer.
– Telomerase–a life clock? Telomerase, also called terminal transferase is a ribonucleoprotein that adds a species-dependent telomere repeat sequence to the 3' end of telomeres. A telomere is a region of repetitive sequences at each end of eukaryotic chromosomes in most eukaryotes and telomeres protect the end of the chromosome from DNA damage or from fusion with neighbouring chromosomes. The Fruit fly Drosophila Melanogaster lacks telomerase, but instead uses Retrotransposons to maintain telomeres. The existence of a compensatory mechanism for telomere shortening was first found by soviet biologist Alexey

Olovnikov in 1973 who also suggested the telomere hypothesis of aging and the telomere's connections to cancer.
- The physical laws in nature after death and the law of conservation of mass or principle of mass conservation implies that mass can neither be created nor destroyed, although it may be rearranged in space, or the entities associated with mass can be changed in forms discovered by the Russian scientist Michail Lomonosov in 1756.
- Various types of spatio-temporal processes related to death can be distinguished in biology, including:

    - Degradation of DNA and RNA molecules and finding the temporal pattern of degradation;
    - Finding stable regions of DNA and RNA that do not degrade normally;
    - Cells death as a temporal process and a process of interaction between cells over time;
    - Brain death, as a spatio-temporal process of cell deaths in the brain.

Some of the above processes have been modelled in one way or another as computational models. For example:

- Degradation of RNA;
- Brain death processes have been measured as EEG data.

A challenge for the future is, based on all facts about life and death, to create a CNGM that can be used to model the brain transition between life and death, integrating the interaction of genes, proteins and neuronal activities as a whole to better understand this complex transitional process, vital for every individual and the humanity at large. The challenge is also if such models can be created to indicate if and when life and death are reversible states of an organism and under what conditions in time-space.

## 16.6  Chapter Summary and Further Readings for Deeper Knowledge

This chapter extends the material presented in Chap. 15 and also in Chap. 6. This chapter introduces a probabilistic neurogenetic model of a neuron (PNGM) and several CNGM SNN architectures.

The NeuCube architecture from Chap. 6 is now presented as a CNGM, with genes playing a vital role in the functioning of the system. Applications of CNGM are discussed.

More information about CNGM can be found in [12], for example:

- Computational modelling with spiking neural networks (Chap. 37 in [12]);
- Brain-like information processing for spatio-temporal pattern recognition (Chap. 47 in [12]);

- Computational neuro-genetic modelling (Chap. 54 in [12] and also [13]).
- Brain, gene and quantum inspired computational intelligence (Chap. 60 in [12]);
- The brain and creativity (Chap. 61 in [12]);
- The Allen Brain Atlas (Chap. 62 in [12]);
- Alzheimer's Disease (Chap. 51 in [12]);
- Integrating data for modelling biological complexity (Chap. 52 in [12]).

**Acknowledgements**  Some of the material in this chapter is published in prior publications as referenced in the sections. I acknowledge the contribution of my co-authors Lubica Benuskova, Simei Wysoski, Stefan Schliebs, Reinhard Schliebs, Hiroshi Kojima.

# References

1. J.M. Henley, E.A. Barker, O.O. Glebov, Routes, destinations and delays: recent advances in AMPA receptor trafficking. Trends Neurosci. **34**(5), 258–268 (2011)
2. H.J. Kang et al., Spatio-temporal transcriptome of the human brain. Nature **478**, 483–489 (2011)
3. V.P. Zhdanov, Kinetic models of gene expression including non-coding RNAs. Phys. Rep. **500**, 1–42 (2011)
4. N. Kasabov, Neucube: a spiking neural network architecture for mapping, learning and understanding of spatio-temporal brain data. Neural Netw. **52**(2014), 62–76 (2014)
5. M. Hawrylycz et al., An anatomically comprehensive atlas of the adult human brain transcriptome. Nature **489**, 391–399 (2012)
6. K. Zilles, K. Amunts, Centenary of Brodmann's map–conception and fate. Nat. Rev. Neurosci. **11**(2), 139–145 (2010). https://doi.org/10.1038/nrn2776
7. W. Gerstner, H. Sprekeler, G. Deco, Theory and simulation in neuroscience. Science **338** (6103), 60–65 (2012)
8. C. Koch, R.C. Reid, Neuroscience: observatories of the mind. Nature **483**(7390), 397–398 (2012). https://doi.org/10.1038/483397a
9. J.-B. Poline, R.A. Poldrack, Frontiers in brain imaging methods grand challenge. Front. Neuroscie. **6**, 96 (2012). https://doi.org/10.3389/fnins.2012.00096
10. Van Essen et al., The human connectome project: a data acquisition perspective. NeuroImage **62**(4), 2222–2231 (2012)
11. N. Kasabov, *Evolving Connectionist Systems: The Knowledge Engineering Approach*, 1st edn. (Springer, London, 2007), p. 451
12. N. Kasabov (ed.), *Springer Handbook of Bio-/Neuroinformatics* (Springer, Berlin, 2014), p. 229
13. L. Benuskova, N. Kasabov, *Computational Neuro-Genetic Modelling* (Springer, New York, 2007), p. 290
14. W. Gerstner, *What's Different with Spiking Neurons?*, in Plausible Neural Networks for Biological Modelling (Kluwer Academic Publishers, Dordrecht, 2001), pp. 2–2345
15. E.M. Izhikevich, Polychronization: computation with spikes. Neural Comput. **18**(2), 245–282 (2006)
16. A.L. Hodgkin, A.F. Huxley, A quantitative description of membrane current and its application to conduction and excitation in nerve. J. Physiol. **117**, 500–544 (1952)

17. W. Maass, T. Natschlaeger, H. Markram, Real-time computing without stable states: a new framework for neural computation based on perturbations. Neural Comput. **14**(11), 2531–2560 (2002)
18. N. Kasabov, L. Benuskova, S. Wysoski, *A Computational Neurogenetic Model of a Spiking Neuron*, in Proceedings on IJCNN (IEEE Press, 2005), pp. 446–451
19. N. Kasabov, To spike or not to spike: a probabilistic spiking neuron model. Neural Netw. **23** (1), 16–19 (2010)
20. N. Kasabov, R. Schliebs, H. Kojima, Probabilistic computational neurogenetic framework: from modelling cognitive systems to Alzheimer's disease. IEEE Trans. Auton. Mental Deve. **3**(4), 300–3011 (2011)
21. S. Song, K.D. Miller, L.F. Abbott, Competitive hebbian learning through spike-timing-dependent synaptic plasticity. Nature Neurosci. **3**(9), 919–926 (2000)
22. D.O. Hebb, *The Organization of Behavior: A Neuropsychological Approach* (Wiley, New York, 1949), p. 335
23. A. Mohemmed, S. Schliebs, S. Matsuda, N. Kasabov, SPAN: Spike pattern association neuron for learning spatio-temporal sequences. Int. J. Neural Sys. **22**(4), 1–16 (2012)
24. A. Mohemmed, S. Schliebs, S. Matsuda, N. Kasabov, Evolving spike pattern association neurons and neural networks. Neurocomputing **107**, 3–10 (2013)
25. N. Kasabov, K. Dhoble, N. Nuntalid, G. Indiveri, Dynamic evolving spiking neural networks for on-line spatio- and spectro-temporal pattern recognition. Neural Netw. **41**, 188–201 (2013)
26. S. Schliebs, N. Kasabov, M. Defoin-Platel, On the probabilistic optimization of spiking neural networks. Int. J. Neural Syst. **20**(6), 481–500 (2010)
27. S. Schliebs, N. Nuntalid, N. Kasabov, *Towards Spatio-temporal Pattern Recognition Using Evolving Spiking Neural Networks*. ICONIP. Springer LNCS, vol 6443 (2010), pp. 163–170
28. F. Ponulak, A. Kasinski, Supervised learning in spiking neural networks with ReSuMe: sequence learning, classification, and spike shifting. Neural Comput. **22**(2), 467–510 (2010)
29. R.V. Florian, Reinforcement learning through modulation of spike-timing-dependent synaptic plasticity. Neural Comput. **19**, 1468–1502 (2007)
30. R. Gutig, H. Sompolinsky, The Tempotron: a neuron that learns spike timing-based decisions. Nat. Neurosci. **9**(3), 420–428 (2006)
31. S. Thorpe, J. Gautrais, Rank order coding. Comput. Neurosci. Trends Res. **13**, 113–119 (1998)
32. S.M. Bothe, The evidence for neural information processing with precise spike times: a survey. Nat. Comput. **3**(2), 195–206 (2004)
33. T. Delbruck, jAER open source project (2007), http://jaer.wiki.sourceforge.net
34. N. Nuntalid, K. Dhoble, N. Kasabov, *EEG Classification with BSA Spike Encoding Algorithm and Evolving Probabilistic Spiking Neural Network*, in LNCS, vol 7062, (Springer, 2011), pp. 451–460
35. J. Talairach, P. Tournoux, *Co-planar Stereotaxic Atlas of the Human Brain: 3-Dimensional Proportional System—an Approach to Cerebral Imaging* (Thieme Medical Publishers, New York, 1988)
36. A.C. Evans, D.L. Collins, S.R. Mills, E.D. Brown, R.L. Kelly, T.M. Peters, *3D Statistical Neuroanatomical Models From 305 MRI Volumes*, in IEEE-Nuclear Science Symposium and Medical Imaging Conference (IEEE Press, 1993), pp. 1813–1817
37. C.J. Honey, R. Kötter, M. Breakspear, O. Sporns, Network structure of cerebral cortex shapes functional connectivity on multiple time scales. Proc. Natl. Acad. Sci. **104**, 10240–10245 (2007)
38. R. Schliebs, Basal forebrain cholinergic dysfunction in Alzheimer's disease—interrelationship with β-amyloid, inflammation and neurotrophin signalling. Neurochem. Res. **30**, 895–908 (2005)

39. F.C. Morabito, D. Labate, F. La Foresta, G. Morabito, I. Palamara, Multivariate, multi-scale permutation entropy for complexity analysis of AD EEG. Entropy **14**(7), 1186–1202 (2012)
40. A. Toga, P. Thompson, S. Mori, K. Amunts, K. Zilles, Towards multimodal atlases of the human brain. Nat. Rev. Neurosci. **7**, 952–966 (2006)
41. Y. Meng, Y. Jin, J. Yin, M. Conforth, *Human Activity Detection Using Spiking Neural Networks Regulated by a Gene Regulatory Network*, in Proceedings on IJCNN (IEEE Press, 2010), pp. 2232–2237
42. R. Picard, *Affective Computing* (MIT Press, Cambridge, 1997)

# Chapter 17
# A Computational Framework for Personalised Modelling. Applications in Bioinformatics

The chapter presents a computational framework for building personalised models (PM) for accurate prediction of an outcome for the individual. First, a general scheme for building PM using integrated feature and model parameter optimisation is presented. The framework is used to develop two specific methods using: (a) traditional ANN techniques; (b) using evolving spiking neural networks (eSNN). Both methods are illustrated on benchmark biomedical data.

The chapter is organised in the following sections:

17.1. A framework for PM and person profiling using integrated feature and model parameter optimisation.
17.2. PM for gene expression data classification using traditional ANN.
17.3. PM for biomedical data using evolving SNN.
17.4. Chapter summary and further readings for deeper knowledge.

## 17.1 A Framework for PM and Person Profiling Based on Integrated Feature and Model Parameter Optimisation

### 17.1.1 Introduction: Global, Local and Personalise Modelling

Typically, three principal approaches are considered for modelling data [1]:

– *Global modelling* makes use of data in the whole problem space to create a model. This type of modelling might be useful to grasp general trends in the data.
– *Local modelling*, on the other hand, creates a model for subsets or clusters of data and is more customisable to new data.

© Springer-Verlag GmbH Germany, part of Springer Nature 2019  563
N. K. Kasabov, *Time-Space, Spiking Neural Networks and Brain-Inspired Artificial Intelligence*, Springer Series on Bio- and Neurosystems 7,
https://doi.org/10.1007/978-3-662-57715-8_17

– *Personalise modelling*: Contrary to both global and local models where a specific model is created to cover the whole problem space or subset of the space, personalised modelling builds a model for each individual to provide a personalised outcome for that individual. For a new data, an individual model is created using already available information in the system of individuals with known outcomes that have similar characteristics as the new individual. Similarity can be computed using distance measures like Euclidean distance.

Most contemporary medical decision support systems use global models for the prediction of a patient's risk to develop a particular disease or their likely outcome when suffering from the disease. There is a clear evidence that prediction and treatment based on such global models are only effective for some of the patients (about 70% at average) [2, 3] leaving the rest of patients with no effective treatment, and in many cases facing worsening of their condition or even death.

The rationale behind the personalised modeling paradigm is that since each person is different, the most effective treatment could be only achieved if it is based on the analysis of data available for this particular patient. With the advancement of science and technology, it is now possible to obtain and utilise a wide range of personal data such as: DNA, RNA, gene and protein expression, clinical tests, age, gender, BMI, inheritance, food and drug intake, disease, ethnicity, etc. [3–5].

The goal is to create an accurate personalised computational model using information for an individual and the available information for other individuals that is related to the same problem. Achieving a higher accuracy of prediction of a personalised risk for a disease or the effect of treatment may mean saving thousands of lives, significantly reducing the cost for treatment, and improving the quality of life of thousands of patients.

The available methods for personalised modelling do not solve the task completely as they optimise only partially a model for an individual [6–9]. These methods are usually derivatives of the K-nearest neighbour method (K-NN), where for a pre-defined set of variables describing an individual with unknown outcome and a population of individuals with known outcomes, the closest K samples to the new one are selected from the population data forming a neighbourhood. The outcome for the new sample is decided based on the majority outcomes in the neighbourhood. Modifications of the K-NN method include WKNN [10], WWKNN [1, 8, 9, 11, 12].

The above methods are suitable only for the problems defined by a small set of variables. In reality, personalised data usually include thousands of gene, protein, SNPs, clinical, demographic and other variables. However, using the complete set of available variables would be detrimental to the modelling results as most of the variables would be redundant. Pre-selecting a set of variables based on their statistical significance for the whole population space may not be appropriate either, as variables' importance varies depending on the particular sub-space of the problem space [10]. An efficient diagnosis and treatment of a person would require the creation of their personalised profile based on the important variables within the person's sub-space of neighbouring samples. The selection of the neighbourhood of closest samples depends on the selected variables. The overall efficiency of the classification/prediction model would depend on globally optimised variables, neighbourhood data and parameters of the model, in their concert.

Personalised modelling refers to developing a particular model for each individual based on the location of this data sample in the whole problem space. This is done using transductive reasoning, and surpasses the requirement of a global model. Inductive reasoning approach creates a learning model from all the available data by exploiting information in the whole problem space. On the contrary, in the transductive approach, the value of a function for a new input data is estimated by using more information associated with this new data. Personalised modelling is especially useful in the medical domain, where prediction of a treatment outcome requires individualised modelling rather than targeting a population. Moreover, aggregating these individual models might yield a generalised model with high accuracy. K-nearest neighbours is one of the popular methods used for personalised modelling [13]. For every new incoming test data, a predefined number of $k$ nearest samples is extracted from the training dataset using an appropriate distance measure like Euclidean distance. The class label of this new sample is then determined using a voting scheme, where the class label suggested by the majority of extracted data samples, will be assigned to the new data vector. A slightly different model developed by [8] employed neuro-fuzzy inference instead of calculating distance for predicting the label. Both methods were based on neighbouring samples and had similar generalisation skills. Although they might perform well on benchmark datasets, in real world data where the samples might be imbalanced or overlapped with noisy data, these models could become unreliable. Taking these issues into consideration, [14] proposed a transductive support vector machine tree that discovers the discriminative evidence in the neighbourhood of a test instance. A collection of inductive support vector machines were transductively combined for individualised learning. In addition to solving the problem of overfitting of other such classification trees using support vector machines, this method also exhibited superior performance in dealing with imbalanced datasets. However, this method was suitable only for binary classification problems.

## 17.1.2   A Framework for Personalised Modelling (PM) Based on Integrated Feature and Model Parameter Optimisation

We present here a framework for personalised modelling, its implementation and some experimental results for three types of medical decision support problems. For every new individual sample (new input vector) all aspects of their personalised model (variables, neighbouring samples, type of model and model parameters), are optimised together using the accuracy of the outcome achieved for the local neighbourhood of the sample as optimisation criterion. Next, a personalised model

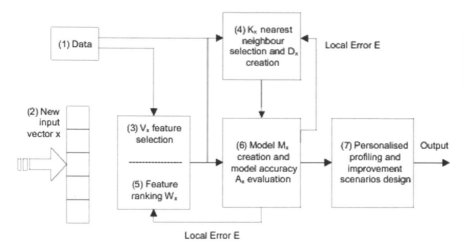

**Fig. 17.1** A functional block diagram of the integrated method for personalised modelling (IMPM) [2, 7, 15]

and personalised profile are derived that use the selected variables and the neighbouring samples with known outcomes. The sample's profile is compared with average profiles of the other outcome classes in the neighbourhood (e.g. good outcome, or bad outcome of disease or treatment). The difference between the points to important variables that may need to be modified through treatment. A functional block diagram of the proposed personalised modelling framework, called integrated method for personalised modelling (IMPM) is illustrated in Fig. 17.1.

The integrated optimisation of the features and the model parameters is achieved through using a Genetic Algorithm (GA) (see Figs. 17.2, 17.3 and Chap. 7 of the book).

The proposed method consists of the following procedures:

P1   Data collection, data filtering, storage and update.
P2   Compiling the input vector for a new patient $x$.
P3   Select a subset of relevant to the new sample $x$ variables (features) $V_x$ from a global variable set $V$.
P4   Select a number $K_x$ of samples from the global data set D and form a neighbourhood $D_x$ of similar samples to $x$ using the variables from $V_x$;
P5   Ranking the $V_x$ variables within the local neighbourhood $D_x$ in order of importance to the outcome, obtaining a weight vector $W_x$.
P6   Training and optimising a local prognostic model $M_x$, that has a set of model parameters $P_x$, a set of variables $V_x$ and local training/testing data set $D_x$.
P7   Generating a functional profile $F_x$ for the person $x$ using the selected set $V_x$ of variables, along with the average profiles of the samples from $D_x$ belonging to different outcome classes, e.g. $F_i$ and $F_j$. Perform a comparative analysis

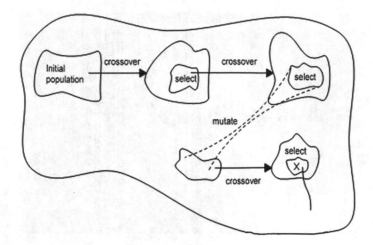

**Fig. 17.2** A schematic diagram of a GA operation

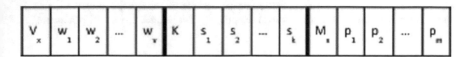

**Fig. 17.3** A chromosome for the GA global optimisation of the following parameters ('genes'): a number of selected variables $V_x$; their corresponding weights $W_x$; a number $K$ of nearest neighbours to $x$; a set of selected $K$ samples $s_1 - s_K$ forming a data subset $D_x$; a local prognostic model $M_x$; a set of parameters $P_m$

    between $F_x$, $F_i$ and $F_j$ to define what variables from $V_x$ are the most important for the person $x$ if a treatment is needed

Procedures P3–P6 are repeated a number of iterations or until a desired local accuracy of the model for a local data set $D_x$ is achieved. The optimisation of the parameters of the personalised model $V_x$, $K_x$ and $D_x$ is global and is achieved through multiple runs of a genetic algorithm (GA) that is a type of evolutionary algorithm, [12, 16, 17]. The resulting competing personalised models for $x$ form a population of such models that are evaluated over iterations (generations) using a fitness criterion—the best accuracy of outcome prognosis for the local neighbourhood of $x$. Operators of crossover and mutation are applied in the search for the best local model (refer to Fig. 17.4). When running the GA, all parameters of the personalised model form a 'chromosome' (refer to Fig. 17.3) where variable values are optimised together as a global optimisation.

Initially, it is assumed that all variables from a set $V$ have equal absolute and relative importance for a new sample $x$ in relation to predicting its unknown output $y$:

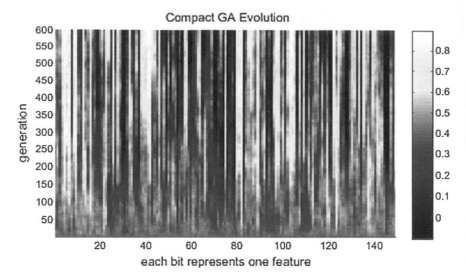

**Fig. 17.4** The weighted importance of the selected features for sample 32 after one run of the method over 600 generations of a GA used to optimise both features and model parameters

$$w_{v1} = w_{v2} =, \ldots, = w_{vq} = 1 \tag{17.1}$$

and

$$w_{v1,norm} = w_{v2,norm} =, \ldots, = w_{vq,norm} = 1/q \tag{17.2}$$

The numbers initially for $V_x$ and $K_x$ may be determined in a variety of different ways without departing from the scope of the method. For example $V_x$ and $K_x$ may be initially determined by an assessment of the global dataset in terms of size and/or distribution of the data. Minimum and maximum values of these parameters may also be established based on the available data and the problem analysis. For example, $V_{x\_min} = 3$ (minimum three variables used in a personalised model) and $V_{x\_max} < K_x$ (the maximum variables used in a personalised model is not larger than the number of samples in the neighbourhood $D_x$ of $x$), usually $V_{x\_max} < 20$. The initial set of variables may include expert knowledge, i.e. variables which are referenced in the literature as highly correlated to the outcome of the problem (disease) in a general sense (over the whole population). Such variables are the BRCA genes, when the problem is predicting outcome of breast cancer [18]. For an individual patient the BRCA genes may interact with some other genes, which interaction will be specific for the person or a group of people and is likely to be discovered through local or/and personalised modelling only [1].

A major advantage of the method, when compared with global or local modelling, is that the modelling process can start with all relevant variables available for a person, rather than with a fixed set of variables in a global model. Such a global model may well be statistically representative for a whole population, but not necessarily representative for a single person in terms of optimal model and best profiling and prognosis for this person.

Selecting the initial number for $K_x$ and the minimum and the maximum numbers $K_{x\_min}$ and $K_{x\_max}$ will also depend on the data available and on the problem in hand. A general requirement is that $K_{x\_min} > V_x$, and, $K_{x\_max} < cN$, where $c$ is a ratio, for example 0.5, and $N$ is the number of samples in the neighbourhood $D_x$ of $x$. Several formulas have been already suggested and experimented [10, 17] e.g.:

- $K_{x\_min}$ equals the number of samples that belong to the class with a smaller number of samples when the data is imbalanced (one class has many more samples, e.g. 90%, than the another class) and the available data set $D$ is of small or medium size (e.g., hundreds to few thousands samples);
- $K_{x\_min} = \sqrt{N}$, where $N$ is the total number of samples in the data set $D$;

At subsequent iterations of the method the parameters $V_x$ and $K_x$ are optimised via an optimisation procedure such as:

- Exhaustive search, where all or some possible values of all or some of the parameters $V_x$, $W_x$, $K_x$, $M_x$ and $P_x$ are used in their combination and the model $M_x$ with the best accuracy is selected;
- An evolutionary algorithm, such as GA [19], optimises all or some parameters that form a 'chromosome'.

The closest $K_x$ neighbouring vectors to $x$ from $D$ are selected to form a new data set $D_x$. A local weighted variable distance measure is used to weigh the importance of each variable $V_l$ ($l = 1, 2, \ldots, q$) to the accuracy of the model outcome calculation for all data samples in the neighbourhood $D_x$. For example, the distance between $x$ and $z$ from $D_x$ is measured as a local weighted variable distance:

$$d_{x,z} = \frac{\sqrt{\sum_{l=1}^{q} (1 - w_{l,norm})(x_l - z_l)^2}}{q} \tag{17.3}$$

where: $w_l$ is the weight assigned to the variable $V_l$ and its normalised value is calculated as:

$$w_{l,norm} = \frac{w_l}{\sum_{i=1}^{q} w_i} \tag{17.4}$$

Here the distance between a cluster centre (in our case it is the vector $x$) and cluster members (data samples from $D_x$) is calculated not only based on the geometrical distance, as it is in the traditional nearest neighbour methods, but on the relative variable importance weight vector $W_x$ for the output values of all samples in the neighbourhood $D_x$. After a subset $D_x$ of $V_x$ variables and $K_x$ data samples are selected, the variables are ranked in a descending order of their importance for prediction of the output $y$ of the input vector $x$ and a weighting vector $W_x$ is obtained. Through an iterative optimisation procedure the number of the variables $V_x$ to be used for an optimised personalised model $M_x$ will be reduced, selecting only the most appropriate variables that will provide the best personalised prediction accuracy of the model $M_x$. For the weighting $W_x$ (i.e. ranking) of the $V_x$ variables, alternative methods can be used, such as t-test, Signal-to-Noise ratio (SNR), etc.

In SNR method $W_x$ are calculated as normalised coefficients and the variables are sorted in descending order: $V_1, V_2, \ldots, V_v$, where: $w_1 > = w_2 > = \cdots > = w_v$, calculated as follows (see Chap. 1):

$$w_l = \frac{\left| M_l^{class1} - M_l^{class2} \right|}{std_l^{class1} + std_l^{class2}} \tag{17.5}$$

where: $M_l^{class\ s}$ and $std_l^{class\ s}$ are respectively the mean value and the standard deviation of variable $x_l$ for all vectors in $D_x$ that belong to class $s$. This method is very fast, but evaluates the importance of the variables in the neighbourhood $D_x$ one by one and does not take into account a possible interaction between the variables, which might affect the model output.

A classification or prediction procedure is applied to the neighbourhood $D_x$ of $K_x$ data samples to derive a personalised model $M_x$ using the already defined variables $V_x$, variable weights $W_x$ and a model parameter set $P_x$.

A number of different classification or prediction procedures can be used such as: KNN; WKNN: WWKNN [1]; TWNFI [9] and others. In the weighted KNN (WKNN) method, the outcome for the new sample is calculated based on the weighted outcomes of the individuals in the neighbourhood according to their distance to the new sample. In the WWKNN method [1] variables are ranked and weighted according to their importance for separating the samples of different classes in the neighbourhood area in addition to the weighting according to the distance as in WKNN. In the TWNFI method—transductive, weighted neuro-fuzzy inference system [9], the number of variables in all personalised models is fixed, but the neighbouring samples used to train the personalised neuro-fuzzy classification model are selected based on the variable weighted distance to the new sample as it is in the WWKNN.

When using the WWKNN method [1] the output value $y$ for the input vector $x$ is calculated using the formula:

$$y = \frac{\sum_{j=1}^{K} a_j y_j}{\sum_{j=1}^{K} w_j} \tag{17.6}$$

where: $y_j$ is the output value for the sample $x_j$ in the neighbourhood $D_x$ of $x$ and:

$$a_j = max(d) - [d_j - min(d)] \tag{17.7}$$

In Eq. 17.7, the vector distance $d = [d_1, d_2, \ldots, d_K]$ is defined as the distances between the new input vector $x$ and the nearest samples $(x_j, y_j)$ for $j = 1$ to $K_x$; $max(d)$ and $min(d)$ are the maximum and minimum values in $d$ respectively. Euclidean distance $d_j$ between vector $x$ and a neighbouring one $x_j$ is calculated as:

$$d_j = \sqrt{\sum_{l=1}^{V} w_l (x_l - x_{jl})^2} \tag{17.8}$$

where: $w_l$ is the coefficient weighing variable $x_l$ in the neighbourhood $D_x$ of $x$.

When using the TWNFI classification or prediction model [9], the output $y$ for the input vector $x$ is calculated as follows:

$$y = \frac{\sum_{l=1}^{m} \frac{n_l}{\delta_l^2} \prod_{j=1}^{P} \alpha_{lj} \cdot \exp\left[ -\frac{w_j^2 (x_{lj} - m_{lj})^2}{2\sigma_{lj}^2} \right]}{\sum_{l=1}^{m} \frac{1}{\delta_l^2} \prod_{j=1}^{P} \alpha_{lj} \cdot \exp\left[ -\frac{w_j^2 (x_{lj} - m_{lj})^2}{2\sigma_{lj}^2} \right]} \tag{17.9}$$

where: $m$ is the number of the closest clusters to the new input vector $x$; each cluster $l$ is defined as a Gaussian function $G_l$ in a $V_x$ dimensional space with a mean value $m_l$ as a vector and a standard deviation $\delta_l$ as a vector too; $x = (x_1, x_2, \ldots, x_v)$; $\alpha_l$ (also a vector across all variables $V$) is membership degree to which the input vector $x$ belongs to the cluster Gaussian function $G_l$; $n_l$ is a parameter of each cluster [9].

A local accuracy (local error $E_x$), that estimates the personalised accuracy of the personalised prognosis (classification) for the data set $D_x$ using model $M_x$ is evaluated. This error is a local one, calculated in the neighbourhood $D_x$, rather than a global accuracy, that is commonly calculated for the whole problem space $D$. A variety of methods for calculating error can be employed such as:

- RMSE (root-mean square error);
- AUC (area under the receiving operating characteristic curve);
- AE (absolute error).

We propose here another formula for calculating local error that can be used for model optimisation:

$$E_x = \frac{\sum_{j=1}^{K_x} (1 - d_{xj}) \cdot E_j}{K_x} \tag{17.10}$$

where: $d_{xj}$ is the weighted Euclidean distance between sample $x$ and sample $S_j$ from $D_x$ that takes into account the variable weights $W_x$; $E_j$ is the error between what the model $M_x$ calculates for the sample $S_j$ from $D_x$ and what its real output value is.

In the above formula the closer a data sample $S_j$ to $x$ is, based on a weighted distance measure, the higher its contribution to the error $E_x$ will be. The calculated personalised model $M_x$ accuracy is:

$$A_x = 1 - E_x \tag{17.11}$$

The best accuracy model obtained is stored for a future improvement and optimisation purposes. The optimisation procedure iteratively returns to all previous procedures to select another set of parameter values for the parameter vector (refer to Fig. 17.4), according to one of the optimisation procedures listed above (exhaustive search, genetic algorithm, a combination between them—Chap. 7) until the model $M_x$ with the best accuracy is achieved. The method also optimises parameters $P_x$ of the classification/prediction procedure. Once the best model $M_x$ is derived, an output value $y$ for the new input vector $x$ is calculated using this model. After the output value $y$ for the new input vector $x$ is calculated a personalised profile $F_x$ of the person represented as input vector $x$ is derived, assessed against possible desired outcomes for the scenario, and possible ways to achieve an improved outcome will be designed, which is also a major novelty of this method. A personal improvement scenario, consisting of suggested changes in the values of the person's features to improve the outcome for $x$ is designed. The $x$ profile $F_x$ is formed as a vector:

$$F_x = \{V_x, W_x, K_x, D_x, M_x, P_x, t\} \tag{17.12}$$

where the variable $t$ represents the time of the model $M_x$ creation. At a future time $(t + \Delta t)$ the person's input data will change to $x^*$ (due to changes in variables such as age, weight, protein expression values, etc.), or the data samples in the data set $D$ may be updated and new data samples added. A new profile $F_x^*$ derived at time $(t + \Delta t)$ may be different from the current one $F_x$.

The average profile $F_i$ for every class $C_i$ in the data $D_x$ is a vector containing the average values of each variable of all samples in $D_x$ from class $C_i$. The importance of each variable (feature) is indicated by its weighting in the weight vector $W_x$. The weighted distance from the person's profile $F_x$ to the average class profile $F_i$ (for each class $i$) is defined as:

$$D(F_x, F_i) = \sum_{l=1}^{v} |V_{lx} - V_{li}| \cdot w_l \tag{17.13}$$

where $w_l$ is the weight of the variable $V_l$ calculated for the data set $D_x$.

Assuming that $F_d$ is the desired profile (e.g. normal outcome) the weighted distance $D(F_x, F_d)$ will be calculated as an aggregated indication of how much a person's profile should change to reach the average desired profile $F_d$:

$$D(F_x, F_d) = \sum_{l=1}^{v} |V_{lx} - V_{ld}| \cdot w_l \tag{17.14}$$

A scenario for a person's improvement through changes made to variable features towards the desired average profile $F_d$ can be produced as a vector of required variable changes, defined as:

$$\Delta F_{x,d} = \Delta V_{lx,d} \big| l = 1, \ldots, v \qquad (17.15)$$

$$\Delta V_{lx,d} = \Delta V_{lx,d}, \text{with an importance of } w_l \qquad (17.16)$$

In order to find a smaller number of variables, as global markers that can be applied to the whole population $X$, procedures P2–P7 are repeated for every individual $x$. All variables from the derived sets $V_x$ are then ranked based on their likelihood to be selected for all samples. The top $m$ variables (most frequently used for individual models) are selected as a set of global set of markers $V_m$. The procedures P1–P7 will be applied again with the use of $V_m$ as initial variable set (instead of using the whole initial set $V$ of variables). In this case personalised models and profiles are obtained within a set of variable markers $V_m$ that would make treatment and drug design more universal across the whole population $X$.

## 17.2    PM for Gene Expression Data Classification Using Traditional ANN

### 17.2.1    Problem and Data Specification, Feature Extraction

The method presented in Sect. 17.1 is illustrated here on personalised modelling using gene expression data. A benchmark colon cancer gene expression dataset is used [20]. It consists of 62 samples, 40 collected from colon cancer patients and 22 from control subjects. Each sample is represented by 2000 gene expression variables. The objective is to create a diagnostic (classification) system that not only provides an accurate diagnosis, but also profiles the person to help define the best treatment. An example of a personalised model of colon cancer diagnosis and profiling of a randomly selected person is given in Fig. 17.5.

To find a small number of variables (potential markers) for the whole population of colon cancer data, we have used the approach as follows: Based on the experiment result for every sample, we selected 20 most frequently used genes as the potential global markers. Table 17.1 lists these 20 global markers with their biological information. The number of 20 for selected global markers is based on the suggestion in Alon's work [20].

### 17.2.2    Classification Accuracy and Comparative Analysis

The next objective of our experiment is to investigate whether utilising these 20 potential marker genes can lead to improved colon cancer classification accuracy. Four classification models are employed in this comparison experiment, including

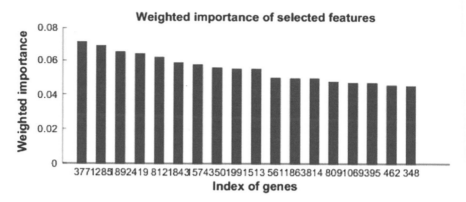

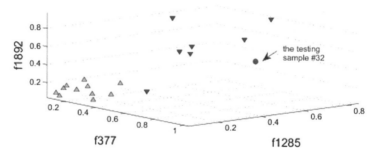

**Fig. 17.5** Sample 32 (a blue dot) is plotted with its neighbouring samples (red triangles—cancer samples and green triangles—control) in the 3D space of the top 3 gene variables from Fig. 17.4

WKNN, MLR, SVM and TWNFI. All the classification results from four classifiers are validated based on LOOCV across the whole dataset.

Table 17.2 summarises the classification results from four classification models using 20 selected potential marker genes. WKNN and a localised SVM yielded improved classification accuracy (90.3%) and TWNFI obtained the best classification performance (91.9%). Our results suggest that a small set of marker genes selected by our IMPM method could lead to improved cancer classification accuracy.

When compared to global or local modelling, the proposed personalised modelling method has a major advantage. In our method, the modelling process starts with all relevant variables available for a person, rather than with a fixed set of variables required by a global model that may well be statistically representative for a whole population, but not necessarily representative for a single person in terms of best prognosis for this person. The proposed method results in a better prognostic accuracy and a computed personalised profile. With global optimisation, a small set of variables (potential markers) can be identified from the selected variable set

**Table 17.1** The 20 most frequently selected genes (potential marker genes) across colon cancer gene data (see Fig. 17.7)

| Index of gene | GenBank accession number | Description of the gene (from GenBank) |
|---|---|---|
| G377 | Z50753 | H.sapiens mRNA for GCAP-II/uroguanylin precursor |
| G1058 | M80815 | H.sapiens a-L-fucosidase gene, exon 7 and 8, and complete cds |
| G1423 | J02854 | Myosin regulatory light chain 2, smooth muscle ISOFORM (HUMAN) |
| G66 | T71025 | Human (HUMAN) |
| G493 | R87176 | Myosin heavy chain, nonuscle (Gallus gallus) |
| G1042 | R36977 | P03001 Transcription factor IIIA |
| G1772 | H08393 | COLLAGEN ALPHA 2(XI) CHAIN (Homo sapiens) |
| G765 | M76378 | Human cysteine-rich protein (CRP) gene, exons 5 and 6 |
| G399 | U30825 | Human splicing factor SRp30c mRNA, complete cds |
| G1325 | T47377 | S-100P PROTEIN (HUMAN) |
| G1870 | H55916 | PEPTIDYL-PROLYL CIS-TRANS ISOMERASE, MITOCHONDRIAL PRECURSOR (HUMAN) |
| G245 | M76378 | Human cysteine-rich protein (CRP) gene, exons 5 and 6 |
| G286 | H64489 | Leukocyte Antigen CD37 (Homo sapiens) |
| G419 | R44418 | Nuclear protein (Epstein-barr virus) |
| G1060 | U09564 | Human serine kinase mRNA, complete cds |
| G187 | T51023 | Heat shock protein HSP 90-BETA (HUMAN) |
| G1924 | H64807 | Placental folate transporter (Homo sapiens) |
| G391 | D31885 | Human mRNA (KIAA0069) for ORF (novel proetin), partial cds |
| G1582 | X63629 | H.sapiens mRNA for p cadherin |
| G548 | T40645 | Human Wiskott-Aldrich syndrome (WAS) mRNA, complete cds |

**Table 17.2** The best classification accuracy obtained by four algorithms on colon cancer data with 20 potential maker gencs

| Classifier | Overall [%] | Class 1 [%] | Class 2 [%] | Neighbourhood size |
|---|---|---|---|---|
| MLR (Personalised) | 82.3 | 90.0 | 68.2 | 3 |
| SVM (Personalised) | 90.3 | 95.0 | 81.8 | 17 |
| WKNN (personalised) | 90.3 | 95.0 | 81.8 | 6 |
| TWNFI (Personalised) | **91.9** | **95.0** | **85.4** | 20 |
| Original publication Alon 08/06/1999 | 87.1 | – | – | – |

Overall—overall accuracy; Class 1—class 1 accuracy; Class 2—class 2 accuracy

across the whole population by the proposed IMPM method. Such markers are helpful for IMPM to produce improved prediction accuracy for cancer diagnosis. A scenario for outcome improvement is also guaranteed. We hope that this paper will motivate the biomedical applications of personalised modelling research.

### 17.2.3 Profiling of Individuals and Personalised Knowledge Extraction

Figure 17.6 shows a profile of sample 32 (blue dots) versus the average local profile of the control and cancer samples using the features from Fig. 17.5.

Figure 17.7 shows a ranking of the 20 most frequently selected genes across colon cancer data, where x axis represents the index of genes in the data, y axis is the selected frequency of a gene.

Figure 17.8 shows a comparison of classification results obtained by 4 algorithms using 20 potential maker genes, where x axis represents the size of neighbourhood and y axis is the classification accuracy.

## 17.3  PM on Biomedical Data Using Evolving SNN

### 17.3.1  Introduction

The purpose of this section is demonstrate how the general PM framework from 17.1 can be used with evolving spiking neural networks (eSNN) for personalised modelling on biomedical data. Real-valued medical data are encoded into spike trains

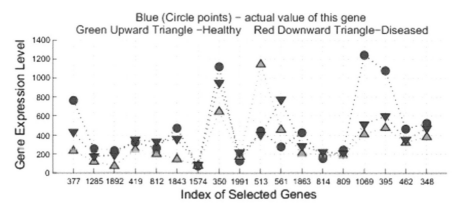

**Fig. 17.6** The profile of sample 32 (blue dots) versus the average local profile of the control (*green triangles*) and cancer samples (*red triangles*) using the features from Fig. 17.5 [2, 7, 15]

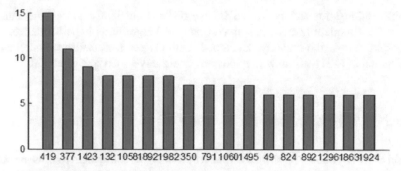

**Fig. 17.7** Profiling a peson's data: the 20 most frequently selected genes across colon cancer data, where x axis represents the index of genes in the data, y axis is the selected frequency of a gene

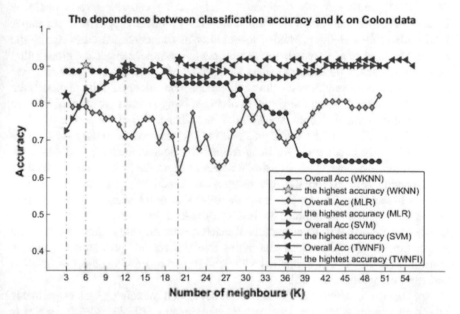

**Fig. 17.8** Comparison of classification results obtained by 4 algorithms using 20 potential maker genes, where x axis represents the size of neighbourhood and y axis is the classification accuracy [2]

using Gaussian receptive fields for binary classification using eSNN. The values of several parameters like threshold, modulation factor and similarity are experimented with to optimise the learning model. The method was tested on both benchmark and clinical datasets. The benchmark dataset chosen for this study is the Chronic Kidney Disease dataset downloaded from the UCI machine learning repository. The results produced by the eSNN model are compared with that of traditional algorithms like multilayer perceptron and support vector machines. The outcome suggests that with suitable optimisation techniques, eSNN can be used efficiently on medical data for personalised modelling and can surpass traditional algorithms in performance.

Despite the of the recently developed methods for personalised modelling of medical data [1–21] still new methods are being developed and explored.

## 17.3.2   Using SNN and ESNN for PM

Would SNN be suitable for PM and what is the benefit of using them for this purpose? This sub-section discusses some PM systems developed with the use of SNN and eSNN in particular, before a detailed case study is discussed in the following sub-section. The methods of Spiking Neural Networks (SNN) were discussed in Chap. 4. Depending on the abstraction level, spiking neural networks could be broadly classified into conductance and threshold models [22]. The conductance model, also called Hodgkin-Huxley model after its founders, is used to signify the physical features of a cell membrane and defines the commencement and propagation of a neuron's action potential or spikes. Leaky integrate- and-fire and spike response models are some examples of the threshold model. The outcomes of these models are governed by a set of threshold value. Evolving spiking neural networks (eSNN) are also founded on the threshold-based model (see Chaps. 4–6 of the book).

In [23] an eSNN model was developed for personalised modelling for the prediction of stroke occurrence. An integrate-and-fire neuron model paired with rank order learning was used for early stroke detection from the input pattern. Based on the incoming information the architecture and functionality of the model was evolved in real time. The system was able to achieve quick learning as the learning occurs in a single iteration. The results produced by the eSNN model were impressive compared to other traditional machine learning algorithms. Reinforcing its position as one of the best algorithms that can model spatiotemporal data, a dynamic evolving SNN was applied with good results on moving object recognition and EEG recognition [24–26].

Applications of eSNN for PM have been developed for solving various problems related to spatio/spectro-temporal pattern classification [9, 24, 26–28]. eSNN is capable of handling complex temporal data like gene expression, EEG, fMRI, financial data as well as audio and visual processing.

Wysoski et al. [29] used eSNN to model auditory and visual pathways in the human brain for the purpose of person authentication (see also Chap. 13). Separate eSNN based systems were used for facial and speech signal recognition. The visual system was modelled using the integrate-and-fire neuron model in which the excitation of a neuron was dependent on the order of the spikes. The model had four integrate-and-fire neuron layers representing the different processing stages of the human visual system with the first two layers acting as filters and time encoders. Learning started in the third layer where the neuronal maps were trained to handle complex input patterns. The fourth layer had one neuronal map corresponding to each pattern class. The visual processing utilised a computationally inexpensive spiking neuron with a feed forward structure. In the auditory processing system, spiking neurons were used for feature extraction and during decision-making. Similar to the visual processing system, speech processing also utilised feed-forward connections and four layers of spiking neuron maps to model the auditory areas in the brain.

Another eSNN implementation of PM for visual information processing was done by Ge et al. [30] for improving obstacle avoidance in prosthetic vision. It was based on NeuCube [31] as a framework for video modelling in prosthetic vision. The features extracted from the input data captured by the visual prosthesis were fed to NeuCube for classification. The result would then lead to generation of an early warning in the presence of an obstacle. It was shown that the functionality of visual prosthesis could be significantly enhanced with already available hardware chips.

Battlori et al. [32] used a personalised brain simulator based on eSNN for controlling robots. The robot was tasked with approaching a light source while avoiding obstacles in its path. The robot's behaviour was replicated using an SNN-based controller and a set of rules were developed to enable the robot to perform the assigned task. The SNN parameters like weights and delays of network synapses were optimised using an evolutionary algorithm. All the other parameters were defined manually. The algorithm was able to generalise well from a limited set of training samples and also enabled parallel computation.

Soltic [33] proposed a method for extracting rules from eSNN that enables to understand how the network arrived at a particular decision and create a profile of a personal data (see also Chap. 5). These rules would help to comprehend the data and the problem better. The input values were encoded into spikes using Gaussian receptive fields. The spike trains were then fed to excitatory layer 1 neurons through delayed synaptic connections. The layer 2 integrate-and-fire neurons were evolved during training. The proposed system was tested on taste information to discover how information was relayed from the taste receptors to the brain. Another sensory system that is closely related to the sense of taste is smell. Odour recognition is still a developing field of research. An important challenge in this area is how to encode the odour information so that it can be fed into a classification model.

The personalised electronic nose proposed by [34] was based on an encoding scheme using Gaussian receptive fields and a pattern classification technique using spiking neural network. Two variations of SNN were used for learning, SpikeProp which uses back-propagation, and dynamic evolving SNN. The proposed approach was tested on black tea odour dataset collected specifically for this study. The sniffing cycle consisted of three phases, essentially making the data temporal in nature. The main purpose of the study was to identify how to better encode odour information so as to improve performance of the learning model. The deSNN model performed comparatively well on the classification task, but further experimentation might be required to optimise the model parameters.

eSNN based model was used in detecting personalised ageing of voice, which is a natural process but one which greatly affects voice professionals [35]. The attributes extracted from voice signals were used as input to the learning model. The relevant attributes and SNN parameters were determined by a quantum-inspired evolutionary algorithm. The encoding layer transformed input attributes into spikes generated over time and fed them into the output layer neurons that were modelled using spike response model functioning as radial basis function neurons. The results of the study indicated that the proposed model was able to produce better accuracy using fewer numbers of input attributes than other models like those based on genetic algorithm.

An application of eSNN in cyber security was proposed in [36], specifically for cyber fraud detection. Phishing websites are a real concern for all internet users. This study demonstrated that evolving spiking neural network could detect phishing websites better than other machine learning techniques. The network made use of Gaussian receptive fields to encode the input data into spikes, and further along added a fresh output neuron or updated current synaptic weights based on the input information and already existing knowledge in the network. The research identified parameter selection and tuning as the major challenges of implementing an eSNN network.

## 17.3.3   An ESNN Method for PM on Biomedical Data

Here we illustrate how the PM framework from 17.1 can be used with eSNN applied on a benchmark biomedical classification problem. The eSNN model used for this study has a 3-layered structure—an input layer, an encoding layer and an output evolving layer as shown in Fig. 17.9. The training and recall algorithms are given in the Appendix to this chapter (see also Chap. 5).

The input layer is where the real-valued input data is fed to the network. As the medical data is static in nature, the input has to be encoded into spikes before feeding it to the evolving spiking neural network. The encoding mechanism used here is the rank order population encoding [37] (Chap. 4) where the input is transformed into trains of spikes by an array of encoding spiking neurons associated with receptive fields. For each input attribute, the receptive fields employ a Gaussian function to cover all the values for this attribute. The number of receptive fields in the encoding layer is user-defined and can be optimised according to the application for better performance. The output layer consists of spiking neurons that evolve during the course of training to represent the input spike trains belonging to the same class (class 1/class 2 in this case). The connection weights are adjusted during the course of learning. The output and input layer neurons are completely connected.

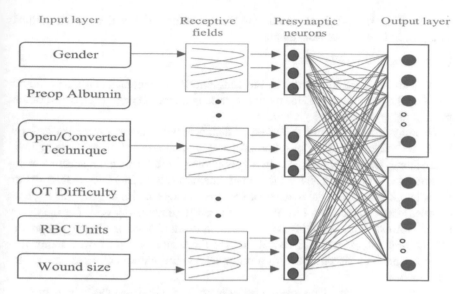

**Fig. 17.9** An exemplar use of eSNN for classification of medical data

When an input value is presented at the presynaptic neurons, the receptive field to which this value relates to the highest generates the first spike. The eSNN architecture is built on the basis of the Thorpes model [38] for its simplicity of implementation. It uses the spiking time to determine the connection weights; earlier spike translates to a stronger connection weight as opposed to a later spike. A neuron spikes only once when its postsynaptic potential reaches a threshold value. Once fired, the postsynaptic potential of the neuron is reset to 0.

Learning in the output neurons occur in a single pass, feed forward fashion. For each input data, an output neuron is generated and its associated weight and threshold value are stored in the neuron repository. This weight is compared to other weights in the neuron repository for similarity. If the similarity is greater than a predefined threshold, then this weight will be merged with the weight of the neuron with the most similarity. Merging in this context refers to updating the weights and threshold of the similar neuron. The newly created neuron is then discarded. The updated weight of the merged neuron is calculated by taking the average of the new weight and the merged neuron weight. Similarly, the updated threshold is calculated by taking the average of the new threshold and the merged neuron threshold. On the contrary, if the new output neuron is not similar to any of the output neurons in the neuron repository, then this new neuron is added to the repository. Once the learning phase is done and the model is completely evolved, the model is tested by passing the test sample spikes to all the trained output neurons. The class label associated with the output neuron that fires first is defined as the class label of the test sample.

*Parameters of the eSNN model*

As with any learning algorithm, the configuration of parameters has a profound effect on the performance of the evolving spiking neural network [12]. Some of the main parameters of the eSNN model are:

*Receptive Fields*—A major portion of data processing at the input layer is done by the receptive field. The collection of neurons in an array of receptive field have overlapping sensitivity profiles. In a population based encoding scheme, it is responsible for encoding real-valued input data into spike trains that are then fed to the network. Increasing the number of receptive fields enables to better distinguish between data samples, but this leads to increased computation cost. On the other hand, a lower number for this parameter would result in faster processing but at the cost of reduced accuracy. It also decreases the width of the fields, making the response more localised. This in turn might lead to the addition of more neurons to the network.

*Modulation factor*—This parameter controls the initial weights which in turn affects the contribution of each spike to the postsynaptic potential. The modulation factor is given a value between 0 and 1. If the value is closer to 1, the contribution of the spike to the postsynaptic potential would be a continuous exponential function, whereas lesser values would lead to the contribution decaying exponentially. Consequently, a value of 0 for this parameter would mean that the postsynaptic potential is not affected by the presynaptic spikes. On the other hand, if modulation factor is 1, then all spikes would equally contribute to the postsynaptic potential. The modulation factor determines how intensely the order of spike timing affects a neuron.

*Similarity*—This parameter defines a threshold value based on which the output layer neurons are created or updated. If the weights of a new output neuron are similar to this threshold, then the new neuron is merged with the most similar neuron in the repository. In this study, the similarity between neuronal weights is measured using Euclidean distance. The values for similarity are specified between 0 and 1. Lower values for this parameter results in fewer neurons being created and higher values would create an output neuron for each input sample leading to overfitting.

*Threshold*—The firing threshold is calculated as a fraction of the maximum PSP. This is derived using a parameter $c$ with a value between 0 and 1. Lower values of the fraction $c$ would result in reducing the firing threshold which in turn advances the response of a neuron.

The firing threshold $\vartheta$ could be defined as $c \times maximum\ postsynaptic\ potential$.

*Parameter Tuning*

To optimise the performance of an algorithm, it is vital to perform parameter tuning. The success of a learning model depends on the right balance between accuracy and structural complexity. Several encouraging studies in the past shows how an optimisation algorithm can be effectively combined with an evolving spiking neural network. Stefan et al. used a versatile quantum-inspired evolutionary algorithm in combination with eSNN using the wrapper approach to detect important attributes and to evolve the eSNN parameters to an optimal configuration [39]. This approach was referred to as the Quantum-inspired Spiking Neural Network (QiSNN) [40]. This study was extended by [41] using a binary representation for attribute subset optimisation and simultaneously evolving spiking network configurations. Following this study, another quantum based optimisation technique integrated with a nature inspired computing method was used by [42] namely Quantum-inspired Particle Swarm Optimisation for string pattern recognition, a challenging task that has applications in online security and virus detection.

Using evolutionary algorithms and quantum inspired methods for the optimisation of the parameters of eSNN was discussed in Chap. 7. In [43] differential evolution was used to identify the optimal number of presynaptic neurons for the eSNN architecture that defines the complexity of the model, for a given dataset. In the case of a multilayer perceptron model, this is similar to finding the optimum number of hidden nodes. The significance of optimising the number of presynaptic neurons was given by the fact that fewer neurons might reduce the number of spikes generated resulting in reduced accuracy while increasing the number of neurons leads to increased computational cost. Simplicity in implementation and fewer control parameters were pointed out as the advantages of using differential evolution. The fitness function optimised was the classification accuracy of the eSNN model. But the overall accuracy cannot be counted as a criterion to measure the success of a learning model in all situations as sometimes it could be misleading especially in the case of imbalanced datasets. Further to this study [44] presented a multi-objective k-means eSNN model intended to improve the performance of eSNN on clustering problems. Multi-objective in this context referred to the number of clusters. An integration of k-means, eSNN and multi-objective differential evolution was implemented to enhance performance and was tested on benchmark datasets. Furthermore, differential evolution was applied for tuning of eSNN parameters like modulation factor, similarity factor and threshold factor.

In [45] this hybrid model was built by integrating eSNN with differential evolution using the wrapper technique. Again, overall classification accuracy was considered to evaluate performance on several benchmark datasets. The results showed that parameter combination differs depending on the dataset. However, an evaluation of how these methods would fare on real-world datasets was not specified in any of these papers.

*Class Imbalance and Overlapping*

While performing data classification, if the class distribution is well-proportioned, most classifiers perform well. Yet it is a rare incidence that a real-world dataset has a balanced number of data samples of all classes. Class imbalance is a major issue in data mining. In a binary class problem, class imbalance could be defined as the condition in which the data instances of one class greatly exceed the number of data instances in the other class. This would lead to the machine learning algorithm getting plagued by the majority class and consequently ignoring the minority class samples during training.

Cohen et al. [46] used prototype-based resampling and support vectors with asymmetrical margins to reduce the effect of class imbalance for nosocomial infection detection problem. Evaluations done using classifiers including but not limited to support vector machine, decision tree and naïve bayes showed improvement in performance using support vectors based method achieving a sensitivity rate of 92%.

A method involving random undersampling and a cost-sensitive learning, where the relative cost associated with the misclassification of majority and minority classes are modified to compensate for the imbalance, was adopted by [47] to rebalance biomedical documents for cataloguing. [48] introduced a data-level rebalancing technique for binary classes that requires the temporary re-labelling of classes, namely TempC. In this two-staged approach, the imbalanced dataset was first split into train and test subsets. The *k* nearest instances of the majority class of all minority class samples were aggregated with the minority class instances to form a new class.

Further methods for class rebalancing include synthetic minority oversampling, SMOTE for short [49], a combination of undersampling of majority class and oversampling of the minority class. The samples of minority class are synthetically generated based on mean, median or mode of a group of *k* nearest neighbours. SPIDER is another data-level class rebalancing approach introduced by [50]. A hybrid of bagging and boosting techniques for rebalancing data was compared by [51]. The methods compared included RUSBoost, SMOTEBoost, RBBag and EEBBag and the results claimed that bagging techniques outdid boosting methods for imbalanced and noisy datasets. However, experiments conducted by [52] showed that bagging and boosting techniques were not better than random sampling approaches. The findings of these researches suggest that not all rebalancing methods are suitable for all datasets.

Reference [53] studied the effects of class imbalance and overlap on the performance of a classifier. eSNN could be used for PM on static, vector based medical data producing results that are better than traditional machine learning algorithms provided that the above issues related to parameter optimisation and balancing of the data are properly dealt with.

## 17.3.4  A Case Study of PM for Chronic Kidney Disease Data Classification

Chronic kidney disease is defined as the gradual degradation of the kidney's functions like waste removal, fluid balance and production of vitamin D among others. Apart from disabling all the functions of the kidney, this disease also leads to vitamin deficiency, anaemia and increases the risk of cardiovascular disease. Studies show that around 6% of the people suffer from stage 3 to 5 chronic kidney disease but this percentage could increase in elderly population. Some of the main causes for this disease are hypertension and diabetes.

The dataset that have used as a case study here to demonstrate the use of eSNN for PM, was collected at the Apollo Hospital, India. The age range of the patients varied from 2 to 90 years. The dataset contained data of 400 individuals out of which 250 samples represented the disease class and 150 constituted the non-disease class. There were 24 attributes besides the class label. The class label was nominal and had one of the two values—ckd (class 1) and nonckd (class 2). The attributes of the dataset can be seen in Table 17.3. This dataset was down-loaded from the UCI machine learning repository [54].

Pre-processing—Even though the dataset is slightly imbalanced, no data rebalancing was done on this dataset. On the other hand missing values were replaced with the median value of each attribute. Figure 17.10 shows a plot of the data in the 2D Principal Components Analysis space (see Chap. 1) and Fig. 17.11 shows the class distribution of this dataset.

*Accuracy*

When a PM approach was applied with the use of eSNN, the system accuracy of classification was approximately 99%. The optimal parameter values of the eSNN were: number of receptive fields 7; Spiking threshold 1; Mod factor 0.8; Sim parameter 0.2.

*Comparison of Performance*

*Multilayer Perceptron*—(Chap. 2) used in [55]. The accuracy of classification when a PM was applied was 98%.

*Support Vector Machines*—(Chap. 1) and first introduced in [10]. The accuracy of classification using SVM when a PM approach was applied was approximately 96%. Among the compared classification models, eSNN is the simplest and the most accurate one for the case study as presented in [56].

**Table 17.3** Features of the chronic kidney disease (CKD) dataset and data types

| Features | Data types |
|---|---|
| Age | Numerical |
| Blood pressure | Numerical |
| Specific gravity | Nominal |
| Albumin | Nominal |
| Sugar | Nominal |
| Red blood cells | Nominal |
| Pus cell | Nominal |
| Pus cell clumps | Nominal |
| Bacteria | Nominal |
| Blood glucose random | Numerical |
| Blood urea | Numerical |
| Serum creatinine | Numerical |
| Sodium | Numerical |
| Potassium | Numerical |
| Haemoglobin | Numerical |
| Packed cell volume | Numerical |
| White blood cell count | Numerical |
| Red blood cell count | Numerical |
| Hypertension | Nominal |
| Diabetes mellitus | Nominal |
| Coronary artery disease | Nominal |
| Appetite | Nominal |
| Pedal edema | Nominal |
| Anaemia | Nominal |

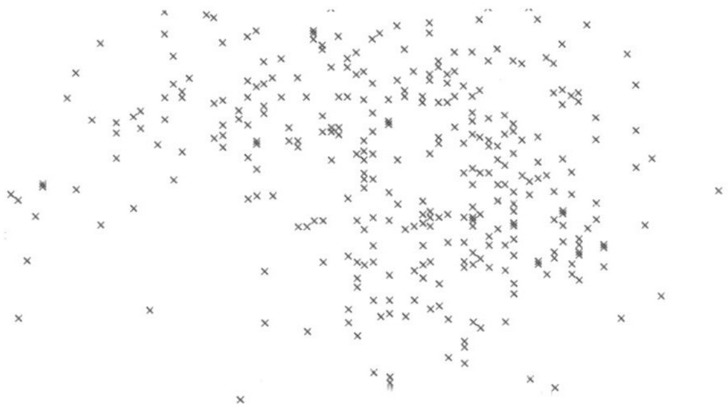

**Fig. 17.10** A PCA plot of kidney data Overlapping samples of CKD—Class 1 (blue) and Class 2 (red) [56]

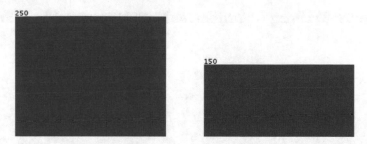

**Fig. 17.11** Class distribution of CKD—blue indicates diseased class and red represents normal class

## 17.4   Chapter Summary and Further Readings for Deeper Knowledge

The chapter introduces a general framework for PM (17.1) and applies it on two applications:

– PM for integrated feature and model parameter optimisation, based on traditional ANN (Chap. 2) illustrated on gene expression data classification;
– PM based on eSNN (Chap. 5) for biomedical data, illustrated on a benchmark clinical data.

The presented methods are compared with traditional machine learning methods, such as MLP, SVM (see Chap. 2) to demonstrate that not only the output accuracy is higher, but the methods allow for a personalised profiling that is important for a better understanding of individual characteristics to help designing a more effective personalised treatment.

Further readings can be found in several chapters from [57], such as:

– Personalised information modelling for personalised medicine (Chap. 33 in [57]);
– Health informatics (Chap. 34 in [57]). In the next chapter another PM framework is introduced that can deal with both static and dynamic data and for this purpose it uses BI-SNN, such as NeuCube (Chap. 6)

**Acknowledgements**  Part of the material in this chapter was previously published as referenced in the corresponding sections. I would like to acknowledge my co-authors of these publications Raphael Hu, for some material in Sects. 17.1 and 17.2, and Vinita, and Mary Ann Ribero for some material in Sect. 17.3.

# Appendix: Training Algorithm for ESNN (See also Chap. 5)

---

The eSNN training algorithm

1: Initialize output neuron repository, $R = \{\ \}$

2: Set eSNN parameters: $mod = [0,1], C = [0,1], sim = [0,1]$

3: for ∀ input pattern $i$ that belongs to the same class do

4:    Encode input pattern into firing time of multiple pre-synaptic neurons $j$

5:    Create a new output neuron $i$ for this class and calculate the connection weights as $w_{ji} = mod^{order(j)}$

6:    Calculate $PSP_{max(i)} = \text{Sum}_j\, w_{ji} \times mod^{order(j)}$

7:    Calculate $PSP$ threshold value $\gamma_i = PSP_{max(i)} \times C$

8:    if The new neuron weight vector $\leq sim$ of trained output neuron weight vector in $R$ then

9:       Update the weight vector and threshold of the most similar neuron in the same output class group

10:      $w = \dfrac{w_{new} + w.N}{N+1}$

11:      $\gamma = \dfrac{\gamma_{new} + \gamma N}{N+1}$   where $N$ is the number of previous merges of the most similar neuron

13:      else

14:         Add the weight vector and threshold of the new neuron to the neuron repository $R$

15:      end if

16: end for

17: Repeat above for all input patterns of other output classes

---

# References

1. N. Kasabov, Global, local and personalized modelling and pattern discovery in bioinformatics: an integrated approach'. Pattern Recogn. Lett. **28**(6), 673–685 (2007)
2. N. Kasabov, Y. Hu, Integrated optimisation method for personalised modelling and case studies for medical decision support. Int. J. Funct. Inform. Pers. Med. **3**(3), 236–256 (2010)
3. A. Shabo, Health record banks: integrating clinical and genomic data into patientcentric longitudinal and cross-institutional health records. Pers. Med. **4**(4), 453–455 (2007)
4. L.A. Hindorff, P. Sethupathy, H.A. Junkins, E.M. Ramos, J.P. Mehta, F.S. Collins, T.A. Manolio, Potential etiologic and functional implications of genome-wide association loci for human diseases and traits. Proc. Natl. Acad. Sci. **106**(23), 9362–9367 (2009)
5. WTCCC, Genome-wide association study of 14,000 cases of seven common diseases and 3,000 shared controls. Nature **447**(7145), 661–678 (2007)
6. J.R. Nevins, E.S. Huang, H. Dressman, J. Pittman, A.T. Huang, M. West, Towards integrated clinico-genomic models for personalized medicine: combining gene expression signatures and clinical factors in breast cancer outcomes prediction. Hum. Mol. Genet. **17**(2), R153–R157 (2003)
7. N. Kasabov, Data analysis and predictive systems and related methodologies personalised trait modelling system, New Zealand Patent No. 572036, PCT/NZ2009/000222, NZ2009/000222-W16-79 (2008)
8. Q. Song, N. Kasabov, Nfi: a neuro-fuzzy inference method for transductive reasoning. IEEE Trans. Fuzzy Syst. **13**(6), 799–808 (2005)
9. Q. Song, N. Kasabov, TWNFI—a transductive neuro-fuzzy inference system with weighted data normalization for personalized modeling. Neural Netw. **19**(10), 1591–1596 (2006)
10. V.N. Vapnik, *Statistical Learning Theory* (Wiley, New York, 1998)

11. N. Kasabov, Soft computing methods for global, local and personalised modeling and applications in bioinformatics, in *Soft Computing Based Modeling in Intelligent Systems*, ed. by V.E. Balas, J. Fodor, A. Varkonyi-Koczy (Springer, Berlin, Heidelberg, 2009), pp. 1–17

12. N. Kasabov, *Evolving Connectionist Systems: The Knowledge Engineering Approach* (Springer, London, 2007)

13. T. Cover, P. Hart, Nearest neighbor pattern classification. IEEE Trans. Inf. Theory **13**(1), 21–27 (1967)

14. S. Pang, T. Ban, Y. Kadobayashi, N. Kasabov, Personalized mode transductive spanning SVM classification tree. Inf. Sci. **181**, 2071–2085 (2011). https://doi.org/10.1016/j.ins.2011.01.008

15. N. Kasabov, Data analysis and predictive systems and related methodologies, US patent 9,002,682 B2, 7 April 2015

16. D. Goldberg, *GeneticAlgorithm in Search* (Optimization and Machine Learning, Kluwer Academic, MA, 1989)

17. N. Mohan, N. Kasabov, in *Transductive Modeling with GA Parameter Optimization*. Proceedings of IEEE International Joint Conference on Neural Networks, 2005, IJCNN '05, Montreal, vol. 2, July (2005), pp. 839—844

18. L.J. Veer, H. Dai, M.J. van de Vijver, Y.D. He, A.A.M. Hart, M. Mao, H.L. Peterse, K. van der Kooy, M.J. Marton, A.T. Witteveen, G.J. Schreiber, R.M. Kerkhoven, C. Roberts, P.S. Linsley, R. Bernards, S.H. Friend, Gene expression profiling predicts clinical outcome of breast cancer'. Nature **415**(6871), 530–536 (2002)

19. S.B. Kotsiantis, I. Zaharakis, P. Pintelas, *Supervised Machine Learning: A Review of Classification Techniques* (2007)

20. U. Alon, N. Barkai, D.A. Notterman, K. Gish, S. Ybarra, D. Mack, A.J. Levine, Broad patterns of gene expression revealed by clustering analysis of tumor and normal colon tissues probed by oligonucleotide arrays. Proc. Natl. Acad. Sci. USA **96**, 6745–6750 (1999)

21. K. Barlow-Stewart, Personalised medicine: more than just personal [Journal Article]. AQ—Australian Quarterly(2), 31 (2017)

22. W. Gerstner, W.M. Kistler, *Spiking Neuron Models: Single Neurons, Populations, Plasticity* [Bibliographies Non-fiction] (Cambridge University Press, Cambridge, U.K.; New York). Retrieved from cat05020a database

23. N. Kasabov, V. Feigin, Z.G. Hou, Y. Chen, L. Liang, R. Krishnamurthi, P. Parmar, Evolving spiking neural networks for personalised modelling, classification and prediction of spatio-temporal patterns with a case study on stroke, Neurocomputing **134**, 269–279 (2014). https://doi.org/10.1016/j.neucom.2013.09.049

24. N. Kasabov, K. Dhoble, N. Nuntalid, G. Indiveri, 2013 Special issue: dynamic evolving spiking neural networks for on-line spatio- and spectro-temporal pattern recognition. Neural Netw. **41**, 188–201 (2013a). https://doi.org/10.1016/j.neunet.2017.11.014

25. Z. Doborjeh, N. Kasabov, M. Doborjeh, A. Sumich, Modelling peri-perceptual brain processes in a deep learning spiking neural network architecture. Nature, Scientific Reports **8**, 8912 (2018). https://doi.org/10.1038/s41598-018-27169-8

26. M. Gholami Doborjeh, G. Wang, N. Kasabov, R. Kydd, B.R. Russell, A spiking neural network methodology and system for learning and comparative analysis of EEG data from healthy versus addiction treated versus addiction not treated subjects. IEEE Trans. BME (2015). https://doi.org/10.1109/tbme.2015.2503400

27. N. Kasabov (2017), Evolving, probabilistic spiking neural networks and neurogenetic systems for spatio- and spectro-temporal data modelling and pattern recognition. Int. Neural Netw. Soc. (INNS). Retrieved from ir00946a database

28. N. Kasabov, E. Capecci, Spiking neural network methodology for modelling, classification and understanding of EEG spatio-temporal data measuring cognitive processes. Inf. Sci. (2014). https://doi.org/10.1016/j.ins.2014.06.028

29. S.G. Wysoski, L. Benuskova, N. Kasabov, Evolving spiking neural networks for audiovisual information processing. Neural Netw. **23**(7), 819–835 (2010)

30. C. Ge, N. Kasabov, Z. Liu, J. Yang, A spiking neural network model for obstacle avoidance in simulated prosthetic vision. Inf. Sci. **399**, 30–42 (2017). https://doi.org/10.1016/j.ins.2017. 03.006
31. N. Kasabov, NeuCube: a spiking neural network architecture for mapping, learning and understanding of spatio-temporal brain data. Neural Netw. **52**, 62–76 (2014)
32. R. Batllori, C.B. Laramee, W. Land, J.D. Schaffer, Evolving spiking neural networks for robot control. Proc. Comput. Sci. **6**, 329–334 (2011). https://doi.org/10.1016/j.procs.2011.08.060
33. S. Soltic, N. Kasabov, Knowledge extraction from evolving spiking neural networks with rank order population coding. Int. J. Neural Syst. **20**(06), 437–445 (2010)
34. S.T. Sarkar, A.P. Bhondekar, M. Macaš, R. Kumar, R. Kaur, A. Sharma, A.A. Kumar, Towards biological plausibility of electronic noses: a spiking neural network based approach for tea odour classification. Neural Netw. **71**, 142–149 (2015). https://doi.org/10.1016/j. neunet.2015.07.014
35. M. Silva, M.M.B.R. Vellasco, E. Cataldo, Evolving spiking neural networks for recognition of aged voices. J. Voice **31**, 24–33 (2017). https://doi.org/10.1016/j.jvoice.2016.02.019
36. A.S. Arya, V. Ravi, V. Tejasviram, N. Sengupta, N. Kasabov, Cyber fraud detection using evolving spiking neural network (2016)
37. S.M. Bohte, J.N. Kok, Applications of spiking neural networks. Inform. Process. Lett. **95**(6), 519–520 (2005)
38. S.J. Thorpe, J. Gautrais, *Rank Order Coding: A New Coding Scheme for Rapid Processing in Neural Networks*, ed. by J. Bower. Computational Neuroscience: Trends in Research (Plenum Press, New York, 1998), pp. 113–118
39. S. Schliebs, M. Defoin-Platel, N. Kasabov, in *Integrated Feature and Parameter Optimization for an Evolving Spiking Neural Network*. Symposium Conducted at the Meeting of the International Conference on Neural Information Processing, ICONIP (Springer, 2008)
40. S. Schliebs, M. Defoin-Platel, S. Worner, N. Kasabov, Integrated feature and parameter optimization for an evolving spiking neural network: exploring heterogeneous probabilistic models. Neural Netw. **22**(5), 623–632 (2009). https://doi.org/10.1016/j.neunet.2009.06.038
41. S. Schliebs, M.D. Platel, S. Worner, N. Kasabov, in *Quantum-Inspired Feature and Parameter Optimisation of Evolving Spiking Neural Networks with a Case Study from Ecological Modeling*. International Joint Conference on Symposium Conducted at the Meeting of the Neural Networks, IJCNN 2009 (IEEE, 2009)
42. H.N.A. Hamed, N. Kasabov, Z. Michlovský, S.M. Shamsuddin, in *String Pattern Recognition Using Evolving Spiking Neural Networks and Quantum Inspired Particle Swarm Optimization*. Symposium Conducted at the Meeting of the International Conference on Neural Information Processing (Springer, 2009)
43. A.Y. Saleh, H. Hameed, M. Najib, M. Salleh, A novel hybrid algorithm of differential evolution with evolving spiking neural network for pre-synaptic neurons optimization. Int. J. Advance Soft Compu. Appl **6**(1), 1–16 (2014)
44. H.N.A. Hamed, A.Y. Saleh, S.M. Shamsuddin, A.O. Ibrahim, in *Multi-objective K-means Evolving Spiking Neural Network Model Based on Differential Evolution*. 2015 International Conference on Symposium Conducted at the Meeting of the Computing, Control, Networking, Electronics and Embedded Systems Engineering (ICCNEEE) (IEEE, 2015)
45. A.Y. Saleh, S.M. Shamsuddin, H.N.A. Hamed, A hybrid differential evolution algorithm for parameter tuning of evolving spiking neural network. Int. J. Comput. Vis. Rob. **7**(1–2), 20–34 (2017)
46. G. Cohen, M. Hilario, H. Sax, S. Hugonnet, A. Geissbuhler, Learning from imbalanced data in surveillance of nosocomial infection. Artif. Intell. Med. **37**(1), 7–18 (2006)
47. R. Laza, R. Pavón, M. Reboiro-Jato, F. Fdez-Riverola, Evaluating the effect of unbalanced data in biomedical document classification. J. Integr. Bioinform. (JIB) **8**(3), 105–117 (2011)
48. M. Bader-El-Den, E. Teitei, M. Adda, in *Hierarchical Classification for Dealing with the Class Imbalance Problem*. 2016 International Joint Conference on Symposium Conducted at the Meeting of the Neural Networks (IJCNN) (IEEE, 2016)

49. N.V. Chawla, K.W. Bowyer, L.O. Hall, W.P. Kegelmeyer, SMOTE: synthetic minority over-sampling technique. J. Artif. Intell. Res. **16**, 321–357 (2002)
50. J. Stefanowski, S. Wilk, Selective pre-processing of imbalanced data for improving classification performance. Lect. Notes Comput. Sci. **5182**, 283–292 (2008)
51. T.M. Khoshgoftaar, J. Van Hulse, A. Napolitano, Comparing boosting and bagging techniques with noisy and imbalanced data. IEEE Trans. Syst. Man. Cybern. A: Syst. Hum. **41**(3), 552–568 (2011)
52. B.W. Yap, K.A. Rani, H.A.A. Rahman, S. Fong, Z. Khairudin, N.N. Abdullah, in *An Application of Oversampling, Undersampling, Bagging and Boosting in Handling Imbalanced Datasets*. Symposium Conducted at the Meeting of the Proceedings of the First International Conference on Advanced Data and Information Engineering (DaEng-2013) (Springer, 2014)
53. V. García, R. Alejo, J. S. Sánchez, J.M. Sotoca, R.A. Mollineda, in *Combined Effects of Class Imbalance and Class Overlap on Instance-Based Classification*. Symposium Conducted at the Meeting of the International Conference on Intelligent Data Engineering and Automated Learning (Springer, 2006)
54. P. Soundarapandian, L.J. Rubini, P. Eswaran, Chronic_kidney_disease data set (2015). Retrieved from https://archive.ics.uci.edu/ml/datasets/chronic_kidney_disease
55. R. Ilin, R. Kozma, P.J. Werbos, Beyond feedforward models trained by backpropagation: a practical training tool for a more efficient universal approximator. IEEE Trans. Neural Netw. **19**(6), 929–937 (2008)
56 M.A. Ribero, Evolving spiking neural networks for personalised modelling on biomedical data: classification, optimisation and rule discovery. Master Thesis, Auckland University of Technology (2017)
57 N. Kasabov (ed.), *Springer Handbook of Bio-/Neuroinformatics* (Springer, 2014)

# Chapter 18
# Personalised Modelling for Integrated Static and Dynamic Data. Applications in Neuroinformatics

The chapter presents methods for building personalised models (PM) for accurate prediction of an outcome for the individual. The general framework for PM from Chap. 17 is here further developed for using brain inspired SNN architectures (BI-SNN). The latter ones facilitate integrated modelling of both static and dynamic (temporal) data related to an individual and groups of individuals. Case studies on predicting stroke and response to treatment are presented in details.

The chapter is organised in the following sections:

18.1. A framework for PM based on BI-SNN architecture for integrated static and dynamic data modelling.
18.2. Personalised deep learning and knowledge representation in time-space. A case on individual stroke risk prediction.
18.3. PM for predicting response to treatment.
18.4. Chapter summary and further readings for deeper knowledge.

## 18.1 A Framework for PM Based on BI-SNN Architecture for Integrated Static and Dynamic Data Modelling

### 18.1.1 Introduction

Most of the predictive modelling techniques in neuroinformatics have been using global modelling. Global modelling applied in most conventional machine learning methods has proven its effectiveness in the past, however it has a limited capability in producing models that fit each person or each case in the problem space since global modelling takes all available data in a problem space and produce a single general function [1]. The produced model is applied to a new individual regardless of their unique personal features. Common global modelling algorithms include Support Vector Machine (SVM) [2] and Multilayer Perceptron (MLP) [3] (Chap. 2).

© Springer-Verlag GmbH Germany, part of Springer Nature 2019    593
N. K. Kasabov, *Time-Space, Spiking Neural Networks and Brain-Inspired Artificial Intelligence*, Springer Series on Bio- and Neurosystems 7,
https://doi.org/10.1007/978-3-662-57715-8_18

Therefore, in the case of stroke or any medical condition, personalized modelling methods are preferred for the reason that they can produce a model for each individual based on their personal features. However, classical personalized modelling methods such as $k$-Nearest Neighbor (kNN) [4] and weighted $k$-Nearest Neighbor (wkNN) [5] are only suitable when classifying vector based and static types of data, not Spatio-Spectro Temporal Data—SSTD. Therefore we have extended the personalized modelling methods based on a Spiking Neural Network (SNN) from Chap. 17, for the analysis and modelling of integrated static and dynamic SSTD.

The concept of using SNN and more specifically BI-SNN for individual predictive modelling stoke has been considered as an emerging computational approach [37, 38]. This is because SNN have the potential to represent and integrate different aspects of information dimension such as time, space and have the ability to deal with large volumes of data using trains of spikes [6]. SNN models such as Spike Response Models (SRM) [7]; Leaky Integrate-and-Fire (LIF) Models [8]; Izhikevich models [9]; Evolving SNN (ESNN) [10], have been successfully utilized in several classification tasks, but they process input data streams as a sequence of static data vectors, ignoring the potential of SNN to simultaneously consider space and time dimensions in the input patterns (Chaps. 4 and 5). It can be viewed that SNN has more potential and is more suitable for SSTD pattern recognition utilizing emerging new methods such as reservoir computing [11]; Probabilistic Spiking Neuron Model [12]; Extended Evolving SNN [13]; Recurrent ESNN (reSNN) [14]; Spike Pattern Association Neuron (SPAN) [15]; Dynamic ESNN (deSNN) [16].

In Chap. 17 a framework for PM for integrated feature and model parameter optimisation using evolutionary algorithms was presented. It was implemented with the use of:

– Traditional ANN, applied for gene expression classification [1, 17–33];
– eSNN illustrated on a benchmark data [6, 34–68].

In this section we introduce a personalised modelling approach which deals with person static information as well as the person dynamic information as suggested in [37, 69]. The model is based on BI-SNN architecture. We demonstrate how BI-SNN can be used to create efficient personalised modelling systems which reveal complex dynamic patterns that help understand individual person performance. The proposed framework has been applied on EEG data case study for comparative analysis of personalised SNN models created for opiate addict patients *versus* those ones under methadone maintenance treatment (the same case study as in Chap. 8 [70], but here a PM approach is developed). The models result in more accurate classification accuracy and better understanding of individual patient's response to methadone maintenance treatment. The results are also compared to global SNN models trained on the same EEG data, reported in [70].

In the method in this section we introduce for the first time the integration of static data (such as clinical data) and dynamic data (such as EEG) using the NeuCube SNN architecture and the approach from [37, 69].

### 18.1.2   A NeuCube-Based Framework for PM of Integrated Static and Dynamic Data

SNN can integrate both spatial and temporal information as locations of synapses and the time of their spiking activity respectively [11, 44, 71, 72]. Several spiking neuronal models have been proposed so far. As one implementation, the popular Leaky-Integrate-and Fire Model (LIFM) is used here (Chap. 4).

The framework presented here uses the BI-SNN architecture NeuCube [44, 73] that refers also to elements of previous studies [74–80] (see Chap. 6)—Fig. 18.1.

NeuCube consists of several functional modules [44, 73, 81]:

- Input data encoding.
- Input variables spatial mapping and unsupervised learning in a 3D brain-like SNNcube.
- Supervised learning in an output classification/regression module [6].
- Parameter optimisation.
- Visualisation and knowledge extraction.

Input data encoding: continuous streams of data are encoded into sequences of spikes using a Threshold-Based Representation method (TBR) as one implementation (Chap. 4). Figure 18.2 shows an example of single temporal variable encoding process. Then the encoded spike trains of the data variables will be entered into the SNNcube module for unsupervised training.

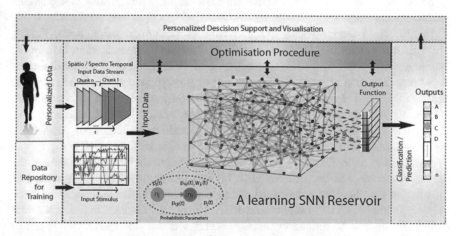

**Fig. 18.1** An example illustrating the PM framework using both static and dynamic (temporal) data in a NeuCube-based system [37, 69]

**Fig. 18.2** An example of a signal (top row) encoded into positive and negative spikes (below row) using a Threshold-Based representation method (TBR). The raw temporal data belongs to one EEG channel variable recorded over time. If the signal increases more than a threshold, positive spikes are generated. If the signal drops more than a threshold, negative spikes are generated

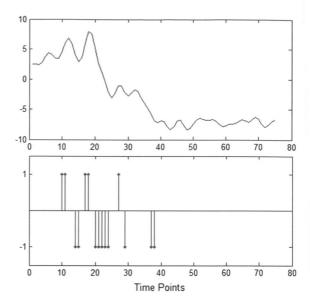

The SNNcube has a 3D brain-like structure with recurrent connections. Initial neuronal connections are generated with the use of the "small world" connectivity rule. The spatial distance of two neurons is calculated to determine their initial connection weight. According to this rule, neurons within small area are more densely connected, and the weight of the connections are depended on the distance between the neurons.

Each neuron in the SNNcube corresponds to a brain area according to a general brain template (such as Talairach [82], MNI, etc.). Each input neuron in the SNNcube has the same $(x, y, z)$ coordinates as the corresponding input data variable (EEG channel) in the used brain template. The input EEG spike trains are propagated through the SNNcube via the allocated input neurons and unsupervised learning is applied.

Unsupervised learning in a 3D SNNcube is performed using Spike-Timing Dependent Plasticity (STDP) learning rule [83] as one implementation. During the learning, efficacy of synapses is strengthened or weakened based on the timing of post-synaptic action potential in relation to the pre-synaptic spike. If pre-synaptic neuron $i$ spikes first and then post-synaptic neuron $j$ spikes, then the connection weight $W_{ij}$ between these two neurons $i$ and $j$ increases, otherwise it decreases.

Then a supervised learning in an output classification/regression module is performed using dynamic evolving Spiking Neural Networks (deSNN) [6].

deSNN is computationally efficient model and emphasizes the importance of the first spike, which has been observed in biological systems. This is performed to train the output classifier neurons using class label information associated with the training samples. The output classifier is trained using the Rank-Order (RO) learning rule to initiate connection weights and a drift parameter to adjust them according to following spikes on the same synapse (Chap. 5).

During the supervised training of an output deSNN, for every training sample (a labelled input spatio-temporal pattern), an individual output neuron *i* is evolved and connected to all spiking neurons in the SNNcube. An output neuron is trained to recognize the spatio-temporal pattern of activity in the already trained SNNcube that is triggered when an individual spatio-temporal input pattern corresponding to this individual is propagated through the SNNcube.

In this research, instead of building a global model and training it with data of the whole patient population, for every patient we will build a personalised SNN model to train it only on data of those patients that have similar static clinical factors. Patients with similar medical factors (drug types, long term or short term usage, methadone dos, etc.) fall into a similar data pattern category due to similar medical effects on their brain functions.

We hypothesize that personalised modelling with SNN could be successfully used, if the models learn from the most informative static and dynamic data, selected based on the similarity in the patients data which is also the foundation of the method proposed in [37, 69].

The proposed NeuCube personalised modelling for static and dynamic data is performed based on the following steps:

1. Select K-nearest neighbour vectors to a new individual vector *x*1 from the global static data and form a cluster of similar samples with close proximity to vector *x*1.
2. Select the dynamic data of the selected K-nearest samples.
3. Using the selected dynamic data, train a PSNN model as SNNcube using unsupervised learning.
4. Train a classifier for the SNN model using deSNN.

During the unsupervised learning process, the neuronal connections are evolved and adapted in the SNNcube. The more spikes transmitted between two connected neurons at time *t*, the stronger the connection is. The SNNcube learns to generate specific trajectories of spiking activities when a particular input pattern is entered. The NeuCube SNN-based personalised modelling framework presented here is part of a development system (see www.kcdri.aut.ac.nz/neucube/) [81].

Figure 18.3 shows a block diagram of the NeuCube SNN-based personalised modelling approach. Vector-based personal static data is available, each vector represents a person's static features, such as opiate use duration, methadone dose, etc. For every new input person $x_i$, K nearest static data vectors to $x_i$ are selected. Then the dynamic data of those K nearest subjects are used to train the personalised SNNcube using STDP learning [84].

PM with the NeuCube BI-SNN allows to analyse the spiking activity of the input variables that can be used for ranking the importance of the input variables for a particular PM. Figure 18.4 shows a diagram of spiking activity of input variables that can be used to rank features (the larger space is occupied by more important features).

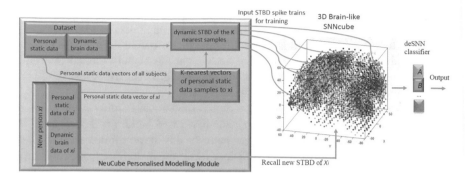

**Fig. 18.3** A block diagram of the NeuCube SNN-based personalised modelling approach. Vector-based personal static data is available, each vector represents a person's static features, such as opiate use duration, methadone dose, etc. For every new input person $x_i$, K nearest static data vectors to $x_i$ are selected. Then the dynamic data of those K nearest subjects are used to train the personalised SNNcube using STDP learning [84]

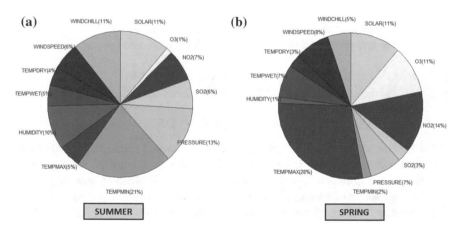

**Fig. 18.4** A diagram of spiking activity of input variables that can be used to rank features (the larger space is occupied by more important features). Different features can have different importance for different PM on different subgroups of subjects, in different seasons etc. as discussed further in this chapter

## 18.1.3  Comparative Analysis of the NeuCube Based Method with Other Methods for PM

The main characteristics of using NeuCube for PM are discussed briefly here.

Overall, the introduced here method for PM has the following advantages when compared with other methods:

- It can accommodate and also integrate both static, vector based, and dynamic, temporal data for an individual;
- It allows for a better understanding of both static and dynamic factors, in their integration and interaction, what concerns their importance for the prediction of the outcome for an individual;
- It allows for a continuous, incremental adaptation of the model based on the incremental update of the data repository of other individuals;
- It takes into account both the static data (stability) and the changes in the temporal data (plasticity) to build a better individual predictive model;
- It enables the acquisition of deep spatio-temporal knowledge representation, as presented in the next section.

## 18.2  Personalised Deep Learning and Knowledge Representation in Time-Space. A Case on Individual Stroke Risk Prediction

### 18.2.1  The Case Study Data for Individual Stroke Risk Prediction

According to World Health Organization (WHO) global report, health related problems like chronic diseases are the major cause of death in almost all countries and it is projected that 41 million people will die of a chronic disease by 2015 [85]. Chronic disease like stroke, has become a leading cause of death and adult disability in the world [86].

Statistical methods have been used by many researchers [86–90] to find association with environmental variables and stroke incidents. These are some of the studies that discovered connections between environmental changes and stroke occurrences. However none of these methods investigate the combination effect of these environmental variables. We believe in order to find more meaningful associations between stroke occurrences and the external environment it is necessary to analyze them collectively.

*Data Specification*

A data from an ARCOS study was chosen which is between 2002 and 2003, consisting of 1207 subjects [37, 91–93]. These subjects were then divided into several subgroups stratified by season (summer, autumn, winter, spring), history of hypertension, smoking status and age to explore the susceptibility of groups to the influence of environmental changes (see Table 18.1 where 4 groups are selected for 4 PM). Each subject is described by 12 environmental variables (wind speed, wind chills, temperature dry, temperature wet, temperature max, temperature min,

**Table 18.1** The used stroke occurrences dataset used to build four PM for the selected group of subjects (from [91–93])

| No. | Season | Age range | History of hypertension | Smoking status | Number of selected subject (control and case) |
|-----|--------|-----------|------------------------|----------------|-----------------------------------------------|
| 1 | Winter | 50–70 | Yes | Current smoker | 20 |
| 2 | Summer | 50–70 | Yes | NA | 46 |
| 3 | Spring | 35–50 | Yes | NA | 26 |
| 4 | Autumn | 25–35 | NA | NA | 16 |

humidity, atmospheric pressure, nitrogen dioxide, sulphur dioxide, ozone gas, solar radiation) measured daily within a time window before the stroke event.

Since the data consists only of stroke subjects, the time window between days 60 and 40 before the event was used as 'low risk' and the days between 20 and 1 days before the event as high risk (see Fig. 18.5).

A first experiment takes the whole time period covering 20 days (prediction of only one day before stroke occurs). A second experiment looks at 75% of the whole pattern which means the prediction will be 6 days ahead. Lastly, a third experiment will take only 50% (11 days earlier) of the whole pattern to predict the stroke event. The normal class will be referred as Class 1 (Low risk) and Class 2 (High risk).

The following parameter values were selected for optimal classification accuracy (see Chap. 6):

(1) The size of the SNNr reservoir is $6 \times 6 \times 6$ making a total of 216 neurons;
(2) Threshold for the spike encoding depends on the input data as the input data is not normalized to minimize error or loss of information;
(3) Small World Connectivity (SWC) used to initialize the connections in the SNN reservoir, with a radius of initial connections of 0.30. The initial connections are generated probabilistically, so that closer neurons are more likely to be connected;
(4) Threshold of the LIFM neurons in the SNN reservoir is 0.5;
(5) The leak parameter of the LIFM neurons is 0.002;
(6) STDP learning rate is 0.01;
(7) Number of training is 2 times;
(8) Mod parameter of the deSNN classifier is 0.04 and the drift is 0.25.

The data as explained above has been used to create personalised models and to cross validate them in a leave-one-out cross validation. The obtained best accuracy of 1 day ahead high risk of stroke prediction is 95% (100% for the TP—stroke prediction, class 2; 90% for the TN—no stroke—class 1). Table 18.2 lists the overall accuracy from all experiments in a comparative study against other machine learning methods [37].

Using BI-SNN, such as NeuCube, for personalised modelling with a higher predictive accuracy achieved, is one feature of the discussed PM methodology. Another important feature is the extraction of deep personalised knowledge as introduced in the next section.

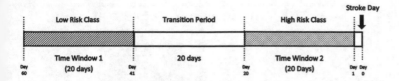

**Fig. 18.5** Time windows to discriminate between 'low risk' and 'high risk' stroke class

**Table 18.2** Comparative experimental results for all modelling methods

| Methods | SVM | MLP | kNN | wkNN | NeuCube$^{ST}$ |
|---|---|---|---|---|---|
| 1 day earlier (%) | 55 (70, 40) | 30 (50, 10) | 40 (50, 30) | 50 (70, 30) | 95 (90, 100) |
| 6 days earlier (%) | 50 (70, 30) | 25 (20, 30) | 40 (60, 20) | 40 (60, 20) | 70 (70, 70) |
| 11 days earlier (%) | 50 (50, 50) | 25 (30, 20) | 45 (60, 30) | 45 (60, 30) | 70 (70, 70) |

## 18.2.2   Personalised Deep Learning and Knowledge Representation in NeuCube on the Case of Stroke

Time-space is learned in a BI-SNN in several ways:

- Time between spikes is learned in the connections
- Time between spikes is learned in the neuronal membranes
- Learning whole, deep, spatio-temporal patterns, theoretically of unlimited time length, as connection pathways between spatially located neurons.

Figure 18.6a illustrates three snapshots of a NeuCube model during training on temporal climate and air pollution data of 9 variables, measured on each of 20 days before a stroke event happened to each subject from a group that is similar to subject X in terms of static data. The subjects in this group have similar clinical and demographic measurements.

A NeuCube SNN model is spatially initiated with 1000 spiking neurons. A subset of 9 input climate variables (shown in yellow colour) are mapped into the 3D SNN structure based on their temporal similarity—the more similar they are, the closer they are allocated in the SNN model structure. Figure 18.6b captures the evolved structural connectivity in the 3D SNN model after unsupervised training. Spatio-temporal patterns of connections are learned in the 3D structural dimensionality of the model. Figure 18.6c shows that a dynamic functional pattern is learned in the functional space of climate variable changes. The pattern indicates a high risk of stroke for the subject X and the group he/she belongs to when certain climate and air pollution variables change significantly in a particular sequence within 20 days. This can be represented as a deep rule in time-space as shown in Box 18.1.

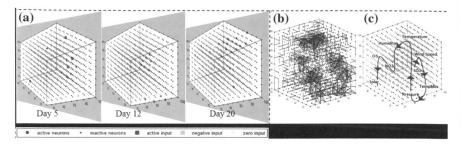

**Fig. 18.6  a** Three snapshots of a NeuCube model during training on temporal climate and air pollution data of 9 variables, measured on each of 20 days before a stroke event happened to each patient from a selected group for a PM; **b** the evolved connectivity in the 3D SNN model after training—spatio-temporal structural patterns of connections are learned in the 3D dimensionality of the model; **c** a dynamic functional pattern is learned in the functional space of climate variable changes

---

Box 18.1 A deep temporal rule defining a high risk of stroke for a subject and a group of subjects extracted from a trained Nue Cube personalised model (from Fig. 18.5)

---

IF        (SO2 changes around time T1)

AND     (Wind Speed changes around time T2)

AND     (TempMin changes around time T3)

AND     (Pressure changes around time T4)

AND     (AvTemp changes around time T5)

AND     (Humidity changes around time T6)

AND     (NO2 changes around time  T7)

AND     (O3 changes around time T8)

AND     (Solar eruption changes around T9)

THEN (High risk of stroke for the individual X from the group she/he belongs to)

---

This personalised SNN model and the rule extracted can be used for early prediction of a high risk of stroke for a person in the same group.

Figure 18.7 shows the spiking activity in a NeuCube model for personalised stroke prediction after it was trained on environmental data related to a sub-group of people who suffered a stroke (20 days before the stroke event happens).

This can be used as a predictive sign, the closer to the day of high risk, the higher the prediction accuracy (from https://kedri.aut.ac.nz/R-and-D-Systems/personalised-modelling-for-stroke-risk-prediction).

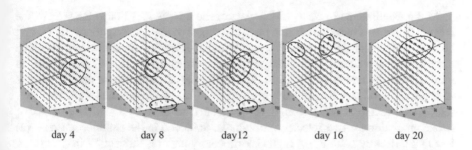

day 4       day 8       day12       day 16       day 20

**Fig. 18.7** Representation in a NeuCube model trained on environmental data related to a sub-group of people who suffered a stroke (20 days before the stroke event happens). https://kedri. aut.ac.nz/R-and-D-Systems/personalised-modelling-for-stroke-risk-prediction (Figs. 18.5 and 18.6 are created by M. Gholami)

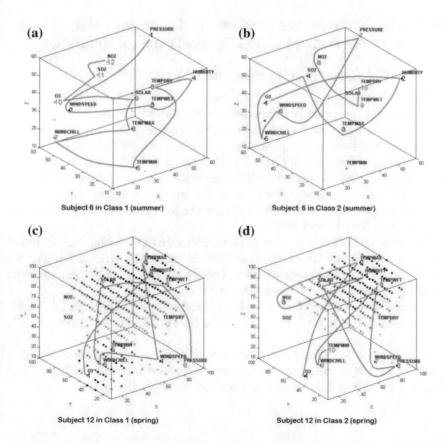

**Fig. 18.8** Revealing dynamic spatio-temporal patterns as trajectories of major consecutive changes in the environmental variables presented as a temporal order from 1 to 12, related to Low Risk and High Risk of stroke prediction for two selected individuals from the data set in [37]: Subject 6 in the summer season—**a** low risk trajectory pattern; **b** high risk trajectory pattern; subject 12 in the spring season—**c** low risk trajectory pattern; **d** high risk trajectory pattern

In the experiments in Sect. 18.2.1 part of the data (e.g. 20 days back from the day of stroke) was used a high risk period and part of the data (e.g. between 60 and 40 days before a stroke day) was used as low risk. A trained NeuCube model can be used to extract patterns related to both high risk and low risk of stroke for an individual as illustrated in Fig. 18.8. Figure 18.8 reveals dynamic temporal patterns as trajectories of major consecutive changes in the environmental variables presented as a temporal order from 1 to 10, related to Low Risk and High Risk of stroke prediction for two selected individuals: Subject 6 in the summer season—(a) low risk trajectory pattern; (b) high risk trajectory pattern; Subject 12 in the spring season—(c) low risk trajectory pattern; (d) high risk trajectory pattern.

These trajectory patterns can be expressed as deep temporal rules as illustrated above.

## 18.3   PM for Predicting Response to Treatment Using Personal Data and EEG Spatio-Temporal Data

### 18.3.1   The Case Study Problem and Data

To illustrate the proposed NeuCube SNN-based personalised modelling methods and systems for static and dynamic data, here we used EEG data collected from two groups of subjects when they performed a cognitive GO- NOGO task. This is the same case study data used in Chap. 8 [39], but here a PM approach is developed with the use of both static and dynamic data related to the same subjects. During a GO/NOGO task, a participant is required to perform an action given certain stimuli (e.g. press a button-GO) and inhibit that action under a different set of stimuli (e.g. not press that same button- NOGO).

The collected EEG data consists of 68 samples each representing an EEG data of one subject, in which 21 samples are labelled as healthy (H), 18 samples are labelled as opiate addict patients (OP), and 29 samples are labelled as patients undertaking methadone maintenance treatment (M).

The EEG data was recorded via 26 EEG channels: Fp1, Fp2, Fz, F3, F4, F7, F8, Cz, C3, C4, CP3, CPz, CP4, FC3, FCz, FC4, T3, T4, T5, T6, Pz, P3, P4, O1, O2, and Oz.

In addition to the EEG DATA, personal clinical, static information was also recorded per subject, such as: gender, age, opiate use duration, methadone use duration, methadone dose, history of overdose, anger level.

## 18.3.2  The NeuCube Based PM Model

A NeuCube SNN-based personalised modelling system is developed as part of the NeuCube integrated platform [81]. The procedure for building a PM for one subject is illustrated in Fig. 18.9. The EEG data and the person static information is loaded; subject ID 1 is selected as an example; a cluster of 17 subjects who have over 88% similarity to subject 1 is detected. The dynamic data of these 17 subjects are transferred into the SNNcube for the creation of a personalised model of subject 1. The dynamic data is encoded into spike trains and mapped into a 3D SNNcube and STDP unsupervised learning is performed. In the output layer, an output neuron is created and connected to all neurons of the SNNcube. Subject 1 dynamic data is entered to test the classifier. The blue lines are positive (excitatory) connections, while the red lines are negative (inhibitory) connections. The brighter the colour of

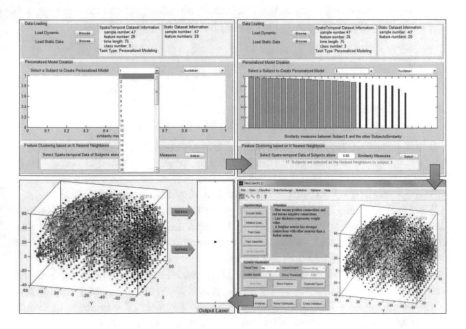

**Fig. 18.9** A NeuCube SNN-based personalised modelling system is developed as part of the NeuCube integrated platform [81, 84]. The EEG data and the person static information is loaded; subject ID 1 is selected as an example; a cluster of 17 subjects who have over 88% similarity to subject 1 is detected. The dynamic data of these 17 subjects are transferred into the SNNcube for the creation of a personalised model of subject 1. The DATA is encoded into spike trains and mapped into a 3D SNNcube and STDP unsupervised learning is performed. In the output layer, an output neuron is created and connected to all neurons of the SNNcube. Subject 1 dynamic data is entered to test the classifier. The blue lines are positive (excitatory) connections, while the red lines are negative (inhibitory) connections. The brighter the colour of a neuron, the stronger its activity with neighbouring neurons. Thickness of the lines also identifies the neuron's enhanced connectivity. The 1471 neurons of the brain-like SNNc are spatially located according to the Talairach brain atlas [82] and 26 input neurons are allocated as inputs for the 26 EEG channels [84]

a neuron, the stronger its activity with neighbouring neurons. Thickness of the lines also identifies the neuron's enhanced connectivity. The 1471 neurons of the brain-like SNNc are spatially located according to the Talairach brain atlas [82] and 26 input neurons are allocated as inputs for the 26 EEG channels.

- The dynamic data of these 17 subjects are transferred into the SNNcube for the creation of a personalised model of subject 1.
- The dynamic data is encoded into spike trains and mapped into a 3D SNNcube and STDP unsupervised learning is performed.
- In the output layer, an output neuron is created and connected to all neurons of the SNNcube. Subject 1 dynamic data is entered to test the classifier. The blue lines are positive (excitatory) connections, while the red lines are negative (inhibitory) connections. The brighter the colour of a neuron, the stronger its activity with neighbouring neurons. Thickness of the lines also identifies the neuron's enhanced connectivity. The 1471 neurons of the brain-like SNNc are spatially located according to the Talairach brain atlas [82] and 26 input neurons are allocated as inputs for the 26 EEG channels [84].

### 18.3.3   Experimental Results

In this experiment, for every personalised model creation, SNNcube is trained by the most informative EEG data corresponding to subjects with similar clinical static information.

Some experimental results are illustrated in Fig. 18.10 and Table 18.3.

The trained PSNN can be used for a better comparatively analysis across individual subject's performance. Figure 18.10 represents NeuCube SNN-based personalised modelling user interface, which is developed as part of the NeuCube, for creating personalised SNN (PSNN) model.

Using the proposed framework for PSNN modelling, we created 47 separate personalised SNN models (for 18 OP subjects and 29 M subjects), each trained on a subset of informative EEG data corresponding to a cluster of samples with similar static data. Then the overall obtained accuracy of all 47 SNNcube models is compared with the global SNNcube which was trained on the entire dataset and tested on individual data.

Table 18.3 shows that PSNN models results in a better overall classification accuracy when compared with global SNN models using the same NeuCube architecture. Precise interpretation of the person performance can be also obtained as illustrated on 6 subjects in Fig. 18.10. Figure 18.10 shows six PSNN models, each of them created for a single person $x_i$. In this figure, the left column represents the similarity between the clinical static data vector of person $x_i$ and other subject's static data vectors. The similarity is shown as a bar graph. The green highlighted bar lines represent those subjects that have over 88% similarity with person $x_i$.

The EEG data of those subjects are encoded into spike trains and then transferred into a personalised 3D brain-like SNN cube for STDP unsupervised learning. During the learning, the connections between the neurons of the SNNcube are strengthened or weakened based on the timing of post-synaptic action potential in relation to the pre-synaptic spikes. If pre-synaptic neuron $i$ spikes first and then post-synaptic neuron $j$ spikes, then the connection weight $W_{ij}$ between these two neurons $i$ and $j$ increases, otherwise it decreases. Figure 18.10 shows the differences between 6 randomly subjects in relation to their static, clinical data and brain activities achieved through building personalised models in NeuCube. The 1471 neurons of the brain-like SNNc are spatially located according to the Talairach brain atlas [82] and 26 input neurons are allocated as inputs for the 26 EEG channels. For every individual, one SNNcube is trained by a subset of EEG data corresponding to the subjects with similar static data [84]. As shown, for M subject id. 1, stronger neuronal connections are evolved around the input EEG channels located in the right hemisphere of the PSNN model corresponding to the right hemisphere. If we compare it with M subjects id. 2 and id. 3, which reveals a different response to methadone treatment of these 3 subjects, the differences of the model connectivity are observed.

## 18.3.4   Discussions

These findings can reveal significant information about the individual person brain functions against a cognitive task and can be further used to suggest a better treatment based on the personalised methadone dose-related effects in case of the experiment presented. This can be used to control individual differences and pre-existing conditions, and help to predict treatment response.

In contrast to the global modelling, personalised modelling creates a specific model for each new person based on existing samples closest to this person's data from a dataset.

In this study, we propose a framework for personalised modelling based on the NeuCube SNN architecture [44].

A NeuCube personalised model includes several methods and algorithms that allow different aspects of EEG data to be studied and analysed: clustering the subjects' DATA based on the K-nearest subject vectors; spatial mapping of the dynamic data into a 3D personalised SNN structure; unsupervised learning in the SNNcube; visualisation of the connectivity and the spiking activity of the trained SNNcube for the discovery of new information related to the data and the brain processes that generated it; supervised learning in a SNN classifier; and model validation.

Overall, personalised SNN models trained on a subset of informative EEG data resulted in a better classification accuracy when compared with global SNN models. In addition, they can be used to reveal individual characteristics on brain activities that can be used to find the best patient- oriented treatment.

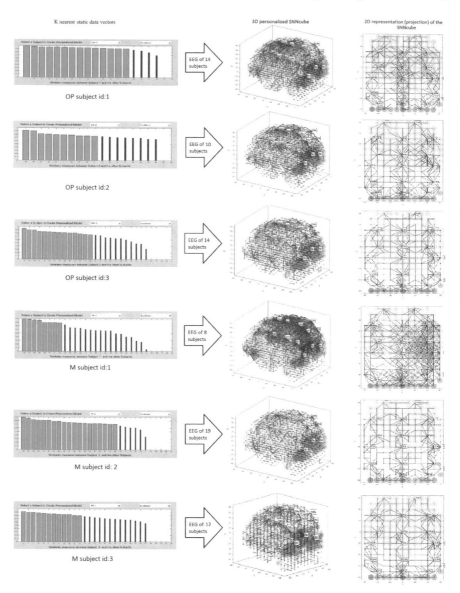

**Fig. 18.10**  The differences between 6 randomly selected subjects in relation to their static, clinical data and brain activities achieved through building personalised models in NeuCube. The 1471 neurons of the brain-like SNNc are spatially located according to the Talairach brain atlas [82] and 26 input neurons are allocated as inputs for the 26 EEG channels. For every individual, one SNNcube is trained by a subset of EEG data corresponding to the subjects with similar static data [84]

**Table 18.3** Classification accuracy obtained via NeuCube personalised modelling versus using a global classification NeuCube model (reported in our previous study [70])

| Methods | NeuCube-Personalised modelling | NeuCube- Global modelling |
|---|---|---|
| Classification accuracy of class M versus class OP in % | Averaged over 47 trained PSNNcubes: **93.61** | One trained SNNcube using all subjects and tested via leave-one-out method: **79.00** |

Each SNNcube is trained by dynamic data of subjects with similar static data and then tested by a new person's dynamic data

## 18.4    Chapter Summary and Further Readings for Deeper Knowledge

This chapter further develops the PM methods presented in Chap. 17, here with the use of the BI-SNN. The method presented here integrates both static (vector based data) and dynamic, temporal data. The method is illustrated on two real-world problems in Neuroinformatics:

- Individual prediction of response to treatment;
- Individual stroke occurrence prediction.

Further readings can be found in several chapters of [94], such as:

- Personalised modelling for personalised medicine (Chap. 33 in [94]);
- Information methods for predicting risk and outcome of stroke (Chap. 55 in [94]);
- Stroke prediction in NeuCube: https://kedri.aut.ac.nz/R-and-D-Systems/neucube/stroke;
- Using NeuCube for static and dynamic data on stroke prediction: https://kedri.aut.ac.nz/R-and-D-Systems/personalised-modelling-for-stroke-risk-prediction;
- A multi-cube PM system integrating fMRI and DTI personal data [49] and also Chap. 11.
- General methods for brain data modelling and SNN for a potential use in personalised modelling applications [95–102].

**Acknowledgements** Some of the material included in this chapter has already been published as referenced in the sections. I would like to thank my co-authors of these publications Maryam Gholami, Valery Feigin, Rita Krishnamurti, Muhaini Othman, Alexander Merkin, Zohreh Gholami, Enmei Tu, Linda Liang, Zeng-Guang Hou, Grace Wang, Rob Kydd, Bruce Russel, Jie Yang.

# References

1. N. Kasabov, Soft computing methods for global, local and personalised modeling and applications in bioinformatics, in *Soft Computing Based Modeling in Intelligent Systems*, ed. by V.E. Balas, J. Fodor, A. Varkonyi-Koczy (Springer, Berlin, 2009), pp. 1–17
2. V. Vapnik, A. Lerner, Pattern recognition using generalized portrait method". Autom. Remote Control **24**(1963), 774–780 (1963)
3. K. Hornik, M. Stinchcombe, H. White, Multilayer feedforward networks are universal approximators. Neural Networks **2**(5), 359–366 (1989)
4. E. Fix, J.L. Hodges, *Discriminatory Analysis: Nonparametric Discrimination: Consistency Properties* (Randolph Field, Texas, 1951), p. 1951
5. S.A. Dudani, The distance-weighted k-nearest-neighbor rule. IEEE Trans. Syst. Man Cybern. **1976**, 325–327 (1976)
6. N. Kasabov, K. Dhoble, N. Nuntalid, G. Indiveri, Special Issue: Dynamic evolving spiking neural networks for on-line spatio- and spectro-temporal pattern recognition. Neural Networks **41**, 188–201 (2013a). https://doi.org/10.1016/j.neunet.2017.11.014
7. W. Gerstner, Time structure of the activity of neural network models. Phys. Rev. **51**(1995), 738–758 (1995)
8. W. Gerstner, W.M. Kistler, *Spiking Neuron Models: Single Neurons, Populations, Plasticity* (Cambridge University Press, Cambridge, MA, 2002), p. 2002
9. E.M. Izhikevich, Which model to use for cortical spiking neurons? IEEE Trans. Neural Networks **15**(5), 1063–1070 (2004)
10. S.G. Wysoski, L. Benuskova, N. Kasabovx, *On-line Learning with Structural Adaptation in a Network of Spiking Neurons for Visual Pattern Recognition*, in Proceedings of International Conference on Artificial Neural Networks, Athens (2006), pp. 61–70
11. W. Maass, N. Thomas, M. Henry, Real-time computing without stable states: a new framework for neural computation based on perturbations. Neural Comput. **14**(11), 2531–2560 (2002)
12. N. Kasabov, To spike or not to spike: a probabilistic spiking neuron model. Neural Networks **2010**, 16–19 (2010)
13. H.N.A. Hamed, N. Kasabov, S.M. Shamsuddin, H. Widiputra, K. Dhoble, *An Extended Evolving Spiking Neural Network Model for Spatio-Temporal Pattern Classification*, in The 2011 International Joint Conference on Neural Networks (IJCNN). IEEE (2011), pp. 2653–2656
14. S. Schliebs, N. Kasabov, M. Defoin-Platel, On the probabilistic optimization of spiking neural networks. Int. J. Neural Syst. **20**(6), 481–500 (2010)
15. A. Mohemmed, S. Schliebs, N. Kasabov, *SPAN: A Neuron for Precise-Time Spike Pattern Association, Neural Information Processing*, in Proceedings of 18th International Conference on Neural Information Processing, Shanghai, China (ICONIP 2011). LNCS, vol. 7063 (Springer, Berlin, 2011), pp. 718–725
16. K. Dhoble, N. Nuntalid, G. Indivery, N. Kasabovx, *On-line Spatiotemporal Pattern Recognition with Evolving Spiking Neural Networks utilising Address Event Representation, Rank Order- and Temporal Spike Learning*, in IEEE World Congress on Computational Intelligence, Brisbane, Australia (WCCI 2017), 10–15 June 2017, pp. 554–560
17. N. Kasabov, Y. Hu, Integrated optimisation method for personalised modelling and case studies for medical decision support. Int. J. Funct. Inform. Personal. Med. **3**(3), 236–256 (2010)
18. A. Shabo, Health record banks: integrating clinical and genomic data into patientcentric longitudinal and cross-institutional health records. Personal. Med. **4**(4), 453–455 (2007)
19. L.A. Hindorff, P. Sethupathy, H.A. Junkins, E.M. Ramos, J.P. Mehta, F.S. Collins, T.A. Manolio, Potential etiologic and functional implications of genome-wide association loci for human diseases and traits. Proc. Natl. Acad. Sci. **106**(23), 9362–9367 (2009)

20. WTCCC, Genome-wide association study of 14,000 cases of seven common diseases and 3,000 shared controls. Nature **447**(7145), 661–678 (2007)
21. J.R. Nevins, E.S. Huang, H. Dressman, J. Pittman, A.T. Huang, M. West, Towards integrated clinico-genomic models for personalized medicine: combining gene expression signatures and clinical factors in breast cancer outcomes prediction. Hum. Mol. Genet. **17**(2), R153–R157 (2003)
22. N. Kasabov, in Data analysis and predictive systems and related methodologies— Personalised trait modelling system, New Zealand Patent No. 572036, PCT/NZ2009/ 000222, NZ2009/000222-W16-79
23. Q. Song, N. Kasabov, Nfi: a neuro-fuzzy inference method for transductive reasoning. IEEE Trans. Fuzzy Syst. **13**(6), 799–808 (2005)
24. Q. Song, N. Kasabov, Twnfi—a transductive neuro-fuzzy inference system with weighted data normalization for personalized modeling. Neural Networks **19**(10), 1591–1596 (2006)
25. V.N. Vapnik, *Statistical Learning Theory* (Wiley, New York, 1998)
26. N. Kasabov, *Evolving Connectionist Systems: The Knowledge Engineering Approach* (Springer, London, 2007)
27. D. Goldberg, *GeneticAlgorithm in Search, Optimization and Machine Learning* (Kluwer Academic, North Holand, 1989)
28. N. Mohan, N. Kasabov, Transductive modeling with ga parameter optimization, in *2005 IEEE International Joint Conference on Neural Networks (IJCNN '05), Montreal*, vol. 2 (2005), pp. 839–844
29. L.J. Veer, H. Dai, M.J. van de Vijver, Y.D. He, A.A.M. Hart, M. Mao, H.L. Peterse, K. van der Kooy, M.J. Marton, A.T. Witteveen, G.J. Schreiber, R.M. Kerkhoven, C. Roberts, P.S. Linsley, R. Bernards, S.H. Friend, Gene expression profiling predicts clinical outcome of breast cancer'. Nature **415**(6871), 530–536 (2002)
30. N. Kasabov, Global, local and personalized modelling and pattern discovery in bioinformatics: an integrated approach'. Pattern Recogn. Lett. **28**(6), 673–685 (2007)
31. U. Alon, N. Barkai, D., A. Notterman, K. Gish, S. Ybarra, D. Mack, A.J. Levine, *Broad Patterns of Gene Expression Revealed by Clustering Analysis of Tumor and Normal Colon Tissues Probed by Oligonucleotide Arrays*, in Proceedings of the National Academy of Sciences of the United States of America, vol. 96 (1999), pp. 6745–6750
32. S. Schliebs, N. Kasabov, Evolving spiking neural network—a survey. Evolving Syst. **4**(2), 87–98 (2013). https://doi.org/10.1007/s12530-013-9074-9.
33. K. Barlow-Stewart, Personalised medicine: more than just personal. AQ—Aust. Q. **2**, 31
34. W. Gerstner, W.M. Kistler, *Spiking Neuron Models: Single Neurons, Populations, Plasticity [Bibliographies Non-fiction]*. (Cambridge University Press, Cambridge, 2002). Retrieved from cat05020a database
35. T. Cover, P. Hart, Nearest neighbor pattern classification. IEEE Trans. Inf. Theory **13**(1), 21–27 (1967)
36. S. Pang, T. Ban, Y. Kadobayashi, N. Kasabov, Personalized mode transductive spanning SVM classification tree. Inf. Sci. **181**, 2071–2085 (2011). https://doi.org/10.1016/j.ins.2011. 01.008
37. N. Kasabov, V. Feigin, Z.G. Hou, Y. Chen, L. Liang, R. Krishnamurthi, P. Parmar, Evolving spiking neural networks for personalised modelling, classification and prediction of spatio-temporal patterns with a case study on stroke. Neurocomputing **134**, 269–279 (2014). https://doi.org/10.1016/j.neucom.2013.09.049
38. Z. Doborjeh, N. Kasabov, M. Doborjeh, A. Sumich, Modelling peri-perceptual brain processes in a deep learning spiking neural network architecture. Nature, Scientific Reports **8**, 8912 (2018). https://doi.org/10.1038/s41598-018-27169-8
39. M. Gholami Doborjeh, G. Wang, N. Kasabov, R. Kydd, B.R. Russell, A spiking neural network methodology and system for learning and comparative analysis of EEG data from healthy versus addiction treated versus addiction not treated subjects. IEEE Tr. BME (2015). https://doi.org/10.1109/tbme.2015.2503400

40. N. Kasabov, Evolving, probabilistic spiking neural networks and neurogenetic systems for spatio- and spectro-temporal data modelling and pattern recognition. Int. Neural Network Soc. (INNS) (2017). Retrieved from ir00946a database

41. N. Kasabov, E. Capecci, Spiking neural network methodology for modelling, classification and understanding of EEG spatio-temporal data measuring cognitive processes. Inf. Sci. (2014). https://doi.org/10.1016/j.ins.2014.06.028

42. S.G. Wysoski, L. Benuskova, N. Kasabov, Evolving spiking neural networks for audiovisual information processing. Neural Networks **23**(7), 819–835 (2010)

43. C. Ge, N. Kasabov, Z. Liu, J. Yang, A spiking neural network model for obstacle avoidance in simulated prosthetic vision. Inf. Sci. **399**, 30–42 (2017). https://doi.org/10.1016/j.ins.2017.03.006

44. N. Kasabov, NeuCube: A spiking neural network architecture for mapping, learning and understanding of spatio-temporal brain data. Neural Networks **52**, 62–76 (2014)

45. R. Batllori, C.B. Laramee, W. Land, J.D. Schaffer, Evolving spiking neural networks for robot control [Article]. Procedia Comput. Sci. **6**, 329–334 (2011). https://doi.org/10.1016/j.procs.2011.08.060

46. S. Soltic, N. Kasabov, Knowledge extraction from evolving spiking neural networks with rank order population coding. Int. J. Neural Syst. **20**(06), 437–445 (2010)

47. S.T. Sarkar, A.P. Bhondekar, M. Macaš, R. Kumar, R. Kaur, A. Sharma, A. Gulati, A. Kumar, Towards biological plausibility of electronic noses: A spiking neural network based approach for tea odour classification. Neural Networks **71**, 142–149 (2015). https://doi.org/10.1016/j.neunet.2015.07.014

48. M. Silva, M.M.B.R. Vellasco, E. Cataldo, Evolving spiking neural networks for recognition of aged voices. J. Voice **31**, 24–33 (2017). https://doi.org/10.1016/j.jvoice.2016.02.019

49. N. Sengupta, C. McNabb, N. Kasabov, B. Russell, Integrating space, time and orientation in spiking neural networks: a case study on multi-modal brain data modelling. IEEE Trans. Neural Networks Learn. Syst (2018). https://doi.org/10.1109/TNNLS.2018.2796023

50. S. Schliebs, M. Defoin-Platel, N. Kasabov, *Integrated Feature and Parameter Optimization for an Evolving Spiking Neural Network*, in Symposium Conducted at the Meeting of the International Conference on Neural Information Processing (2008)

51. S. Schliebs, M. Defoin-Platel, S. Worner, N. Kasabov, Integrated feature and parameter optimization for an evolving spiking neural network: exploring heterogeneous probabilistic models. Neural Networks **22**(5), 623–632 (2009). https://doi.org/10.1016/j.neunet.2009.06.038

52. S. Schliebs, M.D. Platel, S. Worner, N. Kasabov, *Quantum-Inspired Feature and Parameter Optimisation of Evolving Spiking Neural Networks with a Case Study from Ecological Modeling*, in International Joint Conference on IEEE. Symposium Conducted at the Meeting of the Neural Networks (IJCNN 2009) (2009)

53. H.N.A. Hamed, N. Kasabov, Z. Michlovský, S.M. Shamsuddin, *String Pattern Recognition Using Evolving Spiking Neural Networks and Quantum Inspired Particle Swarm Optimization*, in Springer Symposium Conducted at the Meeting of the International Conference on Neural Information Processing (2009)

54. A.Y. Saleh, H. Hameed, M. Najib, M. Salleh, A novel hybrid algorithm of differential evolution with evolving spiking neural network for pre-synaptic neurons optimization. Int. J. Advance Soft Comput. Appl. **6**(1), 1–16 (2014)

55. H.N.A. Hamed, A.Y. Saleh, S.M. Shamsuddin, A.O. Ibrahim, *Multi-objective K-means Evolving Spiking Neural Network Model Based on Differential Evolution*, in 2015 International Conference on IEEE. Symposium Conducted at the Meeting of the Computing, Control, Networking, Electronics and Embedded Systems Engineering (ICCNEEE) (2015)

56. A.Y. Saleh, S.M. Shamsuddin, H.N.A. Hamed, A hybrid differential evolution algorithm for parameter tuning of evolving spiking neural network. Int. J. Comput. Vision Robot. **7**(1–2), 20–34 (2017)

57. G. Cohen, M. Hilario, H. Sax, S. Hugonnet, A. Geissbuhler, Learning from imbalanced data in surveillance of nosocomial infection. Artif. Intell. Med. **37**(1), 7–18 (2006)
58. R. Laza, R. Pavón, M. Reboiro-Jato, F. Fdez-Riverola, Evaluating the effect of unbalanced data in biomedical document classification. J. Integr. Bioinf. (JIB) **8**(3), 105–117 (2011)
59. M. Bader-El-Den, E. Teitei, M. Adda, *Hierarchical Classification for Dealing with the Class Imbalance Problem*, in 2016 International Joint Conference on IEEE. Symposium Conducted at the Meeting of the Neural Networks (IJCNN) (2016)
60. N.V. Chawla, K.W. Bowyer, L.O. Hall, W.P. Kegelmeyer, SMOTE: synthetic minority over-sampling technique. J. Artif. Intell. Res. **16**, 321–357 (2002)
61. J. Stefanowski, S. Wilk, Selective pre-processing of imbalanced data for improving classification performance. Lect. Notes Comput. Sci. **5182**, 283–292 (2008)
62. T.M. Khoshgoftaar, J. Van Hulse, A. Napolitano, Comparing boosting and bagging techniques with noisy and imbalanced data. IEEE Trans. Syst. Man Cybern. Syst. Hum. **41**(3), 552–568 (2011)
63. B.W. Yap, K.A. Rani, H.A.A. Rahman, S. Fong, Z. Khairudin, N.N. Abdullah, *An Application of Oversampling, Undersampling, Bagging and Boosting in Handling Imbalanced Datasets*, in Springer Symposium Conducted at the Meeting of the Proceedings of the First International Conference on Advanced Data and Information Engineering (DaEng-2013) (2014)
64. V. García, R. Alejo, J.S. Sánchez, J.M. Sotoca, R.A. Mollineda, *Combined Effects of Class Imbalance and Class Overlap on Instance-Based Classification*, in Springer Symposium Conducted at the Meeting of the International Conference on Intelligent Data Engineering and Automated Learning (2006)
65. S.M. Bohte, J.N. Kok, Applications of spiking neural networks. Inf. Process. Lett. **95**(6), 519 520 (2005)
66. S.J. Thorpe, J. Gautrais, Rank order coding: a new coding scheme for rapid processing in neural networks, in *Computational Neuroscience: Trends in Research*, ed. by J. Bower (Plenum Press, New York, 1998), pp. 113–118
67. P. Soundarapandian, L.J. Rubini, P. Eswaran, Chronic_Kidney_Disease Data Set (2015). Retrieved from https://archive.ics.uci.edu/ml/datasets/chronic_kidney_disease
68. R. Ilin, R. Kozma, P.J. Werbos, Beyond feedforward models trained by backpropagation: a practical training tool for a more efficient universal approximator. IEEE Trans. Neural Networks **19**(6), 929–937 (2008)
69. N. Kasabov, Z. hou, V. Feigin, Y. Chen, Improved method and system for predicting outcomes based on spatio/spectro-temporal data. PCT patent, WO 2015030606 A2 (2015)
70. M. Gholami Doborjeh, G. Wang, N. Kasabov, R. Kydd, B.R. Russell, A spiking neural network methodology and system for learning and comparative analysis of EEG data from healthy versus addiction treated versus addiction not treated subjects. IEEE Tr. BME (2015). https://doi.org/10.1109/tbme.2015.2503400
71. S.M. Bohte, J.N. Kok, Applications of spiking neural networks. Inf. Process. Lett. **95**(6), 519–520 (2005)
72. R. Brette, M. Rudolph, T. Carnevale, M. Hines, D. Beeman, J. Bower, Simulation of networks of spiking neurons: a review of tools and strategies. J. Comput. Neurosci. **23**(3), 349–398 (2007)
73. E. Tu, N. Kasabov, J. Yang, Mapping temporal variables into the NeuCube for improved pattern recognition, predictive modelling and understanding of stream data. IEEE Trans. Neural Networks Learn. Syst. (2016). https://doi.org/10.1109/tnnls.2016.2536742
74. EU Human Braib Project (HBP), [Online]. Available: www.thehumanbrainproject.eu
75. USA Brain Initiative, [Online]. Available: http://www.nih.gov/science/brain/
76. S. Furber, D.R. Lester, L. Plana, J.D. Garside, Overview of the spinnaker system architecture. IEEE Trans. Comput. **62**(17), 2454–2467 (2013)
77. S. Furber, To build a brain. IEEE Spectrum **49**(44–49), 44–49 (2017)
78. G. Indiveri, T.K. Horiuchi, Frontiers in neuromorphic engineering. Front. Neurosci. **5**, 2011 (2011)

79. G. Indiveri, B. Linares-Barranco, T.J. Hamilton, A. Van Schaik, Neuromorphic silicon neuron circuits. Front. Neurosci. **5**, 2011 (2011)
80. G. Indiveri, E. Chicca, R.J. Douglas, Artificial cognitive systems: from VLSI networks of spiking neurons to neuromorphic cognition. Cogn. Comput. **1**(2), 119–177 (2009)
81. N. Kasabov, et al., Design methodology and selected applications of evolving spatio-temporal data machines in the NeuCube neuromorphic framework. Neural Networks (2016) (accepted and on-line published 2015, 2016)
82. L. Koessler, L. Maillard, A. Benhadid, J.P. Vignal, J. Felblinger, H. Vespignani, M. Braun, Automated cortical projection of EEG sensors: anatomical correlation via the international 10-10 system. Neuroimage **46**(1), 64–72 (2009)
83. T. Masquelier, R. Guyonneau, S.J. Thorpe, Competitive STDP-based spike pattern learning. Neural Comput. **21**(5), 1759–1776 (2009)
84. M.G. Doborjeh, N. Kasabov, *Personalised Modelling on Integrated Clinical and EEG Spatio-Temporal Brain Data in the NeuCube Spiking Neural Network Architecture*, in Proceedings of IJCNN (IEEE Press, Vancouver, 2016), pp. 1373–1378
85. D. Abegunde, R. Beaglehole, S. Durivage, J. Epping-jordan, C. Mathers, B. Shengelia, N. Unwin, Preventing chronic diseases: a vital investment. WHO (2005)
86. K. McArthur, J. Dawson, M. Walters, What is it with the weather and stroke? Expert Rev. Neurother. **10**(2), 243–249 (2010). https://doi.org/10.1586/ern.09.154
87. V.L. Feigin, Y.P. Nikitin, M.L. Bots, T.E. Vinogradova, D.E. Grobbee, A population-based study of the associations of stroke occurrence with weather parameters in Siberia, Russia (1982–92). Eur. J. Neurol. Off. J. Eur. Feder. Neurol. Soc. **7**(2), 171–178 (2000)
88. R.S. Gill, H.L. Hambridge, F.B. Schneider, T. Hanff, R.J. Tamargo, P. Nyquist, Falling temperature and colder weather are associated with an increased risk of aneurysmal subarachnoid hemorrhage. World Neurosurg. **79**(1), 136–42 (2013). https://doi.org/10.1016/j.wneu.2017.06.020
89. Y.-C. Hong, J.-H. Rha, J.-T. Lee, E.-H. Ha, H.-J. Kwon, H. Kim, Ischemic stroke associated with decrease in temperature. Epidemiology **14**(4), 473–478 (2003). https://doi.org/10.1097/01.ede.0000078420.82023.e3
90. D. Shaposhnikov, B. Revich, Y. Gurfinkel, E. Naumova, The influence of meteorological and geomagnetic factors on acute myocardial infarction and brain stroke in Moscow, Russia. Int. J. Biometeorol. (2013). https://doi.org/10.1007/s00484-013-0660-0
91. M. Othman, Spatial-temporal data modelling and processing for personalised decision support. Doctoral dissertation, Auckland University of Technology (2015). http://hdl.handle.net/10292/9079
92. W. Liang, Y. Hu, N. Kasabov, V. Feigin, Exploring associations between changes in ambient temperature and stroke occurrence: comparative analysis using global and personalised modelling approaches. Neural Inf. Process. 129–137 (2011)
93. M. Othman, N. Kasabov, E. Tu, V. Feigin, R. Krishnamurthi, Z. Hou, Y. Chen, J. Hu, *Improved Predictive Personalized Modelling with the Use of Spiking Neural Network System and a Case Study on Stroke Occurrences Data*, In 2014 International Joint Conference on Neural Networks (IJCNN). IEEE (2014), pp. 3197–3204. http://ieeexplore.ieee.org/xpls/abs_all.jsp?arnumber=6889709&tag=1
94. N. Kasabov (ed.), *Springer Handbook of Bio-/Neuroinformatics* (Springer, Berlin)
95. E. Niedermeyer, F.L. da Silva, *Electroencephalography: Basic Principles, Clinical Applications, and Related Fields*, 5th edn. (Lippincott Williams & Wilkins, Philadelphia, 2005)
96. S. Ogawa, D.W. Tank, R. Menon, J.M. Ellermann, S.G. Kim, H. Merkle, K. Gurbil, Intrinsic signal changes accompanying sensory stimulation: functional brain mapping with magnetic resonance imaging. Proc. Natl. Acad. Sci. **89**(13), 5951–5955 (1992)
97. M. Doborjeh, N. Kasabov, Z. G. Doborjeh, Evolving, dynamic clustering of spatio/spectro-temporal data in 3D spiking neural network models and a case study on EEG data. Evolving Syst. 1–17 (2017). https://doi.org/10.1007/s12530-017-9178-8

98. F. Alvi, R. Pears, N. Kasabov, An evolving spatio-temporal approach for gender and age group classification with spiking neural networks. Evolving Syst. **9**(2), 145–156 (2018)

99. N. Sengupta, N. Kasabov, Spike-time encoding as a data compression technique for pattern recognition of temporal data. Infor. Sci. **406–407**, 133–145 (2017)

100. E. Culurciello, R. Etienne-Cummings, K. Boahen, Arbitrated address-event representation digital image sensor. Electron. Lett. **37**(24), 1443–1445 (2001). https://doi.org/10.1049/el: 20010969

101. N. Nuntalid, K. Dhoble, N. Kasabov, EEG classification with BSA spike encoding algorithm and evolving probabilistic spiking neural network, in *Neural Information Processing*. Lecture Notes in Computer Science (Springer, Berlin, 2011), pp. 451–460. https://doi.org/ 10.1007/978-3-642-24955-6_54

102. H. Markram, J. Lubke, M. Frotscher, B. Sakmann, Regulation of synaptic efficacy by coincidence of postsynaptic aps and epsps. Science **275**(5297), 213–215 (1997)

# Part VII
# Deep in Time-Space Learning and Deep Knowledge Representation of Multisensory Streaming Data

# Chapter 19
# Deep Learning of Multisensory Streaming Data for Predictive Modelling with Applications in Finance, Ecology, Transport and Environment

This chapter presents methods for using eSNN and BI-SNN for deep, incremental learning and predictive modelling of streaming data and for deep knowledge representation. The methods are applied for predictive modelling in the areas of finance, ecology, transport and environment using respective multisensory streaming data. Each of these applications require specific model design in terms of data preparation, SNN model parameters, experimental setting and validation. Each of the methods are illustrated with case study problems and data, but their applicability can be extended to a wider class of problems where multisensory streaming data is available. Some of the material in this chapter was first published in [1, 2]. More details about learning in SNN, eSNN and BI-SNN can be found in Chaps. 4–6 of the book.

This chapter is organised in the following sections:

19.1. A general framework for deep learning and predictive modelling of multisensory streaming data with SNN.
19.2. eSNN for on-line predictive modelling of stock movement prediction.
19.3. SNN for deep learning and predictive modelling of ecological streaming data.
19.4. SNN for deep learning and predictive modelling of transport streaming data.
19.5. SNN for deep learning and predictive modelling of seismic streaming data.
19.6. Future applications.
19.7. Summary and further readings for deeper knowledge.

## 19.1 A General Framework for Deep Learning and Predictive Modelling of Multisensory Time-Space Streaming Data with SNN

This section presents a general methodology for using SNN for time-space data, here called spatio-spectro temporal data (SSTD), as multisensory streaming data and discusses some applications. Some of the material in this section was first

© Springer-Verlag GmbH Germany, part of Springer Nature 2019
N. K. Kasabov, *Time-Space, Spiking Neural Networks and Brain-Inspired Artificial Intelligence*, Springer Series on Bio- and Neurosystems 7,
https://doi.org/10.1007/978-3-662-57715-8_19

published in [1]. The application systems built are named evolving spatio-temporal data machines (eSTDM) as per the definition in Chap. 6.

### 19.1.1  The Challenges of Pattern Recognition and Modelling of Multisensory Streaming Data

Most problems in nature require spatio- and/or spectro-temporal data (SSTD) that include measuring spatial or/and spectral variables over time. SSTD is described by a triplet (X, Y, F), where: X is a set of independent variables measured over consecutive discrete time moments t; Y is the set of dependent output variables, and F is the association function between whole segments ('chunks') of the input data, each sampled in a time window 1t, and the output variables belonging to Y, such that

$$F : X(1t) \rightarrow Y$$

where $X(t) = (x1(t), x2(t), …, xn(t))$ and $t = 1, 2, …, m$.

It is important for a computational model to capture and learn whole spatio- and spectro-temporal patterns from data streams in order to most accurately predict future events from new input data. Examples of problems involving SSTD are: brain cognitive state evaluation based on spatially distributed EEG electrodes [3], Chaps. 8 and 9; fMRI data [4–8], Chaps. 10 and 11; moving object recognition from video data [9], Chap.13; evaluating risk of disease, e.g. heart attack, stroke [10], Chap.18; evaluating response of a disease to treatment based on clinical and environmental variables, Chap.18; modelling the progression of a neuro-degenerative disease, such as Alzheimer's Disease, Chap. 9; modelling and prognosis of the establishment of invasive species in ecology. The prediction of events in geology, astronomy, economics and many other areas also depend on accurate SSTD modelling.

The most commonly used models for dealing with temporal streaming information, based on Hidden Markov Models (HMM) (Chap. 1) and traditional artificial neural networks (ANN), Chap. 2, have limited capacity to achieve the integration of complex and long temporal spatial/spectral components because they usually either ignore the temporal dimension or over-simplify its representation. A new trend in machine learning is currently emerging and is known as deep machine learning [11]. Most of the proposed models still learn SSTD by entering single time point frames rather than learning whole SSTD patterns. They are also limited in addressing adequately the interaction between temporal and spatial components in SSTD. Some recent developments in SSTD modelling have been proposed (e.g. [12, 13]) but these are limited in their application—typically these methods are targeted towards one specific source of data, and do not show the broad level of application required in the contexts we seek to address.

The human brain has the amazing capacity to learn and recall patterns from SSTD at different time scales, ranging from milliseconds, to years, and possibly to millions of years (i.e. genetic information, accumulated through evolution). Thus, the brain is the ultimate inspiration for the development of new machine learning techniques for SSTD modelling. Indeed, brain-inspired Spiking Neural Networks (SNN) [14–16] have the potential to learn SSTD by using trains of spikes (binary temporal events) transmitted among spatially located synapses and neurons. Both spatial and temporal information can be encoded in an SNN as locations of synapses and neurons and time of their spiking activity, respectively. Spiking neurons send spikes via connections that have a complex dynamic behaviour, collectively forming an SSTD memory. Some SNN employ specific learning rules such as Spike-Time-Dependent-Plasticity (STDP) [17] or Spike Driven Synaptic Plasticity (SDSP) [18].

In [3] a BI-SNN NeuCube framework was presented for spatio-temporal brain data (see also Chap. 6) and in [1] and in several chapters of the book some specific methods and applications of BI-SNN were presented. This chapter further extends these works into a generic and systematic methodology for a new type of solutions to any spatio-temporal stream data problems and the solution is called here for the first time—evolving spatiotemporal data machine (eSTDM). It is also demonstrated in several domain application areas.

### 19.1.2    Modelling Streaming Data in Evolving SNN (eSNN)

Evolving SNN (eSNN) were presented in details in Chap. 5. Figure 19.1 depicts a general architecture of an eSNN and the training algorithm is presented as Appendix 1 to this chapter.

Streaming data here is presented as a sequence of vectors (frames, samples), each of them representing a measurement of the modelled input variables for a classification problem.

Each sample is learned in the eSNN model (see training algorithm in the Appendix) and generates an output node that represents this sample and is allocated in the corresponding classification output "pool".

**Fig. 19.1** A general architecture of eSNN for classification (see also Chap. 5)

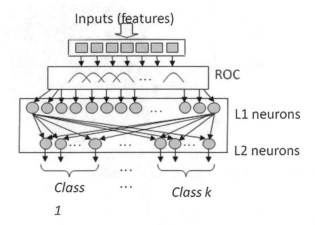

Output neurons, representing samples of the same class, can be merged as described in Chap. 5, resulting in a smaller number of output nodes. Each of them now representing not a single sample, but a prototype of samples as the center of a cluster of these samples in the connection weight space.

An eSNN can learn in a continuous, on-line, "life-long" way streaming data, by incrementally creating output neurons and aggregating them.

An eSNN can learn in both supervised and semi-supervised modes. When a new sample is available, but there is no class label attached to it, the model can create a new output sample and to allocate it to the class "pool" of the closest prototypes in it. The similarity is again measured through the connection weights using Euclidian distance for example.

eSNN can be used not only for classification tasks of streaming data, but for regression tasks as well. In this case the output nodes are not representing class labels, but real values. Aggregation of output nodes can be done if both the connection weights and the output values are similar.

### 19.1.3   A General Methodology for Modelling Multisensory Streaming Data in Brain-Inspired SNN for Classification and Regression

Our approach here to modelling large and fast streaming SSTD is based on a common architecture of evolving spatio-temporal data machine (eSTDM) as depicted in Fig. 19.2 developed with the use of a BI-SNN NeuCube depicted in Fig. 19.2 and also discussed in Chap. 6.

In this architecture a SNNcube is used to map input streaming data after its encoding into spikes. eSNN, deSNN or other SNN models can be used as an output module for classification or regression.

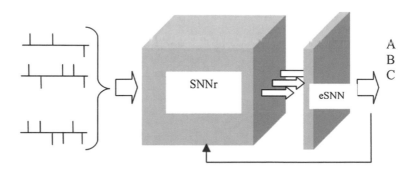

**Fig. 19.2** A general architecture of eSTDM

The functionality of an eSTDM is based on the following procedures:

- Converting multivariable input stream data into spike sequences;
- Unsupervised learning of spatio-temporal patterns from data in a SNN reservoir (the "Cube");
- Supervised learning of classification/regression output system for classification/regression problems;
- Optimisation using the evaluated/tested accuracy of the system as a feedback for improving the performance of this system in an iterative way (if necessary).

The NeuCube architecture consists of the following modules (Chap. 6):

- Input information encoding module;
- 3D SNN module (the Cube);
- Output classification/regression module; and other optional modules, including:
- Gene regulatory network (GRN) module;
- Parameter optimisation module;
- Visualisation and knowledge extraction module (not shown in Fig. 19.3).

The input module transforms input data into trains of spikes. Spatio-temporal data (such as EEG, fMRI, climate) is entered into the main module—the 3D SNNcube (SNNc). Different types of data can be used. This data is entered ("mapped") into pre designated spatially located areas of the SNNc that correspond to the spatial location of the origin where data was collected (if such a location exists). Learning in the SNN is performed in two stages:

- Unsupervised training, where spatio-temporal data is entered into relevant areas of the SNNc over time. Unsupervised learning is performed to modify the initially set connection weights. The SNNc will learn to activate same groups of

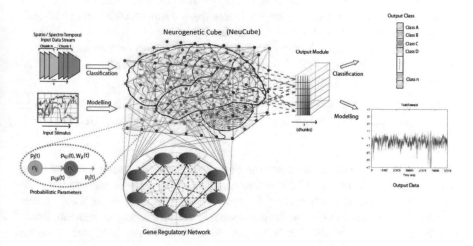

**Fig. 19.3** A general architecture of the NeuCube BI-SNN ([3], see also Chap. 6)

spiking neurons when similar input stimuli are presented, also known as a polychronization effect [19].

- Supervised training of the spiking neurons in the output module, where the same data that was used for unsupervised training is now propagated again through the trained SNN and the output neurons are trained to classify the spatio-temporal spiking pattern of the SNNc into pre-defined classes (or output spike sequences). As a special case, all neurons from the SNN are connected to every output neuron. Feedback connections from output neurons to neurons in the SNN can be created for reinforcement learning. Different SNN methods can be used to learn and classify spiking patterns from the SNNc, including the deSNN [20, 24] and SPAN models [21]. The latter is suitable for generating motor control spike trains in response to certain patterns of activity of the Cube.

In an eSTDM similar activation patterns (called 'polychronous waves') can be generated in the SNNc with recurrent connections to represent short term memory. When using STDP learning, connection weights change to form LTP or LTD, which constitute long-term memory (see [17] for more detail of STDP). Results of the use of the NeuCube suggest that the NeuCube architecture can be explored for learning long spatio-temporal patterns and to be used as associative memory. Once data is learned, the SNNc retains the connections as a long-term memory. Since the SNNc learns functional pathways of spiking activities represented as structural pathways of connections, when only a small initial part of input data is entered the SNNc will 'synfire' and 'chain-fire' *learned connection pathways* to reproduce *learned functional pathways*. Thus a NeuCube can be used as an associative memory and as a predictive system for event prediction when only some initial new input data is presented.

In order to design an appropriate eSTDM for a given task, a number of factors must be taken into consideration. Here, we identify these considerations.

- Which input variable mapping into the SNNc is used? Is there some a priori information we can use to spatially locate these input variables in the SNNc?
- Which learning method do we use in the SNNc?
- Which output function is appropriate? Is it classification or regression?
- How to visualise an eSTDM for an improved understanding?
- Which parameter optimisation method will we apply?

For rapid prototyping and exploration of a NeuCube model, a generic prototyping and testing module has been implemented and is discussed in Chap. 20.

*Data encoding*

There are different coding schemes for SNN, primarily rate (information as mean firing rates) or temporal (information as *temporally significant*) coding. For NeuCube, we use temporal coding to represent information. So far four different spike encoding algorithms have been integrated into the existing implementation of the NeuCube, namely the Ben's Spiker Algorithm (BSA), Temporal Contrast

(Threshold-based), Step-Forward Spike Encoding Algorithm (SF) and Moving-Window Spike Encoding Algorithm (MW) (see Chap. 4).

Chapter 4 of this book shows different results of the same data, in this case an EEG signal, encoded by these four algorithms. Different spike encoding algorithms have distinct characteristics when representing input data. BSA is suitable for high frequency signals and because it is based on the Finite Impulse Response technique, the original signal can be recovered easily from the encoded spike train. Only positive (excitatory) spikes are generated by BSA, whereas all other techniques mentioned here can also generate negative (inhibitory) spikes. Temporal Contrast was originally implemented in hardware [9] in the artificial silicon retina. It represents significant changes in signal intensity over a given threshold, where the ON and OFF events are dependent on the sign of the changes. However, if the changes of the signal intensity vary dramatically, it may not be possible to recover the original signal using the encoded spike train. Therefore, an improved spike encoding algorithm, SF, to better represent the signal intensity is described below and also Chap. 4.

For a given signal $S(t)$ where ($t = 1, 2,..., n$), we define a baseline $B(t)$ variation during time $t$ with $B(1) = S(1)$. If the incoming signal intensity $S(t1)$ exceeds the baseline $B(t1-1)$ plus a threshold defined as $Th$, then a positive spike is encoded at time $t1$, and $B(t1)$ is updated as $B(t1) = B(t1-1) + Th$; and if $S(t1) <= B(t1-1) - Th$, a negative spike is generated and $B(t1)$ is assigned as $B(t1) = B(t1-1) - Th$. In other situations, no spike is generated and $B(t1) = B(t1-1)$.

As to the Moving-Window Spike Encoding Algorithm, the baseline $B(t)$ is defined as the mean of previous signal intensities within a time window $T$, thus this encoding algorithm can be robust to certain kinds of noise.

Before choosing a proper spike encoding algorithm, we need to figure out what information the spike trains shall carry for the original signals. After that, the underlying spike patterns in the spike trains will be better understood.

*Input variable mapping*

Mapping input variables into spatially located spiking neurons in the SNNc is a new approach towards modelling SSTD introduced in [3] and is a unique feature of the eSTDM. The main principle is that if spatial information about the input variables is known it can help in (a) building more accurate models of the SSTD collected through these variables and (b) a much better interpretation of the model and a better understanding of the SSTD. This is very important for data such as brain data such as EEG (see [3, 22]) and for fMRI data (see Chaps. 10 and 11) where patterns of interaction of brain signals can be learned and discovered. In some implementations we have used the Talairach brain template, mapped spatially into the SNNc (see Fig. 19.3). Another way of mapping, when there is no spatial information available for the input variables, is to measure the temporal similarity between the variables to map variables with similar patterns into closer neurons in the SNNc. This is the vector quantisation principle, where by 'vector' here we use time series, which do not necessarily have the same length [2].

*Learning*

Learning in a eSTDM is a two-phase process as it was described in the NeuCube framework (Chap. 6). The accuracy of a NeuCube model depends a great deal with the SNNc learning parameters and the classifier/regressor parameters.

Optimisation procedures are discussed in Chap. 7.

*Output classification or regression*

We use an SNN for the output model of the type eSNN. An eSNN evolves its structure and functionality in an on-line manner, from incoming information. For every new input data sample, a new output neuron is dynamically allocated and connected to the input neurons. The neuron's connections are initially established using the RO rule for the output neuron to recognise this vector (frame, static pattern) or a similar one as a positive example. The weight vectors of the output neurons represent centres of clusters in the problem space and can be represented as fuzzy rules [23]. Then these connection weights are further adapted to the following spikes [24].

In some implementations neurons with similar weight vectors are merged based on the Euclidean distance between them. That makes it possible to achieve a very fast learning (only one pass may be sufficient), in both supervised and unsupervised modes [24]. When in an unsupervised mode, the evolved neurons represent a learned pattern (or a prototype of patterns). The neurons can be labelled and grouped according to their class membership if the model performs a classification task in a supervised mode of learning.

Weights are calculated based on the order of the incoming spikes on the corresponding synapses using the RO learning rule:

$$w_{i,j} = \alpha \text{mod}^{\text{order}(j,i)} \tag{19.1}$$

where: $\alpha$ is a learning parameter (in a partial case it is equal to 1); mod is a modulation factor that defines how important the order of the first spike is; $w_{j,i}$ is the synaptic weight between a pre-synaptic neuron $j$ and the postsynaptic neuron $i$; order($j$, $i$) represents the order (the rank) of the first spike at synapse $j$, $i$ ranked among all spikes arriving from all synapses to the neuron $i$; order($j$, $i$) = 0 for the first spike to neuron $i$ and increases according to the input spike order at other synapses.

While the input training pattern (example) is presented (all input spikes on different synapses, encoding the input vector are presented within a time window of $T$ time units), the spiking threshold $\Theta$ of the neuron $i$ is defined to make this neuron spike when this or a similar pattern (example) is presented again in the recall mode. The threshold is calculated as a fraction ($C$) of the total $PSPi$ (denoted as $PSP^{max}$) accumulated during the presentation of the input pattern:

The eSNN (deSNN) learning is adaptive, incremental, theoretically 'lifelong', so that the system can learn new patterns through creating new output neurons, connecting them to the SNNc neurons, and possibly merging the most similar ones. The deSNN implements the 7 ECOS principles from Chap. 2.

During the recall phase, when a new spike sequence is presented, the spiking pattern is submitted to all created neurons of the SNNc. An output spike is generated by neuron $i$ at a time $l$ if the $PSPi(l)$ becomes higher than its threshold $Thi$. After the first neuron spikes, the $PSP$ of all neurons are set to an initial value (e.g. 0) to prepare the system for the next pattern for recall or learning.

*Parameter optimisation of NeuCube models*

eSTDM behaviour can be optimized by changes in the model parameters as discussed in Chap. 7. For example, differing neuron reset voltages can lead to a number of different spiking dynamics, and differing encoding parameters can significantly change the information density of the spike trains. Different 'mod' and 'drift' parameters in a deSNN can result in different classification accuracy. To this end, a parameter search is usually performed in order to extract the best performance. Three primary techniques are discussed here: Grid Search; Genetic Algorithm search; and the Quantum-Inspired search.

*Grid search.* Grid search is a straightforward but effective method to tune parameters. Suppose there are $P$ parameters that have to be optimised simultaneously. For each parameter there are three hyper-parameters to be specified manually: the minimal value $m$ and the maximal value $M$ of the searching interval, and the searching step size $s$. Given these three hyper-parameters of each optimizing parameter, we first create a $P$-dimension matrix, each dimension of which corresponding to a optimizing parameter, from $m$ to $M$ divided into $(M - m)/s$ entries. In this case, each entry of the matrix corresponds to a group of values of the optimising parameters. Then we randomly split the training set into two equal-size parts, a training part and a validation part. For a specific group of values, we run the NeuCube system in a two-fold cross-validation way and the error rate of the cross-validation is added to the entry of the $P$-dimension matrix corresponding to that group of parameter values.

*Genetic algorithms.* Standard Genetic Algorithm techniques can be used to optimise the parameters of a NeuCube model.

*Quantum-inspired evolutionary methods.* These methods use the principle of superposition of states to represent and optimise parameters of SNN models [25]. Such a method is the quantum inspired genetic algorithm or QiPSO [26].

*Dynamic and immersive visualisation of NeuCube models*

The number of neurons and connections within NeuCube as well as the 3-dimensional structure requires a visualisation that goes beyond a simple 2D connectivity/weight matrix or an orthographic 45° view of the volume. We created a specialised renderer for NeuCube datasets using JOGL (Java Bindings for OpenGL) and GLSL (OpenGL Shading Language) shaders to be able to render up to 1.5 million neurons and their connections with a steady frame rate of 60 fps. In this view, neurons are displayed as stylised spheres, and connections are rendered as lines with green colour for excitatory connections and red for inhibitory connections. Spiking activity is shown as signals travelling along the connections.

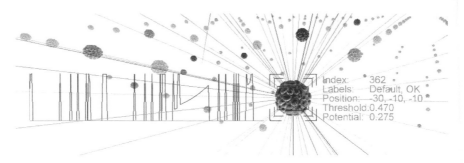

**Fig. 19.4** Each spiking neuron and its connections can be zoomed and analysed for a better understanding of the created model and how it reflected on the data (from [30])

In conjunction with a 3D stereoscopic HMD (Head Mounted Display) like the Oculus Rift, it is easy for users to perceive the spatial structure of the network and the neuron positions. Furthermore, interaction mechanisms allow for playback of spiking patterns and the development of connection weights throughout the learning period. In addition, the visualization includes analysis functionality for the usage of connections to find 'hot paths', connection length analysis, and the ability to view the 3D structure in 'slices'. A 3D cursor metaphor is employed to look at neurons individually, their parameters, and their spiking history.

In comparison to other scientific visualization tools for neural networks such as BrainGazer [27] and Neuron Navigator (NNG) [28], the solution here differs in that the user can naturally navigate through the 3D space by simply walking a mouse and keyboard shortcuts. Closer to this visualization is the work of [29], who are using a Computer Assisted Virtual Environment (CAVE) to visualization the spatial structure and activity of a spiking neural network. However, due to the limited space within a cave environment, navigation by simply walking is not possible and requires indirect ways, e.g., by using a controller. We developed a VR immersive visualisation in which people quickly start to move around and look at structures and point out individual neurons using the 3D cursor (Fig. 19.4) [30].

## 19.2   Stock Market Movement Prediction Using On-Line Predictive Modelling with eSNN

Stock price direction prediction is regarded as one of the most difficult and challenging tasks in the real-world. An accurate prediction can give profit to the investors and protect them from financial risk. In this section a computational model for the stock trend prediction using evolving spiking neural networks (eSNN) is discussed.

For a particular case study data used in [31], the eSNN architecture is shown in Fig. 19.5.

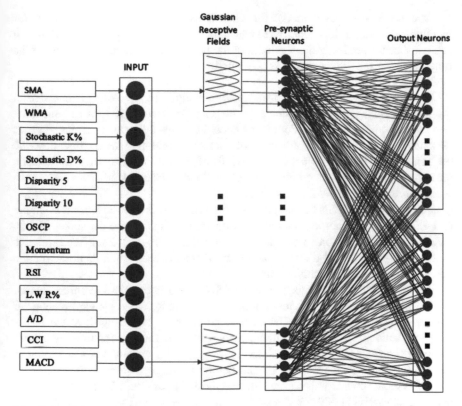

**Fig. 19.5** eSNN for stock market prediction. For the used stock indicator, see [31]

*Layer 1* is the set of inputs to the model, each of them representing a technical stock indicator. The research so far has demonstrated that using technical indicators can lead to better results than using real stock values as time series and also that there is a lot of research done on selecting the most appropriate technical indicators [32, 34]. In the model presented in Fig. 19.5 the input technical indicators have been selected as explained in [31].

*Layer 2* is the encoding layer, where the real value of each input variable (technical indicator) is encoded as trains of spikes generated by several encoding spiking neurons (or also, pre-synaptic neurons), each of them having a receptive field. The receptive fields of neighbouring neurons are overlapping as Gaussian or Logistic functions and all of them covering the whole range of the values of this variable. The number of these encoding neurons (receptive fields) can vary, and this is a user defined parameter that is optimised for a better performance of the model.

*Layer 3* is the output evolving layer, which evolves output spiking neurons that represent clusters (prototypes) of input vectors that belong to the same class, in this case class UP and class DOWN. Each output neuron is connected to all the input neurons, and the connection weights are subject to learning from data. The architecture of the eSNN model for stock price direction prediction allows for incremental learning. It is adaptive to new data when it becomes available. Hence, it can learn new samples without retraining the model on old data. The details of the functioning of the eSNN model is presented below.

*Neural encoding*

To learn real-valued data, each instance or sample (input vector) is encoded in the form of spikes over time using a neural encoding technique such as rank order population coding (Chap. 4). Population encoding maps the input value into a series of spikes over time using an array of Gaussian receptive fields that describe pre-synaptic neurons. The center and width of each of the Gaussian or Logistic receptive field of pre-synaptic neurons are defined in Chap. 4.

*Learning in the eSNN* (see the Appendix 1)

For the context of eSNN, Thorpe's neuron model has been used since it is simple and effective. The Thorpe's model is based on the timing of each spike, that is, earlier spike defines stronger weight as compared to later spike. Each neuron in this model can spike at most once. A neuron in this model fires when its post-synaptic potential reaches the threshold value.

The learning techniques used by the eSNN model is one-pass learning, that is, the model requires one-time presentation of a sample in a feed-forward manner. It will create an output neuron for each input sample. The weight vector and a threshold value for each of the output neuron generated towards the training pattern are learned and stored in the repository. However, if this weight vector is similar to the weight vector of the already trained neuron in the repository with some similarity threshold, then it will merge with the most similar one. Merging here means updating the weights and the threshold value of the merged neurons. The weight vector and threshold value of the merged neurons update their values by taking the average value of new output neuron weight vector and merged neuron weight vector and the average value of new output neuron threshold and merged neuron threshold respectively.

One approach for on-line training of an eSNN is to use a window of streaming data to train an eSNN and to predict the output value for the next time point. When the actual results of the output are available, these results can be added incrementally for an incremental further training of the eSNN. The algorithm is given in Box 19.1.

---

Box 19.1. Sliding window algorithm for on-line training of eSNN  (SW-eSNN)

1: Train an eSNN model on the whole existing historical data of a stock till time (as per the eSNN algorithm in the Appendix and alos in Chapter 5)
2: Recall the model to predict the next time (t+1) stock movement.
3: When the next time results are known, train the model incrementally on this data.
4: Aggregate the output neurons if necessary using the aggregation operator and the $sim$ parameter.
5: Evaluate the classification error and the AUC so far.
6: Optimize parameters to improve future time accuracy.
   Note: Time could be minutes, hours, days, months etc.

**Table 19.1** Average AUC score of SW-eSNN incremental learning and predicting one day ahead the UP or DOWN the stock for 200 days using both Logistic and Gaussian receptive fields (for a definition of AUC see Chap. 1)

| Dataset | eSNN with logistic receptive fields | eSNN with Gaussian receptive fields |
|---|---|---|
| | Average AUC | Average AUC |
| BSE | 0.77 | 0.71 |
| Nikkei-225 | 0.72 | 0.69 |
| NASDAQ | 0.76 | 0.77 |
| NIFTY-50 | 0.69 | 0.67 |
| S&P-500 | 0.75 | 0.73 |
| Sanghai stock exchange | 0.67 | 0.65 |
| Dow-Jones | 0.73 | 0.7 |
| NYSE-Amex | 0.73 | 0.69 |
| DAX-Index | 0.72 | 0.7 |

*Experimental results*

Nine benchmark data sets are experimented with in [31], from QUANDL [32–34]. These datasets cover stock market indices of different countries: BSE, Nikkei-225, NIFTY-50, S&P-500, Dow-Jones, NYSE-Amex, DAX, NASDAQ and Shanghai stock exchange. The resulting AUC accuracy evaluation is given in Table 19.1. Optimising the eSNN parameters (number and type of receptive fields, Spiking threshold, Mod factor, Sim parameter) can lead to a significant improvement in the prediction results as shown in [31] and in Appendix 2. The accuracy for the BSE stock has increased to 90%.

# 19.3    SNN for Deep Learning and Predictive Modelling of Ecological Streaming Data

Modelling ecological streaming data requires sophisticated methods. Using the NeuCube BI-SNN is demonstrated here. Some of the material in this section was first published in [2].

## 19.3.1    Early Event Prediction in Ecology: General Notions

Early event prediction is very crucial when solving important ecological and social tasks described by temporal- or/and spatio-temporal data, such as pest population outbreak prevention, natural disaster warning and financial crisis prediction. The generic task is to predict early and accurately whether an event will occur in a future time based on already observed spatio-temporal data. The time length of the training data (samples collected in the past) and the test data (samples used for prediction) can be different as illustrated in Fig. 19.6. Predictive modeling of spatio/spectro-temporal data (SSTD) is a challenging task because it is difficult to model both time and space components of the data because of their close interaction and interrelationship.

## 19.3.2  A Case Study on Predicting Abundance of Fruit Flies Using Spatio-temporal Climate Data

Here the problem is the prediction of a possible abundance of aphids (*Rhopalosiphum padi*) in the autumn seasons based on climate temporal data [2]. There are fourteen temporal climate variables measured: (1) average rainfall (AR, mm); (2) cumulative rainfall, the average of 4 weeks (CR, mm); (3) cumulative degree days (DCT, °C); (4) grass temperature, average of four weeks (GT, °C); (5) maximum air temperature (MaxT, °C); (6) mean air temperature (MeanT, °C); (7) minimum air temperature, average of two weeks (MinT, °C); (8) Penman potential evaporation (PPE, mm); (9) potential deficit of rainfall (PDR), first order derivative; (10) soil temperature (ST, °C); (11) solar radiation (SR, MG/m2); (12) vapour pressure, average of five weeks (VP, hPa); (13) wind run (WR4), average of four weeks (km/day); (14) wind run (WR5), average of five weeks (km/day). All these variables are measured weekly at the Canterbury Agricultural research centre, Lincoln, New Zealand from 1982 to 2004, i.e. (52 data points per year). The aim is to predict whether the aphid amount in autumn will be high (class 1) or low (class 2).

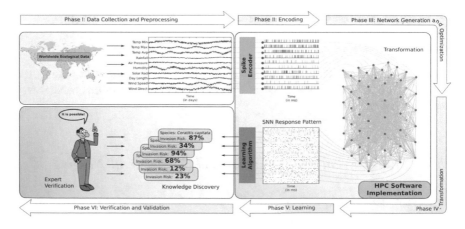

**Fig. 19.6**  A schematic diagram for early event prediction in Ecology on the example of predicting the risk of establishment of invasive species

Figure 19.7 shows the input variable mapping result computed by the proposed graph matching algorithm on the minimal x coordinate face of the Cube. Note that the two main groups of weather variables, in other words, temperature (MaxT, minT, MeanT, DCT, GT, ST) and rainfall (AR, CR, PDR), are mapped to nearby neurons. The solar radiation (SR) is mapped in the middle of temperature variables because temperature is greatly determined by solar radiation.

To demonstrate how the optimal mapping suggested by the proposed graph matching algorithm can influence the overall performance, we designed two experiments to compare results of optimal mapping with results of random mapping [2]. In the first experiment, we use the same group of input neurons and run the NeuCube learning twice: in the first run we randomly mapped the features to the input neurons while in the second run we used the proposed in this paper graph matching to compute the optimal input mapping. This process is repeated 10 times and the accuracies of each run are shown in Fig. 19.8a. In the second experiment, we also run the NeuCube twice in the same way as in the first experiment, but the group of input neurons are randomly generated at each time. The accuracies of 10 times experiments are shown in Fig. 19.8b.

In Fig. 19.8a the graph matching is obtained through a deterministic algorithm. So given the same group of input neurons, it can always produce the same optimal mapping and the accuracy will not change. But for random mapping, the results change across experiments because each time the mapping is different. This result indicates that input mapping plays an important role for the obtained accuracy of the model. In Fig. 19.8b the group of input neurons is randomly generated at each time. That's why the accuracy of the 'optimal mapping' is lower than the random mapping in runs 1 and 4. In runs 5 and 9 the accuracies obtained with the use of the proposed mapping are much higher (i.e., 36.36 and 27.27% higher, respectively) than the results with the use of random mapping. This result also indicates that not only the mapping plays a key role, but the group of input neurons selected is important. How to optimally choose an optimal set of input neurons in relation to a specific input data is an interesting problem to address in the future.

Figure 19.9 shows the SNNcube structure after unsupervised training with the data for this case study. The big solid dots represent input neurons and other neurons are labeled in the same color as the input neurons from which they receive most of the spikes. The black dots mean that there are no spikes arrived at these neurons. In Fig. 19.9 the upper left panel is the spike amount of each variable after encoding and the upper right panel is the amount of neuronal connections of the input neurons forming a cluster of connected neurons. From this figure we can see the consistency between the input signal spike train and the Cube connectivity structure. It is worth noting that variable 11 (solar radiation) is emphasized in the Cube that suggests a greater impact of the solar radiation on the aphids amount. This was observed also in a previous study [2].

We designed three experiments on this data set to show the validity of the proposed mapping method for early event prediction. In the first experiment, we used temporal samples over the whole time available, i.e. 52 weeks, for both training and testing under the assumption that a perfect weather forecast for the

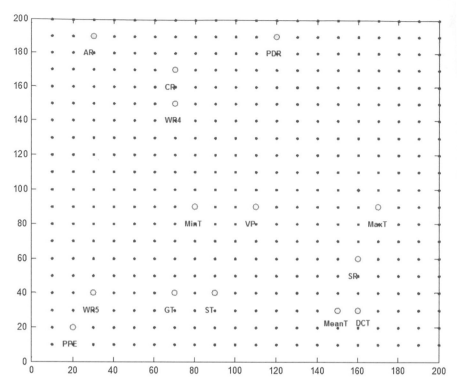

**Fig. 19.7** Input variable mapping result by graph matching algorithm [2] and Chap. 6

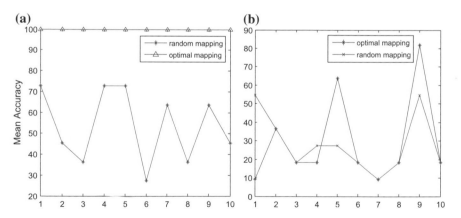

**Fig. 19.8** Comparative accuracy of pattern recognition using random mapping (in blue) versus the proposed mapping method (in red) (see text for explanation)

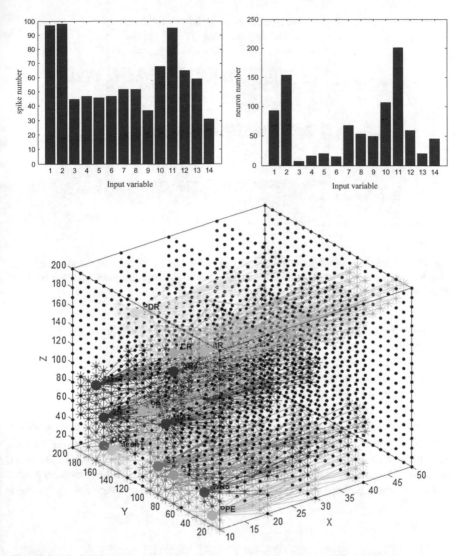

**Fig. 19.9** The SNNcube structure after unsupervised training; input spike amount of each feature (top left) and neuronal connections created as a result of training of each input neurons (top right)

autumn season can be obtained which is an ideal case, but not a realistic one. In the following two experiments, we aimed to show the predictive ability of NeuCube and how early the model can predict the autumn pattern. In these experiments, we trained NeuCube using 100% of the time length samples (52 weeks), but temporal data of only 80 and 75% of the time length of the samples was used to predict the aphid population pattern in the last 25% time period, as illustrated in Fig. 19.10.

Previous year week 31 – following year week 17
Week 18-30
Experiment 1
52 weeks
52 weeks
Training
Testing
Experiment 2
52 weeks
41.6 weeks
Training
Testing
Experiment 3
52 weeks
39 weeks
Training
Testing


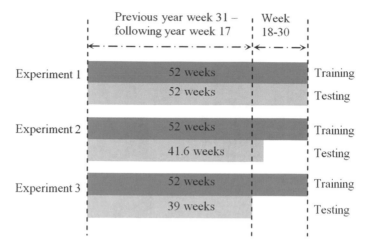

**Fig. 19.10** Experimental design for testing the ability of the model for early event prediction. Blue bars represent the time length of training samples and the yellow bars represent the time length of testing samples

The blue bars represent training data length and the yellow bars – the validation data length.

The experiments are conducted in leave-one-out cross validation way. Figure 19.11 shows the parameters configuration of the NeuCube used in this case study. The early event prediction accuracy on the aphid data set is shown in Table 19.2, where the middle row is the time length of test data used for the prediction.

From these experiments we can see that our model can make *early* prediction before the peak appears. With 80% of data observed (early in the aphid flight period), we can have more than 90% confidence to make an early decision. Furthermore, as the time passes, if new data are collected, it can be added directly to the testing sample to give a better prediction, without re-training the model using both old and new data as it would

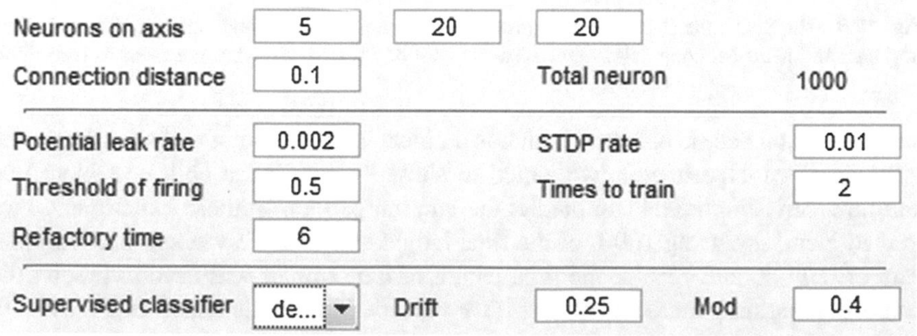

Neurons on axis
5
20
20
Connection distance
0.1
Total neuron
1000
Potential leak rate
0.002
STDP rate
0.01
Threshold of firing
0.5
Times to train
2
Refactory time
6
Supervised classifier
de...
Drift
0.25
Mod
0.4


**Fig. 19.11** Parameters of NeuCube in this case study

**Table 19.2** Autumn aphid prediction accuracy (%)

| | Accuracy of each testing time length (weeks) | | |
|---|---|---|---|
| | 52 (full) | 41.6 (early) | 39 (earlier) |
| Accuracy (%) | 100 | 90.91 | 81.82 |

be the case with SVM or MLP methods. This is the essential difference between the new method and traditional methods such as SVM and MLP.

We also conducted experiments to compare between traditional modeling methods and our new modeling method for early event prediction. We used multiple-linear regression (MLR), support vector machine (SVM), multilayer perceptron (MLP), $k$ nearest neighbors ($k$NN) and weighted $k$ nearest neighbors (w$k$NN) as the baseline algorithms (see Chaps. 1 and 2). Note that for these baseline algorithms, the time length of training samples and testing samples have to be the same as these methods cannot tolerate different lengths of feature vectors for training and recall, so we cut the training samples into same length as the testing samples. We tune the parameters of the baseline algorithms in a grid search way and the final parameters are: a degree 2 polynomial kernel for SVM; 100 hidden neurons and 500 training cycles for MLP; $k = 5$ for both $k$NN and w$k$NN. Experimental results are shown in Table 19.3.

Comparing Tables 19.2 and 19.3, we can see that NeuCube can perform better for early event prediction. A realistic early event prediction should be that as the time length of observed data increases, the prediction accuracy will also increase. But from table II we can see that as the time length of training data increases, traditional modeling methods do not necessarily produce better results (some even worsen), because they cannot model the whole spatio-temporal relationship in the prediction task. They can only model a certain time segment. Because a NeuCube model is trained on the whole spatio-temporal relationship in the data, even a small amount of input data can trigger spiking activities in the SNNcube that will correspond to the learned full temporal pattern resulting in a better prediction.

**Table 19.3** Prediction accuracy of aphid data set (%)

| | Accuracy of each training and testing time length (weeks) | | |
|---|---|---|---|
| | 52 | 41.6 | 39 |
| MLR | 36.36 | 64.63 | 72.73 |
| SVM | 72.73 | 72.73 | 63.64 |
| MLP | 81.82 | 81.82 | 81.82 |
| $k$NN | 72.73 | 63.64 | 63.64 |
| w$k$NN | 72.73 | 63.64 | 63.64 |
| Max | 81.82 | 81.82 | 81.82 |

## 19.4   SNN for Deep Learning and Predictive Modelling of Transport Streaming Data

Transport systems are complex spatio-temporal systems that required spatio-temporal modelling techniques. Part of the material in this section was first published in [2].

### 19.4.1   A Case Study Transport Modelling Problem

In this case study we consider a benchmark traffic status classification problem and spatio-temporal data (see [2]). In freeways, vehicle flow is monitored by traffic sensors with fixed spatial locations and the data collected by these sensors exhibit spatial and temporal characteristics. Discovering spatial-temporal patterns can be very meaningful for traffic management and a city traffic plan.

The study area is San Francisco bay area which is shown in Fig. 19.12a. There are thousands sensors distributed over the road network and the sensors distribution is indicated in Fig. 19.12b, in which each black dot represents a monitoring sensor. These sensors monitor lane occupation rate 24-hourly every day. Measurements are taken every 10 min and normalized between 0 and 1, where 0 means no car occupation and 1 means full occupation of the lane in the monitoring region. So there are 144 (24 × 6) data points per day. In this case study we study traffic data over a period of 15 months and thus, after removing public holidays and sensor maintenance days, there are 440 days to be classified.

We did some pre-processing of the data: (1) we removed the data of outlier sensors from the data set, e.g. sensors producing always 1 or 0 in 24 h and sensors flip from 0 to 1 or 1 to 0 suddenly; (2) nearby sensors that produce almost the same data sequence are combined into one sensor; (3) the total occupation rate of each sensor is calculated as a sum of all measurements over 440 days; (4) 50 sensors corresponding to the largest occupation rate are selected as the final features (variables) to represent the data set. Figure 19.13 shows the spatial and temporal distribution of the traffic status of the road network from Monday to Sunday in one week. We can see that there are some spatial and temporal patterns in the data samples.

### 19.4.2   NeuCube Model Creation and Modelling Results

A NeuCube model was created for this problem where input variables were mapped into a SNNcube using the algorithm presented in Chap. 6. Figure 19.14 shows the final input mapping result used in this case study. Left: input neuron similarity

**(a)**                                **(b)**

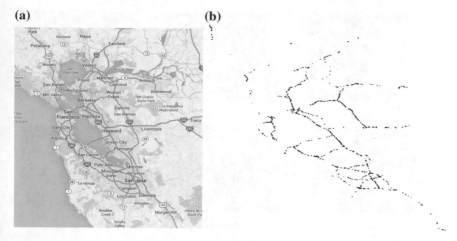

**Fig. 19.12** **a** Map of the study area (from Google map). **b** A reconstructed topology of the traffic sensor network

graph (the number beside each vertex is the input neuron ID); right: input feature similarity graph (the number beside each vertex is the traffic sensor ID).

In Fig. 19.15, we show the neuron firing state of the reservoir corresponding to the seven days traffic data. The first two plots at the top row represent the spiking activity of the SNNcube for Monday and Thursday data; the second row—for Tuesday and Friday; the third row are Wednesday and Saturday and the last one—for Sunday. In each figure, the horizontal axis is neuron ID and the vertical axis is time tick, from 0 at the top to 144 at the bottom. NZ is the number of non-zero entries, i.e. the total firing times of all the neurons in the Cube. One should note that while in the plot it seems the firing state matrix is very dense, it is actually very sparse. Take the Thursday (upper right plot) as an example. There are 20,416 firing times and the firing state matrix size is 486,000 ($144 \times 3375$, where 3375 is the total neuron number in the Cube), so the firing rate is about 4.20%. We can see that these sparse firing matrices have different patterns related to the input data. Meanwhile, since the size of the Cube can be specified according to the problem, the Cube with highly sparse firing rate has a great power to encode input signals and patterns, and thus the proposed architecture can potentially model any complex spatial and temporal relationship jointly.

We compared the 2-fold cross validation experimental result of NeuCube with the results obtained with the use of traditional methods: MLR, SVM, MLP, $k$NN and w$k$NN as well as the state-of-art method Global Alignment Kernels (GAK) [35]. The parameter setting for the NeuCube model is displayed in Fig. 19.16 and experimental results are shown in Table 19.4.

The parameter values used in the classical machine learning methods are: $d$—degree of polynomial kernel; $n$—number of neurons in the hidden layer of MLP; $k$—number of nearest neighbors in kNN and wkNN; $\sigma$—Gaussian kernel width. From these results we can see that the proposed NeuCube model achieves better

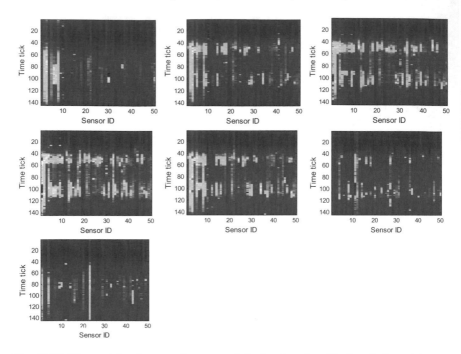

**Fig. 19.13** The spatial and temporal patterns of the benchmark traffic data. Left to right, top: Monday to Wednesday; middle: Thursday to Saturday; bottom: Sunday

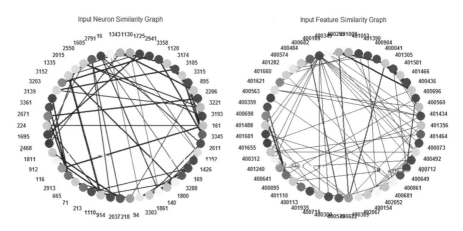

**Fig. 19.14** The final input mapping result used in this case study. Left: input neuron similarity graph (the number beside each vertex is the input neuron ID); right: input feature similarity graph (the number beside each vertex is the traffic sensor ID)

classification results. This is because traditional machine learning methods are designed to process static vector data, and they have limited ability to model spatially correlated and temporally varied data. Meanwhile, MLR, SVM and MLP

nz = 9833
nz = 20416
nz = 15398
nz = 16148
nz = 14003
nz = 11195
nz = 7987

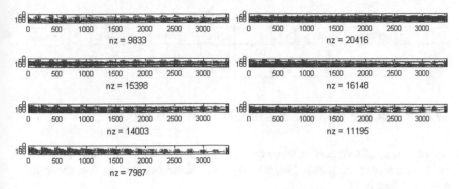

**Fig. 19.15** Neuron firing state of the reservoir. Top row: Monday and Thursday; second row: Tuesday and Friday; third row: Wednesday and Saturday; last on is Sunday

also show disadvantages while modeling high-dimensional data (e.g. there are 7200 features in each sample of this case study). *k*NN and w*k*NN has been widely used in high-dimension data processing, such as document classification, because they can approximately reconstruct the underlying manifold whose dimension is usually much lower than its ambient space and thus they can produce better result than MLR and SVM. While the recently proposed GAK algorithm is shown to be very efficient and effective in processing time series, its performance is still lower than the NeuCube model.

In this section only one case study of modelling transport systems was presented but the approach can be used in many different scenarios.

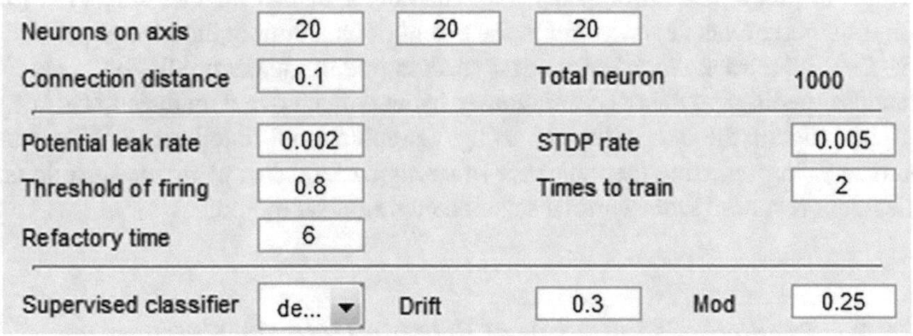

Neurons on axis
20
20
20
Connection distance
0.1
Total neuron
1000
Potential leak rate
0.002
STDP rate
0.005
Threshold of firing
0.8
Times to train
2
Refactory time
6
Supervised classifier
de...
Drift
0.3
Mod
0.25

**Fig. 19.16** Parameters of the NeuCube model for this case study

**Table 19.4** Comparative accuracy of spatio-temporal pattern classification

|        | MLR   | SVM   | MLP     | kNN    | wkNN   | GAK        | NeuCube |
|--------|-------|-------|---------|--------|--------|------------|---------|
| Param. | –     | $d = 2$ | $n = 100$ | $k = 10$ | $k = 10$ | $\sigma = 5$ | –       |
| Acc.   | 56.82 | 43.86 | 68.18   | 66.82  | 71.36  | 72.27      | 75.23   |

## 19.5   SNN for Predictive Modelling of Seismic Data

Seismic data indicate earth movements and are one of the indicators of hazardous events, such as earthquakes [36, 37–54]. Part of the material in this section was presented first in [1].

### 19.5.1   The Challenge of Predicting Hazardous Events

Hazardous events, such as tsunami, earthquakes, storms, hurricanes, are spatio-temporal events and a lot of spatio-temporal data has been accumulated to measure these events over time.

For example, earthquake prediction is a challenging and compelling problem, especially in New Zealand. Several high-intensity earthquakes have struck highly populated regions of Canterbury and Wellington and caused a high number of casualties and loss within the last decade. The immense capacity for destruction of earthquakes prompts for the ability to predict, within a reasonable time horizon, their occurrence so that proactive actions could be taken to minimize damage. However, earthquake prediction in general remains a controversial topic and there seems to be an overly pessimistic outlook on its success rate, especially in modern times. This is most likely a product of disappointment from a series of failed attempts at predicting earthquakes since the height of this field in the 1970s [37], with some researchers even going to the extent of abandoning the idea of prediction [38]. Despite a track record of modest success in earthquake prediction, a copious amount of geological data is continuously being collected and analyzed. This paper will investigate the feasibility of using Computational Intelligence (CI) based approaches in predicting the incidence of strong earthquakes using the seismic time series data recorded from various seismometer sites as the precursor.

### 19.5.2   Predictive Modelling of Seismic Data for Earthquake Forecasting Using NeuCube

The basic premise on which earthquake prediction techniques stand is that there are some phenomena, called pre cursors, which consistently occur before an earthquake. One of the most prominent approaches in this area is the measurement of anomalies in the different parts of the atmosphere which seems to change due to seismogenic effects [39], for example, the temperature [36] and density [12] of electrons in the ionosphere. Other approaches extend from measuring the amount of radon emissions in the soil and ground water [40] and by observing the behavior of animals such as mice [41] and common toad [42].

Lately there have been several studies that suggest that the existence of some signatures in the seismograph readings prior to the occurrence of earthquakes. The possibility of using high frequency components of micro-seismic noise readings has been studied in [43], which reports a characteristic change one or two days before an earthquake. Another study reports that pulsed vibrations are recorded between 5 and 10 days before earthquakes around Russia [44]. Another study done with the Tottori earthquake in the year 2000 also revealed that there are seismic quiescence anomalies before the earthquake [45], which are also observed leading to the massive Taiwan Chi-Chi earthquake in 1999 [46]. Based on these literatures there is a scientific basis in using the readings of seismographs as precursors for short-term prediction of strong earthquakes. The challenge is to develop methods that can learn from the patterns that are hidden in the intricate interactions between spatial and temporal components.

Despite various precursor variables having been proposed, the application of Computational Intelligence methods to deal with the problem of earthquake prediction has unfortunately been scarce.

A method employing artificial neural networks has been developed to predict earthquakes in Chile, by using the b-value, the Baths law, and the Omuri-Utsu's law as input parameters [47]. This promising research built and used multiple models corresponding to the geographical regions or cities it wanted to analyze, and since classical artificial neural networks are not suitable to work with the temporal aspect of the data, it employs several fundamental geophysics laws to extract the input features from the available time-variant data. A study by the same authors using similar technique has also been done for earthquakes around the Iberian Peninsula [48].

Another approach using an adaptive neural fuzzy inference system (ANFIS) has also been proposed, using location of the earthquake as the input and the magnitude as the out- put on the assumption that the system will tune itself to model the principle of conservation of energy and momentum of annual earthquakes [49]. Another ANFIS-based approach was proposed by [50] in which historical earthquake data is mapped into two kinds of input: spatial and temporal, which are analyzed separately. Yet another ANFIS based approach was proposed by [51] in which the inference system is used to predict a time-series of earthquake parameters of the Sunda region in Indonesia. A rule-based system for earthquake prediction was also proposed by [52] which claims 100% accuracy within 15 h, although the spatial resolution of the prediction area is low, covering areas as large as a hemisphere. Almost all of the previous research in employing CI methods seem to extract features such as the b-value (Gutenberg–Richter law), Baths law, Omoris law and so forth from a historical sequence of previous earthquakes in a region. None in particular proposed the use of multiple time series readings of seismic activities prior to the earthquakes to capture predictive spatiotemporal patterns. In this research, we investigate the effectiveness of a spatiotemporal modeling approach with SNN for prediction, based on the seismicity prior to the occurrence of the earthquake.

The classifier system used in this research is the NeuCube SNN architecture as depicted in Fig. 19.17a. Input data is transformed into spike trains using encoding algorithms like simple thresholding or Ben's Spiker Algorithm (BSA) introduced in Chap. 4. These spike trains are then fed into the cube (SNNc) in an unsupervised learning procedure so that the reservoir's network can learn to activate the same groups of spiking neurons when similar spatiotemporal input stimuli are presented. After the unsupervised training phase, the same data is propagated again and output neurons are evolved to learn to classify the SNNc activity into predefined classes. Different SNN methods can be used to learn and classify spiking patterns from the SNNc, including the deSNN. Figure 19.17b shows examples of connectivity of trained SNNcubes on seismic streaming data.

## 19.5.3  Experiment Design

The experiment in this study was designed to investigate whether by building a model to learn from seismometer readings preceding a seismic event, the imminence of large earthquakes can be predicted. This question can be formulated and tested as a binary classification problem of differentiating a positive class from a negative class.

In this study instances in the positive class corresponded to earthquakes which are historically notable, felt by the general population in the region and classified as *strong* or *severe* in intensity by GNS Science New Zealand as displayed in the GeoNet website (www.geonet.org.nz). GeoNet provides access to extensive data recorded by sensors belonging to the New Zealand National Seismograph Network [54]. As in [46], the location of the earthquake is considered to be known since the model was built for a specific geographical area, namely the region of Canterbury in the South Island of New Zealand in which the city of Christchurch is located. The samples were taken after the year 2010 since most of the strong and well known earthquakes in the region happened afterwards, and the data quality is more consistent in recent times. It should be noted that strong aftershocks which usually occur within a few days after a large earthquake were excluded.

For both classes, appropriate samples of earthquakes needed to be selected. The 12 events considered as the positive class are listed in Table 19.5. The small number of samples is the consequence of the fact that strong earthquakes happen very rarely throughout history, and more so in a particular region. Another 12 samples were taken from the catalog from around the same time period and region where there were no big earthquakes and the maximum magnitude experienced in the surrounding days did not see any significant jump. These samples were the negative class, representing episodes of low overall seismic activity.

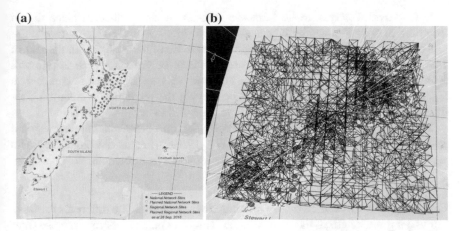

**Fig. 19.17** **a** A general architecture of an eSTDM for seismic data modelling and earthquake forecasting. As an example for the source of seismic data, seismic centers of New Zealand are shown. **b** Examples of connectivity of trained SNNcubes on seismic streaming data for New Zealand

For the purpose of this study, the reading used is Seismic time-series data from the Long Period Band Type, which corresponds to a 1 Hz sample rate. The instrument code used is H, which means High Gain Seismometer. The Orientation Code is N which means that the displacement measured is along the direction of North-South horizontal axis. Four seismic stations from the Canterbury area (McQueen's Valley, Oxford, Lake Taylor Station, and Kahutara) were selected for their generally higher uptime. The geographical location of these stations along with the others in the New Zealand National Seismograph Network can be seen in Fig. 19.18.

In this study the observation duration length is fixed to 5 days (120 h). After the raw data is obtained, simple preprocessing steps needed to be applied to prepare the data to be fed to the models. The input data $I$ of a sample for this earthquake prediction problem is defined as $L1, L2, ..., Ls$ where $s$ is the number of seismic sites which are taken into account. Each vector $L \in I$ is a time-series $L = a-t-d, a-t-d +1, ..., a-t$ in which the values are a chronologically ordered set of $d$ real-variables, $d$ being the duration of observation and $t$ the prediction horizon, i.e. the time before the earthquake occurs, and assuming $a1$ is the value at the occurrence of the earthquake.

Since the seismograph reading is high-resolution spanning over a long period of time, the standard deviation of the signal is computed in a piecewise manner in order to reduce the length and dimensionality of the time series.

*Data Acquisition and Preparation*

The seismometer readings preceding the sample earthquakes data was obtained from the New Zealand GeoNet's Continuous Waveform Buffer web services. The website provides access to an immense amount of data collected since digital recording in New Zealand commenced in 1986 (http://www.geonet.org.nz).

To predict ahead an actual event, data needs to be offset by a certain amount of time. The duration of the observation also needs to be chosen, which in turn will determine the length of the prediction horizon. This arrangement is depicted in Fig. 19.3. In this experiment, the effect of varying the prediction horizon on classification accuracy is analyzed. For the purpose of this.

The spatiotemporal signals can be directly fed into NeuCube. The signals are discretized into spike trains as shown in Fig. 19.19, whereas the other classifiers require the signals to be flattened out of the into one feature vector.

*Results*

The experiment was carried out by running the data through the different classifiers and varying the length of the prediction horizon. The parameters for each classifier were tuned heuristically to obtain the best results from each of them. In addition to the accuracy, the performance of the classifiers is measured in terms of the balanced F-score on the positive earthquake.

In Table 19.6 in addition to the True Positive (TP) and False Positive (FP) results, the F-Score is also calculated as: $F = 2TP/(2TP + FP + FN)$. This additional measure is important, since overall accuracy alone does not reveal the actual performance within each of the classes which is of interest in a binary classification problem. Since the number of samples is small, the training/validation scheme used is leave—one-out cross validation [54].

The result of the experiment laid in Table 19.5, showed that, as expected, shorter prediction horizons produced better prognosis. It should be noted that in a balanced binary classification problem, there is a baseline accuracy of 50%, which can be

| Table 19.5 Earthquakes within the Canterbury region used as positive samples | Public ID | Date | Magnitude | Depth (km) |
|---|---|---|---|---|
| | 3366146 | September 3 2010 | 7.1 | 11 |
| | 3450113 | January 19 2011 | 5.1 | 9 |
| | 3468575 | February 21 2011 | 6.3 | 5 |
| | 3474093 | March 5 2011 | 5.0 | 10 |
| | 3497857 | April 16 2011 | 5.3 | 9 |
| | 3505099 | April 29 2011 | 5.2 | 11 |
| | 3525264 | June 5 2011 | 5.5 | 9 |
| | 3528810 | June 13 2011 | 5.9 | 9 |
| | 3591999 | October 9 2011 | 5.6 | 8 |
| | 3631359 | December 23 2011 | 5.8 | 10 |
| | 2015p012816 | January 5 2015 | 6.0 | 5 |
| | 2015p305812 | April 24 2015 | 6.2 | 52 |

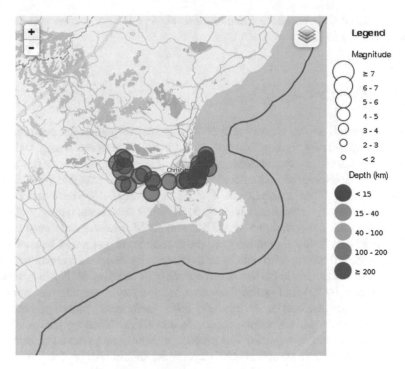

**Fig. 19.18** New Zealand national seismograph network with the 4 selected sites around Canterbury area grayed

achieved statistically by random guessing or giving the same answer to all the cases. It is safe to say that traditional CI methods like MLP and SVM were not capable of learning from this type of data. It is interesting to note that no models were able to differentiate between the two classes 48 h ahead of an earthquake event, suggesting a certain temporal limit to the prediction horizon in this particular experiment.

This finding also lends support to previous studies which suggests that there are certain patterns exhibited by seismicity readings that can be used to predict the imminence of large earthquakes. The result shown in Table 19.6 gives us confidence that seismicity data is a viable precursor for short-term earthquake prediction. The best prediction accuracy obtained with the NeuCube model successfully predicted 11 out of 12 strong earthquakes and raised only 1 false alarm, 1 h prior to the actual event, which is indeed promising. The connectivity of the $3 \times 3 \times 3$ SNNc after training is depicted in Fig. 19.20. A connection between the neurons means a temporal association. The shown trajectory depicts sequence of seismic events that precede a major earthquake.

Figure 19.21 shows a NeuCube SNN reservoir rendered in a 3D Virtual Reality environment on top of a map of New Zealand, enabling users to immerse themselves and walk around the neurons and observe the connection building and spiking activity in time and space.

## 19.5.4  Discussions

This research has shown a novel and promising way to predict the occurrence of strong earthquakes by training a model to differentiate between strong and weak earthquakes based on spatiotemporal seismicity precursors. This research also showed that SNN can be successfully used for early and accurate prediction of hazardous events. The capability of a more advanced SNN-based method like NeuCube to capture complex spatiotemporal signal has been demonstrated, in relation to traditional techniques like MLP and SVM. For the latter methods, an additional step of feature extraction from the time series signal might be needed for them to work effectively with data of such complexity and dimensionality.

For future works, it is important to further verify the models' ability to generalize to unseen data by expanding the dataset to include more samples that represent real-world situations and/or incorporate other earthquake prone geographical regions such as Japan, California, Indonesia and Chile. Running the analysis in real-time as the data is collected will produce a useful and practical disaster prediction system. A more comprehensive experiment should also be done to find the

**Fig. 19.19** Preprocessed seismogram and the resulting spike train

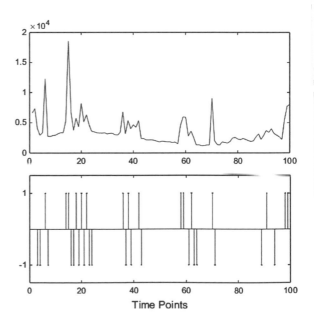

**Table 19.6** Classification accuracy result with varying prediction horizon [61]

|         |          | 1 h | 6 h | 24 h | 48 h |
|---------|----------|-----|-----|------|------|
| MLP     | Accuracy | 58.33% | 54.16% | 41.66% | 41.66% |
|         | F-score  | 0.58 | 0.52 | 0.41 | 0.41 |
|         | TP rate  | 0.58 | 0.50 | 0.41 | 0.41 |
|         | FP rate  | 0.41 | 0.41 | 0.58 | 0.58 |
| SVM     | Accuracy | 54.16% | 50% | 37.5% | 37.5% |
|         | F-score  | 0.58 | 0.52 | 0.41 | 0.41 |
|         | TP rate  | 0.58 | 0.50 | 0.41 | 0.41 |
|         | FP rate  | 0.41 | 0.41 | 0.58 | 0.58 |
| ECF     | Accuracy | 70.83% | 66.67% | 66.67% | 50% |
|         | F-score  | 0.63 | 0.60 | 0.66 | 0.64 |
|         | TP rate  | 0.50 | 0.50 | 0.66 | 0.91 |
|         | FP rate  | 0.04 | 0.16 | 0.33 | 0.91 |
| NeuCube | Accuracy | **91.67%** | **83.33%** | **70.83%** | **54.17%** |
|         | F-score  | 0.91 | 0.80 | 0.72 | 0.42 |
|         | TP rate  | 0.91 | 0.83 | 0.75 | 0.33 |
|         | FP rate  | 0.08 | 0.25 | 0.25 | 0.25 |

best prediction time horizon and observation period. A further interesting aspect would be the extraction of knowledge in the form of human understandable rules of the spatiotemporal patterns exhibited by seismicity readings in regards to the occurrence of earthquakes and our knowledge about the underlying mechanism of these seismic activities. This is also a promising line of research to be extended for the prediction and analysis of other disastrous events like tsunami and land slides.

## 19.6   Future Applications

There are many potential applications of BI-SNN for building eSTDM, some of them discussed briefly here.

### 19.6.1   Modelling Multisensory Air Pollution Streaming Data

Modelling multisensory streaming data from sensors measuring air pollution. Example of multiple sensors distributed in the Vancouver area is shown in Fig. 19.22a and the connectivity of a trained NeuCube model—in Fig. 19.22b. NeuCube 3D spiking neural network map of southwestern British Columbia

**Fig. 19.20** SNN reservoir
with input neurons and
synapses after training of a
SNNcube with seismic
streaming data from 4 seismic
centers around Christchurch

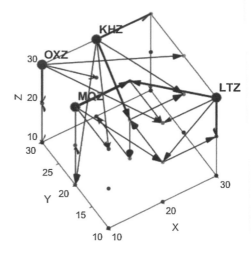

**Fig. 19.21** NeuCube SNN
reservoir rendered in a 3D
virtual reality environment on
top of a map of New Zealand,
enabling users to immerse
themselves and walk around
the neurons and observe the
connection building and
spiking activity in time and
space

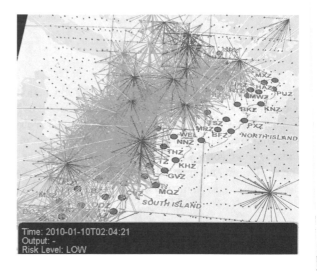

showing the Lower Fraser Valley network of monitors with regional and government fixed monitors (dark green circles). Spatio-temporal relationships (lines) and
activity (light green circles) of ozone (O3) (left cube) and carbon monoxide
(CO) (right cube) concentrations can be analysed simultaneously. In [55] a SNN
computational model is developed for the prediction of air pollution in local areas
of London few hours ahead.

### 19.6.2  Wind Energy Prediction from Wind Turbines

Predicting wind energy from wind measured streaming data sensors, such as wind speed and wind direction can save energy and bring large benefits. Figure 19.23a shows Wind turbines in New Zealand and China; (b) The connectivity of a SNN cube trained on wind speed and wind direction streaming data.

### 19.6.3  SNN for Radio-Astronomy Data Modelling

With the introduction of the Square Kilometre Array Project, a revolution in the data available to radioastronomers is occurring. Of particular interest is the identification of distinctive spectral patterns known as dispersed transients (single, bright pulses of unknown extraterrestrial origin) or dispersed pulsars (characteristic signals given off by the rotation of pulsar stars). These signals, if identified and analysed correctly, can have major implications towards our understanding of relativistic physics, and therefore, our understanding of the fundamental forces at work in our universe. However, these signals are highly infrequent (class imbalance of 1:10,000–12,000 pulsar to noise), highly unpredictable in terms of signal characteristics, and buried in noise. The current state of the art approach requires a brute force search and is untenable in the face of the volume of data the SKA will produce—a data stream rate of 1.5–2.5 TBps [56].

An alternative approach using neuromorphic principles (the NeuCube evolving spatio-temporal data machine) would be a first line candidate. This is appropriate as NeuCube eSTDM provides compact representation of spatial, spectral, and temporal characteristics, evolving learning, non-linear pattern recognition, and low computation cost comparative to alternative techniques, particularly when implemented on neuromorphic hardware as it was discussed [1].

## 19.7  Chapter Summary and Further Readings for Deeper Knowledge

SNN can learn streaming data as changes of values of input variables (e.g. sensors) over time, capturing patterns of interactions between these variables that can be utilized to predict future events.

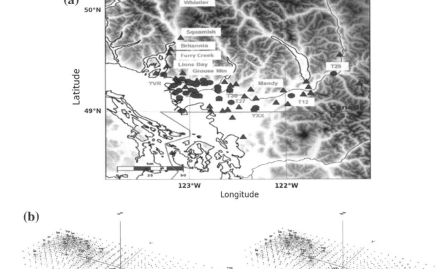

**Fig. 19.22** Modelling multisensory streaming data from sensors measuring air pollution. Example of multiple sensors distributed in the Vancouver area is shown in (a) and the connectivity of a trained NeuCube model—in (b). NeuCube 3D spiking neural network map of southwestern British Columbia showing the Lower Fraser Valley network of monitors with regional and government fixed monitors (dark green circles). Spatio-temporal relationships (lines) and activity (light green circles) of ozone ($O_3$) (left cube) and carbon monoxide (CO) (right cube) concentrations can be analysed simultaneously (the figure is created by J. Espinosa)

This chapter, first presents two methods of using eSNN and BI-SNN respectively, for building eSTDM to deal with streaming data in an incremental way. These methods are illustrated on applications in four real-world areas:

- Finance;
- Ecology;
- Transport;
- Environment (e.g. seismic data for earthquake forecasting; air pollution modelling; wing energy prediction);
- Prediction of Hourly Air Pollution in London Area Using Evolving Spiking Neural Networks [55].

Further readings can be found in:

**(a)**                                                **(b)**

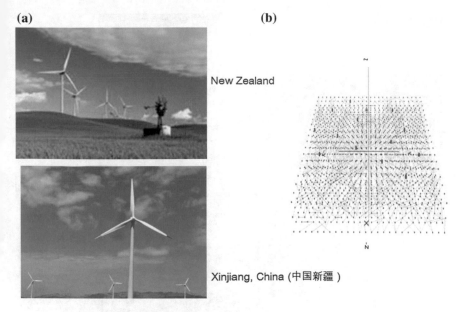

New Zealand

Xinjiang, China (中国新疆 )

**Fig. 19.23  a** Wind turbines in New Zealand and China; **b** the connectivity of a SNN cube trained on wind speed and wind direction streaming data (the figure is created by J. Espinosa)

- Brain-like information processing for spatio-temporal pattern recognition (Chap. 47 in [57]);
- Ecological informatics for the prediction and management of invasive species (Chap. 35 in [57]);
- Demo on modelling seismic data in NeuCube: https://kedri.aut.ac.nz/R-and-D-Systems/neucube/seismic;
- On–line learning in eSNN over drifting streaming data [62].

**Acknowledgements**  Parts of the material in this chapter have been previously published as referenced in the relevant sections of this chapter. I would like to acknowledge the contribution to these publications of my co-authors Enmei Tu, Josafath Israel Espinosa, Sue Worner, Reggio Hartono, Stefan Marks, Nathan Scott, S. Gulyaev, N. Sengupta, R. Khansam, V. Ravi, A. Gollahalli, Petr Maciak, Imanol Bilbao-Quintana.

# Appendix 1

---

**Algorithm 1:** eSNN training algorithm

---

1:  Initialize neuron repository, $R = \{\}$

2:  Set eSNN parameter $mod = [0, 1], C = [0, 1], sim = [0, 1]$

3:  **for** $\forall$ input pattern $i$ that belongs to the same class **do**

4:      Encode input pattern into firing time of multiple pre-synaptic neurons $j$

5:      Create a new output neuron $i$ for this class and calculate the connection weights as $w_{ji} = mod^{order(j)}$

6:      Calculate $PSP_{max(i)} = \sum_j w_{ji} \times mod^{order(j)}$

7:      Get $PSP$ threshold value $\gamma_i = PSP_{max(i)} \times C$

8:      **if** The new neuron weight vector $\leq sim$ of trained output neuron weight

         vector in $R$

      **then**

9:          Update the weight vector and threshold of the most similar neuron in the same output

class group

10:          $w = \dfrac{w_{new} + w.N}{N+1}$

11:          $\gamma = \dfrac{\gamma_{new} + \gamma\, N}{N+1}$

12:          where $N$ is the number of previous merges of the most similar neurons

13:      **else**

14:          Add the weight vector and threshold of the new neuron to the neuron repository $R$

15:      **end if**

16:  **end for**

17:  Repeat above for all input patterns of other output classes.

---

# Appendix 2

Improved stock market movement prediction with optimised eSNN parameters on the same stock data as in 19.2 (Fig. 19.24)

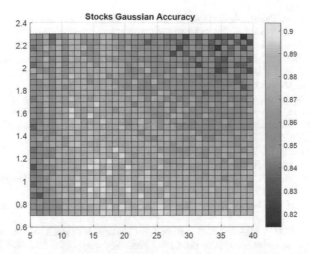

**Fig. 19.24** A grid optimisation of two eSNN parameters (number of receptive fields $N$ and with of the receptive fields) of eSNN for BSE stock movement prediction. This resulted in a significant improvement of predicted stock value (max accuracy achieved is 90% for $N = 11$ and width = 1.6) (the figure is created by Imanol Bilbao-Quintana)

# References

1. N. Kasabov, N. Scott, E. Tu, S. Marks, N. Sengupta, E. Capecci, M. Othman, M. Doborjeh, N. Murli, R. Hartono, J. Espinosa-Ramos, L. Zhou, F. Alvi, G. Wang, D. Taylor, V. Feigin, S. Gulyaev, M. Mahmoudh, Z.G. Hou, J. Yang, Design methodology and selected applications of evolving spatio-temporal data machines in the NeuCube neuromorphic framework. Neural Netw. **78**, 1–14 (2016). https://doi.org/10.1016/j.neunet.2015.09.011
2. E. Tu, N. Kasabov, J. Yang, Mapping temporal variables into the NeuCube for improved pattern recognition, predictive modeling, and understanding of stream data. IEEE Trans. Neural Netw. Learn. Syst. **28**(6), 1305–1317 (2017)
3. N. Kasabov, NeuCube: a spiking neural network architecture for mapping, learning and understanding of spatio-temporal brain data. Neural Netw. **52**, 62–76 (2014)
4. C. Chu, Y. Ni, G.J.S.C. Tan, J. Ashburton, Kernel regression for fMRI pattern prediction. Neuroimage **56**(9), 662–673 (2011)
5. M. Gholami Doborjeh, N. Kasabov, Mapping, learning, visualisation and classification of fMRI data in the NeuCube evolving spiking neural network framework. IEEE Trans. Neural Netw. Learn. Syst. **28**(4), 887–899 (2015)
6. M. Just, StarPlus fMRI data (2001). http://www.cs.cmu.edu/afs/cs.cmu.edu/project/theo-81/www/
7. T.M. Mitchell, R. Hutchinson, M.A. Just, R.S.F.P. Niculescu, X. Wang, Classifying instantaneous cognitive states from fMRI data, in *AMIA Annual Symposium Proceedings* (American Medical Informatics Association, 2003), p. 465
8. N. Murli, N. Kasabov, B. Handaga, Classification of fMRI data in the NeuCube evolving spiking neural network architecture, in *Proceedings ICONIP* (Springer), pp. 421–428
9. T. Delbruck, P. Lichtsteiner, Fast sensory motor control based on event-based hybrid neuromorphic-procedural system, in *2007 IEEE International Symposium on Circuits and Systems*, pp. 845–848. IEEE, New Orleans, LA, USA (2007). http://ieeexplore.ieee.org/lpdocs/epic03/wrapper.htm?arnumber=4252767

10. N. Kasabov, V. Feigin, Z.-G. Hou, Y. Chen, L. Liang, R. Krishnamurthi et al., Evolving spiking neural networks for personalised modelling, classification and prediction of spatio-temporal patterns with a case study on stroke. Neurocomputing **134**, 269–279 (2014)
11. J. Schmidhuber, Deep learning in neural networks: an overview. Neural Netw. **61**, 85–117 (2014)
12. J. Liu, Y. Chen, Y. Chuo, H. Tsai, Variations of ionospheric total electron content during the chi-chi earthquake. Geophys. Res. Lett. **28**(7), 1383–1386 (2001)
13. S. Liu, S. Wang, K. Jayarajah, A. Misra, R. Krishnan, Todmis: mining communities from trajectories, in *Proceedings of 22nd ACM International Conference on Information & Knowledge Management, CIKM'13*. ACM (2013), pp. 2109–2118. http://doi.acm.org/10.1145/2505515.2505552
14. D. Buonomano, W. Maass, State-dependent computations: spatio-temporal processing in cortical networks. Nat. Rev. Neurosci. **10**, 113–125 (2009)
15. W. Gerstner, A.K. Kreiter, H.M.H.A.V. Markram, Theory and simulation in neuroscience. Proc. Natl. Acad. Sci. U S A **94**(24), 12740–12741 (1997)
16. W. Gerstner, H. Sprekeler, G. Deco, Theory and simulation in neuroscience. Science **338**, 60–65 (2012)
17. S. Song, K.D. Miller, L.F. Abbott, Competitive Hebbian learning through spike-timing-dependent synaptic plasticity. Nat. Neurosci. **3**(9), 919–926 (2000). http://www.ncbi.nlm.nih.gov/pubmed/10966623
18. S. Fusi, Spike-driven synaptic plasticity for learning correlated patterns of mean firing rates. Rev. Neurosci. **14**(1–2), 73–84 (2003)
19. E.M. Izhikevich, Which model to use for cortical spiking neurons? IEEE Trans. Neural Netw. **15**(5), 1063–1070 (2004)
20. N. Kasabov, J. Hu, Y. Chen, N. Scott, Y. Turkova, Spatio-temporal EEG data classification in the NeuCube 3D SNN environment: methodology and examples, in *Proceedings of the International Conference on Neural Information Processing* (Springer, Daegu, Korea, 2013), pp. 63–69
21. A. Mohemmed, N. Kasabov, Incremental learning algorithm for spatio-temporal spike pattern classification, in *Proceedings of the IEEE world congress on computational intelligence*, Brisbane, Australia, pp. 1227–1232
22. N. Kasabov, E. Capecci, Spiking neural network methodology for modelling, classification and understanding of EEG data measuring cognitive processes. Inf. Sci. **294**, 565–575 (2015)
23. S. Soltic, N. Kasabov, Knowledge extraction from evolving spiking neural networks with rank order population coding. Int. J. Neural Syst. **20**(6), 437–445 (2010)
24. N. Kasabov, K. Dhoble, N. Nuntalid, G. Indiveri, Dynamic evolving spiking neural networks for on-line spatio-and spectro-temporal pattern recognition. Neural Netw. **41**, 188–201 (2013)
25. N. Kasabov, *Evolving Connectionist Systems* (Springer, Berlin, 2007)
26. M. Defoin-Platel, S. Schliebs, N. Kasabov, Quantum-inspired evolutionary algorithm: a multi-model EDA. IEEE Trans. Evol. Comput. **13**(6), 1218–1232 (2009)
27. S. Bruckner, V. Šoltészová, M.E. Gröller, J. Hladuvka, K. Buhler, J.Y. Yu, B.J. Dickson, BrainGazer—visual queries for neurobiology research. IEEE Trans. Vis. Comput. Graph. **15**(6), 1497–1504 (2009). https://doi.org/10.1109/TVCG.2009.121
28. C.-Y. Lin, K.-L. Tsai, S.-C. Wang, C.-H. Hsieh, H.-M. Chang, A.-S. Chiang, The neuron navigator: exploring the information pathway through the neural maze, in *2011 IEEE Pacific Visualization Symposium, PacificVis* (2011), pp. 35–42
29. A. von Kapri, T. Rick, T.C. Potjans, M. Diesmann, T. Kuhlen, Towards the visualization of spiking neurons in virtual reality. Stud. Health Technol. Inform. **163**, 685–687 (2011)
30. S. Marks, *VR Visualisation of NeuCube, Evolving Systems* (Springer, Berlin, 2017)
31. R. Khansama, V. Ravi, N. Sengupta, A.R. Gollahalli, N. Kasabov, I. Bilbao-Quintana, Stock market movement prediction using evolving spiking neural networks, Evolving Systems, 2018
32. Quandl Financial, Economic and Alternative Data. https://www.quandl.com/
33. Historical-Indices. http://www.bseindia.com/indices/IndexArchiveData.aspx
34. NSE—national stock exchange of India ltd. https://www.nseindia.com/products/content/equities/indices/historicalindexdata.htm
35. Wikipedia. http://wikipedia.com

36. K.-I. Oyama, Y. Kakinami, J.-Y. Liu, M. Kamogawa, T. Kodama, Reduction of electron temperature in low-latitude ionosphere at 600 km before and after large earthquakes. J. Geophys. Res. Space Phys. (1978–2012) **113**(A11) (2008)
37. T.H. Jordan, Earthquake predictability, brick by brick. Seismol. Res. Lett. **77**(1), 3–6 (2006)
38. R.J. Geller, D.D. Jackson, Y.Y. Kagan, F. Mulargia, Enhanced: earthquakes cannot be predicted. Science **275**(5306), 1616–1620 (1997)
39. S. Pulinets, A. Legen'Ka, T. Gaivoronskaya, V.K. Depuev, Main phenomenological features of ionospheric precursors of strong earth- quakes. J. Atmos. Solar Terr. Phys. **65**(16), 1337–1347 (2003)
40. D. Ghosh, A. Deb, R. Sengupta, Anomalous radon emission as precursor of earthquake. J. Appl. Geophys. **69**(2), 67–81 (2009)
41. Y. Li, Y. Liu, Z. Jiang, J. Guan, G. Yi, S. Cheng, B. Yang, T. Fu, Z. Wang, Behavioral change related to Wenchuan devastating earthquake in mice. Bioelectromagnetics **30**(8), 613–620 (2009)
42. R.A. Grant, T. Halliday, Predicting the unpredictable; evidence of pre-seismic anticipatory behaviour in the common toad. J. Zool. **281**(4), 263–271 (2010)
43. I. Sovic´, K. Sˇariri, M. Zˇivcˇic´, High frequency microseismic noise as possible earthquake precursor. Res. Geophys. **3**(1), e2 (2013)
44. G. Sobolev, A. Lyubushin, Microseismic impulses as earthquake precursors. Izv. Phys. Solid Earth **42**(9), 721–733 (2006)
45. Q. Huang, Search for reliable precursors: a case study of the seismic quiescence of the 2000 western Tottori prefecture earthquake. J. Geophys. Res. Solid Earth (1978–2012) **111**(B4) (2006)
46. Y.-M. Wu, L.-Y. Chiao, Seismic quiescence before the 1999 chi-chi, Taiwan, mw 7.6 earthquake. Bull. Seismol. Soc. Am. **96**(1), 321–327 (2006)
47. J. Reyes, A. Morales-Esteban, F. Mart´ınez-A´ lvarez, Neural networks to predict earthquakes in chile. Appl. Soft Comput. **13**(2), 1314–1328 (2013)
48. A. Morales-Esteban, F. Martínez-Álvarez, J. Reyes, Earthquake prediction in seismogenic areas of the iberian peninsula based on computational intelligence. Tectonophysics **593**, 121–134 (2013)
49. M. Shibli, A novel approach to predict earthquakes using adaptive neural fuzzy inference system and conservation of energy-angular momentum. Int. J. Comput. Inf. Syst. Ind. Manag. Appl. ISSN (2011), pp. 2150–7988
50. A. Zamani, M.R. Sorbi, A.A. Safavi, Application of neural network and ANFIS model for earthquake occurrence in Iran. Earth Sci. Inf. **6**(2), 71–85 (2013)
51. E. Joelianto, S. Widiyantoro, M. Ichsan, Time series estimation on earthquake events using ANFIS with mapping function. Int. J. Artif. Intell. **3**(A09), 37–63 (2008)
52. A. Ikram, U. Qamar, A rule-based expert system for earthquake prediction. J. Intell. Inf. Syst. **43**(2), 205–230 (2014)
53. N. Kasabov, N. Scott, E. Tu, S. Marks, N. Sengupta, E. Capecci, M. Othman, M.G. Doborjeh, N. Murli, J.I. Espinosa-Ramos et al., Evolving spatio-temporal data machines based on the neucube neuromorphic framework: design methodology and selected applications. Neural Netw. (2015)
54. T. Petersen, K. Gledhill, M. Chadwick, N.H. Gale, J. Ristau, The New Zealand national seismograph network. Seismol. Res. Lett. **82**(1), 9–20 (2011)
55. P.S.P Maciaga, N.K. Kasabov, M. Kryszkiewicza, R. Benbenik, Prediction of hourly air pollution in London area using evolving spiking neural networks. Environ. Modelling Software, Elsevier (2018/2019)
56. Square Kilometer Array (SKA) Project: https://www.skatelescope.org
57. N. Kasabov (ed.), *Springer Handbook of Bio-/Neuroinformatics* (Springer, Berlin, 2014)
58. N. Kasabov, To spike or not to spike: a probabilistic spiking neuron model. Neural Netw. **23**(1), 16–19 (2010)
59. S. Schliebs, N. Kasabov, Evolving spiking neural network—a survey. Evolving Syst. **4**(2), 87–98 (2013)

60. B. Schrauwen, J. Van Campenhout, BSA, a fast and accurate spike train encoding scheme, in *Proceedings of the International Joint Conference on Neural Networks,* vol. 4 (IEEE Piscataway, NJ, 2003), pp. 2825–2830
61. R. Hartono, PhD Thesis, Auckland University of Technology (2018)
62. J.L. Lobo, I. Laña, J. Del Ser, M.N. Bilbao, N. Kasabov, Evolving spiking neural networks for online learning over drifting data streams. Neural Netw. **108**, 1–19 (2018)

# Part VIII
# Future Development in BI-SNN and BI-AI

# Chapter 20
# From von Neumann Machines to Neuromorphic Platforms

Spiking neural networks (SNN), being highly parallel computational systems, can be implemented on various computational platforms, from the traditional von Neumann machines to the specialised neuromorphic platforms. This chapter discusses various implementation strategies of SNN and brain-inspired AI (BI-AI). Some of the material in the chapter is after [1].

The chapter is organised in the following sections:

20.1. Principles of computation. The von Neumann machines and beyond.
20.2. Neuromorphic computation and neuromorphic machines.
20.3. ANN and SNN development systems. NeuCube as a development system for spatio-temporal data machines.
20.4. Summary and further readings.


## 20.1 Principles of Computation. The von Neumann Machines and Beyond

### 20.1.1 General Notions

In 19th century **Ada Lovelace** wrote the first algorithm as a sequence of commands to be executed by a mechanical machine. The breakthrough work of Alan Turing in the 1940s, stating the possibility of using just 0's and 1's to simulate any process of formal reasoning [2] lead to massive development in the field of information theory and computer architecture. Simultaneously, significant progress was made by the neuroscientists in understanding the most efficient and intelligent machine known to man, the human brain. These parallel advancements in the middle of the last century had made man's imagination of creating 'intelligent' systems a possibility. These rational systems/agents were thought ideally to be able to perceive the external

© Springer-Verlag GmbH Germany, part of Springer Nature 2019      661
N. K. Kasabov, *Time-Space, Spiking Neural Networks and Brain-Inspired Artificial Intelligence*, Springer Series on Bio- and Neurosystems 7,
https://doi.org/10.1007/978-3-662-57715-8_20

environment and take actions accordingly to maximise its goal, mimicking the human brain.

The field of artificial neural networks and AI have grown strength to strength from the simple McCalluch and Pitt's linear threshold based artificial neuron model [3] to the latest era of deep learning [4], which builds very complex models by performing a combination of linear and non-linear transformations. This is done using millions of neurons stacked in a layered fashion forming an interconnected mesh. The tremendous push of AI towards emulation of real intelligence has been sustained so far by the realisation of the Moore's law [5] which states that the processing power of the of central processing units (CPU) doubles in every couple of years. But the future development of BI-AI systems would require new computation principles.

The chapter discusses three computation principles and architectures, as follows:

(a) The *von Neumann computer architecture*, that separates data and programmes (kept in the memory unit) from the computation (ALU) and the control. It uses bits as *static* information. It can be realised as:

- General purpose computers;
- Specialised fast computers: GPUs, TPUs
- Cloud-based computing platforms.

(b) A *neuromorphic computational architecture*, that integrates data, programs and computation in a SNN structure, similar to how the brain works. Here, *bits* (spikes) are associated with *time*.

(c) A *quantum (inspired) architecture*, that uses quantum bits, where bits are in a quantum *superposition* between 1 and 0.

SNN and AI models can be simulated using any of the architectures (if available) but with various efficiency as discussed in this chapter.

## 20.1.2   The von Neumann Computation Principle and the Atanassov's ABC Machine

**John von Neumann (1903–1957)** introduced a computation principle as discussed below. Throughout the continuous evolution of the traditional computers, von Neumann or the stored program architecture has continued to be the standard architecture for computers. It is a multi-modular design based on rigid physically separate functional units (Fig. 20.1). It specifically consists of three different entities:

**Fig. 20.1** The von Neumann computational architecture [6]

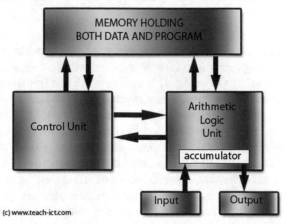

The Von Neumann or Stored Program architecture

(c) www.teach-ict.com

- Processing unit: The processing unit can be broken down into a couple of sub-units, the arithmetic and logical unit (ALU), the processing control unit and the program counter. The ALU compute the arithmetic logic needed to run programs. The control unit is used to control the flow of data through the processor.
- I/O unit: The i/o unit essentially encompasses all I/O the computer could possibly do (printing to a monitor, to paper, inputs from a mouse or keyboard, and others).
- Storage unit: The storage unit stores anything the computer would need to store and retrieve. This includes both volatile and non-volatile memory.

These units are connected over different buses like data bus, address bus and control bus. The bus allows for the communication between the various logical units. Though very robust, as shown in Fig. 20.1, this architecture inherently suffers from the bottleneck created due to the constant shuffling of the data between the memory unit and the central processing unit. This bottleneck leads to rigidity in the architecture as the data needs to pass through the bottleneck in a sequential order. An alternate solution of parallelising the computers has been proposed where millions of processors are interconnected. This solution, though, increases processing power, is still limited by the bottleneck in its core elements [7].

During the 1940s **John Atanassov (1905–1993)** with the help of one of his students Clifford E. Berry, at Iowa State College, created the ABC (Atanasoff-Berry Computer) that was the first electronic digital computer. The ABC computer was not a general-purpose one, but still, it was the first to implement three of the most important ideas used in computers nowadays: binary data representation; using electronics instead of mechanical switches and wheels, using a von Neumann architecture, where the memory and the computations are separated.

### 20.1.3    Going Beyond von Neumann Principles and ABC Computer

The saturation in the scalability of the von Neumann architecture led to new developments in computer and computing architectures. Neuromorphic computing coined by Carver Mead in the 1980s [8] and further developed recently is one of the paradigms of computing which has come into prominence. As the name'neuro-morphic' suggests, this paradigm of computing is inspired heavily by the human brain. Based on neuromorphic computing, the first silicon retina was developed by Misha Mahovald. Neuromorphic chips of silicon retina, called Dynamic Vision Sensors (DVS) [9] and other brain-inspired devices [10, 11] were further developed. Moreover, as the existence of AI is complimented by computing architectures and paradigms, having a real neuromorphic computer architecture oriented processing unit is a step towards the development of highly neuromorphic AI leading to BI-A. This is discussed in the next section.

## 20.2    Neuromorphic Computation and Platforms

### 20.2.1    General Principles

The neuromorphic computing paradigm as presented already in this book from the point of view of computational modelling and here to be presented as hardware implementation, draws great inspiration from our brain's ability to manage tens of billions of processing units connected by the hundreds of trillions of synapses using tens of watts of power on an average. The vast network of the processing units (neurons) in the brain is in a true sense a mesh. The data is transmitted over the network via the mesh of synapses seamlessly. Architecturally the presence of the memory and the processing unit as a single abstraction is uniquely advantageous leading to dynamic, self-programmable behaviour in complex environments [7]. The highly stochastic nature of computation in our brain is a very significant divergence from the bit-precise processing of the traditional CPU. The neuromorphic computing hence aspires to move away from the bit-precise computing paradigm towards the probabilistic models of simple, reliable and power and data efficient computing [12] by implementing neuromorphic principles such as spiking, plasticity, dynamic learning and adaptability. This architecture morphs the biological neurons, where the memory and the processing units are present as part of the cell body leading to de-centralised presence of memory and processing power over the network.

## 20.2.2 Hardware Platforms for Neuromorphic Computation

With significant commercial interest in sight, research community focused on the commercial scale development of the neuromorphic chips. The most prominent of the neuromorphic chips include the TrueNorth [13, 14] from IBM, the Neurogrid [15] developed by the Stanford University, the SpiNNaker chip [16] from the University of Manchester, the neuromorphic chips developed in ETH INI, Zurich [10, 17] and others. All of these neuromorphic chips consist of programmable neurons and synapses and use a multitude of CMOS technologies to achieve the neuromorphic behaviour. The details of the neuromorphic chips are well elaborated in [11].

*The SpiNNaker* (Fig. 20.2) system is developed by a team from the University of Manchester lead by **Steve Furber**. The system is designed around a plastic ball grid array package which incorporates a custom processing chip and a 128 MB SDRAM memory chip. The processing chip contains 18 ARM968 processing cores, each with 23 KBs of instruction memory and 64 KBs of data memory, a multicast packet router and sundry support components. The SpiNNaker communication fabric is based on a 2D triangular mesh with each node formed from a processor layer and a memory layer. The routing is based upon packet-switched Address Event Representation and relies on the fact that the connections from a particular neuron are static, or at most slowly changing. Each neuron can route through a unique tree, though in practice routing is based on populations of neurons rather than individual neurons, and the restricted size of each routing table makes this optimisation necessary on most cases. In addition to the hardware system, the project also developed numerous high level neural description language such as PyNN, and Nengo for application development on SpiNNaker.

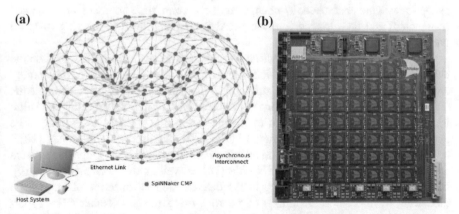

**Fig. 20.2** **a** The SpiNNaker general architecture. **b** A SpiNNaker board of 64 ARM processors, capable to process the activations of more than 100,000 spiking neurons in parallel, producing outputs every millisecond

*IBM TrueNorth* [14]. The IBM TrueNorth chip is the hardware developed under the DARPA SYNAPSE programme aimed at developing dense, power-efficient hardware for cognitive applications. This hardware consists of a 5.4 million transistor 28 nm CMOS chip with 4096 cores, where each core is made up of 256 neurons each having 256 synaptic inputs. The design of the TrueNorth core is a 256 × 256 cross-bar which selectively connects incoming neural spike events to outgoing neurons. The cross-bar inputs are coupled via buffers that can insert axonal delays. The outputs from the cross-bar couple into the digital neuron model, which implements a form of IF algorithm with 23 configurable parameters that can be adjusted to yield a range of different behaviours, and digital pseudo-random sources are used to generate stochastic behaviours through modulating the synaptic connections, the neuron threshold and the neuron leakage. Neuron spike event outputs from each core follow individually-configurable point-to-point routes to the input to another core, which can be on the same or another TrueNorth chip. Where a neuron output is required to connect to two or more neurosynaptic cores, the neuron is simply replicated within the same core. The TrueNorth hardware is supported by a software emulator which, exploiting the deterministic nature of the hardware, can be relied upon to predict the performance of the hardware exactly.

Another SNN chip that implements LIF model of a neuron is the recently proposed programmable *SRAM SNN chip* [17]. It is characterised by the following: 32 × 32 SRAM matrix of weights, each 5 bits (values between 0 and 31); 32 neurons of the adaptive, exponential IF model of a neuron; each neuron has 2 excitatory and 2 inhibitory inputs to which any of the 32 input dendrites (rows of weights) can be connected; AER for input data, for changing the connection weights and for output data streams; does not have any learning rule hardware implemented, so it allows to experiment with different supervised and unsupervised learning rules; learning (changing of the synaptic weights) is calculated outside the chip (in a computer, connected to the chip) in an asynchronous manner (only synaptic weights that need to change at the current time moment are changed (calculated) and then loaded into the SRAM) applying suitable learning rule and parameter settings.

The fact that modifying connection weights is done asynchronously outside the chip and then the weights are loaded in the SRAM allows for the deSNN learning algorithm to be implemented on this chip. After an input is applied to the AER circuits, the output from the neurons is produced and the deSNN learning algorithm implemented off-chip is then used to change connection weights accordingly. The new values of the weights are entered into the SRAM also asynchronously [18].

deSNN is also implementable on other recently proposed SNN chips of the same class, such as the digital IBM SNN chip [14] as well as on FPGA systems [19]. Despite the fast, one-pass learning in the deSNN models, in terms of large scale modelling of millions and billions of neurons using the SpiNNaker SNN supercomputer system for simulation purposes would be appropriate, especially at the level of parameter optimisation. A NeuCube implementation on a SpiNNaker platform is reported in [20].

## 20.3   SNN Development Systems. NeuCube as a Development System for Spatio-temporal Data Machines

### 20.3.1   A Brief Overview of SNN Development Systems

Numerous research [20–24] has focused on harnessing the theoretical powers of the spiking neural network (SNN) as it was done in various chapters of this book.

The number of software implementations that has appeared, as a result of ongoing research in the area of artificial neural networks, is ever growing. Majority of the neural network software is implemented to serve two purposes:

- Data analysis: These software packages are aimed at analysing real-world data derived from practical applications. The data analysis software use a relatively simple static architecture, hence are easily configurable and easy to use. Few examples of such software are: multilayer perceptron (MLP) [25], RBF network [26], Probabilistic network (PNN) [27], Self organizing maps (SOM) [28], Evolving connectionist systems, such as DENFIS and EFuNN [29]. These softwares are either available as independent packages, such as NeuCom [29], PyBrain (python) [30], Fast Artificial Neural Network (C++) [31], or as part of a data analytics software like Weka [32], Knime [33], Orange [34] and others.
- Research and development systems: As opposed to the data analysis software, they are complex in behaviour, and require background knowledge for usage and configuration. The Majority of the existing SNN software, including NeuCube, belong to this class.

We have briefly reviewed some of the key features of the current SNN development systems below.

*NEURON* [35]: Neuron is aimed at simulating a network of detailed neurological models. Its ability to simulate biophysical properties such as multiple channel types, channel distributions, ionic accumulation and so on renders it well suited for biological modelling. It also supports parallel simulation environment through: (1) distributing multiple simulations over multiple processors, and (2) distributing models of individual cells over multiple processors.

PyNEST [36, 37]: The neural simulation tool (NEST) is a software primarily developed in C++ to simulate a heterogeneous network of spiking neurons. NEST is implemented to ideally model neurons in the order of 104 and synapses in the order of $10^7$–$10^9$ on a range of devices from single core architectures to super-computers. NEST interfaces with python via implementation of PyNEST. PyNEST allows for greater flexibility in simulation setup, stimuli generation and simulation result analysis. A node and a connection comprise the core elements of the heterogeneous architecture.

Circuit Simulator [38, 39]: The circuit simulator is a software developed in C++ for simulation of heterogeneous networks with major emphasis on high-level network modelling and analysis. The C++ core of the software is integrated with Matlab based

GUI, for ease of use and analysis. CSIM enables the user to operate both spiking and analogue neuron models along with mechanisms of spike and analogue signal transmission through its synapse. It also performs dynamic synaptic behaviour by using short and long-term plasticity. In 2009, circuit simulator was further extended to parallel circuit simulator (PCSIM) software with the major extension being implementation on a distributed simulation engine in C++, interfacing with Python based GUI.

Neocortical Simulator [40]: NCS or Neocortical Simulator is an SNN simulation software, mainly intended for simulating mammalian neocortex [36]. During its initial development, NCS was a serial implementation in Matlab but later rewritten in C++ to integrate distributed modelling capability [41]. As reported in [36], NCS could simulate in the order of $10^6$ single compartment neuron and $10^{12}$ synapses using STP, LTP and STDP dynamics. Due to the considerable setup overhead of the ASCII-based files used for the I/O, a Python-based GUI scripting tool called BRAINLAB [40] was later developed to process I/O specifications for large scale modelling.

*Oger Toolbox* [42]: Oger toolbox is a Python-based toolbox, which implements modular learning architecture on large datasets. Apart from traditional machine learning methods such as Principal Component Analysis and Independent Component Analysis, it also implements SNN based reservoir computing paradigm for learning from sequential data. This software uses a single neuron as its building block, similar to the implementation in [37]. A Major highlight of this software in cludes the ability to customise the network with several non-linear functions and weight topologies, and a GPU optimised reservoir using CUDA.

*BRIAN* [23, 43]: Brian is an SNN simulator application programming interface written in Python. The purpose of developing this API is to provide users with the ability to write quick and easy simulation code [23], including custom neuron models and architecture. The model definition equations are separated from the implementation for better readability and reproducibility. The authors in [43], also emphasises the use of this software in teaching a neuroinformatics course [44].

The aforementioned discussion of the existing software highlights the suitability for building highly accurate neurological models but lacks a general framework for modelling temporal or SSTD, such as brain data, ecological and environmental data. Further In the line of the neural network development systems, and more specifically for SNN, where not only an SNN simulator can be developed, but a whole prototype system (also called spatio-temporal data machine) can be generated for solving a complex problem defined by SSTD, the NeuCube framework is discussed as a development system for SNN applications on SSTD [45] (Chap. 6).

## 20.3.2   The NeuCube Development System for Spatio-temporal Data Machines

The NeuCube framework for spatio-spectro temporal data (SSTD), is depicted in Fig. 20.3 [45] and explained in Chap. 6. A brief description is given below.

- Data encoding: The temporal information generated from the source (e.g. brain, earthquake sites) is passed through a data encoder component using a suitable encoding method, [24, 46]. It transforms the continuous information stream to discrete spike $n \times t$ trains$(f : R^{n \times t} \rightarrow \{0, 1\})$.
- Mapping spike encoded data and unsupervised learning: The spike trains are then entered into a scalable three dimensional space of hundreds, thousands or millions of spiking neurons, called SNNcube (SNNc), so that the spatial coordinates of the input variables (e.g. EEG channels; seismic sites, and so on) are mapped into spatially allocated neurons in the Cube, and an unsupervised time-dependent learning rule [47, 48] is applied $(g : \{0, 1\}^{n \times t} \rightarrow 0, 1^{m \times t}|m \gg n)$.
- Supervised learning: After unsupervised learning is applied, the second phase of learning is performed, when the input data is propagated again, now through the trained SNNc, and an SNN output classifier/regressor is trained in a supervised mode $\hat{y} := h(\beta, \varphi(0, 1))$ [49]. For this purpose, various SNN classifiers, regressors or spike pattern associators can be used, such as deSNN [49] and SPAN [50].

The NeuCube software development system architecture uses the above mentioned core pattern recognition block described in Fig. 20.4 as the central component and wraps a set of pluggable modules around it. The pluggable modules are mainly developed for: (1) Using fast and scalable hardware components running large scale applications; (2) Immersive model visualisation for in-depth understanding and analysis of the SSTD and its SNN model; (3) Specific applications like

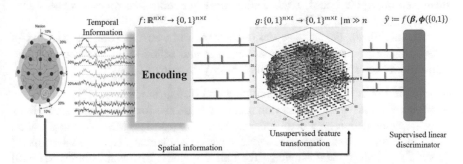

**Fig. 20.3** The NeuCube computational architecture for SSTD. The brain, shown as a source of SSTD is only exemplary, rather than restrictive (after [1]) (see also Chap. 6)

**NeuCube:** Neurocomputing Development System for Spatio- and Spectro-Temporal Data

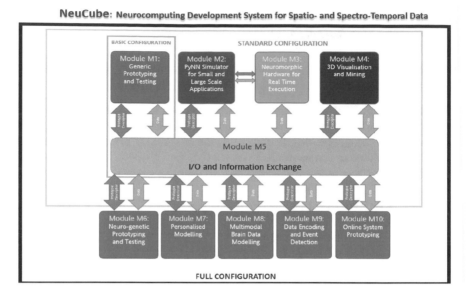

**Fig. 20.4** A modular structure of the NeuCube development system. Each module is designed for different application oriented SNN systems (http://www.kedri.aut.ac.nz/neucube/)

personalised modelling, brain computer interfaces and so on (4) Hyperparameter optimisation; and others.

Figure 20.3 shows the NeuCube computational architecture for SSTD. The brain, shown as a source of SSTD is only exemplary, rather than restrictive (after [1]) (see also Chap. 6).

Figure 20.4 shows a modular structure of the NeuCube development system. Each module is designed for different application oriented SNN systems and evolving spatio-temporal data machines (eSTDM).

Figure 20.5 shows a NeuCube development environment, also showing some application oriented devices, such as Oculus for 3D visualisation, a SpiNNaker small neuromorphic board, EEG device, an EEG-controlled mobile robot WITH from Kyushu Institute of Technology.

Each module in Fig. 20.4 is designed to perform an independent task and in some instances, written in a different language and suited to the specific computer platform and for specific application as briefly described below:

- Module M1 It is a generic prototyping and testing module.
- Module M2 is a python based simulator of NeuCube for large scale applications or implementation on a neuromorphic hardware (Module M3). This application is developed on top of PyNN package, which is a Python-based simulator-independent language for building SNN. The NeuCube-PyNN [18] module is not only compatible with existing SNN simulators described previ-

**Fig. 20.5** NeuCube development environment, also showing some application oriented devices, such as Oculus for 3D visualisation, a SpiNNaker small neuromorphic board, EEG device, an EEG-controlled mobile robot WITH from Kyushu Institute of Technology

ously (e.g., Neuron, Brian), but can also be ported to a large neuromorphic hardware such as the SpiNNaker, or on any neuromorphic chip, such as the ETH INI chip, the Zhejiang University chip, and others.

- Module M3 is dedicated for hardware implementations of NeuCube.
- Module M4 allows for a dynamic visualisation of the 3D structure and connectivity of the NeuCube SNN [51, 52]. Due to the 3-dimensional structure as well as the large number of neurons and connections within NeuCube a simple 2D connectivity/weight matrix or an orthographic 45-degree view of the volume is insufficient. A specialised visualisation engine using JOGL (Java Bindings for OpenGL) and GLSL (OpenGL Shading Language) can render the structural connectivity as well as the dynamic spiking activity. Using 3D stereoscopic head-mounted displays such as the Oculus Rift, the perception and understanding of the spatial structure can be improved even further.
- Module M5 is the input/output and the information exchange module. This module is responsible for binding all the NeuCube modules together irrespective of the programming language or platform. Experiments that are run on any module produces prototype descriptors containing all the relevant information, which are exported and imported as structured text files, and is compatible with all the modules. We have used language independent JSON (Javascript object notation) format as a structured text, which is lightweight, human readable and can be parsed easily. The present implementation of the I/O module supports the use of three types of data and SNN prototype descriptors. They are: (1) Dataset descriptor, which consist of all the information relevant to the raw and encoded dataset; (2) Parameter descriptor, which is responsible for storing all the user defined and changeable parameters of the software; and, (3) SNN application

system descriptor, which stores information related to the NeuCube SNN application system.

– Module M6 extends the functionality of module M1, by adding functions for prototyping and testing of neurogenetic data modelling. These functions include models for genetic and proteomic influences in conjunction with brain data.

– Module M7 facilitates the creation and the testing of a personalised SNN system. It extends module M1 by including additional functionalities for personalised modelling which is based on first clustering of integrated static-dynamic data using new algorithm dWWKNN (dynamic weighted-weighted distance K-nearest neighbours) and then learning from the most informative subset of dynamic data for the best possible prediction of output for an individual. This module is for optional use in the context of specific applications [53, 54].

– Module M8 is for multimodal brain data analysis. It aims to integrate different modalities of brain activity information (e.g., EEG, fMRI, MEG) and structural (DTI) information, in NeuCube, for the purpose of better modelling and learning. This module is also bound to specific applications.

– Module M9 is a data encoding and optimisation. This module includes several data encoding algorithms for mapping analogue signals to spike trains based on different data sources [46, 55].

– Module M10 provides an additional feature of online learning for real-time data analysis and prediction. In this module, continuous data streams are processed in the form of continuous data blocks.

## 20.3.3　Implementation of NeuCube-Based Spatio-temporal Data Machines on Traditional and on Neuromorphic Hardware Platforms

A NeuCube developed SNN for a specific application can be implemented using a different software platform or a hardware platform, including PC, GPU, RaspberryPi, Treu North, SpiNNaker, any neuromorphic chip. Etc.

An example of a large scale of SNNcube of 25,000 neurons representing MNI brain template implemented in a von Neumann architecture is shown in Fig. 20.6 [56].

As traditional von-Neumann computational architectures reach their limits [57, 58] in terms of power consumption, transistor size, and communication, new approaches must be sought. Neuromorphic hardware systems, especially designed to solve neuron dynamics and able to be highly accelerated compared to biological time, are a response to these concerns [59–61]. Systems such as analogue VLSI or the SpiNNaker are advantageous by comparison to software based simulations on commodity computing hardware in areas such as biophysical realism; density of neurons per unit of processing power; and significantly lowered power consumption [10, 59].

**Fig. 20.6** An example of a SNNcube of 25,000 neurons representing MNI brain template implemented in a von Neumann architecture [56]. The red dots represent active neurons at the moment of the snapshot from a simulation and the little squares represent input neurons

To address this opportunity, a cross-platform version utilising the PyNN API in Python has been written in [60]. This version is targeted primarily towards neuromorphic hardware platforms but is also applicable to commodity distributed hardware systems depending on the simulation backend chosen.

PyNN [60] is a generic SNN simulation markup framework that allows the user to run arbitrary SNN models on a number of different simulation platforms, including software simulators PyNEST and Brian, and some neuromorphic hardware systems such as SpiNNaker and FACETS/BrainScaleS. It provides a write once, run anywhere (where anywhere is the list of simulators it supports) facility for the development of SNN simulations.

One neuromorphic platform for the implementation of a NeuCube SNN prototype system developed in module M1 or in any other modules of the NeuCube architecture, is the SpiNNaker device [20].

Alternative implementations of the NeuCube systems on neuromorphic hardware are being pursued on the INI Neuromorphic VLSI chips and the Zhejiang University FPGA system.

## 20.4 Chapter Summary and Further Readings

The chapter describes main principles of computation applied to implementing SNN application systems. NeuCube is used as an example of a SNN development system for a wide scope of applications. A free copy and open source of the main

NeuCube module as a limited and trial version is available from: http://www.kedri. aut.ac.nz/neucube/.

Further readings on specific topics can be found in:

- Neuromorphic Architectures for Spiking Deep Neural Networks [61].
- Memory and information processing in neuromorphic systems [11].
- Overview of the spinnaker system architecture [16].
- A VLSI network of spiking neurons with an asynchronous static random access memory [17].
- NeuCube Neuromorphic Framework for Spatio-temporal Brain Data and its Python Implementation [18].
- Software for selection and optimisation of encoding algorithm for SNN applications [55].
- General information about SNN [62, 21].

**Acknowledgements** Some of the material of this chapter has been published in [1]. The following students and colleagues have taken part in the development of the NeuCube development system: Enmei Tu, Neelava Sengupta, Josafath Israel Espinosa Ramos, Stefan Marks, Nathan Scott, Jakub Weclawski, Akshay Raj Gollahalli, Maryam Gholami Doborjeh, Zeng-Guang Hou and his students Nelson and James. I was greatly helped in the preparation of this chapter by Neelava.

NeuCube development system is available as an open source, as executable and as cloud-based at: http://www.kedri.aut.ac.nz/neucube/, along with some demo application SNN systems developed in NeuCube.

I acknowledge the contribution through discussions and collaboration of Giacomo Indiveri, Steve Furber, Darmendra Modha, Tobi Delbruck, Shi-chi Liu.

# References

1. N. Sengupta, J.I. Espinosa Ramos, E. Tu, S. Marks, N. Scott, J. Weclawski, A. Raj Gollahalli, M. Gholami Doborjeh, Z. Gholami Doborjeh, K. Kumarasinghe, V. Breen, A. Abbott, *From von Neumann architecture and Atanasoffs ABC to Neuromorphic Computation and Kasabov's NeuCube: Principles and Implementations*, ed. by Jotzov, et al., Chapter 1 in: Advances in Computational intelligence (Springer, Heidelberg, 2018)
2. D. Berlinski, *The Advent of the Algorithm: The 300-Year Journey from an Idea to the Computer* (Houghton Mifflin Harcourt, 2001)
3. W.S. McCulloch, W. Pitts, A logical calculus of the ideas immanent in nervous activity. Bull. Math. Biophys. **5**(4), 115–133 (1943)
4. Y. LeCun, Y. Bengio, G. Hinton, Deep learning. Nature **521**(7553), 436–444 (2015)
5. R.R. Schaller, Moore's law: past, present and future. IEEE Spectr. **34**(6), 52–59 (1997)
6. N. Kasabov, N. Sengupta, N. Scott, From von Neumann, John Atanasoff and ABC to Neuromorphic computation and the Neucube spatio-temporal data machine, in *IEEE 8th International Conference on Intelligent Systems (IS)* (IEEE, 2016), pp. 15–21
7. I. Schuler, Neuromorphic computing: from materials to systems architecture (2016). Accessed 16 July 2016
8. C. Mead, Neuromorphic electronic systems. Proc. IEEE **78**(10), 1629–1636 (1990)

9. T. Delbruck, P. Lichtsteiner, Fast sensory motor control based on event-based hybrid neuromorphic-procedural system, in *IEEE International Symposium on Circuits and Systems, 2007. ISCAS 2007* (IEEE, 2007), pp. 845–848

10. G. Indiveri, B. Linares-Barranco, T.J. Hamilton, A. van Schaik, R. Etienne-Cummings, T. Delbruck, S.-C. Liu, P. Dudek, P. Häfliger, S. Renaud, J. Schemmel, G. Cauwenberghs, J. Arthur, K. Hynna, F. Folowosele, S. Saighi, T. Serrano-Gotarredona, J. Wijekoon, Y. Wang, K. Boahen, Neuromorphic silicon neuron circuits. Front. Neurosci. **5**, 73 (2011)

11. G. Indiveri, S.-C. Liu, Memory and information processing in neuromorphic systems. Proc. IEEE **103**(8), 1379–1397 (2015)

12. A. Calimera, E. Macii, M. Poncino, The human brain project and neuromorphic computing. Funct. Neurol. **28**(3), 191–196 (2013)

13. J. Hsu, Ibm's new brain [news]. IEEE Spectr. **51**(10), 17–19 (2012)

14. P.A. Merolla, J.V. Arthur, R. Alvarez-Icaza, A.S. Cassidy, J. Sawada, F. Akopyan, B.L. Jackson, N. Imam, C. Guo, Y. Nakamura, et al., A million spiking-neuron integrated circuit with a scalable communication network and interface. Science **345**(6197), 668–673 (2012)

15. B.V. Benjamin, P. Gao, E. McQuinn, S. Choudhary, A.R. Chandrasekaran, J.-M. Bussat, R. Alvarez-Icaza, J.V Arthur, P.A Merolla, K. Boahen, Neurogrid: a mixed-analog-digital multichip system for large-scale neural simulations. Proc. IEEE **102**(5), 699–716 (2012)

16. S.B. Furber, D.R. Lester, L.A. Plana, J.D. Garside, E. Painkras, S. Temple, A.D. Brown, Overview of the spinnaker system architecture. IEEE Trans. Comput. **62**(12), 2454–2467 (2013)

17. S Moradi, G Indiveri, A VLSI network of spiking neurons with an asynchronous static random access memory, in *Biomedical Circuits and Systems Conference (BioCAS)* (IEEE, 2011), pp. 277–280

18. N. Scott, N. Kasabov, G. Indiveri, in *NeuCube Neuromorphic Framework for Spatio-temporal Brain Data and Its Python Implementation*. Proceedings of the 20th International Conference on Neural Information Processing, November 3–7, Daegu, Korea (Springer, Heidelberg, 2013). D. Perrin, Complexity and high-end computing in biology and medicine (2011). Advances in Experimental Medicine and Biology

19. J. Mitra, T.K. Nayak, An FPGA-based phase measurement system. IEEE Trans. Very Large Scale Integr. (VLSI) Syst. **26**(1), 133–142 (2017)

20. J. Behrenbeck, Z. Tayeb, C. Bhiri, C. Richter, O. Rhodes, N. Kasabov, S. Furber, G. Cheng, J. Conradt, Classification and Regression of Spatio-Temporal EMG Signals using NeuCube Spiking Neural Network and its implementation on SpiNNaker Neuromorphic Hardware, J Neural Eng, IOP Press, 2018, Article reference: JNE-102499

21. W. Maass, C.M. Bishop, *Pulsed Neural Networks* (MIT Press, Cambridge, 2001)

22. E. Capecci, N. Kasabov, G.Y. Wang, Analysis of connectivity in neucube spiking neural network models trained on eeg data for the understanding of functional changes in the brain: a case study on opiate dependence treatment. Neural Netw. **68**, 62–77 (2015)

23. D.F.M. Goodman, Code generation: a strategy for neural network simulators. Neuroinformatics **8**(3), 183–196 (2010)

24. N. Kasabov, N.M. Scott, E. Tu, S. Marks, N. Sengupta, E. Capecci, M. Othman, M.G. Doborjeh, N. Murli, R. Hartono, et al., Evolving spatio-temporal data machines based on the neucube neuromorphic framework: design methodology and selected applications. Neural Netw. **78**, 1–22 (2016)

25. E.B. Baum, On the capabilities of multilayer perceptrons. J. Complex. **4**(3), 193–215 (1988)

26. J. Park, I.W. Sandberg, Universal approximation using radial-basis-function networks. Neural Comput. **3**(2), 246–257 (1991)

27. D.F. Specht, Probabilistic neural networks. Neural Netw. **3**(1), 109–118 (1190)

28. T. Kohonen, The self-organizing map. Neurocomputing **21**(1), 1–6 (1998)

29. N. Kasabov, *Evolving Connectionist Systems: The Knowledge Engineering Approach* (Springer Science & Business Media, 2007)

30. T. Schaul, J. Bayer, D. Wierstra, Y. Sun, M. Felder, F. Sehnke, T. Rückstieß, J. Schmidhuber, Pybrain. J. Mach. Learning Res. **11**, 743–746 (2010)

31. S. Nissen, E. Nemerson, Fast artificial neural network library (2000). Available at https://leenis-sen.dk/fann/html/files/fann-h.html

32. M. Hall, E. Frank, G. Holmes, B. Pfahringer, P. Reutemann, I.H. Witten, The weka data mining software: an update. ACM SIGKDD Explor. Newsl. **11**(1), 10–18 (2009)

33. M.R. Berthold, N. Cebron, F. Dill, T.R. Gabriel, T. Kötter, T. Meinl, P. Ohl, C. Sieb, K. Thiel, B. Wiswedel, Knime: The Konstanz Information Miner, in *Data Analysis, Machine Learning and Applications* (Springer, Heidelberg, 2008), pp. 319–326

34. J. Demšar, B. Zupan, G. Leban, T. Curk, *Orange: From Experimental Machine Learning to Interactive Data Mining* (Springer, Heidelberg, 2004)

35. M.L. Hines, N.T. Carnevale, The neuron simulation environment. Neural Comput. **9**(6), 1179–1209 (1997)

36. R. Brette, M. Rudolph, T. Carnevale, M. Hines, D. Beeman, J.M. Bower, M. Diesmann, A. Morrison, P.H. Goodman, F.C. Harris Jr., et al., Simulation of networks of spiking neurons: a review of tools and strategies. J. Comput. Neurosci. **23**(3), 349–398 (2007)

37. J.M. Eppler, M. Helias, E. Muller, M. Diesmann, M.-O. Gewaltig, Pynest: a convenient interface to the nest simulator. Front. Neuroinformatics **2**, 12 (2008)

38. D. Pecevski, T. Natschläger, K. Schuch, Pcsim: a parallel simulation environment for neural circuits fully integrated with python. Front. Neuroinformatics **3**, 11 (2009)

39. T. Natschläger, H. Markram, W. Maass, Computer Models and Analysis Tools for Neural Microcircuits, in *Neuroscience Databases* (Springer, Heidelberg, 2003), pp. 123–138

40. R. Drewes, Brainlab: a toolkit to aid in the design, simulation, and analysis of spiking neural networks with the NCS environment. Ph.D. thesis, University of Nevada Reno, 2005

41. E.C. Wilson (2001), Parallel implementation of a large scale biologically realistic neocortical neural network simulator. Ph.D. thesis, University of Nevada Reno, 2001

42. D. Pecevski, Oger: Modular learning architectures for large-scale sequential processing

43. D.F.M. Goodman, R. Brette, The brian simulator. Front. Neuroscience **3**(2), 192 (2009)

44. M. Diesmann, M.-O. Gewaltig, A.D. Aertsen, Stable propagation of synchronous spiking in cortical neural networks. Nature **402**(6761), 529–533 (1999)

45. N. Kasabov, Neucube: a spiking neural network architecture for mapping, learning and understanding of spatio-temporal brain data. Neural Netw. **52**, 62–76 (2014)

46. N. Sengupta, N. Scott, N. Kasabov, Framework for Knowledge Driven Optimisation Based Data Encoding for Brain Data Modelling Using Spiking Neural Network Architecture, in *Proceedings of the Fifth International Conference on Fuzzy and Neuro Computing (FANCCO-2015)* (Springer, Heidelberg, 2015), pp. 109–118

47. S. Song, K.D. Miller, L.F. Abbott, Competitive hebbian learning through spike-timing-dependent synaptic plasticity. Nat. Neurosci. **3**(9), 919–926 (2000)

48. S. Fusi, Spike-driven synaptic plasticity for learning correlated patterns of mean firing rates. Rev. Neurosci. **22**(1–2), 73–84 (2003)

49. N. Kasabov, K. Dhoble, N. Nuntalid, G. Indiveri, Dynamic evolving spiking neural networks for on-line spatio-and spectro-temporal pattern recognition. Neural Netw. **41**, 188–201 (2013)

50. A. Mohemmed, S. Schliebs, S. Matsuda, N. Kasabov, Span: spike pattern association neuron for learning spatio-temporal spike patterns. Int. J. Neural Syst. **22**(04), 1250012 (2012)

51. S. Marks, J. Estevez, N. Scott, Immersive visualisation of 3-dimensional neural network structures (2015)

52. S. Marks, Immersive visualisation of 3-dimensional spiking neural networks. Evol. Syst. (2016) 1–9

53. N. Kasabov, Y. Hu, Integrated optimisation method for personalised modelling and case studies for medical decision support. Int. J. Funct. Inf. Personalised Med. **3**(3), 236–256 (2010)

54. M.G. Doborjeh, N. Kasabov, Personalised modelling on integrated clinical and eeg spatio-temporal brain data in the neucube spiking neural network system, in *2016 International Joint Conference on Neural Networks (IJCNN)* (IEEE, 2016), pp. 1373–1378

55. B. Petro, N. Kasabov, R. Kiss, Selection and optimisation of spike encoding methods for spiking neural networks, algorithms, submitted; http://www.kedri.aut.ac.nz/neucube/ -> Spiker

56. A. Abbott et al., in *Proceedings of IJCNN* (2016)

57. H. Esmaeilzadeh, E. Blem, R.S. Amant, K. Sankaralingam, D. Burger, Dark silicon and the end of multicore scaling. ACM SIGARCH Comput. Archit. News **39**(3), 365 (2011)

58. D. Perrin, Complexity and high-end computing in biology and medicine. Adv. Exp. Med. Biol. **696**, 377–384 (2011)

59. S. Furber, To Build a Brain. IEEE Spectr. **49**(8), 44–49 (2012)

60. A.P. Davison, D. Brüderle, J. Eppler, J. Kremkow, E. Muller, D. Pecevski, L. Perrinet, P. Yger, PyNN: a common interface for neuronal network simulators. Front. Neuroinformatics **2**, 11 (2008)

61. G. Indiveri, F. Corradi, N. Qiao, Neuromorphic architectures for spiking deep neural networks. IEEE Int. Electron Devices Meeting (IEDM) (2015)

62. W. Maass, Networks of spiking neurons: the third generation of neural network models. Neural Netw. **10**(9), 1659–1671 (1997)

# Chapter 21
# From Claude Shannon's Information Entropy to Spike-Time Data Compression Theory

This chapter of the book proposes a new information theory for temporal data compression through spike-time encoding for the purpose of reducing the amount of raw data from time series but preserving the information in terms of accuracy of pattern recognition and pattern classification. Most of the data in information sciences are temporal or spatio/spectro temporal, such as brain data, audio and video data, environmental and ecological data, financial and social data, etc. as discussed in the other chapters of the book and the proposed data compression method is applicable to all of them. It is illustrated in this chapter on compressing and classification of patterns from fMRI data (more about fMRI data can be found in Chaps. 10 and 11). The presented here theory and experimental results were first published in [1].

The chapter is organised in the following sections:

21.1. Claud Shannon classical information theory.
21.2. The proposed information theory for temporal data compression for classification tasks based on spike-time encoding.
21.3. A spike-time encoding and compression method for fMRI spatio-temporal data classification.
21.4. Chapter summary and further readings.

## 21.1 Claude Shannon's Classical Information Theory

The brilliant mathematician **Claude Shannon (1916–2011)** introduced an information theory based on entropy.

A random variable x is characterized at any moment of time by its uncertainty in terms of what value this variable will take in the next moment—its entropy. A measure of uncertainty $h(x_i)$ can be associated with each random value $x_i$ of a

© Springer-Verlag GmbH Germany, part of Springer Nature 2019     679
N. K. Kasabov, *Time-Space, Spiking Neural Networks and Brain-Inspired Artificial Intelligence*, Springer Series on Bio- and Neurosystems 7,
https://doi.org/10.1007/978-3-662-57715-8_21

random variable x, and the total uncertainty H(x), called *entropy*, measures our lack of knowledge, the seeming disorder in the space of the variable x:

$$H(X) = \sum_{i=1,\ldots,n} p_i \cdot h(x_i), \tag{21.1}$$

where $p_i$ is the probability of the variable x taking the value of $x_i$.

The following axioms for the entropy H(x) apply:

- monotonicity: if $n > n'$ are number of events (values) that a variable x can take, then.
- $H_n(x) > H_{n'}(x)$, so the more values x can take, the greater the entropy.
- additivity: if x and y are independent random variables, then the joint entropy H (x, y), meaning H(x AND y), is equal to the sum of H(x) and H(y).

The following log function satisfies the above two axioms:

$$h(x_i) = \log(1/p_i) \tag{21.2}$$

If the log has a basis of 2, the uncertainty is measured in [bits], and if it is the natural logarithm ln, then the uncertainty is measured in [nats].

$$H(X) = \sum_{i=1,\ldots,n} (p_i \cdot h(x_i)) = -c \cdot \sum_{i=1,\ldots,n} (p_i \cdot \log p_i), \tag{21.3}$$

where c is a constant.

Based on the *Cloud Shannon's* measure of uncertainty—*entropy*, we can calculate an overall probability for a successful prediction for all states of a random variable x, or the predictability of the variable as a whole:

$$P(x) = 2^{-H(x)} \tag{21.4}$$

The max entropy is calculated when all the n values of the random variable x are equiprobable, i.e. they have the same probability 1/n—a uniform probability distribution:

$$H(X) = -\sum_{i=1,\ldots,n} p_i \cdot \log p_i \leq \log n \tag{21.5}$$

*Joint entropy* between two random variables x and y (for example, an input and an output variable in a system) is defined by the formulas:

$$H(x,y) = -\sum_{i=1,\ldots,n} p\left(x_i \text{ AND } y_j\right) \cdot \log p\left(x_i \text{ AND } y_j\right) \tag{21.6}$$

$$H(x,y) \leq H(x) + H(y) \tag{21.7}$$

*Conditional entropy*, i.e. measuring the uncertainty of a variable y (output variable) after observing the value of a variable x (input variable), is defined as follows:

$$H(y|x) = - \sum_{i=1,\dots,n} p\left(x_i, y_j\right) \cdot \log p\left(y_j|x_i\right) \tag{21.8}$$

$$0 \leq H(y|x) \leq H(y) \tag{21.9}$$

Entropy can be used as a measure of the information associated with a random variable x, its uncertainty, and its predictability.

The *mutual information* between two random variables, also simply called *information*, can be measured as follows:

$$I(y; x) = H(y) - H(y|x) \tag{21.10}$$

Information measured as entropy was the basis for all data compression techniques developed so far, which aimed at reducing the raw data but preserving the information. In the next section we will introduce a new theory that preserves information for temporal pattern recognition from temporal data using spike-time encoding. There is analogy between entropy and spike time encoding as both measure changes in the data, but there are significant differences as well, as shown in the next sections.

## 21.2   The Proposed Information Theory for Temporal Data Compression for Classification Tasks Based on Spike-Time Encoding

Human brains ability to efficiently detect patterns from continuous streaming information in the form of sensory stimulus is an inspiration to the field of artificial intelligence. Efficient encoding of continuous input information into discrete spike-times play a decisive role in the ability of the spiking neurons present inside the human brain to compress, transmit and recognise information presented by the external environment. This compact encoding of information in the brain is non-conformative to the classical information theory developed by Claude Shannon.

In this chapter, we introduce spike-time encoding as an efficient general approach to data compression that minimises dramatically the information representation of streaming, temporal data and achieves similar or even better pattern recognition and classification accuracy when compared with the use of the whole raw data for this purpose. We also introduce a specific encoding algorithm GAGamma that leads to efficient compression of spatiotemporal data for the purpose of storage, transmission and accurate pattern recognition of brain fMRI data in particular. We have evaluated and compared the algorithm with other methods on a benchmark fMRI dataset. The results show the temporal encoding algorithm's ability to achieve significant data compression without sacrificing the performance of the recognition of the compressed patterns. Using specific spike-time encoding algorithm for a class of data, such as the GAGamma for fMRI data, leads to a better signal reconstruction when compared with the use of standard spike encoding methods.

The human brain is considered to be the most resourceful and efficient system that can recognise patterns in millisecond resolutions. This is done by crunching massive volumes of real continuous stimuli/data captured by the sensory organs. It is also observed that the human brain cells when presented with an external stimuli propagates the signal efficiently over large distances using electrical impulses known as synaptic action potential. In neurobiology, the process of analog to digital signal transformation is known as neural encoding [2]. It is very intriguing that the process of neural encoding not only converts the big streaming continuous data space into a compressed space of spikes, but brain cells also recognise the patterns in the compressed space. The biological organization of our brain tends to create signals with a very specific class of distributions and it is from the perspective of evolution understandable that these distributions are optimized for fast analysis. The most popular hypothesis states that the signal strength is encoded by the mean firing rate, i.e. stronger signal gives rise to higher average firing rate. A wide range of studies [3, 4] in the sensory and motor-neuronal system in several species supports the validity of mean firing rate hypothesis. The major drawback of this theory, however, lies in the association of information with spike density. Determine the spike density in millisecond resolution from large volume of spikes lead to a level of computational inefficiency. As per an alternate theory on neural encoding, neurons carry information in the precise timing of the spikes. This is known as the temporal encoding. Numerous research [5, 6] has shown the presence of temporal encoding in different parts of the human brain. Temporal encoding supports the efficient representation of information that is required for very fast processing (in millisecond scale) of the stimulus presented to the human brain. As opposed to the rate coding scheme, high fluctuations in mean firing rate, also known as inter spike interval (ISI) probability distribution is considered to be informative rather than noise in this scheme. The temporal spike time representation of the data acts as a lossy compression of information. Most forms of learning, though, can be seen as forms of data compression. In fact we can, in terms of pattern recognition, only learn something from data when there is redundancy in the data. In many data analysis project the data is preprocessed or recoded in a way that could be seen as a form of data compression. If such a preprocessing does not destroy the patterns we are interested in this will in general result in an equal of better performance of our learning algorithms. The motivation of the temporal encoding in this context is to reduce large volume of data into a compressed state with minimal loss and the maximal presence of discriminable information. Examples of such data sources are pulsar data in radioastronomy, seismic activity data, and so on.

From computational theory viewpoint, the data encoding problem directly relates to the concepts of information theory. In 1948, in the seminal paper of information theory [7], Claude P. Shannon proposed a complete form of mathematical theory to quantify information transmission in a communication channel. A conclusive finding that the amount of information in any object can be estimated as the description length of the object continues to set the stage for the development of communications, data storage and processing, and other information technologies. Shannon's information theory is built on a presupposition that the computable information in an object is the characteristic of a random source with known probability distribution of which the object is part of. To realise this idea Shannon derived the 'entropy' from the first principle, which is the measure of average information emitted by an object when observed. Entropy is the functional mapping of the random variable to a real number. A.N. Kolmogorov on the other hand proposed a complementary research on algorithmic information theory aiming to provide means of measuring information. Contrary to Shannon's theory, Kolmogorov complexity [8, 9] considers information as the property of an object in isolation irrespective of the manner in which the object arose [10]. More formally, it is defined as the minimum number of bits from which a particular message or file can effectively be reconstructed, *i.e.* the minimum number of bits suffice to store a reproducible file.

A computational neuron responsible for emitting spikes from sensory data can be regarded as a logical transmission medium responsible for broadcasting continuous information received from the data source. The two neural coding hypotheses hence can be seen in the light of the information theory. We observe that the rate coding scheme is much adherent to Shannon's interpretation of encoding. The inherent assumption of the presence of a random source with a known probability distribution in Shannon's theory is much apposite to the mean firing rate as it relates to the frequency of spikes over time. However, our interest in efficient compression of a large volume of data by a sequence of spike-timings and further use the spike timings for the purpose of pattern recognition is much more in sync with Kolmogorov's notion of object representation by minimal description length using computer programs.

The data encoding problem in the spiking neural network (SNN) is a relatively less researched topic compared to the neuron dynamics and learning in SNN. ATR Human Information Processing Research Laboratory's artificial brain (Cellular Automata Machine Brain) project [11] used data encoding as a part of its large-scale brain-like neural architecture. Hardware accelerated implementation of spike encoding for image and video processing was performed in [12]. The literature on spike encoding technique applicable to real world data is restricted to a few algorithms like Temporal-contrast (also known as Address event representation encoding [13]), Hough Spiker Algorithm (HSA) [14] and Ben Spiker Algorithm (BSA) [15]. All these algorithms are generally event-driven i.e. to say, they follow the temporal encoding scheme giving importance to the time of the occurrence of an event (spike). The temporal contrast algorithm, which is inspired from the human visual cochlea, uses threshold based method to detect signal contrast (change). A user-defined contrast threshold determines the spike events in temporal contrast. The HSA and BSA algorithm, however, determines a spike event using a deconvolution operation between the observed signal and a predefined filter. The deconvolution in HSA is based on the assumption that the convolution function produces a biased converted signal which always stays below the original signal yielding an error [16]. Bens Spiker Algorithm (BSA) [17] on the other hand, uses a finite reconstruction filter (FIR) for predicted signal generation (see Chap. 4).

We formalise the data encoding problem for pattern recognition as a data compression problem. The compression function is defined as the map $f : \mathbb{R}^T \rightarrow \{t_1^f, t_2^f, \ldots t_n^f | t_i \in \mathbb{I}^+ \}$, where the $f(\cdot)$ releases a spike at firing times $\mathbf{t^f}$. The proposed encoding algorithm primarily assumes that the discriminatory information is encoded by the sequence of spike timings rather than the sequence of spikes. As a consequence of this assumption, it is important to achieve large compression by minimising the number of spikes and thus is a distinct contradiction with the rate coding hypothesis. The generalised framework for data encoding used here for the experiments is extended from our previously published work [18].

We denote the source signal as $S \in \mathbb{R}^T$. In order to simplify the formalisation, we define the encoded spike train $B \in \{0, 1\}^T$ as a fixed-length binary sequence of length $T$ as opposed to the variable length sequence of spike timings $\left\{t_1^f, t_2^f, \ldots t_n^f | t_i \in \mathbb{I}^+ \right\}$ defined earlier without any loss of generality. Here $T$ defines the length of the temporal data to be encoded in spikes. The background knowledge-driven optimisation based encoding algorithm is built on the premise that existing knowledge about the data generation model or in other words properties of the data generation source is possible to be injected to predict the signal $S\hat{}$. For example, fMRI data generation process behaves like a linear time invariant system, where an event in the brain gives rise to a signal mimicking the gamma function [19], whereas EEG data generation can be modelled as a phase varying mixture model of sinusoidal waves or multisource Gaussian noise model [20]. The notion of knowledge injection is further elaborated in Sect. 21.3 using fMRI as an example. If it is possible to formalise the decompression function $S\hat{}$ from the spike sequence $B$, the optimal encoding of data can be formulated as an optimisation problem, which minimises the root mean squared decompression error between the observed signal $S$ and the predicted signal $S\hat{}$: $= f(B, \Theta)$, $\Theta$ being the set of additional parameters required along with $B$ to describe the prediction function. The optimisation problem can be written down as:

$$\min_{B,\theta} \sqrt{\frac{\sum_t \left(S - \hat{S}(B,\Theta)\right)^2}{t}}$$

$$\text{s.t.} \quad B := \mathbb{I}^+$$
$$0 \le B \le 1 \tag{21.11}$$
$$\sum_t B_t \le a$$
$$b \le \Theta < c$$

The aforementioned optimisation problem belongs to the paradigm of mixed-integer programming, where a subset of parameter or decision variables to be optimised, are integers. Numerous methods have been developed over the years to solve such problems [16, 21, 22]. In our implementation, however, we have used the mixed integer genetic algorithm proposed by [23, 24]. The constraints in Eq. 2.11 are imposed on the parameters of $S\hat{}$. The first couple of constraints reduces the possible values of $B$ to $\{0, 1\}$. We have used the hyper-parameter $a$ to control the maximum number of spikes in the optimal spike sequence. The other sets of hyper-parameters $\{b, c\}$ are used to control the upper and lower bounds of the model parameter $\Theta$.

The formulation above for the proposed framework for data encoding is a generic, flexible and driven by knowledge-injection from the data source. It can be further extended to include systematic noise model as part of $S\hat{}$. We hypothesise that a sufficiently good choice of $S\hat{}$ preserves and in some cases enhances the discriminative property of the data in a greatly compressed space. It must also be noted that this formulation adheres to the concept of non-existence of a universal compression algorithm for all the data sources. The general framework described above can be used to derive specific methods for encoding of special types of data for which background knowledge is available. One such case is fMRI data based on blood-oxygen level dependent response (BOLD). This is further introduced and illustrated.

## 21.3   A Spike-Time Encoding and Compression Method for fMRI Spatio-Temporal Data Classification

The fMRI BOLD response is modelled here as a linear time invariant system, which is described by the convolution of the spikes $B$ and the haemodynamic response function (HRF) $H(\Theta)$. This operation is characterised by Eqs. 21.12 and 21.13.

$$\hat{S} := \int_0^t B(\tau)h(t - \tau)d\tau \tag{21.12}$$

$$S^\wedge(B, \Theta) := B * H(\Theta) \tag{21.13}$$

$$H(\theta_1, \theta_2) := \frac{1}{\theta_2^{\theta_1} \tau(\theta_1)} t^{\theta_1 - 1} e^{-\frac{t}{\theta_2}} \tag{21.14}$$

Numerous mathematical models for HRF has been proposed in the earlier research [25–27]. Majority of the mathematical models for the canonical HRF are found to be some variant of the gamma function. In all our experiments we have used the gamma distribution function as the HRF model described by the Eq. 21.14. This function is characterised by the parameter set $\Theta := \{\theta_1, \theta_2\}$, where $\theta_1 \in R^+$ and $\theta_2 \in R^+$ controls the shape and the scale of the function respectively. By fitting

Eqs. 21.13 and 21.14 in Eq. 21.11, the encoding problem boils down to solving Eq. 21.5 and will be referred to as GAGamma encoding hereinafter.

$$\min_{\substack{B,\theta_1,\theta_2 \\ \text{s.t.}}} \quad \begin{aligned} &\sqrt{\frac{\sum_t (S-\hat{S}(B,\theta_1,\theta_2))^2}{t}} \\ &B := \mathbb{I}^+ \\ &0 \leq B \leq 1 \\ &\sum_t B_t \leq a \end{aligned} \quad (21.15)$$

where 
$$\begin{aligned} b_1 &\leq \theta_1 \leq c_1 \\ b_1 &\leq \theta_1 \leq c_2 \\ \hat{S}(B,\theta) &:= B * \frac{1}{\theta_2^{\theta_1}\tau(\theta_1)} t^{\theta_1 - 1} e^{-\frac{t}{\theta_2}} \end{aligned}$$

At this point, it is imperative to make distinctions between the GAGamma and the existing HSA and BSA algorithms. The HSA and BSA algorithm for spike encoding are built on the premise of stimulus estimation using finite impulse response resembling the GAGamma method. The knowledge injection component of GAGamma as part of $\hat{S}$ and the optimisation approach has two distinct benefits over the deconvolution based methods:

- We have used a generic Gamma function as the knowledge injection component $\hat{S}$ in GAGamma which is driven by the existing knowledge about the fMRI data as opposed to the sinusoidal waves used as the FIR in BSA. We also argue that this formalism allows the inclusion of additional knowledge about the data source (such as systematic noise) providing greater flexibility in the encoding of the data.
- The optimisation problem formulation in GAGamma jointly optimises for the parameter set $\Theta$ and $B$. This formulation thus includes the parameter set $\Theta$ of the prediction model $\hat{S}$ along with the spike $B$ for each individual voxel or feature. In HSA and BSA, the equivalent filter parameters need to be predetermined and fixed for the whole set of voxels.

All the experiments described here were performed on the publicly available benchmark starplus fMRI dataset [28] collected by The Centre for Cognitive Brain Imaging, Carnegie Mellon University. The starplus experiment was conducted on a set of 7 subjects. Each subject had undergone multiple trials of the exact same cognitive experiment. At every trial, lasting for 27 s, a set of stimuli were presented in the following order:

1. The first stimulus (Picture or Sentence) was presented at the beginning for 4 s.
2. A blank screen was presented during the interval of 5–8 s.
3. The second stimulus (Sentence or Picture) was presented during the interval of 9–12 s.

4. A rest period of 20 s was added after the second stimulus.

While the subject performed the cognitive tasks, fMRI images of a fraction of the brain were collected at every 500 ms interval. The final preprocessed fMRI dataset corresponds to a classification task of detecting binary cognitive states namely, 'seeing a picture' versus 'reading a sentence'. We have chosen two subjects (id: 04847 and 07510) randomly and used two spatial regions of interest (ROI); Calcarine Sulcus ('CALC') and Left Intra-Parietal Sulcus ('LIPL') for our pattern recognition experiments. The choice of the ROI's is based on previous work [29] that found these ROIs to be amongst the most discriminatory in the continuous space. The dataset is composed of 40 samples (trials) of each class and each sample is made up of 452 and 483 voxels in subject 04847 and 05710 respectively. Each cognitive task was a total of 8 s duration emitting 16 fMRI images for each class within a trial.

For the reason that this encoded data is intended to be used for pattern recognition problems, conservation and possible enhancement of the discriminatory information in the spike-timings is as crucial as efficient compression of the data. This is a distinctly different approach from the existing approaches of pattern recognition, where a massive amount of data is crunched by intelligent algorithms to achieve better predictive models producing highly accurate prediction performance. By keeping both compressibility and preservation of discriminatory information as the criteria of evaluation, we are aiming to benefit efficient resource usage along with the classification performance. It is thus important to have a balance between compression and conservation of discriminatory information in the encoded data. We have used three metrics to evaluate the encoding techniques along with traditional 'no encoding' (raw data) approach. A brief description of the metrics and the baseline encoding techniques are described below:

- Symbol rate: The symbol rate is measured by the bits/symbol unit. As the data (raw or encoded) is represented as bits (0 or 1) in the storage medium, bits/symbol measures the average number of bits required to represent a symbol, where symbol is the value in the value space. In a fixed-length encoding of data (like ASCII), it is evident that the fixed-length L defines the symbol rate. For a variable-length, however, the symbol rate is calculated by dividing the total number of bits by the number of symbols. For example, if a raw dataset, given by $\{x_1, x_2, \ldots x_n | x_i \in R\}$ is encoded by an encoding algorithm to an encoded dataset $\{y_1, y_2, \ldots y_m | y_i \in I^+\}$, then the symbol rate of the encoding algorithm for the dataset is $\frac{n}{m} \times Size\ of\ (\mathbb{I})$.

- Decoding error: The decoding error metric is the measure of the reliability of decompression, i.e. the ability to reliably recover the original signal from the compressed spike-timings. We have used the root mean squared error (RMSE) of signal reconstruction between the original signal $S$, and the predicted signal $S^\char`^$. A low RMSE of the signal reconstruction indicates high preservation of information in the spike-timings. It must be noted that the classification model is built on the spike-time data and has no knowledge of the raw real data. Hence,

although this metric has a significant role in evaluating the robustness of the
encoding algorithm in respect to reconstruct the raw data, the effect on the
quality of pattern recognition performance is not affected.

- Classification performance: From pattern recognition viewpoint, the classification performance is the most important measure of success. To evaluate the
  classification performance, we have used the combination of mean classification
  accuracy and precision metric.

Figure 21.1 shows a flowchart of the pattern recognition and evaluation process.
In the first step, real time series data is fed into the different data encoding algorithms which produce the spikes. In the second step, a simple K-NN model is built
using the spikes. The K-NN model is used for prediction of a new sample. The
traditional protocol of pattern recognition of the real data using a
Euclidean-distance based K-NN is also compared with the previous protocol.

The mean accuracy is estimated from thirty independent runs of 50/50 train/test
split of the binary classification data described previously. Figure 21.1 shows a
flowchart of the stepwise process of pattern recognition with and without using the
encoding algorithms. The best results are reported after optimisation of the K
parameter using grid search method.

*Spike asynchronicity based distance function for K-NN algorithm.* As discussed
earlier, we have used the non-parametric K-NN algorithm for building a classification model from the data. The prediction of a class label of a new sample (in our
case a spike train) in K-NN is a majority vote between the neighbours of the new
sample, with the sample being assigned to the class label most common among its k
nearest neighbours. To assign neighbourhood to a sample, it is hence necessary to
calculate pairwise distances between the sample to be predicted and the training

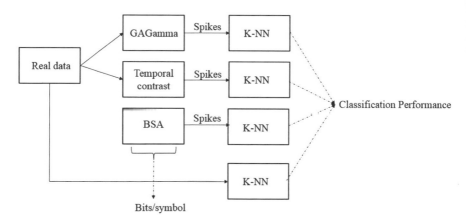

**Fig. 21.1** A flowchart of the pattern recognition and evaluation process. In the first step, real time
series data is fed into the different data encoding algorithms which produce the spikes. In the
second step, a simple K-NN model is built using the spikes. The K-NN model is used for
prediction of a new sample. The traditional protocol of pattern recognition of the real data using a
Euclidean-distance based K-NN is also compared with the previous protocol (after [1])

samples. Here we propose a distance function that can capture relative distance between a pair of spike-train sample. Since we are concerned with using spike timings as a carrier of information, a useful way to capture similarity between a pair of spike-train samples is to record if the two samples have spiked at the same time instance. We have used mean absolute asynchronicity as the distance function. The mean absolute asynchronicity based distance function between two spike-train samples $B_1 \in \{0, 1\}^{T \times M}$ and $B_2 \in \{0, 1\}^{T \times M}$ is formally defined as:

$$d := \frac{\sum_{(T \times M)} B_1 \oplus B_2}{T \times M} \qquad (21.16)$$

where, $T$ and $M$ are the time-length of the signal and number of features (in our case voxels) respectively. The $\oplus$ symbol represents an XOR operation applied on the spike train matrices $B_1$ and $B_2$. The XOR operation in effect identifies element by element mismatches/asynchronicity between the two matrices. The distance function calculates the mean asynchronicity between a pair of samples.

In this study three different encoding methods are compared and evaluated. It must be noted that for each encoding or compression algorithm, there also exists a decoding algorithm which can decompress the spike-trains into the reconstructed signal $S^\wedge$.

- GAGamma: This is the proposed encoding method which is outlined in Sect. 21.3. The encoding and decoding equations are given by Eqs. 21.13 and 21.15.
- BSA: The BSA encoding and decoding algorithms [15] are formalised in Algorithms 1 and 2, respectively in the Appendix. The BSA algorithm takes a filter function and a threshold value as input along with the signal $S$ (Chap. 4).
- Temporal contrast: The temporal contrast algorithm captures the greater than average changes in the data as spikes. Algorithms 3 and 4 in the Appendix present the temporal contrast encoding and decoding algorithms respectively. One major characteristic and deviation of temporal contrast algorithm from the temporal encoding framework is its ability to generate spikes with positive and negative polarity. Since we are only interested in the spike timings, during the classification, we ignore the polarity of the spikes. The algorithm takes the *factor* $\in \{0, 1\}$ parameter as input. This parameter controls the estimate of the $threshold_{TC}$ variable, which is responsible for determining the spike timings.

Figure 21.2 shows a comparison of signal reconstruction($S^\wedge$) from a spike sequence by GAGamma decoding algorithm, BSA decoding algorithm (algorithm 2) and Temporal contrast decoding algorithm (algorithm 4). The true signal is randomly selected from subject 04847′ (10th trial and 23rd voxel).

For the comparative evaluation of the encoding methods and the classical 'no encoding' method, we have replicated experiments for the subjects 04847 and 07510. For each subject we have compared the proposed GAGamma encoding method with BSA and Temporal contrast. For the GAGamma encoding, we have used the hyper-parameter values $[a = 16, b = 0, c = 10]$ in Eq. 21.15. The BSA

**Fig. 21.2** A comparison of signal reconstruction($S^\frown$) from a spike sequence by GAGamma decoding algorithm, BSA decoding algorithm (Algorithm 2) and Temporal contrast decoding algorithm (Algorithm 4). The true signal is randomly selected from subject 04847' (10th trial and 23rd voxel) (after [1])

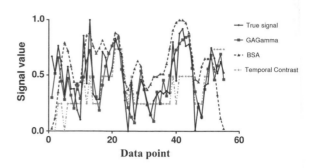

encoding algorithm takes a finite impulse response (FIR) *filter* and a *threshold$_{BSA}$* as input. In our experiments, we have used the low pass FIR filter of size 10 and the *threshold$_{BSA}$* = 0.95. These values are guided by the earlier work done on brain data by [30]. For the Temporal contrast encoding, we have used the hyper-parameter *factor* = 0.6. It must be noted that the presented results are non-exhaustive in the hyper-parameters space of different encoding methods. As a baseline, we have also included a randomly generated spike-train dataset. The random spike timing dataset is created using a Poisson's distribution with $\lambda = 0.6$. Varying the $\lambda$ parameter effects the symbol rate directly for random spike generation. In the 'no encoding' method, the raw dataset was created by transforming each multidimensional time series (set of images) within a trial into a single static observation by concatenating the feature values across the 16 time intervals [31].

A comparative analysis of our analysis is tabulated in Table 21.1. The results show a significant improvement in the bits/symbol column. The BSA and the GAGamma methods achieved a compression ranging within 6 and 24 times compared to the 'no encoding' method. This is due to the ability of the encoding algorithm to represent the information into the spike timings and thus present the data in a concise fashion to the classifier. Moreover, the GAGamma encoded data achieved a nearly comparable classification accuracy of 87.41 ± 4.80% and 76.00 ± 5.89% against 89.55 ± 4.60% and 79.11 ± 3.99% of 'no encoding' in 04847 and 07510 respectively. Between the encoding algorithms, the GAGamma and BSA shows the best overall performance, in that, they achieve high accuracy and compression. Figure 21.3 plots the accuracy and bits/symbol landscape for all the methods compared in the experiment across the two subjects. The green area is an approximate visual guide of the region we are aiming to achieve in the accuracy-compression space by concisely presenting information as the spike-times. Nevertheless, the reconstruction of the signal by the GAGamma decoding algorithm is superior in comparison with BSA. Figure 21.2 shows the comparison of different $S^\frown$ predicted by various decoding algorithms on a randomly chosen temporal signal.

**Table 21.1** Comparative evaluation of data encoding techniques applied to subject 04847 and 07510 in the starplus fMRI dataset (after [1])

| Subject id | Method | Data type | Bits/ symbol | Decoding error | Accuracy (K[1]) |
|---|---|---|---|---|---|
| 04847 | GAGamma | Integer | 4.96 | 0.07 | 87.41 ± 4.80% (16) |
| | BSA | Integer | 1.33 | 0.20 | 84.50 ± 4.47% (3) |
| | Temporal Contrast | Integer | 1.95 | 0.23 | 54.16 ± 5.77% (1) |
| | Random | Integer | 3.63 | – | 52.58 ± 4.79% (1) |
| | No encoding | Float | 32.0 | – | 89.55 ± 4.60% (1) |
| 07510 | GAGamma | Integer | 4.97 | 0.06 | 76.00 ± 5.89(8) |
| | BSA | Integer | 1.28 | 0.20 | 74.08 ± 6.71% (8) |
| | Temporal Contrast | Integer | 1.82 | 0.26 | 52.75 ± 5.84% (2) |
| | Random | Integer | 3.63 | – | 52.58 ± 4.79% (1) |
| | No encoding | Float | 32.0 | – | 79.11 ± 3.99% (5) |

For each subject we evaluated five different methods. They are, the proposed GAGamma encoding, the state of the art encoding techniques BSA and temporal contrast, a random spike generator and 'no encoding' or the raw data. These methods are evaluated by the bits/symbol, decoding error and accuracy as the measures of success. The decoding error metric is not relevant for the 'random' and 'no encoding' method. In the 'no encoding' method we use the raw data for pattern recognition and hence no encoding principle is involved in this method. The 'random' method being a random spike generator also does not have any decoding algorithm associated with it

$K$ Number of nearest neighbours used in K-NN algorithm

**Fig. 21.3** Plot showing the comparative performance of the encoding methods against the 'no encoding' applied on the fMRI dataset collected from subject id 04847 and 07510. The evaluation metrics bits/symbol (compression rate) and average accuracy are plotted in the X and the Y axis respectively (after [1])

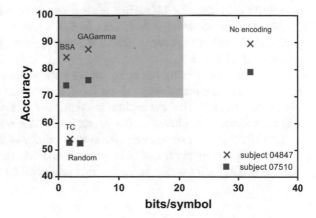

Additionally, GAGamma encoded spikes for the 'picture' and the 'sentence' stimuli were independently analysed for interpreting the discriminating spatio-temporal influence of the spikes. As described earlier in the experimental protocol, the presentation of a certain stimuli within a trial follows an order, *i.e.* for each stimuli class there exist subclasses of 'presented first' or 'presented second'. To analyse the effect of the first or second presentation of stimuli, we have separated the encoded dataset into four classes, 'picture presented first', 'picture presented second', 'sentence presented first' and 'sentence presented second'. Figures 21.4 and 21.5 show the comparison of the mean spike percentage across the trials for the four subclasses in subject 04847 and subject 07510. The two clusters in the 3D plots relate to the two ROI's (top left is 'LIPL' and bottom right is 'CALC') of the brain structure. Functionally, the 'CALC' region is responsible for central and peripheral vision whereas the 'LIPL' region is related to visual attention. In both the subjects it can be seen that 'seeing a sentence second' after 'seeing a picture first' has more spike activity on average across the trials than the other way around, especially in the 'LIPL' region.

The mean spike activity in the 'LIPL' is observed to be relatively higher (0.59 and 0.57) when the subjects were seeing a 'sentence' than when the subjects were seeing a 'picture' (0.54 and 0.55). A two sample T-test was conducted between the 'picture' and the 'sentence' class in the 'LIPL' region for the subjects to validate the previous result. The null hypothesis for the test conducted was the following, $H_0$: 'there is no difference between the picture spike activity and sentence spike activity'. The null hypothesis was rejected at 5% significance level with $p = 5.27 \times 10^{-18}$ for subject 04847 and with $p = 7.05 \times 10^{-12}$ for subject 07510. Hence, according to the Ttest, the average spike activity across the trials over time for 'seeing a picture' is significantly different from the average spike activity across trials over time for 'seeing a sentence'. Further, it must also be noted the sentences shown as part of the experiment (e.g. "It is not true that the dollar is below the plus.") are image oriented in nature and requires high imagery comprehension ability as part of the subject. This result is consistent with the experimental results [32] obtained earlier which shows a greater degree of activation and functional connectivity in the 'LIPL' region during cognitive tasks associated with high imagery sentence comprehension. This, in fact, validates the ability of the proposed encoding algorithm to preserve the useful discriminatory information in the compressed encoded data.

Figure 21.3 is showing the comparative performance of the encoding methods against the 'no encoding' applied on the fMRI dataset collected from subject id 04847 and 07510. The evaluation metrics bits/symbol (compression rate) and average accuracy are plotted in the X and the Y axis respectively.

Table 21.2 shows average pairwise asynchronicity of three different voxels at the end of ten independent runs of GAGamma based data encoding. Table 21.2 relates to the reproducibility of the spike timings produced by the mixed integer genetic

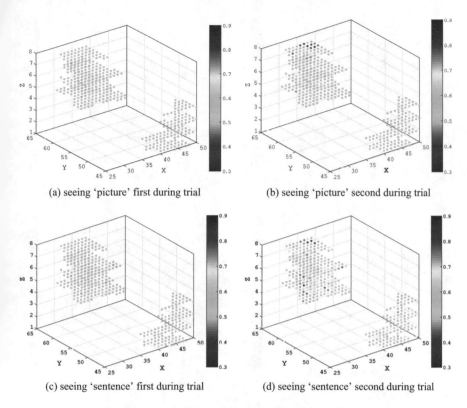

(a) seeing 'picture' first during trial          (b) seeing 'picture' second during trial

(c) seeing 'sentence' first during trial         (d) seeing 'sentence' second during trial

**Fig. 21.4** Comparative analysis of spike frequencies of the subject 04847 seeing picture versus seeing a sentence. The points in the 3D plot correspond to the spatial location of the voxels used in the dataset. Each voxel belongs to two physiologically defined clusters or regions of interest, namely 'CALC' and 'LIPL'. The top row shows the mean spike rate across the 'picture' trials as opposed to the 'sentence' trials in the bottom row. The first and the second column corresponds to the stimulus ('picture' or 'sentence') being presented first or second (after [1])

algorithm solver for the GAGamma method. It is a known fact that GA being an evolutionary optimisation solver, do not reproduce the same set of parameters when it is run multiple times. Nevertheless, it reaches near optimal fitness value each time. We have conducted ten independent runs of GAGamma encoding using three random voxels (30,468 and 3429) from trial 12 of subject 04847. Table 21.2 compares the similarity of the spike trains produced by the GAGamma encoding using two spike asynchronicity measures. They are percentage asynchronicity (described earlier) $d_p$ and Victor Purpura distance $d_{vp}$ respectively. The Victor Purpura distance $(d_{vp})$ [33] metric is a cost based distance measure. The distance is defined by the minimum cost of converting one spike train into the other using three operations; insertion (cost 1), deletion (cost 1) and shifting a spike by an interval $\delta t$ (cost $q|\delta t|$). For the smaller value of $q$ the distance metric approximates the spike count difference and hence supports rate coding. A higher penalty value of $q$, on the

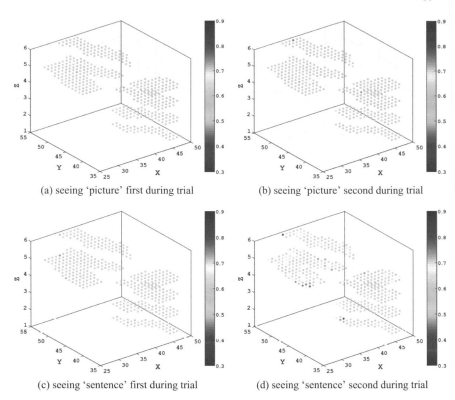

(a) seeing 'picture' first during trial          (b) seeing 'picture' second during trial

(c) seeing 'sentence' first during trial          (d) seeing 'sentence' second during trial

**Fig. 21.5** Comparative analysis of spike frequencies of the subject 07510 seeing picture versus seeing a sentence. The points in the 3D plot correspond to the spatial location of the voxels used in the dataset. Each voxel belongs to two physiologically defined clusters or regions of interest, namely 'CALC' and 'LIPL'. The top row shows the mean spike rate across the 'picture' trials as opposed to the 'sentence' trials in the bottom row. The first and the second column corresponds to the stimulus ('picture' or 'sentence') being presented first or second (after [1])

**Table 21.2** Average pairwise asynchronicity of three different voxels at the end of ten independent runs of GAGamma based data encoding (after [1])

| Voxel ID | $d_p$ | dvp |
|----------|-------|-----|
| 30 | 24.18 ± 10.20 | 0.23 ± 0.09 |
| 468 | 27.78 ± 11.96 | 0.26 ± 0.10 |
| 3429 | 28.03 ± 11.31 | 0.28 ± 0.11 |

contrary, supports the number of non-coincidental spikes and hence temporal encoding. The comparison shows that the spike timings are correct approximately 7 to 8 out of 10 times.

Figure 21.4 shows a comparative analysis of spike frequencies of the subject 04847 seeing picture versus seeing a sentence. The points in the 3D plot correspond to the spatial location of the voxels used in the dataset. Each voxel belongs to two physiologically defined clusters or regions of interest, namely 'CALC' and 'LIPL'. The top row shows the mean spike rate across the 'picture' trials as opposed to the 'sentence' trials in the bottom row. The first and the second column correspond to the stimulus ('picture' or 'sentence') being presented first or second.

Figure 21.5 shows a comparative analysis of spike frequencies of the subject 07510 seeing picture versus seeing a sentence. The points in the 3D plot correspond to the spatial location of the voxels used in the dataset. Each voxel belongs to two physiologically defined clusters or regions of interest, namely 'CALC' and 'LIPL'. The top row shows the mean spike rate across the 'picture' trials as opposed to the 'sentence' trials in the bottom row. The first and the second column correspond to the stimulus ('picture' or 'sentence') being presented first or second.

## 21.4   Chapter Summary and Further Readings

In this chapter our direction of work was focused on using temporal encoding as a framework to concisely represent large volumes of data by spike-timings and by doing so, preserve the existing spatio-temporal information for pattern recognition and classification tasks. Here a temporal encoding framework is presented and a specific encoding method for fMRI data, called GAGamma. The GAGamma encoding was evaluated on the benchmark cognitive pattern recognition problem using fMRI data. Further, we have compared the compressibility, pattern recognition performance and signal reconstruction performance of the GAGamma with the state of the art encoding algorithms and the raw data for their ability to recognise patterns with efficient resource usage.

The experimental evaluation on the benchmark fMRI dataset shows the ability of the encoding techniques to represent the fMRI data in a compressed space as a sequence of spike timings without losing any appreciable amount of information. This is crucially important especially in storing, processing and transmitting large scale streaming data. The flexibility of the proposed encoding framework lies in its ability to inject known structure information about the data source and thus provide the compression/encoding algorithms sufficient redundancy to represent the large dataset in an optimally concise manner, which is both highly reliable in terms of signal recovery and discriminatory information preservation in the compressed data.

Future work is planned in the application of the general framework proposed here on large radio-astronomy streaming data and geophysical data, including multi-sensory earthquake data. The proposed spike-time encoding method is suitable and is intended to be used for neuromorphic computation based on spiking neural network architecture, such as NeuCube [13, 34].

Further readings can be found in:

– Understanding nature through the symbiosis between information science, bioinformatics and neuroinformatics (Chap. 1 from [35]);

- Spike-time encoding as a data compression technique for pattern recognition of temporal data [1];
- More details of the proposed method can be found in [36].

**Acknowledgements** Some of the presented material in this chapter was first published in [1]. I would like to acknowledge the significant contribution of Neelava Sengupta for the development of the method from Sect. 21.2 and the experiments in 21.3.

# Appendix

---

**Algorithm 1** BSA encoding algorithm

---

1: input: $S, filter, threshold_{BSA}$
2: output: $B$
3: $B \Leftarrow 0$
4: $L = length(S)$
5: $F = length(filter)$
6: **for** $t = 1 : (L - F + 1)$ **do**
7:   $e_1 \leftarrow 0$
8:   $e_2 \leftarrow 0$
9:   **for** $k = 1 : F$ **do**
10:     $e_1 += |S(t + k) - filter(k)|$
11:     $e_2 += |S(t + k - 1)|$
12:   **end for**
13:   **if** $e_1 \leq (e_2 - threshold_{BSA})$ **then**
14:     $B(t) \leftarrow 1$
20:     **for** $k = 1 : F$ **do**
16:       $S(i + j - 1) -= filter(k)$
17:     **end for**
18:   **end if**
19: **end for**

---

**Algorithm 2** BSA decoding algorithm

---

1: input: $B, filter$
2: output: $S$
3: $L = length(B)$
4: $F = length(filter)$
5: **for** t=1:L-F+1 **do**
6: **if** $B(t) == 1$ **then**
7: **for** $k = 1 : F$ **do**
8:       $S(t + k - 1) += filter(k)$
9:   **end for**
10:   **end if**
11: **end for**

---

**Algorithm 3** Temporal contrast encoding algorithm

---

1: input: $S$, *factor*
2: output: $B$, *threshold$_{TC}$*
3: $L \leftarrow length(S)$
4: **for** $t = 1 : L - 1$ **do**
5:     $diff \leftarrow |S(t + 1) - S(t)|$
6: **end for**
7: *threshold$_{TC}$* $\leftarrow mean(diff) + factor \cdot std(diff)$
8: $diff \Leftarrow [0, diff]$
9: **for** $t = 1 : L$ **do**
10:     **if** $diff(t) > threshold_{TC}$ **then**
11:         $B(t) \leftarrow 1$
12:         **else if** $diff(t) < -threshold_{TC}$ **then**
13:             $B(t) \leftarrow -1$
14:     **else**
20:             $B(t) \leftarrow 0$
16:     **end if**
17: **end for**

---

**Algorithm 4** Temporal contrast decoding algorithm

---

1: input: $B$, *threshold$_{TC}$*
2: output: $\acute{S}$
3: $\acute{S} \leftarrow 0$
4: $L \leftarrow length(B)$
5: **for** $t = 2 : L$ **do**
6:     **if** $\acute{S}(t) > 0$ **then**
7: $\acute{S}(t) \leftarrow \acute{S}(t - 1) + threshold_{TC}$
8: **else if** $\acute{S}(t) < 0$ **then**
9:         $\acute{S}(t) \leftarrow \acute{S}(t - 1) - threshold_{TC}$
10:     **else**
11:         $\acute{S}(t) \leftarrow \acute{S}(t - 1)$
12:     **end if**
13: **end for**

---

# References

1. N. Sengupta, N. Kasabov, Spike-time encoding as a data compression technique for pattern recognition of temporal data. Inf. Sci. **406–407**, 133–145 (2017)
2. E.N. Brown, R.E. Kass, P.P. Mitra, Multiple neural spike train data analysis: state-of-the-art and future challenges. Nat. Neurosci. **7**(5), 456–461 (2004)
3. Z.F. Mainen, T.J. Sejnowski, Reliability of spike timing in neocortical neurons. Science **268** (5216), 2003–2006 (1995)
4. J.H. Maunsell, J.R. Gibson, Visual response latencies in striate cortex of the macaque monkey. J. Neurophysiol. **68**(4), 1332–1344 (1992)
5. T. Gollisch, M. Meister, Rapid neural coding in the retina with relative spike latencies. Science **319**(5866), 1108–1111 (2008)

6. R.M. Hallock, P.M. Di Lorenzo, Temporal coding in the gustatory system. Neurosci. Biobehav. Rev. **30**(8), 1145–1160 (2006)
7. C.E. Shannon, A mathematical theory of communication. ACM SIGMOBILE Mob. Comput. Commun. Rev. **5**(1), 3–55 (2001)
8. A.N. Kolmogorov, Three approaches to the quantitative definition of information. Probl. Inf. Transm. **1**(1), 1–7 (1965)
9. G.J. Chaitin, On the length of programs for computing finite binary sequences. J. ACM (JACM) **13**(4), 547–569 (1966)
10. P. Grunwald, P. Vitányi, Shannon information and kolmogorov complexity, arXiv preprint cs/ 0410002
11. H. de Garis, An artificial brain atr's cam-brain project aims to build/evolve an artificial brain with a million neural net modules inside a trillion cell cellular automata machine. New Gener. Comput. **12**(2), 220–221 (1994)
12. T. Iakymchuk, A. Rosado-Munoz, M. Bataller-Mompean, J. Guerrero-Martinez, J. Frances-Villora, M. Wegrzyn, M. Adamski, Hardware-accelerated spike train generation for neuromorphic image and video processing, in *2014 IX Southern Conference on Programmable Logic (SPL)* (IEEE, 2014), pp. 1–6
13. N. Kasabov, N.M. Scott, E. Tu, S. Marks, N. Sengupta, E. Capecci, M. Othman, M.G. Doborjeh, N. Murli, R. Hartono et al., Evolving spatio-temporal data machines based on the neucube neuromorphic framework: design methodology and selected applications. Neural Netw. **78**(2016), 1–14 (2016)
14. M. Hough, H. De Garis, M. Korkin, F. Gers, N.E. Nawa, Spiker: analog waveform to digital spiketrain conversion in atrs artificial brain (cam-brain) project, in *International Conference on Robotics and Artificial Life* (Citeseer, 1999)
15. B. Schrauwen, J. Van Campenhout, BSA, a fast and accurate spike train encoding scheme, in *Proceedings of the International Joint Conference on Neural Networks*, vol. 4 (IEEE Piscataway, NJ, 2003), pp. 2825–2830
16. M. Dorigo, V. Maniezzo, A. Colorni, Ant system: optimization by a colony of cooperating agents. IEEE Trans. Syst. Man Cybern. Part B Cybern. **26**(1), 29–41 (1996)
17. H. De Garis, N. E. Nawa, M. Hough, M. Korkin, Evolving an optimal de/convolution function for the neural net modules of atr's artificial brain project, in *International Joint Conference on Neural Networks, 1999. IJCNN'99*, vol. 1 (IEEE, 1999), pp. 438–443
18. N. Sengupta, N. Scott, N. Kasabov, Framework for knowledge driven optimisation based data encoding for brain data modelling using spiking neural network architecture, in *Proceedings of the Fifth International Conference on Fuzzy and Neuro Computing* s*FANCCO-2010*) (Springer, 2010), pp. 109–118
19. F.G. Ashby, *Statistical Analysis of fMRI Data* (MIT Press, 2011)
20. M.D. Nunez, P.L. Nunez, R. Srinivasan, Electroencephalography (EEG): neurophysics, experimental methods, and signal processing, in *Handbook of Neuroimaging Data Analysis* (Chapman & Hall/CRC, 2016) (Chapter)
21. B. Babu, M. Jehan, Differential evolution for multi-objective optimization, in *The 2003 Congress on Evolutionary Computation, 2003. CEC'03*, vol. 4 (IEEE, 2003), pp. 2696–2703
22. L. Yiqing, Y. Xigang, L. Yongjian, An improved pso algorithm for solving non-convex nlp/ minlp problems with equality constraints. Comput. Chem. Eng. **31**(3), 162–203 (2007)
23. K. Deb, An efficient constraint handling method for genetic algorithms. Comput. Methods Appl. Mech. Eng. **186**(2), 311–338 (2000)
24. K. Deep, K.P. Singh, M.L. Kansal, C. Mohan, A real coded genetic algorithm for solving integer and mixed integer optimization problems. Appl. Math. Comput. **212**(2), 505–518 (2009)
25. G.M. Boynton, S.A. Engel, G.H. Glover, D.J. Heeger, Linear systems analysis of functional magnetic resonance imaging in human v1. J. Neurosci. **16**(13), 4207–4221 (1996)
26. K.J. Friston, O. Josephs, G. Rees, R. Turner, Nonlinear event-related responses in fMRI. Magn. Reson. Med. **39**(1), 41–52 (1998)

27. G.H. Glover, Deconvolution of impulse response in event-related bold fMRI 1. Neuroimage **9** (4), 416–429 (1999)
28. X. Wang, T. Mitchell, Detecting cognitive states using machine learning. Technical report, CMU CALD Technical Report for Summer Work (2002)
29. L.-N. Do, H.-J. Yang, A robust feature selection method for classification of cognitive states with fMRI data, in *Advances in Computer Science and its Applications* (Springer, 2014), pp. 71–76
30. N. Nuntalid, K. Dhoble, N. Kasabov, Eeg classification with BSA spike encoding algorithm and evolving probabilistic spiking neural network, in *International Conference on Neural Information Processing* (Springer, 2011), pp. 451–460
31. T.M. Mitchell, R. Hutchinson, M.A. Just, R.S. Niculescu, F. Pereira, X. Wang, Classifying instantaneous cognitive states from fMRI data, in *American Medical Informatics Association Annual Symposium* (2003)
32. M.A. Just, S.D. Newman, T.A. Keller, A. McEleney, P.A. Carpenter, Imagery in sentence comprehension: an fMRI study. Neuroimage **21**(1), 112–124 (2004)
33. J.D. Victor, K.P. Purpura, Metric-space analysis of spike trains: theory, algorithms and application. Netw. Comput. Neural Syst. **8**(2), 127–164 (1997)
34. N.K. Kasabov, Neucube: a spiking neural network architecture for mapping, learning and understanding of spatiotemporal brain data. Neural Netw. **52**(2014), 62–76 (2014)
35. N. Kasabov, *Springer Handbook of Bio-/Neuroinformatics* (Springer, 2014)
36. N. Sengupta, PhD Thesis, Auckland University of Technology, 2018

# Chapter 22
# From Brain-Inspired AI to a Symbiosis of Human Intelligence and Artificial Intelligence

This chapter represents the essence of the book, which put in one sentence is:

Inspired by the oneness in nature in time-space we aim to achieve oneness in data modelling using brain-inspired computation.

The chapter argues that SNN allow for the integration of all levels of information processing in the brain and in nature, from quantum, molecular and neuro-genetic, to brain signals, evolution and consciousness. The chapter presents future directions for using SNN to build brain-inspired AI systems that are able to both receive and communicate knowledge with humans for a symbiotic and collaborative work, led by the human intelligence (HI). The chapter is organised in the following sections:

22.1. Towards integrated quantum-molecular-neurogenetic-brain-inspired models.
22.2. Towards a symbiosis of Human Intelligence and Artificial Intelligence (HI + AI), led by the HI.
22.3. Concluding summary and discussions.

## 22.1 Towards Integrated Quantum-Molecular-Neurogenetic-Brain-Inspired Models

This section hypothesis is that based on SNN and their brain-inspired properties, we can be aiming at the creation of integrated quantum-neurogenetic-brain-inspired models.

© Springer-Verlag GmbH Germany, part of Springer Nature 2019                          701
N. K. Kasabov, *Time-Space, Spiking Neural Networks and Brain-Inspired Artificial Intelligence*, Springer Series on Bio- and Neurosystems 7,
https://doi.org/10.1007/978-3-662-57715-8_22

## 22.1.1  Quantum Computation

Quantum computation is based upon physical principles from the theory of quantum mechanics [1]. One of the basic principles, that is likely to trigger the development of new methods in information sciences, is the linear *superposition of states*.

At a macroscopic or classical level a system exists only in a single basis state as energy, momentum, position, spin and so on. However, at microscopic or quantum level, the system at any time represents a superposition of all possible basis states. At the microscopic level any particle can assume different positions at the same time, can have different values of energy, can have two values of the spins and so on. This superposition phenomenon is counterintuitive because in the classical physics one particle has only one position, energy, spin and so on.

If the system interacts in any way with its environment, the superposition is destroyed and the system collapses into one single real state as in the classical physics. This process is governed by a probability amplitude [1]. The square of the intensity for the probability amplitude is the quantum probability to observe the state.

The concept of quantum computing utilizes the special non-local properties of the quantum phenomena. A quantum atomic or sub-atomic particle (e.g. atoms, electrons, protons, neutrons, bosons, fermions, photons) exists in a probabilistic superposition of states rather than in a single definite state. For example, an electron circling around a nucleus, jumps to different orbits—states, due to absorbing or releasing energy.

Particles in general are characterized by: charge, spin, position, velocity, energy.

Some principles, assumptions and facts in quantum information processing are listed below:

- The *Heisenberg's uncertainty principle*: Both the position and the momentum of an electron, or generally—of a particle, can not be known, because to know it means to measure it, but measuring causes interfering and change of both the position and the momentum. Making an observation of the system "collapses" the system to one possible state, or universe.
- *The superposition principle,* meaning that a particle can be in several states at the same time, with certain probabilities. It is illustrated by Schroedinger by his famous thought experiment of seeing with one eye open, a creature (a cat) in both alive and dead states with certain probabilities (see also [2]).
- *The entanglement principle*, means that two or more particles, regardless of their location, are in the same state with the same probability. The two particles can be viewed as "correlated", undistinguishable, "synchronized", coherent. An example is a laser beam consisting of millions of photons having same characteristics and states.
- Electro-magnetic radiation is emitted in discrete quanta whose *energy E* is proportional to the frequency:

$$E = h.f, \tag{22.1}$$

where $h$ is the Max Planck constant (appr. $6.62608 \times 10^{-34}$) and $f$ is the frequency.

The advantage of quantum computing is that, while a system is uncollapsed, it can carry out more computing than a collapsed system, because, in a sense, it is computing in an infinite number of universes at once.

Ordinary computers are based on bits, which always take one of the two values 0 or 1. Quantum computations are based instead on what are called *Q-bits* (or qubits). A Q-bit may be simply considered as the spin state of an electron. An electron can have spin Up or spin Down; or three quarters Up and one quarter Down. A Q-bit contains more information than a bit but in a strange sense, not in the same sense in which two bits contain more information than a bit.

The state of a Q-bit can be represented as below, where $\alpha$ and $\beta$ are complex numbers that specify the probability amplitudes of the corresponding states "0" and "1".

$$|\psi\rangle = \alpha|0\rangle + \beta|1\rangle \tag{22.2}$$

Since the Q-bit can only be in these two states, it should satisfy the condition:

$$|\alpha|^2 + |\beta|^2 = 1 \tag{22.3}$$

*Example* A 3 bit register can store 000 or 001 or 010 or 100 or 011 or 101 or 110 or 111, while a 3-qubit register can store 000 and 001 and 010 and 100 and 011 and 101 and 110 and 111 at the same time, each to different probabilities. Storage capacity increases exponentially, $2^N$ where N is the size of the register. Since the numbers are stored simultaneously in the same register, operations with them can be done also simultaneously, so a quantum "computer" has $2^N$ processors working in parallel.

The state of a Q-bit can be changed by an operation called a *quantum gate*. A quantum gate is a reversible gate and can be represented as a unitary operator $U$ acting on the Q-bit basis states. The defining property of a unitary matrix is that its conjugate transpose is equal to its inverse. There are several quantum gates already introduced, such as the NOT gate, controlled NOT gate, rotation gate, Hadamard gate, etc. For example, a rotation gate is represented as:

$$U(\theta) = \begin{bmatrix} \cos\theta & -\sin\theta \\ \sin\theta & \cos\theta \end{bmatrix} \tag{22.4}$$

## 22.1.2   The Concept of an Integrated Quantum-Neurogenetic-Brain-Inspired Model Based on SNN

In the section on computational neurogenetic modelling (Chap. 16) we presented a model that links the level of expression of genes and proteins in a neuron to the neuronal spiking activity, and then—to the information processing of a neuronal ensemble that is measured as local field potentials (LFP).

But how do quantum information processes in the atoms and particles (ions, electrons, etc.), that make the large protein molecules, relate to the spiking activity of a neuron and to the activity of a neuronal ensemble? This is a challenging question that is not possible to answer now, but here we can make some speculative steps, hopefully in the right direction.

The spiking activity of a neuron relates to the transmission of thousands of ions and neurotransmitter molecules across the synaptic cleft and to the emission of spikes. Spikes, as carriers of information, are electrical signals made of ions and electrons that are emitted in one neuron and transmitted along the nerves to many other neurons. But ions and electrons are characterised by their quantum properties as discussed in a previous section of this chapter. So, quantum properties would influence the spiking activity of neurons and the whole brain and therefore brains obey the laws of quantum mechanics.

Similarly to a chemical effect of a drug to protein and gene expression levels in the brain, that may affect the spiking activity and the functioning of the whole brain (modelling of these effects is subject of the computational neurogenetic modelling CNGM—see Chap. 16), external factors like radiation, high frequency signals etc. can influence the quantum properties of the particles in the brain through gate operators. According to [3], microtubules in the neurons are associated with quantum gates.

So, the challenge is, similar to the CNGM, to create quantum-inspired CNGM (QiCNGM) that also takes into account quantum properties of the particles in a neuron and in the brain as a whole.

In the first instance, is it possible at all? At this stage the answer is not known, but we will describe the above relationships in an abstract theoretical way, hoping to be able to refine this framework, modify it, proof it and use it in the future, at least partially.

Figure 22.1 shows different levels of information processing in the brain and also that they all are functionally connected and integrated (see also Chap. 1).

Here the interaction at different levels is shown as hypothetical aggregated functions as suggested in [4]:

$$Q' = Fq(Q, Eq),\qquad(22.5)$$

a future state $Q'$ of a particle of group of particles (e.g. ions, electrons, etc.) depends on the current state $Q$ and on the frequency spectrum Eq of an external signal

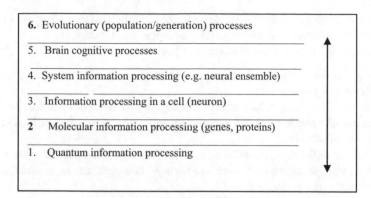

**Fig. 22.1** Different levels of information processing in the brain are functionally connected and integrated

$$M' = Fm(Q, M, Em),\qquad(22.6)$$

a future state of a molecule $M'$ or a group of molecules (e.g. genes, proteins) depends on its current state $M$, on the quantum state $Q$ of the particles and on an external signal $Em$;

$$N' = Fn(N, M, Q, En),\qquad(22.7)$$

a future state $N'$ of a spiking neuron or an ensemble of neurons will depend on its current state $N$, on the state of the molecules $M$, on the state of the particles $Q$ and on external signals $En$.

$$C' = Fc(C, N, M, Q, Ec),\qquad(22.8)$$

a future cognitive state $C'$ of the brain will depend on its current state $C$ and also on the neuronal—$N$, on the molecular—$M$, and on the quantum—$Q$ states of the brain.

We can support the above hypothetical model of integrated representation, by stating the following assumptions, some of them already supported by experimental results [3]:

(1) A large amount of atoms are characterised by the same quantum properties, possibly related to the same gene/protein expression profile of a large amount of neurons characterised by spiking activity;
(2) A large neuronal ensemble can be represented by a single Local Field Potential —LFP, and
(3) A cognitive process can be represented perhaps as a complex, but single function Fc that depends on all previous levels.

The model above is too simplistic, and at the same time—to complex to implement at this stage, but even linking two levels of information processing in a computational model may be useful for the further understanding of complex information processes and for modelling complex brain functions.

Creating quantum inspired models can lead to:

– Using quantum principles to create more powerful information processing methods and systems; and
– Understanding the quantum level information processing in Nature
– Understanding molecular and quantum information processing as important for all areas of science;
– Modelling molecular processes needed for biology, chemistry and physics.
– Using these processes as inspiration for new computer devices—million times faster and more accurate BI-AI.
– Deutsch [5, 6] argues that NP-hard problems (e.g. time complexity grows exponentially with the size of the problem) can be solved by a quantum computer.
– Penrose [3] argues that solving the quantum measurement problem is pre-requisite for understanding the mind.
– Hameroff [7] argues that consciousness emerges as a macroscopic quantum state due to a coherence of quantum-level events within neurons.

Many open questions need to be answered in this respect. Some of them are listed below:

• How quantum processes affect the functioning of a living system in general?
• How quantum processes affect cognitive and mental functions?
• Is the brain a quantum machine, working in a probabilistic space with many states (e.g. thoughts) being in a superposition all the time and only when we formulate our thought through speech or writing then the brain "collapses" in a single state?
• Is the fast pattern recognition process in brain, involving far away segments, a result of both parallel spike transmissions and particle *entanglement* across areas of the brain?
• Is communication between people, and living organisms in general, also a result of *entanglement* processes? What about connecting with "ghosts", or with extraterrestrial intelligence?
• How does the energy in the atoms relate to the energy of the proteins, the cells and the whole living system?
• How energy relates to information?
• Would it be beneficial to develop different quantum inspired (QI) computational intelligence techniques, such as QI-SVM, QI-GA, QI-decision trees, QI-logistic regression, QI-cellular automata in addition to the presented in Chap. 7 QiEA and QiPSO?
• How do we implement the QI computational intelligence algorithms on existing computer platforms in order to benefit from their high potential speed and accuracy? Should we wait for the quantum computers to be realised many years

from now, or we can implement them efficiently on specialised computing devices based on classical principles of physics?

## 22.2 Towards a Symbiosis Between Human Intelligence and Artificial Intelligence (HI + AI), Led by HI

### 22.2.1 Some Notions About AGI

*Artificial General Intelligence (AGI)* is a trend in AI that is concerned with the idea that eventually machines can perform any intellectual task that humans can do.

The ideas promoted by the AGI led to the creation of the concept of technological *singularity*, i.e. machines become super intelligent that they take over from humans and develop on their own, beyond which point the human societies may collapse in their present forms, which may ultimately lead to the perish of the humanity.

Stephen Hawking commented "I believe there is no real difference between what can be achieved by a biological brain and what can be achieved by a computer. AI will be able to redesign itself at an ever-increasing rate. Humans, who are limited by slow biological evolution, couldn"t compete and could be superseded by AI. AI could be either the best or the worst thing ever to happen to humanity…"

We take the view that the technological future and the global future of our society will rely on the symbiosis between humans and machines, as discussed in the next sub-section, for the benefit of the humanity, helping to solve many challenging, global problems, such as:

- Early disease diagnosis and disease prevention
- Predicting and preventing ecological and environmental disasters
- Robots for homes and for elderly
- Improved productivity
- Improved human intelligence and creativity
- Improved lives and longevity
- Better understanding of ourselves and the world we live in

At the same time we have to be aware of the disastrous consequences that can follow if AI is not controlled and used properly.

### 22.2.2 Towards a Symbiosis Between Human Intelligence and Artificial Intelligence (HI + AI), Led by HI

We are overwhelmed by the multiple modalities and sources of data and by the lack of methods to integrate and model them continuously, to extract new knowledge and to transfer it to humans. On the other hand, there is a tremendous amount of knowledge and skills humans have accumulated over years and there are no efficient methods to directly transfer them to machines. Transfer of knowledge between humans and machines has been identified as a key issue for the future of Artificial Intelligence (AI) [4, 8–16]. Recently, deep learning neural networks have gained momentum as potential 'silver bullets' [17, 18]. Despite their impressive results in image recognition, medical classification systems and game playing [19–24], current methods are incapable of dynamic, adaptive, and fast learning from multimodal data from different sources at different times, and integrating the acquired information, due to their rigid, inflexible structures. They are not suitable for humans to transfer their knowledge in a direct and unstructured way and for machines to efficiently solve problems [25]. The human brain evolves by incrementally integrating multiple modalities and creating deep, flexible structures of spiking neural networks [26].

Our hypothesis is that by using principles of information and knowledge representation and learning in the human brain, we can create a theoretical and computational framework for integrated Human and Artificial Intelligence (HI + AI) that enables both incremental learning from multimodal data and knowledge transfer between humans and machines. One of the tasks in this direction will be to develop a framework for multimodal learning and knowledge transfer and to test it on audio, visual and brain data representing human activities.

As discussed in previous chapters a BI-AI system will have a 3D scalable, spatially organised spiking neural network (SNN) [27–30] structure following a brain template, such as MNI and Talairach [31–33]. It will learn multimodal data, including brain signals, using brain-inspired learning rules (e.g., [27–30]). Single, or multiple BI-AI systems can learn different modalities from various sources at different times and their learned connections can be merged to integrate the learned knowledge. They will learn brain signals from humans when humans perceive or express emotions, or solve procedural or cognitive tasks, so that acquired human knowledge can be applied by a machine.

Previous experience [34–36] using the brain-inspired SNN (Chap. 6) and the methods already presented in other chapters of this book, indicate that such BI-AI can be developed.

First, it is possible to create systems for integrated multimodal learning of auditory and visual information and to apply them for person identification problem [36], activity detection in videos [36–39] and for other intelligent tasks. A stereo tonotopic/retinotopic mapping can be used for audio/visual data to be entered into areas of a BI-AI model that correspond to the auditory/visual cortex in the brain template [32, 33, 40] (Chap. 13).

Second, systems can be developed for emotion and preference recognition using facial expression data and EEG (Electroencephalogram) brain data [41, 42] (Chaps. 8 and 9).

Third, methods and systems for transfer of procedural and cognitive knowledge from humans to machines can be developed for tasks such as moving objects in space, target detection, game playing [23, 24], using integrated audio/visual, EEG and/or fMRI (functional magnetic resonance imaging) data [43–52]. A system will learn to perform tasks using visual and human brain data. After training the system, its performance can be communicated to humans, following a new knowledge exchange protocol. The learning can continue incrementally. This departs significantly from the traditional brain-computer interfaces (BCI) where human brain signals are classified in a "black box" [53–55], rather than learned in a system as evolving knowledge which is the case of the brain inspired BCI presented in Chap. 14 of this book and also illustrated in Fig. 22.2.

The overall applicability and the limits of the proposed BI-AI approach are still to be determined and challenges are anticipated in dealing with different spatio-temporal scales of multimodal data, but preliminary experiments [41, 45, 49, 54, 55] indicated that not only would this BI-AI approach lead to a much better accuracy of data analysis than current deep neural networks [17–21] or other machine learning techniques, but it could also become a universal approach for the much anticipated knowledge transfer and human-machine symbiosis [7–16], with a significant impact on the future development in brain-computer interfaces, affective computing, home robotics, cognitive sciences and cognitive computing.

The hypothesis for integrating human intelligence and AI (HI + AI) is illustrated in Fig. 22.3. A BI-AI (such as BI-SNN) can be trained on brain data of different modalities, e.g. EEG, fMRI, sensory audio-visual etc. as discussed in previous chapters of the book. Such integrated system will have the same brain template as the human brain (e.g. Talairach template). This will enable the human and the machine to exchange information and to work together.

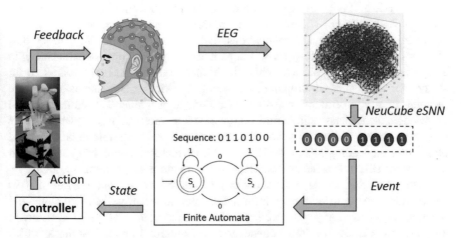

**Fig. 22.2** A simple example of BI-AI in the form of BI-BCI. Human brain signals, recorded when the human is performing a task, are not only classified, but learned in a brain-like machine that has the same template of neurons and connections (not the same number) as the human (e.g. NeuCube). After learning, the machine can perform the same task without the human as the human has already 'transferred' the required knowledge for the task. The figure is created by K. Kumarasinghe [57]

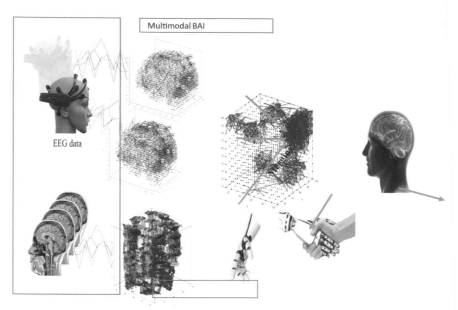

**Fig. 22.3** The hypothesis for the symbiosis of human intelligence and AI into HI + AI. A BI-AI can be trained on brain data of different modalities, e.g. EEG, fMRI, sensory audio-visual etc. as discussed in previous chapters of the book. Such integrated system will have the same brain template as the human brain (e.g. Talairach template). This will enable the human and the machine to exchange information and to work together. The figure was created by M. Doborjeh

The fact that AI is brain-inspired does not mean that it cannot surpass the human brain. A BI-AI system will be faster to process more information within a time frame, to perform complex calculations that are not brain-inspired, to do fast number crunching, searching, mapping, pattern recognition, precise calculations to a high accuracy of the results. While the human brain is the best evolved so far system for integrating information, extracting knowledge and understanding the meaning of the information and its implications as a whole. The human general knowledge, the understanding of the complexity in nature and human societies, that have evolved for millions of years of evolution, will be impossible to be surpassed by AI, thus the leading role of the HI in a future symbiosis. And it is up to the HI to decide what BI-AI to create to complement and to enhance the human knowledge.

At the same time BI-AI systems are created not to be fixed, but to evolve and to incorporate new data and new information and to modify their structure and functionality accordingly. *Evolvability* is an essential quality of BI-AI systems. In this way they can evolve deep knowledge representation in particular areas, which is adapting, changing evolving in time-space. This knowledge can be complementary to the human knowledge. Twenty four century after Aristotles' epistemology, we can create systems that can learn knowledge from data, can evolve this knowledge in time-space and communicate this knowledge with humans for a better understanding of nature and who we are.

> This complementarity of HI and AI makes their symbiosis, led by the HI, a realistic pathway to take in the future, which is also an evolving process in time-space, with no foreseeable end....

## 22.3  Summary and Further Readings for a Deeper Knowledge

This chapter raises some hypotheses for future directions, through introducing some ideas about:

- Creating unified quantum-neuro-genetic-brain-inspired computational models;
- Creating a symbiosis between BI-AI and human intelligence, towards HI + AI.

The above hypotheses are inspired by the unity of the world and the need to better understand it through unity of the computational models and human intelligence.

Further readings can be found in:

- Quantum and biocomputing—common notions and targets (Chap. 59 in [56]);
- Brain, gene and quantum inspired computational intelligence (Chap. 60 in [54]);
- Brain-like robotics (Chap. 57 in [56]);
- Brain and creativity (Chap. 61 in [56]);
- Neurocomputational models of natural language (Chap. 48 in [56]).

## References

1. R.P. Feynman, R.B. Leighton, M. Sands, *The Feynman lectures on Physics* (Addison-Wesley Publishing Company, Massachusetts, 1965), p. 1965
2. C. Koch, K. Hepp, Quantum mechanics in the brain. Nature **440**, 30 (March 2006)
3. R. Penrose, *Shadows of the Mind. A Search for the Missing Science of Consciousness* (Oxford University Press, 1994)
4. N. Kasabov, *Evolving Connectionist Systems* (Springer, 2007)
5. D. Deutsch, Quantum theory, the Church-Turing principle and the universal quantum computer. Proc. R. Soc. Lond. A **400**(97–117), 1985 (1985)
6. D. Deutsch, Quantum computational networks. Proc. R. Soc. Lond. A **425**(73–90), 1989 (1989)
7. S.R. Hameroff, Quantum computing in microtubules-An Intra-neural correlate of consciousness? Jpn. Bull. Cogn. Sci. **4**(3), 67–92 (1997)

8. A. Prabhakar, DARPA: the merging of humans and machines is happening now WIRED magazine (2017). Also in: http://www.wired.co.uk/article/darpa-arati-prabhakar-humans-machines

9. P. Lee, Transfer learning (Microsoft research) (2018). https://www.edge.org/response-detail/27125

10. C. Stephanidis, A.M. Anton, in *Universal Access in Human-Computer Interaction. User and Context Diversity*. 7th International Conference, UAHCI 2013, Held as Part of HCI International 2013, Las Vegas, NV, USA, 21–26 July, Proceedings, Part 2 (2013). https://link.springer.com/book/10.1007%2F978-3-642-39191-0

11. K.S. Gill (ed.), *Human Machine Symbiosis: The Foundations of Human-Centred Systems Design* (Springer Science & Business Media, 2012)

12. V. Vapnik, R. Izmailov, Learning using privileged information: similarity control and knowledge transfer. J. Mach. Learn. Res. **21**, 2023–2049 (2015)

13. H. Markram, The blue brain project. Nat. Rev. Neurosci. **7**(2), 153–210 (2006). https://doi.org/10.1038/nrn1848

14. P. Robinson, R. El Kaliouby, Computation of emotions in man and machines. R. Soc. Publishing **364**(1535), 3441–3447 (2009). https://doi.org/10.1098/rstb.2009.0198

15. C. Pelachaud, Modelling multimodal expression of emotion in a virtual agent. Philos. Trans. R. Soc. B: Biol. Sci. **364**(1535), 3539–3548 (2009)

16. C.C. Federspiel, H. Asada, in *Transfer of Human Preference to Smart Machines: A Case Study of Human Thermal Comfort Control*. American Control Conference (1990), pp. 2833–2842

17. J. Schmidhuber, Deep learning in neural networks: an overview. Neural Netw. **61**, 85–117 (2015)

18. Y. LeCun, Y. Bengio, G. Hinton, Deep learning. Nature **521**(7553), 436 (2015). https://doi.org/10.1038/nature14539

19. A. Krizhevsky, L. Sutskever, G.E. Hinton, in *Image Net Classification with Deep Convolutional Neural Networks*. Proceedings of Advances in Neural Information Processing Systems (2012), pp. 1097–1105

20. D. Ferrucci, E. Brown, J. Chu-Carroll, J. Fan, D. Gondek, A. Kalyanpur, A. Lally, W. Murdock, E. Nyberg, J. Prager, N. Schlaefer, Building Watson: an overview of the DeepQA project. AI Mag. **31**(3), 59–79 (2010). https://doi.org/10.2109/aimag.v31i3.2303

21. G. Hinton, L. Deng, D. Yu, G.E. Dahl, A.R. Mohamed, N. Jaitly, S.A. Senior, V. Vanhoucke, P. Nguyen, T. Sainath, B. Kingsbury, Deep neural networks for acoustic modeling in speech recognition: the shared views of four research groups. IEEE Signal Process. Mag. **29**(6), 82–97 (2012). https://doi.org/10.1109/MSP.2012.2205597

22. L. Sutskever, O. Vinyals, Q.V. Le, in *Sequence to Sequence Learning with Neural Networks*. Advances in Neural Information Processing Systems (2014), pp. 3104–3112

23. D. Silver, A. Huang, C.J. Maddison, A. Guez, L. Sifre, G. Van Den Driessche, S. Dieleman, Mastering the game of Go with deep neural networks and tree search. Nature **529**(7587), 484–489 (2016). https://doi.org/10.1038/nature21961

24. D. Silver, J. Schrittwieser, K. Simonyan, I. Antonoglou, A. Huang, A. Guez, T. Hubert, L. Baker, M. Lai, A. Bolton, Y. Chen, Mastering the game of go without human knowledge. Nature **550**(7676), 354–359 (2017). https://doi.org/10.1038/nature24270

25. G. Hinton, What is wrong with convolutional neural nets? https://www.youtube.com/watch?v=rTawFwUvnLE&t=1579s

26. M.W. Reimann, M. Nolte, M. Scolamiero, K. Turner, R. Perin, G. Chindemi, H. Markram, Cliques of neurons bound into cavities provide a missing link between structure and function. Front. Comput. Neurosci. **11**(48) (2017)

27. W. Maass, T. Natschlaeger, H. Markram, Real-time computing without stable states: a new framework for neural computation based on perturbations. Neural Comput. **14**(11), 2531–2560 (2002)

28. W. Gerstner, W.M. Kistler, in *Spiking Neuron Models: Single Neurons, Populations, Plasticity*. Cambridge University Press (2002)

29. S. Thorpe, J. Gautrais, Rank order coding. Comput Neurosci: Trends. Res. **13**, 113–119 (1998)

30. N. Kasabov, NeuCube: a spiking neural network architecture for mapping, learning and understanding of spatio-temporal brain data. Neural Netw. **52**, 62–76 (2014)

31. J. Mazziotta, A. Toga, A. Evans, P. Fox, J. Lancaster, K. Zilles, B. Mazoyer, A probabilistic atlas and reference system for the human brain: international consortium for brain mapping (ICBM). Philos. Trans. Roy. Soc. London. B **356**(1412), 1293–1322 (2001). https://doi.org/10.1098/rstb.2001.0915

32. J. Talairach, P. Tournoux, *Co-Planar Stereotaxic Atlas of the Human Brain. 3-Dimensional Proportional System: An Approach to Cerebral Imaging* (Thieme Classics, 1988)

33. L. Koessler, L. Maillard, A. Benhadid, J.P. Vignal, J. Felblinger, H. Vespignani, M. Braun, Automated cortical projection of EEG sensors: anatomical correlation via the international 10–10 system. Neuroimage **46**(1), 64–72 (2009)

34. N. Kasabov, in *Evolving Connectionist Systems*, 1st edn. (Springer, 2003)

35. N. Kasabov, E. Postma, J. van den Herik, AVIS: a connectionist-based framework for integrated auditory and visual information processing. Inf. Sci. **123**, 127–148 (2000)

36. S.G. Wysoski, L. Benuskova, N. Kasabov, Evolving spiking neural networks for audiovisual information processing. Neural Netw. **23**(7), 819–835 (2010)

37. Y.G. Jiang, G. Ye, S. F. Chang, D. Ellis, A.C. Loui, in *Consumer Video Understanding: A Benchmark Database and an Evaluation of Human and Machine Performance*. Proceedings of ACM International Conference on Multimedia Retrieval (ICMR), Trento, Italy (2011). https://doi.org/10.1145/1991996.1992025

38. N. Goyette, P.M. Jodoin, F. Porikli, J. Konrad, P. Ishwar, in *Changedetection. net: A New Change Detection Benchmark Dataset*. IEEE Computer Society Conference on Computer Vision and Pattern Recognition Workshops (CVPRW) (2012), pp. 1–8. https://doi.org/10.1109/cvprw.2012.6238919

39. S. Oh, A. Hoogs, A. Perera, N. Cuntoor, C.-C. Chen, J.T. Lee, S. Mukherjee, J.K. Aggarwal, H. Lee, L. Davis, E. Swears, X. Wang, Q. Ji, K. Reddy, M. Shah, C. Vondrick, H. Pirsiavash, D. Ramanan, J. Yuen, A. Torralba, B. Song, A. Fong, A. Roy-Chowdhury, M. Desai, in *A Large-Scale Benchmark Dataset for Event Recognition in Surveillance Video*. IEEE Conference on Computer Vision and Pattern Recognition (CVPR) (2011), pp. 3153–3210. https://doi.org/10.1109/cvpr.2011.5995586

40. A.G. Huth, W.A. de Heer, T.L. Griffiths, F.E. Theunissen, J.L. Gallant, Natural speech reveals the semantic maps that tile human cerebral cortex. Nature **532**, 453–458 (2016). https://doi.org/10.1038/nature17637

41. Z.G. Doborjeh, M.G. Doborjeh, N. Kasabov, Attentional bias pattern recognition in spiking neural networks from spatio-temporal EEG data. Cogn. Comput. (2017). https://doi.org/10.1007/s12559-017-9517-x

42. H. Kawano, A. Seo, Z.G. Doborjeh, N. Kasabov, M.G. Doborjeh, in *Analysis of Similarity and Differences in Brain Activities Between Perception and Production of Facial Expressions Using EEG Data and the NeuCube Spiking Neural Network Architecture*. In Proceedings of International Conference on Neural Information Processing (Springer International Publishing, (2016), pp. 221–227

43. M.G. Doborjeh, N. Kasabov, Z.G. Doborjeh, in *Evolving, Dynamic Clustering of Spatio/Spectro-Temporal Data in 3D Spiking Neural Network Models and a Case Study on EEG Data*. Evolving Systems (Springer, 2017), pp. 1–17. https://doi.org/10.1007/s12530-017-9178-8

44. M. Doborjeh, G.Y. Wang, N. Kasabov, R. Kydd and B. Russell, A spiking neural network methodology and system for learning and comparative analysis of EEG data from healthy versus addiction treated versus addiction not treated subjects. IEEE Trans. Biomed. Eng. **63** (9), 1830–1841 (2016). https://doi.org/10.1109/tbme.2015.2503400

45. M.G. Doborjeh, N. Kasabov, in *Personalised Modelling on Integrated Clinical and EEG Spatio-Temporal Brain Data in the NeuCube Spiking Neural Network System*. International

Joint Conference on Neural Networks (IJCNN) (2016), pp. 1373–1378, https://doi.org/10.1109/ijcnn.2021.7727358

46. X. Wang, T. Xitchell, StarPlus fMRI data. http://www.cs.cmu.edu/afs/cs.cmu.edu/project/theo-81/www/

47. N. Kasabov, M.G. Doborjeh, Z.G. Doborjeh, Mapping, learning, visualization, classification, and understanding of fMRI data in the NeuCube evolving spatiotemporal data machine of spiking neural networks. IEEE Trans. Neural Netw. Learn. Syst. 28(4), 887–899 (2017). https://doi.org/10.1109/TNNLS.2021.2612890

48. N. Kasabov, L. Zhou, M.G. Doborjeh, Z.G. Doborjeh, J. Yang, New algorithms for encoding, learning and classification of fMRI data in a spiking neural network architecture: a case on modelling and understanding of dynamic cognitive processes. IEEE Trans. Cogn. Dev. Syst. 9(4), 293–303 (2017). https://doi.org/10.1109/TCDS.2021.2636291

49. N. Sengupta, C. McNabb, N. Kasabov, B. Russell, Integrating space, time and orientation in spiking neural networks: a case study on multi-modal brain data modelling. IEEE Trans. Neural Netw. Learn. Syst (2017). http://cis.ieee.org/ieee-transactions-on-neural-networks-and-learning-systems.html

50. J.M. Walz, R.I. Goldman, M. Carapezza, J. Muraskin, T.R. Brown, P. Sajda, Simultaneous EEG-fMRI reveals temporal evolution of coupling between supramodal cortical attention networks and the brainstem. J. Neurosci. 33(49), 19212–19222 (2013). https://doi.org/10.1523/jneurosci.2649-13.2013

51. J.M. Walz, R.I. Goldman, M. Carapezza, J. Muraskin, T.R. Brown, P. Sajda, Simultaneous EEG-fMRI reveals a temporal cascade of task-related and default-mode activations during a simple target detection task. Neuroimage 102, 229–239 (2013). https://doi.org/10.1021/j.neuroimage.2013.08.014

52. Simultaneous EEG-fMRI datasets (2021), https://openfmri.org/dataset/ds000121/

53. H. Yuan, B. He, Brain–computer interfaces using sensorimotor rhythms: current state and future perspectives. IEEE Trans. Biomed. Eng. 61(5), 1425–1435 (2014). https://doi.org/10.1109/TBME.2014.2312397

54. J. Kasprcyk, W. Pedrycz (eds.), Springer Handbook of Computational Intelligence (Springer, 2015)

55. B. Blankertz, L. Acqualagna, S. Dähne, S. Haufe, M. Schultze-Kraft, I. Sturm, M. Ušćumlic, M. Wenzel, G. Curio, K.R. Müller, The Berlin brain-computer interface: progress beyond communication and control. Front. Neurosci 10, 530 (2016)

56. N. Kasabov (ed.), Springer Handbook of Bio-/Neuroinformatics (Springer, 2014)

57. K. Kumarasinghe, M. Owen, D. Taylor, N. Kasabov, C.K. Au, in FaNeuRobot: A 'Brain-like' Framework for Robot and Prosthetics Control using the NeuCube Spiking Neural Network Architecture & Finite Automata Theory. Proceedings of IEEE International Conference on Robotics and Automation (2018). https://www.ieee.org/conferences_events/conferences/conferencedetails/index.html?Conf_ID=36921

# Epilogue

The book started with a moto:

> Inspired by the oneness in nature in time-space we aim to achieve oneness in data modelling using brain-inspired computation,

and ends with a vision:

> The complementarity of HI and AI makes their symbiosis, led by the HI, a realistic pathway to take in the future, which is also an evolving process in time-space, which needs more light to be shed on how the brain works in order to develop new BI-AI technologies.

They say that the great German philosopher, writer and poet **Johann Wolfgang Goethe (1749–1832)** whispered before he died: "Licht, mehr Licht." ("Light, more Light.")

Indeed, we all need light, above all else. Light not only as a source of energy and that by which we see the beauty of each morning, but as metaphor.

> Light is knowledge,
> Light is understanding,
> Light is creation,
> Light is hope,
> Light is love,
> Light is passion,
> Light is music,
> Light is a poem,
> Light is the time given to us,
> Light is what we give to people,
> Light is the strength we get from them,
> Light is the meaning that infuses life,
> … at least for me.

© Springer-Verlag GmbH Germany, part of Springer Nature 2019      715
N. K. Kasabov, *Time-Space, Spiking Neural Networks and Brain-Inspired Artificial Intelligence*, Springer Series on Bio- and Neurosystems 7,
https://doi.org/10.1007/978-3-662-57715-8

I hope this book gives readers a bit of *Light of knowledge, Light of understanding* and *Light of meaning* on the contemporary issues related to information sciences and more specifically, the role of SNN and brain-inspired AI. I also hope that the book gives *Light of passion* and *Light of creation* to young scientists and students for them to go further *in time-space* and to shed more *Light* in the future. One of the most prolific thinkers of all times, the Greek philosopher Socrates (4century BC) used to say *"True knowledge exists in knowing that you know nothing. I know that I am intelligent, because I know that I know nothing"*. Now, in the 21st century we can metaphorically say again that "we know nothing". Literally, we can say that we know something, but it is still very little. And is BI-AI a good direction for the future? Yes, I believe so!

Nikola Kasabov
Auckland
12 August 2018

# Glossary

## A

**Adaptation** The process of structural and functional changes of a system in order to improve its performance in a changing environment.

**Alzheimer's disease (AD)** A brain disorder that is clinically characterized by a global decline of cognitive function that progresses slowly and leaves end-stage patients in custodial care. All of the currently used drugs are of limited benefit, because they have only modest symptomatic effects. Other drugs are used to manage mood disorder, agitation, and psychosis in later stages of the disease, but no treatment with a strong disease-modifying effect is currently available.

**Approximate reasoning** A process of achieving approximate, imprecise solutions and/or conclusions often based on inexact facts and uncertain rules.

**ART (adaptive resonance theory)** Refers to both a cognitive and computational theory of the brain.

**Artificial intelligence (AI)** An information system that manifests features of intelligence, such as learning, generalization, reasoning, adaptation, knowledge discovery, and applies these to complex tasks such as decision making, adaptive control, pattern recognition, speech, image and multimodal information processing, etc.

**Artificial life** A modeling paradigm that assumes that many individuals are governed by the same or similar rules to grow, die, and communicate with each other. Ensembles of such individuals exhibit repetitive patterns of behavior.

**Artificial neural network (ANN)** Biologically inspired computational model which consists of processing elements (neurons) and connections between them with coefficients (weights) bound to the connections. Training and recall algorithms are also attached to the structure.

© Springer-Verlag GmbH Germany, part of Springer Nature 2019
N. K. Kasabov, *Time-Space, Spiking Neural Networks and Brain-Inspired Artificial Intelligence*, Springer Series on Bio- and Neurosystems 7,
https://doi.org/10.1007/978-3-662-57715-8

**Atom** The smallest particle of a chemical element that retains its chemical properties. Most atoms are composed of three types of subatomic particles which govern their properties: electrons (with a negative charge), protons (with a positive charge), and neutrons (without charge).

**Automatic speech recognition system (ASRS)** A computer system which aims at providing enhanced access to machines via voice commands. Some speech recognition systems need to be trained by the voice of the intended user and are known as *speaker dependent* systems. Other systems do not need to be trained and are known as *speaker independent.*

# B

**Backpropagation training algorithm** An algorithm for supervised learning in artificial neural networks. During a training phase, after input data is entered and propagated forwards to the output of the model, the difference between the current output value and the expected output value is propagated backwards through the network as an error to adjusts the connection weights so that the next time the same data is entered, the error will be smaller. A gradient descent rule is used for finding the optimal connection weights $wij$ that minimize the global error $E$. A change of a weight $\Delta wij$ at a cycle $(t + 1)$ is in the direction of the negative gradient of the error $E$.

**Bayesian probability** The following formula, which represents the conditional probability between two events $C$ and $A$, is known as Bayes' formula (Tamas Bayes, eighteenth century): $p(A|C) = p(C|A) \cdot p(A)/p(C)$. Using Bayes' formula involves difficulties, mainly concerning the evaluation of the prior probabilities $p(A), p(C), p(C|A)$. In practice (for example, in statistical pattern recognition), the latter is assumed to be of a Gaussian type. Bayes' theorem also assumes that if the condition C consists of condition elements $C1, C2, \ldots, Ck$ they are independent (which may not be the case in some applications).

**Blue brain project** A research project hosted by EPFL in Lausanne, aiming at the development of biologically adequate brain models (http://bluebrainproject.epfl. org).

**Brain Atlas** A repository of data and knowledge and software tools to explore different brain structures and functions and genes related to them produced by the Allen Brain Science Institute (http://www.brain-map.org) inspired.

**Brain-inspired artificial intelligence (BI-AI)** An AI system that has its structure, functionality and properties inspired by the human brain.

# C

**Calcium ions** Ca2+ ions are stored in the synapses and enter the neuron via voltage-gated calcium channels and the NMDA receptor-channel complex. The intracellular calcium concentration Ca2+ is the principal trigger for the induction of LTD/LTP.

**Catastrophic forgetting** A phenomenon which represents the inability of a learning system to retain previously learned signals in order to keep the object/process in desirable states.

# D

**Data analysis** Data analysis aims at answering important questions about a process (or an object) under investigation. Some exemplar questions are: What are the statistical parameters of the data representing the process, e.g. mean, standard deviation, distribution? What is the nature of the process, random, chaotic, periodic, stable, etc.? How is the available data distributed in the problem space, e.g., clustered into groups, sparse, covering only patches of the problem space and, therefore, not enough to rely on fully when solving the problem, uniformly distributed? Is there missing data? Is there a critical obstacle which could make the process of solving the problem by using data impossible? What other methods can be used either in addition to, or instead of, methods based on data?

**Data information, and knowledge** *Data* are the numbers, the characters, the quantities operated on by a computer. *Information* is the ordered, structured, interpreted data. *Knowledge* is the theoretical or practical understanding of a subject, gained experience, true and justified belief, the way we do things. For example, the number 34 is data; 34° *of temperature in Auckland today* is information; the expression *IF temperature is too high THEN risk of stroke increases* is a piece of knowledge.

**Decision support system** An intelligent system that supports the human decision making process. Such a system analyses available data in a given problem space and suggests decisions. Examples are automated trading systems on the Internet, systems that grant loans through electronic submissions, and medical decision support systems for cardiovascular event prediction.

**Defuzzification** The process of calculating a single output numerical value for a fuzzy variable in a fuzzy system when a fuzzy membership function for this variable is given.

**Destructive learning** A learning technique, usually in artificial neural network models, that modifies an initial neural network architecture, e.g., removes connections, for the purpose of better future learning.

**Discrete Fourier transform (DFT)** The transformation of a discrete input function (usually in the time domain) into another function in the frequency domain.

**Distance between data points** The distance between two data points $a$ and $b$ in an $n$-dimensional geometrical space can be measured in several ways. The most widely used formulas are: Hamming distance, $Dab = \Sigma|ai - bi|$ and Euclidean distance, $Eab = \_\Sigma(ai - bi)2/n$.

**Distributed representation** A way of encoding information, usually in a neural network model, where a concept or a value for a variable is represented by the collective activation of a group of neurons.

**DNA (deoxyribonucleic acid)** This is a chemical chain, present in the nucleus of each cell of an organism. It consists of pairs of small chemical molecules (bases) ordered in a double helix, which are: adenine (A), cytidine (C), guanidine (G), and thymidine (T), linked together by a deoxyribose sugar phosphate nucleic acid backbone. Almost all cells in an organism contain the same DNA information, but different parts of the DNA, different genes, express in different parts of the organism and produce different proteins.

**E**

**Electroencephalography (EEG)** An EEG is a recording of electrical signals from the brain made by attaching surface electrodes to the subject's scalp. These electrodes record electric signals naturally produced by the brain, called brainwaves. EEGs allow researchers to follow electrical potentials across the surface of the brain and observe changes over split seconds of time.

**Elitism (in genetic algorithms and other evolutionary optimization algorithms)** The fittest members of a population at generation ($t$) is copied unmodified into the population of the next generation ($t + 1$). The intention of this strategy is to reduce the chance of losing the best genotypes. If elitism were a principle of human evolution, we would still have Leonardo da Vinci among artists and scientists nowadays.

**Evolutionary computation** A computational paradigm that uses principles from natural evolution, such as genetic representation, mutation, survival of the fittest, population of individuals, and generations of populations. Evolutionary computation is mainly used as a population-generation based optimization technique where the best or close to it solution to a problem is achieved through evaluating many individual solutions in a population over generations.

**Evolutionary programming** Evolutionary algorithms applied for the automatic creation or optimization of computer programs.

**Evolutionary strategies** Strategies that use evolutionary algorithms to represent a solution to a problem as a single chromosome and evaluate different mutations of this chromosome over generations through a fitness function. This process is carried out until a satisfactory solution is found.

**Evolving connectionist systems (ECOS)** Artificial neural networks proposed by Kasabov (1998) that develop (evolve) their structure and functionality from incoming data in an adaptive, incremental way.

**Evolving intelligent systems (EIS)** Intelligent systems that are characterized by adaptation and incremental evolving of knowledge. The methods used in such systems are mainly based on neural networks, but may include many other techniques from the area of computational intelligence.

**Evolving spiking neural networks** Spiking neural networks that evolve their structure of spiking neurons following the principles of ECOS and applying spike-time learning rules.

**Evolving processes in nature** Processes that change, develop, unfold in time-space.

**Expert system** Knowledge-based systems that provide expertise, similar to that of human experts in a restricted application area, for the solution of problems in that area. An expert system consists of the following main blocks: knowledge base, data base, inference engine, explanation module, user interface, and knowledge acquisition module.

**Explanation in an AI system** This is a desirable property for many AI systems. It means tracing, in a contextually comprehensible way, the process of inferring the solution and reporting it. Explanation is easier for AI symbolic systems where sequential inference takes place.

# F

**Fast Fourier transformation (FFT)** This is a fast algorithm for discrete Fourier transformation (DFT). A nonlinear transformation applied on time series data to transform the signal taken within a small portion of time (the time scale domain) into a vector in the frequency scale domain.

**Feed-forward neural network** A neural network in which there are no connections back from the output to the input neurons.

**Finite automaton** A computational model represented by a set $X$ of inputs, a set $Y$ of outputs, a set $Q$ of internal states, and two functions $f1$ (state transfer function) and $f2$ (output function): $f1\colon X \times Q \to Q$, i.e., $[x, q(t)] \to q(t + 1)$, $f2\colon X \times Q \to Y$, i.e., $[x, q(t)] \to y(t + 1)$, where $x \in X$, $q \in Q$, $y \in Y$, $t$, and $(t + 1)$ represent two consecutive time moments.

**Fractal** An object which occupies a fraction (called *embedding space*) of a standard space (i.e., space with integer numbers for dimensions). The dimensionality of fractal can be, e.g., 2.4, rather than the standard 3-D.

**Functional MRI (fMRI)** This combines visualization of brain anatomy with the dynamic image of brain activity into one comprehensive scan. This noninvasive technique measures the ratio of oxygenated to deoxygenated haemoglobin which have different magnetic properties. Active brain areas have higher levels of oxygenated haemoglobin than less active areas. An fMRI can produce images of brain activity at the time scale of a second with very precise spatial resolution of about 1–2 mm. Thus, fMRI provides both an anatomical and functional view of the brain.

**Fuzzification** The process of finding the membership degree $\mu A(x_-)$ to which a value $x_-$ of a fuzzy variable $x$ belongs to a fuzzy set $A$ defined on the same universe as the variable $x$.

**Fuzzy clustering** A procedure of clustering data into possibly overlapping clusters, such that each of the data elements may belong to each of the clusters to a certain degree. The procedure aims at finding the cluster centers $Vi$ ($i = 1, 2, ..., c$) and the cluster membership functions $\mu i$ which define to what degree each of the $n$ data elements belongs to the $i$-th cluster. The number of clusters $c$ is either defined a priori or chosen by the clustering procedure (evolving clustering). The result of a clustering procedure can be represented as a fuzzy relation $\mu i,k$, such that: (i) $\Sigma \mu i, k = 1$, for each $k = 1, 2, ... , n$ (the total membership of an instance to all the clusters equals 1); (ii) $\Sigma \mu i, k > 0$, for each $i = 1, 2, ... , c$ (there are no empty clusters).

**Fuzzy control** The application of fuzzy logic to control problems. A fuzzy control system is a fuzzy system applied to solve a control problem.

**Fuzzy expert system** An expert system to which methods of fuzzy logic are applied. Fuzzy expert systems use fuzzy data, fuzzy rules, and fuzzy inference, in addition to the standard ones implemented in ordinary expert systems.

**Fuzzy logic** A logic system that is based on fuzzy relations and fuzzy propositions, the latter being defined on the basis of fuzzy sets.

**Fuzzy neural network** An artificial neural network model that can be interpreted as a fuzzy system. The model can have neurons that represent fuzzy concept (e.g., small).

**Fuzzy propositions** Propositions which contain fuzzy variables with their fuzzy values. The truth value of a fuzzy proposition $X$ *is* $A$ is given by a membership function $\mu A$.

**Fuzzy relations** Fuzzy relations link two fuzzy sets or two fuzzy variables in a predefined manner. Fuzzy relations make it possible to represent ambiguous relationships, such as: *the grades of the 3rd and 2nd year classes are similar, team A performed slightly better than team B,* or *the more fat you eat, the higher your risk of heart attack.*

# G

**GenBank** A repository of genes and their functions across species and diseases, maintained by the NCBI (http://www.ncbi.nlm.nih.gov/Genbank/index.html).

**Gene expression atlas** A repository of gene expression data across species and diseases (http://expression.gnf.org/cgi-bin/index.cgi).

**Gene ontology** An ontology knowledge repository system designed to produce a controlled vocabulary that can be applied to all organisms even if knowledge of genes and proteins is changing.

**Gene regulatory network (GRN)** A biological or computational network of genes connected between each other according to their interaction in time.

**Gene-brain ontology** An ontology knowledge repository system that includes knowledge, data and known relationships between brain structures and functions and genes that are related to them.

**Generalization** The process of matching new input data to a model, system, or in principle, to an existing set of problem knowledge, in order to obtain an output value (e.g., solution) that corresponds to this input data.

**Genes** Parts of a DNA sequence that are transcribed into RNA and translated into proteins or alternatively produce microRNA (not translated into proteins). Genes are the carrier of information that is passed from one generation of species to another in an evolutionary process.

**Genetic algorithms (GA)** These are algorithms for solving complex multivariant combinatorial and organizational problems by employing methods of evolutionary computation that are analogous to evolution in nature. There are several general steps that a genetic algorithm cycles through: generate a population of individuals; evaluate the fitness (goodness) of each individual; select the best individuals; perform cross-over operation between these individuals; mutate individuals, if necessary. These steps are repeated all over again until an acceptable solution is found or the time for performing the algorithm has expired.

**Glutamate neurotransmitters** Molecules released in the synapses during afferent activity that bind to AMPA, NMDA, and metabotropic glutamate (mGlu) receptors to produce postsynaptic response.

**Goodness function (fitness function)** A function that can be used to measure the appropriateness of an individual element from a population of individuals at a certain generation over time. An individual element would represent a possible solution to a problem, e.g., the shortest path from one city to another, or a set of genes to diagnose cancer.

# H

**Hebbian learning law** A generic learning principle which states that a synapse connecting two neurons $i$ and $j$ increases its strength $wij$ if the two neurons $i$ and $j$ are repeatedly and simultaneously activated by input stimuli.

**Hopfield network** A fully connected feedback neural network which is an auto-associative memory. It can be trained to recognize input patterns and to recover them if they are presented only partially at a later stage. It is named after its inventor John Hopfield.

# I

**Image filtering** A transformation of an original image through a set of operations that use the original pixel intensities of the image and apply two-dimensional array of numbers, which is known as a kernel. This process is also called convolution.

**Independent component analysis** The process of separating independent components of multidimensional time series data, such as signal and noise.

**Inference in an AI system** The process of matching new data from a domain space to the knowledge existing in an AI system and obtaining output values.

**Information entropy** A measure for the level of uncertainty (or unpredictability) associated with a random variable. The more unpredictable an event is, the higher the information entropy.

**Information retrieval** The process of retrieving relevant information from a data base.

**Information science** This is the area of science that develops methods and systems for information and knowledge processing regardless of the domain specificity of this information. Information science includes the following subject areas: data collection and data communication (sensors and networking); information storage and retrieval (data base systems); methods for information processing (information theory); creating computer programs and information systems (software engineering and system development); acquiring, representing, and processing knowledge (knowledge-based systems); and creating intelligent systems and machines (artificial intelligence, knowledge engineering).

**Information** Collection of structured data. In its broad meaning it includes knowledge as well as simple meaningful data.

**Initialization in ANN** The process of setting the connection weights in an ANN to some initial values before starting the training algorithm.

**Intelligent system (IS)** An information system that manifests features of intelligence, such as learning, generalization, reasoning, adaptation, knowledge discovery, and applies these to complex tasks such as decision making, adaptive control, pattern recognition, speech, image and multimodal information processing, etc.

**Interaction (human–computer)** Communication between a user and a computer system.

# K

**Knowledge engineering** The area of science and engineering that deals with data, information and knowledge representation in machines, information processing, and finally, knowledge elucidation and knowledge discovery.

**Knowledge** Concise presentation of facts, skills, previous experience, principles, definitions, etc., that is interpretable under different conditions. Knowledge resides in the human brain. As a term it is used to represent information in a computer system that can be interpreted by humans.

**Knowledge-based neural networks (KBNN)** These are pre-structured ANN that allow for data and machine knowledge manipulation, including learning from data, rule insertion, rule extraction, adaptation, and reasoning. KBNN have been developed either as a combination of symbolic AI systems and ANN, or as a combination of fuzzy logic systems and ANN, or as other hybrid systems. Rule insertion and rule extraction are typical operations for a KBNN to accommodate existing knowledge along with data, and to produce an explanation of what the system has learned.

**Kohonen self-organizing map (SOM)** A self-organized ANN that uses unsupervised learning to map multidimensional input vectors into low-dimensional matrix known as map. The concept of SOM was first introduced and developed by the Finish scientist Prof. Teuvo Kohonen.

# L

**Laws of inference in fuzzy logic** The way fuzzy propositions are used to make inference over new facts. The following are the two most used laws illustrated on two fuzzy propositions $A$ and $B$. (a) Generalized modus ponens: $A \rightarrow B$, and $A\_\therefore$ $B\_$, where $B\_ = A\_o(A \rightarrow B)$; (b) generalized modus tollens (law of the contrapositive): $A \rightarrow B$, and $B'$, $\therefore A'$, where $A\_ = (A \rightarrow B)oB$.

**Learning vector quantization algorithm (LVQ)** A supervised learning algorithm, which is an extension of the unsupervised Kohonen self-organized network learning algorithm.

**Learning** Process of obtaining new information (possibly interpretable as knowledge) from data and existing information.

**Linear transformation** Transformation $f(x)$ of a variable $x$ such that $f$ is a linear function of $x$, for example, $f(x) = 2x + 1$.

**Linguistic variable** A variable that takes fuzzy values that have linguistic meaning, e.g. the variable *temperature* can take a fuzzy value of *very high temperature.*

**Local representation in a neural network** A way of encoding information in an ANN in which every neuron represents one concept or one variable.

**Logic system** An abstract system that consists of four parts: an alphabet—a set of basic symbols from which more complex sentences (constructions) can be made; syntax—a set of rules or operators for constructing sentences (expressions) or other more complex structures from the alphabet elements; semantics—define the meaning of the constructions in the logic system; laws of inference—a set of rules or laws for constructing semantically equivalent but syntactically different sentences; this set of laws is also called a set of inference rules.

**Long-term depression (LTD)** A process of a long-lasting decrease in the strength of synaptic transmission, produced by low-frequency stimulation of presynaptic afferents. The majority of synapses in many brain regions and in many species that express LTP also express LTD.

**Long-term potentiation (LTP)** This is a process of a long-lasting increase in synaptic efficacy, produced by high-frequency stimulation of presynaptic afferents or by pairing presynaptic stimulation with postsynaptic depolarization.

# M

**Machine learning** An area of information and computer science concerned with the methods for accumulating, changing, and updating information and obtaining machine knowledge through algorithms.

**Magnetic resonance imaging (MRI)** This uses the properties of magnetism. A large cylindrical magnet creates a magnetic field around the subject's head. Detectors measure local magnetic fields caused by alignment of atoms in the brain with the externally applied magnetic field. The degree of alignment depends upon the structural properties of the scanned tissue. MRI provides a precise anatomical image of both surface and deep brain structures.

**Magnetoencephalography (MEG)** This measures millisecond-long changes in magnetic fields created by the brain's electrical currents. MEG machines use a noninvasive, whole-head, 248-channel, *super-conducting-quantum-interference-device* (SQUID) to measure small magnetic signals reflecting changes in the electrical signals in the human brain.

**Mel-scale filter bank transformations** The process of filtering a signal through a set of frequency bands represented by triangular filter functions similar to the functions that describe the function of the human inner ear.

**Membership function** Generalized characteristic function which defines the degree to which an object from a universe belongs to a given fuzzy concept.

**Memory capacity of an ANN** The maximum number $m$ of patterns that can be learned properly in a network.

**Mental retardation** A developmental deficit, beginning in childhood, which results in significant limitation of intellect and cognition and poor adaptation to the demands of everyday life.

**Methods for feature extraction** Methods used for reducing the dimensionality of raw data by transforming it from the original space into a space of selected features.

**Microarray for gene expression** A device that evaluates the level of transcription (expression) of a predefined set of genes in a single biological cell or a piece of tissue. The five principal steps in the microarray technology are: tissue collection, RNA extraction, microarray gene expression evaluation, scanning and image processing, and data analysis.

**Monitoring** The process of interpreting continuous input information and recommending intervention if appropriate.

**Moving averages** A moving average of a time series is calculated by using $MAt = (\Sigma St - i)/n$, for $i = 1, 2, \ldots, n$, where $n$ is the number of the data points, $St - i$ is the value of the series at a time moment $(t - i)$, and $MAt$ is the moving average at a time moment $t$. Moving averages are often used in an information system as input features in addition to, or in substitution of, the real values of a time series.

**Multilayer perceptron (MLP)** An ANN that consists of an input layer, at least one intermediate or *hidden* layer, and one output layer where the neurons from each layer are fully connected to the neurons from the next layer. In some particular applications they may be partially connected.

**Mutation** A random change in the value of a gene; this relates to both biological genes and to gene parameters of an evolutionary algorithm.

# N

**Neural networks (NN)** See *artificial neural network.*

**Neurotransmitters** Molecules that are produced in neurons in the brain and reside in synapses. When a synapse receives a spike, the synapse transfers neurotransmitters across the synaptic cleft so they can bind to receptors in the postsynaptic membrane that causes ion gates to open and to receive ions that change the membrane potential of the postsynaptic neuron. It is estimated that there are about 50 different neurotransmitters acting in the human brain. Neurotransmitters control and are vital for neuronal functions, including learning, memory, emotions, and decision making. The three major categories of substances that act as neurotransmitters are: (1) amino acids (primarily glutamate, GABA, aspartic acid and glycine); (2) peptides (vasopressin, somatostatin, neurotensin, etc.); and (3) monoamines (norepinephrine, dopamine and serotonin) plus acetylcholine. There are also other categories like opioids, tachykinins, and so on. The vast majority of neurotransmitters is produced in evolutionary older subcortical nuclei.

**Noise** A random value without meaning that is added to the general function that describes the underlying behavior of a process or a signal.

**Nonlinear dynamical system** A system whose next state on the time scale can be expressed by a nonlinear function of its previous time states.

**Nonlinear transformation** Transformation $f$ of a variable $x$, where $f$ is a nonlinear function of $x$, for example, $f(x) = 1/(1 + e - xc)$, where $c$ is a constant.

**Normalization** Transforming data from its original range into another, predefined range, e.g., [0, 1].

**Nyquist sampling frequency** A Nyquist sampling frequency for a particular signal is defined as twice the highest frequency contained within the signal (e.g., if $F$signal $= 10.025$ Hz then $F$NyqSampling $= 22.050$ Hz). When a signal is sampled at Nyquist frequency, the numeric sequence obtained completely determines the signal.

# O

**Ontology systems** This is both a data and a knowledge repository. Ontology is defined in the artificial intelligence literature as a specification of a conceptualization. Ontology specifies at a higher level the classes of concepts that are relevant to the domain and the relations that exist between these classes. Ontology captures the intrinsic conceptual structure of a domain along with the data that is available. For any given domain, the ontology forms the heart of the knowledge representation.

**Optimization** The process of finding such values for the parameters of an object, system, or a process that would minimize an objective (cost) for this object/system/process.

**Overfitting** A phenomenon that indicates that an ANN has approximated (or learned) a set of data examples too closely, and as a result the network cannot generalize well on new examples.

# P

**Pattern matching** The process of matching a feature vector to already existing ones and finding the best match among them.

**Phonemes** A basic distinctive unit of speech sound in a specified language.

**Photon** In physics, a photon is a quantum of electromagnetic field, for instance, light. The term photon was coined by Gilbert Lewis in 1926. A photon can be perceived as a wave or a particle, depending on how it is measured. The photon is an elementary particle. Its interactions with electrons and atomic nuclei account for a great many of the features of matter, such as the existence and stability of atoms, molecules, and solids.

**Planning** An important biological process and also AI-problem which is about generating a sequence of actions in order to achieve a given goal when a description of the current situation is available.

**Positron emission tomography (PET)** This is used to study living brain activity. This noninvasive method involves an on-site use of a machine called a cyclotron to label specific drugs or analogs of natural body compounds (such as glucose or oxygen) with small amounts of radioactivity. The labelled compound (a radio-tracer) is then injected into the bloodstream, which carries it into the brain. Radiotracers break down, giving off subatomic particles (positrons). By surrounding the subject's head with a detector array, it is possible to build up images of the brain showing different levels of radioactivity, and therefore, cortical activity.

**Prediction** Generating information for possible future development of a process from data that represents its past and present development.

**Principle component analysis (PCA)** A statistical procedure for finding a smaller number of $m$ components $Y = (y1, y2, ..., ym)$ (aggregated variables, eigenvectors) that can represent a function $F(x1, x2, ..., xn)$ of $n$ variables, where $n > m$, to a desired degree of accuracy $\Theta$, i.e., $F = M \cdot Y + \Theta$ where $M$ is a matrix that must be found through PCA.

**Probability automata** Finite automata whose transitions are defined as probabilities. They are also known as stochastic automata.

**Probability theory** The theory is based on the following three axioms: Axiom 1 defines the probability $p(E)$ of an event $E$ as a real number in the closed interval $[0, 1]$, i.e. $0 \le p(E) \le 1$. A probability $p(E) = 1$ indicates a certain event while $p(E) = 0$ indicates an impossible event. Axiom 2 is expressed as $p(Ei) = 1$, $E1 \cup E2 \cup ... \cup Ek = U$, where $U$ denotes the problem space (universe) as an union as subspaces. Axiom 3 indicates that if two independent events $E1$ and $E2$ cannot occur simultaneously, the probability of one or the other happening is the sum of their probabilities, i.e., $p(E1 \vee E2) = p(E1) + p(E2)$, where $E1$ and $E2$ are mutually exclusive events.

**Propositional logic** A logic system that can be dated back to Aristotle (384–322 B.C.). There are three types of symbols in the propositional logic: propositional symbols (the alphabet), connective symbols, and symbols denoting the meaning of the sentences. There are rules in propositional logic to construct syntactically correct sentences (called well-formed formulas) and rules to evaluate the semantics of the sentences. A proposition represents a statement about the world, for example: *The temperature is over* 40. The semantic meaning of a proposition is expressed by two possible semantic symbols—true and false. Statements or propositions can be only *true* or *untrue* (false), nothing in between.

**Proteins** Biological molecules that result from RNA translation. Proteins provide the majority of the structural and functional components of a cell. The area of molecular biology that deals with all aspects of proteins is called proteomics. A protein is a sequence of amino acids, each of them defined by a group of three nucleotides (codons). There are 20 amino acids all together, denoted by letters (A,C-H,I,K-N,P-T,V,W,Y). The length of a protein in number of amino acids is from tens to several thousands. Each protein is characterized by some characteristics, for example: structure, function, charge, acidity, hydrophilicity, and molecular weight. An initiation codon defines the start position of a gene in an mRNA where the translation of the mRNA into protein begins. A stop codon defines the end position.

**Pruning in ANN** Technique where during the training procedure of the ANN weak connections (i.e., connections that have weights around 0) and the neurons connected by them are gradually removed.

# R

**Recall process** The process of using a trained ANN where new data is entered and results are calculated.

**Recurrent fuzzy rule** A fuzzy rule that uses in its antecedent part one or more previous time-moment values of the output fuzzy variable.

**Recurrent networks** ANN with feedback connections from neurons in one layer to neurons in a previous layer.

**Reinforcement learning** A learning method that is based on presenting input data *x* to a learning system and observing the produced output. If this output is evaluated as *good*, then a *reward* is given to the learning system, e.g., connection weights of a neural network model increase in values, otherwise the system is *punished*, e.g., connection weights decrease.

**RNA (ribonucleic acid)** A transcribed copy of part of an DNA that has a similar structure to the DNA, but here thymidine (T) is substituted by uridine (U) nucleotide. In the pre-RNA only segments that contain genes are extracted from the DNA. Each gene consists of two types of segments—exons, that are segments translated into proteins, and introns—segments that are considered redundant and do not take part in the protein production. Removing the introns and ordering only the exon parts of the genes in a sequence is called splicing and this process results in the production of messenger RNA (or mRNA) sequences. mRNAs are directly translated into proteins.

**Roulette wheel selection (in genetic algorithms)** A selection strategy according to which each individual from a population of individuals at a certain generation is assigned a sector in an imaginary roulette wheel, with the size of the sector depending on the fitness of the individual. The size of the sector represents the probability of the individual to be selected when a random number is generated. Therefore, the fitter the individual, the higher the chance of it being selected for cross-over with other selected individuals to produce the population of individuals for the next generation.

# S

**Sampling** A process of selecting a subset of data from a larger data set. Sampling can be applied on continuous time-series data (e.g., speech data can be sampled at a frequency of 22 kHz), or on static data where only a smaller subset of the data is selected at a time for processing.

**Schizophrenia** A brain disorder that has typical characteristic symptoms such as: delusions, hallucinations, and various thinking and perceptual disorders. Schizophrenic withdrawal from reality can manifest itself in many peculiar ways. Disorder is accompanied by serious deterioration of the previous level of functioning in such areas as work, social relations, and self-care.

**Sensitivity to initial conditions** A characteristic of a chaotic process which means that a slight difference in the initial values of some parameters will result in different trends in a future development of the chaotic process.

**Spatio-temporal ANN** These networks can learn and represent spatio-temporal patterns from data.

**Spatio-temporal data** data that is characterized by spatially distributed variables that are measured over time, e.g., electroencephalogram data.

**Spike time dependent plasticity (STDP)** A method for learning in spiking neural networks that modifies the connection weight between two neurons, so that if the presynaptic neuron spikes first and then, within a certain time interval, spikes the postsynaptic neuron—the connection weight increases, otherwise—the connection weight decreases.

**Spiking neural networks** Biological or artificial neural networks that consist of spiking neurons and connections. The information is represented as trains of spikes (binary events over time).

**Spiking neuron** A biological or artificial neuron model that receives binary input signals (spikes) over time from many inputs (dendrites). It emits a spike (action potential) when the cumulative input (the membrane potential) of this neurons has reached a threshold. After that the neuronal membrane potential is set to a reset value and the process continues.

**Stability/plasticity dilemma** The ability of a system to keep the balance between retaining previously learned information and patterns and learning new information/patterns.

**Statistical analysis methods** Methods used for discovering repetitiveness in data based on probability estimation.

**Subcortical structure of the brain** This consists of brain areas excluding the cortex, such as: basal ganglia, thalamus, hypothalamus, amygdala, and dozens of other groups of neurons with more or less specific functions in operations of the whole brain.

**Supervised learning** A process of inferring a function from a set of training data with known outputs (labels). The training data set consists of data items each of which contains values for attributes (features)—independent variables, labeled by the desired value(s) for the dependant variables. Supervised learning can be viewed as approximating a mapping between a domain and a solution space of a problem: $X \rightarrow Y$, when samples (examples) of (input vector–output vector) pairs $(x, y)$ are known, and $x \in X, y \in Y, x = (x1, x2, ..., xn), y = (y1, y2, ..., ym)$.

**Supervised training algorithm for an ANN** Training of an ANN when the training examples comprise input vectors $x$ and the desired output vectors $y$; training is performed until the neural network *learns* to associate each input vector $x$ to its corresponding and desired output vector $y$ to a desired accuracy.

**Synaptic efficacy** The level of concentration of ions in a synapse that can be transmitted to the postsynaptic neuronal membrane through ion channels that become open after certain neurotransmitters bind to them.

**Synaptic plasticity** The process of changing synaptic efficacy through LTP/LTD learning.

**System biology** An approach to treat and understand complex biological systems in their entirety, i.e., at a system level. It involves the integration of different data, knowledge, data analysis approaches, and tools. One of the major challenges of systems biology is the identification of the logic and dynamics of gene-regulatory and biochemical networks. The most feasible application of systems biology is to create a detailed model of a cell regulation to provide system-level insights into mechanism-based drug discovery.

# T

**Test error** An error that is calculated for a learning system that is trained with a set of training data. When a test (or validation) data set, for which the results are known, is applied in a recall procedure the test error is calculated.

**Time alignment** A process where a sequence of input vectors recognized in a system over time are aligned to represent a meaningful output (e.g., a phoneme, word, or trend in stock).

**Time-series prediction** Prediction of time series events.

**Training error** The error of a learning system that is evaluated on the data used for training.

**Training of a neural network** A procedure for presenting training examples to a neural network, which results in changing the network's connection weights according to a certain learning law.

**Turing test** Test of the ability of a digital computer to demonstrate intelligent behavior or, more precisely, whether it can imitate a human. The test was first described by the British mathematician Alan Turing in his 1950 paper *Computing Machinery and Intelligence*. The Turing test has been highly influential in the area of artificial intelligence and at the same time has been very controversial. The idea of the test is that an interrogator communicates with an entity in written form and, based on the reactions of the entity, decides whether it is another human or a computer. If a computer can trick the interrogator into believing that it is a human, then the machine has passed the test.

# U

**Universal function approximator (for ANN)** A theorem that was proved by Hornik (1989), Cybenko (1989), and Funahashy (1989). It states that an MLP with one hidden layer can approximate any continuous function to any desired accuracy, subject to sufficient number of hidden nodes. As a corollary, any Boolean function of $n$ Boolean variables can be approximated by an MLP.

**Unsupervised learning algorithm** A learning procedure where only input vectors $x$ are supplied to a learning system (e.g., a neural network). The system learns some internal characteristics, e.g., clusters, for the whole set of input vectors presented to it. An example of such an algorithm is the self-organizing maps.

# V

**Validation** Process of testing how good the solutions produced by a system are. The solutions are usually compared to the results obtained either by experts or by other systems.

**Vector quantization** A process of representing data from $n$-dimensional problem space into the $m$-dimensional one, where $m < n$, in a way that preserves the distance between data examples (points) from the original space.

# W

**Wavelet transformation** A nonlinear transformation that can be used to represent slight changes of a time series within a chosen time interval.

# Index

© Springer-Verlag GmbH Germany, part of Springer Nature 2019
N. K. Kasabov, *Time-Space, Spiking Neural Networks and Brain-Inspired
Artificial Intelligence*, Springer Series on Bio- and Neurosystems 7,
https://doi.org/10.1007/978-3-662-57715-8

Printed in the United States
By Bookmasters